This second edition of a classic book provides an updated look at crystal field theory – one of the simplest models of chemical bonding – and its applications. Crystal field theory can be applied to elements of the first transition series and provides a link between the visible region spectra and thermodynamic properties of numerous rock-forming minerals and gems that contain the elements iron, titanium, vanadium, chromium, manganese, cobalt, nickel or copper. These elements are major constituents of terrestrial planets and significantly influence their geochemical and geophysical properties.

A unique perspective of the second edition is that it highlights the properties of minerals that make them compounds of interest to solid state chemists and physicists as well as to all earth and planetary scientists. This book will be useful as a textbook for advanced students as well as a valuable reference work for all research workers interested in the crystal chemistry, spectroscopy and geochemistry of the transition elements.

MINERALOGICAL APPLICATIONS OF CRYSTAL FIELD THEORY

Second Edition

CAMBRIDGE TOPICS IN MINERAL PHYSICS AND CHEMISTRY

Editors
Andrew Putnis
Robert C. Liebermann

1 *Ferroelastic and co-elastic crystals*
EKHARD K. H. SALJE

2 *Transmission electron microscopy of minerals and rocks*
ALEX C. McLAREN

3 *Introduction to the physics of the Earth's interior*
JEAN-PAUL POIRIER

4 *Introduction to lattice dynamics*
MARTIN DOVE

5 *Mineralogical applications of crystal field theory, Second Edition*
ROGER G. BURNS

MINERALOGICAL APPLICATIONS OF CRYSTAL FIELD THEORY

Second Edition

ROGER G. BURNS
Massachusetts Institute of Technology
Cambridge, Massachusetts

CAMBRIDGE
UNIVERSITY PRESS

CAMBRIDGE UNIVERSITY PRESS
Cambridge, New York, Melbourne, Madrid, Cape Town, Singapore, São Paulo

Cambridge University Press
The Edinburgh Building, Cambridge CB2 2RU, UK

Published in the United States of America by Cambridge University Press, New York

www.cambridge.org
Information on this title: www.cambridge.org/9780521430777

First published 1970
Second edition 1993

A catalogue record for this publication is available from the British Library

Library of Congress Cataloguing in Publication data

Burns, Roger G. (Roger George), 1937–
Mineralogical applications of crystal field theory/Roger G. Burns. – 2nd ed.
p. cm. – (Cambridge topics in mineral physics and chemistry; 5)
Includes bibliographical references and index.
ISBN 0-521-43077-1 (hardback).
1. Mineralogy. 2. Crystal field theory. I. Title. II. Series.
QE364.B87 1993
549–dc20 92–37626 CIP

ISBN-13 978-0-521-43077-7 hardback
ISBN-10 0-521-43077-1 hardback

Transferred to digital printing 2005

Contents

Preface to the first edition — *page* xvii
Preface to the second edition — xxi
1 Introduction — 1
1.1 History of crystal field theory — 4
1.2 Scope of crystal field theory — 5
1.3 Background reading — 6
2 Outline of crystal field theory — 7
2.1 Introduction — 7
2.2 Orbitals — 7
2.2.1 Principal quantum number n — 8
2.2.2 Azimuthal quantum number l — 9
2.2.3 Magnetic quantum number m_l — 9
2.2.4 Spin quantum number m_s — 10
2.2.5 Spin–orbit coupling — 11
2.3 Shape and symmetry of the orbitals — 11
2.4 Transition elements — 14
2.5 Crystal field splitting in octahedral coordination — 15
2.6 Crystal field splittings in other coordinations — 21
2.6.1 Tetrahedral coordination — 21
2.6.2 Cubic and dodecahedral coordinations — 22
2.6.3 Other regular coordination environments — 24
2.7 The 10 *Dq* parameter — 24
2.8 Evaluation of Δ or 10 *Dq* — 27
2.9 Factors influencing values of Δ or 10 *Dq* — 27
2.9.1 Type of cation — 27
2.9.2 Type of ligand — 28
2.9.3 Interatomic distance — 30

2.9.4 Pressure 30
2.9.5 Temperature 31
2.9.6 Symmetry of the ligand environment 32
2.10 Values of CFSE 32
2.11 Crystal fields in non-cubic environments 33
2.12 The Jahn–Teller effect 33
2.13 Stabilization energies in distorted coordination sites 36
2.14 Electronic entropy 40
2.15 Summary 41
2.16 Background reading 43

3 Energy level diagrams and crystal field spectra of transition metal ions 44
3.1 Introduction 44
3.2 Units in absorption spectra 44
3.2.1 Wavelength and energy units 45
3.2.2 Absorption terminology 46
3.3 Examples of crystal field spectra 50
3.4 Energy level diagrams for octahedral environments 50
3.4.1 Spectroscopic terms and crystal field states 50
3.4.2 Tanabe–Sugano diagrams 56
3.4.3 Orgel diagrams 59
3.5 Energy level diagrams for other high-symmetry environments 61
3.6 Energy level diagrams for low-symmetry environments 62
3.7 Selection rules and intensities of absorption bands 64
3.7.1 Laporte or parity selection rule 65
3.7.1.1 Vibronic coupling 67
3.7.2 Spin-multiplicity selection rule 69
3.7.3 Coupling interactions with nearest-neighbour cations 71
3.7.4 Relative intensities of crystal field spectra 71
3.8 Polarization dependencies of absorption bands 72
3.8.1 Electric and magnetic vectors in polarized light 73
3.8.2 Group theoretical interpretation of gillespite polarized spectra 76
3.9 Widths of absorption bands 80
3.9.1 Correlations with energy level diagrams 80
3.9.2 The dynamic Jahn–Teller effect 81
3.9.3 Effects of multiple site occupancies 81
3.9.4 Vibrational interactions 81
3.9.5 Band shape 82
3.10 Ligand field parameters for distorted environments 82
3.11 Summary 84

3.12 Background reading 86

4 Measurements of absorption spectra of minerals 87
4.1 Introduction 87
4.2 Techniques for measuring absorption spectra of minerals 88
4.3 Measurements of intensities of absorption bands 91
4.4 Identification of cation valences and coordination symmetries 93
4.4.1 Valence of titanium in extraterrestrial pyroxenes 93
4.4.2 Stability of Mn^{3+} ions in epidote and andalusite 95
4.4.2.1 Piemontite 95
4.4.2.2 Viridine and kanonaite 99
4.4.3 Coordination symmetry of iron and cobalt in staurolite 100
4.5 Detection of cation ordering in silicate minerals 102
4.6 Concepts of colour and pleochroism 105
4.6.1 Colour 105
4.6.2 Pleochroism and dichroism 105
4.6.3 Visible and invisible pleochroism 108
4.7 Causes of colour and pleochroism in minerals 108
4.7.1 Crystal field transitions 109
4.7.1.1 The colour of ruby 111
4.7.1.2 The 'alexandrite effect' 112
4.7.1.3 Blue colour of tanzanite 113
4.7.1.4 Other minerals coloured by a transition metal 115
4.7.2 Intervalence charge transfer transitions 115
4.7.2.1 Homonuclear intervalence transitions 116
4.7.2.2 Heteronuclear intervalence transitions 116
4.7.2.3 Vivianite 121
4.7.2.4 Glaucophane 124
4.7.2.5 Titaniferous pyroxenes 125
4.7.2.6 Blue sapphire and hibonite 127
4.7.2.7 Kyanite 129
4.7.2.8 Other Fe^{2+}–Fe^{3+} minerals 130
4.7.2.9 Other Fe^{2+}–Ti^{4+} minerals 131
4.7.3 Oxygen → metal charge transfer 132
4.8 Opacity of mixed-valence oxides and silicates 133
4.8.1 Magnetite 136
4.8.2 Ilvaite 137
4.8.3 Other opaque mixed–valence minerals 140
4.8.3.1 Vonsenite 140
4.8.3.2 Deerite 141
4.8.3.3 Cronstedtite 141

4.8.3.4 Laihunite and ferrifayalite 141
4.9 Summary 142
4.10 Background reading 144

5 Crystal field spectra of transition metal ions in minerals 146
5.1 Introduction 146
5.2 Classification of absorption spectra of minerals 148
5.3 Spectra of oxides 149
5.3.1 Periclase 149
5.3.2 Corundum 151
5.3.2.1 Ruby 153
5.3.3 Spinels 155
5.4 Spectra of orthosilicates 155
5.4.1 Garnet group 155
5.4.2 Olivine group 158
5.4.2.1 Background 158
5.4.2.2 Olivine crystal structure 159
5.4.2.3 Crystal field spectra of Fe^{2+} olivines 161
5.4.2.4 Crystal field spectra of Ni^{2+} olivines 165
5.4.2.5 Miscellaneous olivines 168
5.4.3 Silicate spinels 169
5.4.4 Aluminosilicates 172
5.4.4.1 Andalusite 172
5.4.4.2 Kyanite 173
5.4.4.3 Sillimanite 173
5.4.4.4 Yoderite 174
5.4.5 Epidote group 174
5.4.6 Miscellaneous orthosilicates and isostructural minerals 175
5.5 Spectra of chain silicates: Pyroxene group 176
5.5.1 Background 176
5.5.2 Pyroxene crystal structure 176
5.5.3 Calcic pyroxenes 180
5.5.4 Orthopyroxenes 183
5.5.5 Pigeonites 187
5.5.6 Other transition metal-bearing pyroxenes 190
5.6 Spectra of chain silicates: Amphibole group 190
5.6.1 Background 190
5.6.2 Amphibole crystal structure 190
5.6.3 Mg–Fe amphiboles 194
5.6.4 Calcic amphiboles 195
5.6.5 Miscellaneous amphiboles 197

5.7 Spectra of ring silicates 198
5.7.1 Crystal structures 198
5.7.2 Beryl 198
5.7.3 Cordierite 199
5.7.4 Tourmalines 202
5.7.5 Eudialyte 202
5.8 Spectra of layer silicates 204
5.8.1 Crystal structures 204
5.8.2 Muscovites 204
5.8.3 Biotites 204
5.8.4 Miscellaneous layer silicates 205
5.8.4.1 Chlorites 205
5.8.4.2 Chloritoid 205
5.8.4.3 'Garnierite' 206
5.8.4.4 Gillespite 206
5.9 Spectra of framework silicates 206
5.10 Summary of crystal field spectra 208
5.10.1 Ti^{3+} 208
5.10.2 V^{3+} 209
5.10.3 Cr^{3+} 212
5.10.4 Cr^{2+} 214
5.10.5 Mn^{3+} 217
5.10.6 Mn^{2+} 217
5.10.7 Fe^{3+} 222
5.10.8 Fe^{2+} 223
5.10.9 Co^{3+} 227
5.10.10 Co^{2+} 234
5.10.11 Ni^{2+} 234
5.10.12 Cu^{2+} 235
5.11 Summary 239
5.12 Background reading 239
6 Crystal chemistry of transition metal-bearing minerals 240
6.1 Introduction 240
6.2 Interatomic distances in transition metal compounds 240
6.3 Jahn–Teller distortions in crystal structures 243
6.4 Crystal chemistry of spinels 247
6.5 Site occupancies in silicate structures 250
6.6 Measurements of site populations in crystal structures 251
6.6.1 X-ray diffraction techniques 251
6.6.2 Channel enhanced X-ray emission spectroscopy 252

6.6.3 Mössbauer spectroscopy 252
6.6.4 Electron paramagnetic resonance 253
6.6.5 Hydroxyl stretching frequencies in infrared spectra 253
6.6.6 Crystal field spectra 254
6.6.7 X-ray absorption spectra (XANES and EXAFS) 254
6.7 Site occupancies in silicate minerals 254
6.7.1 Olivines 254
6.7.1.1 Fe^{2+} ions 254
6.7.1.2 Ni^{2+} ions 256
6.7.1.3 Other transition metal ions 257
6.7.2 Orthopyroxenes and pigeonites 257
6.7.2.1 Fe^{2+} ions 257
6.7.2.2 Other transition metal ions 258
6.7.3 Amphiboles 258
6.7.4 Biotites 259
6.7.5 Epidotes 259
6.7.6 Other mineral structures 259
6.7.7 Fluorine avoidance in hydroxysilicates 260
6.8 Explanations of cation ordering 260
6.8.1 Cation size 260
6.8.2 Crystal field stabilization energy 262
6.8.3 Site distortion 263
6.8.3.1 Fe^{2+} ions 265
6.8.3.2 Other transition metal ions 266
6.8.4 Covalent bonding 268
6.9 Summary 269
6.10 Background reading 271

7 Thermodynamic properties influenced by crystal field energies 272
7.1 Introduction 272
7.2 Influence of CFSE on thermodynamic data 272
7.2.1 Graphical correlations 272
7.2.2 Correlation with interatomic distances 276
7.3 Ideal solution behaviour in silicate minerals 277
7.3.1 Criteria for ideal solution behaviour 277
7.3.2 Configurational entropy 278
7.3.3 Enthalpy of mixing 281
7.4 Contributions of electronic entropy 284
7.5 Iron: magnesium ratios in coexisting ferromagnesian silicates 288
7.6 Distributions of divalent transition metal ions between coexisting ferromagnesian silicates 290

7.7 Distributions of trivalent transition metal ions in mineral assemblages 292
7.7.1 Ti^{3+} partitioning 292
7.7.2 V^{3+} partitioning 293
7.7.3 Cr^{3+} partitioning 293
7.7.4 Mn^{3+} partitioning 293
7.7.4.1 Influence of CFSE on Al_2SiO_5 phase relationships 294
7.8 Distributions of transition metals between crystals and melts 295
7.8.1 Thermodynamic behaviour of Ni 297
7.9 Summary 298
7.10 Background reading 299

8 Trace element geochemistry: distribution of transition metals in the Earth's crust 300
8.1 Introduction 300
8.2 Trace elements 300
8.3 Trace element distribution rules 301
8.3.1 Background 301
8.3.2 The Goldschmidt Rules 303
8.3.3 Ringwood's modifications of the Goldschmidt Rules 304
8.3.4 Ionic radii and distribution coefficients 305
8.3.5 Discussion 306
8.4 Evaluation of principles governing trace element distribution 307
8.4.1 Ionic radius criterion 307
8.4.2 Melting point criterion 308
8.4.3 Thermodynamic criteria 309
8.5 Igneous geochemistry of the transition elements 312
8.5.1 Background 312
8.5.2 Coordination of transition metal ions in silicate melts 314
8.5.3 Explanation of magmatic differentiation patterns of transition metals 317
8.5.4 The dilemma over nickel 320
8.5.5 Experimental studies of nickel partitioning 322
8.5.6 Origin of chromite layers in stratiform intrusions 324
8.6 Transition element metallogenesis and plate tectonics 325
8.6.1 Enrichments of Ni and Cr in lherzolites 325
8.6.2 Cu and Mn mineralization in porphyry copper systems 327
8.6.3 Nickel in alpine peridotites 328
8.6.4 Chromium in olivine inclusions in diamond 329
8.7 Sedimentary geochemistry of the transition elements 330
8.7.1 Leaching of ions and break-down of silicates 330

8.7.2 Oxidation of transition metal ions in sedimentary processes 335
8.7.3 Authigenic manganese oxides 340
8.7.3.1 Manganese oxides with tunnel structures 340
8.7.3.2 Manganese oxides with layer structures 345
8.7.4 Crystal chemistry of manganese nodules 346
8.7.5 Oxidation state of cobalt in Mn(IV) oxides 348
8.8 Partitioning of transition metal ions during metamorphic processes 349
8.9 Summary 350
8.10 Background reading 352

9 Mantle geochemistry of the transition elements: optical spectra at elevated temperatures and pressures 353
9.1 Introduction 353
9.2 Chemical composition of the Mantle 354
9.3 Mineralogy of the Mantle 355
9.4 Effects of temperature and pressure on optical spectra 360
9.4.1 Background 360
9.4.2 Crystal field spectra 360
9.4.3 Intervalence charge transfer 361
9.4.4 Oxygen → metal charge transfer 361
9.5 Optical spectral data at high *P* and *T* 362
9.5.1 General trends 362
9.5.2 Periclase 369
9.5.3 Corundum 370
9.5.3.1 Ruby 370
9.5.4 Spinels 371
9.5.5 Garnets 371
9.5.6 Olivines 373
9.5.7 Pyroxenes 374
9.5.8 Silicate perovskite 374
9.6 Polyhedral bulk moduli from high pressure spectra 374
9.7 Spin-pairing transitions 379
9.7.1 Background 379
9.7.2 Features of high-spin and low-spin states 379
9.7.3 Evidence for spin-pairing transitions in iron minerals 380
9.7.4 Calculated spin-pairing transition pressures in the Mantle 381
9.7.5 Consequences of spin-pairing transitions in the Mantle 382
9.7.6 Crystal field spectra of low-spin Fe^{2+} in Mantle minerals 383
9.8 Distributions of transition metal ions in the Mantle 383
9.8.1 Changes of coordination number 383
9.8.2 Changes of CFSE 384

9.8.3 Changes of bond-type 384
9.8.4 Changes of oxidation state 385
9.9 Influence of CFSE on phase equilibria in the Mantle 386
9.9.1 The olivine → spinel phase transition 386
9.9.2 Partitioning of iron in post-spinel phases in the Lower Mantle 388
9.10 Geophysical properies of the Earth's interior 389
9.10.1 Radiative heat transport in the Mantle 389
9.10.2 Electrical conduction in the Mantle 391
9.11 Summary 393
9.12 Background reading 395

10 Remote-sensing compositions of planetary surfaces: applications of reflectance spectra 396
10.1 Introduction 396
10.2 Chemical composition of the terrestrial planets 396
10.3 Origin of reflectance spectra 401
10.4 Measurement of telescopic reflectance spectra 404
10.5 Reflectance spectra of ferrous silicates in Moon rocks 406
10.6 The Moon: problems of mineral mixtures 408
10.7 Mercury and the Moon: problems of high temperatures 412
10.8 Venus and Mars: problems of atmospheres and surface weathering products 415
10.8.1 Spectra of ferric oxides 417
10.8.2 Ferric-bearing assemblages on Mars 421
10.8.3 Ferric-bearing assemblages on Venus 422
10.9 Reflectance spectra of meteorites and asteroids 422
10.10 Future measurements and missions 424
10.11 Summary 425
10.12 Background reading 427

11 Covalent bonding of the transition elements 428
11.1 Introduction 428
11.2 Covalency parameters from optical spectra 430
11.2.1 Racah parameters 430
11.2.2 Evaluation of the Racah *B* parameters 432
11.2.3 The nephelauxetic ratio 433
11.2.4 Pressure variations of the Racah *B* parameter 434
11.3 Qualitative molecular orbital diagrams 435
11.3.1 Formation of σ molecular orbitals 435
11.3.2 Formation of π molecular orbitals 438

11.4 π-bond formation in minerals 440
11.4.1 Chalcophilic properties of Fe, Co, Ni and Cu 440
11.4.2 Interatomic distances in pyrites 440
11.5 Element distributions in sulphide mineral assemblages 442
11.6 Computed molecular orbital energy level diagrams 442
11.6.1 The computation procedure 442
11.6.2 MO diagrams for octahedral [FeO_6] clusters 445
11.6.3 MO diagram for the tetrahedral [FeO_4] cluster 449
11.6.4 MO diagrams for low-symmetry coordination clusters 449
11.7 Molecular orbital assignments of electronic spectra 450
11.7.1 Crystal field spectra 450
11.7.2 Oxygen → metal charge transfer transitions 450
11.7.3 Intervalence charge transfer (IVCT) transitions 451
11.7.3.1 Homonuclear IVCT transitions 451
11.7.3.2 Heteronuclear IVCT transitions 454
11.8 Structural stabilities of Mn(IV) oxides 454
11.9 Summary 457
11.10 Background reading 459
Appendices
1 Abundance data for the transition elements 460
2 Isotopes of the transition elements 462
3 Ionic radii of transition metals and related cations 464
4 Nomenclature for atomic states and spectroscopic terms 466
5 Group theory nomenclature for crystal field states 467
6 Correlations between the Schöenflies and Hermann–Mauguin symbols 468
7 Coordination sites in host mineral structures accommodating transition metal ions 470
8 Chemical and physical constants, units and conversion factors 475
References 478
Subject index 523

Preface to the first edition

This book arose from a series of lectures given at the Universities of Cambridge and Oxford during the Spring of 1966. The lectures were based on material compiled by the author between 1961 and 1965 and submitted in a Ph.D. dissertation to the University of California at Berkeley. At the time crystal field theory had become well established in chemical literature as a successful model for interpreting certain aspects of transition metal chemistry.

For many years the geochemical distribution of transition elements had been difficult to rationalize by conventional principles based on ion size and bond type criteria. In 1959 Dr R. J. P. Williams, a chemist at Oxford conversant with the data from the Skaergaard intrusion, was able to present an explanation based on crystal field theory of the fractionation patterns of transition elements during crystallization of basaltic magma. This development led to the author's studying, under the supervision of Professor W. S. Fyfe, other mineralogical and geochemical data of the transition elements which might be successfully interpreted by crystal field theory.

In addition to suggesting applications of crystal field theory to geology, this book reviews the literature on absorption spectral measurements of silicate minerals and determinations of cation distributions in mineral structures. Many of these data have not been published previously. Spectral measurements on minerals have revealed many advantages of silicates as substrates for fundamental chemical studies. First, crystal structures of most rock-forming silicates are known with moderate to high degrees of precision. Secondly, minerals provide a range of coordination symmetries many of which are not readily available to the synthetic inorganic chemist. Thirdly, solid-solution phenomena enable measurements to be made on a phase with a range of composition. These factors commend silicates to further chemical research and theoretical studies.

The book is directed at geologists. mineralogists and chemists seeking an understanding of the role played by transition metals in silicate minerals and in geologic processes. It is not for the mathematically inclined theoretical chemists and physicists. An attempt has been made to cover most references on crystal field theory in geochemistry, absorption spectral measurements of silicate minerals and site population data in mineral structures up to the end of 1968.

The choice of topics is largely governed by the author's interests. Following a brief introduction the crystal field model is described non-mathematically in chapter 2. This treatment is extended to chapter 3, which outlines the theory of crystal field spectra of transition elements. Chapter 4 describes the information that can be obtained from measurements of absorption spectra of minerals, and chapter 5 describes the electronic spectra of suites of common, rock-forming silicates. The crystal chemistry of transition metal compounds and minerals is reviewed in chapter 6, while chapter 7 discusses thermodynamic properties of minerals using data derived from the spectra in chapter 5. Applications of crystal field theory to the distribution of transition elements in the crust are described in chapter 8, and properties of the mantle are considered in chapter 9. The final chapter is devoted to a brief outline of the molecular orbital theory, which is used to interpret some aspects of the sulphide mineralogy of transition elements.

In writing this book I am grateful to several people for advice and suggestions. Professor W. S. Fyfe, through whose suggestion the research at Berkeley originated, has always shown considerable interest, encouragement and offered valuable criticism. At Berkeley, helpful discussions were held with Professor A. Pabst, Dr B. W. Evans, Dr R. G. J. Strens, Dr J. Starkey and Dr J. Wainwright, while Dr G. M. Bancroft, Dr M. G. Clark, Dr A. G. Maddock, Dr E. J. W. Whittaker and Professor J. Zussman have offered valuable suggestions in England. The mistakes that remain are due to the author alone. Several scientists and publishers kindly provided illustrative material. Special thanks are due to the General Electric Company for the colour negative that was used as the basis of plate 2; acknowledgements of the other figures are made in the text. My wife, Virginia, compiled the bibliography. To all these people I owe a debt of thanks.

The greater part of the preparation of this book was undertaken during a period of post-doctoral research at Cambridge University. I am indebted to Dr S. O. Agrell for the invitation to come to Cambridge, and to Professor W. A. Deer for his hospitality and for the facilities placed at my disposal in the Department of Mineralogy and Petrology Thanks are also due to Professor Lord Todd for permission to use facilities in the University Chemical Laboratories. The stay at Cambridge was financed by a bursary from the British Council and by a grant from the Natural Environment Research

Council obtained for me by Dr J. P. L. Long. The research at Berkeley was financed by scholarships from the Royal Commission for the Exhibition of 1851, London; the University of New Zealand; the English Speaking Union (California Branch); and the University of California. Grants from the Petroleum Research Fund of the American Chemical Society and the National Science Foundation are also acknowledged.

Finally, but above all, I am indebted to my wife for the help, encouragement, interest and patience she has shown during the writing of this book. This book is dedicated to her with love as a token of the inspiration I owe to her.

Oxford *Roger G. Burns*
April, 1969

Preface to the second edition

When the first edition of *Mineralogical Applications of Crystal Field Theory* was written during 1968–9, it broke new ground by describing results and suggesting applications of the limited spectroscopic and crystal chemical data then available for transition metal-bearing minerals. The data were derived mainly from visible to near-infrared spectral measurements, together with newly available Mössbauer-effect studies of iron minerals, made principally at ambient temperatures and pressures. The book stimulated considerable interest among subsequent mineral spectroscopists who have developed new and improved methods to study minerals and synthetic analogues under a variety of experimental conditions, including *in situ* measurements made at elevated temperatures and pressures. As a result, the quantity of spectral and crystal chemical data has increased appreciably and may now be applied to a diversity of current new problems involving transition elements in the earth and planetary sciences.

The second edition now attempts to review the vast data-base of visible to near-infrared spectroscopic measurements of minerals containing cations of the first-series transition elements that has appeared during the past 20 years. Several newer applications of the spectral and crystal chemical data are described, including interpretations of remote-sensed reflectance spectra used to identify transition metal-bearing minerals on surfaces of planets. This topic alone warrants the inclusion of a new chapter in the second edition. Many of the classical applications of crystal field theory outlined in the first edition are retained, and each of the original 10 chapters is expanded to accommodate fresh interpretations and new applications of crystal field theory to transition metal geochemistry. An attempt has been made to cover most references to crystal chemistry and visible to near-infrared spectroscopy of transition metal-bearing minerals up to late 1992. Several new illustrations and tables appear throughout the book, and appendices have been added that summarize the abundances, ionic radii, and mineral structures hosting transition metal cations.

The opportunity to write the expanded second edition was created by the award of a Fellowship from the John Simon Guggenheim Memorial Foundation. This Guggenheim Award enabled me to spend several months working on the book during sabbatical leave spent in the Department of Geology and the University of Manchester, England. Here, I deeply appreciate the invitation and facilities provided by Professor David J. Vaughan with whom many inspiring discussions were held.

The book is dedicated to the team of graduate student and colleague investigators with whom I have collaborated in spectral mineralogy research for the past 25 years. Initially, the research was carried out at Cambridge University (1966), Victoria University of Wellington, New Zealand (1967) and Oxford University (1968–70). During the past two decades at the Massachusetts Institute of Technology, however, several new lines of investigation have been developed, results of which are included in this edition. Collaborators included: Rateb Abu-Eid, Kris Anderson, Mike Bancroft, Karen Bartels, Jim Besancon, Janice Bishop, Virginia Burns, Mike Charrette, Mike Clark, Roger Clark, Darby Dyar, Duncan Fisher, Susan Flamm, Mike Gaffey, Colin Greaves, Tavenner Hall, Bob Hazen, Frank Huggins, Bob Huguenin, Tony Law, Irene Leung, Bruce Loeffler, Catherine McCammon, Lucy McFadden, Sandra Martinez, Anne Morawsky, Dan Nolet, Margery Osborne, Kay Parkin, Carle Pieters, Jean Prentice, Virginal Ryan, Martha Schaefer, Ken Schwartz, Dave Sherman, Malcolm Siegel, Bob Singer, Teresa Solberg, Harlan Stockman, D'Arcy Straub, Chien-Min Sung, Matthew Tew, Jack Tossell, David Vaughan, Earle Whipple, Marylee Witner, Sandy Yulke and Valerie Wood. Over the years, stimulating discussions have taken place with John Adams, Mike Bancroft, Jim Bell, Peter Bell, Gordon Faye, Keith Johnson, David Mao, Tom McCord, Dick Morris, George Rossman, Subrata Ghose, Bill White and, in particular, Carle Pieters, Bob Singer, David Sherman, Darby Dyar and David Vaughan. I respect the contributions that all of these people have made to spectral mineralogy of the transition elements. Thanks are due to Carle Pieters, George Rossman and Bob Singer who provided many of the spectra used in the book, and to Fred Frey for helpful discussions about trace element geochemistry. I appreciate the assistance that Jim Besancon and Duncan Fisher provided to prepare some of the crystal structure diagrams. I deeply appreciate all the assistance that Catherine Flack, Conrad Guettler and, in particular, Brian Watts at Cambridge University Press have provided. The research has been supported throughout by grants principally from the National Aeronautics and Space Administration (grant numbers: NGR 22-009-551 and 187; NSG 7604; and NAGW-1078, 2037, 2049 and 2220).

Finally, I deeply appreciate the encouragement and support to complete this undertaking provided by my family Virginia, Kirk, Jonathan and Rachel.

Broomcroft Hall, Department of Geology,
University of Manchester
1992

Roger G. Burns

1

Introduction

The simplicity and convenience of crystal field theory have earned it a place in the 'toolbox' of the (geo)chemist.

F. A. Cotton, G. Wilkinson & P. L. Gauss,
Basic Inorganic Chemistry, 2nd edn, p. 384 (1988)

Crystal field theory is one of several chemical bonding models and one that is applicable solely to the transition metal and lanthanide elements. The theory, which utilizes thermodynamic data obtained from absorption bands in the visible and near-infrared regions of the electromagnetic spectrum, has met with widespread applications and successful interpretations of diverse physical and chemical properties of elements of the first transition series. These elements comprise scandium, titanium, vanadium, chromium, manganese, iron, cobalt, nickel and copper. The position of the first transition series in the periodic table is shown in fig. 1.1. Transition elements constitute almost forty weight per cent, or eighteen atom per cent, of the Earth (Appendix 1) and occur in most minerals in the Crust, Mantle and Core. As a result, there are many aspects of transition metal geochemistry that are amenable to interpretation by crystal field theory.

Cosmic abundance data for the transition elements (Appendix 1) show that each metal of the first transition series is several orders of magnitude more abundant than all of the metals of the second (Y–Ag) and third (Hf–Au) transition series, as well as the rare earth or lanthanide series (La–Lu). Iron, Fe, is by far the predominant transition element, followed by Ni, Cr and Mn. Crustal abundance data also show high concentrations of Fe relative to the other first-series transition elements on the Earth's surface. However, the crustal abundance of Fe is decreased relative to its cosmic abundance as a result of chemical fractionation during major Earth-forming processes involving segregation of the Core, evolution of the Mantle, and chemical weathering of igneous minerals exposed to the atmosphere. Other first-series transition elements have also been

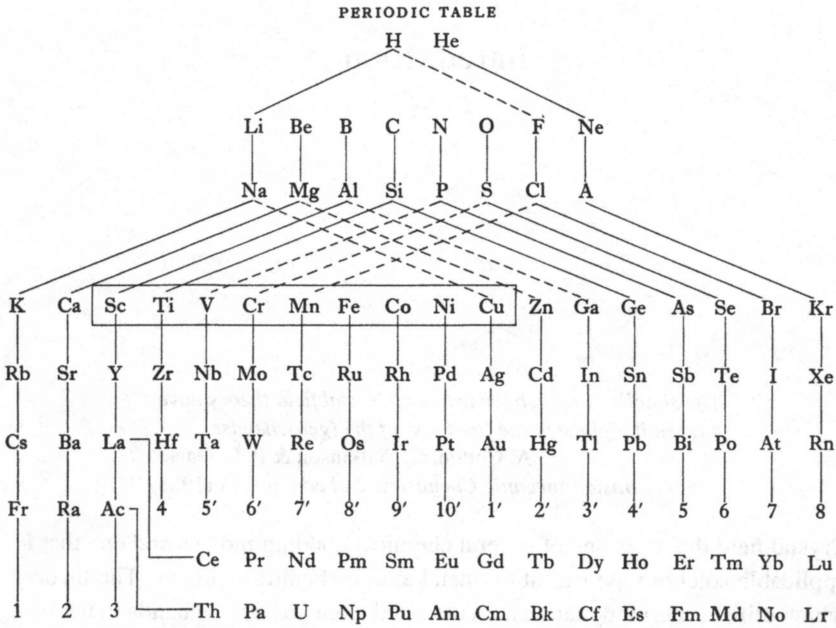

Figure 1.1 The periodic table. Elements of the first transition series are shown enclosed in the block.

subjected to enrichment and depletion processes near the Earth's surface, particularly during ore formation, hydrothermal activity and deep-weathering reactions. Notable examples are the concentrations of Ni in ultramafic rocks, Cu in porphyry calc-alkaline rocks, and Mn in sub-aqueous fissures where deeply eroded igneous rocks, ore deposits, hydrothermal veins and marine sediments outcrop at the Earth's surface. Such enrichment processes are interpretable by crystal field theory. This book is confined to a discussion of the mineralogy and geochemistry of the first-series transition elements.

The chemical properties of the transition metals as a group are more complex than those of other elements in the periodic table due to electrons being located in incompletely filled 3*d* atomic orbitals having different energy levels. Interaction with light causes electrons to be excited between the split 3*d* orbital energy levels, leading to absorption bands in the visible region and causing colour in chemical compounds and minerals containing cations of the first-series transition elements. The origin of the splitting of these 3*d* orbital energy levels may be described by three different models: crystal field theory, ligand field theory and molecular orbital theory. The key feature of all three models is that each transition metal has five 3*d* orbitals having identical energies in a spherical environment such as that of a (hypothetical) gaseous free ion. However, this five-fold degeneracy is removed in a chemical compound or a mineral where the cation is surrounded in a crystal structure by anions (e.g. O^{2-}, OH^-, CO_3^{2-}, SO_4^{2-}, non-bridging oxygens in silicates) or dipolar molecular species (such as H_2O, NH_3, bridging oxygens in silicates), which are referred to collectively as ligands. The relative energies of the 3*d* orbitals depend on the symmetry of the ligand environment defining the geometry of the coordination site.

The crystal field theory treats the transition metal ion in a crystalline environment as though the cation was subjected to purely electrostatic interactions with surrounding anions, which are regarded as point negative charges. The ligand field theory introduces some empirical parameters (such as the Racah *B* and *C* parameters) to account for covalent bonding interactions of the cation with coordinated ligands. The molecular orbital theory focusses on orbital overlap and exchange interactions between the central transition element and its surrounding ligands. All three models make use of the symmetry properties of both the metal orbitals and the configuration of the neighbouring anions or ligands. Consequently, each theory embodies concepts of group theory. Of the three bonding models, crystal field theory is conceptually the simplest and the one most amenable to thermodynamic data derived from optical or visible-region spectral measurements. Such spectrally-derived thermodynamic data or crystal field stabilization energies (CFSE) are useful for *comparing* relative stabilities and geochemical behaviours of cations of Ti, V, Cr, Mn, Fe, Co, Ni, Cu and non-transition metal ions when they occur in similar oxidation states and coordination environments in minerals.

A detailed mineralogical and geochemical study of the transition elements is of interest for several reasons. First, the behaviour of individual transition metal ions in geological processes often differs markedly from those of non-transition metal ions with similar valences and sizes. Second, transition metals are amenable to a variety of experimental techniques which may not be generally applicable to other elements. These include measurements of magnetic susceptibility, optical or electronic absorption spectra, electron spin resonance and, in the case of iron compounds, Mössbauer spectroscopy involving the ^{57}Fe isotope.* Information on the crystal chemistry and speciation of transition metal-bearing phases may also be obtained from other spectroscopic techniques that are also applicable to other elements, including extended X-ray absorption fine structure spectroscopy (EXAFS), X-ray absorption near edge spectroscopy (XANES) and X-ray photoelectron spectroscopy (XPS). Third,

* The Mössbauer effect has also been observed in ^{61}Ni. However, this isotope is very difficult to study and few measurements of geochemical significance have been made on nickel compounds.

interpretations of the behaviour of transition metals as simple and complex ions in chemical systems enable the mineralogical properties and geochemical behaviour of these and other cations in complex geologic media to be better understood. Fourth, mineral structures provide a diversity of coordination sites, making it possible to measure spectra and to determine electronic energy levels of cations in coordination polyhedra having a variety of symmetries, distortions, bond-types, and numbers of nearest-neighbour ligands, as well as different next-nearest-neighbour metal-metal interactions. Such a diversity of coordination environments is not available in more conventional synthetic inorganic salts and organometallic compounds. Finally, the generally high thermal stabilities of most silicate and oxide minerals make it possible to measure optical spectra of these transition metal-bearing phases at elevated temperatures and very high pressures. These features render mineral spectra interesting both to solid state physicists and chemists desiring fundamental information about transition metal ions often in novel bonding situations and to earth and planetary scientists wanting to interpret geochemical and geophysical properties of minerals in the Earth and on surfaces of planets.

1.1 History of crystal field theory

Crystal field theory was developed by H. Bethe in 1929 when he applied group theory and quantum mechanics to electrostatic theory. The theory was utilized by physicists during the 1930s and 1940s and was used mainly to explain magnetic properties and absorption spectra of transition metal and lanthanide compounds. The connection between spectroscopy and thermodynamics was demonstrated by L. E. Orgel in 1952 when he showed that spectroscopically determined energies, termed the crystal field stabilization energy (CFSE), contribute to thermodynamic properties of transition metal compounds. Orgel (1952) showed how crystal field theory could account for trends in the heats of hydration of first-series transition metal ions (see chapter 7, §7.1). Since 1952, crystal field theory has been applied to many aspects of transition metal chemistry. It has been used to interpret and predict crystal chemistry, kinetics and reaction mechanisms, magnetic and spectral properties, and thermodynamic data of transition metal compounds.

Crystal field theory was first applied to a geochemical problem in 1959 when R. J. P. Williams interpreted the relative enrichments of transition metal ions during fractional crystallization of magma in the Skaergaard intrusion (see §8.5). Williams (1959) was able to explain why Cr and Ni were removed from the magma during early crystallization of chromite and olivine, whereas elements like Fe and Mn lagged behind in the melt. Numerous geochemical applications of crystal field theory followed during the 1960s (Burns and Fyfe, 1967a), examples of which formed the basis of the first edition of this book. Subsequently, crystal field theory has been applied to several geochemical, geophysical and mineralogical problems involving surfaces and interiors of the Earth and other terrestrial planets (Burns, 1982, 1985a,b; 1989a).† Many of these topics are discussed in this second edition.

1.2 Scope of crystal field theory

It may, perhaps, be appropriate to justify the name crystal field theory in the book title, rather than ligand field theory or molecular orbital theory which appear to be more rigorous or comprehensive descriptions of chemical bonding. The name 'crystal field theory' refers to the original theory of Bethe (1929) which does not consider the role played by the ligands further than to credit them with producing a steady crystalline field. In the mineral kingdom, it is generally assumed that silicates and oxides of magnesium and aluminium are ionic structures, and that most oxygen atoms carry anionic charges. Much of transition metal geochemistry is concerned with the substitution of host Mg^{2+} and Al^{3+} ions in oxides and silicates by minor amounts of transition metal ions. Thus, the crystal field theory based on electrostatic forces and concepts of symmetry is appropriate to transition metals in most silicate minerals. The use of crystal field theory in the title of this book, therefore, is deliberate.

The major focus of the book is on mineral crystal structures that provide an ordered array of anions forming coordination polyhedra around the central cations. The thermodynamic data underlying many of the geochemical applications described in the first ten chapters are derived from energies of absorption bands in the optical spectra of minerals, which are most simply explained by crystal field theory. Use of experimentally determined energy level data rather than energy separations computed in molecular orbital diagrams is the emphasis of these early chapters.

However, in sulphides and related minerals, the effects of covalent bonding predominate and orbital overlap must be taken into account. Thus, concepts of molecular orbital theory are described in chapter 11 and applied to aspects of the sulfide mineralogy of transition elements. Examples of computed energy diagrams for molecular clusters are also presented in chapter 11. There, it is noted that the fundamental $3d$ orbital energy splitting parameter of crystal field theory, Δ, receives a similar interpretation in the molecular orbital theory.

† A complete list of literature cited will be found in the **Reference** section at the end of the book.

1.3 Background reading

Bethe, H. (1929) Splitting of terms in crystals. (Termsaufspaltung in Kristallen.) *Ann. Phys.*, **3**, 133–206. [*Transl.*: Consultants Bureau, New York.]

Burns, R. G., Clark, R. H. & Fyfe, W. S. (1964) Crystal field theory and applications to problems in geochemistry. In *Chemistry of the Earth's Crust, Proceedings of the Vernadsky Centennial Symposium.* (A. P. Vinogradov, ed.; Science Press, Moscow), **2**, 88–106. [*Transl.*: Israel Progr. Sci. Transl., Jerusalem , pp. 93–112 (1967).]

Burns, R. G. & Fyfe, W. S. (1967a) Crystal field theory and the geochemistry of transition elements. In *Researches in Geochemistry.* (P. H. Abelson, ed.; J. Wiley & Son, New York), **2**, 259–85.

Burns, R. G. (1985) Thermodynamic data from crystal field spectra. In *Macroscopic to Microscopic. Atomic Environments to Mineral Thermodynamics.* (S. W. Kieffer & A. Navrotsky, eds; Mineral. Soc. Amer. Publ.), *Rev. Mineral.*, **14**, 277–316.

Orgel, L. E. (1952) The effects of crystal fields on the properties of transition metal ions. *J. Chem. Soc.*, pp. 4756–61.

Williams, R. J. P. (1959) Deposition of trace elements in a basic magma. *Nature*, **184**, 44.

2

Outline of crystal field theory

Crystal field theory gives a survey of the effects of electric fields of definite symmetries on an atom in a crystal structure.
– – A direct physical confirmation should be obtainable by analysis of the spectra of crystals.

H. A. Bethe, *Annalen der Physik*, 3, 206 (1929)

2.1 Introduction

Crystal field theory describes the origins and consequences of interactions of the surroundings on the orbital energy levels of a transition metal ion. These interactions are electrostatic fields originating from the negatively charged anions or dipolar groups, which are collectively termed ligands and are treated as point negative charges situated on a lattice about the transition metal ion. This is a gross simplification, of course, because sizes of anions or ligands such as O^{2-}, OH^-, H_2O, SO_4^{2-}, etc., are much larger than corresponding ionic radii of cations (Appendix 3). Two effects of the crystalline field are the symmetry and the intensity of the electrostatic field produced by the ligands. The changes induced on the central transition metal ion depend on the type, positions and symmetry of the surrounding ligands.

2.2 Orbitals

The position and energy of each electron surrounding the nucleus of an atom are described by a wave function, which represents a solution to the Schrödinger wave equation. These wave functions express the spatial distribution of electron density about the nucleus, and are thus related to the probability of finding the electron at a particular point at an instant of time. The wave function for each electron, $\Psi(r,\theta,\phi)$, may be written as the product of four separate functions, three of which depend on the polar coordinates of the electron

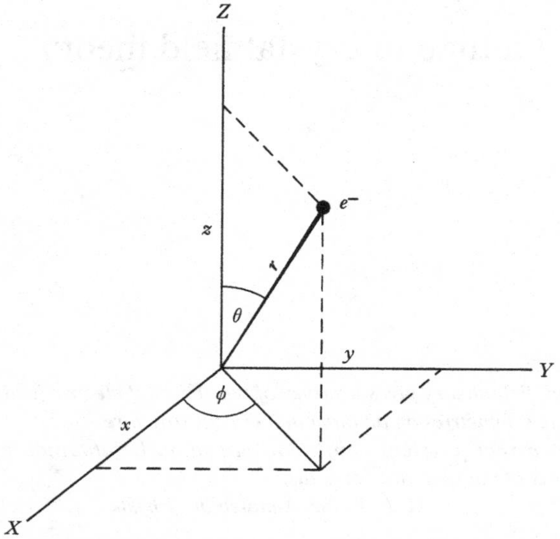

Figure 2.1 Polar coordinates of an electron in space.

illustrated in fig. 2.1. These three functions include the radial function, $R(r)$, which depends only on the radial distance, r, of the electron from the nucleus, and two angular functions $\Theta(\theta)$ and $\Phi(\phi)$ which depend only on the angles θ and ϕ. The fourth function, the spin function, Ψ_s, is independent of the spatial coordinates r, θ and ϕ. Thus, the overall wave function of an electron may be written as the product

$$\Psi(r,\theta,\phi) = R(r)\ \Theta(\theta)\ \Phi(\phi)\ \Psi_s\ . \tag{2.1}$$

Surfaces may be drawn to enclose the amplitude of the angular wave function. These boundary surfaces are the atomic orbitals, and lobes of each orbital have either positive or negative signs resulting as mathematical solutions to the Schrödinger wave equation.

The overall wave function and each of its components are expressed in terms of certain parameters called quantum numbers, four of which are designated n, l, m_l and m_s.

2.2.1 Principal quantum number n

Solutions to the $R(r)$ portion of the Schrödinger equation are expressed as

$$E_n = -\frac{1}{n^2}\frac{2\pi^2\mu^2Z^2e^4}{h^2}\ , \tag{2.2}$$

where E_n is the energy of an electron with charge e; Z is the charge on the nucleus; μ is the reduced mass of the electron (m) and the nucleus (M), with $\mu = Mm/(M + m)$; h is Planck's constant; and n is the principal quantum number relating to the average distance of the electron from the nucleus.

The principal quantum number, n, can have all integral values from 1 to infinity. It is a measure of the energy of an electron. Thus, $n = 1$ is the lowest energy orbital level. As n increases, the energies of the orbitals increase, becoming less negative. In the limit of $n = \infty$, the energy of the electron becomes zero. The electron is no longer bound to the nucleus, so that the atom becomes ionized. There is then a continuum of states with zero bonding energy and an arbitrary kinetic energy. The ionization energy corresponds to the energy difference between the $n = 1$ and $n = \infty$ levels.

2.2.2 *Azimuthal quantum number l*

The azimuthal quantum number or orbital angular momentum quantum number, l, is related to the shape of an orbital and occurs in solutions to the $\Theta(\theta)$ function of the wave equation. It may be regarded as representing the angular momentum of an electron rotating in an orbit, the magnitude of which corresponds to $[l(l + 1)]^{1/2}$ units of $h/2\pi$. Again, l can have only integral values, but its maximum value is limited by the value of n associated with the orbital. Thus, l has the values 0, 1, 2, ..., (n–1). In the first shell, for example, $n = 1$ and there exists only one wave function with $l = 0$. In the second shell $n = 2$, and there are wave functions with $l = 0$ and $l = 1$. Similarly, in the third shell $n = 3$ so that l can take the values 0, 1, and 2; and so on.

Letter symbols are given to the orbitals according to the value of l as follows:

$$\text{for } l = 0, 1, 2, 3, 4, \ldots$$
$$\text{the letter symbol is: } s, p, d, f, g, \ldots \tag{2.3}$$

Thus, s orbitals are states with $l = 0$ and zero orbital angular momentum. The p orbitals with $l = 1$ have orbital angular momenta of $\sqrt{2}$ units of $h/2\pi$, while the d orbitals with $l = 2$ have $\sqrt{6}$ $h/2\pi$ units of orbital angular momenta.

2.2.3 *Magnetic quantum number m_l*

The quantum number, m_l, originating from the $\Theta(\theta)$ and $\Phi(\phi)$ functions of the Schrödinger wave equation, indicates how the orbital angular momentum is oriented relative to some fixed direction, particularly in a magnetic field. Thus, m_l roughly characterizes the directions of maximum extension of the electron cloud in space. The orbiting, negatively charged electron generates a magnetic field. In the presence of an applied magnetic field along one axis, designated by convention as the *z* axis, interactions between the two magnetic fields cause the orbital angular momentum vector to be constrained in specific directions. These directions are quantized such that they can have only integral values along the *z* axis. The quantum number m_l occurs in solutions to both the $\Theta(\theta)$ and $\Phi(\phi)$functions of the Schrödinger equation and can take all integral values from $+l$ to $-l$, or $(2l + 1)$ values in all. For example, for a given principal quantum number *n* and with $l = 0$, there is only one possible orbital, namely one with $m_l = 0$. Thus, there is only one *s* orbital associated with each principal quantum shell. These are designated as 1*s*, 2*s*, 3*s*, 4*s*, etc., orbitals. For a given *n* and with $l = 1$, there are three possible values for m_l: −1, 0 and +1. Hence, there are three different kinds of *p* orbitals for each principal quantum number (except $n = 1$) and they constitute the 2*p*, 3*p*, 4*p*, 5*p*, etc., orbitals. Similarly, there are five different *d* orbitals, since $l = 2$ and m_l can take the values 2, 1, 0, −1, and −2. Thus, the 3*d*, 4*d*, 5*d*, etc., shells each contain five *d* orbitals. The *f* orbitals, designated as 4*f*, 5*f*, etc., occur in sets of 7; *g* orbitals, starting with 5*g*, occur in sets of 9; and so on. The energies of orbitals with the same *n* and *l* values but different values of m_l are the same, except in the presence of a strong electric or magnetic field. Thus, all three *p* orbitals in a given shell have the same energies, as do the five d orbitals. The *p* orbitals are said to be three-fold degenerate and the *d* orbitals are five-fold degenerate.

2.2.4 Spin quantum number m_s

An electron may be visualized as spinning about some axis. Thus, it possesses spin angular momentum. However, since the electron is negatively charged there will be produced a magnetic field associated with its spin. Depending on the sense of rotation, clockwise or anticlockwise, the magnetic field may be in one or the opposite direction. An electron then can have two kinds of spin, characterized by the quantum numbers $m_s = +\frac{1}{2}$ and $m_s = -\frac{1}{2}$. The magnitude of the spin angular momentum is $[s(s + 1)]^{1/2}$ units of $h/2\pi$, where $s = \frac{1}{2}$. Accordingly, for each space orbital characterized by the quantum numbers, *n, l* and m_l, there are two possible arrangements of electron spin generally of the same energy in the absence of a magnetic field. Thus, each orbital can accommodate two electrons which spin in opposite directions. In energy level diagrams these are often referred to as spin-up (or α-spin) and spin-down (or β-spin) configurations.

2.2.5 Spin–orbit coupling

The interaction of the two magnetic fields, one produced by an electron spinning around its axis and the other by its rotation in an orbital around the nucleus, is termed spin-orbit coupling. States with opposed magnetic fields associated with spin and rotation are slightly more stable than those with aligned magnetic fields. Effects of spin–orbit coupling become increasingly important with rising atomic number. They are thus more noticable in compounds of Co and Ni than they are in V- and Cr-bearing phases; spin–orbit interactions are particularly significant in elements of the second and third transition series. Spin–orbit coupling influences electron paramagnetic resonance spectra, and also contribute features to visible-region spectra of transition metal compounds, particularly of Co^{2+} and Ni^{2+}-bearing phases.

2.3 Shape and symmetry of the orbitals

All *s* orbitals are spherically symmetrical because their angular wave functions are independent of θ and ϕ. They are given the group theory symmetry notation a_{1g}; here, symbol *a* indicates single degeneracy, that is, there is one *s* orbital per quantum number; subscript 1 means that the sign of the wave function does not change on rotation about the centre of the atom; and subscript *g* refers to the fact that the wave function does not change sign on inversion through the centre of the atom (German: *gerade* = even). The symbol a_{1g} implies spherical symmetry and *s* orbitals are said to have symmetric wave functions. The spatial properties of the *s* orbitals are shown in fig. 2.2. The balloon picture of the orbital is constructed so that the skin of the balloon includes within it most, perhaps 95 per cent, of the electron density. The sign of the wave function of *s* orbitals is independent of angle in any given spherical shell. Note that for *s* orbitals alone, there is a finite probability that the electron may be at the nucleus. This feature is important in Mössbauer spectroscopy because the *s* electron density at the nucleus influences the value of the chemical isomer shift parameter.

The three *p* orbitals in each shell are mutually perpendicular and depend on the angles θ and ϕ. Individually, they are not spherically symmetrical. The three *p* orbitals in each set are designated by p_x, p_y and p_z, indicating that lobes project along each of the cartesian axes. The signs of the wave functions represented by the two lobes of each orbital are different. Figure 2.2 shows the angular wave functions of the *p* orbitals. The *p* orbitals are shaped like dumbbells and belong to symmetry type t_{1u}: here, *t* represents the three orbitals per principal quantum number, or three-fold degeneracy; subscript 1 again indicates that the sign of the wave function does not change on rotation about the

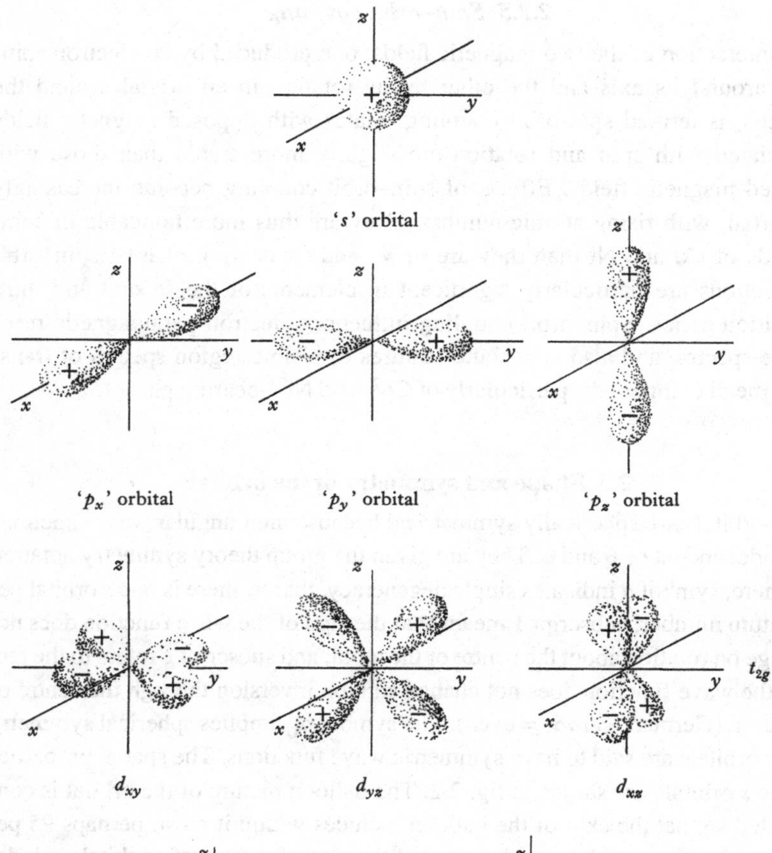

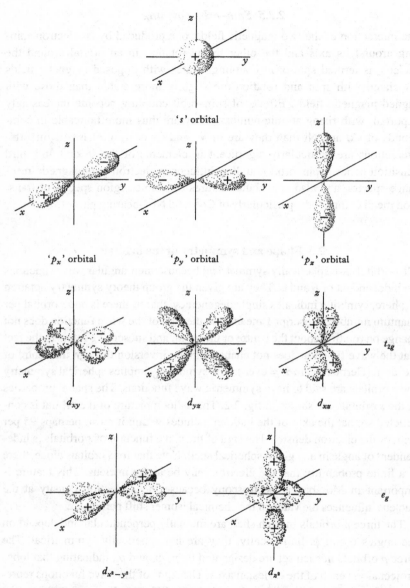

Figure 2.2 Boundary surface of atomic orbitals. The boundaries represent angular distribution probabilities for electrons in each orbital. The sign of each wave function is shown. The d orbitals have been classified into two groups, t_{2g} and e_g, on the basis of spatial configuration with respect to the cartesian axes. (Reproduced and modified from: W. S. Fyfe, *Geochemistry of Solids*, McGraw–Hill, New York, 1964, figure 2.5, p. 19).

cartesian axes; and subscript u shows that the sign of the wave function changes on inversion through the centre of the atom (German: *ungerade* = uneven). Thus, p orbitals are said to be antisymmetric.

The five d orbitals which occur in each shell with principal quantum number 3 or higher are designated by d_{xy}, d_{yz}, d_{xz}, $d_{x^2-y^2}$ and d_{z^2}, and each orbital has four lobes in opposite quadrants. This is illustrated in fig. 2.2. The five d orbitals may be divided into two groups on the basis of their angular distributions. Three of the orbitals, d_{xy}, d_{yz} and d_{xz}, have lobes projecting between the cartesian axes. This group is designated t_{2g}: here, t refers to the three-fold degeneracy; subscript 2 indicates that the sign of the wave function does not change on rotation about the axes diagonal to the cartesian axes; and subscript g again refers to the fact that the wave function does not change sign on inversion (*gerade*). The other two orbitals, $d_{x^2-y^2}$ and d_{z^2}, have lobes directed along the cartesian axes. They are designated e_g, where e stands for the two-fold degeneracy. The d_{z^2} orbital appears to have a different shape from the other four. This difference is only apparent, however. The d_{z^2} orbital represents a linear combination of two orbitals, $d_{z^2-x^2}$ and $d_{z^2-y^2}$, having the same shape as the other d orbitals but which are not independent of them. Note that each of the five d orbitals has a symmetric (g) wave function, like an s orbital. Thus, d and s orbitals are said to have the same parity. However, d orbitals (g) and p orbitals (u) have opposite parities. Also, the plus and minus signs associated with the orbitals illustrated in fig. 2.2 refer to the mathematical signs of the wave functions. The electron density is always positive, however, and is obtained by squaring a wave function. Sometimes, an alternative designation is used for the two groups of d orbitals, namely d_ε and d_γ for the e_g and t_{2g} orbitals, respectively.

In passing, it should be noted that there are seven f orbitals in the $4f$, $5f$, etc., shells. They consist of one orbital, designated f_{xyz} and having a_{2u} symmetry, with eight lobes pointing between the x, y and z axes; three more orbitals, denoted as $f_{z(x^2-y^2)}$, $f_{x(y^2-z^2)}$ and $f_{y(z^2-x^2)}$ with t_{2u} symmetry, in which eight lobes are directed along two of the axes; and three other orbitals labelled as f_{x^3}, f_{y^3} and f_{z^3} with t_{1u} symmetry, that are somewhat analogous to the d_{z^2} orbital by having lobes directed along each cartesian axis and two more regions above and below the plane of the other two axes.

An orbital is capable of accommodating two electrons which, according to the Pauli exclusion principle, must be spinning in opposite directions. Thus, each s orbital may contain two electrons, the three p orbitals a total of six electrons, the five d orbitals up to ten electrons, and the f orbitals a maximum of 14 electrons. When there is an insufficient number of electrons in an atom to completely fill a set of orbitals, the electrons may spread out and occupy singly as many of the orbitals as possible with spins aligned parallel; that is, the electrons spin in the same direction to maximize the exchange energy, in accord with Hund's rules of electronic configurations. Such a distribution minimizes inter-electronic coulombic repulsion and leads to lowest energy states, or ground state configurations, which have the maximum number of unpaired electrons.

2.4 Transition elements

The definition of a transition element, *senso stricto*, is that it is a metal having a partly filled *d* or *f* shell. A broader definition includes also those elements that have partially filled *d* or *f* shells in any one of their commonly occurring oxidation states. Elements of the first transition series have electronic configurations of the general form

$$(1s)^2(2s)^2(2p)^6(3s)^2(3p)^6(3d)^{10-n}(4s)^{1 \text{ or } 2}, \tag{2.4}$$

where $n = 0, 1, 2, ..., 10$. Since the closed shell configuration $(1s)^2(2s)^2(2p)^6(3s)^2(3p)^6$ corresponds to the inert gas argon, Ar, it is termed the argon core and is abbreviated [Ar]. In passing from one element to the next along the first transition series, the 3*d* orbitals are filled progressively. Cations in different oxidation states of the transition elements are formed by the removal of the 4*s* and some, or all, of the 3*d* electrons. Table 2.1 summarizes the electronic configurations of elements of the first transition series in their naturally occurring oxidation states in geochemical environments. Included in table 2.1 are titanium(IV), vanadium(V) and chromium(VI) retaining zero 3*d* electrons, and copper(I) with completely filled 3*d* orbitals. Elements belonging to the second transition series (Y to Ag) and third transition series (from Hf to Au) have partly filled 4*d* and 5*d* orbitals, respectively, whereas the lanthanide elements spanning La to Lu have incompletely filled 4*f* or 5*d* orbitals. Note that on Earth, the most stable oxidation states in minerals occurring in near-surface environments include Ti(IV), V(III), V(V), Cr(III), Cr(VI), Mn(II), Mn(III), Mn(IV), Fe(II), Fe(III), Co(II), Co(III), Ni(II), Cu(II) and Cu(I). On the Moon, however, the oxidation states Fe(II), Mn(II), Ti(III), Ti(IV), Cr(II) and Cr(III) occur in lunar mineral and glass phases. Ferric iron may predominate over Fe(II) in the martian regolith, particularly in bright regions of Mars.

In an isolated transition metal ion, electrons have an equal probability of being located in any one of the five *d* orbitals, since these orbitals have identical energy levels. When a transition metal ion is in a crystal structure, however, the effect of a non-spherical electrostatic field on the five degenerate *d* orbitals is to lower the degeneracy by splitting the *d* orbitals into different energies about the centre of gravity, or baricentre, of the unsplit energy levels. The manner and extent to which the five-fold degeneracy is removed depends on the type, positions and symmetry of ligands surrounding the transition metal.

Table 2.1. *Electronic configurations of the first-series transition elements occurring in minerals*

	Atomic number	Element	Electronic configurations						
			Atom	M(I)	M(II)	M(III)	M(IV)	M(V)	M(VI)
	19	K	[Ar]$4s^1$	[Ar]					
	20	Ca	[Ar]$4s^2$		[Ar]				
first-series transition elements	21	Sc	[Ar]$3d^14s^2$			[Ar]			
	22	Ti	[Ar]$3d^24s^2$		[Ar]$3d^2$	[Ar]$3d^1$	[Ar]		
	23	V	[Ar]$3d^34s^2$		[Ar]$3d^3$	[Ar]$3d^2$	[Ar]$3d^1$	[Ar]	
	24	Cr	[Ar]$3d^54s^1$		[Ar]$3d^4$	[Ar]$3d^3$	[Ar]$3d^2$	[Ar]$3d^1$	[Ar]
	25	Mn	[Ar]$3d^54s^2$		[Ar]$3d^5$	[Ar]$3d^4$	[Ar]$3d^3$		
	26	Fe	[Ar]$3d^64s^2$		[Ar]$3d^6$	[Ar]$3d^5$			
	27	Co	[Ar]$3d^74s^2$		[Ar]$3d^7$	[Ar]$3d^6$			
	28	Ni	[Ar]$3d^84s^2$		[Ar]$3d^8$	[Ar]$3d^7$			
	29	Cu	[Ar]$3d^{10}4s^1$	[Ar]$3d^{10}$	[Ar]$3d^9$				
	30	Zn	[Ar]$3d^{10}4s^2$		[Ar]$3d^{10}$				
	31	Ga	[Ar]$3d^{10}4s^24p^1$			[Ar]$3d^{10}$			
	32	Ge	[Ar]$3d^{10}4s^24p^2$				[Ar]$3d^{10}$		

[Ar] = Argon core, $1s^22s^22p^63s^23p^6$

2.5 Crystal field splitting in octahedral coordination

When a transition metal ion is in octahedral coordination with six identical ligands situated along the cartesian axes, electrons in all five $3d$ orbitals are repelled by the negatively charged anions or dipolar ligands, and the baricentre of the degenerate levels is raised. Since lobes of the e_g orbitals point towards the ligands, electrons in these two orbitals are repelled to a greater extent than are those in the three t_{2g} orbitals that project between the ligands. This difference between e_g and t_{2g} orbitals is illustrated in fig. 2.3 for the $d_{x^2-y^2}$ and d_{xy} orbitals in the x–y plane. Therefore, the e_g orbitals are raised in energy relative to the t_{2g} orbitals, which may be represented by an energy level diagram, as in fig. 2.4. The energy separation between the t_{2g} and e_g orbitals is termed the crystal field splitting and is designated by Δ_o. Alternatively, the symbol 10 Dq utilized in ligand field theory is sometimes used, and $\Delta_o = 10\,Dq$ (see §2.7). An analogous energy separation between t_{2g} and e_g orbital groups exists in the schematic molecular orbital energy level diagram shown in fig. 2.5 which shows 10 Dq (or Δ_o) to be the energy separation between orbitals of the t_{2g} and antibonding e_g^* groups (§11.3.1). The value of Δ_o or 10 Dq is obtained directly, or may be estimated, from spectral measurements of transition metal-bearing phases in the visible to near-infrared region.

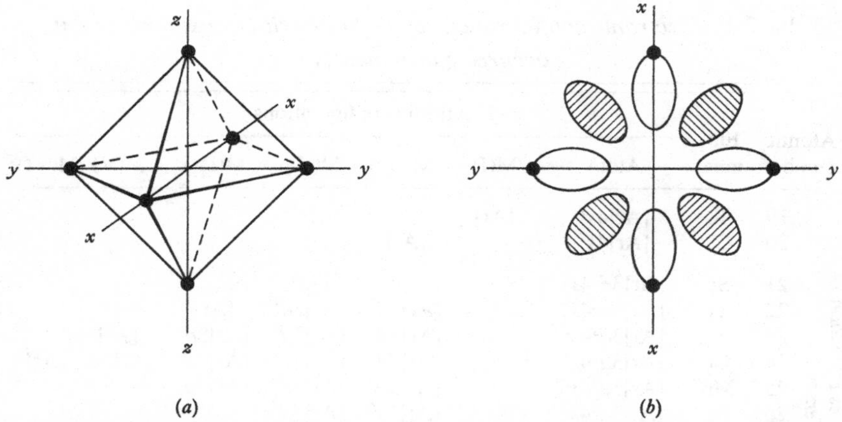

Figure 2.3 Orientations of ligands and d orbitals of a transition metal ion in octahedral coordination. (*a*) Orientation of ligands with respect to the cartesian axes; (*b*) the x–y plane of a transition metal ion in an octahedral crystal field. The d_{xy} orbital is cross-hatched; the $d_{x^2-y^2}$ orbital is open; ligands are black circles.

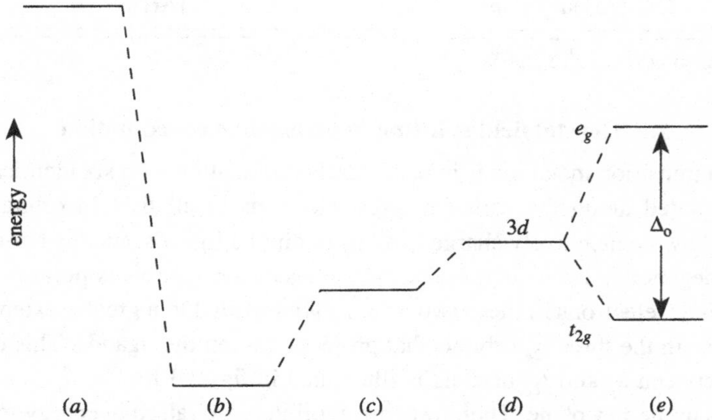

Figure 2.4 Relative energies of a transition metal 3*d* orbitals in octahedral coordination. (*a*) Energy levels of a free cation; (*b*) electrostatic attraction between the cation and anions; (*c*) repulsion between anions and electrons on the cation other than those in 3*d* orbitals; (*d*) repulsion between anions and the 3*d* electrons; (*e*) splitting of 3*d* orbital energy levels in an octahedral crystal field (from Burns, 1985a).

In the crystal field model, the split 3*d* orbital energy levels are assumed to obey a 'centre of gravity' rule. As a result, the three t_{2g} orbitals are lowered by $0.4\Delta_o$ below, and the two e_g orbitals raised by $0.6\Delta_o$ above, the baricentre. This follows from a simple algebraic argument that the energy of six electrons in the three t_{2g} orbitals is compensated by the energy of four electrons in the two e_g

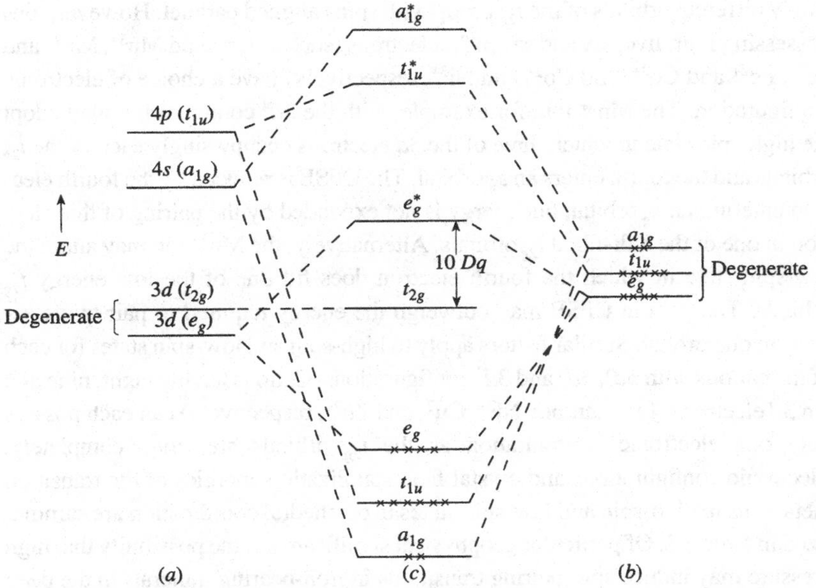

Figure 2.5 Schematic molecular orbital energy level diagram for a transition metal coordination cluster, $[ML_6]$. (*a*) Energy levels of atomic orbitals of the free cation, M; (*b*) energy levels for the six ligands, L, before bonding; (*c*) molecular orbital energy levels for the octahedral $[ML_6]$ cluster.

orbitals. Each electron in a t_{2g} orbital thus stabilizes a transition metal ion by $0.4\Delta_o$, whereas every electron in an e_g orbital diminishes the stability by $0.6\Delta_o$. The resultant nett stabilization energy is termed the crystal field stabilization energy and is designated by CFSE.

The distribution of 3*d* electrons in a given transition metal ion is controlled by two opposing tendencies. First, repulsion and exchange interactions between electrons causes them to be distributed over as many of the 3*d* orbitals as possible with parallel spins. This is in accordance with Hund's first rule. Secondly, the effect of crystal field splitting is to cause electrons to populate 3*d* orbitals having the lowest energy. These two opposing tendencies lead to high-spin and low-spin electronic configurations in certain transition metal ions. The distinction between the high-spin and low-spin states is an experimental one based on, for example, interatomic distances, magnetic susceptibility, visible-region and various X-ray spectroscopic measurements and, in the case of iron, the chemical isomer shift parameter determined by Mössbauer spectroscopy.

Consider the elements of the first transition series in octahedral coordination. Ions with one, two or three 3*d* electrons (for example, Ti^{3+}, V^{3+} and Cr^{3+}, respectively) each can have only one electronic configuration, and the electrons occupy singly different orbitals of the t_{2g} group with spins aligned parallel. However, ions possessing four, five, six and seven $3d$ electrons (such as Cr^{2+} and Mn^{3+}, Mn^{2+} and Fe^{3+}, Fe^{2+} and Co^{3+}, and Co^{2+} and Ni^{3+}, respectively) have a choice of electronic configuration. The Mn^{3+} ion, for example, with the $3d^4$ configuration may adopt the high-spin state in which three of the $3d$ electrons occupy singly each of the t_{2g} orbitals and the fourth enters an e_g orbital. The CFSE is reduced by the fourth electron entering an e_g orbital, but energy is not expended by the pairing of this electron in one of the half-filled t_{2g} orbitals. Alternatively, the Mn^{3+} ion may attain the low-spin state in which the fourth electron does fill one of the low energy t_{2g} orbitals. The gain in CFSE may outweigh the energy required to pair two electrons in one orbital. Similar factors apply to high-spin and low-spin states for each of the cations with $3d^5$, $3d^6$ and $3d^7$ configurations. Cations having eight, nine and ten $3d$ electrons (for example, Ni^{2+}, Cu^{2+} and Zn^{2+}, respectively) can each possess only one electronic configuration as the t_{2g} orbitals are filled completely. Electronic configurations and crystal field stabilization energies of the transition metal ions in high-spin and low-spin states in octahedral coordination are summarized in table 2.2. Of particular geophysical significance is the possibility that high pressure may induce spin-pairing transitions in iron-bearing minerals in the deep in the Earth's Mantle. This aspect is discussed in chapter 9 (§9.7)

It can be seen from table 2.2 that transition metal ions with $3d^3$, $3d^8$ and low-spin $3d^6$ configurations acquire significantly higher CFSE's in octahedral coordination than other cations. Therefore, ions such as Cr^{3+}, Mn^{4+}, Ni^{2+}, and Co^{3+} are expected to show strong preferences for octahedral coordination sites. Cations with $3d^0$, $3d^{10}$ and high-spin $3d^5$ configurations, such as Ca^{2+}, Zn^{2+}, Mn^{2+} and Fe^{3+}, receive zero CFSE in octahedral coordination.

If there is competition between pairing energy and crystal field stabilization energy, a cation may be unstable towards oxidation and reduction. For example, the Mn^{3+} ($3d^4$) ion is readily oxidized to the Mn(IV) oxidation state and is easily reduced to the Mn^{2+} ion. In addition, the Mn^{3+} ion disproportionates in aqueous solution:

$$2\,Mn^{3+}{}_{(aq)} + 2\,H_2O_{\,(l)} \rightarrow Mn^{2+}{}_{(aq)} + MnO_{2\,(s)} + 4\,H^+{}_{(aq)} \tag{2.5}$$

$$\Delta G^\circ_{298} = -109 \text{ kJ mole}^{-1}$$

$$2\,Mn(OH)_{3\,(s)} \rightarrow Mn(OH)_{2\,(s)} + MnO_{2\,(s)} + 2\,H_2O_{\,(l)}. \tag{2.6}$$

$$\Delta G^\circ_{298} = -42 \text{ kJ mole}^{-1}$$

Table 2.2. *Electronic configurations and crystal field stabilization energies of transition metal ions in octahedral coordination*

Number of 3*d* electrons	Cation	High-spin state: Electronic configuration t_{2g}			High-spin: e_g		High-spin: Unpaired electrons	High-spin: CFSE	Low-spin state: Electronic configuration t_{2g}			Low-spin: e_g		Low-spin: Unpaired electrons	Low-spin: CFSE
0	Ca^{2+}, Sc^{3+}, Ti^{4+}						0	0						0	0
1	Ti^{3+}	↑					1	$\frac{3}{5}\Delta_0$	↑					1	$\frac{3}{5}\Delta_0$
2	Ti^{2+}, V^{3+}	↑	↑				2	$\frac{4}{5}\Delta_0$	↑	↑				2	$\frac{4}{5}\Delta_0$
3	V^{2+}, Cr^{3+}, Mn^{4+}	↑	↑	↑			3	$\frac{6}{5}\Delta_0$	↑	↑	↑			3	$\frac{6}{5}\Delta_0$
4	Cr^{2+}, Mn^{3+}	↑	↑	↑	↑		4	$\frac{3}{5}\Delta_0$	⇅	↑	↑			2	$\frac{3}{5}\Delta_0$
5	Mn^{2+}, Fe^{3+}	↑	↑	↑	↑	↑	5	0	⇅	⇅	↑			1	$\frac{10}{5}\Delta_0$
6	Fe^{2+}, Co^{3+}, Ni^{4+}	⇅	↑	↑	↑	↑	4	$\frac{3}{5}\Delta_0$	⇅	⇅	⇅			0	$\frac{13}{5}\Delta_0$
7	Co^{2+}, Ni^{3+}	⇅	⇅	↑	↑	↑	3	$\frac{4}{5}\Delta_0$	⇅	⇅	⇅	↑		1	$\frac{9}{5}\Delta_0$
8	Ni^{2+}	⇅	⇅	⇅	↑	↑	2	$\frac{6}{5}\Delta_0$	⇅	⇅	⇅	↑	↑	2	$\frac{6}{5}\Delta_0$
9	Cu^{2+}	⇅	⇅	⇅	⇅	⇅	1	$\frac{3}{5}\Delta_0$	⇅	⇅	⇅	⇅	↑	1	$\frac{3}{5}\Delta_0$
10	Cu^{+}, Zn^{2+}, Ga^{3+}, Ge^{4+}	⇅	⇅	⇅	⇅	⇅	0	0	⇅	⇅	⇅	⇅	⇅	0	0

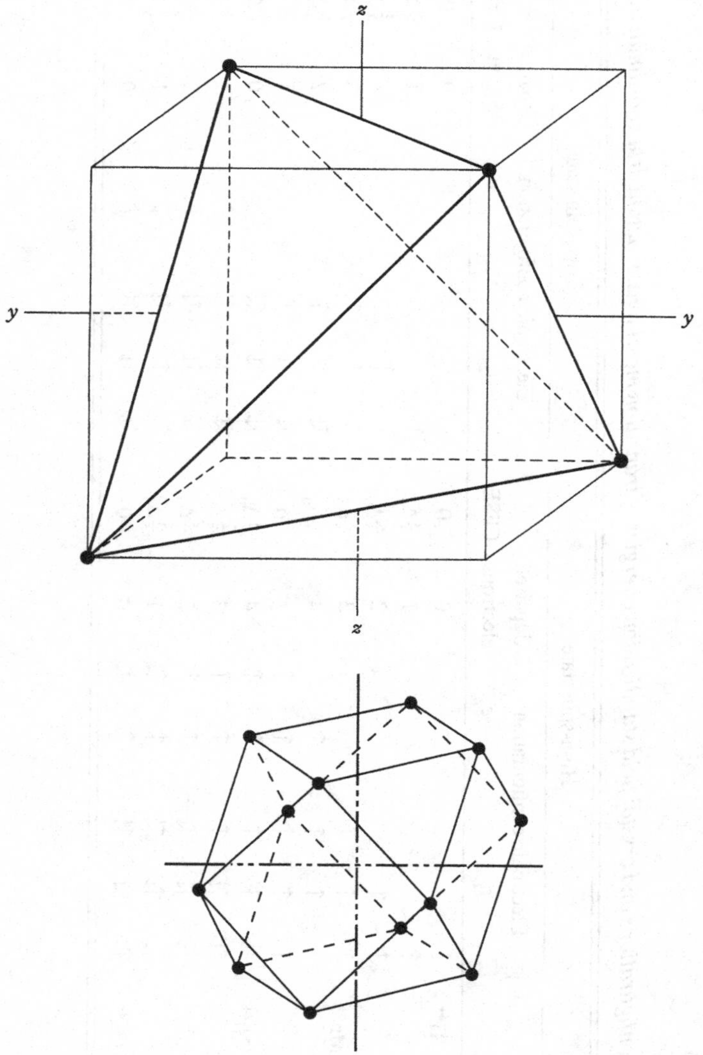

Figure 2.6 Arrangements of ligands about a transition metal ion in (*a*) tetrahedral and cubic coordinations, and (*b*) dodecahedral or cuboctahedral coordination. In tetrahedral coordination, the ligands may be regarded as lying at alternate vertices of a cube. In cubic coordination the ligands are situated on all eight vertices.

The Mn^{2+} ion, $(t_{2g})^3(e_g)^2$, formed in these processes possesses zero CFSE but has the stable electronic configuration with a maximum of five unpaired electrons spinning in the same direction. The Mn(IV) oxidation state, $(t_{2g})^3$, has a very high CFSE. Unstable cations such as Mn^{3+} may be stabilized in distorted

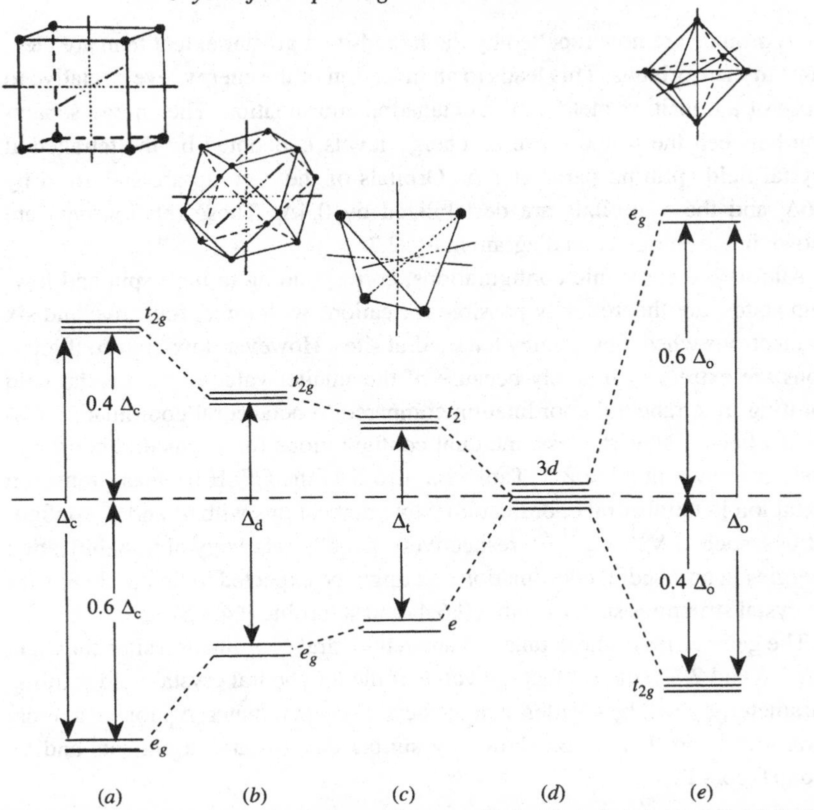

Figure 2.7 Crystal field splittings of transition metal 3*d* orbitals in (*a*) cubic (8-fold); (*b*) dodecahedral (12-fold); (*c*) tetrahedral (4-fold); (*d*) spherical; and (*e*) octahedral (6-fold) coordinations (from Burns, 1985a).

environments, however. This property, which is the result of the Jahn–Teller effect, is discussed in §2.12, §4.4.2 and §6.3.

2.6 Crystal field splittings in other coordinations

2.6.1 Tetrahedral coordination

In structures containing tetrahedral coordination sites, the ligands may be regarded as lying on alternate vertices of a cube with the transition metal ion at the centre. This arrangement is shown in fig. 2.6*a*. A tetrahedron lacks a centre of symmetry so that by group theory nomenclature the two groups of orbitals are designated by t_2 (for the d_{xy}, d_{yz} and d_{zx} orbitals) and e (for the $d_{x^2-y^2}$ and d_{z^2} orbitals) with the subscript g omitted. In tetrahedral coordination, electrons in the t_2 orbitals are now repelled by the ligands to a greater extent than are electrons in the e orbitals. This leads to an inversion of the energy levels relative to those of a transition metal ion in octahedral coordination. The energy separation between the t_2 and e orbital energy levels is denoted by the tetrahedral crystal field splitting parameter Δ_t. Orbitals of the e group are stabilized by $0.6\Delta_t$ and the t_2 orbitals are destabilized by $0.4\Delta_t$. These relationships are shown in the energy level diagram in fig. 2.7.

Alternative electronic configurations, corresponding to high-spin and low-spin states, are theoretically possible for cations with three, four, five and six $3d$ electrons when they occupy tetrahedral sites. However, low-spin configurations are extremely unlikely because of the smaller value of the crystal field splitting in tetrahedral coordination compared to octahedral coordination discussed below. Nevertheless, the dual configurations for tetrahedral coordination are shown in table 2.3. Table 2.3 also lists the CFSE for each transition metal ion in tetrahedral coordination. Note that cations with d^2 and d^7 configurations, such as V^{3+} and Co^{2+}, respectively, acquire relatively high stabilization energies in tetrahedral coordination and might be expected to favour these sites in crystal structures such as spinel (§6.4) and staurolite (§4.4.3).

The geometries of the octahedral and tetrahedral coordination sites shown in figs 2.3 and 2.6*a* suggest that the value of the tetrahedral crystal field splitting parameter, Δ_t, will be smaller than the octahedral parameter, Δ_o, for each transition metal ion. It may be shown by simple electrostatic arguments and by group theory that

$$\Delta_t = -\frac{4}{9}\Delta_o \qquad (2.7)$$

when cations, ligands, and metal-ligand distances are identical in the two coordinations. The implication of the negative sign in eq. (2.7) is that the relative stabilities of the two groups of d orbitals are reversed in the two coordinations. Thus, the d_{z^2} and $d_{x^2-y^2}$ orbitals are less stable in octahedral coordination but are the more stable group in tetrahedral coordination relative to the d_{xy}, d_{yz} and d_{zx} orbitals.

2.6.2 Cubic and dodecahedral coordinations

In body-centred cubic coordination, the eight ligands surrounding a transition metal ion lie at the vertices of a cube (cf. fig. 2.6*a*.). In one type of dodecahedral coordination site found in the ideal perovskite structure (cf. fig. 9.3), the 12 nearest-neighbour anions lie at the vertices of a cuboctahedron illustrated in fig. 2.6*b*. The relative energies of the e_g and t_{2g} orbital groups in these two centrosymmetric coordinations are identical to those of the e and t_2 orbital groups

Table 2.3. *Electronic configurations and crystal field stabilization energies of transition metal ions in tetrahedral coordination*

Number of 3*d* electrons	Cation	High-spin state: Electronic configuration *e*	High-spin state: Electronic configuration t_2	High-spin state: Unpaired electrons	High-spin state: CFSE	Low-spin state: Electronic configuration *e*	Low-spin state: Electronic configuration t_2	Low-spin state: Unpaired electrons	Low-spin state: CFSE
0	Ca^{2+}, Sc^{3+}, Ti^{4+}			0	0			0	0
1	Ti^{3+}	↑		1	$\frac{3}{5}\Delta_t$	↑		1	$\frac{3}{5}\Delta_t$
2	Ti^{2+}, V^{3+}	↑ ↑		2	$\frac{6}{5}\Delta_t$	↑ ↑		2	$\frac{6}{5}\Delta_t$
3	V^{2+}, Cr^{3+}, Mn^{4+}	↑ ↑	↑	3	$\frac{4}{5}\Delta_t$	↑↓ ↑		1	$\frac{9}{5}\Delta_t$
4	Cr^{2+}, Mn^{3+}	↑ ↑	↑ ↑	4	$\frac{3}{5}\Delta_t$	↑↓ ↑↓		0	$\frac{13}{5}\Delta_t$
5	Mn^{2+}, Fe^{3+}	↑ ↑	↑ ↑ ↑	5	0	↑↓ ↑↓	↑	1	$\frac{10}{5}\Delta_t$
6	Fe^{2+}, Co^{3+}, Ni^{4+}	↑↓ ↑	↑ ↑ ↑	4	$\frac{3}{5}\Delta_t$	↑↓ ↑↓	↑ ↑	2	$\frac{3}{5}\Delta_t$
7	Co^{2+}, Ni^{3+}	↑↓ ↑↓	↑ ↑ ↑	3	$\frac{6}{5}\Delta_t$	↑↓ ↑↓	↑↓ ↑ ↑	3	$\frac{6}{5}\Delta_t$
8	Ni^{2+}	↑↓ ↑↓	↑↓ ↑ ↑	2	$\frac{4}{5}\Delta_t$	↑↓ ↑↓	↑↓ ↑ ↑	2	$\frac{4}{5}\Delta_t$
9	Cu^{2+}	↑↓ ↑↓	↑↓ ↑ ↑↓	1	$\frac{3}{5}\Delta_t$	↑↓ ↑↓	↑↓ ↑↓ ↑	1	$\frac{3}{5}\Delta_t$
10	Cu^{+}, Zn^{2+}, Ga^{3+}, Ge^{4+}	↑↓ ↑↓	↑↓ ↑↓ ↑↓	0	0	↑↓ ↑↓	↑↓ ↑↓ ↑↓	0	0

in tetrahedral coordination, so that the electronic configurations and relative crystal field stabilization summarized in table 2.3 are also applicable to transition metal ions in cubic and dodecahedral coordinations. The cubic crystal field splitting parameter, Δ_c, however, is larger than Δ_t. Electrostatic and group theory arguments, again based on the same cation, similar ligands and identical metal-ligand distances in the two coordinations, lead to the following relationships between the cubic and octahedral crystal field splitting parameters

$$\Delta_c = -\frac{8}{9}\Delta_o \,. \qquad (2.8)$$

Similarly, for dodecahedral (Δ_d) and octahedral coordinations,

$$\Delta_d = -\frac{1}{2}\Delta_o \,. \qquad (2.9)$$

These relative crystal field splittings are portrayed schematically in fig. 2.7.

2.6.3 Other regular coordination environments

Several regular, non-distorted coordination environments providing coordination numbers ranging from 3 to 12 exist in chemical compounds of the transition elements, many being represented in mineral crystal structures discussed in later chapters. Relative energy levels corresponding to identical metal–ligand distances in each of these coordination environments are summarized in table 2.4, together with mineralogical examples.

2.7 The 10 *Dq* parameter

In an octahedral coordination site, the electrostatic field produced by the six ligands (represented as point negative charges and interacting with an electron in the vicinity of the central cation) is expressed by the potential

$$V_{oct} = \frac{6Z_L e}{R} + \frac{35 Z_L e}{4R^5}\left(x^4 + y^4 + z^4 - \frac{3}{5}r^4\right), \qquad (2.10)$$

where ($Z_L e$) is the charge on the ligands separated by a distance R from the cation, and x, y, z and r are the polar coordinates of the electron (fig. 2.1). The first and largest term of this expression, ($6Z_L e/R$), is the constant potential contributed by electrons of the argon core (table 2.1). This term contributes the

Table 2.4. *Relative energy levels of 3d orbitals in crystal fields of different symmetries**

Coordination number	Symmetry of site	d_{z^2}	$d_{x^2-y^2}$	d_{xy}	d_{xz}	d_{yz}	Examples
3	triangular	−0.321	0.546	0.546	−0.386	−0.386	(carbonate ion)
4	tetrahedral	−0.267	−0.267	+0.178	+0.178	+0.178	spinel
4	square planar	−0.528	1.228	0.228	−0.464	−0.464	gillespite, eudialyte
5	trigonal bipyramid	0.707	−0.082	−0.082	−0.272	−0.272	hibonite, andalusite
5	square pyramid	0.086	0.914	−0.086	−0.457	−0.457	S_N1 reactions, eudialyte
6	octahedral	0.600	0.600	−0.400	−0.400	−0.400	many Fe–Mg and Al silicates
6	trigonal prism	0.096	−0.584	−0.584	0.536	0.536	troilite
7	pentagonal bipyramid	0.493	0.282	0.282	−0.528	−0.528	S_N2 reactions
8	cube	−0.534	−0.534	0.356	0.356	0.356	garnet
8	square antiprism	−0.534	−0.089	−0.089	0.356	0.356	diopside M2
12	dodecahedron (cuboctahedron)	0.300	0.300	−0.200	−0.200	−0.200	perovskite

*Expressed as fractions of the octahedral crystal field splitting parameter, Δ_o.

major share to thermodynamic properties of a transition metal compound and influences the baricentre of the 3d orbital energy levels (fig. 2.4). However, since the ($6Z_L e/R$) term is spherically symmetrical, it does not participate in the splitting of the 3d orbital energy levels. Hence, it is usually neglected in crystal field theory, being unimportant when *energy differences* are being considered.

The second term in eq. (2.10) may be expressed by

$$V'_{\text{oct}} = D\,(x^4 + y^4 + z^4 - \frac{3}{5}r^4)\,, \qquad (2.11)$$

where constant, D, is given by

$$D = \frac{35Z_L e}{4R^5}\,. \qquad (2.12)$$

The potential, V'_{oct}, interacting with 3d electrons in the t_{2g} and e_g orbitals leads to the following expressions for their respective energies:

$$E(t_{2g}) = -\,4\,Dq \qquad (2.13)$$

and

$$E(e_g) = +\,6\,Dq\,, \qquad (2.14)$$

where

$$q = \frac{2e}{105}\int_0^\infty r_{3d}^2(r)r^4r^2\mathrm{d}r = \frac{2e<r^4>}{105}\,, \qquad (2.15)$$

so that the energy difference between the t_{2g} and e_g orbitals is

$$E(e_g) - E(t_{2g}) = 10\,Dq\,, \qquad (2.16)$$

which is equivalent to the crystal field splitting parameter Δ_o.

In eq. (2.15), $<r^4>$ is the mean value of the fourth power of the radial distance of a 3d orbital from the nucleus. Although accurate estimates of $<r^4>$ are not yet available, its value is assumed to be approximately constant for cations of similar valence in the same transition series. Combining eqs (2.12) and (2.15), the product of D and q is $(Z_L e^2/6R^5)<r^4>$, so that

$$\Delta_o = 10Dq = \frac{Z_L e^2}{6R^5}<r^4> = \frac{Q<r^4>}{R^5}\,. \qquad (2.17)$$

where Q is a constant.

The inverse fifth-power dependency of crystal field splitting on metal–oxygen distance expressed in eq. (2.17) is of fundamental importance in transition metal geochemistry, particularly in mineral physics at high pressures and interpretations of visible to near-infrared spectra of minerals. Thus, the $\Delta \propto R^{-5}$ relationship, eq. (2.17), is referred to frequently in later chapters.

2.8 Evaluation of Δ or 10 *Dq*

The magnitude of the crystal field splitting parameter, Δ or 10 *Dq*, may be estimated by two independent methods. The conventional way for evaluating Δ is from positions of absorption bands in visible-region spectra of transition metal compounds. The energy required to excite an electron from one 3*d* orbital to another 3*d* orbital of higher energy corresponds to radiation in the visible and near-infrared regions of the electromagnetic spectrum. Absorption of visible radiation by such intra-electronic transitions is the most common cause of colour in transition metal compounds and minerals. Accordingly, crystal field spectra are commonly referred to as optical spectra, alluding to colour variations of transition metal-bearing phases that are visible to the eye. Crystal field spectra are also called *d–d* spectra and, more generally, electronic absorption spectra, although the latter designation also includes interelectronic or charge transfer transitions between adjacent ions. The theory of crystal field spectra is described in chapter 3, while applications to colours of minerals are discussed in chapter 4.

The second method for estimating the value of Δ is from plots of thermodynamic data for series of similar compounds of transition elements. This method is discussed further in chapter 7.

2.9 Factors influencing values of Δ or 10 Dq

Certain generalizations may be made about factors that influence the magnitude of Δ or 10 *Dq*, values of which depend on individual cations, types of anions or ligands, interatomic distances, pressure and temperature variations, and symmetries of ligand environments.

2.9.1 Type of cation

Values of Δ_o are generally higher for trivalent cations than for corresponding divalent ions. This is demonstrated by the data summarized in table 2.5 for divalent and trivalent transition metal ions in simple oxide structures and in hydrated environments. For example, Δ_o values for Fe^{2+} and Fe^{3+} ions in aqueous solution are 10,400 cm^{-1} and 13,700 cm^{-1}, respectively.

There is a general sequence of Δ_o values which increase in the order

$$Mn^{2+} < Ni^{2+} < Co^{2+} < Fe^{2+} < V^{2+} < Fe^{3+} < Cr^{3+} < V^{3+} < Co^{3+} < Mn^{4+}. \quad (2.18)$$

This sequence is particularly well characterized for fluoride complexes of high-spin cations of the first-series transition elements (Allen and Warren, 1971). Moreover, between successive transition metal series, values of Δ_o increase by about thirty to fifty per cent. For example, in hydrated cations of the first and second transition series, Δ_o for $[Cr(H_2O)_6]^{3+}$ and $[Mo(H_2O)_6]^{3+}$ are 17,400 cm^{-1} and 26,110 cm^{-1}, respectively.

2.9.2 Type of ligand

Ligands coordinated to transition metal ions may also be arranged in order of increasing Δ. This order is called the spectrochemical series, reflecting colour variations in chemical compounds of individual cations with different ligands. Thus, for Cr^{3+} and Co^{3+} cations in octahedral coordination with different ligands, the order of increasing Δ_o is

$$I^- < Br^- < Cl^- < SCN^- < F^- < \text{urea} = OH^- < CO_3^{2-} = \text{oxalate} < O^{2-} < H_2O < \text{pyridine} < NH_3 < \text{ethylene diamene} < SO_3^{2-} < NO_2^- < HS^- < S^{2-} < CN^-. \quad (2.19)$$

Such an order is difficult to rationalize in terms of electrostatic energies embodied in the simple point-charge model of crystal field theory. For example, the charged O^{2-} anion precedes the dipolar H_2O molecule in the spectrochemical series. Note that the spectrochemical series differs from the nephelauxetic series discussed in chapter 11 (eq. 11.6), which is a measure of the degree of covalent bonding.

Transition metal ions with $3d^4$ to $3d^7$ configurations in coordination with ligands at the beginning of the spectrochemical series generally have high-spin configurations, whereas low-spin states exist in complexes with ligands at the end of the series. Thus, fluoride compounds of the first transition series contain cations in high-spin states, whereas low-spin states exist in cyanide complexes. The crossover point from high-spin to low-spin configuration varies from one cation to another, and may be ascertained from magnetic susceptibility measurements, interatomic distances in crystal structure refinements and information from spectroscopic techniques such as X-ray photoelectron (XPS), X-ray absorption (EXAFS, XANES) and Mössbauer spectroscopies. Most cations of the first transition series have high-spin configurations in oxide structures at

Table 2.5. *Crystal field splittings in transition metal-bearing oxides*

Electronic Structure		Cation			Electronic configuration	Δ corundum (cm^{-1})	Δ periclase (cm^{-1})	Δ aqueous (cm^{-1})	CFSE of hexahydrate (Δ_o)	CFSE of hexahydrate (cm^{-1})	CFSE of hexahydrate (kJ/g.ion)	Pairing energy* (cm^{-1})
	$3d^0$		Sc^{3+}	Ti^{4+}	[Ar]							
	$3d^1$		Ti^{3+}		$(t_{2g})^1$	19,100	11,360	18,950	$0.4\Delta_o$	= 7,580	= −90.7	
	$3d^2$		V^{3+}		$(t_{2g})^2$	18,730		19,100	$0.8\Delta_o$	= 15,280	= −182.8	
	$3d^3$	V^{2+}			$(t_{2g})^3$			12,600	$1.2\Delta_o$	= 15,120	= −180.9	
	$3d^3$		Cr^{3+}		$(t_{2g})^3$	18,150	16,200	17,400	$1.2\Delta_o$	= 20,880	= −249.9	
	$3d^3$			Mn^{4+}	$(t_{2g})^3$			21,000				
hs	$3d^4$	Cr^{2+}			$(t_{2g})^3(e_g)^1$			13,900	$0.6\Delta_o$	= 8,340	= −99.8	20,425
hs	$3d^4$		Mn^{3+}		$(t_{2g})^3(e_g)^1$	19,370		21,000	$0.6\Delta_o$	= 12,600	= −150.8	25,215
hs	$3d^5$	Mn^{2+}			$(t_{2g})^3(e_g)^2$			7,800	0	0	0	23,825
hs	$3d^5$		Fe^{3+}		$(t_{2g})^3(e_g)^2$	14,300		13,700	0	0	0	29,875
hs	$3d^6$	Fe^{2+}			$(t_{2g})^4(e_g)^2$		10,200	9,400	$0.4\Delta_o$	= 3,760	= −45.0	19,150
ls	$3d^6$		Co^{3+}		$(t_{2g})^6$			18,600	$2.4\Delta_o$	= 44,640	= −534.2	23,625
hs	$3d^7$	Co^{2+}			$(t_{2g})^5(e_g)^2$	12,300	9,000	9,300	$0.8\Delta_o$	= 7,440	= −89.1	20,800
ls	$3d^7$		Ni^{3+}		$(t_{2g})^6(e_g)^1$	18,000						
	$3d^8$	Ni^{2+}			$(t_{2g})^6(e_g)^2$	10,700	8,500	8,500	$1.2\Delta_o$	= 10,200	= −122.1	
	$3d^9$	Cu^{2+}			$(t_{2g})^6(e_g)^3$			13,000	$0.6\Delta_o$	= 7,800	= −93.3	
	$3d^{10}$	Zn^{2+}	Ga^{3+}	Ge^{4+}	$(t_{2g})^6(e_g)^4$							

* Pairing energies for high-spin (hs) to low-spin (ls) configurations in field-free cations are from Huheey (1983), p.380. The values are 15–30 % smaller when cations are chemically bound in coordination sites as a result of the nephelauxetic effect (§11.2.3).

atmospheric pressures. The Co^{3+} and Ni^{3+} ions are exceptions, however, and usually have low-spin configurations $(t_{2g})^6$ and $(t_{2g})^6(e_g)^1$, respectively, in oxide structures at atmospheric pressure.

In many transition metal-bearing minerals, oxygen is the most common ligand coordinated to cations in silicates, oxides, carbonates, phosphates and sulphates, for example. However, the oxygen bond-type varies from free O^{2-} anions in oxides and some silicates (e.g., Al_2SiO_5 polymorphs, epidote, wadsleyite or β–Mg_2SiO_4), to OH^- ions in oxyhydroxides and many silicates (e.g., amphiboles, micas, epidote, tourmaline), to non-bridging Si–O^- in most silicates (e.g., garnets, olivines, pyroxenes), to bridging Si–O–Si atoms in pyroxenes and amphiboles, to H_2O in a variety of hydrated minerals. Such differences of oxygen bond-type may produce small variations of Δ and influence the visible-region spectra, colours of rock-forming minerals and cation site occupancies in crystal structures. Such effects are discussed in later chapters.

2.9.3 Interatomic distance

The value of Δ is influenced by the distances between the transition metal and surrounding ligands, as formulated in eq. (2.17). This inverse fifth-power relationship indicates that Δ changes rapidly with variations of metal-ligand distances. For example, in a ferromagnesian silicate such as olivine, $(Mg,Fe)_2SiO_4$, the replacement of smaller Mg^{2+} ions (octahedral ionic radius, r_{oct} = 72 pm) by larger Fe^{2+} ions (r_{oct} = 77 pm), which causes a decrease of lattice parameters and bond-lengths in the crystal structure, leads to a decrease of Δ in the crystal field spectra of forsterite-fayalite series (§5.4.2.3). Ruby is coloured red, and not green as eskolaite (Cr_2O_3) is, due to the compression of Cr^{3+} (61 pm) replacing Al^{3+} ions (57 pm) in the corundum structure.

2.9.4 Pressure

At elevated pressures, the inverse fifth-power relationship expressed by eq. (2.17) leads to

$$\frac{\Delta_P}{\Delta_0} = \left(\frac{R_0}{R_P}\right)^5 , \tag{2.20}$$

where Δ_P, Δ_0 and R_P, R_0 are the crystal field splittings and average cation to oxygen distances for a transition metal ion in a mineral structure at high and ambient pressures, respectively. Since $R = 1/V^3$, where V is volume,

$$\Delta \propto V^{-5/3}, \text{ or } \ln \Delta = -\frac{5}{3} \ln V \tag{2.21}$$

Differentiation of eq. (2.21) gives

$$\frac{d\Delta}{\Delta} = \frac{5}{3}\frac{dV}{V} = \frac{5}{3}\frac{d\rho}{\rho} = \frac{5}{3}\frac{dP}{\kappa} , \tag{2.22}$$

where ρ is the density and κ the bulk modulus or incompressibility of the coordination polyhedron. κ is related to the compressibility, β, of the coordination polyhedron by

$$\kappa = \frac{1}{\beta} = -V\frac{dP}{V} , \tag{2.23}$$

Rewriting eq. (2.22) gives

$$\frac{d\Delta}{dP} = \frac{5}{5}\frac{\Delta}{\kappa} , \tag{2.24}$$

from which site incompressibilities or polyhedral bulk moduli of transition metal-bearing minerals may be estimated from Δ derived from high pressure crystal field spectral measurements. This application of high pressure spectroscopy is discussed in chapter 9 (§9.6).

Another application of eq. (2.20) concerns spin-pairing transitions in cations with $3d^4$ to $3d^7$ configurations. According to eq. (2.17), a small contraction of metal-ligand distance caused, for example, by increased pressure leads to a large increase in Δ. Such sensivity of Δ to contraction of bond lengths raises the possibility of spin-pairing transitions; that is, a change from a high-spin to low-spin configuration, in some transition metal ions. Such pressure-induced transitions are discussed further in chapter 9 in connection with possible electronic-induced phase changes in minerals deep in the Lower Mantle (§9.7).

2.9.5 Temperature

The temperature variation of Δ may be expressed by

$$\frac{\Delta_T}{\Delta_0} = \left(\frac{V_0}{V_T}\right)^{5/3} = [1 = \alpha(T - T_0)]^{-5/3} , \tag{2.25}$$

where Δ_T, Δ_0 and V_T, V_0 are crystal field splittings and molar volumes at elevated temperatures (T) and room temperature (T_0), respectively; α is the volume coefficient of thermal expansion. Generally, $V_T > V_0$, so that decreased values of Δ might be expected at elevated temperatures. However, other factors contribute to temperature-induced variations of absorption bands and some of these are discussed in chapters 3 and 10 (§3.9.4 and §10.7).

2.9.6 Symmetry of the ligand environment

As noted earlier, Δ depends on the symmetry of the ligands surrounding a transition metal ion. The relationships expressed in eqs (2.7), (2.8) and (2.9) for crystal field splittings in octahedral, tetrahedral, body-centred cubic and dodecahedral coordinations are summarized in eq. (2.26)

$$\Delta_o:\Delta_c:\Delta_d:\Delta_t = 1:-\frac{8}{9}:-\frac{1}{2}:-\frac{4}{9} \qquad (2.26)$$

and are shown schematically in fig. 2.7. These splittings which apply to regular or high-symmetry coordination polyhedra are further modified in non-cubic coordination environments such as those listed in table 2.4, as well as distorted coordination sites described later (§2.11).

2.10 Values of CFSE

The data in table 2.5 illustrate some of the generalizations enunciated in §2.9 regarding relative values of the crystal field splitting parameter, Δ. Crystal field stabilization energies (CFSE's) estimated from the Δ_o values of the hexahydrated ions are also calculated in table 2.5. The spectrum of Cr^{3+} in aqueous solution, for example, yields a value for Δ_o of 17,400 cm^{-1}. The CFSE of the Cr^{3+} ion, $(t_{2g})^3$, is $1.2\Delta_o$, or 20,880 cm^{-1}. This may be converted to joules by applying the conversion factor, 1 cm^{-1} = 11.966 J (see Appendix 8) giving −249.9 kJ/(g. ion) as the CFSE of Cr^{3+} in aqueous solution. The corresponding CFSE for Cr^{3+} in the corundum structure is (1.2 x 18,150 x 11.966) or −260.6 kJ/(g.ion). In Cr^{3+}–doped periclase the CFSE is about −233 kJ/(g.ion). These values for octahedrally coordinated Cr^{3+} ions, which are among the highest for transition metal ions with high-spin configurations in oxide structures, profoundly influence the geochemical behaviour of chromium (e.g., §8.5.3). Similarly, the unique configuration of low-spin Co^{3+}, $(t_{2g})^6$, also results in this trivalent cation having a very high CFSE, which has important consequences in sedimentary geochemistry (§8.7.5). Among divalent cations, Ni^{2+} has the highest CFSE. Such a relatively high electronic stability in octahedral sites has important implications in crystal chemical and geochemical properties of nickel (e.g., §8.5). The relatively high CFSE values for Cr^{3+}, Co^{3+} and Ni^{2+} contrast to the zero CFSE's of Mn^{2+} and Fe^{3+} resulting from their unique high-spin $3d^5$ configurations. Cations with zero 3*d* electrons (for example, Mg^{2+}, Ca^{2+}, Sc^{3+}, Ti^{4+}) and ten 3*d* electrons (for example, Cu^{1+}, Zn^{2+}, Ga^{3+}, Ge^{4+}) also have zero CFSE. Such differences of relative CFSE's profoundly influence the crystal chemical and geochemical properties of the cations, and are highlighted in later chapters.

2.11 Crystal fields in non-cubic environments

The discussion so far has been concerned with crystal field splittings within transition metal ions when they are surrounded by identical ligands in regular octahedral, tetrahedral, cubic and dodecahedral coordinations. Such high-symmetry coordination sites possibly exist in aqueous solutions, fused salts and silicate melts. However, coordination polyhedra are rarely so regular in crystal structures of most transition metal-bearing phases. In silicate minerals in particular, oxygen atoms frequently occur at the vertices of distorted polyhedra, and metal-oxygen distances are not identical within a coordination site (see Appendix 7). Furthermore, ligands may not be identical or equivalent in the coordination polyhedra. Thus, certain silicates (e.g., micas, amphiboles, tourmaline, topaz) contain appreciable amounts of OH^-, F^- and Cl^- anions, in addition to oxygens linked to silicon atoms, surrounding the cation coordination sites. As noted earlier (§2.9.2), the bond-type of oxygen ligands may vary within a site. Thus, minor differences in crystal field splittings may occur between O^{2-}, OH^-, Si–O^-, Si–O–Si, and H_2O groups. These factors, and their impact on cation site occupancies, are discussed in §6.8.

Further resolution of the 3*d* orbital energy levels takes place within a transition metal ion when it is located in a low-symmetry site, including non-cubic coordination environments listed in table 2.4 and polyhedra distorted from octahedral or cubic symmetries. As a result, the simple crystal field splitting parameter, Δ, loses some of its significance when more than one energy separation occurs between 3*d* orbitals of the cation.

2.12 The Jahn–Teller effect

Distortions of coordination polyhedra in the crystal structures of certain transition metal compounds are to be expected for theoretical reasons known as the Jahn–Teller effect. Jahn and Teller (1937) proved that if the ground state or lowest energy level of a molecule is degenerate, it will distort spontaneously to a lower symmetry so as to remove the degeneracy and make one energy level more stable. For example, if one of the 3*d* orbitals is completely empty or completely filled while another of equal energy is only half-filled, the environment about the transition metal ion is predicted to distort spontaneously to a different geometry in which a more stable electronic configuration is attained by making the half-filled orbital lower in energy.

The proof of the Jahn–Teller theorem lies in group theory and quantum mechanics. However, the origin of the distorting forces may be illustrated by considering the Mn^{3+} ion in octahedral coordination with oxygen. The Mn^{3+} ion has the high-spin configuration, $(t_{2g})^3(e_g)^1$, in which each t_{2g} orbital is occupied by one 3*d* electron and the fourth electron may occupy either the $d_{x^2-y^2}$ or d_{z^2} orbital (fig. 2.8). If the four oxygen atoms in the *x*–*y* plane move towards the central Mn^{3+} ion and, simultaneously, the two oxygen atoms along the *z* axis move away, the solitary e_g electron will favour the d_{z^2} orbital in which repulsion by the oxygen ions is smaller than in the $d_{x^2-y^2}$ orbital. Therefore, the e_g orbital group is separated into two energy levels with the d_{z^2} orbital becoming the the more stable. At the same time the t_{2g} orbital group is split into two energy levels, the d_{xz} and d_{yz} orbitals becoming more stable than the d_{xy} orbital. The four Mn–O distances in the *x*–*y* plane become smaller than the two distances along the *z* axis. The resulting coordination polyhedron and relative energies of the 3*d* orbitals are shown schematically in fig. 2.8*a*.

A converse situation exists whereby the two oxygen ions along the *z* axis may move closer to the Mn^{3+} ion (fig. 2.8*b*). This results in the stabilization of the $d_{x^2-y^2}$ orbital relative to the d_{z^2} orbital, and shorter Mn–O distances along the *z* axis compared to the *x*–*y* plane. In either of the tetragonally distorted environments shown in fig. 2.8 the Mn^{3+} ion becomes more stable relative to a regular octahedral coordination site. In most minerals, however, the Mn^{3+} ion occurs in an axially elongated octahedron (see table 6.1).

Transition metal ions most susceptible to large Jahn–Teller distortions in octahedral coordination in oxide structures are those with $3d^4$, $3d^9$ and low-spin $3d^7$ configurations, in which one or three electrons occupy e_g orbitals. Thus, the Cr^{2+} and Mn^{3+}, Cu^{2+}, and Ni^{3+} ions, respectively, are stabilized in distorted environments, with the result that compounds containing these cations are frequently distorted from type-structures. Conversely, these cations may be stabilized in distorted sites already existing in mineral structures. Examples include Cr^{2+} in olivine (§8.6.4) and Mn^{3+} in epidote, andalusite and alkali amphiboles (§4.4.2). These features are discussed further in chapter 6.

Uneven electron distributions also exist in the t_{2g} orbitals of some transition metal ions, including those with $3d^1$, $3d^2$, and high-spin $3d^6$ and $3d^7$ configurations, suggesting that the t_{2g} orbital group may also undergo Jahn–Teller splitting. However, van Vleck (1939) proved and examples have confirmed that Jahn–Teller distortions are small when there is degeneracy in the t_{2g} orbital group. For example, the lower-level splittings of t_{2g} orbitals, δ_1, depicted in fig. 2.8 are about 10 cm^{-1} to 100 cm^{-1} when cations such as Ti^{3+}, V^{3+}, Fe^{2+} and Co^{2+} occur in structures (e.g., MgO) and environments (e.g., aqueous solutions) providing regular octahedral coordination sites. In contrast, the upper-level split-

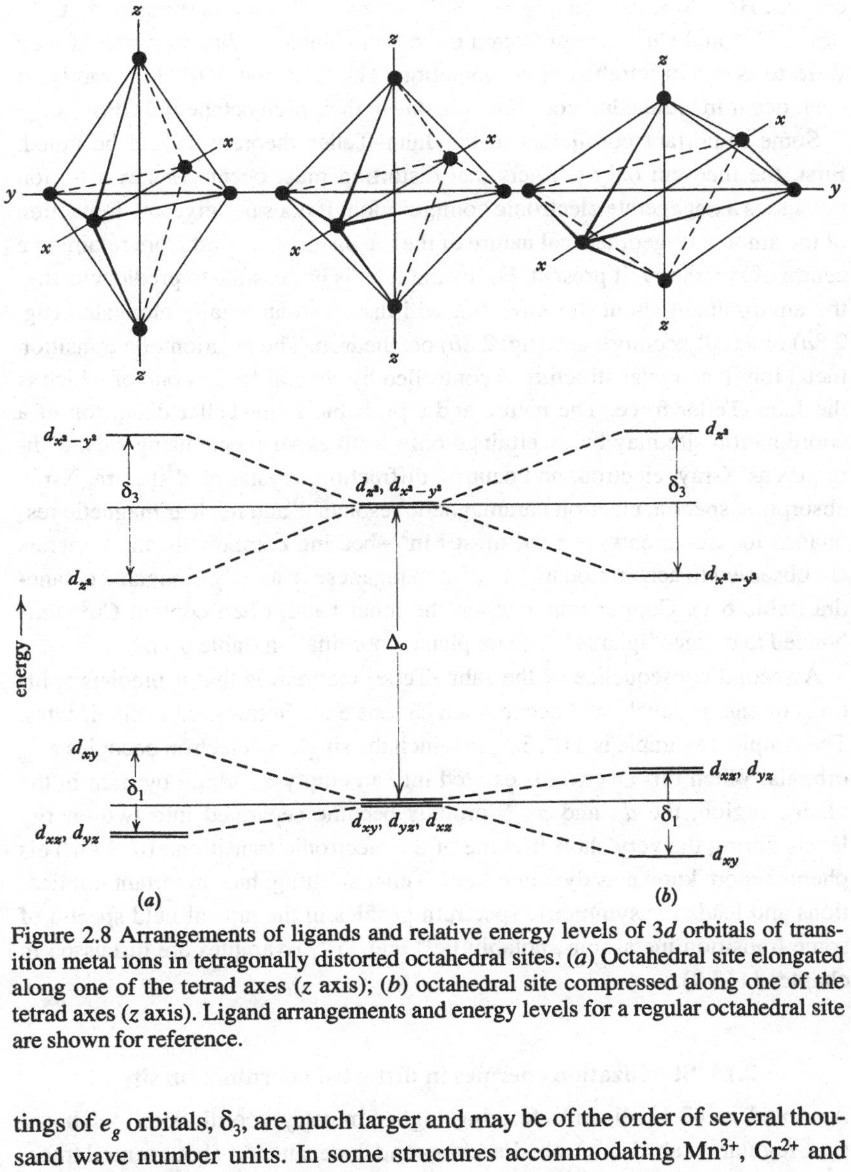

Figure 2.8 Arrangements of ligands and relative energy levels of 3*d* orbitals of transition metal ions in tetragonally distorted octahedral sites. (*a*) Octahedral site elongated along one of the tetrad axes (*z* axis); (*b*) octahedral site compressed along one of the tetrad axes (*z* axis). Ligand arrangements and energy levels for a regular octahedral site are shown for reference.

tings of e_g orbitals, δ_3, are much larger and may be of the order of several thousand wave number units. In some structures accommodating Mn^{3+}, Cr^{2+} and Cu^{2+} ions in octahedral sites, energy separations between the d_{z^2} and $d_{x^2-y^2}$ orbitals (δ_3) may be comparable to Δ_o values.

Jahn–Teller distortions are also predicted for certain transition metal ions in tetrahedral coordination. The electronic configurations predicted to undergo no Jahn–Teller distortions are the high-spin $3d^2$, $3d^5$, $3d^7$ and low-spin $3d^4$

cations. However, ions having $3d^3$, $3d^4$, $3d^8$ and $3d^9$ configurations (e.g., Cr^{3+}, Mn^{3+}, Ni^{2+} and Cu^{2+}) are predicted to be more stable in distorted sites if they were to occur in tetrahedral coordination. The Cr^{3+} and Mn^{3+} ions rarely, if ever, occur in tetrahedral coordination, due to their high octahedral CFSE's.

Some important corollaries to the Jahn–Teller theorem should be noted. First, the theorem only predicts that distortion must occur whenever an ion possesses a degenerate electronic configuration. It does not give any indication of the amount or geometrical nature of the the distortion, apart from retaining a centre of symmetry, if present. For example, it is impossible to predict whether the environment about the Mn^{3+} ion will distort to an axially elongated (fig. 2.8*a*) or axially compressed (fig. 2.8*b*) octahedron. The position of a transition metal ion in a crystal structure is controlled by several factors one of which is the Jahn–Teller force. The nature and a probable Jahn–Teller distortion of a coordination site may be ascertained only from experiments using such techniques as X-ray, electron, and neutron diffraction, crystal field spectra, X-ray absorption spectra, electron paramagnetic resonance and nuclear magnetic resonance measurements. In fact, most Mn^{3+}–bearing compounds and minerals are observed to accommodate trivalent manganese in axially elongated octahedra (table 6.1). Copper minerals, on the other hand, often contain Cu^{2+} ions bonded to oxygen ligands in square planar coordination (table 6.1).

A second consequence of the Jahn–Teller theorem is that it predicts splittings of energy levels will occur when cations exist in transient excited states. The simplest example is Ti^{3+}, $3d^1$, in which the single $3d$ electron occupies a t_{2g} orbitals. When this electron is excited into an empty e_g orbital by light in the visible region, the d_{z^2} and $d_{x^2-y^2}$ orbitals become separated into two energy levels during the very short lifetime of the electronic transition (10^{-15} s). This phenomenon, known as dynamic Jahn–Teller splitting, has important implications and leads to asymmetric spectrum profiles in the crystal field spectra of some transition metal ions, notably Fe^{2+} and Ti^{3+}. Examples are discussed in chapter 3 (§3.9).

2.13 Stabilization energies in distorted coordination sites

As noted in §2.11, ligands forming high-symmetry coordination polyhedra (i.e., regular octahedra, tetrahedra, cubes and dodecahedra) about central transition metal ions are rare. Such highly idealized coordinations, nevertheless, do exist in the periclase (octahedra), cubic perovskite (octahedra, dodecahedra) and spinel (tetrahedra) structures. The more important rock-forming oxide and silicate minerals provide, instead, low-symmetry coordination environments. These include trigonally distorted octahedra in the corundum, spinel and garnet structures, the rhombic distorted dodecahedra in garnets, and a wide variety of deformed octahedra and six-coordinated sites in ferromagnesian silicates, including the olivine, pyroxene and amphibole structures illustrated later (figs 5.8, 5.16 and 5.18; see also Appendix 7). Non-cubic symmetry also exists in coordination sites when different arrangements of oxygen ligands surround a cation, such as $Si–O^-$ and OH^- anions in amphiboles and micas. In micas, for example, the *trans*–M1 and *cis*–M2 sites differ with respect to the positions of the two hydroxyl ions in the coordination polyhedron surrounding the cation in octahedral sites (fig. 5.21). A similar distinction exists between the *trans*–M3 and *cis*–M1 sites in the amphibole structure (fig. 5.18). The occurrence of transition metal ions in such low-symmetry sites leads to further resolution of the $3d$ orbitals into additional energy levels which influence electronic configurations and modify estimates of CFSE's.

For example, in a tetragonally distorted octahedron elongated along the tetrad axis, such as that shown in fig. 2.8a, the t_{2g} and e_g orbital groups are each resolved into two additional energy levels. The energy separations between the resolved orbitals of the t_{2g} and e_g groups are denoted by δ_1 and δ_3 respectively. If the split energy levels obey centre of gravity rules about the t_{2g} and e_g baricentres, the d_{xz} and d_{yz} orbitals are stabilized by $\delta_1/3$ and the d_{z^2} orbital by $\delta_3/2$, respectively. In a site compressed along the tetrad axis, on the other hand, the d_{xy} and $d_{x^2-y^2}$ orbitals are the more stable ones of the t_{2g} and e_g groups, respectively. In table 2.6 are listed the stabilization energies of each transition metal ion in regular and tetragonally distorted octahedral sites, together with the configurations of the most stable site predicted for each transition metal ion. Ions that are most stable in regular octahedral sites are those with $3d^3$, $3d^8$ and low-spin $3d^6$ configurations. The remaining cations attain increased stability in distorted coordination sites.

Transition metal ions in crystal structures may also occur in coordination sites possessing symmetries other than regular and tetragonally distorted octahedral and tetrahedral. Some of the geometries of low-symmetry coordination polyhedra occurring in silicates include trigonally distorted octahedra, orthorhombic and skewed octahedra, distorted cubes, square planar and, rarely, linear and square antiprisms. Transition metal ions in each of these coordination sites have their $3d$ orbitals split into several energy levels and examples are shown schematically in fig. 2.9. In each configuration, the z electronic axis, by convention, is designated as the axis of highest symmetry and usually corresponds to the axis of distortion of an octahedron. Other examples of distorted coordination polyhedra are portrayed in later chapters.

Distorted coordination sites in oxides and silicates have several important consequences in transition metal geochemistry and crystal chemistry. First,

Table 2.6. *Stabilization energies of transition metal ions in tetragonally distorted octahedral sites in oxides*

Electronic structure		Cation			Electronic configuration	Stabilization energy: Octahedron elongated along tetrad axis	Stabilization energy: Octahedron compressed along tetrad axis	Configuration of the most stable six-coordinated site
	$3d^0$	Ca^{2+}	Sc^{3+}	Ti^{4+}	[Ar]	-	-	octahedron
	$3d^1$		Ti^{3+}	V^{4+}	$(t_{2g})^1$	$\frac{2}{5}\Delta_o + \frac{1}{3}\delta_1$	$\frac{2}{5}\Delta_o + \frac{2}{3}\delta_1$	compressed octahedron
	$3d^2$		V^{3+}	Cr^{4+}	$(t_{2g})^2$	$\frac{4}{5}\Delta_o + \frac{2}{3}\delta_1$	$\frac{4}{5}\Delta_o + \frac{1}{3}\delta_1$	elongated octahedron
	$3d^3$	V^{2+}	Cr^{3+}	Mn^{4+}	$(t_{2g})^3$	$\frac{6}{5}\Delta_o$	$\frac{6}{5}\Delta_o$	octahedron
hs	$3d^4$	Cr^{2+}	Mn^{3+}		$(t_{2g})^3(e_g)^1$	$\frac{2}{5}\Delta_o + \frac{1}{2}\delta_3$	$\frac{2}{5}\Delta_o + \frac{1}{2}\delta_3$	deformed octahedron
ls	$3d^4$	Cr^{2+}	Mn^{3+}		$(t_{2g})^4$	$\frac{8}{5}\Delta_o + \frac{1}{3}\delta_1$	$\frac{8}{5}\Delta_o + \frac{2}{3}\delta_1$	compressed octahedron
hs	$3d^5$	Mn^{2+}	Fe^{3+}		$(t_{2g})^3(e_g)^2$	0	0	octahedron
ls	$3d^5$	Mn^{2+}	Fe^{3+}		$(t_{2g})^5$	$2\Delta_o + \frac{2}{3}\delta_1$	$2\Delta_o + \frac{1}{3}\delta_1$	elongated octahedron
hs	$3d^6$	Fe^{2+}	Co^{3+}		$(t_{2g})^4(e_g)^2$	$\frac{2}{5}\Delta_o + \frac{1}{3}\delta_1$	$\frac{2}{5}\Delta_o + \frac{2}{3}\delta_1$	compressed octahedron
ls	$3d^6$	Fe^{2+}	Co^{3+}		$(t_{2g})^6$	$\frac{12}{5}\Delta_o$	$\frac{12}{5}\Delta_o$	octahedron
hs	$3d^7$	Co^{2+}	Ni^{3+}		$(t_{2g})^5(e_g)^2$	$\frac{4}{5}\Delta_o + \frac{2}{3}\delta_1$	$\frac{4}{5}\Delta_o + \frac{1}{3}\delta_1$	elongated octahedron
ls	$3d^7$	Co^{2+}	Ni^{3+}		$(t_{2g})^6(e_g)^1$	$\frac{9}{5}\Delta_o + \frac{1}{2}\delta_3$	$\frac{9}{5}\Delta_o + \frac{1}{2}\delta_3$	deformed octahedron
	$3d^8$	Ni^{2+}			$(t_{2g})^6(e_g)^2$	$\frac{6}{5}\Delta_o$	$\frac{6}{5}\Delta_o$	octahedron
	$3d^9$	Cu^{2+}			$(t_{2g})^6(e_g)^3$	$\frac{3}{5}\Delta_o + \frac{1}{2}\delta_3$	$\frac{3}{5}\Delta_o + \frac{1}{2}\delta_3$	deformed octahedron
	$3d^{10}$	Zn^{2+}	Ga^{3+}	Ge^{4+}	$(t_{2g})^6(e_g)^4$	-	-	octahedron*

hs and ls are high-spin and low-spin configurations, respectively; Δ_o Crystal field splitting parameter in octahedral coordination; δ_1 Splitting of t_{2g} orbital group; δ_3 Splitting of e_g orbital group.
* Zn^{2+} occurs in tetrahedral coordination in most oxide structures.

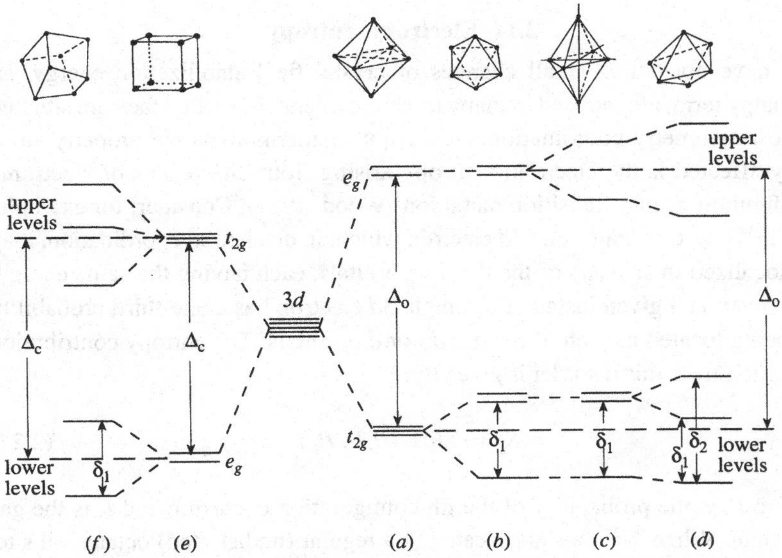

Figure 2.9 Relative energies of 3*d* orbital energy levels of a transition metal ion in low-symmetry distorted sites. *(a)* Regular octahedron (e.g., periclase); *(b)* trigonally distorted octahedron (e.g., corundum, spinel, approx. olivine M2 site); *(c)* tetragonally distorted octahedron (e.g., approx. olivine M1 site); *(d)* highly distorted six-coordinated sited (e.g., pyroxene M2 site); *(e)* regular cube; *(f)* distorted cube (e.g., triangular dodecahedral site of garnet).

distorted coordination polyhedra induce cation ordering in many mineral structures. If there is a variety of sites available to a cation during mineral crystallization, the cation will favour that site which accommodates it best and bestows greatest electronic stability. Site preferences of the transition metals are discussed in chapter 6 (§6.7). Second, unusual cation valences may be stabilized in distorted sites. Examples including Cr^{2+} in lunar olivines, Ti^{3+} in some pyroxenes found in carbonaceous chondrites and mare basalts from the Moon, and Mn^{3+} in the epidote and andalusite structures. These and other examples are described in §4.4.1, §4.4.2 and §6.2. Third, low-symmetry environments contribute to visual pleochroism in many non-cubic minerals when viewed in polarized light under the polarizing microscope. These effects on colours of minerals and gems are discussed in chapter 4 (§4.7). Finally, different splittings of the 3*d* orbitals, particularly for Fe^{2+} ions in highly distorted sites in pyroxenes and olivine, produce absorption bands at different energies in the near-infrared region and enable each mineral to be identified in remote-sensed spectra of planetary surfaces. Examples are considered in chapter 10.

2.14 Electronic entropy

We have noted that small changes of crystal field stabilization energy, an enthalpy term, are induced in many transition metal ions when they are situated in low-symmetry coordination sites. Another thermodynamic property similarly affected is the electronic entropy arising from differences of electronic configurations of a transition metal ion (Wood, 1981). Consider, for example, the Ti^{3+} ion. It contains one $3d$ electron which, in octahedral coordination, may be localized in any one of the three t_{2g} orbitals, each having the same energy. However, at a given instant, this single $3d$ electron has a one-third probability of being located in each of the d_{xy}, d_{yz} or d_{zx} orbitals. The entropy contribution resulting from this disorder is given by

$$S_{\mathrm{el}} = -R\,\Sigma\,(P_i \ln P_i)\,, \qquad (2.27)$$

where P_i is the probability of the ith configuration occurring and R is the gas constant. When Ti^{3+} ions are located in a regular (undistorted) octahedral site, there are three configurations each with a one-third probability of occurrence, so that per gram ion

$$S_{\mathrm{el}}\,(\mathrm{oct}) = -\,3R\left(\frac{1}{3}\ln\frac{1}{3}\right) = 9.13\ \mathrm{J\ (deg.\ g.ion)^{-1}}. \qquad (2.28)$$

Similarly, if the Ti^{3+} ions were to occur in tetrahedral coordination, the single $3d$ electron could occupy either one of the two e orbitals, d_{z^2} or $d_{x^2-y^2}$, and the electronic entropy would be

$$S_{\mathrm{el}}(\mathrm{tet}) = -\,2R\left(\frac{1}{2}\ln\frac{1}{2}\right) = 5.76\ \mathrm{J\ (deg.\ g.ion)^{-1}}. \qquad (2.29)$$

For Cr^{3+} ions in octahedral coordination, there is just one unique electronic configuration in which its three $3d$ electrons occupy singly each of the t_{2g} orbitals. With probability of occurrence equal to one, the electronic entropy of Cr^{3+} is given by

$$S_{\mathrm{el}}(\mathrm{oct}) = -\,R\,(1\ln 1) = 0. \qquad (2.30)$$

The presence of a Ti^{3+} ion in a distorted octahedral site would also yield a zero electronic entropy term. This results from removal of the three-fold degeneracy of t_{2g} orbitals in the low-symmetry environment. Other effects of electronic entropy on thermodynamic properties of transition metal-bearing minerals are discussed in chapter 7 (§7.4).

2.15 Summary

Chapter 2 outlines the crystal field theory of chemical bonding, which is appropriate to ionic structures such as silicate and oxide minerals. In this electrostatic model, anions and dipolar groups (ligands) are regarded as point negative charges, and the effects of the surrounding ligands on the orbital energy levels of a transition metal cation are examined.

Orbitals. Atomic orbitals represent the angular distribution of electron density about a nucleus. The shapes and energies of these amplitude probability functions are obtained as solutions to the Schrödinger wave equation. Corresponding to a given principal quantum number, for example $n = 3$, there are one $3s$, three $3p$ and five $3d$ orbitals. The s orbitals are spherical, the p orbitals are dumb-bell shaped and the d orbitals crossed dumb-bell shaped. Each orbital can accomodate two electrons spinning in opposite directions, so that the d orbitals may contain up to ten electrons.

Transition elements. Elements of the first transition series are characterized by having incompletely filled $3d$ orbitals in one or more of their common oxidation states. The series includes scandium, titanium, vanadium, chromium, manganese, iron, cobalt, nickel and copper, which have electronic configurations of the form $(1s)^2(2s)^2(2p)^6(3s)^2(3p)^6(3d)^{10-n}(4s)^{1 \text{ or } 2}$.

Crystal field splitting, Δ. In the absence of magnetic and electric fields, the five $3d$ orbitals of an isolated transition metal ion have the same energy, that is they are five-fold degenerate, but have different spatial orientations. Lobes of the d_{z^2} and $d_{x^2-y^2}$ orbitals (the e_g group) project along the cartesian axes, whereas lobes of the d_{xy}, d_{yz} and d_{zx} orbitals (the t_{2g} group) are directed between these axes. When a transition metal ion is in octahedral coordination in a structure, the effect of the six ligands situated along the cartesian axes is to repel electrons in e_g orbitals to a greater extent than electrons in t_{2g} orbitals. The octahedral crystal field splitting parameter, Δ_o, is the induced energy separation between the t_{2g} and e_g orbital groups. The energy level of the t_{2g} group is lowered by $0.4\Delta_o$ and the e_g orbital group raised by $0.6\Delta_o$ relative to the hypothetical level of unresolved $3d$ orbitals.

Crystal field stabilization energy, CFSE. Each electron in a t_{2g} orbital stabilizes a transition metal ion in octahedral coordination by $0.4\Delta_o$, whereas every electron in an e_g orbital destabilizes it by $0.6\Delta_o$. The crystal field stabilization energy, CFSE, represents the algebraic sum of these factors. Cations may have either high-spin configurations in which electrons with spins aligned occupy singly as many t_{2g} and e_g orbitals as possible, or low-spin configurations in which electrons fill t_{2g} orbitals. Favourable exchange interactions, reduced interelectronic repulsion and unnecessary pairing energies of two electrons in one orbital contribute to the stabilities of high-spin states. Low-spin states lead to a substantial increase in CFSE. Magnetic susceptibilities, interatomic distances and a variety of spectroscopic measurements may identify the spin state of each transition metal ion. Apart from Co^{3+} and Ni^{3+} ions, all transition metal cations exist in high-spin states in oxide structures on the Earth's surface. The ions Cr^{3+}, Ni^{2+} and Co^{3+} acquire high CFSE in octahedral coordination and Mn^{2+} and Fe^{3+} receive zero CFSE in high-spin states.

Evaluation of Δ. Energy separations between resolved 3*d* orbital energy levels correspond to visible and near-infrared radiation. Measurements of absorption spectra of transition metal compounds and minerals are used to obtain Δ and to evaluate the CFSE of the ions. Crystal field splittings may also be estimated from plots of thermodynamic data for the first-series transition metal compounds. The magnitude of Δ depends on:

(i) the symmetry of the coordinated ligands. Ratios of crystal field splittings in octahedral, body-centred cubic, dodecahedral and tetrahedral sites (where the cation is coordinated to 6, 8, 12 or 4 ligands, respectively) are: $\Delta_o : \Delta_c : \Delta_d : \Delta_t = 1 : -\frac{8}{9} : -\frac{1}{2} : -\frac{4}{9},$ where the negative signs indicate that d_{z^2} and $d_{x^2-y^2}$ orbitals become the more stable set relative to the other three d orbitals;

(ii) the valence of the cation, values for M^{3+} ions being higher than those for M^{2+} ions;

(iii) the nature and type of ligand coordinated to the cation. Thus, Δ_o for six CN^- anions is considerably larger than that for six H_2O ligands, with the result that cyanide complexes contain transition metal ions in low-spin states;

(iv) the interatomic distance, R, between metal and ligand. The $\Delta \propto R^{-5}$ relationship shows that small decreases in R lead to large increases of Δ;

(v) pressure, the effect of which is to shorten interatomic distances at elevated pressures, thereby increasing the CFSE of most transition metal cations; and

(vi) temperature, because thermal expansion may lower Δ.

The Jahn–Teller effect and stabilization energies in distorted coordination sites. Transition metal ions rarely occur in high-symmetry coordination sites in silicate minerals, which show various distortions from regular octahedral, tetrahedral, cubic and dodecahedral symmetries. The Jahn–Teller theorem states that cations with $3d^4$, $3d^9$ and low-spin $3d^7$ configurations, which have uneven distributions of electrons in e_g orbitals, should spontaneously distort their environments, thereby attaining increased electronic stability in low-symmetry sites. This explains why compounds of Cr^{2+}, Mn^{3+}, Cu^{2+} and Ni^{3+} are usually distorted from type-structures. In low-symmetry coordination sites the simple crystal field splitting parameter loses some of its significance because more than one energy separation between $3d$ orbital energy levels must be considered. Lowering of one or two orbital energy levels results in certain transition metal ions acquiring increased stabilization in distorted coordination sites. This leads to the possibility of relative enrichment of some cations in certain sites in silicate structures during mineral formation, stabilization of unusual oxidation states, pleochroism in non-cubic minerals, and specific profiles for Fe^{2+} silicates in remote-sensed spectra of planetary surfaces.

Electronic entropy. As a result of unequal electron occupancies of degenerate t_{2g} orbitals, the Ti^{3+} and Fe^{2+} ions in octahedral or tetrahedral sites, for example, may have large electronic entropies compared to zero values for Cr^{3+}. Electronic entropies decrease at elevated temperatures and are smaller when cations are located in distorted sites.

2.16 Background reading

Ballhausen, C. J. (1962) *Introduction to Ligand Field Theory.* (McGraw–Hill, New York), ch. 1.

Burns, R. G. (1985) Thermodynamic data from crystal field spectra. In *Macroscopic to Microscopic. Atomic Environments to Mineral Thermodynamics.* (S. W. Kieffer & A. Navrotsky, eds; Mineral. Soc.Amer. Publ.), *Rev. Mineral.*, 14, 277–316.

Burns, R. G. & Fyfe, W. S. (1967) Crystal field theory and the geochemistry of transition elements. In *Researches in Geochemistry, vol 2.* (P. H. Abelson, ed.; J. Wiley, New York), pp. 259–85

Cotton, F. A. & Wilkinson, G. (1988) *Advanced Inorganic Chemistry, 4th edn.* (Interscience Publ., New York), ch. 20.

Cotton, F. A., Wilkinson, G., & Gaus, P. L. (1988) *Basic Inorganic Chemistry, 2nd edn.* (J. Wiley & Sons, New York), chs 2 & 23.

Huheey, J. E. (1983) *Inorganic Chemistry: Principles of Structure and Reactivity, 3rd edn.* (Harper & Row, New York), ch. 9.

Orgel, L. E. (1966) *An Introduction to Transition Metal Chemistry: Ligand–Field Theory, 2nd edn.* (Methuen, London), chs 2 & 3.

Phillips, C. S. G. & Williams, R. J. P. (1966) *Inorganic Chemistry.* (Clarendon Press, Oxford), ch. 24.

3

Energy level diagrams and crystal field spectra of transition metal ions

A variety of selection rules derived from quantum mechanics governs the intensity of the various types of absorption phenomena.
– – The rules can be bent when ions get together.
G. R. Rossman, *Rev. Mineral.*, 18, 214 (1988)

3.1 Introduction

In the previous chapter it was shown how electrostatic fields produced by anions or negative ends of dipolar ligands belonging to coordination sites in a crystal structure split the 3*d* orbitals of a transition metal ion into two or more energy levels. The magnitude of these energy separations, or crystal field splittings, depend on the valence of the transition metal ion and the symmetry, type and distances of ligands surrounding the cation. The statement was made in §2.8 that separations between the 3*d* orbital energy levels may be evaluated from measurements of absorption spectra in the visible to near-infrared region. The origins of such crystal field spectra, also termed *d–d* spectra and optical spectra, are described in this chapter. Later chapters focus on measurements and applications of crystal field spectra of transition metal-bearing minerals.

3.2 Units in absorption spectra

When light is passed through a compound or mineral containing a transition metal ion, it is found that certain wavelengths are absorbed, often leading to coloured transmitted light. One cause of such absorption of light is the excitation of electrons between the split 3*d* orbital energy levels. Measurements of the intensity of light incident on and transmitted through the transition metal-bearing phase produces data for plotting an absorption spectrum. An absorption spectrum shows the amount of radiation absorbed or transmitted at each wavelength or energy. The absorption scale forms the ordinate and the abscissa is an energy or wavelength scale.

3.2.1 Wavelength and energy units

The position of an absorption band is measured on a wavelength scale which may be calibrated in ångstroms (Å), nanometre (nm) or micron (μm) units. Ångstrom units were most commonly used in early mineral spectroscopy literature, including the first edition of this book. However, in current spectral mineralogy research, absorption spectra are often plotted on nanometre scales, whereas micron units are commonly employed in reflectance spectra and remote-sensing applications (chapter 10). The relationship between these wavelength units is

$$1.0\ \mu m = 1{,}000\ nm = 10{,}000\ Å = 10^{-6}\ m\,. \tag{3.1}$$

Some spectra are recorded on a wavenumber scale, which is inversely proportional to wavelength. The most common wavenumber unit is the reciprocal centimetre, which is related to wavelength units as follows:

$$1.0\ \mu m = 10^{-4}\ cm = 10^{4}\ cm^{-1} \text{ or } 10{,}000\ cm^{-1}. \tag{3.2}$$

Thus, 1,000 nm corresponds to 10,000 cm^{-1}, 400 nm to 25,000 cm^{-1} and 2,000 nm to 5,000 cm^{-1}. Wavenumbers are sometimes equated with kaysers, ***K***, where

$$1\ \boldsymbol{K} = 1\ cm^{-1}, \text{ and } 10{,}000\ cm^{-1} = 10\ k\boldsymbol{K}\,. \tag{3.3}$$

The advantage of the wavenumber scale is that it is linearly proportional to other energy units. Some of the relationships and conversion factors (see Appendix 8) are as follows:

$$1\ cm^{-1} = 11.966\ J = 2.859\ cal = 1.24 \times 10^{-4}\ eV. \tag{3.4}$$

Convenient bench-marks in visible to near-infrared spectroscopy are summarized in table 3.1. Note that the Å, nm and cm^{-1} units are used interchangeably in this chapter and elsewhere throughout the book. Note, too, that the wavelength range 400 to 2,000 nm (or 4,000 to 20,000 Å, or 25,000 to 5,000 cm^{-1}) corresponds to energies of approximately 300 to 60 kJ. Energies of this magnitude are comparable to changes of enthalphies and free energies in many chemical reactions.

Table 3.1. *Wavelength and energy units used in crystal field spectra of minerals*

Wavelength units			Energy units			Colour
microns (μm)	nanometres (nm)	ångstroms (Å)	wave-numbers (cm^{-1})	electron volts (eV)	kilojoules (kJ)	
0.3	300	3,000	33,000	4.09	394.7	ultraviolet
0.4	400	4,000	25,000	3.10	299.0	violet
0.5	500	5,000	20,000	2.48	239.4	green
0.6	600	6,000	16,667	2.07	199.3	orange
0.7	700	7,000	14,286	1.77	170.9	red
0.8	800	8,000	12,500	1.55	149.5	near-infrared
1.0	1,000	10,000	10,000	1.24	119.6	↓
1.25	1,250	12,500	8,000	0.99	95.7	
1.5	1,500	15,000	6,667	0.83	79.7	
1.75	1,750	17,500	5,714	0.71	68.3	
2.0	2,000	20,000	5,000	0.62	59.8	mid-infrared
2.5	2,500	25,000	4,000	0.50	47.8	↓

3.2.2 Absorption terminology

A variety of constants have been used to express absorption. All of these constants are based on the following general equation

$$\log \frac{I_0}{I} = a\, b, \tag{3.5}$$

where: I_o is the intensity of the incident light,
I is the intensity of the emergent light,
a is the absorption constant
b is a constant depending on the conditions of the measurement.

The ratio I_0/I is obtained directly from a spectrophotometric measurement, and the value of a is then calculated from eq. (3.5) to yield the desired absorption constant. The numerous absorption constants found in the literature arise from the choice of quantities incorporated in the constant b. Some of the terms most commonly used to express absorption in minerals are summarized in table 3.2. Note that optical densities (*O.D.*), representing the direct output from many spectrophotometers, lack specificity about sample thickness and element concentrations. Absorption coefficients (α) indicate that sample thicknesses have been measured or estimated. Molar extinction coefficients (ε) require chemical analytical data as well as knowledge of sample thicknesses.

Table 3.2. *Absorption units used in crystal field spectra of minerals*

Intensity term	Symbol	Equation*	Units
Transmission	T	$T = (I/I_0)$	–
Absorption	A	$A = 1 - T = 1 - (I/I_0)$	–
Optical density or Absorbance	$O.D.$	$O.D. = \log(I/I_0)$	–
Absorption coefficient or Extinction coefficient	α	$\alpha = \log(I/I_0)/d$	cm^{-1}
Molar extinction coefficient	ε	$\varepsilon = \log(I/I_0)/C\,d$	litre $mole^{-1}$ cm^{-1} or litre $(g.ion)^{-1}$ cm^{-1}
Oscillator strength	f	$f \approx (\text{const})\ \varepsilon_{max}\ \Delta\upsilon_{1/2}$	–

* I_0 and I are the intensities of incident and emergent light, respectively; d is the thickness, in cm, of the material or optical path length in the medium; C is the concentration of the absorbing species in moles per litre or gm ions per litre; $\Delta\upsilon_{1/2}$ is the width of an absorption band (expressed as wavenumber units) at half peak-height at which the molar extinction coefficient, ε, is a maximum.

The assumption is made that absorption bands have Gaussian shapes, that is, they fit an expression of the form

$$\varepsilon = \varepsilon_0 \exp[-C(\upsilon - \upsilon_0)^2], \tag{3.6}$$

where C is a constant, υ_0 is the wavenumber at the centre of the band, and ε_0 is the molar extinction coefficient there. The area under the absorption band is approximately equal to $\varepsilon_0\Delta\upsilon_{1/2}$, where $\Delta\upsilon_{1/2}$ is the full width of the band at half peak-height in wavenumber units. This area also appears in approximations of the oscillator strength, f, which is related to the probability of an electronic transition, discussed later (§3.7.1).

A fundamental relationship used in electronic absorption spectroscopy is the Beer–Lambert law which states that the amount of light absorbed is proportional to the number of absorbing molecules or ions through which the light passes. The Beer–Lambert law is formulated as

$$\log \frac{I_0}{I} = \varepsilon\, C\, d, \tag{3.7}$$

where ε is the molar extinction coefficient in units such as litre $mole^{-1}$ cm^{-1}
C is concentration expressed in moles/litre or gram ions/litre
d is the path length measured in centimetres.

The concentration, C, of a transition metal ion in solid-solution in a silicate mineral may be calculated from the mole fraction, X, and the molar volume, V, of the silicate phase by the equation

$$C = \frac{X}{V}. \tag{3.8}$$

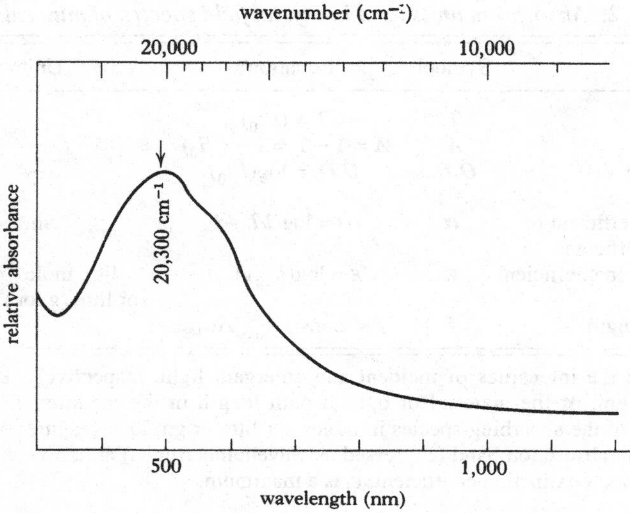

Figure 3.1 Absorption spectrum of the hexahydrated Ti^{3+} ion in an aqueous solution of caesium titanium(III) sulphate.

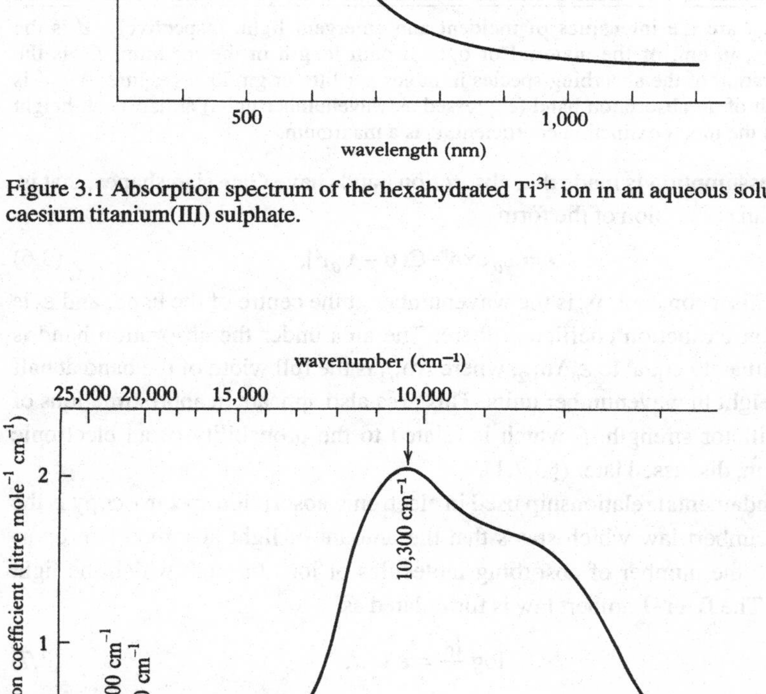

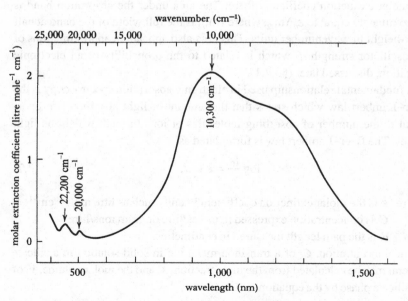

Figure 3.2 Absorption spectrum of the hexahydrated Fe^{2+} ion in an aqueous solution of iron(II) ammonium sulphate

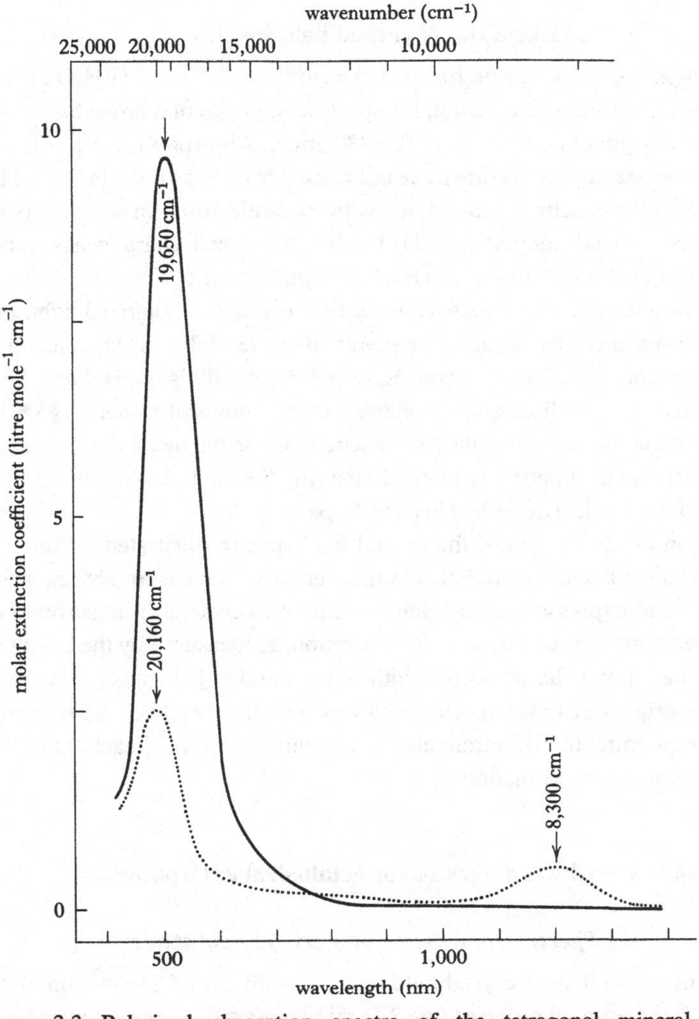

Figure 3.3 Polarized absorption spectra of the tetragonal mineral gillespite, $BaFe^{2+}Si_4O_{10}$, measured with light polarized parallel and perpendicular to the *c* crystallographic axis (see §3.8.1). —— **E**∥*c* spectrum; ······· **E**⊥*c* spectrum.

Methods for calculating molar extinction coefficients of minerals are outlined in chapter 4 (§4.3). The importance of the Beer–Lambert law, eq. (3.7), is that the molar extinction coefficient of an absorption band should be independent of the concentration of the absorbing species. Deviations from this law originating from cation ordering are discussed in chapter 4.

3.3 Examples of crystal field spectra

The absorption spectrum profile for the hexahydrated Ti^{3+} ion, $[Ti(H_2O)_6]^{3+}$, is shown in fig. 3.1. For this $3d^1$ cation, the spectrum consists of a broad band centred at a wavelength of about 493 nm (20,280 cm^{-1}). Absorption spectra of ions with more than one $3d$ electron are more complex. This is illustrated in fig. 3.2 by the spectrum of the hexahydrated Fe^{2+} ion with six $3d$ electrons. In addition to the dominant broad band centred around 1,000 nm, several sharp peaks occur between 400 and 500 nm for the $[Fe(H_2O)_6]^{2+}$ ion. Further complexities arise in the spectra of transition metal ions in anisotropic crystals in polarized light. For example, the spectra of the tetragonal mineral gillespite, $BaFe^{2+}Si_4O_{10}$, shown in fig. 3.3 differ when the electric vector, **E**, of polarized light is transmitted along the two different crystallographic directions of this uniaxial mineral (§3.8.1). Even larger variations exist in polarized spectra of transition metal ions in biaxial minerals, such as Fe^{2+} in orthorhombic olivine (fig. 5.8) and Mn^{3+} in monoclinic epidote crystals (fig. 4.3) described in later chapters.

Absorption bands in each of the crystal field spectra illustrated in figs 3.1, 3.2 and 3.3 have three characteristic features: energy, as measured by the position of the band expressed in wavelength (e.g., nanometres or ångstroms) or wavenumber (cm^{-1}) units; intensity of absorption, as measured by the height or area of the band; and sharpness or width of the band at half maximum peak-height. The origins of these spectral features may be correlated with energy levels corresponding to differerent electronic configurations of each transition metal ion in various coordination sites.

3.4 Energy level diagrams for octahedral environments

3.4.1 Spectroscopic terms and crystal field states

The effect of an octahedral crystal field on the $3d$ orbitals of a transition metal ion is to split the original group of five $3d$ orbitals into two levels, a lower-level group of three t_{2g} orbitals and an upper-level group of two e_g orbitals separated by an energy Δ_o (see fig. 2.4). Illustrated in fig. 3.4 are the energy levels of the Ti^{3+} ion before and after its single $3d$ electron is excited from a t_{2g} orbital to an e_g orbital when the cation is in octahedral coordination with six ligands. The energy level diagram shows that the energy separation between the two electronic states increases with increasing strength of the crystal field. The two crystal field states are derived from one energy level for the gaseous or field-free Ti^{3+} ion, designated as the ^2D spectroscopic term. The lowest energy state, or ground state, of the Ti^{3+} ion in an octahedral crystal field can have only one

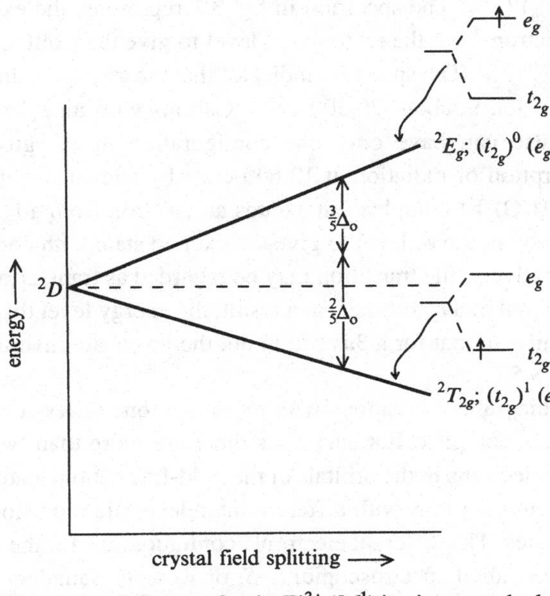

Figure 3.4 Energy level diagram for the Ti^{3+} ($3d^1$) ion in an octahedral crystal field.

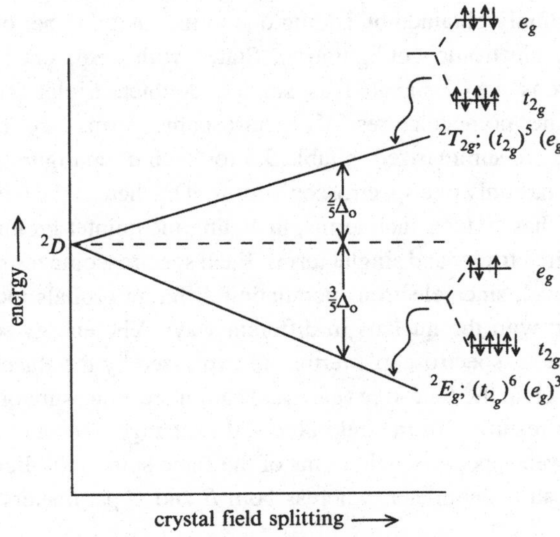

Figure 3.5 Energy level diagram for the Cu^{2+} ($3d^9$) ion in an octahedral crystal field.

configuration, $(t_{2g})^1(e_g)^0$. The spectrum in fig. 3.1 represents the excitation of the single 3*d* electron from the t_{2g} to the e_g level to give the excited state with configuration $(t_{2g})^0(e_g)^1$. The spectrum indicates that the average value of Δ_o for the hydrated Ti^{3+} ion is about 20,300 cm^{-1}. Cations with nine 3*d* electrons, such as Cu^{2+}, also can have only one configuration in the ground state, $(t_{2g})^6(e_g)^3$. Absorption of radiation at 12,600 cm^{-1} by aqueous solutions containing the $[Cu(H_2O)_6]^{2+}$ complex ion excites an electron from a t_{2g} orbital to the hole or vacancy in the e_g level to give the excited state with configuration $(t_{2g})^5(e_g)^4$. Alternatively, the transition may be regarded as transfer of a hole in the unfilled e_g orbital to a t_{2g} orbital. As a result, the energy level diagram for a $3d^9$ cation is similar to that for a $3d^1$ cation but the levels are inverted. This is illustrated in fig. 3.5.

Energy level diagrams for cations with more than one or less than nine 3*d* electrons are more complex. For such ions there are more than two ways of arranging the 3*d* electrons in the orbitals of the field-free cation, leading to several electronic configurations with different interelectronic repulsion energies for the excited states. The different electronic configurations for the isolated or field-free ions are called spectroscopic, *L–S*, or Russell–Saunders terms and are designated by symbols such as ^2D, ^3F, ^6S, etc. The capital letters denote the total atomic orbital angular momenta ($L = \Sigma m_l$) summed over all of the 3*d* electrons, the nomenclature being similar to those for individual *s, p, d* and *f* electrons given in §2.2.2. The superscript numerals denote the spin multiplicities, *r*, where $r = (2\Sigma m_s + 1)$ and $m_s = \frac{1}{2}$ for each unpaired electron. Thus, the spin-multiplicity is simply obtained by adding one to the total number of unpaired electrons in an electronic configuration. States with zero, one, two, etc., unpaired electrons are designated as singlet, doublet, triplet, etc., states, respectively. The nomenclatures of spectroscopic terms are defined in Appendix 4 and are summarized in table 3.3 for each d^n configuration. A $3d^1$ ion such as Ti^{3+} has only one spectroscopic term, ^2D, whereas the Fe^{2+} ion with six 3*d* electrons has sixteen such terms, including the quintet ground term ^5D and several excited triplet and singlet terms. Each spectroscopic term has a different energy level, since electrons populating different orbitals interact with one another and with the nucleus in different ways. The energy separations between the various spectroscopic terms are expressed by the Racah *B* and *C* parameters* which in the field-free (gaseous) cation are a measure of interelectronic repulsion resulting from coulombic and exchange interactions. Energy separations between spectroscopic terms of the same spin multiplicity usually contain the Racah *B* parameter, whereas both *B* and *C* parameters appear in

* Occasionally, the Condon-Shortley parameters F_2 and F_4, are used instead, and are related to the Racah parameters by $B = F_2 - 5F_4$ and $C = 35F_4$

Table 3.3. *Spectroscopic terms arising from each 3d^n configuration for transition metal ions*

Electronic configuration	Cations							Spectroscopic terms*
	M^{1+}	M^{2+}	M^{3+}	M^{4+}	M^{5+}	M^{6+}	M^{7+}	
[Ar]	K^+	Ca^{2+}						^1S
[Ar]$3d^0$			Sc^{3+}	Ti^{4+}	V^{5+}	Cr^{6+}	Mn^{7+}	^1S
[Ar]$3d^{10}$	Cu^+	Zn^{2+}	Ga^{3+}	Ge^{4+}				
[Ar]$3d^1$			Ti^{3+}	V^{4+}				^2D
[Ar]$3d^9$		Cu^{2+}						
[Ar]$3d^2$			V^{3+}					^3F, ^3P, ^1G, ^1D, ^1S
[Ar]$3d^8$		Ni^{2+}						
[Ar]$3d^3$			Cr^{3+}	Mn^{4+}				^4F, ^4P, ^2H, ^2G, ^2F, ^2D, ^2D
[Ar]$3d^7$		Co^{2+}	Ni^{3+}					
[Ar]$3d^4$		Cr^{2+}	Mn^{3+}					^5D, ^3H, ^3G, ^3F, ^3F, ^3D, ^3P, ^3P, ^1I, ^1G, ^1G, ^1F, ^1D, ^1D, ^1S, ^1S
[Ar]$3d^6$		Fe^{2+}	Co^{3+}					
[Ar]$3d^5$		Mn^{2+}	Fe^{3+}					^6S, ^4G, ^4F, ^4D, ^4P, ^2I, ^2H, ^2G, ^2G, ^2F, ^2F, ^2D, ^2D, ^2D, ^2P, ^2S

* The ground term for each electronic configuration is listed first.

expressions for energy differences between terms having different spin multiplicities (Lever, 1984, p. 126). Although these parameters could be obtained by accurate calculations of the coulomb and exchange integrals (Ballhausen, 1962; Lever, 1984), they are determined empirically from measurements of atomic spectra. The use of Racah parameters as a measure of covalent bonding is discussed in chapter 11 (§11.2).

The energy level diagram for Ti^{3+} in fig. 3.4 shows the manner by which the ^2D spectroscopic term is resolved into two different levels, or crystal field states, when the cation is situated in an octahedral crystal field produced by surrounding ligands. In a similar manner the spectroscopic terms for each $3d^n$ configuration become separated into one or more crystal field states when the transition metal ion is located in a coordination site in a crystal structure. The extent to which each spectroscopic term is split into crystal field states can be obtained by semi-empirical calculations based on the interelectronic repulsion Racah *B* and *C* parameters derived from atomic spectra (Lever, 1984, p. 126).

Table 3.4. *Crystal field states arising from free ion spectroscopic terms of transition metals in octahedral coordination*

Free-ion spectroscopic terms	Crystal field states
S	A_{1g}
P	T_{1g}
D	$T_{2g} + E_g$
F	$A_{2g} + T_{1g} + T_{2g}$
G	$A_{1g} + E_g + T_{1g} + T_{2g}$
H	$E_g + T_{1g} + T_{1g} + T_{2g}$
I	$A_{1g} + A_{2g} + E_g + T_{1g} + T_{2g} + T_{2g}$

Bethe, in 1929, used symmetry arguments to determine the qualitative nature of the orbital splittings for transition metal ions in various coordination symmetries. Bethe (1929) demonstrated how it is possible, using the methods of group theory, to determine just what crystal field states will result when an ion of any given electronic configuration is introduced into a crystal field with a definite symmetry. Bethe's calculations showed that for octahedral environments there are collectively only five types of crystal field states that can arise from the various spectroscopic terms of field-free transition metal ions when they are placed in an octahedral crystal field. These crystal field states are summarized in table 3.4 (see also Appendix 5). Each crystal field state is given a group theory symbol notating the symmetry properties of its electronic configuration, by analogy with the wave functions of electrons in individual s (a_{1g}), p (t_{1u}) and d (t_{2g} and e_g) orbitals described earlier (§2.3). Thus, the ^2D spectroscopic term of Ti^{3+} is resolved into the $^2T_{2g}$ and ^2E_g crystal field states when the cation is present in an octahedral site in a crystal structure. This symbolism for crystal field states is analogous to that used for $3d$ orbitals which are separated into t_{2g} and e_g groups when the cation occurs in octahedral coordination. The symbol $^2T_{2g}$ represents the symmetry of the ground-state configuration $(t_{2g})^1(e_g)^0$ and highlights the symmetry of the three-fold degeneracy (symbol T_2) of the single $3d$ electron in three equivalent t_{2g} orbitals in the centrosymmetric (subscript g) octahedral site. The spin-multiplicity of Ti^{3+} is denoted by the superscript 2 (doublet) in the $^2T_{2g}$ representation. When the single $3d$ electron of Ti^{3+} is induced to occupy one of the e_g orbitals by absorption of light in the visible region (*e.g.*, at the wavelength 493 nm in fig. 3.1), the excited state is represented by ^2E_g to designate the symmetry of the two-fold degeneracy (symbol E) of the $(t_{2g})^0(e_g)^1$ electronic configuration. The $^2T_{2g} \rightarrow {}^2E_g$

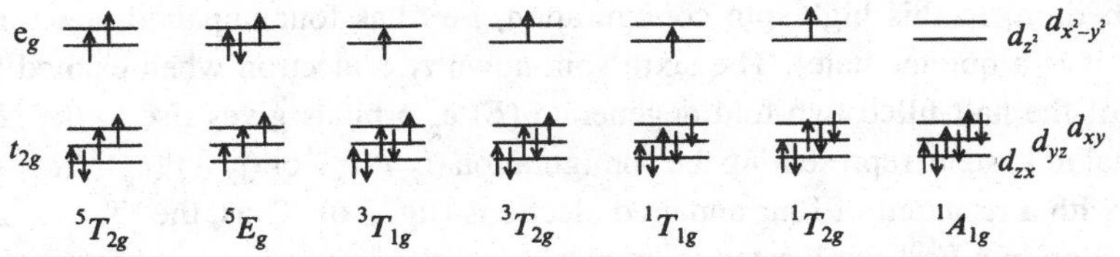

Figure 3.6. Electronic configurations of the ground state and some of the excited crystal field states of the Fe^{2+} ($3d^6$) ion in octahedral coordination.

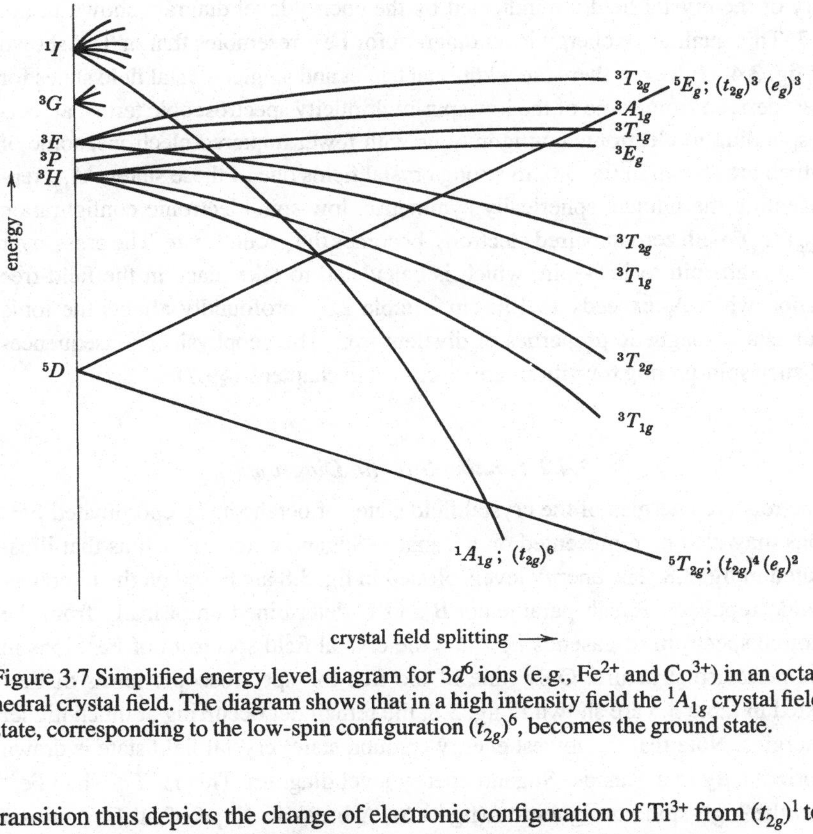

Figure 3.7 Simplified energy level diagram for $3d^6$ ions (e.g., Fe^{2+} and Co^{3+}) in an octahedral crystal field. The diagram shows that in a high intensity field the $^1A_{1g}$ crystal field state, corresponding to the low-spin configuration $(t_{2g})^6$, becomes the ground state.

transition thus depicts the change of electronic configuration of Ti^{3+} from $(t_{2g})^1$ to $(e_g)^1$ in which there is no change in the number of unpaired electrons.

The Fe^{2+} ion, $3d^6$, in octahedral coordination in a silicate has the ground state configuration $(t_{2g})^4(e_g)^2$ or alternatively $(t_{2g}\uparrow)^3(t_{2g}\downarrow)^1(e_g\uparrow)^2$ depicted in fig. 3.6. This configuration is represented by $^5T_{2g}$, in which five of the $3d$ electrons spinning, say, clockwise (or spin-up) occupy singly each of the $3d$ orbitals and the sixth electron with anticlockwise spin (or spin-down) pairs up in one of the three-fold degenerate (symmetry T_2) t_{2g} orbitals in the centrosymmetric (g)

octahedron. In this high-spin configuration, Fe^{2+} has four unpaired electrons (i.e., it is a quintet state). The sixth spin-down $t_{2g}\downarrow$ electron when excited to one of the half-filled two-fold degenerate (*E*) e_g orbitals gives rise to the ^5E_g crystal field state, representing the configuration $(t_{2g})^3(e_g)^3$ or $(t_{2g}\uparrow)^3(e_g\uparrow)^2(e_g\downarrow)^1$, still with a resultant of four unpaired electrons (fig. 3.6). Thus, the $^5T_{2g} \rightarrow {}^5E_g$ transition in a Fe^{2+} ion located in an octahedral site results in no change in the number of unpaired electrons. The relative energies of the two quintet crystal field states of Fe^{2+} derived from the ^5D spectroscopic term with increasing intensity of the crystal field are indicated by the energy level diagram shown in fig. 3.7. This qualitative energy level diagram for Fe^{2+} resembles that of Ti^{3+} shown in fig. 3.4. However, there are additional triplet and singlet crystal field states for Fe^{2+} derived from some of the low spin-multiplicity spectroscopic terms, all corresponding to electronic configurations with fewer unpaired electrons, some of which are shown in fig. 3.6. In strong crystal fields one of these states, $^1A_{1g}$, representing the unique, spherically symmetric, low-spin electronic configuration $(t_{2g})^6(e_g)^0$ with zero unpaired electrons, becomes the ground state. The cross-over from high-spin to low-spin, which is calculated to take place in the field-free cation when Δ_o exceeds 19,150 cm^{-1} (table 2.5), profoundly affects the ionic radius and magnetic properties of divalent iron. The geophysical consequences of such spin-pairing transitions are discussed in chapter 9 (§9.7).

3.4.2 Tanabe–Sugano Diagrams

The relative energies of the crystal field states of octahedrally coordinated Fe^{2+} ions may also be represented on a Tanabe–Sugano diagram such as that illustrated in fig. 3.8. The energy levels plotted in fig. 3.8 are based on the interelectronic repulsion Racah parameters *B* and *C* determined empirically from the atomic spectrum of gaseous Fe^{2+} and the crystal field spectrum of Fe^{2+} ions in the periclase structure. Only nine of the free ion spectroscopic terms of Fe^{2+} listed in table 3.3 are shown in fig. 3.8, the remainder occurring at much higher energies. Note that the lowest energy (ground state) crystal field state is drawn horizontally in a Tanabe–Sugano energy level diagram. This is $^5T_{2g}$ when Fe^{2+} has the high-spin configuration (fig. 3.6). A break in slope of all Fe^{2+} crystal field states occurs beyond the vertical dashed line in fig. 3.8 when the $^1A_{1g}$ level, corresponding to the low-spin configuration (fig. 3.6), becomes the ground state. This occurs when the strength of the crystal field in a coordination site becomes sufficiently high to induce spin-pairing in Fe^{2+} ions (§2.5 and §9.7). Transitions shown by the vertical dotted line in fig. 3.8 correspond to features in the spectrum of the hexahydrated Fe^{2+} ion illustrated in fig. 3.2, including the $^5T_{2g} \rightarrow {}^5E_g$ transition responsible for the broad absorption band near 10,000 cm^{-1}.

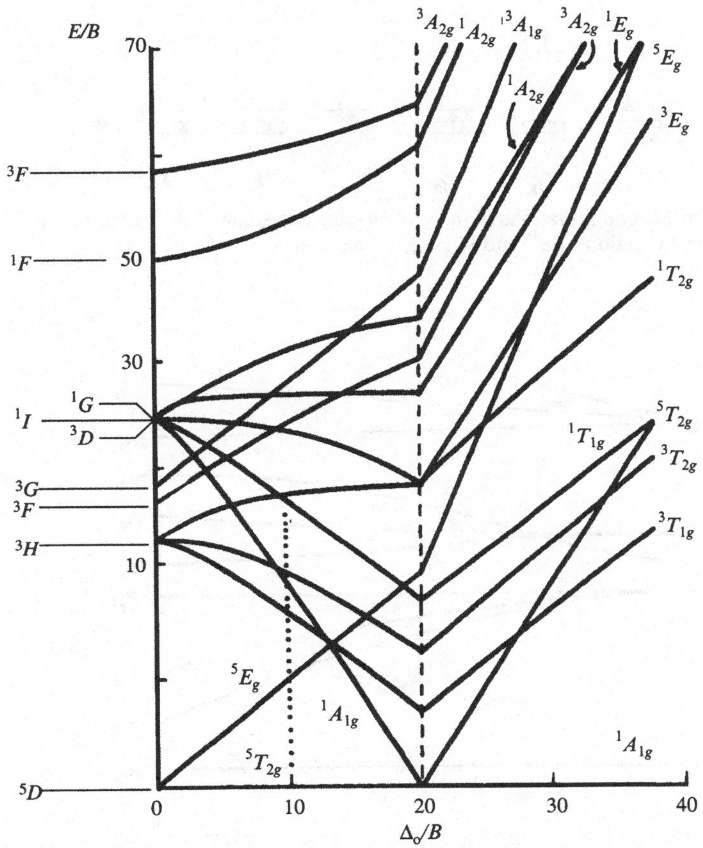

Figure 3.8 Tanabe–Sugano energy level diagram for a $3d^6$ ion in an octahedral crystal field. Note that some of the highest energy triplet and singlet crystal field states listed in table 3.3 are not shown in the diagram.

This transition results in no change in the number of unpaired electrons in Fe^{2+} and is referred to as a spin-allowed transition. The sharp peaks at 20,000 cm^{-1} and 22,200 cm^{-1} in fig. 3.2 correspond to transitions to the triplet states $^3T_{1g}$ and $^3T_{2g}$, respectively (cf. fig. 3.6), having lower spin-multiplicities than the quintet $^5T_{2g}$ ground state. Such transitions leading to fewer unpaired electrons in excited crystal field states are termed spin-forbidden transitions.

Comparable energy level diagrams to that shown for Fe^{2+} in fig. 3.8 may be constructed for each transition metal cation from spectroscopic terms of the field-free ion. The Tanabe–Sugano diagrams for other octahedrally coordinated transition metal ions are discussed later in chapter 5 (§5.10)

Cations with $3d^5$ configurations have distinctive energy level diagrams and crystal field spectra which are worthy of special attention. The electronic con-

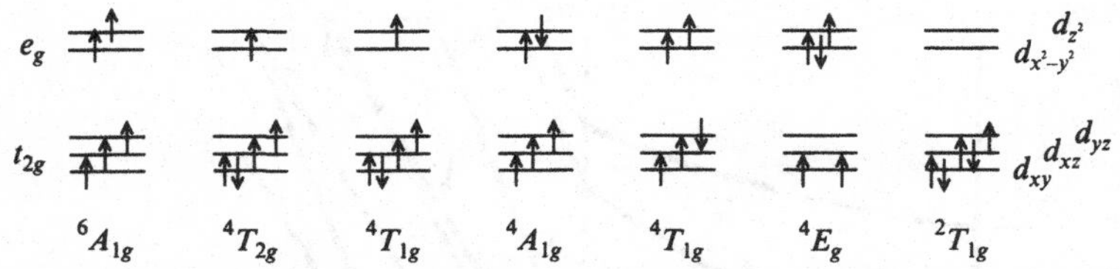

Figure 3.9 Electronic configurations of the ground state and some of the excited crystal field states of the $3d^5$ cations, Fe^{3+} and Mn^{2+}, in octahedral coordination.

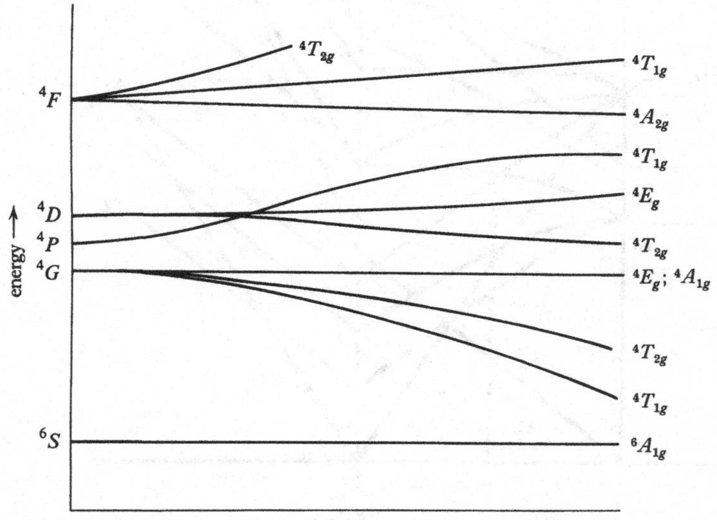

Figure 3.10 Partial energy level diagram for the Fe^{3+} or Mn^{2+} ions with $3d^5$ configurations in high-spin states in an octahedral crystal field. Only sextet and quartet spectroscopic terms and crystal field states are shown. Note that the same energy level diagram applies to the cations in tetrahedral crystal fields (with g subscripts omitted from the state symbols for the acentric coordination site).

figurations of Fe^{3+} and Mn^{2+} illustrated in fig. 3.9 are such that in the ground state the five $3d$ electrons occupy singly each of the three t_{2g} and the two e_g orbitals. In this half-filled high-spin state, $(t_{2g}\uparrow)^3(e_g\uparrow)^2$, represented by $^6A_{1g}$ (where A_{1g} again represents an unique spherically symmetric electronic configuration), there are five unpaired electrons (i.e. a sextet state) with spins aligned parallel (spin-up) when Fe^{3+} ions occupy centrosymmetric octahedral sites. In the energy level diagram shown in fig. 3.10, the $^6A_{1g}$ level represents the ground state of Fe^{3+} in an octahedral crystal field. All excited states of Fe^{3+}, some of which are depicted in fig. 3.9, correspond to electronic configurations with fewer unpaired electrons and, hence, lower spin-multiplicities (i.e. they are either quartets or doublets) than the sextet ground state. As a result, all electronic transitions between the 3*d* orbitals of the Fe^{3+} and Mn^{2+} ions are spin-forbidden, leading to weak spectral features in their absorption spectra. Energy levels of the $^4T_{1g}$ and $^4T_{2g}$ crystal field states shown in fig. 3.10 correspond to electronic configurations derived from $(t_{2g})^4(e_g)^1$ or $(t_{2g}\uparrow)^3(t_{2g}\downarrow)^1(e_g\uparrow)^1$. The difference between these two excited crystal field states is related to different interaction energies between the solitary electron remaining in one of the e_g orbitals and the paired electron in one of the t_{2g} orbitals. Thus, less repulsion occurs when these electrons occupy, say, the d_{z^2} and d_{xy} orbitals ($^4T_{1g}$), compared to the d_{xy} and $d_{x^2-y^2}$ orbitals ($^4T_{2g}$). Other distributions such as ($d_{x^2-y^2}$ + d_{yz},) or ($d_{x^2-y^2}$ + d_{xz},) and (d_{xz} +d_{z^2}) or (d_{yz} + d_{z^2}) account for the three-fold degeneracies represented by the $^4T_{1g}$ and $^4T_{2g}$ states, respectively.

3.4.3 Orgel diagrams

Certain qualitative similarities exist between the energy level diagrams for the various $3d^n$ configurations when only crystal field states with the highest spin-multiplicity are considered. There are essentially only three distinct types of energy level diagrams for transition metal ions, corresponding to crystal field states derived from the field-free *S, D,* and *F* spectroscopic ground terms, respectively. Such features, first recognised by Orgel (1952), are summarized in the Orgel diagrams shown in figs 3.11 and 3.12. As might be inferred from fig. 3.4 (Ti^{3+}) and fig. 3.6 (Fe^{2+}), octahedral cations with $3d^1$ and $3d^6$ configurations have the same type of diagram, which in inverted form correspond to the energy level diagram for $3d^4$ (Mn^{3+}) and $3d^9$ (Cu^{2+}) ions (cf. fig. 3.5). These cations, each possessing a *D* spectroscopic ground term, are all represented by the Orgel diagram shown in fig. 3.11. Similarly, octahedrally coordinated cations with $3d^2$ (V^{3+}) and $3d^7$ (Co^{2+}) configurations have similar Orgel diagrams, which correspond to the inverted diagrams for $3d^3$ (Cr^{3+}) and $3d^8$ ions (Ni^{2+}). The Orgel diagram linking all cations with *F* spectroscopic ground terms is shown in fig. 3.12. Curvature of crystal field states with T_{1g} symmetry derived from the spectroscopic *F* and *P* terms is a consequence of the non-crossing rule of group theory. Thus, states with the same symmetry may interact with one another, whereby the lower energy state becomes stabilized and the higher energy one destabilized the closer they approach each other. This non-crossing behaviour of the $T_{1g}(F)$ and $T_{1g}(P)$ states is a manifestation of molecular orbital bonding interactions and influences the Racah *B* parameter, which is used as a measure of covalent bond character in transition metal compounds described in chapter 11 (§11.2). The third type of Orgel diagram, that

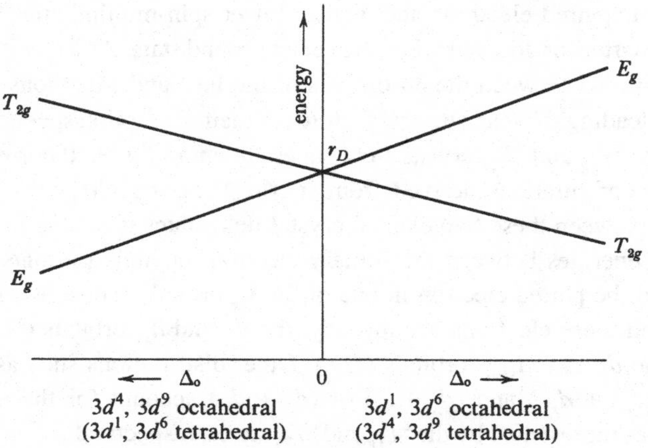

Figure 3.11 Orgel diagram for transition metal ions possessing ^rD spectroscopic terms in octahedral crystal fields of increasing intensity. The right-hand side applies to $3d^1$ (e.g., Ti^{3+}) and $3d^6$ (e.g., Fe^{2+}) cations and the left-hand side to $3d^4$ (e.g., Mn^{3+}) and $3d^9$ (e.g., Cu^{2+}) cations in octahedral coordination. The diagram in reverse also applies to the cations in tetrahedral, cubic and dodecahedral coordinations.

for $3d^5$ cations (Fe^{3+} and Mn^{2+}), corresponds to the Tanabe–Sugano diagram already illustrated in fig. 3.10.

It must be stressed that although qualitative features exist in Orgel diagrams for the three groups of transition metal ions, they are not identical for any pair of cations since the energy separations between the crystal field states of individual ions are different. For example, although qualitative similarities exist in the energy level diagrams for octahedral $3d^1$ and $3d^6$ ions, the energy separations for Ti^{3+} are larger than those for Fe^{2+}, which influence the positions of absorption bands in their respective crystal field spectra (compare figs 3.1 and 3.2.). A similar situation exists for the octahedral $3d^3$ and $3d^8$ cations (Cr^{3+} and Ni^{2+}) and the $3d^2$ and $3d^7$ cations (V^{3+} and Co^{2+}).

The Orgel diagrams illustrated in figs 3.11 and 3.12 indicate that, for electronic transitions between crystal field states of highest spin-multiplicities, one absorption band only is expected in the spectra of $3d^1$, $3d^4$, $3d^6$ and $3d^9$ cations in octahedral coordination, whereas three bands should occur in the spectra of octahedrally coordinated $3d^2$, $3d^3$, $3d^7$ and $3d^8$ ions. Thus, if a crystal structure is known to contain cations in regular octahedral sites, the number and positions of absorption bands in a spectrum might be used to identify the presence and valence of a transition metal ion in these sites. However, this method of cation identification must be used with caution. Multiple and displaced absorption bands may occur in the spectra of transition metal ions situated in low-symmetry distorted coordination sites.

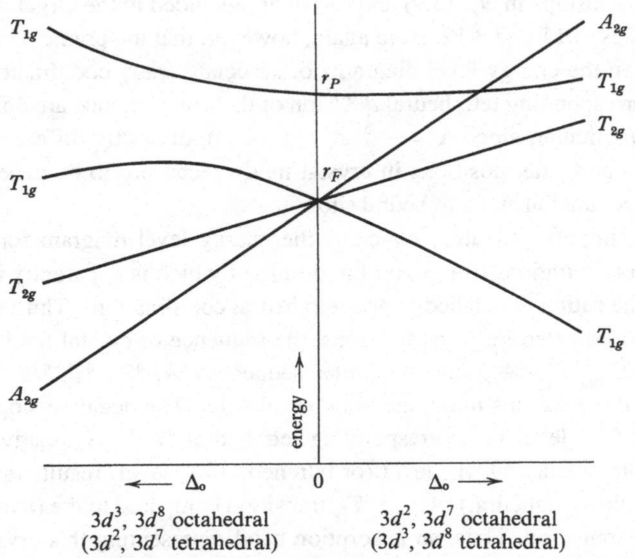

Figure 3.12 Orgel diagram for transition metal ions possessing ^rF and ^rP spectroscopic terms in octahedral crystal fields of increasing intensity. The right-hand side applies to $3d^2$ (e.g., V^{3+}) and $3d^7$ (e.g., Co^{2+}) cations and the left-hand side to $3d^3$ (e.g., Cr^{3+}) and $3d^8$ (e.g., Ni^{2+}) cations in octahedral coordination. The diagram in reverse also applies to the cations in tetrahedral, cubic and dodecahedral coordinations.

3.5 Energy level diagrams for other high-symmetry environments

Energy level diagrams for cations in tetrahedral and body-centred cubic coordinations are related to those for octahedrally coordinated ions. Since the crystal field splittings of $3d$ orbitals in tetrahedral and eight-fold cubic coordinations are opposite to that for octahedral coordination (§2.6), energy levels for cations in tetrahedral, cubic and dodecahedral environments correspond to inverted diagrams for the corresponding ions in octahedral coordination, except for cations with $3d^5$ configurations. These correlations lead to the following general relationships between energy level diagrams:

$$3d^n_{\text{octahedral}} \text{ is the same as } 3d^{10-n}_{\text{tetrahedral}},\ 3d^{10-n}_{\text{cubic}} \text{ and } 3d^{10-n}_{\text{dodecahedral}} \quad (3.9)$$

and

$$3d^n_{\text{tetrahedral}},\ 3d^n_{\text{cubic}},\ 3d^n_{\text{dodecahedral}} \text{ are the same as } 3d^{10-n}_{\text{octahedral}}. \quad (3.10)$$

For $3d^5$ ions, the same diagram applies for octahedral, tetrahedral, cubic and dodecahedral coordination environments.

The relationships in eqs (3.9) and (3.10) are included in the Orgel diagrams shown in figs. 3.11 and 3.12. Note again, however, that the predicted similarities between the energy level diagrams of an octahedrally coordinated d^n ion and the corresponding tetrahedral d^{10-n} ion of the same valence are only qualitative. In particular, since $\Delta_t = -\frac{4}{9}\Delta_o$ (eq. (2.7)), distinctly different energy separations and peak positions in crystal field spectra are to be expected for cations in octahedral and tetrahedral sites.

Another important feature concerns the energy level diagram for cations with $3d^5$ configurations (fig. 3.10), the profile of which is apparently identical for the same cation in octahedral and tetrahedral coordinations. Thus, for octahedrally coordinated Fe^{3+} or Mn^{2+} ions, the sequence of crystal field states is $^6A_{1g}$, $^4T_{1g}$, $^4T_{2g}$, ^4E_g, $^4A_{1g}$, etc. A similar sequence, $^6A_1$, $^4T_1$, $^4T_2$, ^4E, $^4A_1$, etc., applies to these cations in tetrahedral coordination. The negative slope of the octahedral $^4T_{1g}$ level (or corresponding tetrahedral $^4T_1$ level) relative to the ground state octahedral $^6A_{1g}$ level (or tetrahedral $^6A_1$ level) results in a *lower* energy for the (octahedral) $^6A_{1g} \rightarrow {}^4T_{1g}$ transition compared to the (tetrahedral) $^6A_1 \rightarrow {}^4T_1$ transition. Thus, an absorption band representing this crystal field transition in tetrahedrally coordinated Fe^{3+} ions occurs at a *higher* energy than that for Fe^{3+} ions in octahedral coordination, apparently in contradiction to eq. (2.7). A similar situation applies to the (octahedral) $^6A_{1g} \rightarrow {}^4T_{2g}$ and (tetrahedral) $^6A_1 \rightarrow {}^4T_2$ transitions, but not to transitions involving the (octahedral) $^4A_{1g}$,^4E_g or (tetrahedral) $^4A_1$,^4E levels. This is because the energy separation between the $^6A_{1g}$ and $^4A_{1g}$,^4E_g levels is relatively unaffected by changes of crystal field strength. Such factors are apparent in crystal field spectral data of Mn^{2+}-and Fe^{3+}-bearing minerals summarized in tables 5.14, 5.15 and 10.2.

3.6 Energy level diagrams for low-symmetry environments

Further resolution of the 3*d* orbital energy levels takes place, as illustrated in fig. 2.9, when the symmetry of the environment about a transition metal is lowered from octahedral symmetry by either distortion of the coordination site or uneven distribution of the ligands forming the coordination polyhedron. Additional electronic configurations arise from the different arrangements of electrons in the resolved orbitals, each having a different energy and symmetry. Therefore, several additional crystal field states of lower degeneracies may now exist for each $3d^n$ configuration. The symmetry of each crystal field state is again designated by group theory notation.

To designate point symmetries of regular and distorted coordination polyhedra in transition metal-bearing phases, Schöenflies symbols are assigned to spectroscopic states, by convention, rather than the Hermann–

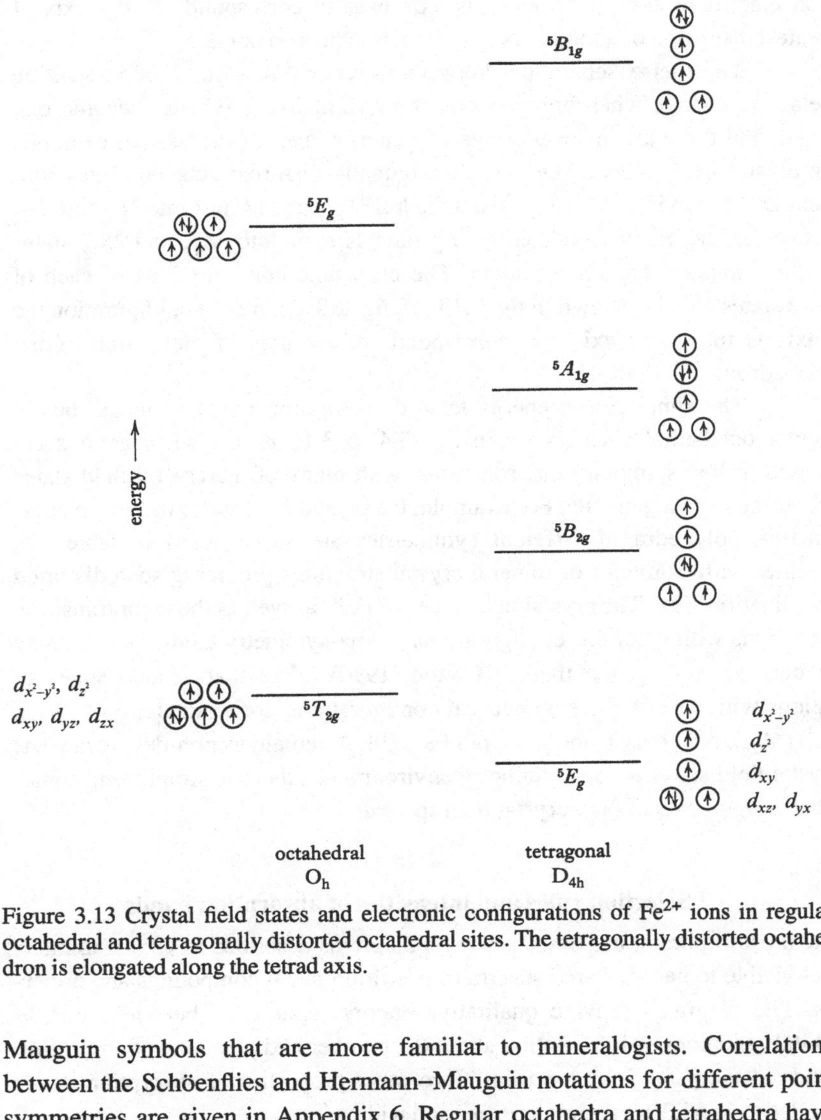

Figure 3.13 Crystal field states and electronic configurations of Fe^{2+} ions in regular octahedral and tetragonally distorted octahedral sites. The tetragonally distorted octahedron is elongated along the tetrad axis.

Mauguin symbols that are more familiar to mineralogists. Correlations between the Schöenflies and Hermann–Mauguin notations for different point symmetries are given in Appendix 6. Regular octahedra and tetrahedra have Schöenflies point symmetries O_h and T_d, respectively, while D_{4h} and C_{3v} may represent tetragonally distorted and trigonally distorted octahedra, respectively. Other examples of Schöenflies symbols denoting low-symmetry coordination sites in minerals are contained in Appendix 7, which also summarizes information about crystal structures of many transition metal-bearing phases. By convention, the major symmetry axis is taken to be the z electronic axis of each electronic configuration. This axis usually corresponds to the axis of greatest distortion of a regular octahedron, tetrahedron or cube.

The energy separations between the crystal field states of a transition metal ion change when high-symmetry coordination polyhedra become distorted, and these are again portrayed by energy level diagrams. For example, the crystal field states of Fe^{2+} in a tetragonally distorted octahedral environment are shown in fig. 3.13. The octahedral $^5T_{2g}$ state is split into ^5E_g and $^5B_{2g}$ states and the excited octahedral ^5E_g state is split into $^5A_{1g}$ and $^5B_{1g}$ states in the symmetry D_{4h} environment. The electronic configurations of each of these states are also shown in fig. 3.13 (cf. fig. 2.8.*a*). In each configuration the *z* axis is the tetrad axis and corresponds to the axis of elongation of the octahedron.

The simplicity of energy level diagrams for transition metal ions in regular octahedral sites shown in figs 3.4 to 3.12 is lost when cations are located in low-symmetry environments, with many of the crystal field states becoming non-degenerate. For example, the crystal field states of Fe^{2+} in coordination polyhedra of different symmetries are summarized in table 3.5, together with examples of mineral crystal structures providing such distorted coordination sites. The crystal field states of Fe^{2+}, as well as those for transition metal ions with other $3d^n$ configurations in low-symmetry environments, may be derived from group theory (Cotton, 1990). Note that ground states of cations with spherically symmetrical configurations, for example, Fe^{3+} ($^6A_{1g}$), Cr^{3+} ($^4A_{2g}$), Ni^{2+} ($^3A_{2g}$), and low-spin Fe^{2+} ($^1A_{1g}$), remain as non-degenerate (*A*) crystal field states in low-symmetry environments, thereby simplifying somewhat assignments of their crystal field spectra.

3.7 Selection rules and intensities of absorption bands

The most important use of energy level diagrams described in §3.5 is to interpret visible to near-infrared spectra of transition metal compounds and minerals. The diagrams provide qualitative energy separations between split $3d$ orbitals and convey information about the number and positions of absorption bands in a crystal field spectrum. Two other properties of absorption bands alluded to in §3.3 are their intensities and widths.

Several factors affect intensities of crystal field spectra. In addition to enhancement by increased temperature and pressure discussed in chapter 9 (§9.4), intensities of absorption bands depend on: first, the spin-state or number of unpaired electrons possessed by a transition metal ion; second, whether or not the cation is located at the centre of symmetry of a coordination site; and third, interactions with next-nearest-neighbour cations.

Table 3.5. *Crystal field states of Fe^{2+} in coordination sites of different symmetries*

Symmetry of crystal field	Symmetry notation	Crystal field states	Mineral examples
octahedral	O_h	$T_{2g} + E_g$	magnesiowüstite; ≈ orthopyroxene M1
tetrahedral	T_d	$E + T_2$	spinel; ≈ staurolite
tetragonal (elongated oct.)	D_{4h}	$E_g + B_{2g} + A_{1g} + B_{1g}$	≈ olivine M1
tetragonal (compressed oct.)	D_{4h}	$B_{2g} + E_g + B_{1g} + A_{1g}$	≈ amphibole M3; ≈ biotite M1
square planar	D_{4h}	$A_{1g} + E_g + B_{2g} + B_{1g}$	≈ gillespite; ≈ eudialyte
trigonal (compressed oct.)	D_{3d}	$A_{1g} + E_g + E_g$	silicate spinel (γ-phase)
trigonal (compressed oct.)	C_{3v}	$A_1 + E + E$	≈ corundum; ≈ olivine M2
trigonal (compressed oct.)	C_3	$A + E + E$	ilmenite; corundum
orthorhombic (distorted cube)	D_2	$A + A + B_1 + B_2 + B_3$	almandine garnet
orthorhombic	C_{2v}	$A_2 + B_1 + B_2 + A_1 + A_1$	amphibole M4; ≈ orthopyroxene M2
monoclinic	C_{2h}	$B_g + B_g + A_g + A_g + A_g$	amphibole M3; biotite M1
monoclinic	C_2	$A + B + B + A + A$	calcic pyroxene M1; amphibole M1; biotite M2
monoclinic	C_s	$A' + A' + A' + A'' + A''$	olivine M2
triclinic	C_i	$A_g + A_g + A_g + A_g + A_g$	olivine M1
triclinic	C_1	$A + A + A + A + A$	orthopyroxene M1

Intensities of absorption bands are governed by probabilities of electronic transitions between the split 3*d* orbital energy levels. The probabilities are expressed by selection rules, two of which are the spin-multiplicity selection rule and the Laporte selection rule.

3.7.1 *Laporte or parity selection rule*

An electronic transition takes place through the interaction of the electric field of incident electromagnetic radiation with a component of the dipole moment of the absorbing atomic or molecular system. Such transitions usually involving light in the visible region of the spectrum can occur only between states that differ in parity; that is, one state must have a symmetric (*g*) wave function and the other an antisymmetric (*u*) wave function.

The probability of a transition, P_x, which influences the intensity of an absorption band, depends on the square of the transition moment, Q, expressed as

$$\sqrt{P_x} = Q \propto \int_0^{\infty} \Psi_g \, \mu_x \, \Psi_e \, dx, \tag{3.11}$$

where Ψ_g and Ψ_e are the wave functions of the ground and excited states, respectively, and μ_x is the dipole moment component of light polarized along the x axis. The significance of eq. (3.11) is that the probability of an electronic transition is zero unless Ψ_g and Ψ_e differ in parity.

The transition moment, Q, is related to the oscillator strength, f, which is a measure of intensity of an absorption band (table 3.2), by

$$f = 1.085 \times 10^{11} \, G \upsilon Q, \tag{3.12}$$

where υ is the frequency of the band in wavenumber units (cm^{-1}) and G is unity for a transition between non-degenerate states. Otherwise, $G = 1/\Omega$ for states of degeneracy Ω, and the oscillator strength is summed over all pairs of possible transitions.

The Laporte selection rule may also be expressed in the form $\Delta l = \pm 1$, for orbitally allowed transitions, where l is the azimuthal or orbital angular momentum quantum number (§2.2.2). In §2.2 it was noted that orbitals, being probability amplitude functions of electron distributions about the nucleus of an atom, have mathematical positive and negative signs associated with them as solutions to the Schrödinger wave equation. The *s* and *d* orbitals have similar signs because they are centrosymmetric (designated as subscript *g* in group theory notation), as do the *p* and *f* orbitals which lack inversion symmetry (designated as subscript *u*). Thus, *s* and *d* orbitals have identical parities, as do *p* and *f* orbitals, but *d* and *p* orbitals have opposite parities. Therefore, electronic transitions between *d* and *p* orbitals are allowed by the Laporte selection rule, whereas those between *d* and *s* orbitals are forbidden. In particular, electron transfer between two different *d* orbitals is forbidden. Thus, the excitation of an electron between a t_{2g} orbital and an e_g orbital of a transition metal ion in octahedral coordination is not allowed by the Laporte selection rule.

The Laporte selection rule is weakened, or relaxed, by three factors: first, by the absence of a centre of symmetry in the coordination polyhedron; second, by mixing of *d* and *p* orbitals which possess opposite parities; and third, by the interaction of electronic 3*d* orbital states with odd-parity vibrational modes. If the coordination environment about the cation lacks a centre of symmetry, which is the case when a cation occupies a tetrahedral site, some mixing of *d* and *p* orbitals may occur because they have symmetry operations in common in a tetrahedral environment. It may happen that a transition involving electron transfer from one 3*d* orbital to another embodies a small amount of transfer via *p* orbitals embraced in covalent bonding with surrounding ligands. Since transitions between *d* and *p* orbitals are allowed, a mechanism is available for the normally Laporte-forbidden transition of an electron between *e*-type and t_2-type 3*d* orbitals in a tetrahedrally coordinated transition metal ion. This leads to an absorption band in a spectrum, the intensity of which is proportional to the extent of mixing of *d* and *p* orbitals which, in turn, is inflenced by covalent bonding in a transition metal–ligand coordination cluster (chapter 11). In an octahedral site, however, mixing of the transition metal 3*d* and 4*p* orbitals is not possible when the cation lies at the centre of symmetry of the site because these orbitals have few symmetry operations in common in an octahedral environment. The overall result is that absorption bands originating from crystal field transitions within a given cation in an octahedral environment may be one-hundred times less intense than those within the cation tetrahedrally coordinated to the same ligands. One manifestation of this difference appears in crystal field spectra of basaltic glasses, for example, in which an absorption band at 1,800 nm originating from one per cent Fe^{2+} in tetrahedral sites would have an intensity comparable to an absorption band at 1,050 nm contributed by ninety-nine per cent Fe^{2+} in octahedral sites in the glass structure (Nolet *et al.*, 1979; Dyar and Burns, 1981).

3.7.1.1 Vibronic coupling

Another mechanism by which Laporte-forbidden transitions may occur, even in cations located in centrosymmetric (octahedral) sites, is through vibronic coupling, which involves coupling of *vib*rational and elec*tronic* wave functions with opposite parities. Electronic transitions between ground and excited states involve vibrational levels of the two states. The energy diagram illustrated in fig. 3.14 shows the potential energy surfaces of the ground and excited electronic states drawn as a function of one of the normal vibrational modes of a transition metal–ligand coordination cluster. In the simplest case illustrated in fig. 3.14, an electronic transition from the lowest vibrational mode of the ground electronic state takes place to several vibrational modes of the excited electronic state. According to a condition known as the Frank–Condon principle, an electronic transition between two energy states takes place in such a short time interval ($\sim 10^{-15}$ s) that the nuclei remain almost stationary during the transition. At the instant of an electronic transition, the excited state may have the same nuclear geometry as the ground state but may be highly excited vibrationally. However, relaxation to the ground vibrational level of the

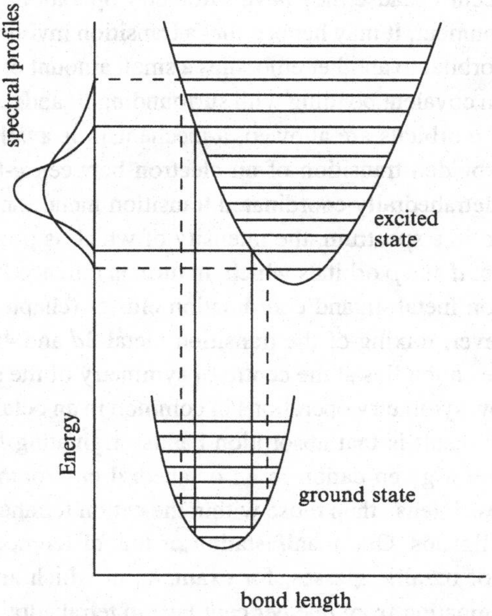

Figure 3.14 Potential energy diagram showing how an electronic transition takes place between vibrational levels of the ground and excited states. The illustration also demonstrates how the width and asymmetry of an absorption band changes at elevated temperature as a result of increased thermal population of vibrational levels of the ground electronic state (— low temperature; – – – elevated temperature). (Modified from Hitchman, 1985, figs 8 and 18.)

excited electronic state also occurs less rapidly (~10^{-12} s). In the course of molecular vibrations between the transition metal ion and ligands surrounding it in octahedral coordination, some of the vibrational modes cause the cation to be offset periodically from its equilibrium centrosymmetric position. Electronic transitions between different $3d$ orbital energy levels may then take place via vibrational sublevels with opposite parities, such as two of the acentric vibrational modes of an octahedron illustrated in fig. 3.15. Note that Greek letters are used to designate symmetries of vibrational modes (*e.g.*, α_{1g}, ε_u, τ_{1u}, etc.) in order not to confuse them with the symmetry notation of electron orbitals (*e.g.*, a_{1g}, t_{1u}, etc.) and crystal field states (e.g., A_{1g}, E_g, T_{1g}, etc.,) which are represented by Arabic symbols. Since molecular vibrations are thermally activated, intensification of crystal field transitions by the vibronic coupling mechanism is expected, and often observed, in high temperature crystal field spectra. Increased temperature also broadens absorption bands by the same mechanism and is discussed later (§3.9.4).

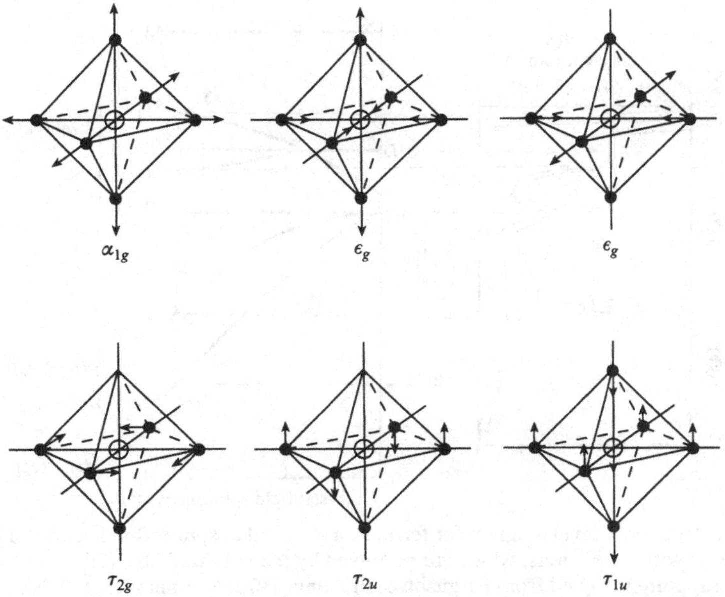

Figure 3.15 Vibrational modes of an octahedron. Note the acentric modes τ_{1u} and τ_{2u} which aid vibronic coupling.

3.7.2 Spin-multiplicity selection rule

The spin-multiplicity selection rule relates to changes in the number of unpaired electrons when they are excited between split 3*d* orbital energy levels within a cation. The selection rule states that the total number of unpaired electrons on an atom must remain unchanged during an electronic transition. Accordingly, spin-allowed transitions are distinguished in intensity from very weak spin-forbidden transitions in a crystal field spectrum, and depend on the electronic configurations of each cation. The simplest example of a spin-allowed transition is Ti^{3+}, $3d^1$, the single 3*d* electron of which in octahedral coordination occupies one of the t_{2g} orbitals in the ground state. The $^2T_{2g} \rightarrow {}^2E_g$ transition, which is portrayed by the absorption band at 493 nm in fig. 3.1 and results from a change of electronic configuration within Ti^{3+} from $(t_{2g})^1$ to $(e_g)^1$, is spin-allowed because there is no change in the number of unpaired electrons. A similar spin-allowed transition in octahedral Fe^{2+}, $^5T_{2g} \rightarrow {}^5E_g$, is responsible for the broad band centred near 1,000 nm in fig. 3.2.

In contrast to these spin-allowed transitions in Ti^{3+} and Fe^{2+}, only spin-forbidden transitions are possible in $3d^5$ cations such as Fe^{3+} and Mn^{2+} when they are coordinated to oxygen ligands. The electronic structures of these cations are such that in their ground-state configurations, $(t_{2g})^3(e_g)^2$ represented by

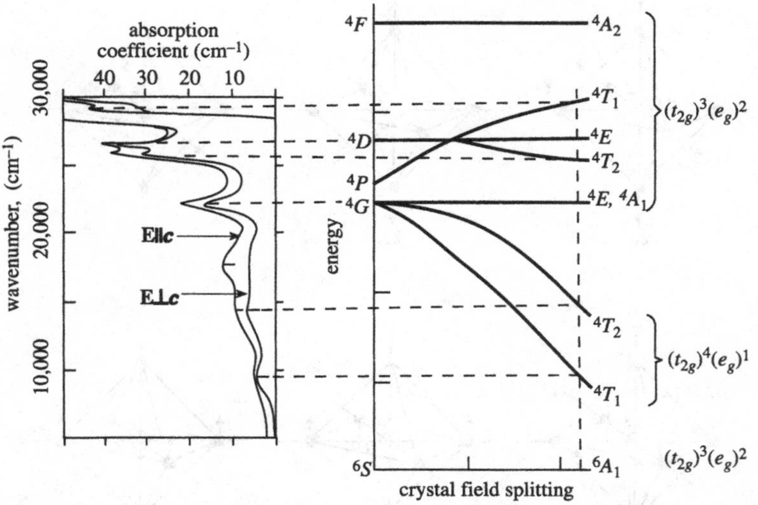

Figure 3.16 Energy level diagram for ferric iron matched to spin-forbidden crystal field transitions within Fe^{3+} ions, which are portrayed by the polarized absorption spectra of yellow sapphire (adapted from Ferguson & Fielding, 1972; Sherman, 1985a). Note that the unassigned band at ~17,600 cm^{-1} represents a paired transition within magnetically coupled Fe^{3+} ions located in adjacent face-shared octahedra in the corundum structure.

$^6A_{1g}$, there are five unpaired electrons when the cations occupy centrosymmetric octahedral sites. All possible crystal field transitions within Fe^{3+} and Mn^{2+} are to excited states with electronic configurations containing fewer unpaired electrons (fig. 3.9). Such transitions, because they are spin-forbidden, result in relatively low intensity absorption bands compared to Ti^{3+} and Fe^{2+}. The crystal field spectra of Fe^{3+}-bearing silicates and oxides, for example, such as the absorption spectra of yellow sapphire illustrated in fig. 3.16, contain two weak, broad bands in the vicinity of 9,500 cm^{-1} and 14,500 cm^{-1} (representing transitions from the $^6A_{1g}$ ground state to the $^4T_{1g}(G)$ and $^4T_{2g}(G)$ excited states, respectively), as well as superimposed sharp peaks at approximately 22,000 cm^{-1} (representing the $^6A_{1g} \rightarrow {}^4E_g, {}^4A_{1g}(G)$ transition) and other peaks located at higher energies in the ultraviolet region. The intensities of these spin-forbidden transitions in Fe^{3+}-bearing silicates are generally one or two orders of magnitude lower than the spin-allowed transitions within Fe^{2+} ions.

Spin-forbidden transitions are possible in other cations with two or more $3d$ electrons (except Cu^{2+}, $3d^9$). Thus, in oxygen ligand environments containing Fe^{2+} ions in octahedral coordination, spin-forbidden transitions also occur in the crystal field spectra, often appearing as weak, sharp peaks in the visible region, in addition to the broad spin-allowed bands located in the near-infrared around 1,000 nm. For example, in the spectra of Fe^{2+} in the hydrated $[Fe(H_2O)_6]^{2+}$ ion (fig. 3.2), the two peaks identified near 20,000 cm^{-1} and 22,200 cm^{-1} represent the spin-forbidden transitions $^5T_{2g} \rightarrow {}^3T_{1g}$ and $^5T_{2g} \rightarrow {}^3T_{2g}$, respectively (compare fig. 3.7). Similar weak spin-forbidden peaks occur in the visible region in crystal field spectra of many of the ferromagnesian silicates described in chapter 5.

3.7.3 Coupling interactions with nearest-neighbour cations

Intensities of spin-forbidden crystal field bands, particularly those of ferric oxide minerals such as crystalline hematite (see fig. 10.2), may be enhanced by magnetic coupling of electron spins on next-nearest-neighbour Fe^{3+} ions when they occupy adjacent sites in a crystal structure. Spectroscopic selection rules for coupled Fe^{3+}–Fe^{3+} pairs differ from those for isolated individual Fe^{3+} ions (Sherman and Waite, 1985). Furthermore, additional transitions corresponding to simultaneous excitations within two adjacent Fe^{3+} ions may be present in visible-region spectra (Ferguson and Fielding, 1972). In the spectra of yellow sapphire shown in fig. 3.16, for example, another absorption band occurs at 18,690 cm^{-1} in addition to the crystal field transitions within individual Fe^{3+} ions at 9,450 cm^{-1} ($^6A_1 \rightarrow {}^4T_1$), 14,350 cm^{-1} ($^6A_1 \rightarrow {}^4T_2$) and 22,730 cm^{-1} ($^6A_1 \rightarrow {}^4E,{}^4A_1$). The additional feature at 18,690 cm^{-1} represents the pair excitation $^6A_1 + {}^6A_1 \rightarrow {}^4T_1 + {}^4T_1$ between two Fe^{3+} ions located in adjacent face-shared octahedra in the corundum structure (see fig. 5.4). Similar intensity enhancement of Fe^{2+} crystal field transitions occurs in optical spectra of mixed-valence Fe^{2+}–Fe^{3+} and Fe^{2+}–Ti^{4+} minerals (Smith, 1977, 1978b; Amthauer and Rossman, 1984; Mattson and Rossman, 1987a,b, 1988), including vivianite, babingtonite, tourmalines and other minerals described in chapters 4 and 5.

3.7.4 Relative intensities of crystal field spectra

Taking into account all of the factors influencing intensities of crystal field spectra discussed so far, the following generalizations may be made. Transitions of 3*d* electrons within cations in octahedral coordination are expected to result in relatively weak absorption bands. Intensification occurs if the cation is not centrally located in its coordination site. In tetrahedral coordination, the intensities of crystal field transitions should be at least one-hundred times larger than those in octahedrally coordinated cations. Spin-forbidden transitions are usually about one-hundred times weaker than spin-allowed transitions in centrosymmetric, octahedrally coordinated cations, but become

Table 3.6. *Relative intensities of absorption bands in transition metal-bearing minerals*

Type of electronic transition	Molar extinction coefficient, ε*	Mineral examples	Reference
	Crystal Field Transitions		
1. Spin-forbidden, Laporte-forbidden, centrosymmetric site	10^{-3} to 1	Fe^{3+} in garnet	fig. 4.9
2. Spin-forbidden, Laporte-forbidden, acentric site	10^{-1} to 10	Fe^{3+} in feldspars (tetrahedral)	fig. 5.22
		Fe^{3+} in epidote M3 (distorted oct)	fig. 4.9
3. Spin-forbidden, Laporte-forbidden, pair-enhancement	1 to 10	Fe^{3+} in corundum, hematite	fig. 3.16 fig. 10.2
4. Spin-allowed, Laporte-forbidden, centrosymmetric site	1 to 10	Fe^{2+} in olivine M1, gillespite	figs 5.9 & 4.3
	10 to 50	Mn^{3+} in andalusite M1	§4.4.2.2 & §5.4.4.1
5. Spin-allowed, Laporte-forbidden, acentric site	10 to 15	Fe^{2+} in olivine M2	fig. 5.9
	10 to 100	Fe^{2+} in pyroxene M2, amphibole M4	fig. 5.15
	50 to 300	Mn^{3+} in epidote M3	fig. 4.2
	Electron Transfer Transitions		
6. Spin-allowed metal → metal intervalence charge transfer	10^2 to 10^3	$Fe^{2+} \rightarrow Fe^{3+}$ in glaucophane	fig. 4.16
		$Fe^{2+} \rightarrow Ti^{4+}$ in blue sapphire	fig. 4.17
7. Spin- and Laporte-allowed oxygen → metal charge transfer	10^3 to 10^5	$O \rightarrow Fe^{3+}$ in biotite and hornblende	§4.7.3

* ε has units litre $mole^{-1}$ cm^{-1} or litre $(g.ion)^{-1}$ cm^{-1}

intensified by magnetic interactions with cations in adjacent sites, especially in face-shared and edge-shared octahedra. Spin-forbidden transitions may also be intensified when cations occupy acentric coordination sites. Transitions allowed by both the spin-multiplicity and Laporte selection rules, such as oxygen → metal charge transfer transitions described later (§4.7.3 and §11.7.2), may be ten-thousand times more intense than crystal field transitions in octahedrally coordinated cations. Relative intensities of crystal field transitions are summarized in table 3.6.

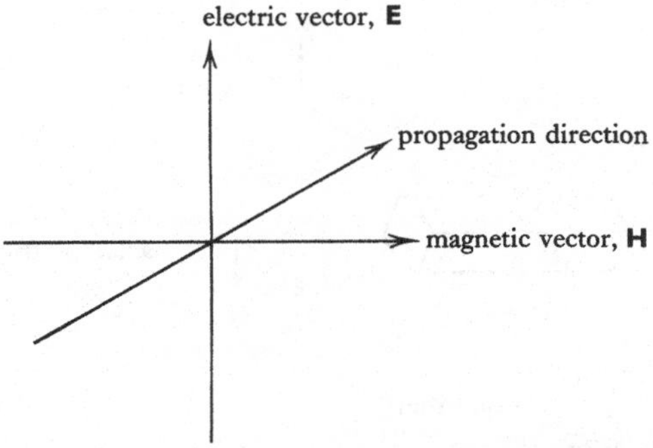

Figure 3.17 Orientations of the electric vector and magnetic vector with respect to the propagation direction of linearly polarized electromagnetic radiation.

3.8 Polarization dependencies of absorption bands

The crystal structures of transition metal compounds and minerals have either cubic or lower symmetries. The cations may occur in regular octahedral (or tetrahedral) sites or be present in distorted coordination polyhedra in the crystal structures. When cations are located in low-symmetry coordination environments in non-cubic minerals, different absorption spectrum profiles may result when linearly polarized light is transmitted through single crystals of the anisotropic phases. Such polarization dependence of absorption bands is illustrated by the spectra of Fe^{2+} in gillespite (fig. 3.3) and of Fe^{3+} in yellow sapphire (fig. 3.16).

Gillespite, $BaFe^{2+}Si_4O_{10}$, belongs to the tetragonal system and the crystal structure contains a coordination site closely approximating D_{4h} symmetry. The four coplanar oxygen atoms that are nearest to Fe are bonded to Si and lie at the corners of a square surrounding the Fe^{2+} cations. As a result of Fe^{2+} ions being in square planar sites, the crystal field spectra of gillespite differ for light polarized parallel and perpendicular to the optic axis (*c* axis). Similarly, since corundum is a trigonal mineral and trivalent cations occupy a trigonally distorted octahedron (point symmetry C_3), different spectrum profiles are obtained for Fe^{3+} in yellow sapphire when light is polarized parallel and perpendicular to the *c* axis (fig. 3.16). In order to account for the different spectrum profiles of these and other non-cubic minerals, it is necessary to examine the nature and properties of polarized light.

3.8.1 Electric and magnetic vectors in polarized light

Electromagnetic radiation has associated with it an electric vector **E** and a magnetic vector **H**. In linearly polarized light these vectors are mutually per-

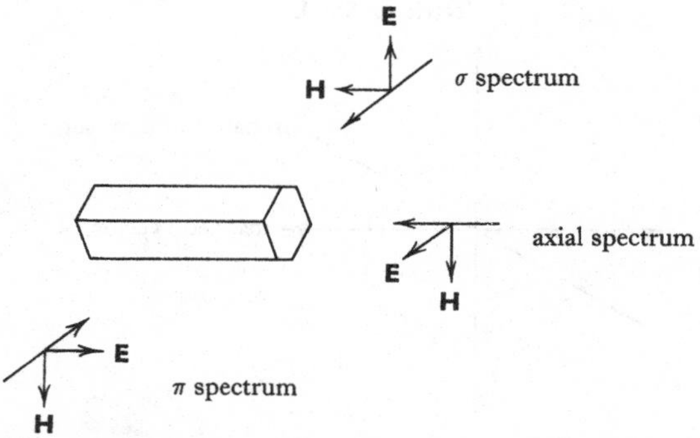

Figure 3.18 Three ways of measuring the polarized absorption spectra of a uniaxial crystal.

pendicular and are each perpendicular to the direction of propagation of light through a crystal. The orientations of the electric vector, magnetic vector and propagation directions are shown in fig. 3.17. Electromagnetic radiation may induce a transition by either a magnetic dipole or an electric dipole mechanism. In the visible region, electronic transitions between orbitals are dominated by electric dipole interactions. Contributions from magnetic dipole interactions are negligible, except when they help intensify weak spin-forbidden transitions as in Fe^{3+}-bearing minerals such as yellow sapphire (fig. 3.16), sulphates (Rossman, 1974; 1975) and oxyhydroxides (Sherman and Waite, 1985).

When light is passed through a polarizer such as a calcite Nicol prism, the electric vector is parallel to the plane of polarization. The electric vector transforms in different ways when the polarized light is transmitted through a single crystal of a mineral, depending on its crystal symmetry.

In cubic crystals no polarization of light occurs and the electric vector is identical in all crystallographic directions. Therefore, one spectrum only is observed in cubic crystals. Three types of spectrum are theoretically possible in uniaxial crystals, depending on the orientations of the electric and magnetic vectors with respect to the crystallographic axes (McClure, 1959). These are shown in fig. 3.18. In the axial spectrum the light propagation direction is along the optic axis corresponding, for example, to the *c* axis of gillespite and corundum. In the so-called σ spectrum (McClure, 1959), the propogation direction and electric vector are perpendicular to the optic axis, whereas the propagation direction and magnetic vector are perpendicular to the optic axis in the π spectrum. In electric dipole transitions the axial and σ spectra become identical so that in minerals of the tetragonal, trigonal and hexagonal systems only two spectrum profiles may be distinguished. These correspond to spectra obtained with the ordinary and extraordinary rays of polarized light, and are designated as the ω (or $\mathbf{E}\perp c$) and ε (or $\mathbf{E}\| c$) spectra, respectively. Examples of such polarized spectra include those for gillespite and yellow sapphire illustrated in figs 3.3 and 3.16.

Six types of spectra are theoretically possible in minerals of the orthorhombic, monoclinic and triclinic systems (McClure, 1959). However, for electric dipole transitions only three spectra are usually distinguished. These are the α, β and γ spectra obtained when light is polarized along each of the three indicatrix axes, which in orthorhombic minerals such as olivine and orthopyroxene correspond to the three crystallographic axes. The majority of the spectra of minerals described in chapters 4 and 5 consist of polarized spectra measured in the three mutually perpendicular directions corresponding to α, β and γ polarized light.

When light impinges on a non-cubic mineral the electric vector transforms into different components, the symmetry properties of which depend on the direction of propagation of light through the crystal and the symmetry of the crystal structure. In cubic crystals, the electric vector is identical for light polarized along each of the three tetrad axes. For an octahedal site in a cubic crystal, the electric vector has group theory symmetry T_{1u}; for a tetrahedral site, it is represented by T_2 symmetry. In a tetragonal crystal with D_{4h} symmetry such as gillespite, the electric vector has A_{2u} symmetry when light is polarized along the *c* axis and E_u symmetry for light polarized perpendicular to the *c* axis. In a trigonal crystal with C_3 symmetry such as yellow sapphire, the electric vector has either A_1 ($\mathbf{E}\| c$) or E ($\mathbf{E}\perp c$) symmetry. In crystals of lower symmetries, the electric vector is represented by different group theory symmetries when polarized along each of the three optical indicatrix axes.

Methods of group theory are used to ascertain whether electronic transitions will take place between crystal field states of a transition metal ion for light polarized along each electronic axis. This axis may or may not coincide with a crystallographic or optical indicatrix axis of the host crystal. Intensities of absorption bands in polarized light hinge upon expressions for the transition probability such as that in eq. (3.10). Not only must the transition probability, P_x, be non-zero and have a finite value, implying that the wave functions for the ground state (Ψ_g) and excited state (Ψ_e) have opposite parities, but the product of group theory representations of quantities in the integral in eq. (3.10) must be totally symmetric, that is, it must possess A_1 or A_{1g} symmetry. Stated in group theory terminology, the transition probability (P_x) is determined from the symmetry of the ground state and excited state by constructing the direct product involving the irreducible representations of the wave functions of the two states (Γ_g and Γ_e, respectively) and the electric dipole vector (Γ_μ). If the direct product, $\Gamma_g \times \Gamma_\mu \times \Gamma_e$, contains the A_1 or A_{1g} representation, the transition is allowed. If the A_{1g} (or A_1) representation is absent from the

Figure 3.19 The crystal structure of gillespite projected down the *c* axis. Large spheres: Ba^{2+}; small spheres: Fe^{2+} in square planar coordination; tetrahedra: linked $[SiO_4]$.

factorized direct product, the transition is forbidden. Such direct product calculations are simplified by the use of character tables, which provide a shorthand notation of the symmetry properties of the crystal field states and electric vectors in coordination sites with a particular symmetry. An elementary account of such calculations is given by Cotton (1990).

3.8.2 Group theoretical interpretation of gillespite polarized spectra

The polarized spectra of gillespite, $BaFe^{2+}Si_4O_{10}$, shown in fig. 3.3 serve to illustrate these group theoretical methods for assigning absorption bands. This rare tetragonal layered silicate mineral containing Fe^{2+} ions in square planar coordination with oxygen ions has figured prominently in developments of spectral mineralogy, having played a key role in understanding 3*d* orbital energy levels and predicting pressure-induced spin-pairing transitions in Fe^{2+}–bearing minerals in the Earth's interior (§9.7.3). The platey habit of gillespite is particularly favourable for measuring and assigning absorption bands in its polarized crystal field spectra.

The gillespite structure is illustrated in fig. 3.19. The planar $[FeO_4]$ group, strictly speaking, has C_4 symmetry. However, the Fe^{2+} ion is only 0.3 pm out of

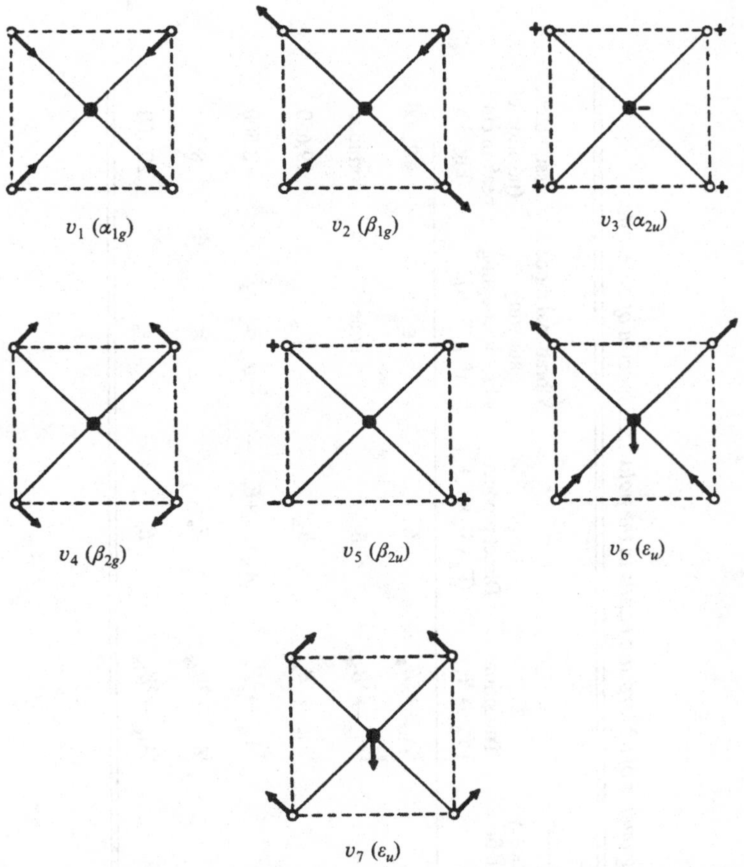

Figure 3.20 Vibrational modes, υ_1 to υ_7, of atoms in a square planar site such as the [FeO_4] coordination environment in gillespite (from Hitchman, 1985). Note the acentric α_{2u}, β_{2u} and ε_u modes that facilitate electronic transitions within Fe^{2+} ions by vibronic coupling.

the plane of the four coordinating oxygen atoms, with diagonal O–Fe–O angles of 178° and all four Fe–O distances equal to 200 pm. Therefore, the [FeO_4] group very closely approximates D_{4h} symmetry. In addition, the square planar site is aligned perpendicular to the optic axis. Thus, the z electronic axis of the 3d orbitals coincides with the c crystallographic axis. A square planar coordination site may be regarded as one of the limiting cases of tetragonal distortion of an octahedron (cf. figs 2.8 and 3.13) in which two ligands along the z electronic axis are completely removed. Since repulsion of electrons by surrounding ligands is minimal along the z axis, orbitals projecting in this direction are more stable than those with lobes in the x–y plane. Therefore, the order

Table 3.7. *Assignments of absorption bands in the polarized spectra of gillespite*

Spectrum	Electric vector: Polarization	Electric vector: Group theory representation (Γ_μ)	Transition ($\Psi_g \rightarrow \Psi_e$)	Direct product (Γ_g x Γ_μ x Γ_e)	Vibrational mode allowing vibronic coupling (Fig. 3.20)	Assignment (position of peak in cm^{-1}) (Fig. 3.3)
ε	parallel to *c* (E‖*c*)	A_{2u}	$^5A_{1g} \rightarrow {}^5E_g$	E_u	ε_u	≈ 2,500
			$^5A_{1g} \rightarrow {}^5B_{2g}$	B_{1u}	none	forbidden
			$^5A_{1g} \rightarrow {}^5B_{1g}$	B_{2u}	β_{2u}	19,650
ω	perpendicular to *c* (E⊥*c*)	E_u	$^5A_{1g} \rightarrow {}^5E_g$	A_{2u}+B_{2u}+E_u	α_{2u},β_{2u} ,ε_u	≈ 2,500
			$^5A_{1g} \rightarrow {}^5B_{2g}$	E_u	ε_u	8,300
			$^5A_{1g} \rightarrow {}^5B_{1g}$	E_u	ε_u	20,160

of increasing energy of the 3*d* orbitals is: $d_{z^2} < d_{xz} = d_{yz} < d_{xy} < d_{x^2-y^2}$. The ground state of Fe^{2+} in gillespite has an electronic configuration $(d_{z^2})^2(d_{xz})^1(d_{yz})^1(d_{xy})^1(d_{x^2-y^2})^1$ and symmetry $^5A_{1g}$. Excited states are ^5E_g, $^5B_{2g}$ and $^5B_{1g}$. Thus, electronic transitions between the $^5A_{1g}$ ground state and each of these excited states appear likely. However, since each crystal field state is centrosymmetric, such transitions are forbidden by the Laporte selection rule (§3.7.1). Transitions may be allowed by vibronic coupling, however. The symmetries of the vibrational modes of the square [FeO_4] group illustrated in fig. 3.20 are α_{1g}, α_{2u}, β_{1g}, β_{2g}, β_{2u} and two with ε_u symmetry. Thus, they include three types (α_{2u}, β_{2u} and ε_u) having acentric vibrational modes.

To account for the polarization dependencies of the absorption bands in the polarized spectra of gillespite in fig. 3.3, it is necessary to consider the symmetry properties of the electric vector of light in a tetragonal (symmetry D_{4h}) crystal, which are A_{2u} (parallel) and E_u (perpendicular) relative to the optic axis (or *c* crystallographic axis) coinciding exactly with the *z* electronic axis of iron in the square planar site in gillespite. Evaluations of direct products involving the symmetry properties of the crystal field ground state (A_{1g}), with each excited state (E_g, B_{2g} or B_{1g}), each component of the electric dipole moment (A_{2u} or $\mathbf{E}_u$) and the various acentric vibrational modes (α_{2u}, β_{2u} or ε_u) through the use of character tables (e.g., Cotton, 1990) leads to the interpretation of the polarized spectra summarized in table 3.7.

The assignments summarized in table 3.7 reveal that all electronic transitions in gillespite are allowed through the mechanism of vibronic coupling *except* the transition $^5A_{1g} \rightarrow {}^5B_{2g}$ in light polarized along the optic axis (**E**||*c* spectrum) for which no vibrational mode exists to assist this electronic transition. This accounts for the absence of an absorption band at about 1,200 nm (8,300 cm^{-1}) in the **E**||*c* spectrum of gillespite shown in fig. 3.3. Note that the low energy transition $^5A_{1g} \rightarrow {}^5E_g$ leads to an absorption band beyond the wavelength range shown in the spectra in fig. 3.3. Studies of the crystal field spectra of cuprorivaite or Egyptian blue, $CaCuSi_4O_{10}$ (Clark and Burns, 1967; Ford and Hickman, 1979), and synthetic $CaCrSi_4O_{10}$ (Belsky *et al.*, 1984), which are isostructural with gillespite and contain Cu^{2+} or Cr^{2+} ions in the square planar coordination site, lead to estimates of $\leq$2,500 cm^{-1} for the $^5A_{1g} \rightarrow {}^5E_g$ transition in gillespite.

The interpretation of the crystal field spectra of gillespite is simplified by the high symmetry of this tetragonal mineral and by the coincidence of the electronic axes of Fe^{2+} in the square planar site with the electric vectors of polarized light, due to the perfect alignment of the [FeO_4] plane perpendicular to the *c* axis. Detailed spectral analyses of most other non-cubic transition metal-bearing minerals become increasingly difficult when electronic axes within a coordination site do not coincide with optic axes of polarized light. This occurs in most silicate minerals, including orthopyroxene, the spectra of which have been interpreted, nevertheless, by group theory (e.g., Goldman and Rossman, 1977a; Rossman, 1988). Because of such misalignments in silicates, a small component of an electronic transition which may be allowed along one electronic axis that is inclined to the light polarization axes often causes the absorption band to occur with different intensities in all three polarized spectra.

3.9 Widths of absorption bands

Absorption bands in crystal field spectra are not sharp lines. Instead, as the spectra illustrated in figs 3.1, 3.2 and 3.3 show, they contain rather broad envelopes approximating gaussian profiles which at half peak-height may have full widths ranging from <100 cm^{-1} to 1,000–2,000 cm^{-1}. Several factors lead to broadened absorption bands and they are discussed below.

3.9.1 Correlations with energy level diagrams

Energy level diagrams, which provide a measure of relative energies of crystal field transitions, also account for relative widths of absorption bands in crystal field spectra of transition metal compounds. The widths are related to the slope differences of crystal field states in an Orgel or Tanabe–Sugano energy level diagram. Atoms in a structure are in continual thermal vibration about their mean positions. As cations and surrounding ligands vibrate about their structural positions through vibrational modes such as those illustrated in figs 3.15 and 3.20, metal–ligand distances vary so that Δ oscillates about a mean energy corresponding to the average position of the atoms. If the energy separation between the ground state and an excited state is a sensitive function of Δ, the energy difference will vary appreciably over the range through which Δ varies in response to vibrations of the metal and ligand atoms. Electronic transitions between the two states will result in a broad absorption band. If, however, there is little variation in the energy separation between the ground and excited states with fluctuating Δ, an electronic transition between these states will lead to a sharp absorption band.

These features are particularly well illustrated by crystal field spectra of $3d^5$ cations, such as those of Fe^{3+} ions in yellow sapphire illustrated in fig. 3.16. The energy separations between the $^6A_{1g}$ ground state and the $^4T_{1g}$ and $^4T_{2g}$ excited states of Fe^{3+} converge with increasing Δ but the energy separation between the $^6A_{1g}$ and ^4E_g,$^4A_{1g}$ levels shown in fig. 3.10 remains constant. As a result, crystal field spectra of Fe^{3+} in yellow sapphire (fig. 3.16), Fe^{3+} in epidote (fig. 4.9) and Mn^{2+} in tephroite (fig. 4.8*a*) typically show two broad bands at longer wavelengths and one or two sharp peaks in the visible region.

Broadening of absorption bands is commonly found for electronic transitions between two crystal field states arising from the same spectroscopic term. For example, the $^5T_{2g}$ and ^5E_g states are both derived from the ^5D ground term of Fe^{2+} (figs 3.7 and 3.8). The energy separation between the two crystal field states increases with increasing Δ_o. As a result, the $^5T_{2g} \rightarrow {}^5E_g$ transition in Fe(II) compounds is represented by a broad absorption band. This partially accounts for the broad band centred around 10,000 cm^{-1} in the spectrum of the hydrated Fe^{2+} ion in fig. 3.2. A similar explanation applies to the $[Ti(H_2O)_6]^{3+}$ spectrum shown in fig. 3.1.

3.9.2 The dynamic Jahn–Teller effect.

Another factor contributing to the asymmetry and breadth of absorption bands in crystal field spectra of transition metal ions is the dynamic Jahn–Teller effect, particularly for dissolved hexahydrated ions such as $[Fe(H_2O)_6]^{2+}$ and $[Ti(H_2O)_6]^{3+}$, which are not subjected to static distortions of a crystal structure. The degeneracies of the excited ^5E_g and ^2E_g crystal field states of Fe^{2+} and Ti^{3+}, respectively, are resolved into two levels during the lifetime of the electronic transition. This is too short to induce static distortion of the ligand environment even when the cations occupy regular octahedral sites as in the periclase structure. A dual electronic transition to the resolved energy levels of the E_g excited states causes asymmetry and contributes to the broadened absorption bands in spectra of most Ti(III) and Fe(II) compounds and minerals (cf. figs 3.1, 3.2 and 5.2).

3.9.3 Effects of multiple site occupancies

Broadening of absorption bands also results from the superposition of close-spaced bands originating from a single transition metal ion located in a low-symmetry coordination site, or from the cations being distributed over two or more structurally similar sites. Each of these factors is significant in crystal field spectra of silicate minerals, the structures of which often contain transition metal ions in two or more distorted sites. Even in crystal structures providing a single crystallographic position, different next-nearest-neighbour distributions resulting from atomic substitution in minerals may prevent a specific transition metal ion from being in identical coordination environments, again contributing to the breadth of absorption bands.

3.9.4. Vibrational interactions

Implicit in the foregoing discussion is that increased temperature, which increases thermal motions of atoms in a crystal structure, also contributes to the broadening of absorption bands in a crystal field spectrum. Since the crystal field splitting is proportional to the inverse fifth-power of the metal–ligand distance, eq. (2.17), energy separations between crystal field states are very sensitive to changes of interatomic distances induced by thermal expansion of a crystal structure, eq. (2.25).

Thermal broadening of an electronic transition results from the population of additional vibrational levels of the electronic ground state illustrated by the potential energy diagram in fig. 3.14. Conversely, spectral features may be narrowed, and better resolution of absorption bands achieved, by performing crystal field spectral measurements at low temperatures. Under these conditions, vibrational peaks may contribute to fine structure observed on electronic absorption bands of transition metal-bearing phases, particularly in low temperature spectra.

3.9.5 Band shape

As noted earlier (§3.2.2), absorption bands are assumed to have gaussian shape, eq. (3.6). However, vibrational interactions may modify their shapes and contribute to asymmetric profiles in crystal field spectra. At low temperatures when few vibrational modes are active in the electronic ground state, absorption bands may be comparatively narrow, as indicated by the potential energy level diagram in fig. 3.14. At elevated temperatures, a Boltzmann distribution results in increased populations of higher energy vibrational modes so that excitation of electrons to higher energy crystal field states takes place over a wider range of vibrational modes. Due to anharmonicity of potential energy curves, absorption bands broaden more towards lower frequencies, thereby modifying the gaussian shape and causing asymmetry towards long wavelength tails of absorption bands. The position of the absorption maximum may also change with temperature as an increased number of higher energy vibrational modes become occupied.

3.10 Ligand field parameters for distorted environments

The ligand field parameter for cations in a regular octahedral environment, 10 *Dq*, expressed by eq. (2.17),

$$\Delta_0 = 10Dq = \frac{Z_L e^2}{6R^5} < r^4 > \qquad (2.17)$$

and embracing the radial integral term, $<r^4>$, must be modified in low-symmetry environments to take into account angular distortions and variations of

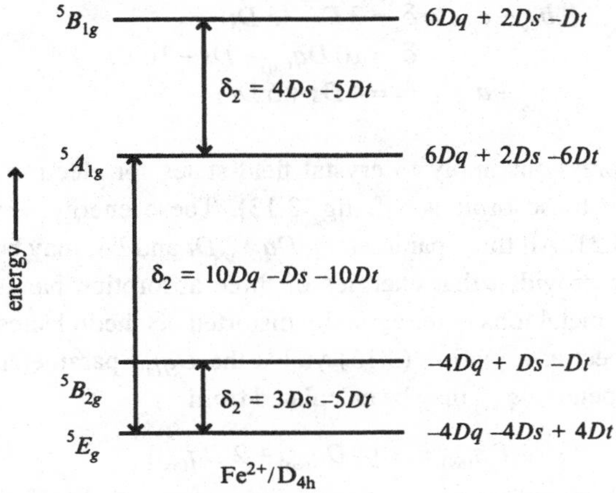

Figure 3.21 Ligand field energy separations for Fe^{2+} ions in a tetragonally elongated octahedron (compare fig. 3.13). The inverted diagram applies to Mn^{3+} in a similar D_{4h} environment.

metal–ligand distances (R) in deformed coordination polyhedra or uneven arrangements of different ligands surrounding the central cation. Such asymmetries require the specification of an additional parameter, C_p, (Gerloch and Slade, 1973), defined as

$$Cp = \frac{Z_L e^2}{7R^3} < r2 >. \tag{3.13}$$

A particularly common type of distortion involves the lowering of symmetry of a ligand coordination polyhedron from O_h to D_{4h}. In such a tetragonally distorted octahedron, the Dq parameter has two values resulting from different metal–ligand distances represented by the four equatorial (eq) and two axial (ax) bonds. Two additional radial integral splitting parameters, Ds and Dt, are used to express the orbital energies of the individual $3d$ orbitals (Ballhausen, 1962, p. 100; Gerloch and Slade, 1973; Lever, 1984, p. 19)

$$Ds = Cp_{(eq)} - Cp_{(ax)} \tag{3.14}$$

$$Dt = \tfrac{4}{7}\,[Dq_{(eq)} - Dq_{(ax)}]\ . \tag{3.15}$$

The Dt parameter is a measure of the difference between the equatorial and axial Dq parameters. In an elongated octahedron, the energy separations between pairs of $3d$ orbital energy levels are expressed by

$$\begin{aligned} b_{2g} - e_g &= \delta_1 = 3\,Ds - 5\,Dt \\ a_{1g} - e_g &= \delta_2 = 10\,Dq_{(eq)} - Ds - 10\,Dt \\ b_{1g} - a_{1g} &= \delta_3 = 4\,Ds + 5\,Dt\,. \end{aligned} \tag{3.16}$$

Analogous expressions apply to crystal field states for electronic configurations involving these orbitals (cf. fig. 3.13). These energy separations are shown in fig. 3.21. All three parameters, $Dq_{(eq)}$, Dt and Ds, may be determined experimentally provided that energies of three absorption bands originating from transition metal ions in tetragonally distorted octahedral sites occur in the crystal field spectra. Equation (3.15) yields the $Dq_{(ax)}$ parameter, so that the mean Dq parameter, $Dq_{(av)}$ may be calculated from

$$Dq_{(av)} = \tfrac{1}{6}\,[4\,Dq_{(eq)} + 2\,Dq_{(ax)}]\,. \tag{3.17}$$

This method of evaluating ligand field parameters for transition metal ions in tetragonally distorted octahedra is particularly applicable to minerals containing Mn^{3+} and Cr^{2+}, as well as Fe^{2+} ions in some very distorted octahedra, because their crystal field spectra enable δ_1, δ_2 and δ_3 (eq. 3.16)) to be determined experimentally. Examples are discussed in chapter 5 particularly when crystal field spectra of Mn^{3+}-bearing minerals are described (§5.10.5).

3.11 Summary

Chapter 3 describes the theory of electronic spectra of transition metal ions. The three characteristic features of absorption bands in a spectrum are position or energy, intensity of absorption and width of the band at half peak-height. Positions of bands are commonly expressed as wavelength (micron, nanometre or ångstrom) or wavenumber (cm^{-1}) units, while absorption is usually displayed as absorbance, absorption coefficient (cm^{-1}) or molar extinction coefficient [litre $(g.ion)^{-1}$ cm^{-1}] units.

Spectroscopic terms. Each isolated ion gives rise to several spectroscopic terms, each term representing an energy level resulting from a different arrangement of electrons in atomic orbitals. Spectroscopic terms are designated by letter symbols, ^{r}L, denoting the total atomic orbital angular momenta of unpaired electrons (L) and the spin multiplicity (r)which are calculated by adding one to the number of unpaired electrons. Energy separations between spectroscopic terms are determined from atomic spectra of gaseous ions.

Crystal field states and positions of absorption bands. The spectroscopic terms are split into one or more crystal field states when a transition metal ion is located in a coordination site in a crystal structure. Each crystal field state represents a different electronic configuration of $3d$ orbitals. Crystal field states are labelled by group theory notation summarizing the symmetry properties of the electronic configurations and the spin multiplicities. Energy level diagrams, including Orgel and Tanabe–Sugano diagrams, show how anions or ligands coordinated to the cation influence interelectronic coulombic and exchange interactions between the $3d$ electrons. For crystal field states with highest spin multiplicities, energy level diagrams for cations with d^n and d^{10-n} configurations show qualitative similarities, with one diagram representing the inverted form of the other. The diagrams also bear an inverse relationship for cations in octahedral and tetrahedral (or cubic and dodecahedral) sites. As a result, there are essentially only three types of energy level diagrams, correlating with three distinct absorption spectrum profiles, for the various $3d$ electronic configurations in high-symmetry environments. One type applies to $3d^5$ cations, a second to $3d^1$, $3d^6$, $3d^4$ and $3d^9$ cations, and the third to cations with $3d^2$, $3d^7$, $3d^3$ and $3d^8$ configurations. When a transition metal ion is situated in a distorted coordination site, or the ligands are unevenly distributed about the cation, further resolution of the $3d$ orbital energy levels takes place. The number and positions of absorption bands in visible to near-infrared spectra reflect energy separations between electronic configuration differing in symmetry in the non-cubic coordination environments.

Selection rules and intensities of absorption bands. The probability of electronic transitions between different crystal field states, which influences intensities of absorption bands in crystal field spectra, may be deduced from selection rules. The spin-multiplicity selection rule specifies that transitions may take place only between ground and excited states having the same number of unpaired electrons. Spin-allowed transitions occur in crystal field spectra of most transition metal ions except cations with high-spin $3d^5$ configurations. Thus, spin-forbidden transitions, observed in minerals containing Mn^{2+} and Fe^{3+} ions, produce very weak absorption bands, which may gain intensity by spin–orbit coupling of electrons and magnetic interactions between neighbouring cations. The Laporte selection rule specifies that electronic transitions between orbitals of the same type and quantum number, such as two $3d$ orbitals, are forbidden. However, crystal field transitions involving electrons in t_{2g} and e_g orbitals may take place and gain intensity whenever cations occur in coordination environments, such as tetrahedral sites, that lack a centre of symmetry. In centrosymmetric octahedral sites, absorption bands gain intensity by vibronic coupling when electronic transitions proceed via acentric vibrational modes of the cation and surrounding ligands.

Polarization dependence of absorption bands. The occurrence of transition metal ions in low-symmetry coordination sites leads to crystallographic dependencies of absorption spectra for non-cubic minerals in polarized light. Interpretations of such polarized spectra using methods of group theory hinge upon considerations of symmetry properties of ground and excited crystal field states as well as the electric vectors of polarized light transmitted through the crystal. For centrosymmetric cations, symmetries of vibrational modes must also be taken into account. The polarized spectra of gillespite, $BaFeSi_4O_{10}$, a tetragonal mineral with Fe^{2+} ions in square planar coordination, illustrate group theoretical methods for analysing absorption bands in visible to near-infrared spectra.

Widths of absorption bands. Thermal vibrations of atoms produce fluctuating energy separations between 3*d* orbital energy levels, leading to broadened and asymmetric absorption bands, particularly at elevated temperatures. Sharp peaks in visible-region spectra of Mn^{2+}- and Fe^{3+}-bearing minerals are due to transitions between crystal field states of constant energy separation. Superposition of two or more bands representing transitions to closely spaced energy levels of transition metal ions either located in a single low-symmetry site or distributed over several structurally similar sites may also broaden absorption bands.

3.12 Background reading

Ballhausen, C. J. (1962) *Introduction to Ligand Field Theory.* (McGraw–Hill, New York), 298 pp.

Burns, R. G. (1985) Electronic spectra of minerals. In *Chemical Bonding and Spectroscopy in Mineral Chemistry.* (F. J. Berry & D. J. Vaughan, eds; Chapman & Hall, London), pp. 63–101.

Cotton, F. A. (1990) *Chemical Applications of Group Theory, 3rd edn.* (Wiley–Interscience, New York), 579 pp.

Figgis, B. N. (1966) *Introduction to Ligand Fields.* (Interscience Publ., New York), 351 pp.

Gerloch, M. & Slade, R. C. (1973) *Ligand–Field Parameters.* (Cambridge Univ. Press, Cambridge), 235 pp.

Henderson, B. & Imbusch, G. F. (1989) *Optical Spectroscopy of Inorganic Solids.* (Oxford Univ. Press, Oxford), ch. 2 & 3.

Hitchman, M. A. (1985) Chemical information from the polarized crystal spectra of transition metal complexes. In *Transition Metal Chemistry, vol. 9.* (G. A. Melson & B. N. Figgis, eds; Marcel Dekker, Inc., New York), pp. 1–223.

Lever, A. B. P. (1984) *Inorganic Electronic Spectroscopy, 2nd edn.* (Elsevier, Amsterdam), 863 pp.

McClure, D. S. (1959) Electronic spectra of molecules in crystals. Part II: Spectra of ions in crystals. *Solid State Phys.*, 9, 399–425.

Marfunin, A. S. (1979) *Physics of Minerals and Inorganic Materials.* (Springer–Verlag, New York), chs 1 & 2.

Orgel, L. E. (1966) *An Introduction to Transition-Metal Chemistry: Ligand-Field Theory, 2nd edn.* (Methuen, London), ch. 6.

4

Measurements of absorption spectra of minerals

Since ferrous iron usually colors minerals green, and ferric iron yellow or brown, it may seem rather remarkable that the presence of both together should give rise to a blue color, as in the case of vivianite. – – Other instances may perhaps be discovered, should this subject ever be investigated as it deserves to be.

E. T. Wherry, *Amer. Mineral.*, 3, 161 (1918)

4.1 Introduction

In order to apply crystal field theory to geologic processes involving transition metal ions, it is necessary to have crystal chemical information and thermodynamic data for these cations in mineral structures. In §2.8, it was noted that the principal method for obtaining crystal field splittings, and hence stabilization energies of the cations, is from measurements of absorption spectra of transition metal compounds at wavelengths in the visible and near-infrared regions. The origins of such absorption bands in crystal field spectra were discussed in chapter 3. The focus of this chapter is on measurements of absorption spectra of minerals, with some applications to fundamental crystal chemical problems.

When minerals occur as large, gem-sized crystals, it is comparatively easy to obtain absorption spectra by passing light through natural crystal faces or polished slabs of the mineral. However, most rock-forming minerals are present in assemblages of very small crystals intimately associated with one another, leading to technical problems for measuring spectra of minerals *in situ*. In addition, several of the transition elements occur in only trace amounts in common minerals, making spectral features of individual cations difficult to resolve, especially in the presence of more abundant elements such as iron, titanium and manganese which also absorb radiation in the visible to near-infrared region. Mineral spectroscopists are continually searching for exotic minerals, preferably in the form of large gem quality crystals, containing unusually high concentrations of the less abundant transition elements for absorption spectral measurements. Discoveries of such rare minerals, including gem varieties of corundum (ruby), beryl (emerald and morganite), garnet (tsavorite) and zoisite (tanzanite), have considerably broadened the crystal field spectral data-base for several cations, notably Cr^{3+} and V^{3+}, in a variety of crystal structures. In other cases, synthetic analogues and extraterrestrial minerals have enabled visible-region spectra to be measured of inherently unstable cations, including Ti^{3+}, Mn^{3+} and Cr^{2+}.

There are several approaches for obtaining spectral data for low-abundance transition metal ions, rare minerals and crystals of small dimensions. Data for a transition element in its chemical compounds, such as hydrates, aqueous solutions, molten salts or simple oxides, may be extrapolated to minerals containing the cation. Such data for synthetic transition metal-doped corundum and periclase phases used to describe principles of crystal field theory in chapter 2, appear in table 2.5, for example. There is a growing body of visible to near-infrared spectral data for transition metal-bearing minerals, however, and much of this information is reviewed in this chapter and the following one. These results form the data-base from which crystal field stabilization energies (CFSE's) of most of the transition metal ions in common oxide and silicate minerals may be estimated.

Measurements of electronic absorption spectra in the visible region not only lead to the evaluation of CFSE's, but they also provide useful information about the crystal chemistry of transition metal ions in the crystal structures and causes of colour and pleochroism of minerals. In this chapter, techniques for measuring absorption spectra of minerals are briefly described and some general applications of the optical spectra to basic crystal chemical properties, such as colour and pleochroism, are discussed. These examples also amplify many of the features of crystal field spectra outlined in chapter 3.

4.2 Techniques for measuring absorption spectra of minerals

The optimum method for obtaining absorption spectra of minerals is to use chemically analysed, gem-size crystals. A major problem encountered in measurements of solid-state spectra, though, is the adverse effect of light scattering by the sample which may be alleviated by using highly polished plates when large crystals are available. A typical method is to mount a crystal having a minimum diameter of about 1 mm (1,000 microns) and thickness ranging from 10 to 100 microns into a holder fitted with an appropriate aperture (Rossman, 1975a). The crystal thickness depends on the specific cation, its concentration in the mineral, and the symmetry of the coordination site occupied by the cation. The holder may then be placed in the cell compartment of a double-beam spectrophotometer. For measurements of polarized absorption spectra, various types of polarizers have been used, including calcite Nicol, Glan–Thompson or air-gap prisms, sheets of Polaroid film (e.g., HN42 for the visible and HR for the near-infrared) and stacked AgI plates. This technique, using oriented single crystals and pinhole-size apertures, has provided a plethora of mineral absorption spectra (Rossman, 1988).

Silicate minerals most frequently occur as very small crystals in rock, however, and it is often impossible to prepare polished plates of individual minerals for spectral measurements. Various methods have been devised to obtain absorption spectra for powdered minerals separated from crushed rock by heavy-liquid and magnetic separation techniques (Faye, 1971a). A powdered specimen may be placed between opal scattering discs or embedded in pressed KBr or CsI pellets. A simple but effective method for overcoming light scattering in minerals is to mount between glass slides a finely divided specimen (particle size 10 to 45 microns) as a paste in a transparent oil matching the average refractive index of the mineral. Since powders contain grains mainly in random orientations, measurements made on pulverized samples give average spectra. A few minerals, including amphiboles, micas, gillespite and clay silicates with acicular or platey habits, may give spectrum profiles that closely match polarized spectra of single crystals due to effects of preferred orientation arising from a unidirectional alignment of the crystallites. In diffuse reflectance spectral measurements, powdered mineral samples are used with an integrating sphere accessory, particularly in research to calibrate telescopic remotely sensed spectra of planetary surfaces. Examples of such reflectance spectral measurements are described in chapter 10.

In the development of mineral spectroscopy, microscopes have figured prominently in spectral measurements of small crystals. One early method for measuring polarized spectra of fine-grained minerals contained in rock utilized polarizing microscopes equipped with universal stage attachments (Burns, 1966b). The mineral specimen in the form of a rock thin-section or single crystal mounted in a transparent cement is placed between the glass hemispheres of a three-axis universal stage attached to a polarizing microscope. The microscope is then mounted in the sample chamber of the double-beam spectrophotometer so that radiation from the light-source is transmitted through the polarizer of the microscope before impinging on the specimen and entering the objective. An identical microscope arrangement is placed in the reference beam of the spectrophotometer except that a glass slide containing only the

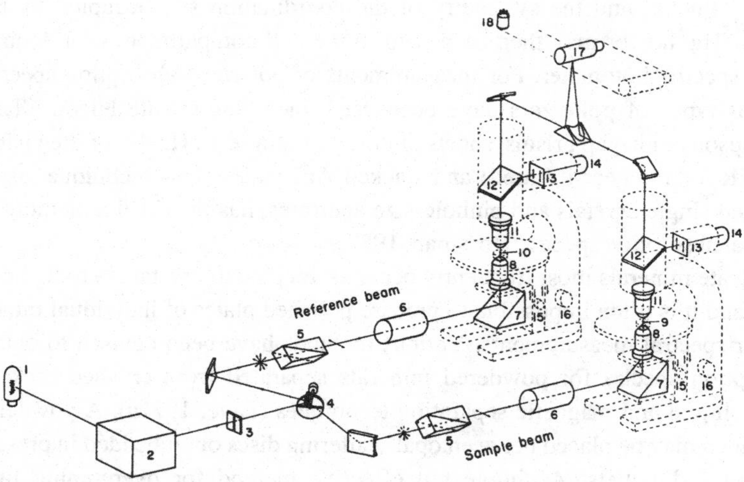

Figure 4.1 Schematic view of a microscope spectrophotometric system for measuring the polarized absorption spectra of minerals. 1. Tungsten–halogen light source; 2. prism-grating primary and intermediate slit system of the spectrophotometer; 3. exit slit; 4. beam chopper and curved mirrors; 5. calcite Nicol polarizing prisms; 6. auxiliary lens system; 7. mirror; 8. condensing lens; 9. mineral crystal or rock thin section; 10. reference mount; 11. objective lens; 12. half-mirror; 13. analyser; 14. viewing occular; 15. substage adjustment; 16. stage adjustment; 17. photomultiplier tube; 18. PbS solid-state detector (from Bell and Mao, 1972a; Mao and Bell, 1973a).

transparent cement is used. A schematic view of one such apparatus for spectral measurements by the microscope technique is shown in fig. 4.1 (Bell and Mao, 1972a; Mao and Bell, 1973a; Abu–Eid, 1976).

Such a microscope technique, which was used to obtain many of the spectra illustrated in chapters 4 and 5, has numerous advantages for absorption spectral measurements of minerals. First, only very small amounts of material are required. A single crystal of the mineral should cover the field of view of the microscope and this may be controlled by choice of a suitable objective lens. Second, the use of rock thin-sections enables spectral measurements to be made on individual minerals without having to extract single crystals from the rock matrix. Third, the calcite Nicol prism of the petrographic microscope provides a convenient method for transmitting polarized light through the specimen, thereby enabling polarized spectra to be measured. Fourth, the use of a universal stage attachment enables a crystal to be oriented accurately in any desired manner and permits the measurement of polarized absorption spectra along any crystallographic axis or extinction direction in the crystal. One disadvantage of microscope spectrophotometry, however, is that while accurate measurements of positions (energies) of absorption bands are possible, quantitative estimates of intensities are subject to error due to the fact that convergent light, necessarily introduced by the condenser of the petrographic microscope, mixes the polarization components of the various absorption bands (Goldman and Rossman, 1979). Effects of light convergence are most evident in spectra taken with high power objective lenses and especially for samples which have a highly absorbing band in one polarization only. Resulting absorption bands may be flat-topped and have lowered peak-height intensities, displaying characteristics similar to those caused by stray light in a spectrophotometer. Another disadvantage is that the microscope technique is often restricted to spectral measurements in the wavelength region 400 to 2,500 nm. Beyond these limits absorption by the glass optics of the microscope system may become appreciable.

The development of microscope spectrophotometers during the past two decades has greatly extended the range of spectral measurements of very small crystals. In fact, the evolution of microscope absorption spectroscopy has closely paralleled developments of high-pressure diamond anvil cells (Bell and Mao, 1969, 1972a; Mao and Bell, 1973a, 1978; Abu–Eid, 1976; Langer and Abu–Eid, 1977; Mao, 1976; Sung, 1976; Langer and Frentrup, 1979; Frentrup and Langer, 1982; Smith and Langer, 1982a,b; Langer, 1988), with the result that pressure variations of mineral spectra are now routinely studied. These methods allow spectral measurements to be made from crystal areas with minimum diameters of 30 nm and smaller (Langer and Frentrup, 1979). The stability of transparent diamonds to temperature also has been exploited by using a laser beam to heat mineral specimens under pressure, so that it is becoming increasingly feasible to measure mineral spectra at high pressures and temperatures by microscope spectrophotometery. Results of such spectra are described in chapter 9.

4.3 Measurements of intensities of absorption bands

In §3.2 it was noted that a variety of intensity units appear in published mineral spectra, ranging from optical densities obtained directly from a spectrophotometer to calculated extinction coefficients (table 3.2). The Beer–Lambert Law,

$$\log \frac{I_0}{I} = \varepsilon\, C\, d, \qquad (3.7)$$

so fundamental to electronic absorption spectroscopy, utilizes the molar extinction coefficient, ε. The value of ε is readily calculated for a solution of a dissolved transition metal compound of known concentration (C) contained in a cell of specific thickness (d). However, computations of ε are somewhat more complicated for transition metal-bearing minerals.

As noted in eq. (3.8),

$$C = \frac{X}{V}, \tag{3.8}$$

so the concentration, C, of a transition metal ion in solid-solution in a silicate mineral may be calculated from the mole fraction, X, and the molar volume, V, of the silicate phase. For example, the molar volume of an olivine containing 12 mole per cent Fe_2SiO_4 (Fa_{12}) is 0.0441 litre mole^{-1}. Thus, the value of C is (0.12/0.0441) or 2.79 moles per litre of Fe_2SiO_4 or 5.58 g. ions per litre of Fe^{2+}. The molar extinction coefficient, ε, of an absorption band at 1,050 nm with optical density 1.26 measured on a crystal of thickness 0.055 cm is [1.26/(2.79 x 0.055)] or 8.2 litre mole^{-1} cm^{-1} based on Fe_2SiO_4. However, since half the Fe^{2+} ions occur in the olivine M2 site and they, alone, are responsible for the CF band at 1,050 nm, the value of ε is, in fact, 16.4 litre (g. ion)$^{-1}$ cm^{-1} for olivine Fe^{2+}/M2 site cations.

An alternative method for calculating the molar concentration of a transition metal ion in solid-solution is as follows (Rossman, 1980, 1988). The forsterite Fa_{12} with 11.95 wt per cent FeO has a specific gravity of 3.35. One litre of the crystal weighs 3,350 grams, of which 11.95 wt per cent or 400.3 g is FeO. The formula weight of FeO is 71.85, so that the number of moles of FeO in one litre of crystal is 400.3/71.85 or 5.57 moles. There is one mole of Fe in FeO so that the concentration of Fe^{2+} in the forsterite Fa_{12} is 5.57 g. ions per litre, which is then used to calculate the ε value of 16.4 litre (g. ion)$^{-1}$ cm^{-1}.

According to the Beer–Lambert law, eq. (3.7), the molar extinction coefficient of an absorption band should be independent of the total concentration of the absorbing species. Deviations from this law may arise whenever a transition metal ion is distributed over two or more coordination sites in a structure having different configurations of surrounding anions. Non-constancy of molar extinction coefficients may be indicative of cation ordering in a structure. In Mg^{2+}–Fe^{2+} olivines, values of ε are virtually unchanged across the forsterite–fayalite series in accord with the negligible cation ordering found in Mg^{2+}–Fe^{2+} olivines (§6.7.1.1). In the orthopyroxene series, on the other hand, values of ε for crystal field bands vary with total iron content, which correlates with cation ordering in Mg^{2+}–Fe^{2+} pyroxenes (§6.7.2.1). Other consequences of cation ordering are compositional variations of spectrum profiles manifested in differences of peak positions and intensities across a solid-solution series. Examples described later include crystal field spectra of Mg–Fe^{2+} orthopyroxenes (fig. 5.15), Mg–Ni olivines (fig. 5.12) and Mn^{3+}-bearing epidotes and andalusites (§4.4.2).

4.4 Identification of cation valences and coordination symmetries

The valence and coordination symmetry of a transition metal ion in a crystal structure govern the relative energies and energy separations of its 3*d* orbitals and, hence, influence the *positions* of absorption bands in a crystal field spectrum. The *intensities* of the absorption bands depend on the valences and spin states of each cation, the centrosymmetric properties of the coordination sites, the covalency of cation–anion bonds, and next-nearest-neighbour interactions with adjacent cations. These factors may produce characteristic spectra for most transition metal ions, particularly when the cation occurs alone in a simple oxide structure. Conversely, it is sometimes possible to identify the valence of a transition metal ion and the symmetry of its coordination site from the absorption spectrum of a mineral.

4.4.1 Valence of titanium in extraterrestrial pyroxenes

The Ti(III) oxidation state is rare in terrestrial minerals compared to Ti(IV) due to the comparatively high redox conditions associated with mineral-forming processes on Earth. Nevertheless, in Fe–Ti minerals, such as subsilicic garnets of the melanite–schorlomite series, [$Ca_3(Fe,Ti)_2(Si,Fe)_3O_{12}$], the presence of Ti^{3+} ions has been inferred from wet chemical analyses, crystal field and Mössbauer spectral measurements (Burns, 1972a; Huggins *et al.*, 1977; Schwartz *et al.*, 1980). Trivalent titanium is more prevalent in extraterrestrial materials, including titaniferous pyroxenes and glasses from the Moon and in certain meteorites (Burns *et al.*, 1972a; Bell and Mao, 1972b; Dowty and Clark, 1973; Sung *et al.*, 1974; Bell *et al.*, 1976). However, the presence of iron produces ambiguous assignments of the optical spectra of these Fe–Ti phases due to overlapping absorption bands in the visible region arising from spin-forbidden transitions in Fe^{2+} ions (cf. fig. 3.2) and $Fe^{2+} \rightarrow Ti^{4+}$ intervalence charge transfer (IVCT) transitions discussed later (§4.7.2.5).

Trivalent titanium has been positively identified by optical spectral measurements of a green calcic pyroxene from the meteorite that fell near Pueblo de Allende, Mexico, in 1969. The chemical analysis of this titanian pyroxene (Dowty and Clark, 1973) revealed it to be an iron-free subsilicic diopside (fassaite) containing coexisting Ti^{3+} and Ti^{4+} ions and having the chemical formula $Ca_{1.01}Mg_{0.38}Ti^{3+}{}_{0.34}Ti^{4+}{}_{0.14}Al_{0.87}Si_{1.26}O_6$.

The polarized spectra of the pyroxene from the Allende meteorite illustrated in fig. 4.2 (Mao and Bell, 1974a) consist of three broad bands centred near 480 nm (20,830 cm^{-1}), 610 nm (16,390 cm^{-1}) and 666 nm (15,000 cm^{-1}). There has been much debate about assignments of these bands (Dowty and Clark, 1973;

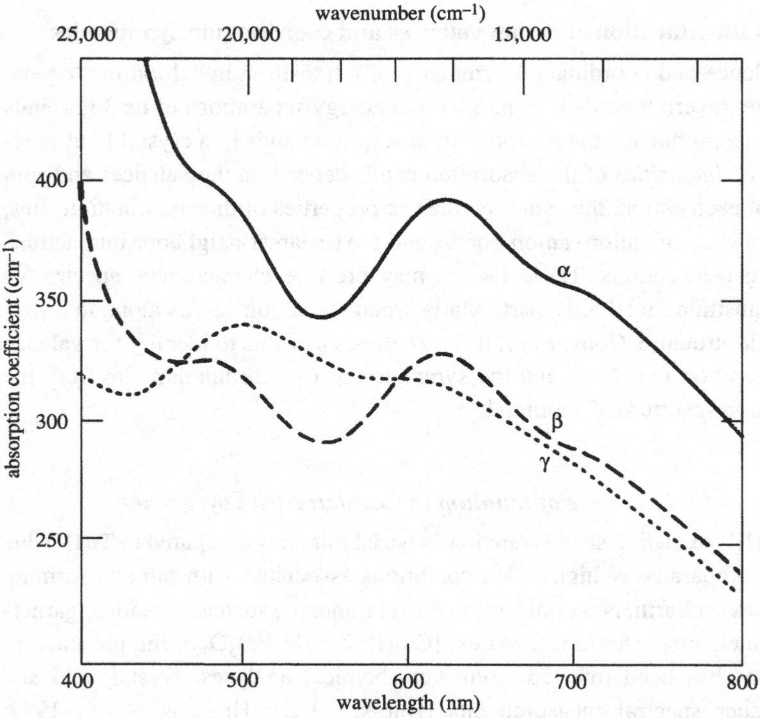

Figure 4.2 Polarized absorption spectra of the titanian subsilicic diopside (fassaite) from the Allende meteorite (from Mao and Bell, 1974a). — α spectrum; - - - - β spectrum; .... γ spectrum. The pyroxene contains 0.34 Ti^{3+} ions per formula unit.

Burns and Huggins, 1973; Mao and Bell, 1974a; Dowty, 1978; Strens *et al.*, 1982). Pressure-induced shifts of the 480 nm and 610 nm bands to lower wavelengths indicate that they originate from crystal field transitions in Ti^{3+} ions located in the distorted M1 octahedral site of the calcic pyroxene structure described later (fig. 5.13). On the other hand, increased intensification and negligible shift of the 666 nm band at high pressures suggest its origin to be a $Ti^{3+} \rightarrow Ti^{4+}$ IVCT transition between Ti atoms 315 pm apart in edge-shared M1 octahedra in the pyroxene structure. Positive identifications of Ti^{3+} ions have also been made from optical spectral measurements of synthetic Ti^{3+}-doped Al_2O_3 crystals (McClure, 1962; Townsend, 1968) described later (fig. 4.16c) and silica glasses approximating lunar basaltic compositions (Bell *et al.*, 1976; Nolet *et al.*, 1979), all devoid of iron.

Optical spectral measurements thus provide a method for detecting, and possibly estimating the proportions of, Ti^{3+} ions in minerals which is difficult to do when iron is also present. During wet chemical analysis, Ti^{3+} ions are oxidized to the Ti(IV) state by coexisting Fe^{3+} ions. As a result, Fe^{2+}/Fe^{3+} ratios deduced from chemical analyses of Ti-garnets, for example, are in error (Howie and Woolley, 1968; Burns, 1972a; Schwartz *et al.*, 1980). The occurrences of Ti^{3+} ions in other silicates have been suggested from optical spectral measurements of garnets (Manning, 1967a,b), micas (Faye, 1968a,b), tourmalines (Faye *et al.*, 1968), terrestrial titanaugites (Chesnokov, 1957), several pyroxenes from mare basalts (Burns *et al.*, 1972a; Sung *et al.*, 1974), kosmochlor in the Toluca meteorite (Abs-Wurmbach *et al.*, 1985) and orange glass spherules from the regolith of the Moon (Vaughan and Burns, 1973). However, the coexistence of Fe^{2+} and Ti^{4+} cations in many of these phases may produce complications from intervalence charge transfer transitions described later (§4.7.2). These transitions are very intense and occur at similar energies in the visible region as the Ti^{3+} crystal field bands (Burns, 1981, 1991; Strens *et al.*, 1982; Mattson and Rossman, 1988).

4.4.2 Stability of Mn^{3+} ions in epidote and andalusite

4.4.2.1 Piemontite

The Mn^{3+} ion, $3d^4$, is relatively unstable and, as noted in §2.5, Mn(III) compounds readily undergo oxidation, reduction and disproportionation reactions forming more stable Mn(II) and Mn(IV) compounds. One mineral that stabilizes the Mn^{3+} ion is the manganiferous epidote, piemontite, $Ca_2(Al,Fe,Mn)Al_2Si_3O_{12}(OH)$, in which substantial amounts of manganese replace Fe^{3+} and Al^{3+} ions. The stability of Mn^{3+} cations in the epidote structure may be assessed from energy separations deduced from polarized absorption spectra of piemontite (Burns and Strens, 1967; Xu et al., 1982; Smith et al., 1982; Kersten et al., 1987), and the configurations of the coordination sites in the epidote structure (Dollase, 1968, 1969; Gabe et al., 1973; Stergiou and Rentzeperis, 1987).

The three polarized spectra of piemontite shown in fig. 4.3 are distinctive with absorption bands of different intensities occurring in each spectrum at approximately 450 nm (22,200 cm^{-1}), 550 nm (18,200 cm^{-1}) and 830 nm (12,000 cm^{-1}) (Burns and Strens, 1967; Smith *et al.*, 1982; Kersten *et al.*, 1987). The molar extinction coefficients of these bands are one order of magnitude higher than those in absorption spectra of other Mn(III) compounds, indicating that Mn^{3+} ions are situated in an acentric coordination site in the epidote structure. Moreover, extinction coefficients of corresponding bands in the polarized spectra of a suite of manganiferous epidotes vary with composition,

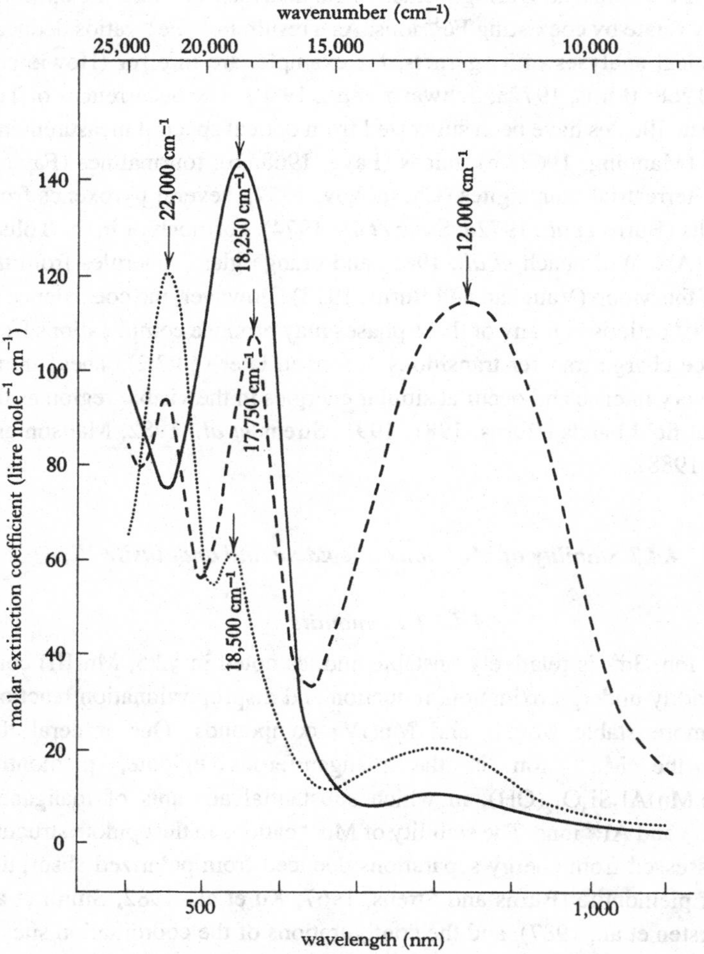

Figure 4.3 Polarized absorption spectra of piemontite. ······· α spectrum; – – – – β spectrum; —— γ spectrum. The epidote mineral, containing 0.625 Mn^{3+} ions per formula unit, is from St. Marcel, Piemonte, Italy [Optic orientation: $\alpha : c = 5^\circ$; $\beta = b$; $\gamma : a = 30^\circ$]

indicating that the Mn^{3+} ions are not located entirely in one site (Burns and Strens, 1967; Smith *et al.*, 1982). This is borne out by crystal structure refinements of a piemontite (Dollase, 1969) from the same locality at St. Marcel, Italy, that yielded the polarized absorption spectra illustrated in fig. 4.3.

The epidote structure illustrated in fig. 4.4 contains three positions of six-fold coordination. The M3 coordination polyhedron is acentric with irregular metal–oxygen and oxygen–oxygen distances due to compression of the site along one

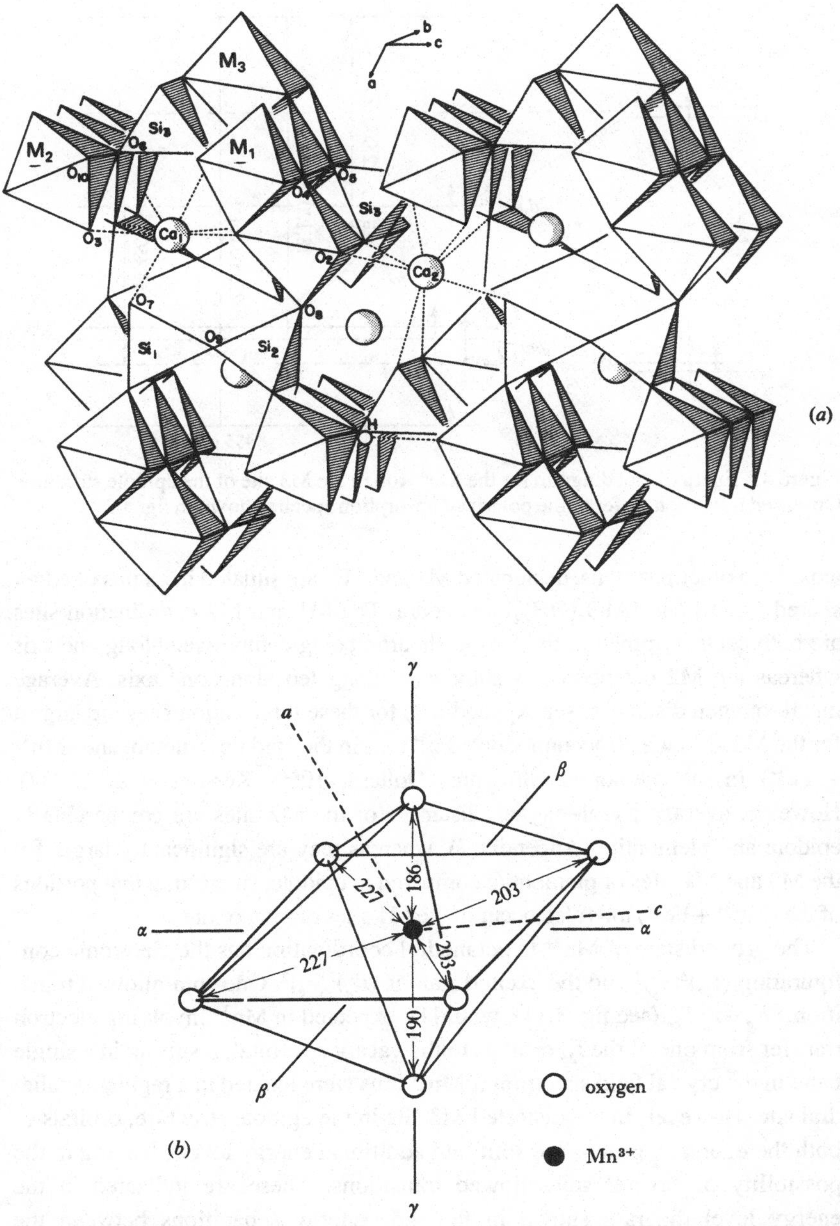

Figure 4.4 Crystal structure of epidote. (*a*) Linkages of the coordination polyhedra viewed along the *b* axis; (*b*) configuration of the M3 coordination site accommodating Mn^{3+} ions in the epidote structure. The figure shows the orientations of the crystallographic and indicatrix axes (from Dollase, 1969). Metal–oxygen distances are pm.

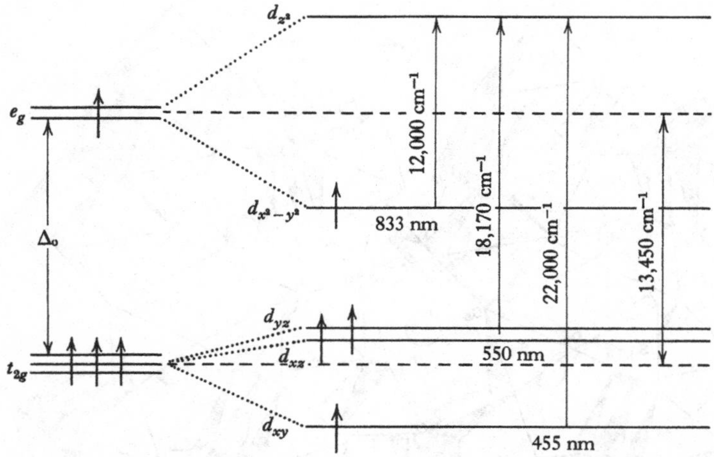

Figure 4.5 Energy level diagram for the Mn^{3+} ion in the M3 site of the epidote structure. Observed transitions refer to the polarized absorption spectra shown in fig. 4.3.

axis. Two other positions, designated M1 and M2, are situated in chains of edge-shared [AlO_6] and [$AlO_4(OH)_2$] octahedra. The M1 and M2 coordination sites are both centrosymmetric, the M1 octahedron being compressed along one axis whereas the M2 octahedron is somewhat elongated along one axis. Average metal–oxygen distances (see Appendix 7) for these three cation sites are largest for the M3 sites which accommodate Fe^{3+} ions in the epidote structure and (Mn^{3+} + Fe^{3+}) in the piemontite structure (Dollase, 1969; Kersten *et al.*, 1987). However, average metal–oxygen distances of the M2 sites are comparable in epidote and piemontite (Appendix 7), whereas they are significantly larger for the M3 and M1 sites of piemontite compared to epidote, suggesting that portions of the (Mn^{3+} + Fe^{3+}) ions also occur in the M1 sites of piemontite.

The ground state of Mn^{3+} in octahedral coordination has the electronic configuration $(t_{2g})^3(e_g)^1$ and the excited state is $(t_{2g})^2(e_g)^2$. One spin-allowed transition, $^5E_g \rightarrow {}^5T_{2g}$ (see fig. 3.11), would be expected in Mn^{3+} involving electron transfer from one of the t_{2g} orbitals to the vacant e_g orbital, resulting in a single band in the crystal field spectrum if Mn^{3+} ions were located in a regular octahedral site. However, in the distorted M3 site in the epidote structure, orbitals of both the e_g and t_{2g} groups are split into additional energy levels, leading to the possibility of several spin-allowed transitions. These are indicated in the energy level diagram shown in fig. 4.5. Energy separations between the resolved 3*d* orbitals derived from the polarized spectra of piemontite illustrated in fig. 4.3 provide a value of Δ_o = 13,450 cm^{-1} for the crystal field splitting parameter, assuming that split 3*d* orbital energy levels obey centre of gravity rules about the t_{2g} and e_g baricentres. The CFSE of a high-spin $3d^4$ ion such as Mn^{3+} in an undistorted octahedral site would normally amount to $0.6\Delta_o$ (table 2.2) or 8,070 cm^{-1} (96.6 kJ/g.ion) if the epidote M3 site had regular octahedral symmetry. However, due to distortion of the M3 site, the Mn^{3+} ion in piemontite acquires a CFSE of about 167.8 kJ/g.ion, of which 71.8 kJ/g.ion (6,000 cm^{-1}) results from the stabilization of one orbital of the e_g group. Thus, the M3 site of the epidote structure considerably stabilizes the Mn^{3+} ion, which is normally unstable in octahedral coordination.

A qualitative explanation for the polarization dependence of the absorption bands is as follows. According to eq. (4.1) the probability, P_x, of a transition is proportional to the square of the integral

$$P_x \propto \left[\int_0^{\infty} \Psi_{\mathrm{g}} \mu_x \Psi_{\mathrm{e}} \mathrm{d}x \right]^2 , \tag{4.1}$$

where Ψ_{g} and Ψ_{e} are the wave functions of the ground and excited states, respectively, and μ_x is the dipole moment component along x. In the spectra of piemontite, Ψ_{e} is the d_{z^2} orbital. Transitions between the d_{xy} or $d_{x^2-y^2}$ and d_{z^2} orbitals (see fig. 4.5) lead to intense absorption when radiation is polarized in the plane of the d_{xy} and $d_{x^2-y^2}$ orbitals, accounting for the bands at 450 nm and 830 nm in the α and β spectra. Similarly, transitions between the d_{xz} or d_{yz} and d_{z^2} orbitals lead to absorption when a component of the polarized radiation interacts with electrons in the d_{xz} or d_{yz} orbitals. These produce the peaks at 540 to 560 nm in the α, β and γ spectra.

4.4.2.2 Viridine and kanonaite

The polarized spectra of Mn^{3+}-bearing andalusites, including kanonaite ($MnAlSiO_5$) and viridine [$(Al,Fe,Mn)_2SiO_5$], show features in common with the spectra of piemontite illustrated in fig. 4.3. Three band systems again dominate the crystal field spectra of andalusites containing Mn^{3+} ions (Abs-Wurmbach *et al.*, 1981; Smith *et al.*, 1982; Langer, 1988; Rossman, 1988); they are centred near 700 nm, 450 nm and 425 nm with different relative intensities in each polarized spectrum. The polarization dependencies of the absorption bands and the stability of the Mn^{3+} ion may again be correlated with its occurrence in a distorted six-coordinated site in the andalusite structure. Crystal structure refinements (Abs-Wurmbach *et al.*, 1981; Weiss *et al.*, 1981) have shown that the major portion of Mn^{3+} ions occur in the acentric tetragonally elongated M1 octahedra of the andalusite structure, the deformation of which increases with increasing Mn^{3+} content of andalusite (Abs-Wurmbach *et al.*, 1983; Weiss *et al.*, 1981). An important consequence of the high CFSE of

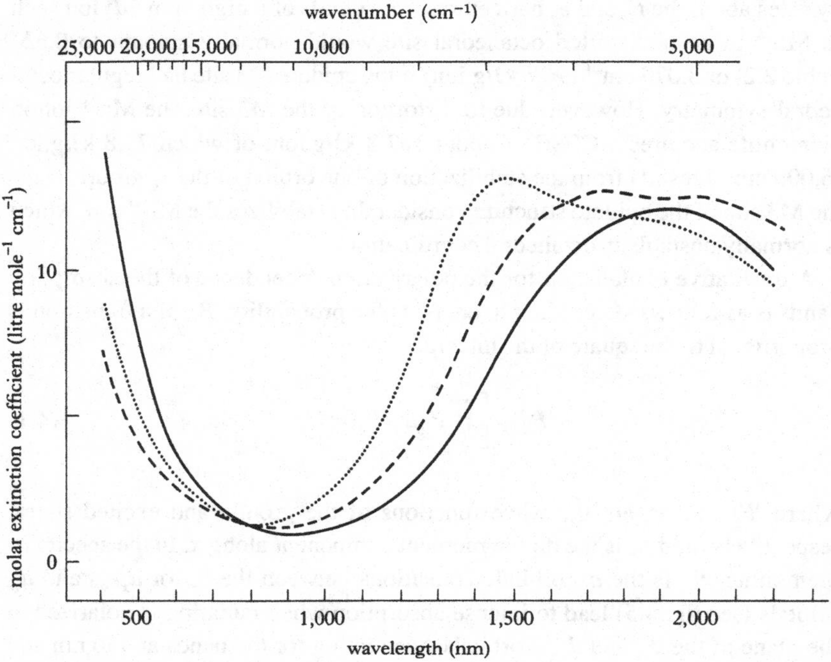

Figure 4.6 Polarized absorption spectra of staurolite. – – – – α spectrum; ······ β spectrum; —— γ spectrum. The staurolite, containing 1.42 Fe^{2+} ions per formula unit, is from Pizzo Forno, Tessin, Switzerland. [Optic orientation: α = *b*; β = *a*; γ = *c*.]

Mn^{3+}, estimated to be about 198 kJ/g.ion in the andalusite M1 site, is that it increases the stability field of viridines in the pressure–temperature phase diagram for the system Al_2SiO_5–$(Al_{1-x}Mn_x)_2SiO_5$ where $x < 0.2$ (Abs-Wurmbach *et al.*, 1981; Langer, 1988). The influence of CFSE on the stability field of Mn^{3+} andalusite is discussed in chapter 7 (§7.7.4.1).

4.4.3 Coordination symmetry of iron and cobalt in staurolite

The crystal field spectra of Fe^{2+} ions surrounded by oxygen in regular octahedral sites normally contain absorption bands centred near 1,000 nm or 10,000 cm^{-1} (see fig. 3.2). By changing from octahedral to tetrahedral coordination, absorption bands for tetrahedral Fe^{2+} ions would, according to eq. (2.7), be predicted to occur at ($^4/_9$ x 10,000) or 4,444 cm^{-1} (2,250 nm) if the iron–oxygen distances remain identical in the two coordinations. Tetrahedrally coordinated Fe^{2+} ions in spinel, $MgAl_2O_4$, for example, produce an absorption band near 4,830 cm^{-1} (2,070 nm) (§5.3.3).

Staurolite, $(Fe,Mg,Co)_2Al_9(SiO_4)_4O_6(OH)_2$, also contains the major portion of Fe^{2+} ions in tetrahedral sites. Its polarized absorption spectra shown in fig. 4.6 consist of a very broad, intense band spanning the region 1,400 to 2,200 nm (7,000 to 4,550 cm^{-1}). The position of such bands may be correlated with the occurrence of Fe^{2+} ions in a slightly distorted tetrahedron, the average metal–oxygen distance of which is about 201 pm. The location of the absorption bands in the staurolite spectra at higher energies than that predicted by eq. (2.7) is due to the shorter Fe–O distances in the staurolite structure compared to those in the six-coordinated sites in ferromagnesian silicates, which are of the order of 216 pm (Appendix 7).

The polarized spectra of lusakite, the cobaltian staurolite from Zambia (Cech *et al.*, 1981), which are illustrated in fig. 4.7 together with (inset) the spectra of the hexahydrated Co^{2+} and tetrahedral Co^{2+} cations in spinel, show that Co^{2+} ions also occur in tetrahedral coordination in the staurolite structure. The profiles and intensities of the triple-peak absorption band between 500 nm and 650 nm are characteristic of tetrahedral Co^{2+} ions.

Spectral characteristics of tetrahedral Fe^{2+} ions may be shown coincidently by Fe^{2+} ions in very distorted coordination polyhedra. For example, bands at 4,420 cm^{-1} in the spectra of calcic pyroxenes (§5.5.3) and at 5,400 cm^{-1} in orthopyroxene spectra (fig. 5.15*a*) were once erroneously assigned to tetrahedral Fe^{2+} ions in the pyroxene structure. Instead, they originate from Fe^{2+} ions concentrated in the very distorted pyroxene M2 site (Bancroft and Burns, 1967a). In another example, bands at 4,100 to 4,300 cm^{-1} initially assigned to tetrahedral Ni^{2+} ions in garnierite were subsequently shown to be O–H vibrational features (Faye, 1974). The presence of other absorption bands located around 9,000 cm^{-1}, 16,000 cm^{-1} and 25,000 cm^{-1} confirmed that octahedral Ni^{2+} ions only are present in this clay silicate assemblage (Faye, 1974; Manceau and Calas, 1985). However, despite the evidence from optical and X-ray absorption spectra for octahedrally coordinated Ni^{2+} ions in phyllosilicates, Ni^{2+} ions have continued to be misassigned to tetrahedral coordination in phyllosilicates as a result of incorrect assignments of absorption bands (Tejedor-Tejedor *et al.*, 1983). Similar misinterpretations of absorption spectra of chrome diopsides have led to the suggestion that low-spin tetrahedrally coordinated Cr^{3+} ions occur in the pyroxene structure (§5.10.4). Such a cation species is extremely unlikely for two reasons. First, the tetrahedral crystal field splitting is too small to induce spin-pairing in any transition metal ion (§2.6.1), and second, the very high CFSE of Cr^{3+} in octahedral coordination (table 2.5) precludes its occupancy in tetrahedral sites. Such examples, whereby tetrahedral Fe^{2+}, Ni^{2+} and Cr^{3+} ions have been misidentified in silicate minerals, indicate that deductions about cation coordination symmetries based on positions of absorption bands in optical spectra sometimes need to be approached with caution.

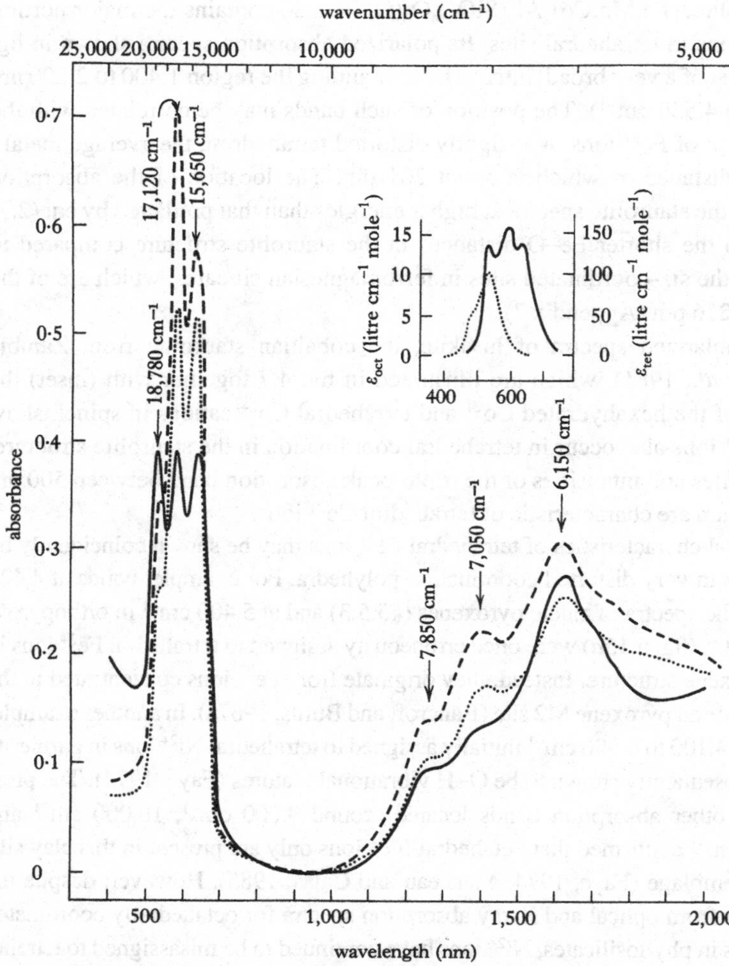

Figure 4.7 Polarized absorption spectra of lusakite. ······ α spectrum; – – – – β spectrum; —— γ spectrum. The staurolite contains 0.97 Co^{2+} ions per formula unit, and is from Lusaka, Zambia. [Optic orientation: $\alpha = b$; $\beta = a$; $\gamma = c$]
Inset: Absorption spectra of cobalt(II) in octahedral and tetrahedral coordination in aqueous solution (······) and spinel (——), respectively.

4.5 Detection of cation ordering in silicate minerals

The decrease of molar extinction coefficients of absorption bands in the polarized spectra of piemontites with increasing Mn^{3+} ion contents (§4.4.2), which is contrary to the Beer–Lambert law, eq. (3.7), indicates that Mn^{3+} ions are not located entirely in one site of the epidote structure (§4.4.2.1). Most of the manganese occurs in the M3 sites. However, crystal structure refinements indicate that some of the Mn^{3+} ions, and perhaps some of the Fe^{3+} ions as well, are also present in the M1 sites (Dollase, 1969). More accurate measurements and calibrations of intensities of the absorption bands (e.g., Smith *et al.*, 1982) could possibly be used to yield quantitative estimates of the proportions of Mn^{3+} ions in the structural sites of epidote.

Intracrystalline Fe^{2+}–Mg^{2+} distributions in natural and synthetic orthopyroxenes have been determined from intensities of absorption bands in their polarized spectra (Goldman and Rossman, 1977a; Steffen *et al.*, 1988). Molar extinction coefficients of crystal field bands centred at 10,500 to 11,000 cm^{-1} and 4,900 to 5,400 cm^{-1} originating from Fe^{2+} ions located in pyroxene M2 sites (§5.5.4) enabled the iron contents to be estimated from the Beer–Lambert law equation, eq. (3.7).

Cation ordering in silicate minerals may also be indicated by changes of spectrum profiles with changing chemical composition of a mineral. Such an effect is illustrated by manganiferous olivines of the fayalite–tephroite series, $(Fe,Mn)_2SiO_4$, the spectra of which are shown in fig. 4.8. The spectra of iron-rich manganiferous olivines are similar to those of fayalite described later (fig. 5.9). However, pronounced changes in the γ spectrum profiles take place with increasing Mn^{2+} ion content of the olivine. The intense band at 1,078 nm in the fayalite spectrum migrates to longer wavelengths with decreasing Fe^{2+} contents of Mn–Fe olivines, which is a reversal of trends shown in the forsterite–fayalite series (fig. 5.10) and reflects the larger octahedral ionic radius of Mn^{2+} (83 pm) relative to Fe^{2+} (78 pm). At the same time, prominent shoulders appear on both the long and short wavelength sides of the 1,080 to 1,120 nm band, becoming comparable in intensity in the spectrum of the tephroite $(Mn_{0.668}Fe_{0.311}Mg_{0.021})_2SiO_4$. Additional weak, sharp peaks at about 410 nm, 440 nm and 585 nm emerge and gain intensity with increasing Mn^{2+} content.

The intense band at 1,080 to 1,120 nm in the γ spectra of Fe–Mn olivines originates from Fe^{2+} ions in the acentric M2 positions of the olivine structure (fig. 5.8). The reduced intensity of this band in manganiferous olivines shows that Mn^{2+} ions enter preferentially, and displace Fe^{2+} ions from, the M2 positions of the olivine structure. The larger Mn^{2+} ions are apparently accommodated more readily in the larger M2 site, in which the average metal–oxygen distance is slightly larger than the M1 site (fig. 5.8; Appendix 7). Such Mn^{2+}–Fe^{2+} ordering in silicates is difficult to show by X-ray crystal structure refinements due to the similar scattering factors of iron and manganese. The Mn^{2+} preference for olivine M2 sites first demonstrated by absorption spectroscopy (Burns, 1970) has been confirmed subsequently by neutron diffraction and Mössbauer spectroscopy measurements of Fe–Mn olivines (Ballet *et al.*, 1987; Annersten *et al.*, 1984).

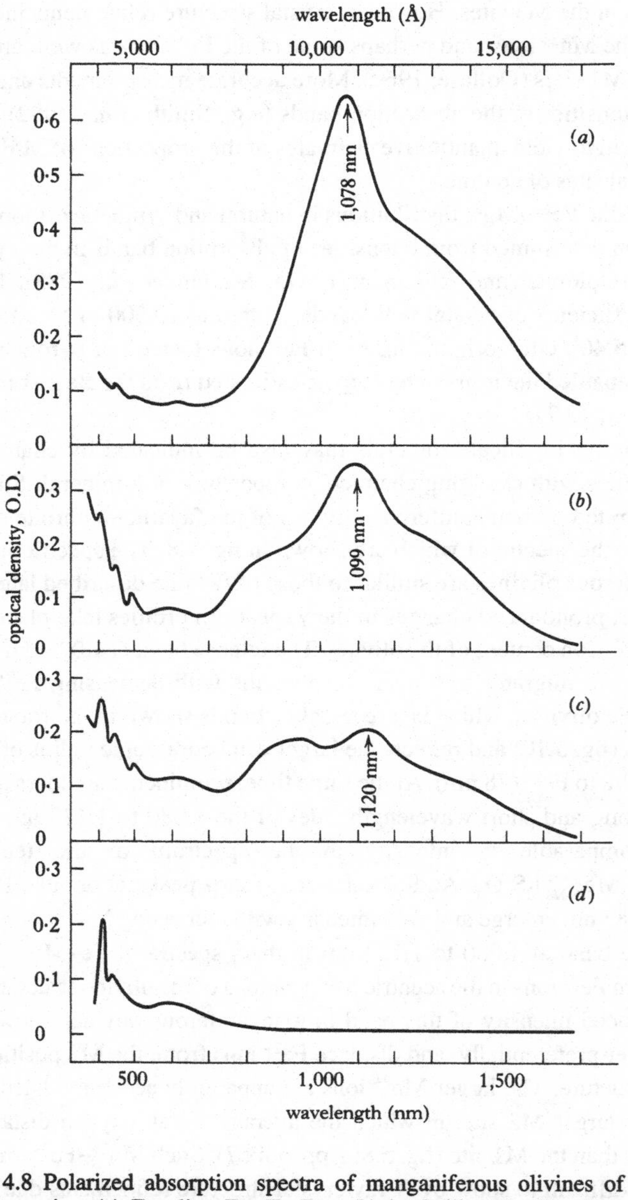

Figure 4.8 Polarized absorption spectra of manganiferous olivines of the fayalite–tephroite series. (a) Fayalite, $Fa_{96}Te_3Fo_1$; (b) knebellite, $Fa_{69}Te_{26}Fo_5$; (c) tephroite, $Fa_{31}Te_{67}Fo_2$; (d) tephroite, $Fa_1Te_{94}Fo_5$. The reduced relative intensity of the absorption band at 1,078 to 1,120 nm is due to substitution of Fe^{2+} by Mn^{2+} in M2 positions of the olivine structure. [γ spectra; γ = *a*.]

Other examples discussed later where changes of spectrum profiles across a solid–solution series correlate with cation ordering in the crystal structure include Ni–Mg olivines (Hu *et al.*, 1990), in which Ni^{2+} ions are strongly ordered in the M1 sites (§5.4.2.4), and Mg–Fe^{2+} orthopyroxenes mentioned earlier where strong enrichment of Fe^{2+} ions occurs in the very distorted M2 sites (§5.5.4).

4.6 Concepts of colour and pleochroism

Perhaps the most widely recognized influence that transition metal ions have on properties of minerals concerns the colours produced in many gems. Indeed, many varietal names of mineral species are synonymous with a specific cation in the gem. Notable examples are ruby, emerald and alexandrite which imply Cr^{3+}-bearing corundum, beryl and chrysoberyl, respectively, and the contemporary vanadium-bearing garnet (tsavorite) and zoisite (tanzanite) gem minerals. Other examples include yellow and blue sapphires (Fe^{3+} and Fe–Ti corundums, respectively) and aquamarine and morganite (beryls coloured by Fe and Mn, respectively). Such spontaneous mineral recognition based largely on the property of colour alone has spawned numerous reviews of the causes of colour of minerals, gems and birthstones (Loeffler and Burns, 1976; Burns, 1983; Nassau, 1978, 1980, 1983; Fritsch and Rossman, 1987, 1988a,b).

4.6.1 Colour

The term 'colour' may have different connotations, depending on the user and the context in which it is used. To the artist, colour may mean a pigment; to the psychologist, a perception; to the physiologist, a response of the nervous system; to the mineralogist, a property for identifying a mineral in hand-specimen or in thin-section; and to the physicist, radiation in the visible region of the electromagnetic spectrum.

The visible region consists of electromagnetic radiation the wavelengths of which lie between 400 nm and 700 nm (table 3.1). However, the visible region constitutes only a very small portion of the electromagnetic spectrum. Radiation of many wavelengths outside the visible region may be absorbed or scattered by matter without optical effects.

4.6.2 Pleochroism and dichroism

Pleochroism is the property of a non-cubic crystal whereby certain wavelengths of polarized light are absorbed in different amounts in different crystallographic directions. Such differential absorption may produce different

Table 4.1. *Origins of colours in gems and birthstones**

Month	Gem (mineral)	Formula	Colour	Chromophore	Origin of colour
January	garnet				
	e.g. rhodolite (pyrope–almandine)	$(Mg,Fe)_3Al_2Si_3O_{12}$	red	Fe^{2+}	Crystal field transition in Fe^{2+} in distorted cubic (eight–coordinated) site.
	e.g. demantoid (andradite–grossular)	$Ca_3(Fe^{3+},Al)_2Si_3O_{12}$	green	Fe^{3+}	Crystal field transition in Fe^{3+} in octahedral site
February	amethyst (quartz)	SiO_2	purple	colour centre plus Fe	colour centre involving trapped electron and transient Fe^{4+}.
March	aquamarine (beryl)	$Be_3Al_2Si_6O_{18}$	blue	$Fe^{2+} + Fe^{3+}$	$Fe^{2+} \rightarrow Fe^{3+}$ intervalence charge transfer transition between cations in octahedral and channel sites.
April	diamond	C	colourless (pale yellow, blue)	colour centre plus N or B	When coloured, due to N (yellow) or B (blue) Colour developed and manifested by heat and radiation.
May	emerald (beryl)	$Be_3Al_2Si_6O_{18}$	green	Cr^{3+}	Crystal field transitions in Cr^{3+} in trigonally distorted octahedral site.
June	alexandrite (chrysoberyl)	Al_2BeO_4	red/green	Cr^{3+}	Crystal field transitions in Cr^{3+} concentrated in non-centrosymmetric distorted six-coordinated site.
	moonstone (feldspar)	$(Ca,Na)(Si,Al)_4O_8$	iridescent	diffraction	Diffraction by exsolution lamellae of Ca-rich and Na-rich feldspars.

July	ruby (corundum)	Al_2O_3	red	Cr^{3+}	Crystal field transitions in Cr^{3+} in trigonally distorted octahedral site.
August	peridot (olivine)	$(Mg,Fe^{2+})_2SiO_4$	yellow-green	Fe^{2+}	Crystal field transitions in Fe^{2+} in two distorted six-coordinated sites.
September	sapphire (corundum)	Al_2O_3	blue	$Fe^{2+} \rightarrow Ti^{4+}$	$Fe^{2+} \rightarrow Ti^{4+}$ intervalence charge transfer transition between cations in face-shared octahedra.
October	opal	$SiO_2.nH_2O$	opalescent	diffraction	Diffraction by spheres of silica.
	tourmaline e.g. rubellite (elbaite)	$Na(Li,Al)_3Al_6(BO_3)_3(Si_6O_{18})(OH,F)_4$	pink	Mn^{2+}	Crystal field transitions in Mn^{2+} and Fe^{2+} in distorted octahedral sites
	e.g. indicolite (schorl)	$NaFe^{2+}{}_3Al_6(BO_3)_3(Si_6O_{18})(OH,F)_4$	blue	Fe^{2+}	
November	topaz	$Al_2SiO_4(OH,F)_2$	colourless (pale yellow, pink,blue, green)	colour centres (heat, radiation)	When coloured, due to colour centres induced by radiation and heat.
December	zircon	$ZrSiO_4$	colourless (red–brown, blue)	colour centres (heat, radiation)	When coloured, due to colour centres induced by radioactive U and Th substituting for Zr
	turquoise	$CuAl_6(PO_4)_4(OH)_8.4H_2O$	blue	Cu^{2+}	Crystal field transitions in Cu^{2+} coordinated to $4OH^-$ and $2H_2O$ in distorted octahedral site.

* *From*: Burns, R. G. (1983) Colours of gems. *Chemistry in Britain*, no. 12, 1004–7.

colours when the crystal is viewed in polarized transmitted light. Pleochroism, or dichroism when it occurs in uniaxial crystals, may be observed in minerals belonging to all crystal systems except the cubic system. This fact provides one clue to the cause of pleochroism, that it originates within cations in low-symmetry environments.

4.6.3 Visible and invisible pleochroism

Useful as it is for noticing subtle changes of radiation in the visible region, the eye has a very limited range of detection. Factors leading to pleochroism in the visible region and, therefore, producing different colours in polarized light, may also operate outside the visible region and not be seen. This is illustrated by comparing the spectra of piemontite in fig 4.3 with those of olivine in fig. 5.9. Each polarized spectrum is distinctive in the two minerals. The pleochroism is seen in piemontite, which has the pleochroic scheme α = lemon yellow, β = amethyst, and γ = bright red, because radiation in the visible region is affected. However, the pleochroism is not seen in olivine because the differential absorption takes place outside the visible region. The invisible pleochroism of olivine illustrates the importance of spectrophotometric measurements for evaluating pleochroism quantitatively within and beyond the visible region.

4.7 Causes of colour and pleochroism in minerals

Most minerals owe their colour and pleochroism to the presence of transition metals or lanthanide elements in their crystal structures, either as major constituents or in trace amounts. Examples involving gems and birthstones are contained in table 4.1. Apart from physical effects such as internal scattering and reflexion phenomena causing iridescence in opal and some feldspars (e.g., moonstone), the most common cause of colour in minerals is absorption of radiation through electronic processes (Loeffler and Burns, 1976; Nassau, 1978, 1980, 1983; Burns, 1983; Fritsch and Rossman, 1987, 1988 a,b). Such absorption in the visible region produces coloured transmitted and reflected light. Electronic processes leading to absorption in the ultraviolet, visible and short-wave infrared regions spanning the wavelength range 300 nm to 1,000 nm, include: first, crystal field transitions within individual cations of the transition elements; second, electron transfer between adjacent transition metal ions, or intervalence charge transfer transitions; and third, charge transfer involving transition metal ions and surrounding anions. In addition, there are electron transfer transitions induced by crystal structure imperfections, as well as band gap transitions exhibited by minerals with semiconductor and metallic properties. These processes are described elsewhere (Nassau, 1983; Fritsch and Rossman, 1978b).

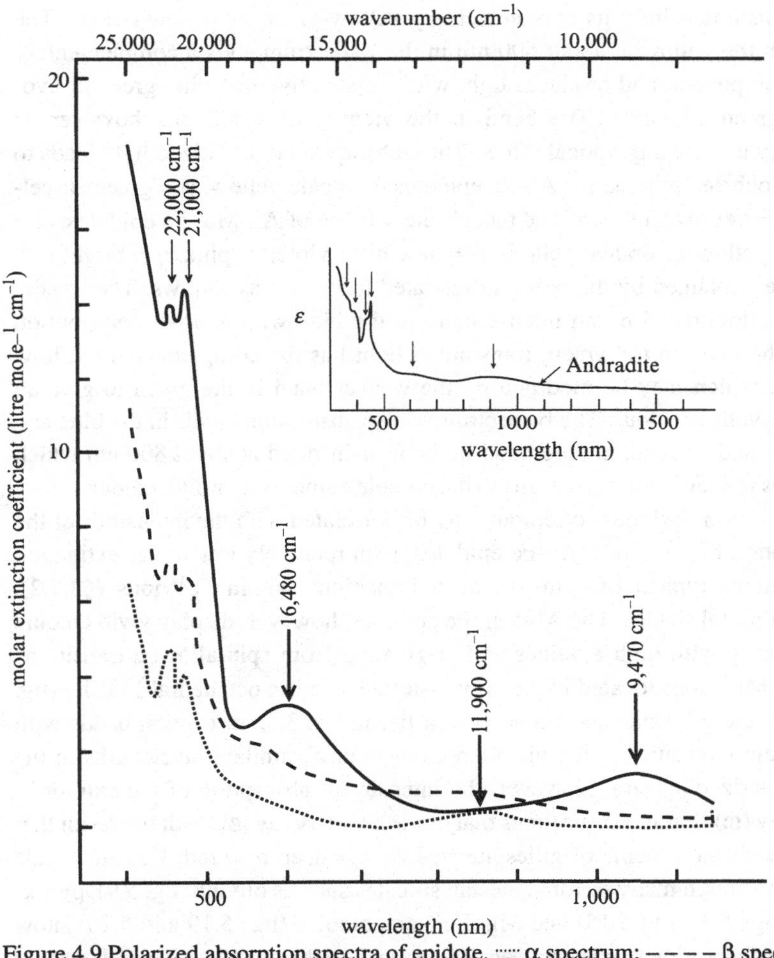

Figure 4.9 Polarized absorption spectra of epidote. ······ α spectrum; – – – – β spectrum; —— γ spectrum. The epidote, from Berkeley, California, contains 0.864 Fe^{3+} ions per formula unit. [Optic orientation: $\alpha : c = 10°$; $\beta = b$; $\gamma : a = 35°$.]
Inset: crystal field spectrum of andradite from Val Maleneo, Italy, contains 0.995 Fe^{3+} ions per formula unit.

4.7.1 Crystal field transitions

Absorption of light may excite electrons between *d* or *f* orbitals producing coloured transmitted light if orbital energy separations lie in the visible region. The positions of the absorption bands determine the actual colours observed as illustrated by minerals of the epidote group, the spectra of which are shown in fig. 4.3 (Mn^{3+} in piemontite) and fig. 4.9 (Fe^{3+} in epidote). Absorption in the violet and blue regions dominate the spectra of Al–Fe epidotes (fig. 4.9), giving transmitted light its complementary yellow-green or yellow colour. The band in the yellow at about 600 nm in the γ spectrum adds a complementary blue component, and produces light with a distinctive pistachio green or avocado green character. The band in the vicinity of 1,000 nm, however, is unlikely to have any optical effect. Thus, absorption of polarized light leads to the pleochroic scheme for Al–Fe epidotes: α = pale yellow, β = greenish yellow, γ = yellowish green. The pleochroic scheme of Al–Mn–Fe epidotes: α = lemon yellow or orange yellow, β = amethyst, violet or pink, γ = bright red, may be explained by the spectra illustrated in fig. 4.3 as follows. The γ spectrum is dominated by an intense band in the blue with a small contribution from the band in the green; transmitted light has the complementary yellow colour, which may be modified by the weaker band in the green to give an orange-yellow colour. The β spectrum shows absorption bands in the blue and yellow, and a broad, intense band in the near-infrared at about 800 nm which absorbs red radiation and results in the complex amethyst or pink colour.

The hue or vividness of colour may be correlated with the intensities of the absorption bands. Thus, Al–Fe epidotes, with relatively low molar extinction coefficients typical of spin-forbidden transitions within Fe^{3+} ions (§3.7.2), exhibit pastel shades. The Al–Mn–Fe epidotes, however, display vivid colours correlating with high ε values and originating from spin-allowed transitions within Mn^{3+} ions located in the very distorted acentric octahedral M3 site (fig. 4.4). In the gillespite spectra shown earlier in fig. 3.3, absorption bands with moderate intensities in the visible region occur at similar wavelengths in the two polarized spectra. However, the intensity of absorption of the extraordinary ray (ω) is almost four times that of the ordinary ray (ε), with the result that the pleochroic scheme of gillespite is: $\mathbf{E}\|c$, ε = deep rose red, $\mathbf{E}\perp c$, ω = pale pink. Although many ferromagnesian silicates such as olivine (fig. 5.9), pyroxenes (figs 5.14 and 5.15) and Mg–Fe^{2+} amphiboles (figs 5.19 and 5.20) show intense absorption bands between 800 nm and 1,300 nm, these minerals are not strongly coloured. The characteristic green colours of peridot, the gem olivine, and, indeed, most Fe(II) compounds containing Fe^{2+} ions octahedrally coordinated to oxygen ligands, result from masking of red radiation by tails of absorption bands emerging from maxima located in the near-infrared region. The absorption spectra of almandine (fig. 5.6), in which Fe^{2+} ions occur in eight-coordination with oxygen ions at the vertices of a distorted cube (fig. 5.5), again show several bands in the infrared resulting from Fe^{2+} ions in a low-symmetry environment. However, because garnet is a cubic mineral, the spectra are identical for all orientations of polarized light transmitted through almandine crystals, so that no pleochroism occurs. The deep red colour of almandine results from maximum transmission of light in the red region

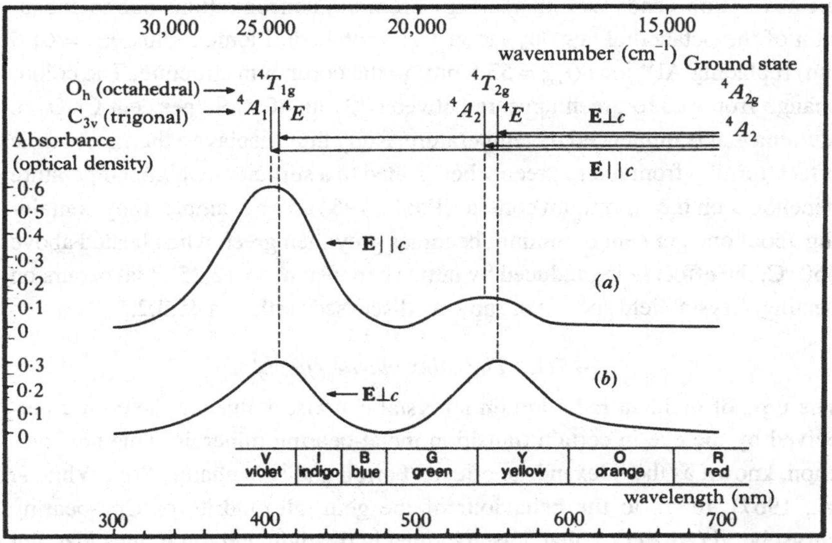

Figure 4.10 Polarized absorption spectra of ruby (from Burns, 1984). The ruby formula is $(Al_{0.998}Cr_{0.002})_2O_3$, and the spectra originate from crystal field transitions within Cr^{3+} replacing Al^{3+} ions in trigonally distorted octahedral sites in the trigonal corundum structure. Consequently, the spectra differ slightly for light polarized (*a*) parallel (E||*c*) and (*b*) perpendicular (E⊥*c*) to the *c* crystallographic axis. The group theoretical assignments of the absorption bands are also indicated. [Reproduced from: *Chemistry in Britain*, 1984, p. 1004]

located between intense spin-allowed Fe^{2+} crystal field bands in the near-infrared and the weak, sharp spin-forbidden peaks situated on the absorption edge of the oxygen → Fe^{2+} charge transfer band located in the ultraviolet.

4.7.1.1 The colour of ruby

The spectra of ruby, $(Al,Cr)_2O_3$, illustrated in fig. 4.10 show Cr^{3+} crystal field bands near 18,000 cm^{-1} and 24,500 cm^{-1} with regions of minimum absorption occurring around 21,000 cm^{-1} (blue) and beyond 16,000 cm^{-1} (red). The 'window' in the red dominates the human eye's perception of ruby's 'pigeon blood' colour in transmitted light. On the other hand, the optical spectrum of eskolaite or chromia, Cr_2O_3, has its intense spin-allowed transitions in Cr^{3+} centred at 16,600 cm^{-1} and 21,700 cm^{-1} and a 'window' of minimum absorption located near 20,000 cm^{-1} (green), resulting in its green colour (Neuhaus, 1960). Numerous studies have been made of the compositional variations of the optical spectra of Al_2O_3–Cr_2O_3 solid-solution series (Neuhaus, 1960; Poole, 1964; Reinen, 1969), which demonstrate red shifts (movements to lower energies) of the absorption bands with increasing chromium concentrations due to expansion of the octahedral sites by larger Cr^{3+} (octahedral ionic radius, r_{oct} = 61.5 pm) replacing Al^{3+} ions (r_{oct} = 53.5 pm) in the corundum structure. The colour change from red to green appears between 20 and 40 mole per cent Cr_2O_3 in alumina and is influenced by temperature. Ruby thus displays a thermochromic effect, turning from red to green when heated to a sufficiently high temperature dependent on the chromium content (Poole, 1964). For example, ruby containing about one per cent chromium becomes grey then green when heated above 450 °C, the effect being induced by lattice expansion, eq. (2.25), that occurs on heating. Crystal field spectra of ruby are discussed further in §5.3.2.1.

4.7.1.2 The 'alexandrite effect.'

The type of incident radiation on a crystal can also influence the colour perceived by the eye in certain transition metal-bearing minerals. This phenomenon, known as the 'alexandrite effect' (Farrell and Newnham, 1965; White *et al.*, 1967), refers to the behaviour of the gem, alexandrite or Cr^{3+}-bearing chrysoberyl (Al_2BeO_4), that appears green in daylight and red in incandescent light. The 'alexandrite effect' is thus defined as the property of a solid to change its apparent colour when viewed under different light sources. Although the effect is best known in alexandrite, other examples have been described and include Cr^{3+}- and V^{3+}-bearing corundum, spinel and garnet (White *et al.*, 1967; Schmetzer and Gübelin, 1980; Schmetzer *et al.*, 1980), Sm^{2+}- and Y^{3+}-bearing fluorite (Schmetzer *et al.*, 1980), Nd^{3+}-bearing monazite (Bernstein, 1982) and coquimbite (Rossman, 1975).

The positions and intensities of absorption bands in the optical spectra of alexandrite are polarization dependent (Farrell and Newnham, 1965) due to the occurrence of Cr^{3+} ions in distorted octahedral sites in the orthorhombic chrysoberyl structure, which is isostructural with olivine (fig. 5.10). The optical spectra of alexandrite are bimodal, by analogy with ruby (fig. 4.10), with absorption maxima occurring in the violet-blue and yellow-orange regions and with 'windows' of comparably low absorption in the red and green regions. Because daylight is equally rich in all wavelengths in the visible region, alexandrite transmits equal proportions of red and green light. However, because the eye is most sensitive to green light, alexandrite is perceived to be green. A green colour also dominates when alexandite is viewed in fluorescent light. However, under incandescent light, which is richer in low-energy red wavelengths, alexandrite transmits more red light swamping out the green so that the gem appears red.

The optical spectra of other colour-changing materials exhibiting the 'alexandrite effect', including chromium- and vanadium-bearing garnets and corundums and rare earth-bearing fluorites and monazite listed earlier, have features in common with those of alexandrite (Schmetzer, 1978; Schmetzer *et al.*, 1980). Thus, absorption maxima located between 580 and 560 nm (17,240 to 17,860 cm^{-1}) are flanked by minima between 665 and 625 nm (15,040 to 16,000 cm^{-1}) and between 520 and 470 nm (19,600 to 21,280 cm^{-1}). Crystals that are green in daylight generally change to red in artificial light, while crystals that are bluish in daylight change to reddish-violet. A notable example of a synthetic phase displaying the 'alexandrite effect' is V^{3+}-doped corundum, optical spectra of which (McClure, 1962; White *et al.*, 1967) are remarkably similar to those of the mineral alexandrite with respect to the positions and intensities of the absorption bands and the flanking 'windows' of minimum absorption. Indeed, V^{3+}-doped corundums are used widely as imitations of the rare gem alexandrite. Because Cr^{3+}-doped Al_2O_3 crystals also show red shifts of absorption maxima and minima with increasing Cr^{3+} concentration, the 'alexandrite effect' is displayed by some natural rubies, too, that contain significant concentrations of both vanadium and chromium (Schmetzer and Bank, 1980a,b; Schmetzer *et al.*, 1980).

4.7.1.3 Blue colour of tanzanite

Heat-induced oxidation can produce spectacular changes of colour and pleochroism of minerals, particularly by the enhancement of blue colours of some cordierite and corundum specimens due to the partial oxidation of Fe^{2+} to Fe^{3+}. Tanzanite, the vanadium-bearing gem-quality zoisite, $Ca_2(Al,V)_3Si_3O_{12}(OH)$, also undergoes striking colour changes after heat treatment. The natural crystals containing only 0.012 V^{3+} ions per formula unit show a variety of colours in reflected light, with the amethystine colour predominating (Hurlbut, 1969). However, when the zoisite is heated in air, it becomes an intense sapphire-blue colour, and in this form tanzanite is a popular gem. The pleochroism also changes as a result of heating in air. Before heating, α = red-violet; β = deep blue; γ = yellow-green. After heating, α = violet-red; β and γ = deep blue. To maximize the blue colour in the gem stone, crystals are cut and mounted perpendicular to the α direction (*b* crystallographic axis).

The colour changes induced in tanzanite may be correlated with the polarized spectra illustrated in fig. 4.11 of the vanadium-bearing zoisite crystals before and after heat treatment (Faye and Nickel, 1970b). The most conspicuous changes occur in the γ spectra. Here, the intense band at 22,500 cm^{-1} in the unheated zoisite (absorbing blue-violet light and transmitting some yellow-green light) disappears and is replaced by an intense absorption band at 27,000 cm^{-1}. Absorption bands located near 13,400 cm^{-1}, 16,800 cm^{-1} and 19,000 cm^{-1} in the β and γ spectra are unchanged in the heated zoisite. It is the

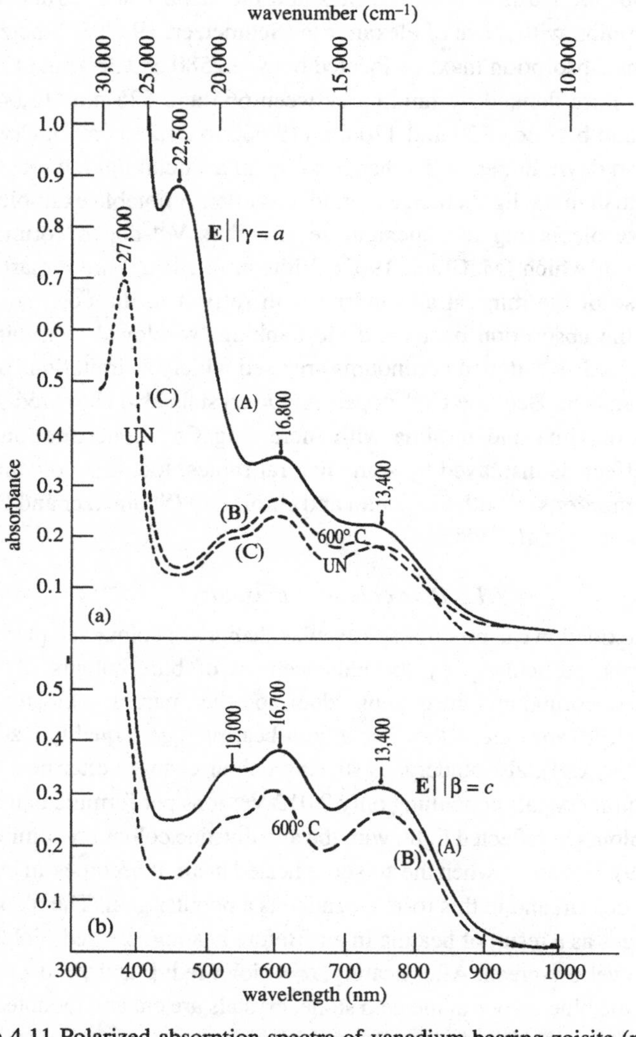

Figure 4.11 Polarized absorption spectra of vanadium-bearing zoisite (modified from Faye and Nickel, 1969). The zoisite from Tanzania contains 0.012 V^{3+} ions per formula unit. (*a*) Spectrum (A): γ-spectrum of unheated zoisite; Spectrum (B): γ-spectrum of zoisite (now tanzanite) after heating to 600 °C Spectrum (C): unpolarized spectrum of tanzanite. (*b*) Spectrum (A): β-spectrum of unheated zoizite; Spectrum (B): β-spectrum of tanzanite after heating to 600 °C. [Optic orientation: α = *b*; β = *c*; γ = *a*.]

minimum occurring around 22,700 cm^{-1}, already present in the β spectrum of the unheated zoisite but induced in the γ spectrum by heat, that is responsible for the blue colour of tanzanite in γ- and β-polarized light. In the α spectrum, minima around 25,000 cm^{-1} and beyond 15,000 cm^{-1} are accentuated by heat, accounting for the violet-red colourations of tanzanite in α-polarized light.

The origin of these colour changes in tanzanite are caused by changes of oxidation state of vanadium ions in the zoisite structure. Evidence from electron spin resonance (Tsang and Ghose, 1971) indicates the presence of V^{2+} and V^{3+} ions substituting for Ca^{2+} and Al^{3+} ions in the unheated zoisite and the possible formation of tetravalent vanadium as V^{4+} or vanadyl (VO^{2+}) ions in tanzanite after heating in air. The spectral signatures of these ions are discussed later (§5.10.2).

4.7.1.4 Other minerals coloured by a transition metal

In addition to many of the gems listed in table 4.1, other examples of minerals that owe their colours to crystal field bands (intra-electronic transitions between 3*d* orbitals within individual cations) include the following:

- Ti^{3+}: fitzroyite (Ti phlogopite), some meteoritic and lunar pyroxenes
- V^{3+}: goldmanite (garnet), tsavorite (garnet), tanzanite (zoisite)
- Cr^{3+}: uvarovite (garnet), kammererite (chlorite), tawmawite (epidote)
- Mn^{3+}: piemontite (epidote), thulite (zoisite), viridine (andalusite), blanfordite (pyroxene), juddite and winchite (amphiboles)
- Mn^{2+}: rhodonite, pyroxmangite and bustamite (pyroxenoids), johannsenite (pyroxene), spessartine (garnet), tirodite (amphibole)
- Fe^{3+}: epidote, andradite, vesuvianite, sinhalite
- Fe^{2+}: many green ferromagnesian silicates such as peridot, diopside and actinolite, gillespite,
- Co^{2+}: lusakite (staurolite), erythrite (arsenate)
- Ni^{2+}: liebenbergite (olivine), garnierite (clay silicates)
- Cu^{2+}: dioptase, cuprorivaite (Egyptian blue), malachite, azurite, turquoise.

Other examples are listed by Fritsch and Rossman (1988b).

Cations of the lanthanide elements also produce colours in some minerals through intra-electronic transitions within 4*f* orbitals (Adams, 1965; Bernstein, 1982). Absorption bands are usually sharp and weak, leading to pastel shades. Examples of such coloured minerals are monazite, bastnaesite, rhabdophane, xenotime, gadolinite, and certain apatites, calcites, scheelites and fluorites. As noted earlier, some rare earth-bearing minerals, notably fluorite and monazite, also display the alexandrite effect (Berstein, 1982; Schmetzer *et al.*, 1980).

4.7.2 Intervalence charge transfer transitions

In crystal field spectra, electrons excited between split 3*d* orbital energy levels remain on individual transition metal ions, although magnetic interactions between adjacent cations may result in paired crystal field excitations within each cation. When a 3*d* electron is *transferred* between neighbouring cations in adjacent coordination sites, there are momentary changes of valences during the lifetime of the electronic transition (10^{-15} s). This process is termed a metal–metal or cation–cation intervalence charge transfer (IVCT) transition (Smith and Strens, 1976; Burns, 1981, 1991; Amthauer and Rossman, 1984; Mattson and Rossman, 1987a,b, 1988; Sherman, 1987a,b; 1990).

Two different situations exist, depending on whether the same element (homonuclear) or two different elements (heteronuclear) are involved in the electron transfer.

4.7.2.1 *Homonuclear intervalence transitions* involve single elements in different oxidation states. For example, electron transfer between Fe^{2+} and Fe^{3+} ions located in adjacent octahedral sites A and B, may be represented by:

$$Fe^{2+}{}_{A}[(t_{2g}\uparrow)^3(t_{2g}\downarrow)^1(e_g\uparrow)^2] + Fe^{3+}{}_{B}[(t_{2g}\uparrow)^3(e_g\uparrow)^2] \rightarrow Fe^{3+}{}_{A}[(t_{2g}\uparrow)^3(e_g\uparrow)^2] + Fe^{2+}{}_{B}[(t_{2g}\uparrow)^3(t_{2g}\downarrow)^1(e_g\uparrow)^2] , \quad (4.2)$$

in which the electron transferred during the transition occupies the Fe^{2+} spin-down t_{2g} orbital (i.e. the sixth 3*d* electron of Fe^{2+} filling one of the t_{2g} orbitals in the ground state). There is no change in the number of unpaired electrons in eq. (4.2), so that the $Fe^{2+} \rightarrow Fe^{3+}$ IVCT transition is spin-allowed (§3.7.2)

Similarly, if Ti^{3+} and Ti^{4+} ions are involved, the electron transfer is represented by:

$$Ti^{3+}{}_{A}[(t_{2g}\uparrow)^1] + Ti^{4+}{}_{B}[(t_{2g})^0] \rightarrow Ti^{4+}{}_{A}[(t_{2g})^0] + Ti^{3+}{}_{B}[(t_{2g}\uparrow)^1] . \quad (4.3)$$

Other possible examples of homonuclear IVCT transitions include $Mn^{2+} \rightarrow Mn^{3+}$ and $Cr^{2+} \rightarrow Cr^{3+}$ IVCT transitions.

4.7.2.2 *Heteronuclear intervalence transitions* involve different cations. An example involving the octahedral Fe^{2+} and Ti^{4+} cation pair, may be depicted by

$$Fe^{2+}{}_{A}[(t_{2g}\uparrow)^3(t_{2g}\downarrow)^1(e_g\uparrow)^2] + Ti^{4+}{}_{B}[(t_{2g})^0] \rightarrow Fe^{3+}{}_{A}[(t_{2g}\uparrow)^3(e_g\uparrow)^2] + Ti^{3+}{}_{B}[(t_{2g}\downarrow)^1]. \quad (4.4)$$

Other examples are $Mn^{3+} \rightarrow Fe^{3+}$, $Mn^{2+} \rightarrow Fe^{3+}$ and $Fe^{2+} \rightarrow Mn^{3+}$ IVCT.

Examples of silicate and oxide minerals showing IVCT transitions are summarized in table 4.2. Most IVCT transitions take place between octahedrally coordinated cations, although examples involving octahedral–tetrahedral (e.g., cordierite) and cubic–tetrahedral (e.g., garnet) pairs are known. An IVCT transition most commonly occurs between cations located in edge-shared coordination polyhedra, although several examples involving face-shared octa-

Table 4.2. *Intervalence charge transfer energies from optical spectra of mixed-valence minerals*

Mineral	M–M distances* (pm)	IVCT energy (cm^{-1})	Molar absorptivity ($M^{-1} cm^{-1}$)	Half-width (cm^{-1})	Sources of data	Symbol in fig. 4.17 [see p.130]
			$Fe^{2+} \rightarrow Fe^{3+}$ IVCT			
vivianite	285	15,870			fig.4.12; [1–5]	Vv
lazulite	271^F	14,970	120	5,250	[4,6]	Lz
rockbridgeite	271^F	13,100	110		[4,6]	Rk
babingtonite	330, 337	14,710	60	5,500	[4,6]	Ba
ilvaite	283; 301; 303	12,300	150		[4]	Il
		14,500				
andalusite	265.8; 290.1	10,900			[8]	
		13,400				
kyanite	275–278	16,500			[9–16]	Ky
yoderite	290	13,800			[17,18]	Yo
aquamarine		16,100			[19–22]	
cordierite	274^T	17,500			[2,12–14,23–25]	
osumilite	274^T	15,480			[26,27]	
tourmaline	Mg–Mg = 304	18,900			[2,12–14,28–33]	
	Mg–Al = 297					
	Al–Al = 280					
orthopyroxene	M1–M1 = 315	14,500			fig.5.15*b*;	Or
	M1–M2 = 308;327				[34,35]	
augite	M1–M1 = 311	13,700	160		fig.5.13;	Aa
	M1–M2 = 321				[4,36]	
glaucophane	M1–M2 = 309	16,130		6,600	fig.4.15;	Gl
	M1–M1 = 322				[6,12,34,37]	
	M1–M3 = 310					
	M2–M3 = 331	18,520		7,000	[6]	

Table 4.2 *continued*

Mineral	M–M distances* (pm)	IVCT energy (cm^{-1})	Molar absorptivity ($M^{-1} cm^{-1}$)	Half-width (cm^{-1})	Sources of data	Symbol in fig. 4.17 [see p. 130]
riebeckite	cf. glaucophane	15,000			[38–40]	
(crocidolite)		to 18,000				
actinolite	cf. glaucophane	15,120			fig.5.20; [34,41]	
(hornblende)		13,740				
biotite	310; 311	13,650			[1,2,12–14,42–43]	AN
(phlogopite–		16,400				
annite)						
chloritoid		16,300			[2,44]	CH
chlorite		14,300			[1,12–14]	
ellenbergerite	273^F	12,500		5,000	[45]	EL
taramellite	339	14,290			[46]	TA
euclase	294	14,930		4,600	[6]	EU
sapphire	279;	11,500			fig.4.16*b*,*d*;	SA
	265^F	9,700			[12–14,34,47–50]	
calculated	293.7	10,570			[51]	CALC
			$Fe^{2+} \rightarrow Ti^{4+}$ IVCT			
sapphire	279;	17,400			fig.4.16*b*,*d*;	
	265^F	14,000			[12–14,34,47–50]	
hibonite	263^F	14,000			[52]	
tourmaline	304	24,100		4,000	[46]	
omphacite	290	15,040			[53,54]	
fassaite	313	20,600			[55]	
(Angra dos Reis)						
neptunite	314; 328	24,100	225	9,000	[46]	
taramellite	339	21,740	450–1,300	9,000	[46]	

traskite		22,730	75	7,000	[46]
ellenbergerite	273^F	19,500	1,600	6,500	[56]
calculated	290.6	18,040			[57]
			$Ti^{3+} \rightarrow Ti^{4+}$ IVCT		
Al_2O_3/Ti^{3+}	265^F	12,500			fig.4.16*c*; [50]
andalusite	265.8, 290.1	20,800			[58]
fassaite (Allende)	315	15,000			fig.4.2; [59]
			$Mn^{2+} \rightarrow Mn^{3+}$ IVCT		
yoderite	290				[18]
			$Mn^{2+} \rightarrow Ti^{4+}$ IVCT		
tourmaline	304	30,770	450	7,000	[60]
			$Mn^{2+} \rightarrow Fe^{3+}$ IVCT		
calculated	298	14,900			[61]
			$Fe^{3+} \rightarrow Mn^{3+}$ IVCT		
babingtonite	330	18,020			[4,62]
calculated	284	17,800			[61]

* edge-sharing octahedra, except: T = edge-sharing octahedra–tetrahedra; F = face-sharing octahedra.

Sources of data : [1] Faye (1968b); [2] Faye, Manning & Nickel (1968); [3] Townsend & Faye (1970); [4] Amthauer & Rossman, (1984); [5] Mao (1976); [6] Mattson & Rossman (1987a); [7] Güttler, Salje & Ghose (1989); [8] Langer, Hålenius & Fransolet (1984); [9] White & White (1967); [10] Faye & Nickel (1969b); [11] Faye (1971b); [12] Smith & Strens (1976); [13] Smith (1977); [14]Smith (1978b); [15] Ghera, Graziani & Lucchesi (1986); [16] Parkin, Loeffler & Burns (1977); [17] Abu-Eid, Langer & Seifert (1978); [18] Langer, Smith & Hålenius (1982); [19] Samoilovich, Isinober & Dunin-Barkovskii (1971); [20] Goldman, Rossman & Parkin (1978); [21] Blak, Isotani & Watanabe (1982); [22] Price, Vance, Smith, Edgar & Dickson (1976); [23] Pollak (1976); [24] Goldman, Rossman & Dollase (1977); [25] Vance & Price (1984); [26] Faye (1972); [27] Goldman & Rossman (1978); [28] Wilkins, Farrell & Naiman (1969); [29] Townsend (1970); [30] Burns (1972a); [31] Faye, Manning, Gosselin & Tremblay (1974); [32] Smith (1978a); [33] Mattson & Rossman (1987a); [34] Burns (1981); [35] Steffen, Langer & Seifert (1988); [36] Burns & Huggins (1973); [37] Bancroft & Burns (1969); [38] Littler & Williams (1965); [39] Manning & Nickel (1969); [40] Faye & Nickel (1969a); [41] Goldman & Rossman (1977b); [42] Robbins & Strens (1972); [43] Kleim & Lehmann (1979); [44] Hålenius, Annersten & Langer (1981); [45] Chopin & Langer (1988); [46] Mattson & Rossman (1988); [47] Townsend (1968); [48] Eigenmann & Gunthard (1971); [49] Ferguson & Fielding (1972); [50] Burns & Burns (1984a); [51] Sherman (1987a); [52] Burns & Burns (1984b); [53] Abu-Eid (1976); [54] Strens, Mao & Bell (1982); [55] Mao, Bell & Virgo (1977); [56] Chopin & Langer (1988); [57] Sherman (1987b); [58] Faye & Harris (1969); [59] Mao & Bell (1974a); [60] Rossman & Mattson (1986); [61] Sherman (1990); [62] Burns (1991).

hedra (e.g., blue sapphire, lazulite, rockbridgeite, ellenbergerite and, perhaps, some meteoritic hibonites) are known. Intervalence charge transfer transitions are facilitated by short metal–metal interatomic distances and are strongly polarization dependent. Thus, when light is polarized along the cation–cation axis, intense absorption bands located principally in the visible region appear in optical spectra, generally at energies different from those of crystal field transitions within individual cations.

The breadths of IVCT bands are significantly larger than those of CF bands. In fact, large widths at half peak-height are considered to be a diagnostic property of IVCT transitions (Mattson and Rossman, 1987a). Intensities of the spin-allowed IVCT transitions, which may be 1 to 3 orders of magnitude higher than those of spin-allowed CF transitions, depend on populations of Fe^{2+}–Fe^{3+}, Ti^{3+}–Ti^{4+} or Fe^{2+}–Ti^{4+} cation pairs in adjacent sites rather than concentrations of individual cations. As a result, molar extinction coefficients are determined from products of the concentrations of the two interacting cations (Smith and Strens, 1976). In mixed-valence silicate and oxide minerals with high concentrations of Fe^{2+}–Fe^{3+} or Fe^{2+}–Ti^{4+} pairs, electron transfer between adjacent cations may become sufficiently extensive to induce electron delocalization throughout the structure. Optically induced IVCT band intensities then become immeasurable and thermally activated electron delocalization (electron hopping) occurs, accounting for the opacity, high electrical conductivity, and metallic properties of minerals such as magnetite, ilmenite, ilvaite, deerite, riebeckite, cronstedtite, etc. (Burns, 1981, 1991; Sherman, 1987a,b). Such opaque Fe^{2+}–Fe^{3+} and Fe^{2+}-Ti^{4+} minerals are discussed later in the chapter (§4.8). Electron hopping in mixed-valence Fe^{2+}–Fe^{3+}–Ti^{4+} minerals is blocked by atomic substitution of non-transition metal ions such as Mg^{2+} and Al^{3+} in adjacent coordination sites. Opacities are thereby reduced, resulting in dark blue, green, or red colourations exhibited by vivianite, glaucophane, babingtonite, blue sapphire, hibonite, neptunite, taramellite, etc. Increased temperature decreases the intensities of intervalence transitions, in contrast to crystal field transitions which usually increase in intensity at elevated temperatures. Rising pressure, on the other hand, intensifies both IVCT and CF transitions, but in contrast to CF spectra, pressure-induced shifts of absorption bands are small in IVCT spectra. Thus, CF and IVCT transitions often may be distinguished by measuring optical spectra of minerals at different pressures and temperaures (§9.5.1). However, IVCT transitions have been shown to enhance intensities of CF bands by pair-coupling mechanisms (Smith, 1977, 1978b, 1980; Amthauer and Rossman, 1984; Mattson and Rossman, 1987a,b, 1988). Examples include vivianite (§4.7.2.3), blue sapphire (§4.7.2.6) and kyanite (§4.7.2.8) discussed later.

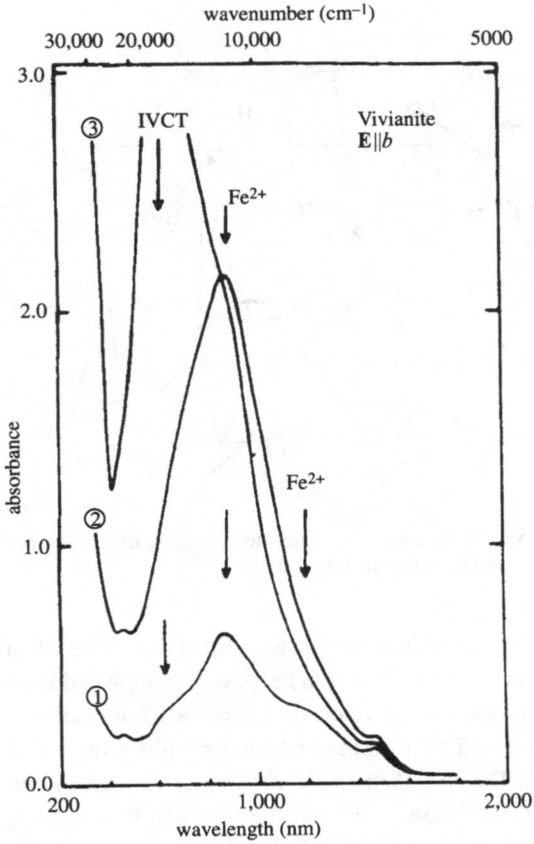

Figure 4.12 Polarized absorption spectra of a vivianite single crystal measured in three zones of increasing oxidation. (1) Nearly colourless; (2) light blue; and (3) dark blue. The spectra were measured with light polarized along the *b* axis corresponding to the Fe^{2+}–Fe^{3+} vector of edge-shared $[FeO_6]$ octahedra. Note the intensification of the Fe^{2+} crystal field bands at 1,200 nm and 800 nm by the $Fe^{2+} \rightarrow Fe^{3+}$ IVCT at 630 nm (from Amthauer and Rossman, 1984).

4.7.2.3 Vivianite

Vivianite, $Fe^{2+}_3(PO_4)_2.8H_2O$, is the classic example of a mineral showing an intervalence charge transfer transition (Wherry, 1918; Burns, 1981). Vivianite has a diagnostic indigo-blue colour and a well characterized $Fe^{2+} \rightarrow Fe^{3+}$ IVCT absorption band in the polarized spectra illustrated in fig. 4.12 and is the datum with which electron interaction parameters for other minerals are compared. The chemical formula of vivianite is not indicative of a mixed-valence compound. However, the pale-green colour of newly cleaved vivianite crystals or freshly

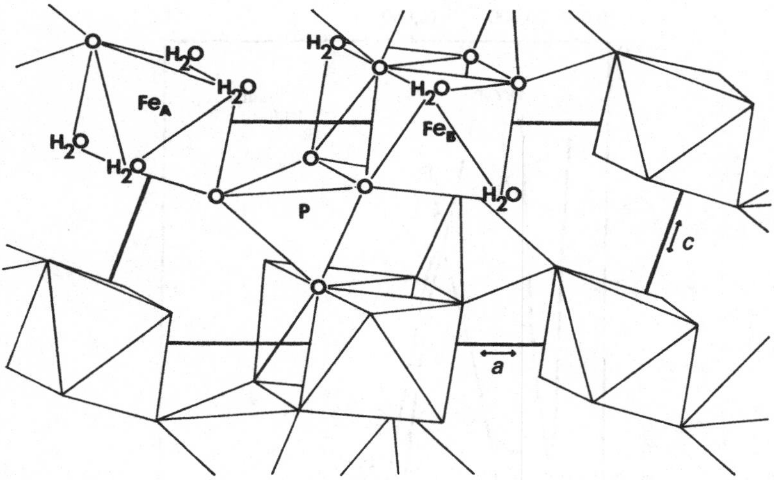

Figure 4.13 The vivianite crystal structure viewed along the *b* axis. Note the dimers of edge-shared octahedra forming the Fe_B site.

precipitated ferrous phosphate suspensions rapidly turn blue when exposed to air due to oxidation of some Fe^{2+} ions to Fe^{3+} (McCammon and Burns, 1980). Such intense blue colourations are atypical of pure Fe(II) or Fe(III) compounds containing only Fe^{2+} or Fe^{3+} ions in octahedral coordination with oxygen ligands (Wherry, 1918). The crystal structure of vivianite illustrated in fig. 4.13 contains two distinct octahedral sites, designated as Fe_A and Fe_B. The Fe_A octahedra are isolated, whereas pairs of Fe_B octahedra share a common edge across which adjacent Fe atoms are separated by only 285 pm along the *b* axis. The blue colour of partially oxidized vivianite results from an intense absorption band located at about 15,800 cm^{-1} (630 nm) when light is polarized parallel to the *b* axis coinciding with the Fe^{2+}–Fe^{3+} vector between these cations occupying adjacent Fe_B octahedral dimers. Measurements of the $Fe^{2+} \rightarrow Fe^{3+}$ IVCT band in vivianite have proven to be difficult due to the great intensity of this transition. The spectra illustrated in fig. 4.12 obtained from three areas of a colour-zoned single crystal of vivianite (Amthauer and Rossman, 1984) show overlapping absorption bands centred at about 1,470 nm (representing a water stretching mode), 1,200 nm and 880 nm (due to Fe^{2+} CF transitions), 630 nm (originating from the $Fe^{2+} \rightarrow Fe^{3+}$ IVCT), and 450 nm (attributed to a spin-forbidden transition in Fe^{2+}). The spectrum of the nearly colourless zone (1) in fig. 4.12 shows absorption by Fe^{2+} with just a minor indication of an IVCT interaction. The spectrum of the light blue region (2) shows intensification of the IVCT band at 630 nm but the Fe^{2+} CF band

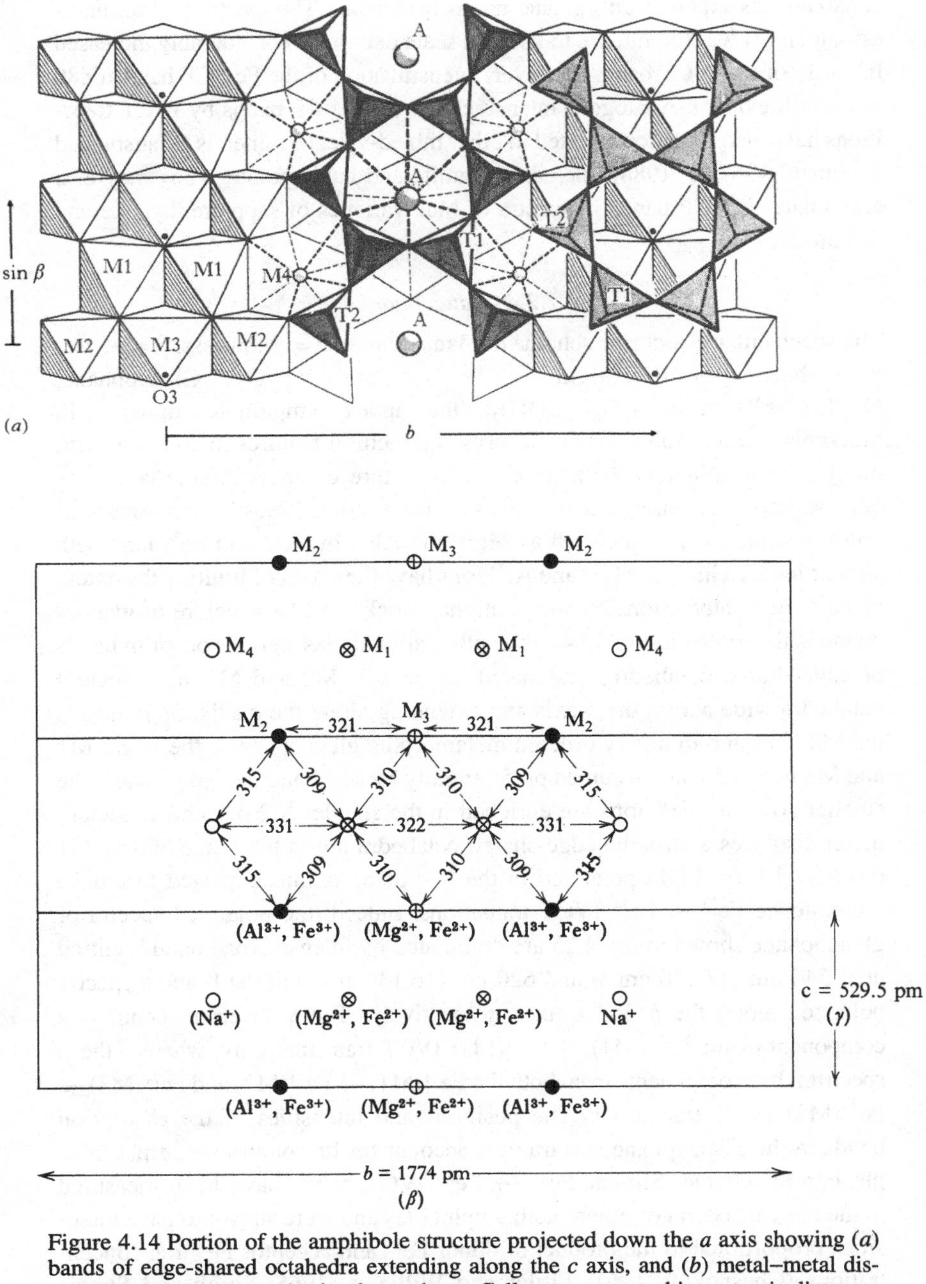

Figure 4.14 Portion of the amphibole structure projected down the *a* axis showing (*a*) bands of edge-shared octahedra extending along the *c* axis, and (*b*) metal–metal distances and site occupancies in glaucophane. Cell parameters and interatomic distances from Papike and Clark (1968).

at 880 nm has experienced a greater intensity increase. The spectrum of the darkest region (3) was too intense to keep on scale due to the considerably increased intensity of the IVCT band. However, intensification of the Fe^{2+} CF band at 880 nm is still evident. Analogous intensification of Fe^{2+} CF bands by IVCT transitions have also been documented in other mixed-valence minerals (Mattson and Rossman, 1987a,b, 1988), including lazulite and babingtonite. They may also accentuate Fe^{2+} CF bands in spectra of blue varieties of sapphire, kyanite and andalusite.

4.7.2.4 Glaucophane

The spectacular pleochroic scheme of glaucophane, α = colourless, β = purple, γ = blue, has its origin in $Fe^{2+} \rightarrow Fe^{3+}$ IVCT. Glaucophane, $Na_2(Mg,Fe^{2+})_3(Al,Fe^{3+})_2Si_8O_{22}(OH)_2$, the index amphibole mineral in blueschist metamorphic rocks, has crystal structural features in common with many silicate minerals: first, its crystal structure contains infinite chains of edge-shared octahedra; and second, extensive atomic substitution occurs of non-transition metal ions, such as Mg^{2+} and Al^{3+}, by Fe^{2+} and Fe^{3+} ions with similar ionic radii. The Mg^{2+} and Al^{3+} ions have the effect of limiting the extent of Fe^{2+}–Fe^{3+} interactions by substitutional blocking. The structure of glaucophane is illustrated in fig. 4.14. As in other amphiboles, cations occur in bands of edge-shared octahedra, designated as the M1, M2 and M3 sites, several octahedra wide across the *b* axis and extending along the *c* axis. Sodium fills the M4 sites and in highly ordered metamorphic glaucophanes, the larger M1 and M3 octahedra are occupied preferentially by Mg^{2+} and Fe^{2+} ions, while the smaller Al^{3+} and Fe^{3+} ions are enriched in the smaller M2 octahedra. Metal–metal distances across the edge-shared octahedra are in the range 300 to 331 pm (fig. 4.14*b*). Light polarized in the *b*–*c* plane is thus expected to induce intervalence $Fe^{2+} \rightarrow Fe^{3+}$ IVCT transitions. Indeed, the polarized spectra of glaucophane shown in fig. 4.15 are dominated by intense broad bands centred near 540 nm (18,520 cm^{-1}) and 620 nm (16,130 cm^{-1}) in the β and γ spectra polarized along the *b* and *c* axes, respectively. The γ spectrum contains a component of the Fe^{2+}(M1) $\rightarrow$ Fe^{3+}(M2) IVCT transition only, whereas the β spectrum has components from both the Fe^{2+}(M1) $\rightarrow$ Fe^{3+}(M2) and Fe^{2+}(M3) $\rightarrow$ Fe^{3+}(M2) IVCT transitions. The positions and intensities of the absorption bands in the glaucophane spectra thus account for its colourless–purple–blue pleochroic scheme. Similar $Fe^{2+} \rightarrow Fe^{3+}$ IVCT bands have been measured in the optical spectra of other alkali amphiboles and were shown to have intensities proportional to the product of donor Fe^{2+} and acceptor Fe^{3+} ion concentrations (Chesnokov, 1961; Littler and Williams, 1965; Smith and Strens, 1976).

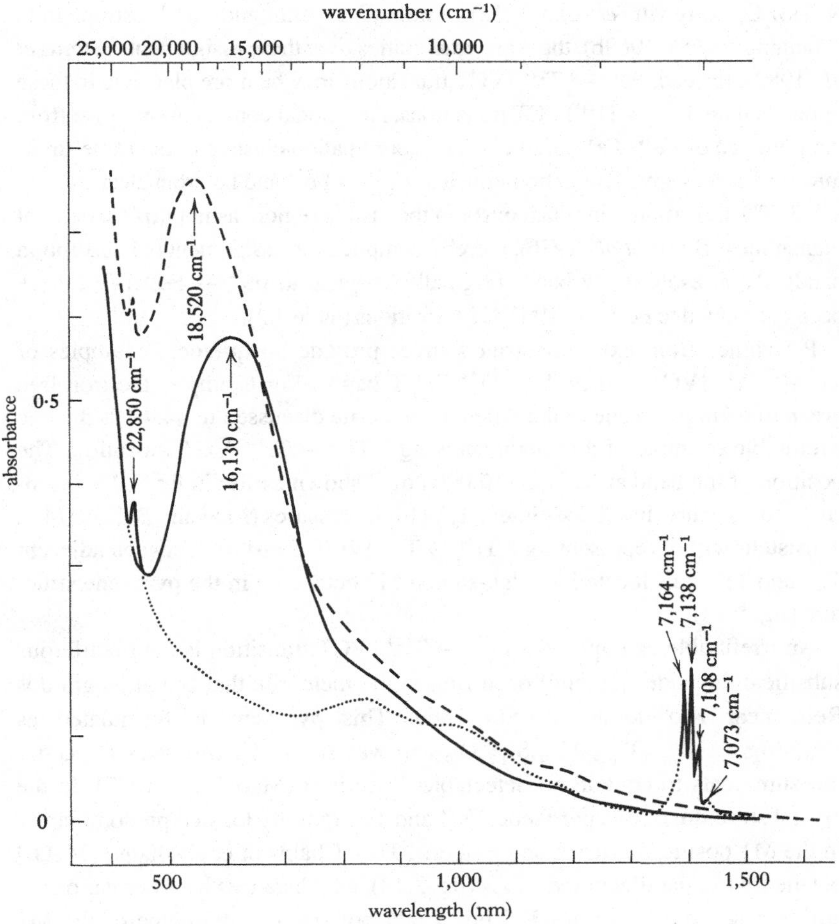

Figure 4.15 Polarized absorption spectra of glaucophane. ········ α spectrum; – – – – β spectrum; ——— γ spectrum. The bands centred between 540 nm and 620 nm in the β and γ spectra represent charge transfer transitions between neighbouring Fe^{2+} and Fe^{3+} ions in the glaucophane structure. The glaucophane, containing 1.10 Fe^{2+} and 0.53 Fe^{3+} ions per formula unit, is from Tiberon, California. [Optic orientation: $\alpha : a = 10°$; $\beta = b$; $\gamma : c = 4°$]

4.7.2.5 Titaniferous pyroxenes

The Ti(IV) oxidation state is more stable in terrestrial minerals than the Ti(III) state so that the existence of Fe^{2+}–Fe^{3+}–Ti^{4+} assemblages are much more common than are Ti^{3+}–Ti^{4+} assemblages. Thus, although $Ti^{3+} \rightarrow Ti^{4+}$ IVCT transitions have been suggested to occur in several titaniferous silicates, including melanite garnets (Moore and White, 1971), andalusite (Faye and Harris, 1969), synthetic

$NaTiSi_2O_6$ (Prewitt *et al.*, 1972), tourmaline, clintonite and astrophyllite (Manning, 1968a, 1969b), there are ambiguities over these assignments (Strens *et al.*, 1982). Instead, $Fe^{2+} \rightarrow Ti^{4+}$ IVCT transitions may be more plausible in these minerals than $Ti^{3+} \rightarrow Ti^{4+}$ IVCT transitions. Additional complications arise from the presence of Fe^{2+}–Fe^{3+} and Fe^{2+}–Ti^{4+} coordination clusters in the same structure. These may give rise to homonuclear $Fe^{2+} \rightarrow Fe^{3+}$ and heteronuclear $Fe^{2+} \rightarrow Ti^{4+}$ IVCT transitions simultaneously in the visible region, as in purple terrestrial titanaugites (Burns *et al.*, 1976), thereby complicating assignments of absorption bands. As a result, many bands originally assigned to $Fe^{2+} \rightarrow Fe^{3+}$ IVCT have been reassigned to $Fe^{2+} \rightarrow Ti^{4+}$ IVCT transitions (table 4.2).

Pyroxenes from extraterrestrial sources provide unequivocal examples of $Ti^{3+} \rightarrow Ti^{4+}$ IVCT and $Fe^{2+} \rightarrow Ti^{4+}$ IVCT bands. For example, the iron-free green titanian pyroxene in the Allende meteorite discussed in §4.4.1 is the one irrefutable example of a mineral showing a $Ti^{3+} \rightarrow Ti^{4+}$ IVCT transition. The position of the band at 666 nm (15,000 cm^{-1}) shown earlier in fig. 4.2 is insensitive to pressure, but it does intensify at high pressures (Mao and Bell, 1974a), consistent with it representing a $Ti^{3+} \rightarrow Ti^{4+}$ IVCT transition between adjacent Ti^{3+} and Ti^{4+} ions located in edge-shared M1 octahedra in the pyroxene structure (fig. 5.13).

An irrefutable example of a $Fe^{2+} \rightarrow Ti^{4+}$ IVCT transition is the titaniferous subsilicic diopside (fassaite) occurring in the meteorite that fell at Angra dos Reis, near Rio de Janeiro, in 1869. This pyroxene is formulated as $Ca_{0.97}Mg_{0.58}Fe^{2+}_{0.22}Ti_{0.06}Al_{0.43}Si_{1.79}O_6$, and was shown by Mössbauer spectral measurements to contain no detectable ferric iron (Mao *et al.*, 1977). In the crystal structure of this pyroxene, Fe^{2+} and Ti^{4+} ions are located predominantly in the M1 positions (Hazen and Finger, 1977). Chains of edge-shared [$M1O_6$] octahedra run parallel to the *c* axis (fig. 5.13), and distances between cations in adjacent M1 octahedra are 315 pm. Light polarized in the plane of the M1 cations gives rise to an intense broad absorption band centred around 20,600 cm^{-1} (Bell and Mao, 1976; Mao *et al.*, 1977), which may be unambiguously assigned to a $Fe^{2+} \rightarrow Ti^{4+}$ IVCT transition, and produces deep pink colouration in the pyroxene. Pressure not only intensifies this band, but also results in a systematic shift of it to lower energies, so that at 5.2 GPa (52 kb) it is centred at 19,200 cm^{-1} (Hazen *et al.*, 1977a). Although such a pressure-induced shift for the $Fe^{2+} \rightarrow Ti^{4+}$ IVCT transition contrasts with the negligible shifts observed for both the $Ti^{3+} \rightarrow Ti^{4+}$ IVCT transition at 15,000 cm^{-1} in the Allende pyroxene (fig. 4.2) and the $Fe^{2+} \rightarrow Fe^{3+}$ IVCT transition at 15,800 cm^{-1} in vivianite (fig. 4.12) (Bell and Mao, 1974; Mao, 1976), the intensification of the 19,200 cm^{-1} band at elevated pressures in the Angra dos Reis pyroxene spectra is consistent with trends observed for other IVCT transitions.

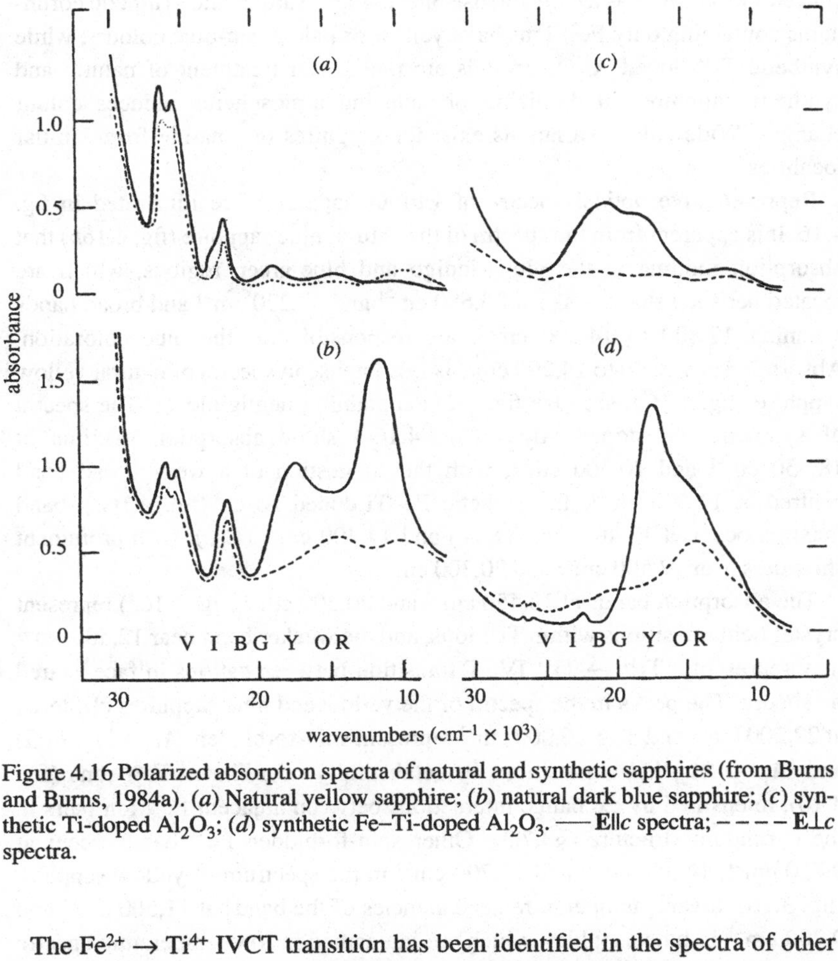

Figure 4.16 Polarized absorption spectra of natural and synthetic sapphires (from Burns and Burns, 1984a). (*a*) Natural yellow sapphire; (*b*) natural dark blue sapphire; (*c*) synthetic Ti-doped Al_2O_3; (*d*) synthetic Fe–Ti-doped Al_2O_3. —— **E**||*c* spectra; – – – – **E**⊥*c* spectra.

The $Fe^{2+} \rightarrow Ti^{4+}$ IVCT transition has been identified in the spectra of other titanian pyroxenes, including samples from the Moon (Burns *et al.*, 1976; Dowty, 1978) and, perhaps, a blue titaniferous omphacite from Labrador (Curtis *et al.*, 1975). An intense broad band in this omphacite centred around 15,000 cm^{-1} and showing a pressure-induced shift to lower energies was originally assigned to $Fe^{2+} \rightarrow Fe^{3+}$ IVCT (Abu-Eid, 1976) but later it was re-assigned as a $Fe^{2+} \rightarrow Ti^{4+}$ IVCT transition (Strens *et al.*, 1982).

4.7.2.6 Blue sapphire and hibonite

The cause of the dark blue colour of sapphire has been widely investigated (Burns and Burns, 1984a). Analyses of natural sapphires, coupled with crystal growth studies, established that both Fe and Ti must be present but in very low concentrations to produce the intense blue colour. Natural and synthetic corundums containing only Fe^{3+} ions have yellow or pale green-blue colours, while synthetic Ti^{3+}-doped Al_2O_3 crystals are pink. Heat treatment of natural and synthetic sapphires in oxidizing or reducing atmospheres induces colour changes. Wide colour variations exist for sapphires originating from similar localities.

Representative optical spectra of various sapphires are illustrated in fig. 4.16. It is apparent from the spectra of the natural blue sapphire (fig. 4.16*b*) that absorption minima in the violet-indigo and blue-green regions, which are located between sharp peaks at 25,680 cm^{-1} and 22,220 cm^{-1} and broad bands spanning 17,800 to 14,200 cm^{-1}, are responsible for the blue coloration. Absorption at 17,800 to 14,200 cm^{-1} is less intense in spectra of natural yellow sapphire (fig. 4.16*a*; see also fig. 3.21) containing negligible Ti. The spectra of synthetic Ti^{3+}-doped Al_2O_3 (fig. 4.16*c*) show absorption maxima at 18,450 cm^{-1} and 20,300 cm^{-1}, with the suggestion of a weak broad band centred at 12,500 cm^{-1}. In synthetic Fe–Ti-doped Al_2O_3 (fig. 4.16*d*), band maxima occur at 17,400 cm^{-1} ($\mathbf{E}\perp c$) and 14,400 cm^{-1} ($\mathbf{E}\| c$), with prominent shoulders near 12,500 cm^{-1} and 20,300 cm^{-1}.

The absorption bands at 18,450 cm^{-1} and 20,300 cm^{-1} (fig. 4.16*c*) represent crystal field transitions within Ti^{3+} ions, and the weaker band near 12,500 cm^{-1} may represent a $Ti^{3+} \rightarrow Ti^{4+}$ IVCT transition between cations in face-shared octahedra. The peaks in the spectra of the yellow and blue sapphires clustered at 22,200 cm^{-1} and near 26,000 cm^{-1} represent spin-forbidden ${}^6A_1 \rightarrow {}^4A_1,{}^4E(G)$ and ${}^6A_1 \rightarrow {}^4A_2,{}^4E(D)$ transitions in octahedrally coordinated Fe^{3+} ions (fig. 3.10), intensified by exchange interactions between adjacent Fe^{3+} ion pairs in the corundum structure (§3.7.3). Other spin-forbidden Fe^{3+} bands occur at 9,450 cm^{-1}, 14,350 cm^{-1} and 18,700 cm^{-1} in the spectrum of yellow sapphire (fig. 3.16). Inverse temperature dependencies of the bands at 11,500 cm^{-1} and 9,700 cm^{-1} in blue sapphire, which are intensified in low temperature spectra (Smith and Strens, 1976; Smith, 1977), were originally assigned to $Fe^{2+} \rightarrow Fe^{3+}$ IVCT transitions taking place perpendicular and parallel to the *c* axis, respectively. However, by analogy with vivianite in which Fe^{2+} CF bands are intensified by a $Fe^{2+} \rightarrow Fe^{3+}$ IVCT transition, the bands at 11,500 cm^{-1} and 9,700 cm^{-1} in blue sapphires may represent, instead, CF transitions within Fe^{2+} ions substituting for Al^{3+} ions in the corundum structure. The broad bands around 17,400 cm^{-1} and 14,200 cm^{-1} also intensify in low-temperature spectra, indicating that they represent $Fe^{2+} \rightarrow Ti^{4+}$ IVCT transitions across edge-shared ($\mathbf{E}\perp c$) and face-shared ($\mathbf{E}\| c$) octahedra, respectively. The loss of intensity of these bands when synthetic Fe–Ti-doped Al_2O_3 and some natural blue sapphires are heated in air may be attributed to oxidation of Fe^{2+} to Fe^{3+} ions and the removal of $Fe^{2+} \rightarrow Ti^{4+}$ IVCT transitions near 17,400 cm^{-1} and 14,300 cm^{-1}. The latter two bands confine the 'windows' of minimum absorption to the blue and indigo-violet region of the spectrum, resulting in dark blue sapphires. The absence of titanium and/or the conversion of Fe^{2+} to Fe^{3+} ions eliminates the possibility of $Fe^{2+} \rightarrow Ti^{4+}$ IVCT transitions, resulting in yellow or pale-blue sapphires.

Corundum is rarely found in meteorites. Instead, blue-coloured hibonite, $CaAl_{12}O_{19}$, occurs in refractory inclusions of several carbonaceous chondrites. Such blue hibonites are enriched in vanadium and also contain significant amounts of Mg, Ti and traces of Fe. The hibonite structure contains Al^{3+} ions in tetrahedral and trigonal bipyramidal (five-fold) coordination sites, as well as in three distinct octahedral sites (Burns and Burns, 1984b). One of these octahedral sites, corresponding to the Al3 position in the hibonite structure, consists of face-shared octahedra with cations separated by only 263 pm. Several assignments have been proposed for the two absorption bands measured in optical spectra of blue hibonites at 400 nm (25,000 cm^{-1}) and 700 nm (14,100 cm^{-1}) (Ihinger and Stolper, 1986), including crystal field transitions within V^{3+} and Ti^{3+} ions located in the trigonal bipyramidal site (Burns and Burns, 1984b). However, the band at 14,100 cm^{-1} could also originate from a $Fe^{2+} \rightarrow Ti^{4+}$ IVCT transition involving these cations in the face-shared Al3 octahedra of hibonite, by analogy with blue sapphire and ellenbergerite (Chopin and Langer, 1988).

4.7.2.7 Kyanite

The origin of the characteristic blue colour of bladed crystals of kyanite, Al_2SiO_5, has been extensively investigated. The major feature in visible-region spectra of blue kyanites is an intense polarization-dependent band at 16,500 cm^{-1} (Faye and Nickel, 1969; White and White, 1967). The 16,500 cm^{-1} band was originally assigned to a $Fe^{2+} \rightarrow Fe^{3+}$ IVCT transition (Faye and Nickel, 1969; Faye, 1971b) between Fe^{2+} and Fe^{3+} ions located in chains of edge-shared $[AlO_6]$ octahedra in the kyanite structure, in which Al–Al distances are 270 to 288 pm. A shoulder at 12,500 cm^{-1} was attributed to a spin–allowed crystal field transition in octahedrally coordinated Fe^{2+} ions. Mössbauer spectroscopy subsequently confirmed that octahedral Fe^{2+} and Fe^{3+} ions coexist in blue kyanites (Parkin *et al.*, 1977). However, reported correlations between intensity of the blue colour and Ti content of kyanite (White and White, 1967; Rost and Simon, 1972) led to the suggestion that the 16,500 cm^{-1} is due, instead, to the $Fe^{2+} \rightarrow Ti^{4+}$ IVCT transition (Smith and Strens, 1976). A coupled substitution of Fe^{2+} and Ti^{4+} for two Al^{3+} ions was assumed, since this mechanism maintains local charge balance within the crystal structure. It also

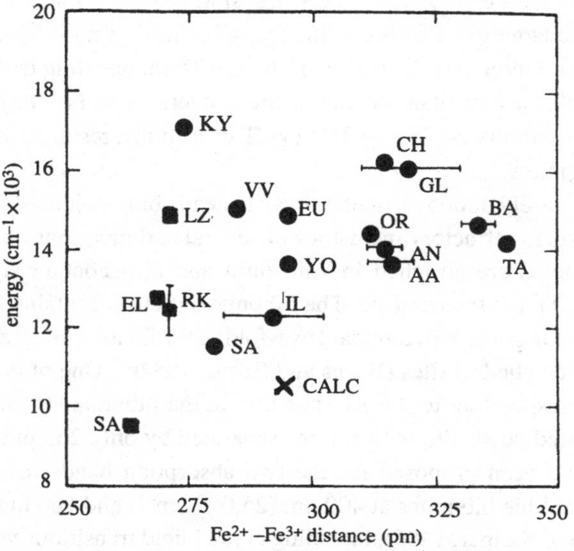

Figure 4.17 Correlations between transition energies of $Fe^{2+} \rightarrow Fe^{3+}$ IVCT bands and metal–metal distances of several mixed-valence minerals (modified from Mattson & Rossman, 1987a). Circles: edge-shared octahedra; squares: face-shared octahedra; cross: calculated from molecular orbital energy level calculations (§11.7.3; Sherman, 1987a). The key to the symbols is given in table 4.2, p. 117.

allows for a high probability of Fe^{2+}–Ti^{4+} couples to exist in adjacent sites, which is necessary to explain the intensity of absorption with rather low concentrations of Ti. The shoulder at 11,500 to 12,500 cm^{-1} intensifies at low temperatures, leading to this feature being reassigned to a $Fe^{2+} \rightarrow Fe^{3+}$ IVCT transition (Smith and Strens, 1976). However, by analogy with intensity enhancement of CF bands by IVCT transitions observed in vivianite (Amthauer and Rossman, 1984) and, perhaps, blue sapphire (§4.7.2.6), a CF transition within Fe^{2+} ions is favoured in accord with the evidence from Mössbauer spectroscopy of coexisting Fe^{2+} and Fe^{3+} ions in blue kyanite (Parkin *et al.*, 1977). The assignment of the 16,500 cm^{-1} band to $Fe^{2+} \rightarrow Fe^{3+}$ IVCT is also preferred (table 4.2).

4.7.2.8 Other Fe^{2+}–Fe^{3+} minerals

For a variety of minerals listed in table 4.2, $Fe^{2+} \rightarrow Fe^{3+}$ IVCT transitions have been assigned to a range of energies spanning 9,700 cm^{-1} to 18,500 cm^{-1}. These data enable two widely held beliefs to be examined: first, that the charge transfer energies should decrease with decreasing separation between the interacting cations; and, second, that IVCT transitions involving cations in face-shared octahedra should occur at lower energies than those involving cations in edge-shared octahedra. The data plotted in fig. 4.17 suggest that IVCT transitions between face-shared octahedra do, indeed, occur at lower energies, but there is a poor correlation between $Fe^{2+} \rightarrow Fe^{3+}$ IVCT energies and internuclear distances. As a result, band energies are considered not to be definitive criteria of IVCT transitions (Mattson and Rossman, 1987a). The one spectral feature that appears to distinguish IVCT from CF electronic transitions is the width of the absorption bands which is generally much broader for IVCT transitions than for CF transitions. Another useful criterion for distinguishing the two types of electronic transitions may be their relative responses to increased pressure (Burns, 1982). Thus, at elevated pressures, CF bands generally intensify and migrate to higher energies whereas greatly intensified IVCT bands show negligible pressure-induced energy changes. Such trends are discussed in chapter 9 (table 9.3).

4.7.2.9 Other Fe^{2+}–Ti^{4+} minerals

Other silicate minerals, in common with pyroxene from the Angra dos Reis meteorite, also show intense red colours attributed to $Fe^{2+} \rightarrow Ti^{4+}$ IVCT. These include taramellite and neptunite, for example. The high Fe and Ti contents of taramellite, $Ba_4Fe^{2+}Fe^{3+}{}_2TiSi_8O_{24}(OH)_4$, combined with features of its crystal structure, facilitate assignments of $Fe^{2+} \rightarrow Ti^{4+}$ IVCT bands. The cations occur in single chains of edge-shared octahedra parallel to the *a* axis, in which nearest-neighbour cations are 339 pm apart. Light polarized along these chains produces a broad, intense band at 21,740 cm^{-1} (460 nm) attributable to $Fe^{2+} \rightarrow Ti^{4+}$ IVCT (Mattson and Rossman, 1988). Other bands at 14,300 cm^{-1} (700 nm) representing $Fe^{2+} \rightarrow Fe^{3+}$ IVCT and at 900 nm and 1,150 nm attributed to Fe^{2+} CF transitions increase significantly in intensity, whereas the $Fe^{2+} \rightarrow Ti^{4+}$ IVCT band at 460 nm shows negligible temperature intensification.

In the crystal structure of neptunite, $KNa_2LiFe^{2+}Ti_2Si_8O_{24}$, Fe and Ti occur in two interconnecting chains of edge-shared octahedral chains in which nearest cations are 314 pm and 328 pm apart and the interaction vectors have substantial components in all three optical directions. An ordered arrangement exists within the chains so that Fe^{2+} ions alternate with Ti^{4+} ions. The optical spectra of neptunite show a very broad and intense band centred at 24,100 cm^{-1} (415 nm) which lies well outside the range of Fe^{2+} CF transitions and can be unequivocally assigned to $Fe^{2+} \rightarrow Ti^{4+}$ IVCT (Mattson and Rossman, 1988). Other examples are listed in table 4.2.

Ellenbergerite, $(Mg,Fe,Ti)_2Mg_6Al_6Si_8O_{25}(OH)_{10}$, discovered as recently as 1986 in high-pressure metamorphic rocks, represents another mineral besides blue sapphire and hibonite in which $Fe^{2+} \rightarrow Ti^{4+}$ IVCT transitions may involve cations in face-shared octahedra (Chopin and Langer, 1988). The crystal structure of ellenbergerite consists of two types of octahedral chains running parallel to the *c* axis. In one chain, face-shared M2 octahedra run along six-fold screw axes; in the other, zig-zag double chains of stacked edge-sharing pairs of distorted face-shared $M1_1$ and $M1_2$ octahedra occur on two-fold screw axes. These single and double chains are interconnected by six [SiO_4] tetrahedra. The octahedral double chains are fully ordered with Al and Mg occupying the smaller $M1_1$ and $M1_2$ octahedra of each edge-shared pair, and interatomic distances are 284 pm and 300 pm. The single octahedral chains are only partially filled by (Mg+Fe) and (Ti+Zr) in nearly equal amounts leaving about one-third of the sites vacant. Interatomic distances across face-shared octahedra in these chains are 273 pm. Optical spectra of ellenbergerite show an intense broad band lying between 18,500 cm^{-1} and 19,500 cm^{-1} when light is polarized along the *c* axis. This absorption band has been assigned to a $Fe^{2+} \rightarrow Ti^{4+}$ IVCT transition (Chopin and Langer, 1988) involving cations in the single chains of face-shared octahedra.

4.7.3 Oxygen → metal charge transfer

Electronic transitions between nearest-neighbour oxygen ligands and the central transition metal ion in a coordination site are induced by high-energy ultraviolet light. However, absorption edges of bands located in the ultraviolet may extend into the visible region and contribute to optical spectra. Such oxygen → metal charge transfer (OMCT) transitions are responsible for the intense colours of compounds and minerals containing high-valence transition metal ions with electronic configurations devoid of 3*d* electrons and, therefore, CF and IVCT transitions. Examples include permanganates such as $KMnO_4$ containing Mn(VII), chromates such as crocoite, $PbCrO_4$, and other Cr(VI)-bearing minerals, and vanadates consisting of vanadinite, $Pb_5(VO_4)_3Cl$, and other V(V)-containing minerals (Abu-Eid, 1976; Mao, 1976). Several common ferromagnesian silicate minerals containing transition metal ions in lower oxidation states have dark-brown colours attributable to OMCT transitions, notable examples being biotites and hornblendes. These intense colours may be interpreted by results obtained from molecular orbital energy level calculations described in chapter 11 (§11.7.2).

The molecular orbital energy level calculations have yielded relative energies of OMCT transitions (Loeffler *et al.*, 1974, 1975; Sherman, 1985a,b; Sherman and Waite, 1985). Peak maxima are centred well into the ultraviolet region. However, since OMCT transitions are fully allowed by both the

Laporte and spin-multiplicity selection rules (§3.7) and have intensities 10^3 to 10^4 times higher than those of crystal field transitions (table 3.6), their absorption edges may extend well into the visible region and overlap crystal field spin-allowed and spin-forbidden peaks.

For cations most frequently encountered in terrestrial and lunar minerals, OMCT energies involving transition metal ions are calculated or observed to decrease in the order

$$Cr^{3+} > Ti^{3+} > Fe^{2+} > Fe^{3+}_{tet} > Fe^{3+}_{oct} > Ti^{4+} . \qquad (4.5)$$

As a result, silicate minerals containing significant amounts of Ti^{4+} and Fe^{3+} cations in their crystal structures are most strongly affected by OMCT absorption edges, accounting for the dark-brown colours of titaniferous oxyamphiboles (hornblendes and kaersutites) and biotites occurring in igneous rocks.

With rising temperatures and pressures, absorption edges of OMCT bands show red-shifts, absorbing increasing amounts of the visible region in spectra measured at elevated temperatures and high pressures (Burns, 1982). The importance of such effects in radiative heat transfer and electrical conductivity in the Earth's Mantle is discussed in chapter 9 (§9.10).

4.8 Opacity of mixed-valence oxides and silicates

Several mixed valence Fe^{2+}–Fe^{3+} oxide and silicate minerals which are coloured black are so opaque that it is extremely difficult to measure their absorption spectra in the visible region. Examples include magnetite, vonsenite, ilvaite, deerite, riebeckite, cronstedtite and laihunite. Common features of all of these minerals are: first, they approximate Fe^{2+}–Fe^{3+} end-member compositions and possess minor Mg^{2+} and Al^{3+} contents; and second, their crystal structures contain cations in infinite chains or bands of edge-shared octahedra. The close proximity of Fe^{2+} and Fe^{3+} ions in adjacent octahedra enables electron hopping to occur between the neighbouring cations throughout the crystal structure, leading to thermally activated delocalization of electrons (Burns, 1991). These delocalized or itinerant electrons cause the opaque oxides and silicates to exhibit metallic properties, such as high electrical and thermal conductivities. The Fe^{2+}–Fe^{3+} minerals are amenable to Mössbauer spectral measurements and have been studied extensively by this technique.

One consequence of thermally activated electron delocalization behaviour is that techniques such as Mössbauer spectroscopy, which might be expected to distinguish between discrete Fe^{2+} and Fe^{3+} valences, instead often detect Fe cations in intermediate oxidation states. Such cation species originate when electrons are transferred between Fe^{2+} and Fe^{3+} ions in neighbouring sites more rapidly than the lifetime of the Mössbauer transition (10^{-7} s between the ground state and 14.41 keV excited nuclear energy level of individual ^{57}Fe nuclei). The Mössbauer spectrum profiles of opaque mixed-valence Fe minerals show significant temperature variations due to thermally activated electron delocalization between Fe cations. This is one property of mixed-valence compounds. Other features are summarized in table 4.3.

The classification scheme for mixed-valence compounds in table 4.3 (Robin and Day, 1967; Day, 1976) is based on structural similarities of coordination sites occupied by cations with different valences and the consequent ease or difficulty of transferring an electron from one site to another. When the two coordination sites are very different (e.g., Fe^{3+}-bearing octahedra and Fe^{2+}-bearing cubes in the garnet structure; see fig. 5.5), the transfer of electrons is difficult so that the properties of the mineral essentially correspond to those of the individual cations in the two sites. These minerals are termed Class I compounds (Day, 1976). If the two sites are identical, the electron may be transferred from one site to the other with no expenditure of energy. Thus, if a crystal structure contains continuous arrays of edge-shared or face-shared coordination polyhedra, the mineral may have metallic properties. Because the valences of the individual cations are 'smeared out', the characteristic properties of the single valence states may not be found. Minerals of this type are called Class IIIB compounds. Magnetite and ilvaite are classic examples of Class IIIB compounds. If the structure is not completely continuous, or atomic substitution of Mg^{2+} or Al^{3+} block the Fe^{2+}–Fe^{3+} interactions, electron delocalization takes place only within finite clusters of equivalent cations. Although the individual cation properties of the mineral may not be seen the structure does not conduct electrons. This type of substance is called a Class IIIA compound, and mineral examples include glaucophane, blue sapphire, kyanite and slightly oxidized vivianite.

Intermediate between these two extremes are minerals classified as Class II compounds in which the two sites are similar but distinguishable; that is, both are octahedral sites, but with slightly different metal–oxygen distances, ligand orientation or bond-type. Examples include the amphibole M1, M2 and M3 sites (figs 4.14 and 5.18), the mica *trans*-M1 and *cis*-M2 sites (fig. 5.21) and babingtonite (Burns and Dyar, 1991). Such materials still exhibit properties of cations with discrete valences, but they have low energy IVCT bands and may be semiconductors.

Some mixed-valence Fe^{2+}–Fe^{3+} minerals, because they possess high electrical conductivities at elevated temperatures, may be important on Venus (Burns and Straub, 1992). As a result, magnetite, ilvaite, laihunite, oxyhornblendes

Table 4.3. *Physical properties of mixed-valence minerals*

Property	Class I	Class II	Class III A	Class IIIB
Nature of sites in crystal structure	Vastly different, e.g., cube-octahedron; octahedron-tetrahedron	Similar, e.g., octahedra with slightly different bond lengths or ligands	Identical, finite clusters, e.g., edge-shared octahedral dimers	Identical, continuous chains, e.g., chains of edge-shared octahedra
Optical absorption spectroscopy	No intervalence transitions in visible region	One mixed-valence transition in visible region; absorption bands intensify at elevated pressures and low temperatures	One or more mixed-valence transitions in visible region; temperature lowers intensity of absorption bands, pressure intensifies	Opaque; metallic reflectivity in visible region
Electrical conductivity	Insulator; conductivity < 10^{-10} ohm^{-1} m^{-1}	Semiconductor; conductivity 10 to 10^{-6} ohm^{-1} m^{-1}	Insulator or low-conductivity semiconductor	Metallic conductor; conductivity 10^4 to 10^8 ohm^{-1} m^{-1}
Magnetic properties	Diamagnetic or paramagnetic to very low temperatures	Magnetically dilute; either ferromagnetic or antiferromagnetic at low temperatures	Magnetically dilute	Pauli paramagnetism or ferromagnetic with high Curie temperature
Mössbauer spectra	Spectra of constituent Fe cations; discrete Fe^{2+} and Fe^{3+} ions	Spectra of constituent Fe cations; discrete Fe^{2+} and Fe^{3+} ions	Electron delocalized species may contribute to spectra; higher contributions at elevated temperatures and pressures	Electron delocalized species contribute to spectra; higher contributions at elevated temperatures and pressures
Mineral examples	Garnet; cordierite and osumilite; aquamarine (beryl); perovskite A & B sites; pyroxene M1 & M2	Amphibole M1 & M2; biotite M1 & M2; tourmaline Mg & Al; ludwigite M3 and M4; lazulite Mg & Al; babingtonite	vivianite FeB; lazulite Al; riebeckite (amphibole) M1; tourmaline Mg; sapphire (corundum); hibonite Al3; rockbridgeite; babingtonite	magnetite; ilvaite; deerite; laihunite M1; cronstedtite [also: pyroxene M1; tourmaline Al; kyanite M1]

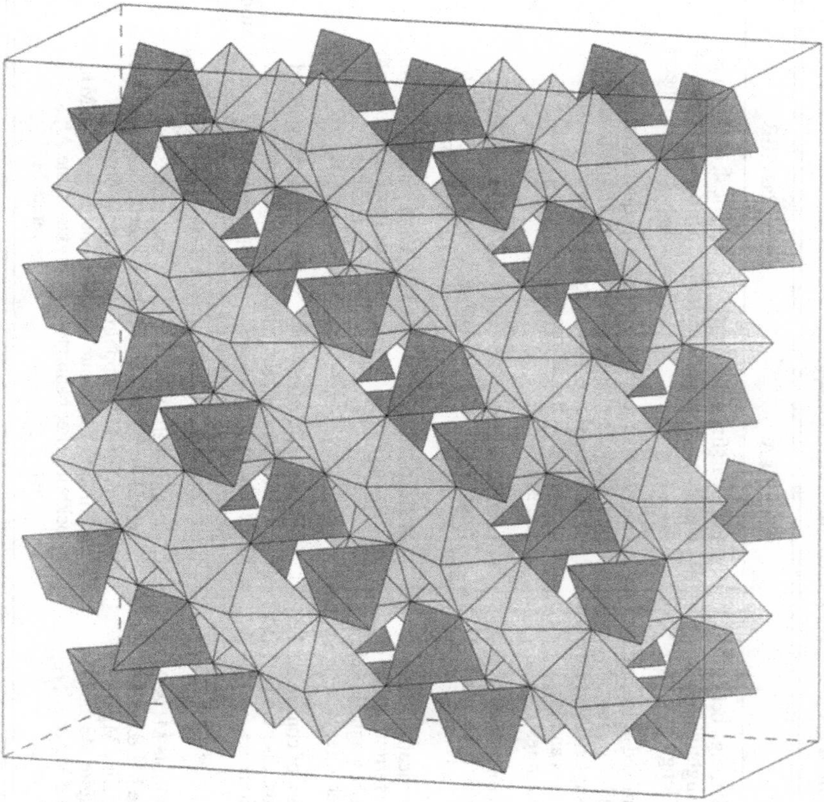

Figure 4.18 The spinel crystal structure adopted by magnetite, $Fe^{2+}Fe^{3+}{}_2O_4$. Note the three–dimensional infinite chains of edge-shared [FeO_6] octahedra extending along [110] directions, which accommode Fe^{2+} and Fe^{3+} ions separated by 297 pm. Fe^{3+} ions also occur in isolated tetrahedra linking the octahedral chains.

and oxybiotites may contribute to high radar-reflectivities emitted from Venusian mountainous regions (§10.8.3).

4.8.1 Magnetite

Just as vivianite is regarded as the simplest example of a mineral with isolated clusters of Fe^{2+}–Fe^{3+} octahedra showing IVCT transitions, so too is magnetite considered to be the classic example of a structure-type with infinite chains of Fe^{2+} –Fe^{3+} octahedra exhibiting electron delocalization. Magnetite, Fe_3O_4 or $Fe^{2+}Fe^{3+}{}_2O_4$, has the spinel structure illustrated in fig. 4.18 with an *inverse* cation distribution (§6.4). Thus, half the Fe^{3+} ions occupy isolated tetrahedral sites, while Fe^{2+} and the remaining Fe^{3+} ions occur in octahedral coordination in the three-dimensional infinite single chains of edge-shared octahedral projecting along the [110] axes of the cubic unit cell. Above 119 K, electron delocalization occurs in magnetite between cations in octahedral sites which are separated by only 297 pm, resulting in opacity and high electrical and thermal conductivities. Mössbauer spectra of magnetite show two cation species: tetrahedral Fe^{3+} ions and intermediate valence $Fe^{2.5+}$ ions in octahedral sites. At 119 K, the Verwey transition temperature, a phase change occurs which doubles the unit cell, leads to magnetic ordering of Fe^{2+} and Fe^{3+} in octahedral sites, and produces distinguishable Fe^{2+} and Fe^{3+} cations in octahedral sites with attendant drop in electrical conductivity and changes in the Mössbauer spectra (Verwey and Haayman, 1941; Verble, 1974). The Verwey transition in magnetite has been interpreted as an ionic order–disorder transition. A similar phase change occurs in ilvaite.

4.8.2 Ilvaite

Many of the attributes of electron delocalization in Class IIIB mixed-valence compounds (table 4.3) are displayed by ilvaite, $CaFe^{2+}{}_2Fe^{3+}Si_2O_8(OH)$. Evidence for thermally activated electron hopping between Fe^{2+} and Fe^{3+} ions at and above room temperature was suggested by several Mössbauer spectral studies prior to 1980 (see references in Burns, 1981). Such behaviour of electrons influences the crystal structure and optical, electrical and magnetic properties of ilvaite which have since been investigated in great detail. Thus, interactions between Fe^{2+} and Fe^{3+} in neighbouring sites in ilvaite cause a sequence of two magnetic and a coupled electronical and displacive phase transition. It is now known that between 333 K and 343 K, ilvaite changes from monoclinic to orthorhombic symmetry. This is accompanied by thermally activated electron delocalization between 300 K and 400 K. Below about 120 K ilvaite becomes antiferromagnetically ordered.

The basic ilvaite structure (Takéuchi *et al.*, 1983; Ghose *et al.*, 1984; Finger and Hazen, 1987) illustrated in fig. 4.19 consists of a framework of infinite double chains of edge-shared $[FeO_6]$ octahedra parallel to the *c* axis which contain equal numbers of Fe^{2+} and Fe^{3+} ions. These sites are crystallographically equivalent in the high temperature orthorhombic phase (space group *Pnam*), and are designated as the Fe(A) sites (alternatively labelled the M1 or 8*d* sites). Additional Fe^{2+} octahedra in sites designated as the Fe(B) sites (also called the M2 or 4*c* sites) are attached to the Fe(A) sites alternating above and below the double chains of edge-shared octahedra. The double chains are linked to one another by seven-coordinated Ca^{2+} ions and by the sorosilicate $[Si_2O_7]$ groups. The Fe(A) sites occupied by both Fe^{2+} and Fe^{3+} ions are nearly regular octahe-

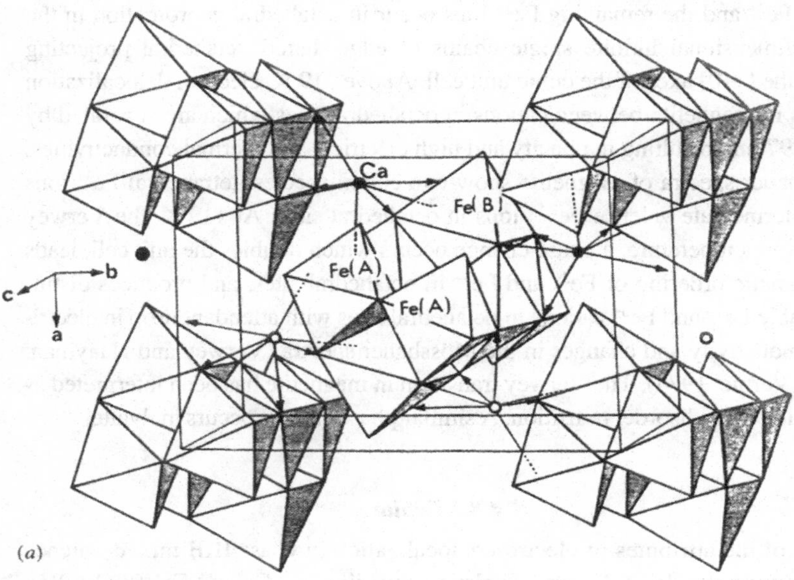

(*a*)

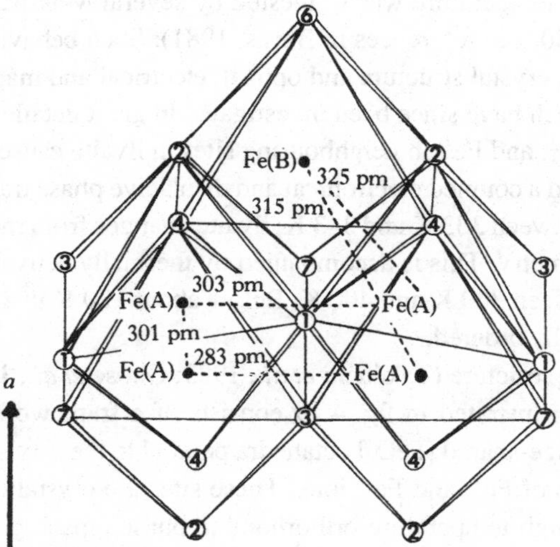

(*b*)

Figure 4.19 The ilvaite crystal structure showing (*a*) the arrangement of double chains of edge-shared Fe(A) octahedra extending along the *c* axis, which are flanked by Fe(B) octahedra and are linked by Ca^{2+} ions and Si_2O_7 groups (not shown). (*b*) Projection of one unit of the double chain of Fe(A) octahedra showing Fe–Fe interatomic distances across shared edges (from Nolet and Burns, 1979).

dra with an average metal–oxygen distance of 208.1 pm (ranging from 200.6 pm to 217.2 pm), whereas the more distorted Fe(B) sites occupied by Fe^{2+} ions alone have an average metal–oxygen distance of 219.0 pm. The Fe(A)–Fe(A) distances between edge-shared octahedra along the chains are 283 pm and 303 pm parallel to the *c* axis and across the chains they are 301 pm in the *a–b* plane. Corresponding Fe(A)–Fe(B) distances range from 315 pm to 325 pm.

Decreasing temperature results in a phase transition at about 335 K from the orthorhombic (*Pnam*) to monoclinic ($P2_1/a$) space group. Ordering of ferrous and ferric iron occurs within the Fe(A) sites so that Fe^{2+} and Fe^{3+} ions alternate in sites now labelled as Fe(A1) and Fe(A2) (alternatively M11 and M12), respectively, with identical valences existing within planes perpendicular to the *c* axis. In the monoclinic phase, average metal–oxygen distances in the cation sites are: Fe(A1) = 210.3 pm (range 201.7 to 222.2 pm); Fe(A2) = 205.7 pm (range 197.6 to 213.3 pm); and Fe(B) = 216.8 pm (range 196.6 to 228.7 pm). At lower temperatures, antiferromagnetic ordering occurs between adjacent double chains of Fe(A) sites, resulting in a magnetic transition at 123 ± 1 K, followed by an antiferromagnetic ordering of Fe^{2+} in attached Fe(B) sites at 50 ± 3 K (Ghose *et al.*, 1984; Coey *et al.*, 1984; Ghosh *et al.*, 1987). Neutron diffraction studies of ilvaite at low temperatures (Ghose *et al.*, 1984) show that at 80 K, the Fe^{2+} and Fe^{3+} spins on the Fe(A) sites along one of the infinite *c* axis chains are parallel and antiparallel to those along the adjoined edge-sharing centrosymmetric related chain. The spin vectors are all perpendicular to the plane of these chains, i.e. almost parallel to the *b* axis. At 5 K, this order is maintained, but the Fe^{2+} ions in the Fe(B) sites order antiferromagnetically as well, again almost parallel to the *b* axis.

Evidence in support of electron delocalization in ilvaite comes from Mössbauer spectral measurements, which show significant temperature-induced changes of positions and intensities of the peaks. One interpretation of the variable spectrum profiles is that the positions and intensities of peaks originating from localized Fe^{2+} and Fe^{3+} ions and delocalized Fe^{2+}–Fe^{3+} species do not change abruptly with temperature, but instead their relative intensities change as a result of increased participation of Fe(A) site Fe^{2+} and Fe^{3+} ions in forming Fe^{2+}–Fe^{3+} species (Nolet and Burns, 1979; Ghazi–Bayat *et al.*, 1987, 1989). Another interpretation of the Mössbauer spectra is that strong relaxation effects cause line broadening and deviations from Lorentzian shape due to electron hopping (Litterst and Amthauer, 1984). The activation energy for the relaxation process was estimated to be 0.11 eV indicating that at elevated temperatures the small barrier for electron hopping is exceeded. A comparable activation energy (0.13 eV) was also deduced from resistivity measurements of ilvaite (Coey *et al.*, 1984).

The relaxation model proposed for the Mössbauer spectra of ilvaite (Litterst and Amthauer, 1984) does not provide conclusive information about electron transport behaviour significantly faster than the Mössbauer time-window (10^{-7} s). Optical spectroscopy, on the other hand, acting on a timescale of 10^{-15} s may be more informative about transport properties related to the electron order–disorder processes of ilvaite (Güttler *et al.*, 1989). However, the opacity of ilvaite makes measurements of its optical absorption spectra extremely difficult (Amthauer and Rossman, 1984; Güttler *et al.*, 1989). Nevertheless, crystals polished to a thickness of 10 to 20 microns become transparent with intense green to dark-brown pleochroism. The optical spectra of these thinned ilvaite crystals with light polarized along the *c* axis show a prominent shoulder located near 12,300 cm^{-1} (813 nm) attributed to $Fe^{2+} \rightarrow Fe^{3+}$ IVCT situated on a broad, intense band centred at about 8,260 cm^{-1} (1,080 nm), which was assigned to a Fe^{2+} CF transition intensified by the IVCT band (Amthauer and Rossman, 1984). More recent measurements (Güttler *et al.*, 1989) suggest that the IVCT band may be located at 14,500 cm^{-1} (690 nm).

4.8.3 Other opaque mixed-valence minerals

4.8.3.1 Vonsenite

Oxyborate minerals belonging to the ludwigite–vonsenite series, $(Mg,Fe^{2+})_2Fe^{3+}BO_5$, are of interest because most magnesian ludwigites are completely opaque, in contrast to other mineralogical Mg–Fe^{2+} solid-solution series which are transparent even at high Fe concentrations. In the crystal structure of ludwigite, Mg and Fe cations occupy four crystallographically distinct six-coordinated sites, labelled M1 through M4, with Fe^{2+} and Fe^{3+} ions preferentially occupying the M3 and M4 sites, respectively. Extensive edge-sharing of the cation octahedra occurs within bands parallel to the *c* axis and also in the zig-zag chains running along the *a* axis. Metal–metal distances are as short as 300 to 312 pm across edge-shared octahedra parallel to the *c* axis, and only 275 pm between M3 (Fe^{2+}) and M4 (Fe^{3+}) sites in the zig-zag chains. The structure is thus strongly conducive to electron delocalization between adjacent Fe^{2+} and Fe^{3+}, even in Mg-rich ludwigites.

The Mössbauer spectra of specimens spanning the ludwigite–vonsenite series show temperature and compositional variations (Burns and Burns, 1982; Swinnea and Steinfink, 1983). These variations resemble those observed in the ilvaite Mössbauer spectra and may be attributed to electron hopping between Fe^{2+}(M3) and Fe^{3+}(M4) cations. Charge delocalization in vonsenite becomes possible over the three-dimensional structure via edge-shared octahedra parallel to the *c* axis.

4.8.3.2 Deerite

Another black mixed-valence mineral exhibiting large temperature variations of its Mössbauer spectral profiles is deerite, $Fe^{2+}{}_6Fe^{3+}{}_3O_3Si_6O_{17}(OH)_5$. The crystal structure of this chain silicate mineral resembles amphibole (fig. 4.14) but contains bands of edge-shared octahedra, six octahedra wide, extending along the *c* axis. There are nine distinct crystallographic positions containing Fe, which may be grouped into three sets of three virtually equivalent octahedra. Extensive edge-sharing of these octahedra produces Fe–Fe distances ranging from 310 pm to 331 pm. The low-temperature Mössbauer spectra show that discrete Fe^{2+} and Fe^{3+} ions exist in deerite, but by room temperature more than 20 per cent of the iron cations take part in thermally activated electron delocalization across edge-shared octahedra (Amthauer *et al.*, 1980). This correlates with the high electrical conductivity of deerite for which the activation energy is only 0.043 eV (Pollak *et al.*, 1981).

4.8.3.3 Cronstedtite

Another conducting opaque silicate is the serpentine-group mineral cronstedtite, $(Fe^{2+},Fe^{3+})_3(Si,Fe^{3+})_2O_5(OH)_4$, which also undergoes thermally activated charge transfer (Coey *et al.*, 1982). Sheets of edge-shared octahedra with Fe–Fe distances around 310 pm occur in this black, mixed-valence Fe^{2+}–Fe^{3+} phyllosilicate in which cations are octahedrally coordinated in identical sites to four OH^- anions and two *cis*-oxygens belonging to the sheets of $[SiO_4]$ tetrahedra. Some ferric iron also replaces Si in tetrahedral coordination. Electron hopping in cronstedtite is indicated by its high electrical conductivity parallel to the sheets (activation energy 0.25 eV) and by the rapid relaxation effects seen in its Mössbauer spectra particularly above 200 K (Coey *et al.*, 1982).

Ferric-rich biotites, particularly annites, also are often opaque but their Mössbauer spectra do not portray evidence of electron delocalization behaviour (Dyar and Burns, 1986) at ambient temperatures. Although Fe–Fe distances in annite, 310 to 311 pm, resemble those in cronstedtite, electron delocalization may be restricted in biotites by the presence of cations in two different octahedral sites, the *cis*-M2 and *trans*-M1 sites (fig. 5.21), as well as by cation vacancies in the octahedral sites.

4.8.3.4 Laihunite and Ferrifayalite

Laihunite, $Fe^{2+}Fe^{3+}{}_2Si_2O_8$, a relatively new Fe^{2+}–Fe^{3+} silicate mineral first discovered in mainland China in 1976, is commonly microscopically twinned and topotactically intergrown with magnetite and fayalite, giving rise to mixed-mineral assemblages called ferrifayalite. The opacity of laihunite is suggestive of electron delocalization which is limited, however, due to ordered vacancies in the crystal structure. Although optical spectra have not been reported, several Mössbauer spectral measurements have been made of laihunite and ferrifayalites which have a bearing on Fe^{2+}–Fe^{3+} interactions in this nesosilicate.

The crystal structure of laihunite is based on that of the olivine mineral fayalite (fig. 5.8). Laihunite, however, contains ordered vacancies that result in finite clusters of three edge-shared [FeO_6] octahedra. In pure fayalite, Fe^{2+} ions occur in two distorted octahedral sites, designated as Fe(1) and Fe(2). Linear chains of edge-shared [Fe(1)O_6] octahedra extend parallel to the *c* axis with Fe(1)–Fe(1) distances of about 304 pm. Each [Fe(1)O_6] octahedron shares opposite edges with two [Fe(2)O_6] octahedra which are in a serrated or zig-zag configuration relative to the chains of [Fe(1)O_6] octahedra. The Fe(1)–Fe(2) distances are about 318 pm. The serrated chains are linked to adjacent chains at different levels along the *a* axis by corner-sharing of [Fe(2)O_6] octahedra. In the laihunite structure, ordered vacancies occur in the [Fe(1)O_6] octahedra, giving superstructures with 2*c* and 3*c* repeats of the basic olivine structure. Ferric ions preferentially occupy the Fe(2) sites which also contain some Fe^{2+} ions, but most of the ferrous iron is located in Fe(1) sites alternating with the vacancies. The average structure of the laihunite–3M polymorph thus contains triple clusters of Fe(2)–Fe(1)–Fe(2) octahedra with Fe^{2+} and Fe^{3+} ions 318 pm apart.

The Mössbauer spectra of laihunite and ferrifayalites (Kan and Coey, 1985; Schaefer, 1985) consist of two quadrupole doublets with Mössbauer parameters consistent with discrete Fe^{3+} and Fe^{2+} ions. Low-temperature measurements indicate that ferric ions order magnetically below 160 K, whereas Fe^{2+} ions begin to order magnetically below 100 K. Analysis of the 4.2 K spectrum suggests that there is a distortion of the structure at low temperatures which leads to inequivalent Fe(1) and Fe(2) sites each with relative populations of 2:1, analogous to that observed in magnetite below the Verwey transition. Thermally activated electrical conduction measured in the range 290 to 500 K provided an activation energy of 0.53 eV for laihunite (Kan and Coey, 1985) which is significantly higher than those of deerite (0.043 eV), ilvaite (0.13 eV), and cronstedtite (0.25 eV), indicating that the ordered vacancies in the Fe(1) sites prevent electron delocalization in laihunite.

4.9 Summary

Chapter 4 outlines methods for measuring absorption spectra of minerals in the visible to near-infrared region, and describes applications of electronic spectral measurements to some fundamental crystal chemical problems, including causes of colour and pleochroism of minerals.

Techniques for measuring mineral spectra. When minerals are available as large single crystals, spectral measurements may be made on highly polished plates of the mineral mounted in a holder with a suitable aperture which is placed in the light beam of a spectrophotometer. Otherwise, measurements are made either on powdered minerals extracted from rock or on crystals in rock-matrix using microscope accessories. Such microscope techniques enable polarized absorption spectra to be measured on accurately oriented small crystals often in conventional petrographic thin sections of rock. Microscope absorption spectroscopy is particularly important in high-pressure measurements of minerals enclosed in diamond anvil cells.

Applications of spectral measurements in crystal chemistry. Each transition metal ion in a particular oxidation state or coordination symmetry produces a characteristic crystal field spectrum. Thus, optical spectra of minerals may serve to identify cations in unusual oxidation states, as well as the symmetry of coordination sites accommodating the cations. For example, spectra of staurolites confirm the presence of tetrahedral Fe^{2+} and Co^{2+} ions; titanian pyroxenes from the Moon and some carbonaceous chondrites contain Ti^{3+} ions; and Mn^{3+} ions exist in manganiferous epidotes and andalusites. The existence of these relatively unstable cations results from their occurrences in distorted six-coordinated sites. Such low-symmetry sites in the pyroxene and alumino-silicate structures not only bestow additional CFSE's on Ti^{3+} and Mn^{3+} relative to those acquired by these cations in regular octahedral sites, but they also produce polarization-dependent absorption spectra in the host mineral crystals if they are non-cubic. Site preferences and cation ordering in silicate structures may be indicated by changes of spectrum profiles and non-constancy of extinction coefficients with variations of chemical compositions of the minerals. Thus, ordering of Mn^{2+} in Fe–Mn olivines, Fe^{2+} in Mg–Fe pyroxenes, Ni^{2+} in Mg–Ni olivines and Mn^{3+} in epidotes and andalusites are revealed in their crystal field spectra.

Colour and pleochroism. Colour and pleochroism may be evaluated quantitatively from measurements of optical spectra. Colour arises from absorption of light in the visible region, 400 to 700 nm. Absorption in the ultraviolet and infrared regions may not be seen by the eye. Pleochroism in biaxial minerals (or dichroism in uniaxial minerals) is caused by differential absorption of polarized radiation in different crystallographic directions and is seen as colour variations when light in the visible region is affected. Spectrophotometric measurements of polarized absorption spectra reveal the invisible pleochroism in the near-infrared region not seen in most ferromagnesian silicate minerals.

Causes of colour and pleochroism. The majority of silicates owe their colours to the presence of transition metal and lanthanide ions in the crystal structures, which absorb radiation when electrons are excited between 3*d* or 4*f* orbitals within individual cations. The occurrence of these cations in coordination sites distorted from octahedral or tetrahedral symmetries causes pleochroism in non-cubic minerals. Intervalence charge transfer transitions (IVCT) are common in silicate minerals containing transition metals that can exist in two or more oxidation states. $Fe^{2+} \rightarrow Fe^{3+}$ and $Fe^{2+} \rightarrow Ti^{4+}$ IVCT between neighbouring cations in a crystal structure may be facilitated through suitably orientated overlapping 3*d* orbitals when light is polarized along certain crystallographic directions, leading to pleochroism. Common mineralogical examples include $Fe^{2+} \rightarrow Fe^{3+}$ IVCT in vivianite, glaucophane and kyanite, and $Fe^{2+} \rightarrow Ti^{4+}$ IVCT in sapphire and lunar pyroxenes. Oxygen → metal charge transfer transitions contribute to red-brown colours observed in many augites, hornblendes, biotites and staurolites.

Opacity of mixed-valence minerals. The opacities of many end-member Fe^{2+}–Fe^{3+} oxide and silicate minerals result from electron hopping between neighbouring cations when they are located in infinite chains or bands of edge-shared octahedra in the crystal structures. Opaque minerals such as magnetite, ilvaite, deerite, cronstedtite, riebeckite and laihunite owe their relatively high electrical conductivities to thermally activated electron delocalization, contributing to intermediate valence states of iron cations which may be detected by Mössbauer spectroscopy.

4.10 Background reading

Amthauer, G.. & Rossman, G. R. (1984) Mixed valence of iron in minerals with cation clusters. *Phys. Chem. Minerals*, 11, 37–51.

Burns, R. G. (1981) Intervalence transitions in mixed-valence minerals of iron and titanium. *Ann. Rev. Earth Planet. Sci.*, 9, 345–83.

Burns, R. G. (1983) Colours of gems. *Chem. Britain*, (12), 1004–7.

Burns, R. G. (1991) Mixed valency minerals: influences of crystal structures on optical and Mössbauer spectra. In *Mixed Valence Systems: Applications in Chemistry, Physics and Biology.* (K. Prassides, ed.; Plenum Press, New York), *NATO ASI–C Series, Math. Phys. Sci.*, C343, 175–200.

Fritsch, E. & Rossman, G. R. (1987, 1988) An update on color in gems. Parts 1, 2 and 3. *Gems & Gemology*, 23, 126–39; 24, 3–15 and 81–102.

Loeffler, B. M. & Burns, R. G. (1976) Shedding light on the color of gems and minerals. *Amer. Sci.*, 64, 636–47.

Marfunin, A. S. (1979a) *Physics of Minerals and Inorganic Materials* (Springer-Verlag, New York), 340 pp.

Marfunin, A. S. (1979b) *Spectroscopy, Luminescence and Radiation Centers in Minerals* (Springer-Verlag, New York), 352 pp.

Nassau, K. (1978) The origins of color in minerals. *Amer. Mineral.*, 63, 219–29.
Rossman, G. R. (1988) Optical spectroscopy. In *Spectroscopic Methods in Mineralogy and Geology*. (F. C. Hawthorne, ed.; Mineral Soc. Amer. Publ.), *Rev. Mineral.*, 18, 207–43.

5

Crystal field spectra of transition metal ions in minerals

The end products of the analysis (measurement and interpretation of mineral absorption spectra by crystal field theory) are some parameters that can be correlated with structural properties.

K. L. Keester & W. B. White, *Proc. 5th IMA Meeting (Cambridge, 1966)*, p. 22 (1968)

5.1 Introduction

In the previous chapter it was shown how measurements of polarized absorption spectra in the visible to near-infrared region can provide information on such crystal chemical problems as oxidation states of transition metal ions, coordination site symmetries and distortions, cation ordering and the origins of colour and pleochroism of minerals. Much attention was focused in chapter 4 on energies of intervalence charge transfer transitions appearing in electronic absorption spectra of mixed-valence minerals.

Perhaps a more fundamental application of crystal field spectral measurements, and the one that heralded the re-discovery of crystal field theory by Orgel in 1952, is the evaluation of thermodynamic data for transition metal ions in minerals. Energy separations between the 3*d* orbital energy levels may be deduced from the positions of crystal field bands in an optical spectrum, making it potentially possible to estimate relative crystal field stabilization energies (CFSE's) of the cations in each coordination site of a mineral structure. These data, once obtained, form the basis for discussions of thermodynamic properties of minerals and interpretations of transition metal geochemistry described in later chapters.

As noted in previous chapters, several complexities arise in crystal field spectra of silicate minerals making it difficult, if not impossible in some cases, to estimate accurately energy separations between all 3*d* orbitals and to calculate CFSE's of individual transition metal ions in every site of some minerals. One difficulty is that coordination sites in silicate structures usually possess low symmetries, causing 3*d* orbitals to be resolved into several energy levels and resulting in many bands in an absorption spectrum. Some of the energy separations between the resolved 3*d* orbitals are so small and correspond to energies in the mid- and far-infrared regions, as to be hidden among vibrational and rotational absorption bands. Secondly, rigorous interpretation of the spectra of cations in the low-symmetry coordination sites by group theory is not always possible due to non-coincidence of electronic axes with indicatrix axes of polarized light. A third difficulty is the occurrence of a cation in two or more distinct sites that produce multiple, overlapping absorption bands in a crystal field spectrum. Resolving such closely spaced absorption bands not only is impossible even at low temperatures but also presents a problem of assigning the bands to the transition metal in the correct coordination site. A fourth problem is the presence of small amounts of other transition metal ions that may have profound effects on optical spectra by producing interfering absorption bands, perhaps with enhanced intensity. Thus, as noted in §4.7.2, traces of Fe^{3+} or Ti^{4+} ions in ferromagnesian silicates may produce intense intervalence charge transfer (IVCT) bands in a spectrum through electronic transitions between adjacent Fe^{2+} and Fe^{3+} or Ti^{4+} ions in the structure which obscure low-intensity CF bands within the Fe^{3+}, Fe^{2+} and Ti^{3+} ions, themselves, as well as in less abundant cations. Such IVCT absorption bands must be identified before crystal field splittings and stabilization energies can be estimated from an optical spectrum.

Despite these problems, energy levels of transition metal-bearing minerals have been deduced with various degrees of accuracy from crystal field spectra. The present chapter surveys spectral measurements of many common rock-forming minerals containing individual transition metal ions. The spectra are correlated with the configurations of coordination sites accommodating the cations in the crystal structures. These spectral and structural data form the basis for estimating CFSE's of cations of the transition elements, particularly for Fe^{2+}, Cr^{3+}, Mn^{3+}, V^{3+}, Ti^{3+} and Ni^{2+}, in their host mineral structures. However, due to inherent assumptions about some of the energy levels, the estimated CFSE's are only semi-quantitative for some minerals. Nevertheless, the data are useful for *comparing* crystal chemical properties of individual cations in different minerals and assessing geochemical distributions of several transition elements in mineral assemblages, which are the topics of later chapters.

5.2 Classification of absorption spectra of minerals

Many features of absorption spectra depend on the electronic configurations of individual transition metal ions and the geometries of coordination sites in the host structures. These attributes suggest several schemes for classifying crystal field spectral measurements of minerals. First, the classification may be based on structure-types of specific mineral groups containing a variety of transition elements. This format is most commonly adopted in standard mineralogical reference texts and in recent reviews of mineral spectroscopy (Burns, 1985a,b; Rossman, 1988); it is the principal classification scheme chosen here. Secondly, a classification may be based on the spectra of individual cations in a variety of mineral structures (e.g. Rossman, 1980, 1984; Hofmeister and Rossman, 1983). This method for classifying the spectra of transition metal compounds is adopted in most chemistry texts and reviews (e.g., McClure, 1959; Dunn, 1960; Ballhausen 1962; Cotton and Wilkinson, 1988; Marfunin, 1979; Lever, 1984) and is utilized at the end of the chapter to summarize the spectral data for each cation. A third method is to classify spectra according to the coordination number and symmetry of the cations. This enables small variations of absorption bands to be correlated with changes of site distortion and bond-type of ligands surrounding a given transition metal ion in a crystal structure. This approach has been used, for example, to correlate data from X-ray absorption spectra (XANES and EXAFS: Waychunas *et al.*, 1983, 1986) and Mössbauer spectroscopy (Burns and Solberg, 1990).

A review of the literature shows that there is a vast amount of crystal field spectral data for iron, the major transition metal in silicate and oxide minerals. The focus of this chapter, therefore, is mainly on ferromagnesian silicates. However, there is also a significant amount of information for chromium-, vanadium- and manganese-bearing minerals. The data are more sporadic for other cations. The optical spectra of the transition metal-bearing minerals enable semi-quantitative estimates to be made of the relative CFSE's of Fe^{2+}, Cr^{3+}, Mn^{3+}, V^{3+}, Ti^{3+}, Ni^{2+} and Co^{2+} in many mineral structures. Note, however, that Mn^{2+} and Fe^{3+} in high-spin states acquire zero CFSE in oxides and silicates. The crystal field spectra of Mn(II) and Fe(III) minerals are described separately later in the chapter (§5.10.6 and §5.10.7).

In order to evaluate CFSE's from the spectra, correlations of the absorption bands are made with the geometries of the coordination sites. Structural data for most of the minerals described in this chapter are summarized in Appendix 7. In the following sections details about relevant crystal structures precede descriptions of the crystal field spectra of many of the mineral groups.

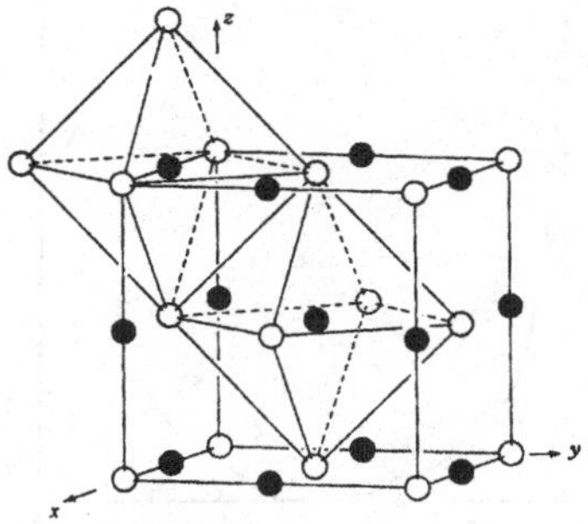

Figure 5.1 Crystal structure of periclase. Note that undistorted octahedra containing Fe^{2+} ions share all edges with other $[MgO_6]$ octahedra throughout the structure. Filled circles: cations; open circles: O^{2-} ions.

5.3 Spectra of oxides

5.3.1 Periclase

Perhaps the most highly symmetric coordination environment available to a transition metal ion in a mineral is the periclase structure illustrated in fig. 5.1. The structure consists of a cubic close-packed array of O^{2-} anions providing regular octahedral coordination sites (point symmetry O_h), all of which are occupied by Mg^{2+} ions in MgO. The $[MgO_6]$ octahedra are undistorted and Mg^{2+} ions are centrosymmetric in the coordination site with Mg–O distances all equal to 210.5 pm. Each octahedron shares all of its edges with 12 adjacent $[MgO_6]$ octahedra, so that next-nearest-neighbour Mg–Mg distances are 298 pm. Substitution of larger Fe^{2+} (r_{oct} = 78 pm) for smaller Mg^{2+} (r_{oct} = 72 pm) leads to a small expansion of the cubic unit cell parameter from a_0 = 421.2 pm in MgO to 424.0 pm in the magnesiowüstite $Mg_{0.75}Fe_{0.25}O$ (denoted as $Wü_{25}$). The Ni^{2+} (r_{oct} = 70 pm) and Co^{2+} (r_{oct} = 73.5 pm) cations are accomodated in transition metal-doped MgO crystals without much change in the unit cell parameters. Smaller trivalent cations such as Cr^{3+} (r_{oct} = 61.5 pm) and Ti^{3+} (r_{oct} = 67 pm), however, find themselves in comparatively large sites in MgO leading to the likelihood of relatively low crystal field splitting parameters compared to hexahydrated cations (cf. table 2.5). Because the cubic periclase structure contains cations in regular centrosymmetric octahedral sites, pleochroism is not observed in transition metal-doped MgO crystals; thus, positions and intensities of absorption bands in optical spectra are unaffected by the direction of propogation of polarized light through the crystals. As a result, the visible-region spectra of transition metal-bearing periclases are relatively simple and easily interpreted.

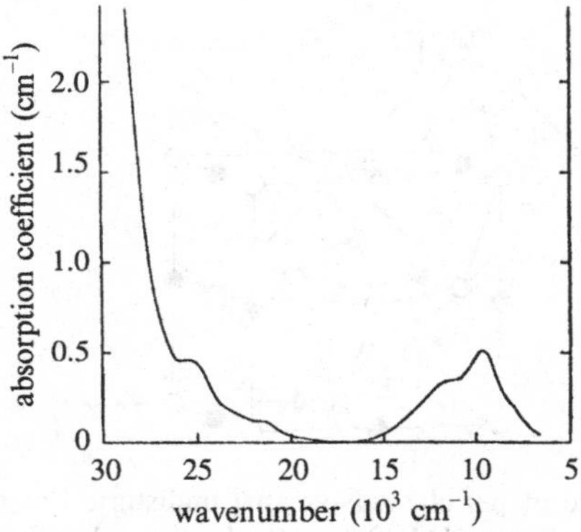

Figure 5.2 Crystal field spectrum of magnesiowüstite, $(Mg_{0.74}Fe_{0.26})O$ (from Goto *et al.*, 1980).

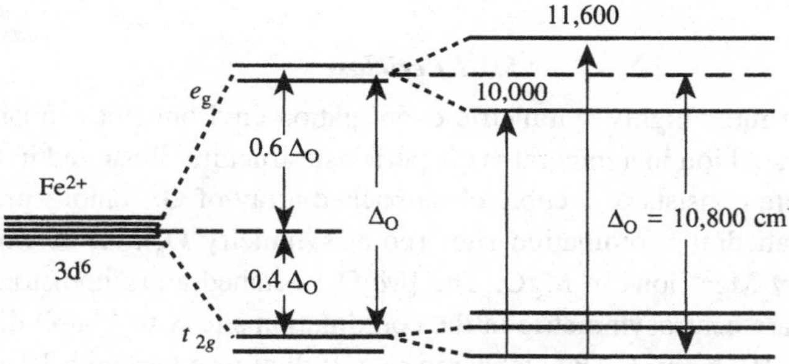

Figure 5.3 $3d$ orbital energy level diagram for Fe^{2+} ions in the periclase structure (from Burns, 1985a). Observed transitions in the spectrum of magnesiowüstite (fig. 5.2) are indicated.

For example, the absorption spectrum of magnesiowüstite $Wü_{26}$ illustrated in fig. 5.2 (Goto *et al.*, 1980) may be assigned to the ${}^5T_{2g} \rightarrow {}^5E_g$ crystal field transition representing the excitation of electrons between the Fe^{2+} $3d$ orbital energy levels shown in fig. 5.3. The two absorption maxima occurring at 10,000 cm^{-1} and 11,600 cm^{-1} are attributed to splitting of the upper level e_g orbitals during the electronic transition within Fe^{2+}. This is a consequence of the dynamic Jahn–Teller effect described in §3.9.2. Thus, the cation octahedra are not actually distorted when Fe^{2+} is in the excited ${}^5E_g$ state $(t_{2g})^3(e_g)^3$ because the life-time of the crystal field transition ($\approx 10^{-15}$ s) is considerably smaller than frequencies of vibrational modes of the $[FeO_6]$ octahedra. In other minerals discussed later, Fe^{2+} ions occur in statically distorted octahedral sites so that the upper level e_g orbitals are already separated into two discrete energy levels in the ground state. In the magnesiowüstite spectrum (fig. 5.2), peaks in the

Table 5.1. *Crystal field parameters for transition metal ions in periclase structures*

Oxide (mineral)	Cation	Δ_o (cm^{-1})	CFSE (cm^{-1})	CFSE (kJ/g.ion)	Sources of data
MgO	Ti^{3+}	11,300	4,544	−54.4	[1]
MgO	Cr^{3+}	16,200	19,440	−232.6	[1,2]
MnO (manganosite)	Mn^{2+}	9,790	0	0	[3,4]
MgO	Fe^{2+}	10,800	4,320	−51.7	[5–11]
MgO	Co^{2+}	9,000	7,200	−86.2	[12–14]
CoO	Co^{2+}	8,500	6,800	−81.4	[3]
MgO	Ni^{2+}	8,500	10,200	−122.1	[1,15–17]
NiO (bunsenite)	Ni^{2+}	8,750	10,500	−125.6	[18]

Sources of data : [1] Drickamer & Frank (1973); [2] Low (1957); [3] Pratt & Coelho (1959); [4] Keester & White (1968); [5] Blazey (1977); [6] Goto, Ahrens, Rossman & Syono (1980); [7] Manson, Gourley, Vance, Sengupta & Smith (1976); [8] Mao (1976); [9] Modine, Sonder & Weeks (1977); [10] Shankland, Duba & Woronow (1974); [11] Smith (1977); [12] Low (1958b); [13] Pappalardo, Wood & Linares (1961b); [14] Reinen (1970); [15] Low (1958a); [16] Schmiitz-Dumont, Gossling & Brokopf, (1959);[17] Pappalardo Wood & Linares (1961a); [18] Rossman, Shannon & Waring (1981).

visible region at 21,300 cm^{-1} and 26,200 cm^{-1} originate from the spin-forbidden transitions $^5T_{2g} \rightarrow {}^3T_{1g}(H)$ and $^5T_{2g} \rightarrow {}^3T_{2g}(H)$, respectively (see figs 3.6 and 3.7). The value of Δ_o for Fe^{2+} in the periclase structure estimated from the spectrum shown in fig. 5.2 is 10,800 cm^{-1}, so that the CFSE amounts to 4,320 cm^{-1}. Spectral data and crystal field parameters for other transition metals ions in the periclase structure are summarized in table 5.1.

5.3.2 Corundum

The corundum structure illustrated in fig. 5.4 is based on a hexagonal closest-packed lattice of O^{2-} anions, in which two-thirds of the octahedral sites are filled by Al^{3+} cations. Pairs of Al^{3+} ions are stacked along the *c* axis, and each Al^{3+} ion is surrounded by a trigonally distorted octahedron of O^{2-} ions (point symmetry C_3). These pairs of adjacent Al^{3+} ions share three oxygens located at the vertices of an equilateral triangle between the two cations, and Al–O distances are 196.9 pm to these three oxygens. The other three oxygens of each $[AlO_6]$ octahedron, which also surround unfilled octahedral sites above and below the Al^{3+} ion pairs parallel to the *c* axis, lie closer to each Al^{3+} and these

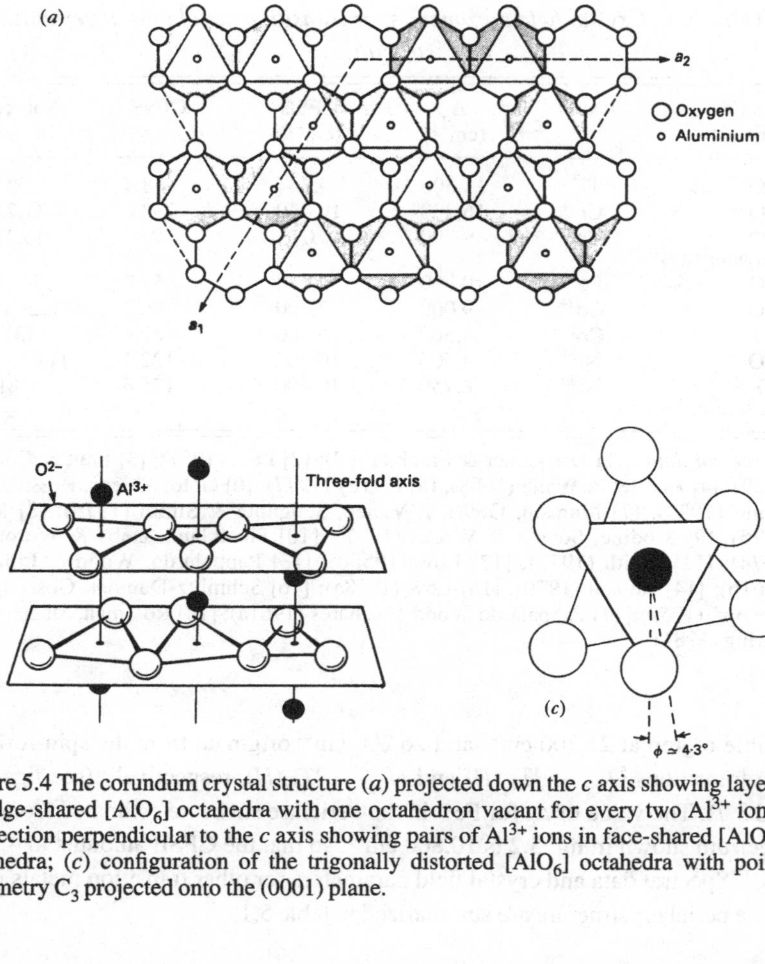

Figure 5.4 The corundum crystal structure (*a*) projected down the *c* axis showing layers of edge-shared $[AlO_6]$ octahedra with one octahedron vacant for every two Al^{3+} ions; (*b*) section perpendicular to the *c* axis showing pairs of Al^{3+} ions in face-shared $[AlO_6]$ octahedra; (*c*) configuration of the trigonally distorted $[AlO_6]$ octahedra with point symmetry C_3 projected onto the (0001) plane.

three Al–O distances are 185.6 pm. Therefore, the Al^{3+} cations are not centrally located in the $[AlO_6]$ octahedra. The Al–Al distance between pairs of face-shared $[AlO_6]$ octahedra is only 265 pm parallel to the *c* axis. Each $[AlO_6]$ octahedron also shares edges with three neighbouring $[AlO_6]$ octahedra, and the Al atoms lie in puckered planes approximately normal to the *c* axis. The Al–Al distances across edge-shared octahedra are 279 pm at about 80° to the *c* axis. Although all Al^{3+} sites are crystallographically equivalent in the corundum structure, there are two magnetically inequivalent sites when occupied by Fe^{3+} ions in hematite (§10.8.1). All cations in a given (0001) plane of edge-shared octahedra perpendicular to the *c* axis are equivalent but magnetically inequivalent to cations in adjacent (0001) planes separated by the face-shared octahedra.

Most of the cations substituting for Al^{3+} ions (r_{oct} = 53.5 pm) in transition metal-bearing Al_2O_3 have significantly larger octahedral ionic radii (e.g., Ti^{3+}, 67 pm; V^{3+}, 64 pm; Cr^{3+}, 61.5 pm; Fe^{3+}, 64.5 pm), exceptions being low-spin Co^{3+} (54.5 pm) and Ni^{3+} (56 pm). Since these trivalent cations occupy highly compressed sites in the corundum structure, absorption bands in crystal field spectra of transition metal-doped Al_2O_3 crystals generally occur at higher energies relative to the hexahydrated cations so that crystal field splittings and stabilization energies summarized in table 5.2 are generally higher in the corundum structure than for the hydrated cations (table 2.5). In addition, the occurrence of cations in acentric distorted octahedral sites in the trigonal corundum structure leads to dichroism, and hence polarization-dependencies of positions and intensities of absorption bands, in visible-region spectra (McClure, 1962).

The optical spectra of natural and synthetic transition metal-doped corundums have been extensively studied both theoretically and experimentally. Data reviewed elsewhere (Burns and Burns, 1984a; Burns, 1985a) are summarized in table 5.2. The Δ_o and CFSE values are only approximate because the ground-state splittings, δ_1, of the t_{2g} orbital group (cf. fig. 2.9) are unknown but could amount to 1,000 cm^{-1} for some cations in corundum.

5.3.2.1 Ruby.

Because of its importance as a gem and in ruby-laser technology, the spectra of Cr^{3+}–doped Al_2O_3 have been studied under a variety of experimental conditions. The optical spectra of ruby single crystals measured in polarized light at ambient temperature and atmospheric pressure illustrated earlier (fig. 4.10) show two broad intense bands near 18,000 cm^{-1} and 25,000 cm^{-1}. These represent spin-allowed transitions to the excited states $^4T_{2g}(F)$ and $^4T_{1g}(F)$ (see fig. 5.24), which are resolved into (^4E + $^4A_2$) and (^4E + $^4A_1$) states, respectively, because Cr^{3+} ions in the corundum structure occupy trigonally distorted octahedral sites that lack a centre of symmetry. The weak peaks near 14,430 cm^{-1} and 15,110 cm^{-1} originate from spin-forbidden transitions to excited states derived from $^2E_g(G)$ and $^2T_{1g}(G)$. These doublet states are further split by spin-orbit coupling interactions, the $^2E_g(G)$ state producing the ruby R_1 (14,418 cm^{-1}) and R_2 (14,447 cm^{-1}) levels so important in laser technology. The polarized spectra illustrated in fig. 4.10 portray the dichroism expected for Cr^{3+} in the trigonally distorted site, with small differences of band maxima and intensities for light polarized parallel and perpendicular to the *c* axis. In the $\mathbf{E}\|c$ spectrum, band maxima are at 18,450 cm^{-1} ($^4A_2 \rightarrow {}^4A_1(^4T_{2g})$) and 25,200 cm^{-1} ($^4A_2 \rightarrow {}^4A_2(^4T_{1g})$), while in the $\mathbf{E}\perp c$ spectrum they occur at 18,000 cm^{-1} ($^4A_2 \rightarrow {}^4E(^4T_{2g})$) and 24,400 cm^{-1} ($^4A_2 \rightarrow {}^4E(^4T_{1g})$). Strictly speaking, the $^4A_2 \rightarrow {}^4A_1$ transition in the $\mathbf{E}\|c$ spectrum is symmetry-forbidden by group theoretical

Table 5.2. *Crystal field parameters for transition metal ions in corundum structures*

Oxide (mineral)	Positions of absorption bands (cm^{-1}) (**E**∥*c*; **E**⊥*c*)	Δ_o (cm^{-1})	CFSE (cm^{-1})	CFSE (kJ/g.ion)	Sources of data
Ti^{3+} Al_2O_3	18,450 (∥,⊥); 20,300 (∥,⊥)	19,100	8,870	−106.2	[1–3]
V^{3+} Al_2O_3	17,510 (∥); 24,930 (∥)	18,730	14,980	−179.3	[1,4–7]
Cr^{3+} Al_2O_3 (ruby)	18,450 (∥); 25,200 (∥) 18,000(⊥); 24,400(⊥)	18,150	21,780	−260.6	[1,7]
Cr_2O_3 (eskolaite)	16,670; 21,750	16,670	19,920	−238.4	[8–11]
Mn^{4+} Al_2O_3	21,300	21,300	25,560	−305.9	[12]
Mn^{3+} Al_2O_3	18,750 (⊥); 20,600 (∥)	19,370	11,620	−139.1	[1]
Fe^{3+} Al_2O_3 (yellow sapphire)	9,450 ; 14,350 (⊥) 18,700 (∥, ⊥); 22,220 (∥, ⊥) 25,600 (∥, ⊥); 26,700 (∥, ⊥)	14,300	0	0	[3,13–20]
Fe_2O_3 (hematite)	11,600; 18,500; 22,200	14,000	0	0	[21,22]
Co^{3+} Al_2O_3	15,740 (∥); 23,170 (∥) 15,380 (⊥); 22,800 (⊥)	17,300	24,220*	−290.0*	[1,23]
Co^{2+} Al_2O_3	9,100(⊥); 21,400 (∥) 22,800 (∥, ⊥); (7,700; 16,000; 17,200; 18,300: spin-forbidden)	12,300	9,840	−117.8	[23]
Ni^{3+} Al_2O_3	16,800 (⊥); ~20,000 (∥) 16,300 (⊥); ~20,000 (⊥)	18,000	14,400*	−172.3*	[1,23]
Ni^{2+} Al_2O_3	18,000(∥, ⊥); 25,000 (∥, ⊥) (16,200; 22,700: spin-forbidden)	10,700	12,840	−153.6	[23]

* Low-spin cations. The CFSE's for these cations are corrected for electron-pairing energies which approximate the Δ_o values.

Sources of data : [1] McClure (1962); [2] Townsend (1968); [3] Eigenmann, Kurtz & Gunthard (1972); [4] Macfarlane (1964); [5] White, Roy & Crichton (1967); [6] Rahman & Runciman (1971); [7] Macfarlane (1963); [8] Neuhaus (1960); [9] McClure (1963); [10] Poole (1964); [11] Reinen (1969); [12] Geschwind, Kisliuk, Klein, Remeika & Wood (1962); [13] Lehmann & Harder (1970); [14] Lehmann & Harder (1971); [15] Faye (1971b); [16] Eigenmann & Gunthard (1971); [17] Eigenmann, Kurtz & Gunthard (1971); [18] Ferguson & Fielding (1971); [19] Ferguson & Fielding (1972); [20] Krebs & Maisch (1971); [21] Morris, Agresti, Lauer, Newcomp, Shelfer & Murali (1989); [22] Sherman & Waite (1985); [23] Muller & Gunthard (1966).

selection rules (McClure, 1962); the presence of the absorption band at 18,450 cm^{-1} is attributed to vibronic coupling (§3.7.1.1). Such vibronic coupling is accentuated by increased temperature, accounting for increased intensity of the crystal field bands at elevated temperatures (Poole, 1964; McClure, 1959; Parkin and Burns, 1980).

5.3.3 Spinels

A cubic close-packed array of oxygen ions exists in the spinel structure, in which one-eighth of the tetrahedral voids and one-half of the octahedral interstices are occupied by cations. The arrangement of the atoms is such that perpendicular to each 3-fold axis, layers occupied by cations in octahedral coordination alternate with other layers in which the tetrahedral and octahedral sites are filled in the ratio of 2:1. The crystal structure illustrated earlier (fig. 4.18) shows chains of edge-shared octahedra running along [110] directions interspersed with isolated tetrahedral sites. The tetrahedral sites are regular (point symmetry T_d), whereas the octahedral sites are trigonally distorted (point symmetry D_{3d}).

Optical spectral measurements of natural and synthetic oxide spinel minerals include spinel itself, $MgAl_2O_4$, containing tetrahedral (Fe^{2+}, Cr^{2+}) and octahedral (Cr^{3+}, V^{3+}) ions (Poole, 1964; Gaffney, 1973; Sviridov *et al.*, 1973; Mao and Bell, 1975a; Shankland *et al.*, 1974). An absorption band at 4,830 cm^{-1} originating from tetrahedral Fe^{2+} ions results in a CFSE of 2,900 cm^{-1}. Octahedrally coordinated Cr^{3+} in a lunar chromite absorbs at 17,700 cm^{-1} and 22,400 cm^{-1} (Mao and Bell, 1975a) indicating an octahedral CFSE of 21,240 cm^{-1}. Spectral measurements have also demonstrated the existence of tetrahedrally coordinated Cr^{2+} ions in natural chromites (Mao and Bell, 1975a). Synthetic Mg–Cr-spinels yielded absorption features at 6,250 and 6,670 cm^{-1} attributed to tetrahedral Cr^{2+} ions (Ulmer and White, 1966; Greskovitch and Stubican, 1966).

5.4 Spectra of orthosilicates

5.4.1 Garnet group

The garnet structure hosts transition metal ions in sites having three different coordination symmetries: eight-fold (triangular dodecahedron approximating a distorted cube, point symmetry D_2), six-fold (trigonally distorted octahedron, point symmetry C_{3i}) and four-fold (regular tetrahedron, point symmetry T_d) (Novak and Gibbs, 1971). In silicate garnets, site occupancies of the first series transition elements mainly involve divalent cations in the larger eight-coordinated sites and trivalent cations in the smaller octahedral sites (Meagher, 1982), although octahedral Fe^{2+} and tetrahedral Fe^{3+} ions also occur in titaniferous melanites and schorlomites (Burns, 1972a; Schwartz *et al.*, 1980). In the structures of pyrope and almandine, the eight-fold coordination polyhedra approximate distorted cubes shown in fig. 5.5, in which there are two sets of four identical metal–oxygen distances. In almandine, the Fe^{2+}–O distances are 222.0 pm and 237.8 pm (mean 229.9 pm), while in pyrope the Mg–O distances

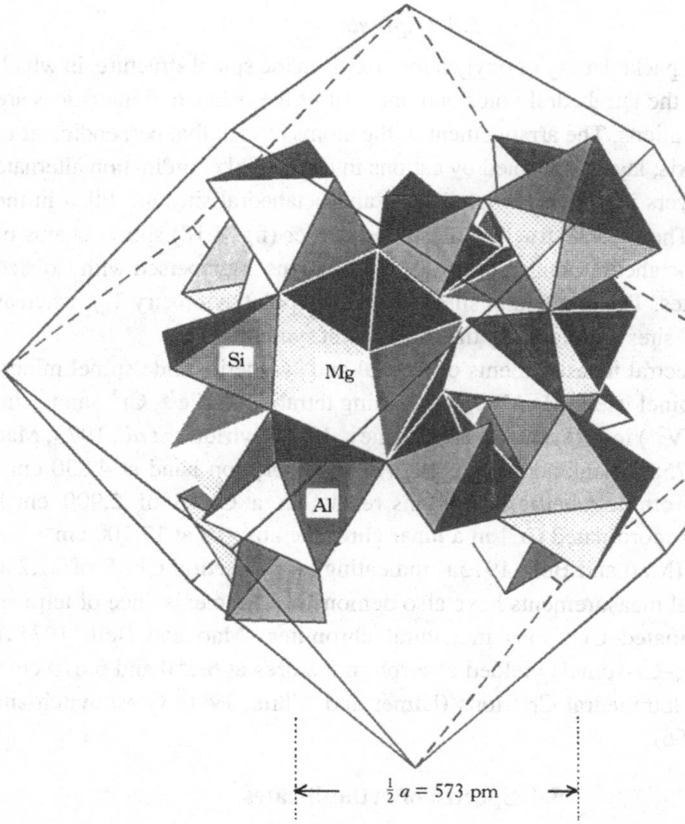

Figure 5.5 The crystal structure of garnet (pyrope) projected into a dodecahedron. Note the configuration of the eight-fold coordinated Mg sites (distorted cube) which share edges with $[AlO_6]$ octahedra and $[SiO_4]$ tetrahedra.

are 219.7 pm and 234.3 pm (mean 227.0 pm). The divalent cations are not located at the centre of symmetry of the deformed cube site. The octahedral site in the garnet structure is centrosymmetric and only slightly distorted, so that all six metal–oxygen distances are identical. They range from Al^{3+}–O distances of 188.7 pm and 192.4 pm in pyrope and grossular, respectively, to Fe^{3+}–O = 202.4 pm in andradite, Cr^{3+}–O = 198.5 pm in uvarovite, V^{3+}–O = 198.8 in goldmanite, and $(Al,Cr)^{3+}$–O = 190.5 in chrome-pyrope.

Because garnet belongs to the cubic system, crystal field spectra are identical for all orientations of polarized light. The absorption spectra of garnets of the pyrope–almandine series have been discussed extensively (e.g., Grum-Grzhimailo, 1954; Clark, 1957; Manning, 1967a,b; Burns, 1968b; Lyubutin and

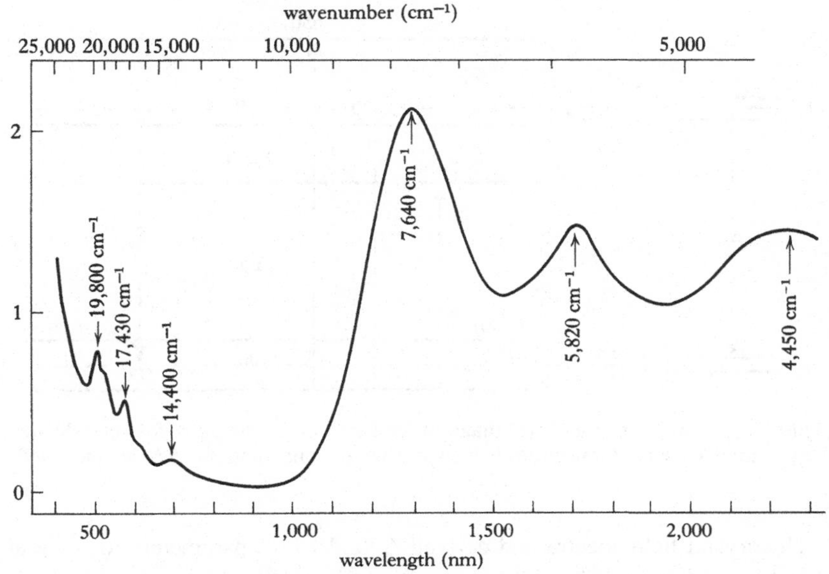

Figure 5.6 Crystal field spectrum of almandine. The garnet with 70 mole per cent $Fe_3Al_2Si_3O_{12}$ is from Fort Wrangell, Alaska.

Dodokin, 1970; White and Moore, 1972; Runciman and Sengupta, 1974; Huggins, 1975; Marshall and Runciman, 1975; Evans and Sergent, 1975; Amthauer, 1976; Newman *et al.*, 1978; Smith and Langer, 1983). The spectrum illustrated in fig. 5.6 for an almandine with 70 mole per cent $Fe_3Al_2(SiO_4)_3$, contains three bands near 4,400 cm^{-1}, 5,800 cm^{-1} and 7,600 cm^{-1}, corresponding to transitions to the upper t_{2g} orbital energy levels (fig. 2.9), together with a number of spin-forbidden peaks in the visible region. The splitting of the lower-level e_g orbitals, δ_3, cannot be obtained from the visible to near-infrared spectra, but was estimated to be about 1,100 cm^{-1} from calculations based on temperature variations of Mössbauer spectra of almandine (Lyubutin and Dodokin, 1970; Huggins, 1975). The energy level diagram for Fe^{2+} in the distorted cubic site in almandine shown in fig. 5.7 leads to an estimated value of 5,420 cm^{-1} for Δ_c and a CFSE of 3,530 cm^{-1}, calculated from $(0.6\Delta_c + 0.5\delta_3)$. The spectrum of a pyrope with 20 mole per cent $Fe_3Al_2(SiO_4)_3$, having absorption bands at 4,540 cm^{-1}, 6,060 cm^{-1} and 7,930 cm^{-1}, yields larger values, Δ_c = 5,630 cm^{-1} and CFSE = 3,650 cm^{-1}, which are consistent with the smaller metal–oxygen distances in pyrope. Ranges of Δ_c and CFSE of Fe^{2+} in the garnet structure over the composition range Alm_1–Alm_{100}, are estimated to be

$$\Delta_c = 5{,}710 \text{ to } 5{,}300 \text{ cm}^{-1}; \text{ and CFSE} = 3{,}700 \text{ to } 3{,}460 \text{ cm}^{-1}. \quad (5.1)$$

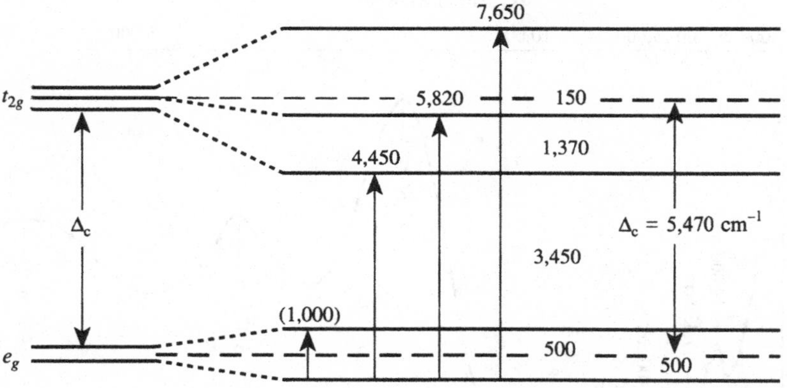

Figure 5.7 3*d* orbital energy level diagram for Fe^{2+} ions in the eight-fold coordinated site in garnet. Observed transitions in the spectrum of almandine (fig. 5.6) are indicated.

The crystal field spectra and derived Δ_o and CFSE parameters for several garnets containing octahedrally coordinated trivalent transition metal ions are summarized in table 5.3. The values of Δ_o and CFSE reflect the variations of metal–oxygen distances in the garnet structures.

5.4.2 Olivine group

5.4.2.1 Background

Olivines of the forsterite–fayalite series, $(Mg,Fe)_2SiO_4$, occurring in igneous rocks, granulite facies metamorphic terranes, beach sands and sedimentary placer deposits, meteorites and many lunar samples, have been the focus of numerous spectroscopic investigations (e.g., Clark, 1957; Grum-Grzhimailo, 1958, 1960; Farrell and Newnham, 1965; White and Keester, 1966; Burns, 1966b, 1968b, 1970, 1974; Bell and Mao, 1969; Mao and Bell, 1971; Runciman *et al.*, 1973a, 1974; Smith and Langer, 1982a,b; King and Ridley, 1987). The popularity of this orthorhombic silicate to mineral spectroscopists may be attributed to several factors: first, the availability of large euhedral, homogeneous crystals such as gem peridot from which highly polished, crystallographically oriented sections may be easily cut; second, the importance of olivine (α-phase) and denser isochemical structure-types in the Earth's Upper Mantle (e.g., spinel or γ-phase, and the 'modified spinel' or β-phase); and third, the influence of olivines on geophysical properties of the Earth's interior, including radiative transfer of heat, electrical conductivity, and tectonic and seismic features associ-

Table 5.3 *Crystal field parameters for transition metal ions in garnets*

Garnet	Positions of absorption bands (cm^{-1})	Δ_o (cm^{-1})	CFSE (cm^{-1})	CFSE (kJ/g.ion)	Sources of data
		V^{3+} garnets			
pyrope	18,500; 14,800	17,500	14,000	−167.5	[1,2]
grossular (tsavorite)	16,500; 23,800	17,670	14,140	−169.2	[1–3]
goldmanite	16,000; 22,500	17,100	13,680	−163.7	[1,2]
		Cr^{3+} garnets			
pyrope	17,606; 24,272	17,606	21,130	−252.8	[4]
pyrope	17,450; 24,240	17,450	20,940	−250.6	[5]
pyrope	17,790; 24,150	17,790	21,350	−255.5	[5]
grossular	16,500; 22,290	16,500	19,800	−236.9	[5]
uvarovite	16,667; 22,727	16,667	20,000	−239.3	[6]
uvarovite/9.7GPa	17,820; 23,840	17,820	21,385	−255.9	[6]
uvarovite	16,529; 24,814	16,529	19,835	−237.3	[7]
uvarovite	16,600; 23,100	16,600	19,920	−238.4	[8]
uvarovite	16,260; 22,720	16,260	19,510	−233.5	[6]
		Mn^{3+} garnets			
grossular	17,000; 20,400	18,150	10,890	−130.3	[9]
calderite	19,200; 21,000	19,800	11,880	−142.2	[10]
		Fe^{3+} garnets			
grossular	13,111; 22,865; 23,592; 27,040	12,800	0	0	[4]
andradite	12,453; 16,650; 22,701; 24,000; 27,000	12,600	0	0	[4,11]

Sources of data : [1] Schmetzer & Ottemann (1979); [2] Schmetzer (1982); [3] Schmetzer, Berdesinski & Traub (1975); [4] Moore & White (1972); [5] Amthauer (1976); [6] Abu-Eid (1976); [7] Neuhaus (1960); [8] Manning (1969c); [9] Frentrup & Langer (1981); [10] Langer & Lattard (1984); [11] Parkin & Burns (1980).

ated with the olivine → spinel phase change. These aspects of olivine spectra are discussed in chapter 9.

5.4.2.2 Olivine crystal structure

From a crystal chemical standpoint, the olivine structure is particularly interesting because it contains divalent cations in two distinct six-fold coordination sites having contrasting symmetries. Consequently, crystal field spectra of several transition metal-bearing olivines have been studied in attempts to detect cation ordering, including evidence for site preferences of Mn^{2+} (§4.5) and Ni^{2+} (Hu *et al.*, 1990), and to estimate CFSE data for Fe^{2+} and Ni^{2+} in each site in the olivine structure.

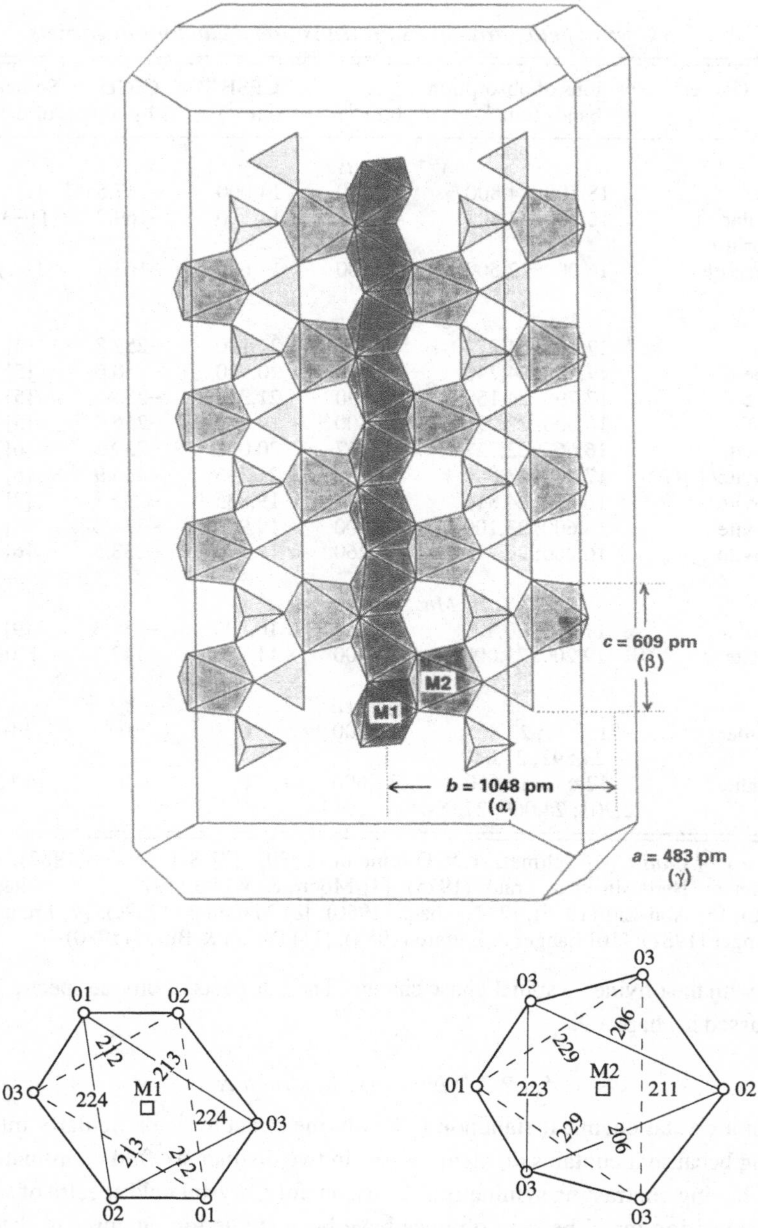

Figure 5.8 The crystal structure of olivine. (*a*) The structure projected onto (100) showing serrated chains of octahedra running parallel to the *c* axis; (*b*) oxygen coordination polyhedra projected about the M1 and M2 positions. Metal–oxygen distances in each coordination site are indicated. Cell parameters and interatomic distances are for fayalite (from Smyth & Bish, 1988).

Numerous refinements of the crystal structure of olivines have been made (see Brown, 1982; Princivalle, 1990). Like garnet, olivine contains discrete [SiO_4] tetrahedra so that, in contrast to pyroxenes discussed later, all oxygen ligands are bonded to just one Si atom in the olivine structure. These non-bridging oxygen atoms surround cations in M1 and M2 positions, coordination polyhedra about which form serrated bands of edge-shared octahedra running parallel to the *c* axis, as illustrated in fig. 5.8. The two six-coordinated sites of olivine are both distorted from octahedral symmetry. Projections of the coordination polyhedra illustrated in fig. 5.8 show that the slightly smaller centrosymmetric M1 site is elongated along the O_3–O_3 axis, which is not perpendicular to the equatorial plane, itself showing rhombic distortion. Although the point symmetry of the M1 position is C_i, the M1 coordination site crudely approximates a tetragonally elongated octahedron. The slightly larger M2 coordination site is also distorted and is acentric. The point symmetry of the M2 position is C_s. However, based on metal–oxygen distances in the M2 site, three of which are significantly longer than the other three, the M2 polyhedon crudely resembles a trigonally distorted octahedron such as that in the corundum structure. Crystal structure refinements of a series of Mg–Fe olivines (Birle *et al.*, 1968; Brown, 1982) show that average metal–oxygen distances in fayalite (M1 site = 216 pm; M2 site = 219 pm) decrease in peridot, Fa_{10} (M1 site = 210 pm; M2 site = 214 pm). In contrast to orthopyroxenes, which show pronounced Mg–Fe^{2+} cation ordering (§6.7.2), the distribution of Fe^{2+} ions between the olivine M1 and M2 site is more random. The site occupancies, however, are influenced slightly by temperature, composition and oxygen fugacity (Princivalle, 1990; Ottonello *et al.*, 1990).

5.4.2.3 Crystal field spectra of Fe^{2+} olivines

Differences between the two cation environments in the olivine structure produce contrasting crystal field transitions in Fe^{2+} ions which may be readily distinguished in spectra measured in the near-infrared region, such as those for fayalite illustrated in fig. 5.9. Because olivine is an orthorhombic mineral, it produces three different spectrum profiles when light polarized along each of the three crystallographic directions is transmitted through oriented single crystals. The α and β polarized spectra contain broad bands with absorption maxima near 11,000 cm^{-1} and 8,000 cm^{-1}. The γ spectrum consists of an intense band centred at 9,300 cm^{-1} in fayalite, with a prominent shoulder around 8,000 cm^{-1}. Several weak, sharp peaks occur in the visible region of each spectrum, with intense absorption edges emerging from the ultraviolet region, particularly in β-polarized light. Similar spectrum profiles are shown for other olivines in the forsterite–fayalite series which is indicative of only

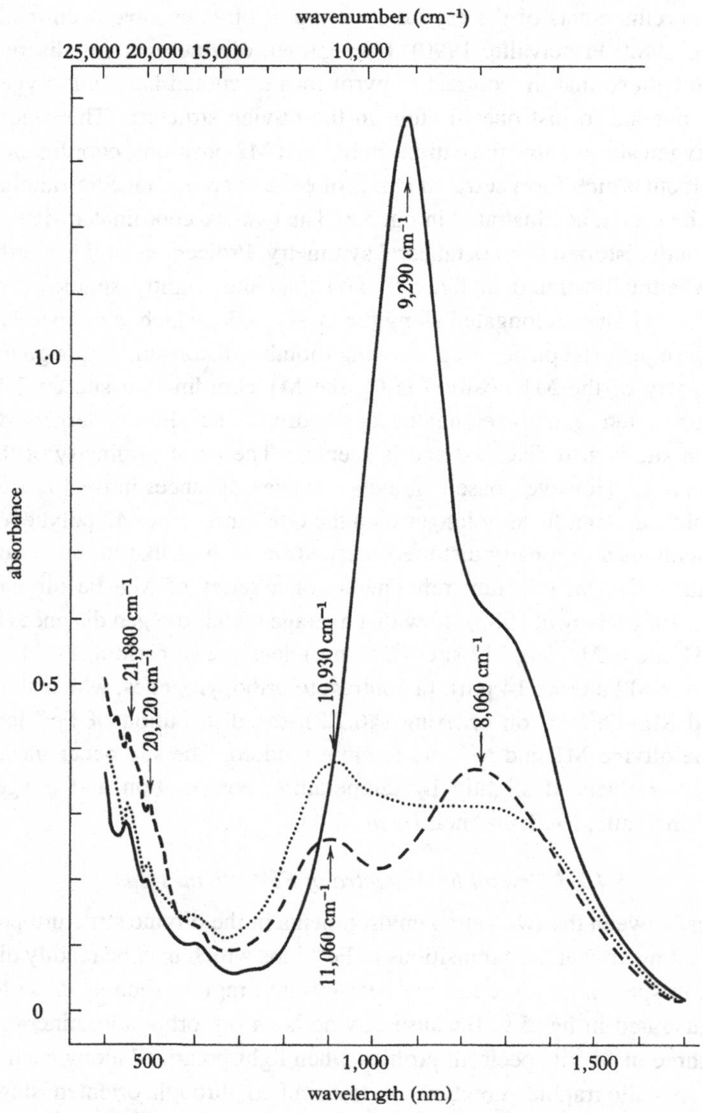

Figure 5.9 Polarized absorption spectra of fayalite. ······· α spectrum; – – – – β spectrum; —— γ spectrum (thickness of crystals: 80 μm). [Optic orientation: α = *b*; β = *c*; γ = *a*.] The fayalite, with 96 mole percent Fe_2SiO_4, is from Rockport, Massachusetts.

minor Mg^{2+}–Fe^{2+} ordering in the M1 and M2 sites, in marked contrast to orthopyroxene spectra described later (§5.5.4). However, absorption maxima of all bands show compositional variations and shift to shorter wavelengths

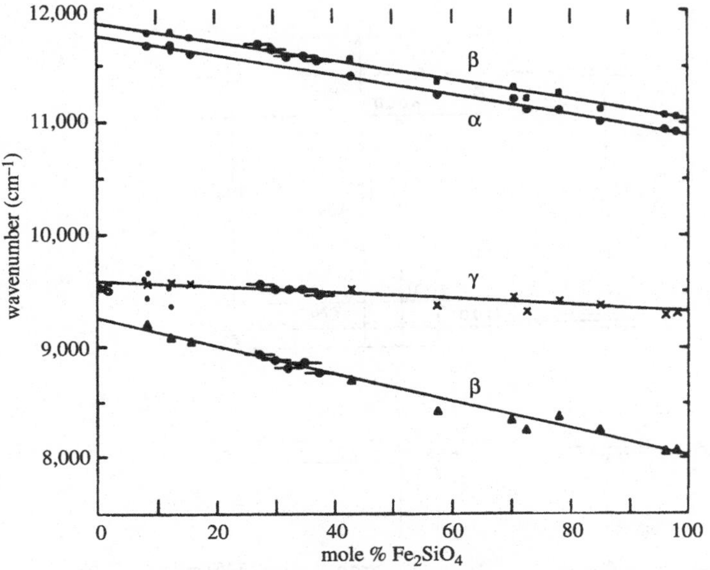

Figure 5.10 Compositional variations of peak maxima in the polarized spectra of Mg–Fe olivines (from Burns *et al.*, 1972b).

with decreasing iron content (Burns, 1970; Hazen *et al.*, 1977). Figure 5.10 shows that these shifts are linear with respect to changing Mg^{2+}–Fe^{2+} compositions when expressed in wavenumber units (cm^{-1}), reflecting diminished metal–oxygen distances of the M1 and M2 sites in Mg-rich olivines. Note that the extreme polarization dependence of the Fe^{2+} crystal field bands in the near-infrared region is not detected by the eye, so that the pleochroism of olivines is invisible (§4.6.3). The weakly pleochroic green colouration of olivines results from absorption of red radiation as absorption bands intensify in the 800–1,700 nm region (fig. 5.9).

Energy level diagrams for Fe^{2+} ions in the M1 and M2 sites derived from the spectra of fayalite (fig. 5.9) are shown in fig. 5.11. The absorption bands around 11,000 cm^{-1} and 8,000 cm^{-1} are assigned to the split upper levels of the e_g orbitals of Fe^{2+}/M1 site cations, while the band centred near 9,300 cm^{-1} represents a transition to the e_g orbitals of Fe^{2+}/M2 site ions. Debate has centred on assignments of absorption bands in olivine spectra (Runciman *et al.*, 1973a, 1974; Burns, 1970, 1974; Hazen *et al.*, 1977b), which has now been resolved (Gonschorek, 1986) in favour of the energy level diagrams shown in fig. 5.11. Uncertainties exist, however, about energy splittings of the lower-level t_{2g} orbitals for Fe^{2+} in the two sites. From measurements of the temperature dependency of the Mössbauer quadrupole splitting parameter of fayalite,

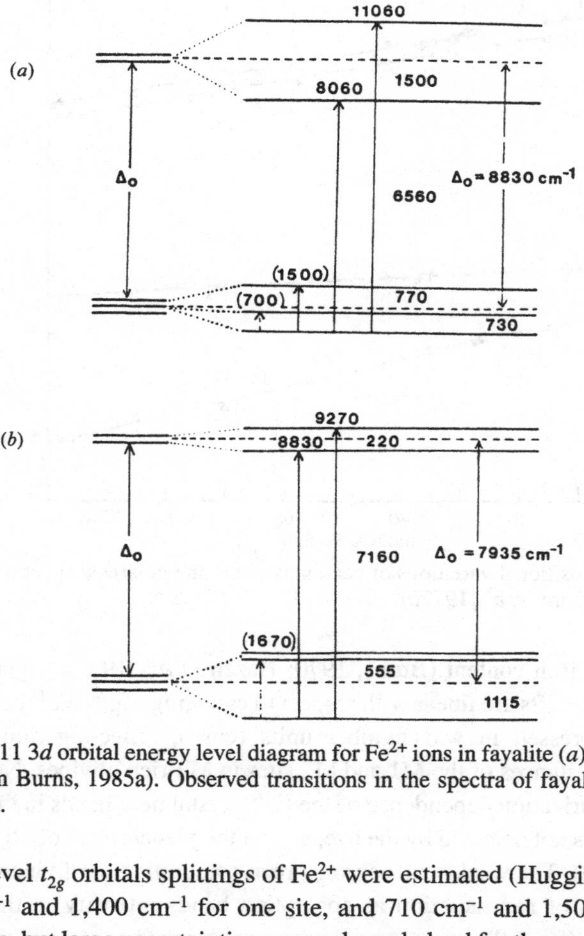

Figure 5.11 3*d* orbital energy level diagram for Fe^{2+} ions in fayalite (*a*) M1 site; (*b*) M2 site (from Burns, 1985a). Observed transitions in the spectra of fayalite (fig. 5.9) are indicated.

lower-level t_{2g} orbitals splittings of Fe^{2+} were estimated (Huggins, 1976) to be 620 cm^{-1} and 1,400 cm^{-1} for one site, and 710 cm^{-1} and 1,500 cm^{-1} for the other site, but large uncertainties were acknowledged for these values. The values of 700 cm^{-1} and 1,500 cm^{-1} were used to construct the energy level diagram for the M1 site (fig. 5.11*a*). The polarized absorption spectra of peridot (Fa_{10}) indicated an absorption band at 1,668 cm^{-1} (Runciman *et al.*, 1973a), the intensity of which suggested an origin from Fe^{2+} ions in the acentric M2 sites. The value of 1,670 cm^{-1} was used as one of the lower-level t_{2g} orbital splittings for the M2 site. Transitions to the upper-level e_g orbitals of Fe^{2+} ions in olivine M2 sites were also suggested to occur at 9,540 cm^{-1} and 9,100 cm^{-1} (Runciman *et al.*, 1973a), indicating a small splitting of about 440 cm^{-1}. This value was used to construct the energy level diagram for the M2 site (fig. 5.11*b*).

Compositional variations of the olivine spectra across the forsterite–fayalite series (fig. 5.10) indicate ranges of Δ_o and CFSE of Fe^{2+} ions to be

$$\text{M1 site: } \Delta_o = 9{,}760 \text{ to } 8{,}830 \text{ cm}^{-1}; \text{ CFSE} = 4{,}700 \text{ to } 4{,}250 \text{ cm}^{-1}, \quad (5.2)$$

$$\text{M2 site: } \Delta_o = 8{,}250 \text{ to } 7{,}930 \text{ cm}^{-1}; \text{ CFSE} = 4{,}420 \text{ to } 4{,}280 \text{ cm}^{-1}, \quad (5.3)$$

for the composition range Fa_1 to Fa_{100}. Note that in the first edition of this book, an error occurred in computed data for the Fe^{2+}/M1 site cations (Walsh *et al.*, 1974; Ottonello *et al.*, 1990). Subsequent estimates of the Δ_o and CFSE values have varied (Burns and Sung, 1978; Burns, 1985a) due mainly to different assumptions made about splittings of lower-level t_{2g} energy levels in Fe^{2+} ions in the two cation sites of olivine.

For the pseudo-tetragonally elongated M1 octahedral site in the olivine structure, an alternative method for obtaining the CFSE of Fe^{2+} ions in this site is to evaluate the ligand field splitting parameters in the equatorial plane and along the axial direction using eqs (3.14) to (3.16). The crystal field spectral data for the fayalite M1 site (see fig. 5.11*a*) yield values for 10 $Dq_{(ax)}$ and 10 $Dq_{(eq)}$ of 8,172 cm^{-1} and 9,327 cm^{-1}, respectively. The mean 10 Dq value of 8,942 cm^{-1} and the CFSE of 4,133 cm^{-1} compare favourably with Δ_o = 8,830 cm^{-1} and CFSE = 4,250 cm^{-1} for Fe^{2+} in the fayalite M1 sites determined by the baricentric method (fig. 5.11*a*).

5.4.2.4 Crystal field spectra of Ni^{2+} olivines

Crystal field spectra have been described for a number of individual Mg^{2+}–Ni^{2+} olivine compositions (e.g., Wood, 1974; Rossman *et al.*, 1981; Rager *et al.*, 1988; Hu *et al.*, 1990). When Ni^{2+} ions are present in a regular octahedral site, the lowest energy crystal field transition, $^3A_{2g} \rightarrow {}^3T_{2g}$ (see fig. 5.29), provides a direct measure of Δ_o from which the CFSE of Ni^{2+} is estimated as $1.2\Delta_o$. However, in the two low-symmetry distorted octahedral sites in the olivine structure, the degeneracy of the $^3T_{2g}$ crystal field state, as well as other higher energy states of Ni^{2+}, is removed so that several absorption bands appear in the polarized absorption spectra, raising problems of peak assignments to Ni^{2+} ions in the olivine M1 and M2 sites. From diffuse reflectance spectra of powdered Mg–Ni olivines, CFSE values of Ni^{2+} were estimated to be 9,550 cm^{-1} (M1 site) and 8,990 cm^{-1} (M2 site) for the forsterite $(Mg_{0.95}Ni_{0.05})_2SiO_4$ (Wood, 1974). Polarized spectra of single crystals of the forsterite $(Mg_{0.98}Ni_{0.02})_2SiO_4$ (Rager *et al.*, 1988), yielded a CFSE of 10,800 cm^{-1} for Ni^{2+} assumed to occupy the M1 site. Polarized spectra of single crystals of synthetic Ni_2SiO_4 and $(Ni_{0.57}Mg_{0.43})_2SiO_4$ (Rossman *et al.*, 1981), led to estimated CFSE's of Ni^{2+} in the M1 and M2 sites of Mg–Ni olivines in the ranges 9,620 to 9,430 cm^{-1} and 8,100 to 8,360 cm^{-1}, respectively (Burns, 1985a).

A detailed investigation of the polarized spectra of a wide compositional

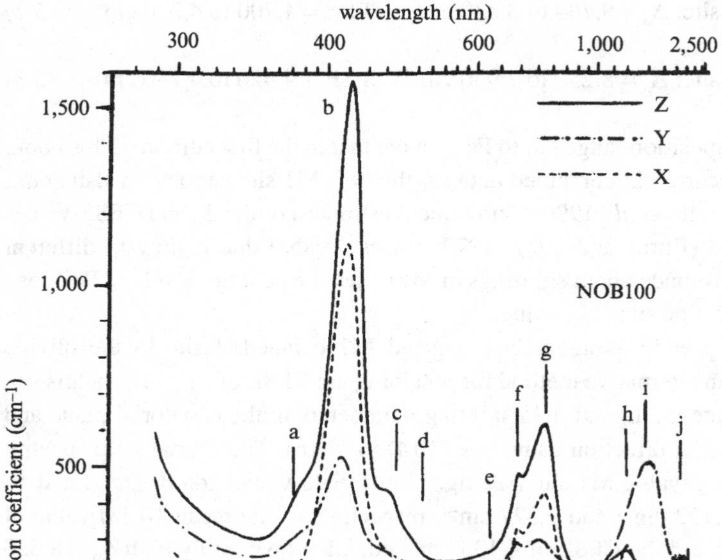

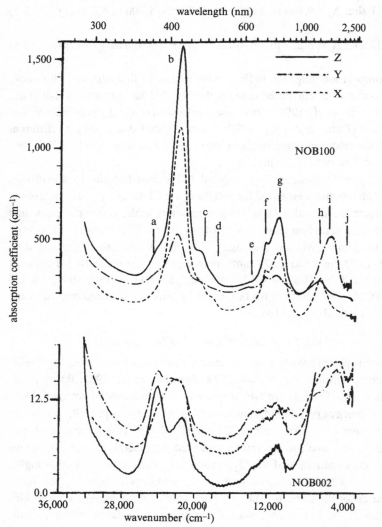

Figure 5.12 Polarized absorption spectra of Ni^{2+}-bearing olivines (modified from Hu *et al.*, 1990). Upper set of spectra: Ni_2SiO_4; lower set: $(Mg_{0.98}Ni_{0.02})_2SiO_4$. Band positions are listed in table 5.4. Note the increased intensity of bands a, e and h with decreasing Ni content originating from Ni^{2+} ions enriched in the centrosymmetric M1 sites.

range of synthetic $(Mg_{1-x}Ni_x)_2SiO_4$ olivine crystals (where x_{Ni} = 1.00, 0.75, 0.31, 0.07 and 0.04) demonstrated that the spectrum profiles are strongly influenced by cation ordering and temperature (Hu *et al.*, 1990), the two effects being manifestations of the site symmetries of Ni^{2+} ions in the M1 and M2 coordination polyhedra.

Table 5.4 *Assignments of absorption bands in the crystal field spectra of Mg^{2+}–Ni^{2+} olivines**

Peak	Position (cm^{-1})	Assignment	Site
a	25,100–26,200	$^3A_2 \rightarrow {}^3T_1\ (P)$	M1
b	23,400–24,000	$^3A_2 \rightarrow {}^3T_1\ (P)$	M2
c	21,000–21,800	$^3A_2 \rightarrow {}^1A_1\ (G)$	M1
d	20,000–20,800	$^3A_2 \rightarrow {}^1T_2\ (D)$	M2
e	15,400–15,900	$^3A_2 \rightarrow {}^3T_1\ (F)$	M1
f	13,900–14,300	$^3A_2 \rightarrow {}^1E\ (D)$	M2
g	12,600–13,000	$^3A_2 \rightarrow {}^3T_1\ (F)$	M2
h	8,700–9,000	$^3A_2 \rightarrow {}^3T_2\ (F)$	M1
i	7,200–8,000	$^3A_2 \rightarrow {}^3T_2\ (F)$	M2
j	6,000–7,000	$^3A_2 \rightarrow {}^3T_2\ (F)$	M1

* Modified from Hu, Langer & Boström (1990).

Reproduced in fig. 5.12 are polarized crystal field spectra of two end-member Mg^{2+}–Ni^{2+} olivines with compositions corresponding to pure Ni_2SiO_4 (liebenbergite) and forsterite $(Mg_{0.98}Ni_{0.02})_2SiO_4$. The spectra clearly show pronounced compositional variations of peak intensities attributable to changing Ni^{2+} occupancies of the acentric M2 sites of the Mg–Ni olivines and the strong cation ordering of Ni^{2+} in the centrosymmetric M1 sites. Intensities of crystal field transitions within cations occupying centrosymmetric sites such as the olivine M1 site are most affected by the temperature-sensitive vibronic coupling mechanism (§3.7.1.1). Thus, measurements of optical spectra at low temperatures facilitated assignments of absorption bands in the crystal field spectra of the Ni^{2+} olivines given in table 5.4 that are based on the Tanabe–Sugano energy level diagram for a $3d^8$ cation illustrated later in fig. 5.29.

Crystal field parameters derived from the polarized absorption spectra of Mg–Ni olivines were expressed by the empirical equations (Hu *et al.*, 1990)

$$\text{M1 site: } \Delta_o = [9{,}500 + (461\ x_{Ni})];\ \text{CFSE} = [11{,}384 + (574\ x_{Ni})], \quad (5.4)$$

$$\text{M2 site: } \Delta_o = [7{,}368 + (514\ x_{Ni})];\ \text{CFSE} = [8{,}828 + (621\ x_{Ni})], \quad (5.5)$$

where x_{Ni} is the mole fraction of Ni_2SiO_4.

The value of Δ_o, and hence the calculated CFSE, for Ni^{2+} ions in the olivine M1 sites appears to be anomalously high compared to values acquired in other octahedral sites in oxide structures, including bunsenite and MgO (table 5.1), Ni_2SiO_4 spinel (§5.4.3), and other phases listed later (table 5.19). The discrepency results from incorrect assignments of bands i and j listed in table 5.4

based on peak positions in polarized spectra of Mg–Fe^{2+} olivines. The crystal field spectra of fayalite show that the Fe^{2+}/M2 site band at 9,300 cm^{-1} is located between the Fe^{2+}/M1 site bands at 11,000 cm^{-1} and 8,000 cm^{-1}. By analogy, in the spectra of Mg–Ni^{2+} olivines, Ni^{2+}/M2 site cations are responsible for band i which lies between bands h and j originating from Ni^{2+}/M1 site cations. These re-assignments of absorption bands in Mg–Ni^{2+} olivine spectra result in revised ranges of Δ_o and CFSE for the forsterite–liebenbergite series

$$\text{M1 site: } \Delta_o = 7{,}430 \text{ to } 7{,}780 \text{ cm}^{-1}; \text{ CFSE} = 8{,}920 \text{ to } 9{,}340 \text{ cm}^{-1}, \quad (5.4a)$$

$$\text{M2 site: } \Delta_o = 7{,}210 \text{ to } 7{,}490 \text{ cm}^{-1}; \text{ CFSE} = 8{,}660 \text{ to } 8{,}980 \text{ cm}^{-1}. \quad (5.5a)$$

These values still conform with other estimates (Wood, 1974; Burns, 1985a; Hu *et al.*, 1990) indicating that Ni^{2+} ions receive a higher CFSE in the olivine M1 site, accounting for the strong cation ordering of Ni^{2+} ions in this slightly smaller coordination site (§6.7.1.2).

The interelectronic repulsion Racah *B* parameters were also derived from the crystal field spectra of Mg–Ni olivines. (Hu *et al.*, 1990). The compositional variations given by

$$B_{\text{Ni(M1)}} = 882 - 127\, x_{\text{Ni}}; \text{ and } B_{\text{Ni(M2)}} = 950 - 92\, x_{\text{Ni}} \quad (5.6)$$

decrease with increasing Ni occupancies of the sites, indicating an increase of covalent bond character in the Ni-rich olivines. The Racah *B* parameters are also smaller for the M1 site than for the M2 site, pointing to stronger covalency of the M1 site (cf. §6.8.4).

Compositional variations of intensity of the absorption band at 23,400 to 24,000 cm^{-1} attributed to Ni^{2+}/M2 site cations have provided site occupancy data for the Mg^{2+}–Ni^{2+} olivines (Hu *et al.*, 1990). The results obtained from crystal field spectra showing strong cation ordering of Ni^{2+} ions in the olivine M1 sites are in good agreement with estimates from crystal structure refinements (e.g. Boström, 1987; Ottonello *et al.*, 1989) described later (§6.7.1.2).

5.4.2.5 Miscellaneous olivines

Crystal field spectra of a chromium-bearing forsterite yielded bands at 23,500 and 16,900 cm^{-1} (Scheetz and White, 1972), indicating a CFSE of 20,280 cm^{-1} for Cr^{3+} in the olivine structure. Additional broad, asymmetric bands centred at 11,800 and 6,400 to 6,700 cm^{-1} were attributed to crystal field transitions in Cr^{2+} (Scheetz and White, 1972; Burns, 1974). Although EPR measurements of forsterite show slight enrichments of Cr^{3+} in the M1 sites (Rager, 1977), polarized absorption spectra measured at 77 K of a forsterite doped with 2 x 10^{-4} per cent Cr produced peaks at 13,500, 15,100 and 17,500 cm^{-1} which were attributed to Cr^{3+} in the non-centrosymmetric M2 sites (Rager and Weiser, 1981). Spectra of synthetic Co_2SiO_4 in α-polarized light show features near 6,500 cm^{-1}, 8,330 cm^{-1}, 12,500 cm^{-1} and 20,000 cm^{-1} (Rossman, 1988). These yield approximate values of Δ_o = 7,210 cm^{-1} and CFSE = 6,270 cm^{-1} for Co^{2+}, assuming that cations located in the acentric M2 sites dominate the crystal field spectra. Reflectance spectra of synthetic Co_2SiO_4 with bands at 13,000 cm^{-1} and 18,000 cm^{-1} yielded an estimated value for Δ_o of 7,200 cm^{-1} (Goodgame and Cotton, 1961). Spectra of Mn^{2+}–bearing olivines and tephroite (Keester and White, 1968; Burns, 1970; Weeks *et al.*, 1974) illustrated in fig. 4.8 and discussed earlier (§4.5) show that Mn^{2+} ions are enriched in the M2 sites. Table 5.5 summarizes crystal field spectral data for transition metal-bearing olivines.

5.4.3 Silicate spinels

Although only a few transition metal-bearing silicate spinels have been synthesized, spectral data exist for γ-Fe_2SiO_4 (ringwoodite) and γ-Ni_2SiO_4 and were used to interpret the olivine → spinel phase change (Mao and Bell, 1972b; Burns and Sung, 1978) described later (§9.9.1). The transition metal ions in these silicate spinels occupy the trigonally distorted octahedral sites (point symmetry D_{3d}). The crystal field spectra are simplified by showing no polarization dependencies for the cubic spinel phase.

The crystal field spectrum of Fe_2SiO_4 spinel contains a broad, slightly asymmetric band centred around 11,430 cm^{-1} (Mao and Bell, 1972b; Burns and Sung, 1978), leading to the crystal field parameters for Fe^{2+} of

$$\Delta_o = 10{,}760\ cm^{-1};\ \text{and CFSE} = 4{,}970\ cm^{-1}. \qquad (5.7)$$

These values are significantly higher than the values estimated for Fe^{2+} in the olivine M1 and M2 sites (eqs (5.2) and (5.3)).

The reflectance spectrum of Ni_2SiO_4 spinel contains intense bands with absorption maxima of 9,150 cm^{-1}, 14,780 cm^{-1} and 22,550 cm^{-1} at atmospheric pressure (Yagi and Mao, 1977). These led to initial estimates of the crystal field parameters for Ni^{2+} in silicate spinel of Δ_o = 9.,150 cm^{-1} and CFSE = 10,980 cm^{-1}. However, there is also a prominent shoulder in the reflectance spectra around 8,000 cm^{-1} attributable to trigonal distortion of the octahedral site in the spinel structure (Burns, 1985a). This led to revised estimates for Ni^{2+} in Ni_2SiO_4 spinel of

$$\Delta_o = 8{,}400\ cm^{-1};\ \text{and CFSE} = 10{,}080\ cm^{-1}, \qquad (5.8)$$

Table 5.5. *Crystal field parameters for transition metal-bearing olivines*

Mineral or synthetic phase	Cation site	Positions of absorption bands (cm^{-1})	Δ_o (cm^{-1})	CFSE (cm^{-1})	Sources of data
forsterite–fayalite $(Mg_{1-x}Fe_x)_2SiO_4$ x_{Fe}: 0.08 to 0.98	M1	9,090 to 8,060; 11,830 to 11,050	9,670 to 8,830	4,600 to 4,250	[1] §5.4.2.3
	M2	9,570 to 9,280	8,210 to 7,930	4,400 to 4,280	
fayalite–tephroite $(Mn_{1-x}Fe_x)_2SiO_4$ x_{Fe}: 0.31 to 0.98	M1	8,200 to 8,060; 10,990 to 11,050	8,900 to 8,830	4,280 to 4,250	[1] fig.4.8
	M2	8,930 to 9,300	7,480 to 7,930	4,090 to 4,280	
$(Mg_{1-x}Ni_x)_2SiO_4$ $x_{Ni} = 0.05$ (reflectance)	M1	8,400; 14,700; 25,500; 21,300	8,700	9,550	[2]
	M2	6,600; 12,800; 23,400	7,680	8,990	
liebenbergite (Ni_2SiO_4)	M1,M2	α spectrum: 7,970; 12,690; 14,140; 20,000; 21,100; 23,810; β spectrum: 6,750; 7,290; 12,920; 14,160; 16,100; 21,100; 24,100 γ spectrum: 8,020; 9,090; 12,720; 14,120; 21,190; 23,640	8,360	9,620 (M1) (M2)	[3,4]
Ni^{2+} forsterite $x_{Ni} = 0.02$	M1	9,000; 14,000; 25,000	8,500	10,200	[5]
forsterite–liebenbergite x_{Ni}: 0.02 to 1.0	M1	8,700 to 9,000; 15,400 to 15,900; 25,100 to 26,200; (21,000 to 21,800)	7,430 to 7,780	8,920 to 9,340	[6] §5.4.2.4 fig.5.12
	M2	6,000 to 7,000; 7,200 to 8,000; 12,600 to 13,000; 23,400 to 24,000; 20,000 to 20,800	7,210 to 7,490	8,660 to 8,980	

Co_2SiO_4	M1,M2	6,500; 8,330; 12,500; 20,000 (3 peaks)	7,210	6,270	[7]
Co_2SiO_4 (reflectance)	M1,M2	13,000; 18,000	7,200	6,260	[8,9]
$CaCoSiO_4$	M1		7,500	6,500	[8]
Cr^{3+} forsterite	M1,M2	16,900; 23,500	16,900	20,280	[10]
Cr^{2+} forsterite	M1	6,400 to 6,700; 11,800			[10]
fayalite–tephroite $(Fe_{1-x}Mn_x)_2SiO_4$ x_{Mn}: 0.3 to 1.0	M2,M1	17,180 to 16,670; 22,350 to 22,730; 24,420 to 24,210		0	[1] fig.4.8
tephroite (Mn_2SiO_4)	M2,M1	17,390; 22,680; 24,630; 27,700; 28,570	8,750	0	[11]

Sources of data : [1] Burns (1970); [2] Wood (1974); [3] Rossman, Shannon & Waring (1981); [4] Burns (1985a); [5] Rager, Hosoya & Weiser, (1988); [6] Hu, Langer & Boström (1990); [7] Rossman (1988); [8] Reinen (1970); [9] Goodgame & Cotton (1961); [10] Scheetz & White (1972); [11] Keester & White (1968).

which, again, are significantly higher than values estimated for Ni^{2+} in the M1 and M2 sites of the olivine structure (eqs (5.4a) and (5.5a)).

5.4.4 Aluminosilicates

Trivalent transition metal ions substituting for Al^{3+} occur in several aluminosilicate minerals, including epidotes, yoderite and the Al_2SiO_5 polymorphs andalusite, kyanite and sillimanite, often producing spectacular colours and pleochroism. As a result, numerous optical spectral measurements have been made of several natural and synthetic Al_2SiO_5 minerals (Kerrick, 1990).

5.4.4.1 Andalusite

In the andalusite structure, Al^{3+} ions occur in six-coordinated M1 and five-coordinated M2 sites (Winter and Ghose, 1979). Crystal structure refinements of viridine, $(Al,Fe,Mn)_2SiO_5$, show that transition metal ions are concentrated in the distorted octahedral sites (point symmetry C_2), although results from Mössbauer spectral measurements suggest that some Fe^{3+} ions may also occupy the trigonal bipyramidal M2 sites (Abs-Wurmbach *et al.*, 1981). In kanonaite, $(Mn,Al)_2SiO_5$, most of the Mn^{3+} ions occupy the M1 sites, the tetragonal distortion of which increases with increasing Mn^{3+} concentration (Abs-Wurmbach *et al.*, 1981; Weiss *et al.*, 1981). A small proportion of Mn^{3+} ions may also be present in the five-coordinated M2 sites (Weiss *et al.*, 1981).

Optical spectra of Mn^{3+}-bearing andalusites, including natural viridines, kanonaite and synthetic $(Mn_xAl_{1-x})_2SiO_5$, have been studied extensively (Abs-Wurmbach *et al.*, 1977, 1981; Hålenius, 1978; Kai *et al.*, 1980; Smith *et al.*, 1982; Rossman, 1988; Langer, 1988). The spectra of kanonaite resemble those of piemontite (fig. 4.3) by displaying pronounced polarization-dependencies of absorption bands in the visible region which are responsible for the strong pleochroism of Mn^{3+}-bearing andalusites. Energy level diagrams derived from the polarized spectra (Abs-Wurmbach *et al.*, 1981; Langer, 1988) show a pronounced increase of upper-level splitting of e_g orbital levels with rising Mn^{3+} content of the $(Al,Mn)_2SiO_5$ phases (cf. fig. 4.5), correlating with the large tetragonal distortion of the M1 octahedral site in kanonaite due to the Jahn–Teller effect (§2.12). In fact, Jahn–Teller distortion of the M1 octahedron in kanonaite is surpassed by very few Mn^{3+} octahedra in minerals and compounds whose structures have been refined (Weiss *et al.*, 1981). Examples of such Mn(III) minerals listed in table 6.1 are discussed later (§6.3).

In some andalusites, $Fe^{2+} \rightarrow Fe^{3+}$ IVCT transitions have been identified (table 4.2) involving cations in the chains of edge-shared M1 octahedra paralleling the *c* axis (Langer *et al.*, 1984). An intense broad band at 20,800 cm^{-1} in the E||*c* spectrum of andalusite originally attributed to $Ti^{3+} \rightarrow Ti^{4+}$ IVCT (Faye and Harris, 1969) more likely represents a $Fe^{2+} \rightarrow Ti^{4+}$ IVCT transition (Burns, 1981; Strens *et al.*, 1982).

5.4.4.2 Kyanite

In the kyanite structure, Al^{3+} ions occupy four octahedral sites, including chains of edge-shared M1 and M2 octahedra and isolated M3 and M4 octahedra located between these chains. Some green kyanites owe their colours to crystal field transitions in Fe^{3+} ions (Faye and Nickel, 1969; Parkin *et al.*, 1977). Other green kyanites containing Cr^{3+} ions yield strongly polarization-dependent crystal field spectra (Rossman, 1988; Langer, 1988), which have been assigned to Cr^{3+} ions located in M1 and M2 octahedra (both having point symmetry C_1) of this triclinic mineral. Crystal field parameters derived by averaging energies of the bands centred around 16,600 cm^{-1} in the three polarized spectra yielded a CFSE of 19,800 cm^{-1} for Cr^{3+} in the kyanite structure (Langer, 1988). Analogous polarized absorption spectra are shown by vanadium-bearing kyanites (Schmetzer, 1982).

In more typical blue-coloured kyanites, the spectra of which were discussed earlier (§4.7.2.7), two overlapping absorption bands occur, one with a maximum at 16,670 cm^{-1} and the other producing a prominent shoulder around 11,760 cm^{-1} (White and White, 1967; Faye and Nickel, 1969; Parkin *et al.*, 1977; Ghera *et al.*, 1986). The 16,670 cm^{-1} band has been assigned to charge transfer involving either $Fe^{2+} \rightarrow Ti^{4+}$ and/or $Fe^{2+} \rightarrow Fe^{3+}$ IVCT transitions (Smith and Strens, 1976; Burns, 1981). Since Mössbauer spectral measurements demonstrated the presence of ferrous iron in blue kyanites (Parkin *et al.*, 1977), the band centred around 11,760 cm^{-1} may represent a CF transition within Fe^{2+} ions (Faye and Nickel, 1969) which, by analogy with vivianite (fig. 4.12), is intensified by the IVCT transition (Amthauer and Rossman, 1984; Mattson and Rossman, 1987a). Such an assignment of the kyanite 11,760 cm^{-1} band yields a CFSE of about 4,700 cm^{-1} for Fe^{2+} ions in the kyanite structure.

5.4.4.3 Sillimanite

In the sillimanite structure, half of the Al^{3+} ions occur in single chains of edge-shared M1 octahedra (point symmetry C_i) running parallel to the *c* axis. These chains are linked laterally by double chains of tetrahedra parallel to the *c* axis and are occupied alternately by silicon and the remaining Al^{3+} ions in the M2 tetrahedra (point symmetry C_s). Mean metal–oxygen distances are M1–O =191.2 pm and M–O = 176.4 pm for octahedrally and tetrahedrally coordinated Al^{3+} ions, respectively.

The spectra of yellow sillimanites have been attributed to Fe^{3+} ions in octahedral and tetrahedral sites (Hålenius, 1979; Rossman *et al.*, 1982). The spectra of a green Cr-bearing sillimanite yielded bands at 16,130 and 23,640 cm^{-1} attributed to Cr^{3+} ions in the octahedral M1 sites. Samples of blue sillimanite yielded bands at 16,810 cm^{-1} and 11,960 cm^{-1} in light polarized along the *c* axis (Rossman *et al.*, 1982). The bands resemble those measured in blue kyanite and appear to represent $Fe^{2+} \rightarrow Fe^{3+}$ and/or $Fe^{2+} \rightarrow Ti^{4+}$ IVCT transitions. However, the 11,960 cm^{-1} band in sillimanite may represent a CF transition in the Fe^{2+} ions located in octahedral M1 sites, by analogy with kyanite (§4.7.2.7 and §5.4.4.2) and corundum (§4.7.2.6).

5.4.4.4 Yoderite

The crystal structure of yoderite, the ideal formula of which is $A_8Si_4O_{18}(OH)_2$ where A = Mg, Al, Fe^{2+}, Fe^{3+}, Mn^{2+} and Mn^{3+}, contains three cation sites including distorted M1 octahedra with average metal–oxygen distance of 196.6 pm forming chains parallel to the *c* axis (Higgins *et al.*, 1982), These chains are interconnected by [SiO_4] tetrahedra and two five-coordinated trigonal bipyramidal sites, M2 and M3, with average metal–oxygen distances of 193.5 pm and 187.0 pm, respectively. The polarized spectra of dark blue and purple yoderites contain three major bands centred around 16,900 cm^{-1}, 18,600 cm^{-1} and 20,600 cm^{-1} and a fourth less intense band at 25,500 cm^{-1}. Originally, only the two bands at 16,700 cm^{-1} and 20,600 cm^{-1} were resolved (Abu–Eid *et al.*, 1978) and were assigned to two different $Mn^{2+} \rightarrow Mn^{3+}$ IVCT transitions involving pairs of cations located in edge-shared M1–M1 octahedra and M2–M2 bipyramids, respectively. These assignments were based on polarization dependencies, large widths, and extremely high intensities of the two absorption bands. In a later study (Langer *et al.*, 1982), two additional bands were resolved at 18,600 cm^{-1} and 25,500 cm^{-1} leading to the suggestion that they are the ones representing the two $Mn^{2+} \rightarrow Mn^{3+}$ IVCT transitions. The bands at 16,700 cm^{-1} and 20,600 cm^{-1} were assigned to CF transitions in Mn^{3+} ions located in the M2 or M3 trigonal bipyramidal sites.

5.4.5 Epidote group

The optical spectra of Fe^{3+}-, Mn^{3+}- and vanadium-bearing epidotes were described earlier in chapter 4 (figs 4.3, 4.9 and 4.11). The monoclinic epidote crystal structure illustrated there (fig. 4.4) shows the presence of three octahedral sites, two of which are equivalent in the orthorhombic zoisite structure. The centrosymmetric M1 and M2 sites form chains of edge-shared octahedra running parallel to the *c* axis. Calcium in two sites having eight- and nine-fold coordination, silicon in isolated [SiO_4] and linked [Si_2O_7] tetrahedral groups, and trivalent cations in acentric, very distorted M3 sites occur between the chains of edge-shared M1 and M2 octahedra. Ferric and Mn^{3+} ions are concentrated in the distorted M3 octahedra, resulting in intense, polarization dependent absorption bands in Al–Fe and Al–Mn epidotes (Burns and Strens, 1967). Other studies of polarized spectra Al–Mn^{3+} epidotes (Langer *et al.*, 1976; Langer and Abu-Eid, 1977; Smith *et al.*, 1982; Kersten *et al.*, 1987; Langer, 1988) have generally confirmed the spectrum profiles illustrated in fig. 4.3 and the peak assignments represented by the energy level diagram in fig. 4.5. Compositional variations of the molar extinction coefficients, however, suggest that some of the Mn^{3+} ions may occur in a second octahedral site in piemontite, perhaps the M2 site indicated by crystal structure refinements (Dollase, 1969). However, intensities of the Mn^{3+}/M2 site CF bands are expected to be low due to the centrosymmetric properties of the M2 site, causing them to be obscured by the very intense Mn^{3+}/M3 site CF bands. The optical spectra of piemontite have also been measured at high pressures (Abu-Eid, 1976; Langer, 1988; table 9.2) and at low temperatures (Smith *et al.*, 1982).

Crystal field spectra for other transition metal-bearing epidotes include those of Cr^{3+} in tawmawite (Burns and Strens, 1967) and zoisite (Schmetzer and Berdesinski, 1978), as well as the vanadium-bearing gem zoisite, tanzanite (Faye and Nickel, 1970b; Tsang and Ghose, 1971) illustrated in fig. 4.11 and described in §4.7.1.3. Several assignments of these absorption bands have been proposed with major contributions from Cr^{3+} or V^{3+} ions in the very aistorted M3 sites. In tanzanite, however, multiple site occupancies and oxidation states of vanadium have been proposed and these are discussed later (§5.10.2).

5.4.6 Miscellaneous orthosilicates and isostructural minerals

The polarized spectra of staurolite, $Fe_2Al_9Si_4O_{23}(OH)$, accomodating tetrahedrally coordinated Fe^{2+} ions (point symmetry C_m; mean Fe–O = 200.8 pm) and consisting of absorption bands spanning the 5,000 to 7,000 cm^{-1} region (Bancroft and Burns, 1967a; Dickson and Smith, 1976), are illustrated in fig. 4.6 and discussed in §4.4.3. These bands completely mask any contributions from Fe^{2+} ions that might be present in centrosymmetric octahedral sites in the staurolite structure. The Δ_t and CFSE parameters of tetrahedral Fe^{2+} ions in staurolite are estimated to be about 5,300 cm^{-1} and 3,700 cm^{-1}, respectively. The spectra of the cobaltian staurolite, lusakite, illustrated in fig 4.7, indicates that tetrahedrally coordinated Co^{2+} ions have Δ_t and CFSE values of about 6,500 cm^{-1} and 7,800 cm^{-1}, respectively.

Published spectra are available for various types of topazes, including Cr^{3+}-bearing (Petrov *et al.*, 1977) and irradiated (Aines and Rossman, 1986) specimens. Spectra of Cr^{3+} in vesuvianite (Manning, 1968b, 1976, Manning and Tricker, 1975), euclase and mullite (Neuhaus, 1960), have also been described. Polarized spectra of minerals isostructural with olivine include those of chrysoberyl, sinhalite and alexandrite (Farrell and Newnham, 1965; White *et al.*, 1967; Schmetzer *et al.*, 1980), the complex colour of alexandrite being discussed in §4.7.1.2.

5.5 Spectra of chain silicates: Pyroxene group

5.5.1 Background

The electronic spectra of pyroxenes have been studied more extensively than those of any other group of minerals (Rossman, 1980, 1988). Interest in the crystal field spectra of pyroxenes may be attributed to this mineral being the major ferromagnesian silicate in mafic igneous rocks, granulite facies metamorphic rocks and many meteorites. Pyroxenes also predominate in basaltsic rocks from the Moon and are responsible for the most conspicuous features observed in remote-sensed reflectance spectral profiles of the Moon's surface and certain asteroids (Burns, 1989a). The paragenesis, cleavages and crystal habits of pyroxenes often result in small crystallites being present in meteorites, lunar, igneous and metamorphic rocks. The rapid development of pyroxene absorption spectroscopy has been due, in part, to the exploitation of microscope accessories in spectrophotometers, and also to the availability of specimens with exotic chemical compositions in meteorites and in rocks brought back from the Moon.

The pyroxene structure is also of considerable interest to mineral spectroscopists because, like olivine, it again contains distinguishable coordination sites yielding distinctive Fe^{2+} crystal field spectra. In contrast to olivine, however, Fe^{2+} ions in pyroxenes show strong intracrystalline cation ordering, so that there are major compositional variations of visible to near-infrared spectra.

5.5.2 Pyroxene crystal structure

In pyroxenes, cations occur in two crystallographic positions designated as the M1 and M2 sites. The crystal structure illustrated in fig. 5.13 contains single chains of corner-sharing [SiO_4] tetrahedra running parallel to the *c* axis which are linked by bands of cations located in the M1 and M2 positions. Each [SiO_4] tetrahedron shares two of its oxygens with adjacent tetrahedra and these bridging oxygens, together with non-bridging oxygens in each tetrahedron, constitute the oxygen ligands of coordination polyhedra surrounding cations in the

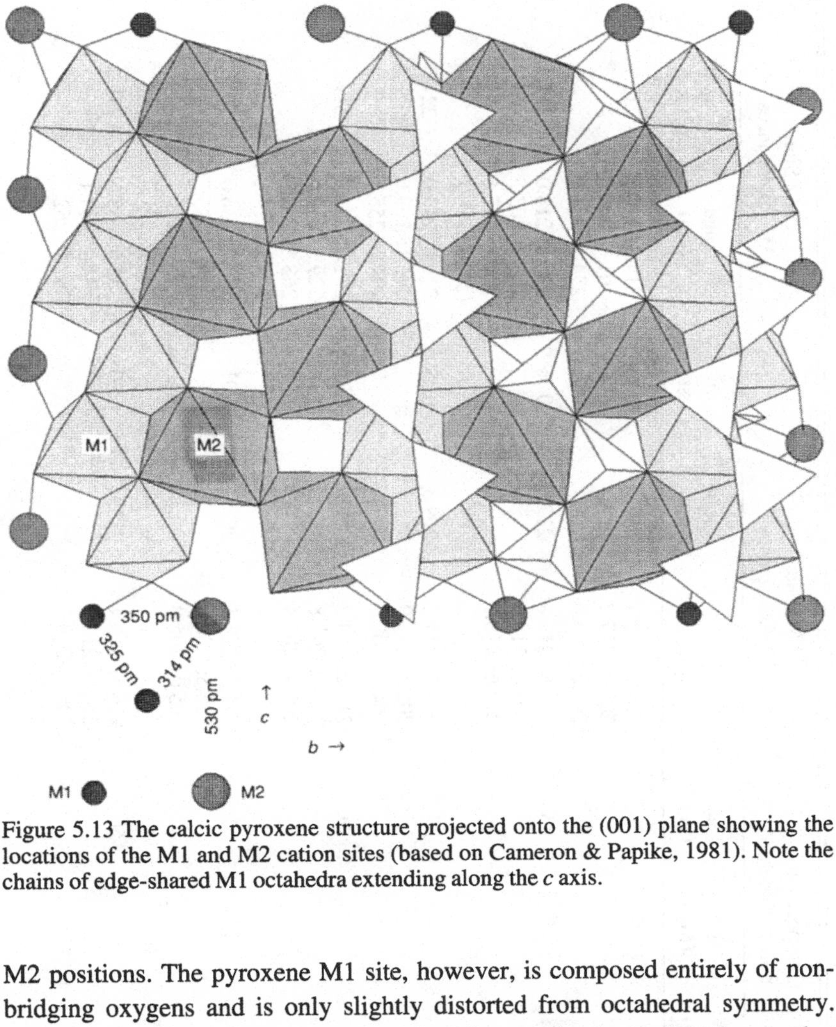

Figure 5.13 The calcic pyroxene structure projected onto the (001) plane showing the locations of the M1 and M2 cation sites (based on Cameron & Papike, 1981). Note the chains of edge-shared M1 octahedra extending along the *c* axis.

M2 positions. The pyroxene M1 site, however, is composed entirely of non-bridging oxygens and is only slightly distorted from octahedral symmetry. Cations are not centrally located in the M1 coordination polyhedon so that there are ranges of cation–oxygen distances within the pyroxene M1 sites (point symmetry C_1). Chains of edge-shared M1 octahedra zig-zag along the *c* axis and M1–M1 interatomic distances are in the range 311 to 315 pm.

The pyroxene M2 site, on the other hand, is very distorted and metal–oxygen distances span wide ranges as demonstrated by the data summarized in table 5.6 chosen for a variety of extraterrestrial pyroxenes. Distances from the M2 cation to the four non-bridging oxygens are considerably shorter than those to oxygens bridging two silicon atoms. Moreover, distances to, and numbers

Table 5.6. *Crystal chemical data for pyroxenes*

Pyroxene	Composition mole per cent			M1 Site M1–O distances		Fe in site	M2 Site M2–O distances		Fe in site
	Wo (Ca)	En (Mg)	Fs (Fe)	range (pm)	mean (pm)		range (pm)	mean (pm)	
enstatite (Luna 20) [1]	7	60	33	204.2; 215.2 202.8; 217.2 207.2; 205.6	208.7	0.13	214.4; 210.5 201.0; 206.7 234.5; 251.0	219.7	0.41
ferrosilite (Earth) [2]	2	13	85	207.6; 217.8 208.6; 218;6 210.6; 210.6	212.3	0.76	216.1; 213.0 199.7; 203.5 244.4; 257.6 (293.0;343.0)	222.6	0.94
pigeonite (Apollo 11) [3]	9	54	37	205.7; 214.2 205.2; 217.9 204.4; 208.7	209.3	0.33	216.8; 213.0 205.1; 199.6 237.6; 255.7	221.3	0.59
augite (Apollo 14) [4]	40	32	28	214.2; 214.2 207.8; 207.8 205.4; 205.4	209.1	0.39	231.0; 231.0 229.5; 229.5 263.1; 263.1 274.1; 274.1	249.6	0.13
diopside (Earth) [5]	49	50	1	211.5; 211.5 206.5; 206.5 205.0; 205.0	207.7	0.01	236.0; 236.0 235.3; 235.3 256.1; 256.1 271.7; 271.7	249.8	0
hedenbergite (synthetic) [6]	50	0	50	216.4; 216.4 214.0; 214.0 208.7; 208.7 271.9; 271.9	213.0	1.00	235.5; 235.5 234.1; 234.1 262.7; 262.7	251.0	1.00

fassaite (Allende) [7]	50	19	24 (Ti)	212.0; 212.0 206.4; 206.4 199.3; 199.3	205.9	0.34 (Ti^{3+}) 0.14 (Ti^{4+})	239.9; 239.9 239.9; 239.9 254.6; 254.6 265.8; 265.8	250.0	0
fassaite (Angra dos Reis) [8]	44	30	11 (Fe) 3 (Ti)	211.5; 211.5 205.1; 205.1 201.6; 201.6	206.1	0.21 (Fe^{2+}) 0.06 (Ti^{4+})	236.3; 236.3 234.9; 234.9 268.7; 268.7 255.6; 255.6	248.9	0.02

Sources of data: [1] orthopyroxene K–7, Luna 20: $Ca_{0.083}Mn_{0.008}Fe_{0.501}Mg_{1.389}Cr_{0.022}Ti_{0.009}Al_{0.054}Si_{1.946}O_6$ (Ghose & Wan, 1973);
[2] iron-rich orthopyroxene, granulite, Greenland: $Ca_{0.038}Mg_{0.259}Fe_{1.680}Mn_{0.003}Ti_{0.004}Al_{0.018}Si_{1.992}O_6$ (Burnham, Ohashi, Hafner & Virgo, 1971);
[3] pigeonite 10003,38, Apollo 11: $Ca_{0.163}Fe_{0.718}Mg_{1.050}Ti_{0.026}Al_{0.060}Si_{1.983}O_6$ (Clark, Ross & Appleman, 1971);
[4] augite 14310, Apollo 14; $Ca_{0.773}Na_{0.005}Fe_{0.541}Mg_{0.633}Ti_{0.033}Al_{0.061}Si_{1.946}O_6$ (Takeda, Miyamoto & Reid, 1974);
[5] diopside, New York; $Ca_{0.99}Na_{0.02}Mg_{1.01}Fe_{0.01}Al_{0.01}Si_{1.97}O_6$ (Clark, Appleman & Papike, 1969);
[6] hedenbergite, synthetic $CaFeSi_2O_6$ (Clark, Appleman & Papike, 1969);
[7] titanian subsilicic diopside (fassaite), Allende meteorite: $Ca_{1.01}Mg_{0.38}Ti^{3+}{}_{0.34}Ti^{4+}{}_{0.14}Al_{0.87}Si_{1.26}O_6$ (fig. 4.2; Dowty & Clark, 1973) ;
[8] titaniferous subsilicic diopside (fassaite), Angra dos Reis meteorite: $Ca_{0.97}Fe_{0.21}Mg_{0.59}Ti_{0.06}Al_{0.45}Si_{1.79}O_6$ (Hazen & Finger, 1977).

of, bridging oxygens forming the M2 coordination polyhedron vary with pyroxene structure-type and composition (table 5.6; Appendix 7). In calcic clinopyroxenes of the diopside–hedenbergite series, $Ca(Mg,Fe)Si_2O_6$, Ca^{2+} ions fill the M2 positions and the eight nearest-neighbour oxygen ligands forming the M2 coordination polyhedron comprise four bridging and four non-bridging oxygens. In orthopyroxenes, the M2 polyhedron consists of four non-bridging oxygens and only two bridging oxygens. Cations in M2 polyhedra of pigeonite have similar nearest-neighbour oxygen ligands to those forming the orthopyroxene M2 sites, but an additional bridging oxygen further away could be regarded as belonging to the pigeonite M2 coordination polyhedron making it a seven-coordinated site. In orthopyroxenes and pigeonites, Fe^{2+} ions show strong preferences for the M2 coordination polyhedron and are enriched in this site relative to the M1 site. In augites, Ca vacancies in M2 positions are filled preferentially by Fe^{2+} ions. It is the Fe^{2+} ions in the M2 sites of orthopyroxenes, pigeonites, and subcalcic clinopyroxenes that dominate the crystal field spectra of pyroxenes.

5.5.3 Calcic pyroxenes

Before discussing the dominant Fe^{2+}/M2 site crystal field spectra of pyroxenes, it is necessary to identify the locations of absorption bands originating from Fe^{2+} ions in the less distorted M1 sites. Crystal field spectra of stoichiometric hedenbergite, $CaFeSi_2O_6$, consist of two broad bands centred at 10,200 cm^{-1} and 8,475 cm^{-1} (Rossman, 1980; Straub *et al.*, 1991), from which the crystal field parameters are estimated to be

$$\Delta_o = 9{,}100\ cm^{-1};\ \text{and CFSE} = 3{,}890\ cm^{-1}, \tag{5.10}$$

assuming a lower-level splitting of only 500 cm^{-1} for the t_{2g} orbitals of Fe^{2+} ions located in the relatively undistorted M1 site. These values remain almost constant over the diopside–hedenbergite series. Traces of Fe^{3+} ions usually present in terrestrial calcic pyroxenes may be responsible for an additional band located between 12,000 cm^{-1} and 14,000 cm^{-1} assigned to a $Fe^{2+} \rightarrow Fe^{3+}$ IVCT transition (§4.7.2; table 4.2) and resulting from electron transfer between iron cations in edge-shared M1 octahedra approximately 312 pm apart.

When slight deficiencies of calcium occur in calcic pyroxenes, Fe^{2+} replacing Ca^{2+} ions in the M2 positions produce additional bands centred around 9,600 cm^{-1} and 4,400 cm^{-1} (White and Keester, 1966; Burns and Huggins, 1973; Hazen *et al.*, 1978). These two bands may be more intense than the Fe^{2+}/M1 site CF bands located near 10,200 cm^{-1} and 8,500 cm^{-1}, even in spectra of subcalcic hedenbergites, such as that portrayed in fig. 5.14, even though

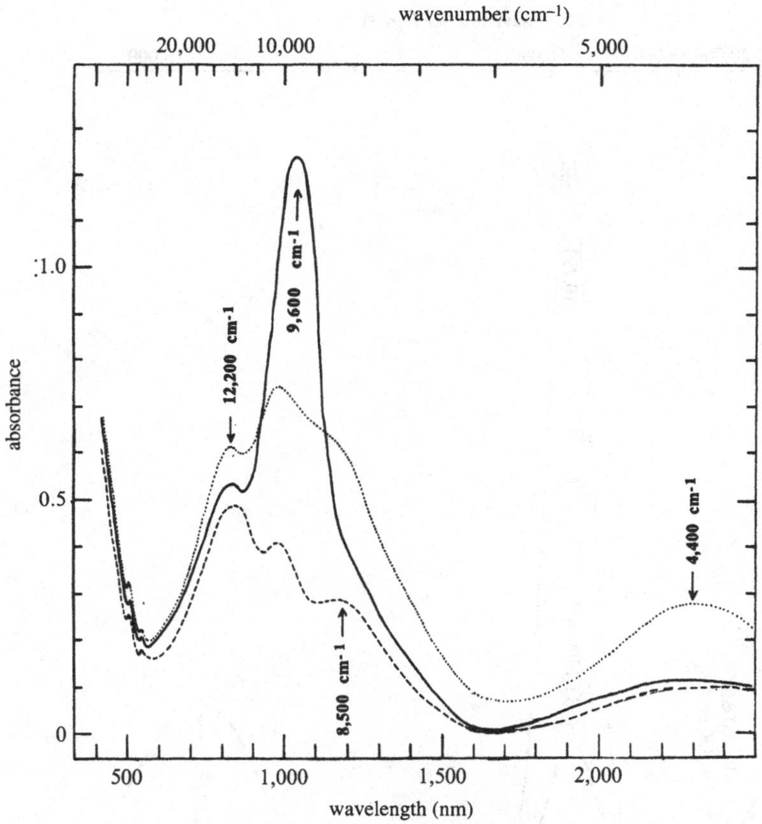

Figure 5.14 Polarized absorption spectra of hedenbergite. ······· α spectrum; – – – – β spectrum; —— γ spectrum. The hedenbergite, containing 88 mole per cent $CaFeSi_2O_6$ and 1.5 wt per cent Fe_2O_3, is from the Quebec Iron Formation. [Optic orientation: α : *a* = 34°; β = *b*; γ : *c* = 48°.]

higher concentrations of Fe^{2+} ions are present in the M1 sites. The high intensities of the Fe^{2+}/M2 site CF bands at 9,600 cm^{-1} and 4,400 cm^{-1} are consequences of the highly distorted, non–centrosymmetric properties of the M2 coordination polyhedron. As noted earlier, the absorption band centred around 12,200 cm^{-1} in fig. 5.14 represents a $Fe^{2+} \rightarrow Fe^{3+}$ IVCT transition between neighbouring Fe^{2+} and Fe^{3+} ions in the pyroxene structure.

In augites, Fe^{2+} occupancies of the M2 sites increase as the pyroxene becomes increasingly subcalcic. Significant replacement of larger Ca^{2+} by smaller Fe^{2+} ions leads to slight contraction of the M2 site (table 5.6), which causes the dominant Fe^{2+} CF bands located near 10,000 cm^{-1} (wavelength of 1 micron) and 5,000 cm^{-1} (2 microns) to move to shorter wavelengths (Adams,

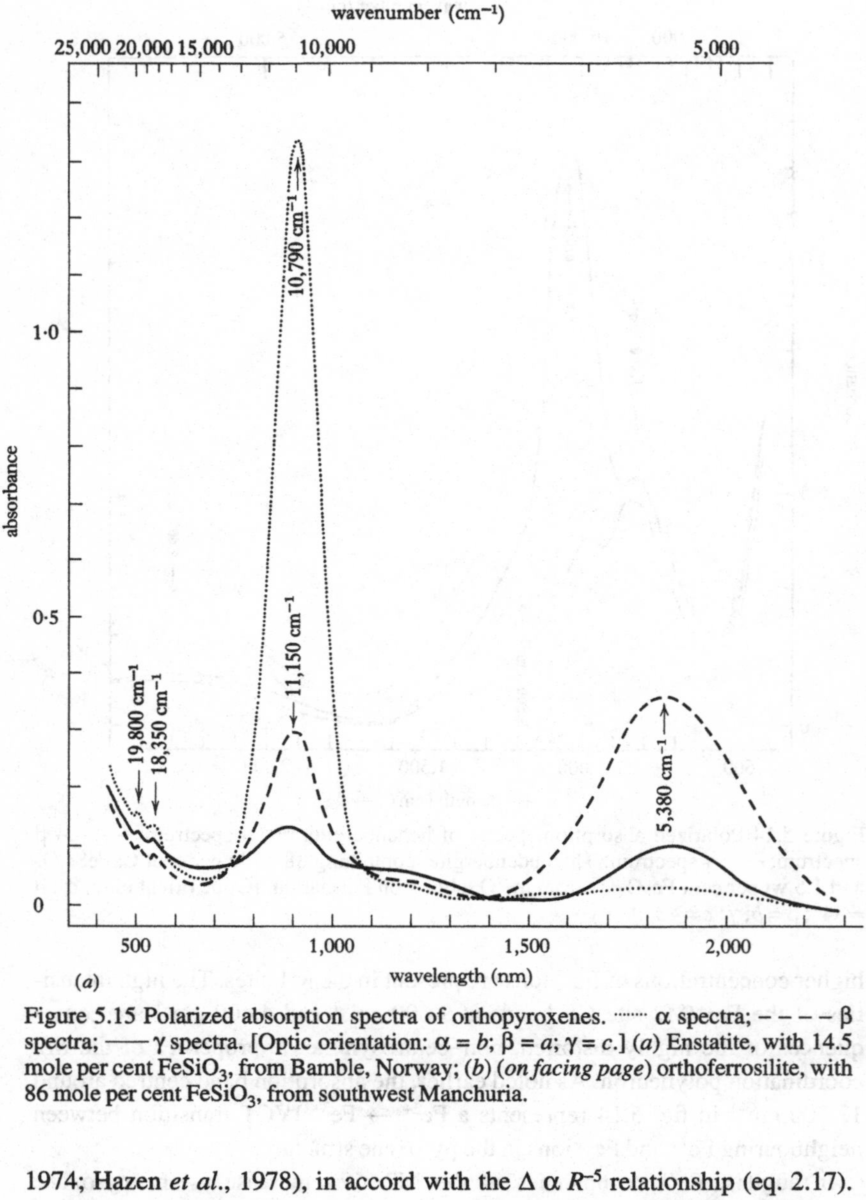

Figure 5.15 Polarized absorption spectra of orthopyroxenes. ······· α spectra; – – – – β spectra; —— γ spectra. [Optic orientation: α = *b*; β = *a*; γ = *c*.] (*a*) Enstatite, with 14.5 mole per cent $FeSiO_3$, from Bamble, Norway; (*b*) (*on facing page*) orthoferrosilite, with 86 mole per cent $FeSiO_3$, from southwest Manchuria.

1974; Hazen *et al.*, 1978), in accord with the $\Delta \propto R^{-5}$ relationship (eq. 2.17). The positions of the '1 micron' and '2 micron' bands (cf. fig. 10.5) provide a powerful means of identifying pyroxene structure-types and compositions on surfaces of terrestrial planets by telescopic reflectance spectral measurements discussed in chapter 10.

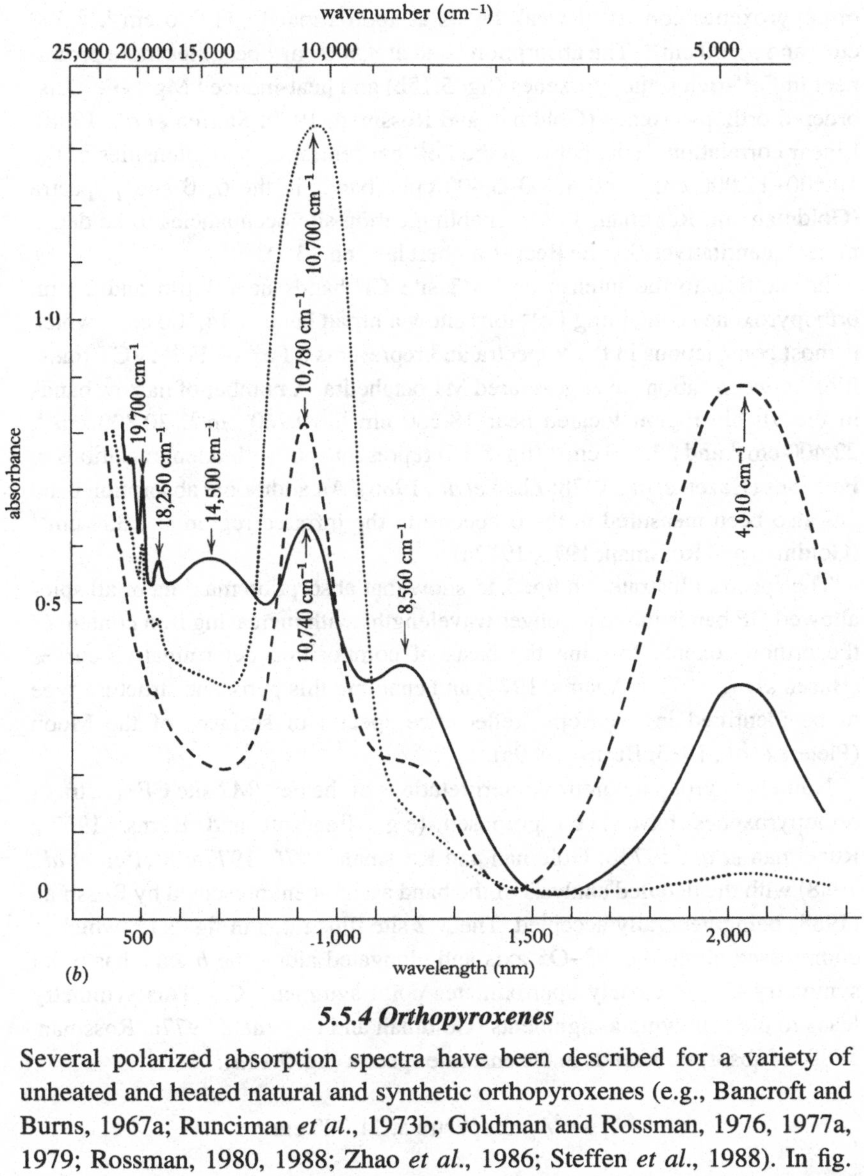

(*b*)

5.5.4 Orthopyroxenes

Several polarized absorption spectra have been described for a variety of unheated and heated natural and synthetic orthopyroxenes (e.g., Bancroft and Burns, 1967a; Runciman *et al.*, 1973b; Goldman and Rossman, 1976, 1977a, 1979; Rossman, 1980, 1988; Zhao *et al.*, 1986; Steffen *et al.*, 1988). In fig. 5.15, representative spectra are illustrated of two orthopyroxenes, enstatite $Fs_{14.5}$ and ferrosilite $Fs_{86.4}$. The α spectra show a very intense band centred near 10,700 cm^{-1}. The β spectra consist of two bands located near 11,100 cm^{-1} and between 5,400 to 4,900 cm^{-1}, with a shoulder around 8,500 cm^{-1} becoming well developed in Fe^{2+}-rich orthopyroxenes. The γ spectra of Mg^{2+}-rich orthopyroxenes consist of weak bands at approximately 11,000 cm^{-1}, 8,500 cm^{-1} and 5,300 cm^{-1}. The absorption band at 8,500 cm^{-1} becomes more prominent in Fe^{2+}-rich orthopyroxenes (fig. 5.15b) and heat-induced Mg^{2+}–Fe^{2+} disordered orthopyroxenes (Goldman and Rossman, 1979; Steffen *et al.*, 1988). Linear correlations exist between the Fe^{2+} concentration and intensities of the 10,500–11,000 cm^{-1} and 4,900–5,400 cm^{-1} bands in the α, β and γ spectra (Goldman and Rossman, 1979), enabling cation site occupancies to be determined quantitatively by the Beer–Lambert law, eq. (3.7).

In addition to the intense Fe^{2+}/M2 site CF bands near 1 μm and 2 μm, orthopyroxenes containing Fe^{3+} ions show a broad band at 14,500 cm^{-1} which is most conspicuous in the γ spectra and represents a $Fe^{2+} \rightarrow Fe^{3+}$ IVCT transition between cations in edge-shared M1 octahedra. A number of narrow bands in the visible region located near 18,250 cm^{-1}, 19,760 cm^{-1}, 20,800 cm^{-1}, 22,400 cm^{-1} and 23,470 cm^{-1} (fig. 5.15) represent spin-forbidden transitions in Fe^{2+} ions (Hazen *et al.*, 1978; Zhao *et al.*, 1986). An additional absorption band has also been measured in the α spectra in the infrared region at 2,350 cm^{-1} (Goldman and Rossman, 1976, 1977a).

The spectra illustrated in fig. 5.15 show that absorption maxima of all spin-allowed CF bands move to longer wavelengths with increasing iron content of the orthopyroxene, forming the basis of composition determinative curves (Hazen *et al.*, 1977b; Adams, 1974) and enabling this pyroxene structure-type to be identified in telescopic reflectance spectra of surfaces of the Moon (Pieters *et al.*, 1985; Burns, 1989a).

Numerous group theoretical interpretations of the Fe^{2+}/M2 site CF spectra of orthopyroxenes have been proposed (e.g., Bancroft and Burns, 1967a; Runciman *et al.*, 1973b; Goldman and Rossman, 1976, 1977a; Steffen *et al.*, 1988) with the detailed analysis of the band assignment presented by Rossman (1988) being generally accepted. The M2 site illustrated in fig. 5.16, which is compressed along the O2–O5 axis and elongated along the *b* axis, has point symmetry C_1 but closely approximates point symmetry C_{2v}. This symmetry leads to the following assignments (Goldman and Rossman, 1977a; Rossman, 1988) of absorption bands in the enstatite spectra (fig. 5.15*a*):

$$^5A_1 \rightarrow {}^5A_1\text{: } \alpha \text{ spectra at } 11{,}000 \text{ cm}^{-1}$$

$$^5A_1 \rightarrow {}^5B_1\text{: } \beta \text{ spectra at } 5{,}400 \text{ cm}^{-1}$$

$$^5A_1 \rightarrow {}^5B_2\text{: } \gamma \text{ spectra at } 2{,}350 \text{ cm}^{-1}$$

$$^5A_1 \rightarrow {}^5A_2\text{: forbidden, predicted at } 354 \text{ cm}^{-1}. \qquad (5.11)$$

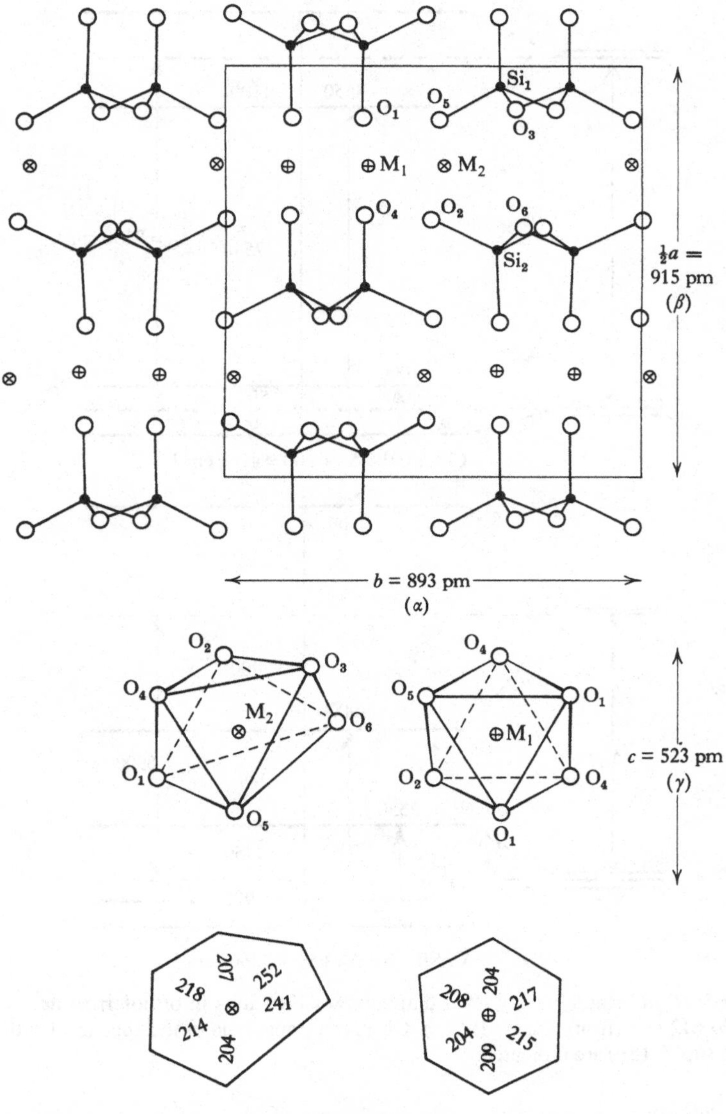

Figure 5.16 The orthopyroxene crystal structure. The figure shows the structure projected onto (001). Oxygen coordination polyhedra [(100) projections] and metal–oxygen distances in each site are indicated (pm). ⊕ M1; ⊗ M2; • Si. Atomic coordinates and cell parameters from Ghose (1965).

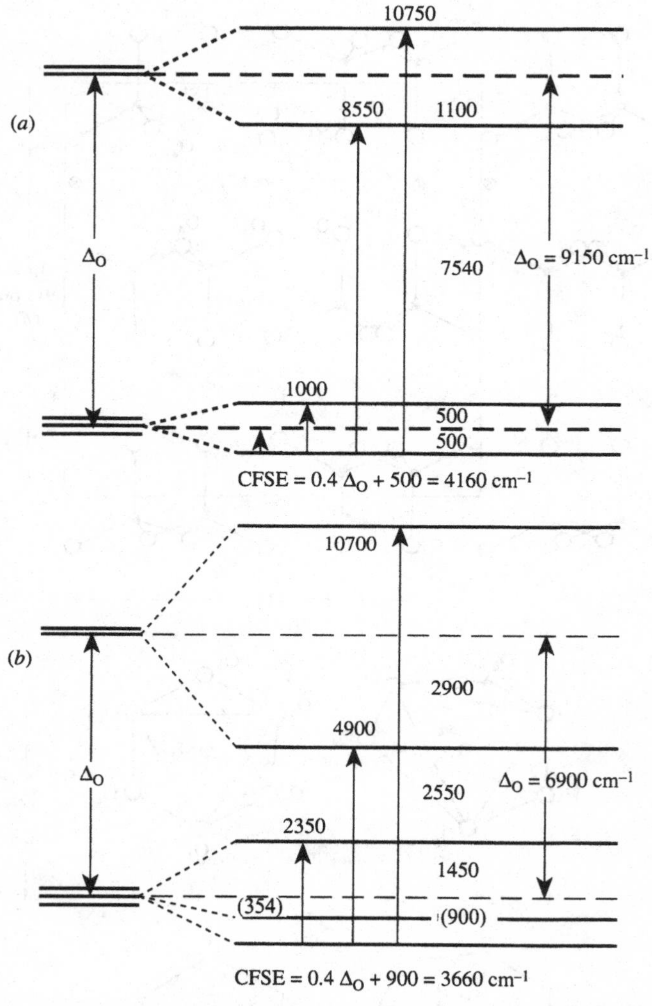

Figure 5.17 3*d* orbital energy level diagrams for Fe^{2+} ions in orthoferrosilite. (*a*) M1 site; (*b*) M2 site (from Burns, 1985a). Observed transitions in the spectra of orthoferrosilite (fig. 5.15*b*) are indicated.

These assignments of the crystal field bands may be used to construct the 3*d* orbital energy level diagrams illustrated in fig. 5.17 for Fe^{2+} ions in the M1 and M2 sites of ferrosilite, $Fs_{86.4}$. The polarized absorption spectra of this ferrosilite (fig. 5.15*b*) show that two of the M2 site Fe^{2+} bands are centred near 10,700 cm^{-1} and 4,900 cm^{-1}. The lower-level splittings of 2,350 cm^{-1} and 354 cm^{-1} listed in eq. (5.11) for enstatite are assumed to apply to ferrosilite. This information is embodied in the energy level diagram for the M2 site shown in fig. 5.17*b*. For the M1 site, one of the Fe^{2+} CF bands lies at 8,560 cm^{-1} (fig. 5.15*b*, which was confirmed in heating experiments to induce Mg^{2+}–Fe^{2+} disordering (Goldman and Rossman, 1977a; Steffen *et al*., 1988). The breadth of the intense band located around 10,700 cm^{-1} indicates that Fe^{2+} CF transitions in the M1 and M2 sites overlap there, again confirmed by the controlled heating experiments (Steffen *et al*., 1988). This information is embodied in the energy level diagram for the M1 site shown in fig. 5.17*a*. The splitting of the lower level t_{2g} orbitals is unknown, and is assumed to have a small value of 500 cm^{-1} since the M1 coordination polyhedron is not significantly distorted from octahedral symmetry (fig. 5.16). The approximate values of Δ_o and CFSE for Fe^{2+} in the two sites of orthopyroxenes in the composition range Fs_{14}–Fs_{86} are estimated to be

$$\text{M1 site: } \Delta_o = 9{,}650 \text{ to } 9{,}400 \text{ cm}^{-1}; \text{ CFSE} = 4{,}200 \text{ to } 4{,}010 \text{ cm}^{-1}, \quad (5.12)$$

$$\text{M2 site: } \Delta_o = 7{,}300 \text{ to } 6{,}900 \text{ cm}^{-1}; \text{ CFSE} = 3{,}820 \text{ to } 3{,}660 \text{ cm}^{-1}. \quad (5.13)$$

The crystal field splitting estimated for Fe^{2+} in the enstatite M2 site is in reasonable agreement with the Δ_o value of 6,522 cm^{-1} calculated by a theoretical point charge model (Goldman and Rossman, 1977a; Rossman, 1988).

5.5.5 *Pigeonites*

The polarized spectra of pigeonites are similar to those of orthopyroxenes (Sung *et al*., 1977). The crystal chemical data summarized in table 5.6 show that the relative enrichments of Fe^{2+} ions in pigeonite M2 sites are not as high as those in orthopyroxene M2 sites. This factor, and the slightly larger M2–oxygen distances in pigeonites, result in the CF bands for Fe^{2+}/M2 site cations being located at slightly longer wavelengths than for orthopyroxenes with comparable total iron contents. For example, in a pigeonite with composition Fs_{57}, intense absorption bands occur at 10,360 cm^{-1} in the α spectrum, moderately intense bands are present at 10,400 cm^{-1} and 4,880 cm^{-1} in the β spectrum, and weak bands occur at 10,750 cm^{-1} and 8,550 cm^{-1} in the γ spectrum. Assuming lower level t_{2g} orbital splittings to be similar to those for orthopyroxenes, the crystal field splittings and stabilization energies for Fe^{2+} in pigeonite are estimated to be

$$\text{M1 site: } \Delta_o = 9{,}400 \text{ cm}^{-1}; \text{ CFSE} = 4{,}000 \text{ cm}^{-1}, \quad (5.14)$$

$$\text{M2 site: } \Delta_o = 6{,}650 \text{ cm}^{-1}; \text{ CFSE} = 3{,}560 \text{ cm}^{-1}. \quad (5.15)$$

Table 5.7. *Crystal field parameters for transition metal ions in chain silicates*

Mineral	Site occupancy	Positions of absorption bands (cm^{-1})	Δ_o (cm^{-1})	CFSE (cm^{-1})	Sources of data
		Fe^{2+} pyroxenes & pyroxenoids			
diopside–hedenbergite	M1	8,475; 10,200	9,100	3,890	§5.5.3
enstatite–ferrosilite	M1	≈8,500; ≈10,700	9,460 to 9,150	4,280 to 4,160	§5.5.4
	M2	≈5,400; ≈11,000	7,300 to 6,900	3,820 to 3,660	§5,5,4
babingtonite	Fe(1)	7,720; 10,580	8,600	3,900	(§4.7.2.8)
rhodonite	M4	4,250; 6,700; 8,000; 9,750			[1–3]
		Ni^{2+} pyroxenes			
Ni^{2+} enstatite	M1	6,900; 12,700; 24,500	6,900	8,280	[4,5]
Ni^{2+} enstatite	M1	7,200; 12,970; 24,510	7,200	8,640	[4,6–8]
	M2	6,500; 12,440; 21,550	6,500	7,800	
Ni^{2+} diopside	M1	8,400; 13,800; 24,500	8,400		[4,5]
		Co^{2+} pyroxenes			
Co^{2+} enstatite	M1,M2	6,250; 8,200; 15,600; 20,200	8,370	6,700	[4,5]
Co^{2+} diopside	M1	5,200; 7,900; 15,400; 19,100	8,020	6,420	[4,5]
		Cr^{2+} pyroxenes			
Cr^{2+} diopside	M1	9,200; 11,900; 13,300 13,500; 17,150	9,000	7,900	[9,10]
Cr^{2+} diopside	M1	9,170; 13,550; 17,150 22,830			[11]
		Cu^{2+} pyroxene			
Cu^{2+} enstatite	M2	7,280; 12,200			[6]
		Mn^{3+} & Mn^{4+} pyroxenes			
Mn^{3+} enstatite	M1	18,520	18,520	11,110	[6]
blanfordite (Mn^{3+} omphacite)	M1	7,700 to 8,500; 16,800 17,300; 19,000	13,650	12,200	[12]
Mn^{4+} spodumene	M1	22,800 to 23,150; 26,400 to 27,100	23,000	27,600	[13]
		Cr^{3+} pyroxenes			
Cr^{3+} enstatite	M1	15,480; 22,620	15,480	18,580	[4,6]
Cr^{3+} diopside	M1	16,129; 22,969	16,129	19,360	[14]
Cr^{3+} diopside	M1	15,500; 22,000	15,500	18,600	[9]
Cr^{3+} diopside	M1	15,900; 21,740	15,900	19,080	[11]
Cr^{3+} diopside	M1	15,420; 22,040	15,420	18,500	[15]

Table 5.7 *continued*

Mineral	Site occupancy	Positions of absorption bands (cm^{-1})	Δ_o (cm^{-1})	CFSE (cm^{-1})	Sources of data
Cr^{3+} omphacite	M1	15,520; 22,300	15,520	18,650	[15]
Cr^{3+} aegerine augite	M1	15,392; 21,925	15,392	18,470	[16]
Cr^{3+} jadeite	M1	15,740; 22,620	15,740	18,890	[15]
kosmochlor (ureyite)	M1	15,600; 22,000	15,600	18,720	[5]
syn. kosmochlor	M1	15,458; 21,742	15,458	18,544	[16]
kosmochlor	M1	15,200; 21,917	15,200	18,240	[16]
ureyite–jadeite	M1	15,580; 22,170	15,580	18,700	[15]
Cr^{3+} spodumene	M1	16,250; 23,510	16,250	19,500	[13]
Cr^{3+} spodumene	M1	15,920; 23,500	15,920	18,100	[15]
		V^{3+} pyroxenes			
V^{3+} diopside	M1	15,700	16,700	13,360	[17]
V^{3+} spodumene	M1	16,000; 23,500	17,210	13,770	[17]
		Ti^{3+} pyroxenes			
Ti^{3+} diopside	M1	16,500; 21,000	18,470	7,900	[18–20]
syn. $NaTiSi_2O_6$	M1	15,600; (22,000)	18,770	8,000	[19,21]
Ti^{3+} kosmochlor	M1	22,000	<22,000	<8,800	[16]
		Fe^{2+} amphiboles			
cummingtonite–grunerite	M4	≈10,000; 4,200	6,300 to 6,000	3,520 to 3,400	§5.6.3
	M1,M2,M3	10,100; 8,500	9,250 to 9,050	3,950 to 3,850	
actinolite	M4	9,710; 4,050	5,890	3,350	§5.6.4
	M1,M2,M3	(10,200; 8,470)	(9,100)	(3,890)	
glaucophane	M1,M3	10,000; 8,600	9,100	3,870	§5.6.5
		V^{3+} amphibole			
tremolite	M1,M2,M3	15,500; 22,500	16,650	13,320	[17]
		Cr^{3+} amphibole			
tremolite	M1,M2,M3	16,310; 23,530	16,310	19,570	[14]
		Mn^{3+} amphiboles			
juddite (riebeckite)	M2	5,800 to 6,800; 16,500; 18,500	11,925	11,870	[12]
winchite	M2	16,300; ≈18,000	11,640	11,700	[12]

Sources of data: [1] Manning (1969a); [2] Gibbons, Ahrens & Rossman (1974); [3] Marshall & Runciman (1975); [4] Burns (1985a); [5] White, McCarthy & Scheetz (1971); [6] Rossman (1980); [7] Rossman, Shannon & Waring (1981); [8] Rossman (1988); [9] Mao, Bell & Dickey (1972); [10] Burns (1975a); [11] Ikeda & Yagi (1982); [12] Ghose, Kersten, Langer, Rossi & Ungretti (1986); [13] Cohen & Janezic (1983); [14] Neuhaus (1960); [15] Khomenko & Platonov (1985); [16] Abs-Wurmbach, Langer & Oberhänsli (1985); [17] Schmetzer (1982); [18] Dowty & Clark (1973); [19] Burns & Huggins (1973); [20] Mao & Bell (1974a); [21] Prewitt, Shannon & White (1972).

5.5.6 Other transition metal-bearing pyroxenes

Optical spectral measurements have been made on a variety of natural and synthetic pyroxenes containing a number of transition metal ions (Rossman, 1980). Crystal field splitting and stabilization energy parameters derived from these spectra are summarized in table 5.7.

5.6 Spectra of chain silicates: Amphibole group

5.6.1 Background

The crystal field spectra of iron-bearing amphiboles have posed problems of interpretation mainly on account of closely overlapping bands which originate from Fe^{2+} ions in two or more coordination sites each with different ligand-types and symmetries (Hawthorne, 1983). There are also interferences from Fe^{3+} ions participating in $Fe^{2+} \rightarrow Fe^{3+}$ IVCT transitions (Hawthorne, 1981b). Amphiboles may accommodate a variety of transition metal ions in several of the octahedral sites, as well as Fe^{3+} in the $[SiO_4]$ tetrahedra, while hydroxyl groups may be replaced by other anions such as F^-, Cl^- and O^{2-} (Hawthorne, 1981a). As a result of the complex crystal chemistry, assignments of absorption bands in optical spectra of amphiboles have been somewhat controversial, particularly for calcic amphiboles.

5.6.2. Amphibole crystal structure

The amphibole structure, such as that of cummingtonite illustrated in fig. 5.18, consists of infinite double chains of $[Si_4O_{11}]$ groups providing bridging and non-bridging oxygen atoms of coordination polyhedra surrounding the cations (Hawthorne, 1981a). The silicate chains are linked by bands of cations in edge-shared octahedra extending along the *c* axis. There are three positions of six-fold coordination, designated as the M1, M2 and M3 sites, and one of six- to eight-fold coordination which is designated as the M4 site. The cation site multiplicities are in the ratio M1 : M2 : M3 : M4 = 2 : 2 : 1 : 2. Cations in the M1 and M3 positions are each coordinated to four non-bridging oxygen ions and two hydroxyl ions. The OH groups are in *cis* arrangement in the M1 sites and *trans* in the M3 site. Six non-bridging oxygen ions surround cations in the M2 positions. Cations in the M4 positions, which are occupied by Mg^{2+} in anthophyllite, Fe^{2+} in grunerite, Ca^{2+} in calcic amphiboles and Na^+ in alkali amphiboles, are surrounded by four non-bridging oxygens and two to four bridging oxygens. The amphibole M4 site thus resembles the pyroxene M2 site. The oxygen polyhedra surrounding the M1, M2 and M3 positions approximate regular octahedra (M1 and M3), or are only slightly distorted from octahedral

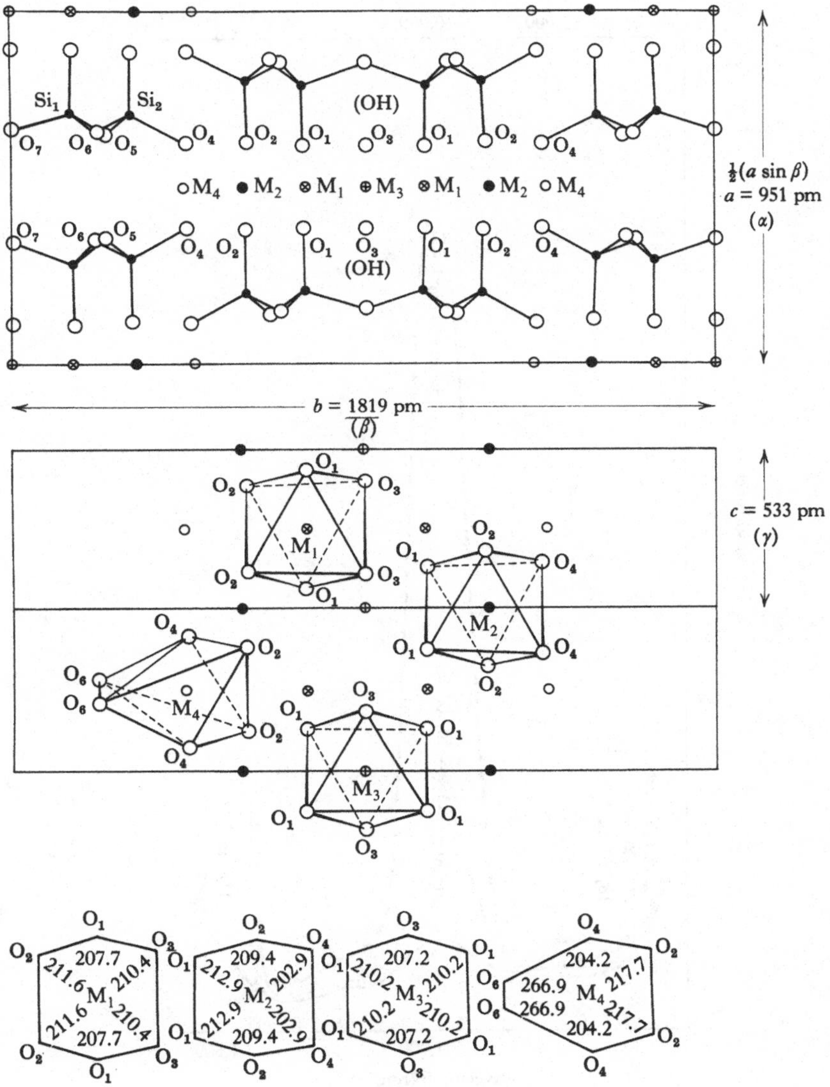

Figure 5.18 The crystal structure of Mg–Fe amphibole (cummingtonite). The figure shows the structure projected onto (001), and a (100) projection of each oxygen coordination polyhedron about a metal position. Metal–oxygen distances in each coordination site are indicated (pm). ⊗ M1; • M2; ⊕ M3; ○ M4. Atomic coordinates and cell parameters from Ghose (1961).

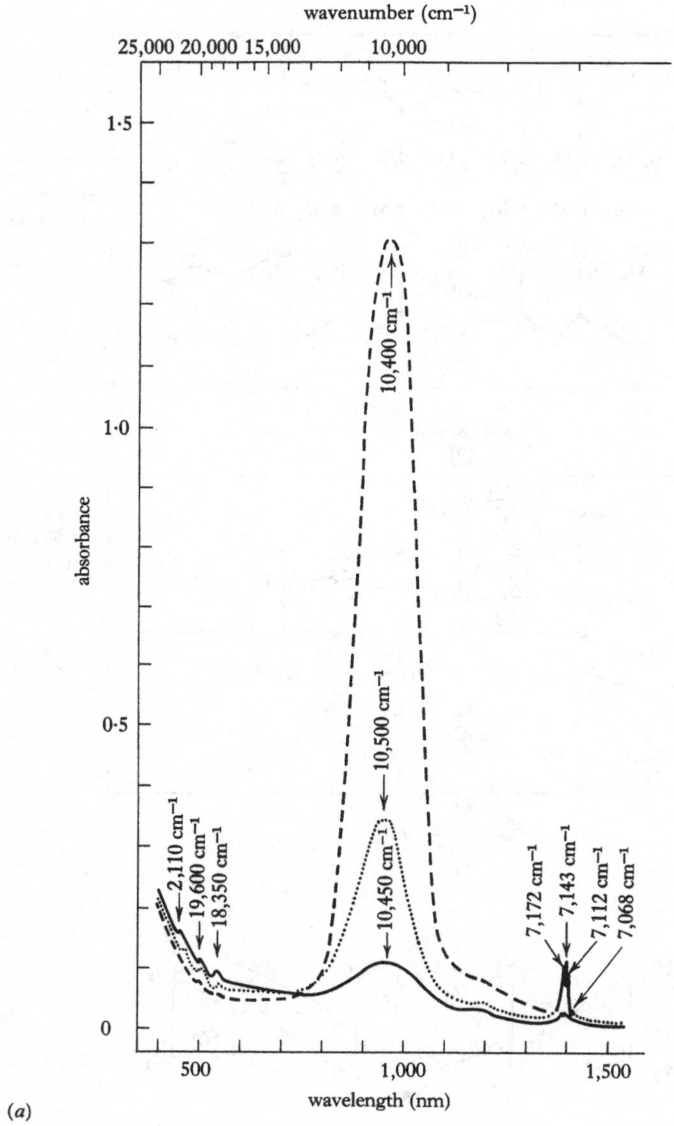

Figure 5.19 Polarized absorption spectra of Mg–Fe amphiboles. ······· α spectra; – – – – β spectra; —— γ spectra. [Optic orientation: α : *a* = 4°; β = *b*; γ : *c* = 15°.] (*a*) Cummingtonite, with 35.8 mole per cent $Fe_7Si_8O_{22}(OH)_2$; (*b*) (*on facing page*) grunerite, with 95.3 mole per cent $Fe_7Si_8O_{22}(OH)_2$. Both specimens, 30 μm thick, are from the Quebec Iron Formation.

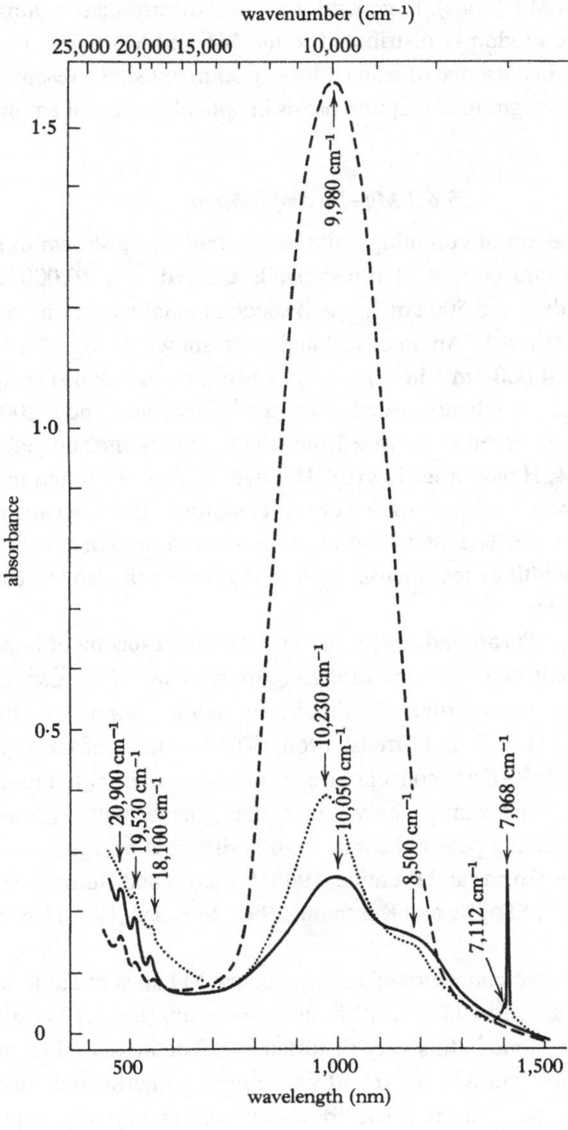

(*b*)

symmetry (M2), and the cations are either centrosymmetric (M3 site) or nearly so (M1 and M2 sites). However, the M4 coordination site is very distorted and cations do not lie at the centre of symmetry of the site. A variety of X-ray crystal structure refinements and spectroscopic measurements have shown that Fe^{2+} ions enter preferentially any vacancies in the M4 positions (except in competition with Mn^{2+} ions). In general, Fe^{2+} ions discriminate against the M2 positions and are randomly distributed in the M1 and M3 positions of many amphiboles. The occurrence of iron cations in so many sites present problems of resolving and assigning absorption bands in optical spectra of amphiboles.

5.6.3 Mg–Fe amphiboles

The polarized spectra of cummingtonite and grunerite are shown in fig. 5.19. The α and γ spectra consist of broad bands centred near 10,000 cm^{-1} and prominent shoulders at 8,500 cm^{-1}. The β spectra contain a very broad band at 10,000 to 10,400 cm^{-1}. An intense band (not shown in fig. 5.19) is also observed around 4,000 cm^{-1} in α spectra of Mg–Fe amphiboles (Hawthorne, 1981b). Similar intense bands near 10,800 cm^{-1} (β spectra) and 4,200 cm^{-1} (α spectra) are also observed in polarized spectra of anthopyllite and gedrite (Mao and Seifert, 1974; Hawthorne, 1981b). The two spectra illustrated in fig. 5.19 indicate that all bands migrate to longer wavelengths with increasing iron content of the amphibole. The intensities of the bands and shoulders in the α and γ spectra and the width of the intense band in the β spectra also increase in the spectra of grunerites.

Several sharp peaks around 7,100 cm^{-1} occur in the α spectra of most amphiboles and represent first overtone stretching frequencies of oxygen–hydrogen bonds of OH^- anions coordinated to different cation assemblages in the M1 and M3 positions (Burns and Strens, 1966). The positions of these peaks are unaffected by Mg^{2+}–Fe^{2+} concentrations in the cummingtonite–grunerite series, but their number and relative intensities change with increasing iron content. Corresponding peaks occur at slightly different energies in sodic and calcic amphiboles (Burns and Prentice, 1968; Bancroft and Burns, 1969; Burns and Greaves, 1971; Skogby and Rossman, 1991; Mustard, 1992) (see figs 4.15 and 5.20).

Since Fe^{2+} ions are concentrated in the acentric M4 sites of cummingtonite, the intense band around 10,000 cm^{-1} in the β spectrum (fig. 5.19*a*) arises from absorption by Fe^{2+} ions in this very distorted site. The increased intensity and width of absorption bands in the α and γ spectra of grunerite and the broadening of the intense band in the β spectra result from increased occupancies of Fe^{2+} ions in the more regular M1, M2 and M3 octahedral sites.

There is insufficient spectral resolution to determine the energy separations between all 3*d* orbitals of each Fe^{2+} ion in all four sites of the cummingtonite structure. Approximate energy separations may be obtained from correlations with the spectra of orthopyroxenes. The cummingtonite M4 and orthopyroxene M2 sites resemble one another in the nature of the oxygen ligands and the distortions of coordination polyhedra. Furthermore, the direction of β-polarized light in cummingtonite is related crystallographically to that of α-polarized light in orthopyroxene: both lie in the plane and across bands of cations in edge-shared octahedra in the structures. Therefore, the intense bands at 10,000 to 10,400 cm^{-1} in the β spectra of cummingtonite and grunerite, as well as the bands at 4,200 to 4,000 cm^{-1} in these and other Mg^{2+}–Fe^{2+} amphiboles (Mao and Seifert, 1974; Hawthorne, 1981b), have similar origins to the bands at 11,000 to 10,700 cm^{-1} and 5,400 to 4,900 cm^{-1} in the crystal field spectra of orthopyroxenes (fig. 5.15); both originate from Fe^{2+} ions in very distorted, acentric M4 (amphibole) or M2 (pyroxene) sites. Using estimates of 1,000 cm^{-1} and 2,000 cm^{-1} for the lower-level t_{2g} orbital splitting of Fe^{2+} ions in amphibole M4 sites (Goldman and Rossman, 1977b), the calculated crystal field parameters are approximately

$$\text{M4 site: } \Delta_o = 6{,}300 \text{ to } 6{,}000\ cm^{-1};\ \text{CFSE} = 3{,}520 \text{ to } 3{,}400\ cm^{-1} \quad (5.16)$$

for the composition range 35–100 mole per cent $Fe_7Si_8O_{22}(OH)_2$ in the cummingtonite–grunerite series.

The breadth of the band near 10,100 cm^{-1} in the α and γ spectra of grunerite and the shoulder at 8,500 cm^{-1} are attributed to absorption by Fe^{2+} ions in the M1, M2 and M3 positions. Assuming a lower-level splitting of 500 cm^{-1} for the t_{2g} orbitals of Fe^{2+} ions located in these relatively undistorted octahedral sites, an energy level diagram similar to that for the orthopyroxene M1 site (fig. 5.16) may be constructed leading to the approximate crystal field parameters

$$\text{M1} \approx \text{M2} \approx \text{M3 sites: } \Delta_o = 9{,}250 \text{ to } 9{,}050\ cm^{-1};$$
$$\text{CFSE} = 3{,}950 \text{ to } 3{,}850\ cm^{-1} \quad (5.17)$$

for the cummingtonite–grunerites series. Note that the CFSE values decrease with increasing Fe^{2+} ion contents of the M4 and (M1,M2,M3) sites for Mg^{2+}–Fe^{2+} amphiboles

5.6.4 Calcic amphiboles

The optical spectra of calcic amphiboles, with approximate formulae $Ca(Mg,Fe)_5Si_8O_{22}(OH)_2$, have been described in several papers (e.g., White and Keester, 1966; Burns, 1968b; Goldman and Rossman, 1977b, 1982; Aldridge *et al.*, 1982; Mustard, 1992). Polarized spectra of a typical actinolite are shown in fig. 5.20. The α spectrum consists of broad bands spanning the 800–1,200 nm wavelength range and several sharp peaks near 1,400 nm (7,100 cm^{-1}) representing overtone vibrations of O–H bonds of hydroxyl groups coor-

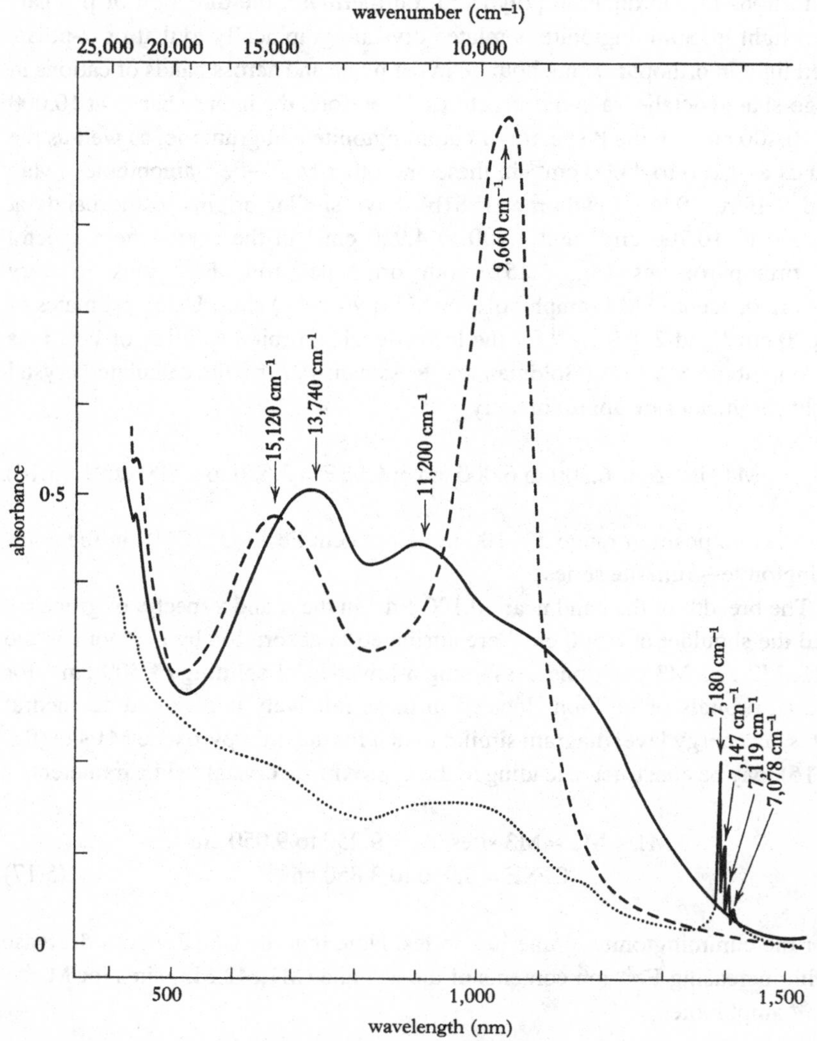

Figure 5.20. Polarized absorption spectra of a calcic amphibole. ········ α spectra; – – – – β spectra; —— γ spectra. [Optic orientation: α : *a* = 5°; β = *b*; γ : *c* = 10°.] The actinolite, containing 10 mole per cent $Ca_2Fe_5Si_8O_{22}(OH)_2$ and 1.40 wt per cent Fe_2O_3 is from California.

dinated to different cation assemblages in the M3 and M1 sites (Burns and Strens, 1966). In several calcic amphiboles, an additional band superimposed on vibrational combination peaks near 4,050 cm^{-1} (not shown in fig. 5.20) has been identified in α spectra (Goldman and Rossman, 1977b, 1982). The β spectrum contains an intense band at 9,660 cm^{-1} and a less intense band of variable intensity around 15,100 cm^{-1}. The γ spectrum shows absorption bands at 11,200 cm^{-1} and 13,740 cm^{-1}, as well as a prominent inflexion at about 10,000 cm^{-1}. The bands at 15,120 cm^{-1} (β spectrum) and 13,740 cm^{-1} (γ spectrum) represent $Fe^{2+} \rightarrow Fe^{3+}$ IVCT transitions between neighbouring Fe^{2+} and Fe^{3+} ions in edge-shared octahedra parallel to the *b* and *c* axes (cf. figs 4.14 and 4.15). The remaining absorption bands represent superimposed crystal field transitions within Fe^{2+} ions located in the M1, M2, M3 and M4 sites of the actinolite structure.

The two bands at 9,710 cm^{-1} (β spectra) and 4,050 cm^{-1} (α spectra) reported for calcic amphiboles (Goldman and Rossman, 1977b; 1982) appear to originate from small amounts of Fe^{2+} ions filling calcium vacancies in the very distorted, acentric M4 sites of subcalcic actinolites, although this assignment for amphiboles containing a surplus of Ca^{2+} and Na^{+} ions to fill M4 positions has been debated (Aldridge *et al.*, 1982; Goldman and Rossman, 1982). If these are, indeed, Fe^{2+}/M4 site CF bands, energy level calculations analogous to those for the cummingtonite M4 and orthopyroxene M2 sites would lead to estimated crystal field parameters of Δ_o = 5,890 cm^{-1} and CFSE = 3,350 cm^{-1} for Fe^{2+} in actinolite M4 sites. The remaining spectral profiles of actinolites are too complex to deduce separate energy level diagrams for Fe^{2+} ions in the individual M1, M2 and M3 sites of calcic amphiboles. However, similarities between the actinolite M2 and diopside M1 sites suggest that the crystal field parameters are comparable for these two sites. Also, the configurations of the actinolite M1 and M3 sites resemble the corresponding sites in cummingtonite (fig. 5.18) so that the Δ_o and CFSE parameters in eq. (5.17) are expected to be similar for Fe^{2+} ions in these octahedral sites of calcic and Mg–Fe amphiboles.

5.6.5 Miscellaneous amphiboles

The polarized spectra of alkali amphiboles such as those of glaucophane illustrated in fig. 4.15, have been studied extensively (Chesnokov, 1961; Littler and Williams, 1965; Bancroft and Burns, 1969; Faye and Nickel, 1970a; Smith and Strens, 1976) and are dominated by $Fe^{2+} \rightarrow Fe^{3+}$ IVCT bands (§4.7.2). However, Fe^{2+} crystal field bands occurring at 10,000 cm^{-1} and 8,600 cm^{-1} are distinguishable in fig. 4.15, indicating that glaucophane provides CFSE's comparable to those of Fe^{2+} ions in M1 and M3 octahedral sites of other amphiboles, eq. (5.17).

Crystal field spectra of other transition metal-bearing amphiboles, including those for V^{3+}- and Cr^{3+}-bearing tremolites (Neuhaus, 1960; Hawthorne, 1981b; Schmetzer, 1982), are summarized in table 5.7. Absorption bands in polarized spectra of the Mn^{3+}-bearing alkali amphiboles winchite and juddite originate from cations in the M2 sites (Ghose *et al.*, 1986; Langer, 1988), which provide Jahn–Teller stabilization for Mn^{3+} ions in the amphibole structure (§6.3).

5.7 Spectra of ring silicates

5.7.1 Crystal structures

A common structural feature of minerals such as beryl ($Be_3Al_2Si_6O_{18}$), cordierite ($Mg_2Al_4Si_5O_{18}$), and tourmaline [e.g., schorl, $(Na,Ca)(Fe^{2+},Mg)_3(Al,Fe^{3+})_6(BO_3)_3Si_6O_{18}(OH,F)_4$] is that they each contain six-membered rings of linked [SiO_4] tetrahedra that contribute non-bridging oxygen ligands to coordination polyhedra surrounding cations in octahedral and tetrahedral sites. In beryl, the [Si_6O_{18}] rings are arranged in layers in the basal plane and are linked laterally and vertically by Be^{2+} and Al^{3+} ions in tetrahedral and octahedral coordinations, respectively, and are stacked one above the other along the (hexagonal) *c* axis. In cordierite, [Si_5AlO_{18}] rings form channels along the (orthorhombic) *c* axis, and the rings are linked by Mg^{2+} and Al^{3+} in octahedral (M) and tetrahedral (T_2) sites. In tourmaline, the [Si_6O_{18}] rings are connected to spirals of edge-shared [AlO_6] octahedra extending along the (trigonal) *c* axis and are further linked to trimers of edge-shared [MgO_6] octahedra and triangular BO_3^{3-} groups. Hydroxyl anions also constitute the oxygen coordination polyhedra about the octahedral Al and Mg sites of tourmaline. Atomic substitution is prevalent in each of these ring silicates, particularly in the octahedral Mg sites of tourmaline and Al sites of beryl. In addition, the channels within the stacked silicate rings may accommodate a variety of ion and molecular species, including alkali metal cations, anions such as OH^-, F^-, and CO_3^{2-}, molecular species such as H_2O and CO_2, inert gases (He, Ar), actinide elements and their decay products, and perhaps transition metal ions themselves, including Fe^{2+} and Fe^{3+}. These channel constituents of ring silicates interfere with and complicate interpretations of CF bands in visible to near-infrared spectra. Crystal field spectral data for ring silicates are summarized in table 5.8 and discussed in the following sections.

5.7.2 Beryl

Although there is general agreement that the broad band centred at 16,100 cm^{-1} causing the blue colour of aquamarine represents a $Fe^{2+} \rightarrow Fe^{3+}$ IVCT transition (table 4.2), uncertainty exists over the assignments of two other bands at 12,200 cm^{-1} and 10,300 cm^{-1} (Farrell and Newnham, 1967; Wood and Nassau, 1968;

Samoilovich *et al.*, 1971; Price *et al.*, 1976; Loeffler and Burns, 1976; Parkin *et al.*, 1977; Goldman *et al.*, 1978; Blak *et al.*, 1982). The 12,200 cm^{-1} band, together with another one resolved around 4,760 cm^{-1} amidst vibrational features, was assigned to Fe^{2+} ions in a channel site (Goldman *et al.*, 1978). A later interpretation assigned the 12,200 cm^{-1} band to Fe^{2+} replacing Be^{2+} ions in tetrahederal sites (Price *et al.*, 1976). The assignment adopted here is that the two bands at 12,200 cm^{-1} and 10,300 cm^{-1} originate from CF transitions in Fe^{2+} ions replacing Al^{3+} ions in the octahederal sites (point group D_3; mean Al–O = 190.6 pm) of the beryl structure. They yield Fe^{2+} crystal field parameters of $\Delta_o = 11{,}000$ cm^{-1} and CFSE = 4,650 cm^{-1}.

In optical spectra of emerald (Neuhaus, 1960; Poole, 1964; Wood and Nassau, 1968; Schmetzer and Bank, 1981), Cr^{3+} CF bands are located near 16,130 cm^{-1} and 23,530 cm^{-1} and are assigned to cations in octahedral sites. Similar bands for octahedral V^{3+} ions in beryl occur around at 16,000 cm^{-1} and 23,800 cm^{-1} (Beckwith and Troup, 1973; Schmetzer, 1982; Ghera and Lucchesi, 1987). Spectral features of pink and red beryls in the region 18,000–20,000 cm^{-1} (Wood and Nassau, 1968) may originate from crystal field transitions in Mn^{3+} ions in morganite.

5.7.3 Cordierite

Numerous investigations of cordierite (Farrell and Newnham, 1967; Faye *et al.*, 1968; Smith and Strens, 1976; Pollak, 1976; Smith, 1977; Goldman and Rossman, 1978b; Vance and Price, 1984) have identified several absorption bands centred around 10,000 cm^{-1} and a broad band at 17,700 cm^{-1}. The latter occurring in the β- and γ-polarized light (**E**||*b* and **E**||*a* spectra, respectively) is assigned to a $Fe^{2+} \rightarrow Fe^{3+}$ IVCT transition involving Fe^{2+} and Fe^{3+} ions located in adjacent octahedral and tetrahedral sites, respectively (table 4.2). From combined Mössbauer and optical spectral measurements, it was suggested (Goldman *et al.*, 1977) that peaks at 10,200 cm^{-1} and 8,700 cm^{-1} in α-polarized light (**E**||*c* spectrum) originate from Fe^{2+} replacing Mg^{2+} ions in octahedral sites (point group C_2: mean Mg–O = 211.0 pm). Bands centred near 10,600 cm^{-1} in the β and γ spectra were assigned to Fe^{2+} ions located in the channel sites. It was also suggested that some Fe^{3+} ions occur in the channel sites, leading to an alternative assignment of the 17,700 cm^{-1} band, that it originates from a $Fe^{2+} \rightarrow Fe^{3+}$ IVCT transition involving octahedral Fe^{2+} ions and channel-site Fe^{3+} ions. In spectral studies of the related mineral, osumilite, $K(Mg,Fe)_2(Al,Fe^{3+})(Si,Al)_{12}O_{30}$, (Faye, 1972; Goldman and Rossman, 1978b), bands at 10,280 cm^{-1} and 7,020 cm^{-1} (γ spectrum) were assigned to Fe^{2+} in structural octahedral sites, while bands in the 10,200 to 10,400 cm^{-1} (α and β

Table 5.8. *Crystal field parameters for transition metal ions in ring, layer and framework silicates*

Mineral	Site or symmetry	Positions of absorption bands (cm^{-1})	Δ (cm^{-1})	CFSE (cm^{-1})	Sources of data
V^{3+} beryl	Al	16,000; 23,800	17,240	13,790	[1]
V^{3+} beryl	Al	16,450; 23,260	17,600	14,080	[2,3]
		15,300; 25,000	16,500	13,200	
emerald (Cr^{3+} beryl)	Al	16,130; 23,530	16,130	19,360	[4,5]
morganite (Mn^{3+} beryl)	Al	(18,020); 18,520; 20,200; (28,170)			[6]
heliodore (Fe^{3+} beryl)	Al	12,300; 17,500; 23,600; 26,500		0	[7]
		14,200; 20,000			
aquamarine (Fe^{2+} in beryl)	Al	12,200; 10,300	11,000	4,650	[8]
	channel	4,760; 12,200			
cordierite (Fe^{2+})	Mg/oct	10,200; 8,700	9,150	3,900	[9]
Fe^{2+} tourmaline	Al	9,500; 14,500	11,000	4,900	[10]
	Mg	7,900; 13,200	10,000	4,500	
V^{3+} tourmaline	Al	16,700; 22,700	17,370	13,900	[1]
		16,400; 23,900			
Cr^{3+} tourmaline	Al	17,000; 24,000	17,000	20,400	[11]
Mn^{3+} tourmaline	Al	9,000; 13,500; 18,400; 22,700	13,900	12,840	[12]
V^{3+} kornuperine	M4	15,200; 23,300	16,420	13,140	[1]
Cr^{3+} kornuperine	M4	17,300; 23,000	17,300	20,760	[1]
eudialyte (Fe^{2+})	sq.planar	7,150; 18,900			[13]
	tetrag.pyr.	4,000; 10,900			
Ti^{3+} phlogopite (fitzroyite)	M1,M2	22,000 to 24,000	23.000	9,200	[14]
Mn^{3+} phlogopite (manganophyllite)	M2	14,850; 19,050; 21,400; (23,800)	12,800 (14,760)	15,105 (14,060)	§5.8.3
Fe^{3+} phlogopite	tet	19,200; 20,300; 22,700; 25,000		0	[15,16]
biotite (Fe^{2+})	M1,M2	11,900; 8,900	10,000	4,400	§5.8.3
V^{3+} roscoelite	oct	16,200; 23,500	17,400	13,920	[1]
V^{3+} muscovite	M2	16,950; 24,000	18,130	14,505	[1]
Cr^{3+} muscovite	M2	16,200; 23,800	16,200	19,440	[17]
Mn^{3+} muscovite	M2	13,200; 18,250; 21,850	13,450 (16,370)	14,670 (16,420)	[18]
Fe^{3+} muscovite	M2	11,600; 14,700; 21,500; 26,600; 27,400		0	[14]
	C_1	17,200; 19,650; 22,600			[18–22]
Cr^{3+} clinochlore	oct	18,200; 25,300	18,200	21,840	[17]
kammererite	oct	18,450; 25,000	18,450	22,140	[4]
Cr^{3+} chlorite	oct	18,020; 25,640	18,020	21,625	[23]

Table 5.8 *continued*

Mineral	Site or symmetry	Positions of absorption bands (cm^{-1})	Δ (cm^{-1})	CFSE (cm^{-1})	Sources of data
Cr^{3+} smectite	oct	16,800; 24,000	16,800	20.160	[17]
Cr^{3+} stichtite	oct	18,500; 25,300	18,500	22,200	[17]
Mn^{3+} montmorillonite	M1,M2	10,400; 18,800; 20,600; 22,400	15,400 (16,301)	14,440 (14,980)	[24]
nontronite (Fe^{3+})	oct	10,692; 15,775; 22,400; 26,509; 27,067	15,050	0	[25,26]
chlorite (Fe^{2+})	Mg/oct	11,500; 9,500	10,200	4,300	§5.8.4.1
chloritoid (Fe^{2+})	Mg/oct	10,900; 8,000	9,000	4,050	§5.8.4.2
garnierite (Ni clay silicates)	oct	9,100; 15,200; 26,300; (13,000)	9,100	10,910	[27–29]
Ni^{2+} talc (kerolite)	oct	8,950; 17,325; 25,800; 13,868	8,950	10,740	[27]
Ni^{2+} lizardite	oct	8,845	8,845	10,615	[27]
nepouite (Ni^{2+} serpentine)	oct	9,120	9,120	10,945	[27]
gillespite (Fe^{2+})	sq.planar	20,000; 8,300	11,600	6,125	[30]
$CaCrSi_4O_{10}$	sq. planar	14,925; 19,570; 22,075	14,925	18,330	[31]
cuprorivaite	sq. planar	12,740; 16,130; 18,520	12,740		[30]
$BaCuSi_4O_{10}$	sq. planar	12,900; 15,800; 18,800	12,900		[32]
plagioclase (Fe^{2+})	Ca/oct	8,500; 4,500	5,730	3,550	[33]
orthoclase (Fe^{3+})	tet	20,700; 22,650; 24,000; 26,500		0	[34,35]
quartz (Fe^{3+})	tet	18,700; 20,200; 22,500; 24,800; 27,300		0	[36]

Sources of data : [1] Schmetzer (1982); [2] Beckwith & Troup (1973); [3] Ghera & Lucchesi (1987); [4] Neuhaus (1960); [5] Parkin & Burns (1980); [6] Wood & Nassau (1968); [7] Low & Dvir (1960); [8] Goldman, Rossman & Parkin (1978); [9] Goldman, Rossman & Dollase (1977); [10] Faye, Manning, Gosselin & Tremblay (1974); [11] Manning (1968a); [12] Manning (1973); [13] Pol'shin *et al* (1991)[14] Faye (1968a); [15] Faye & Hogarth (1969); [16] Hogarth, Brown & Pritchard (1970); [17] Calas, Manceau, Novikoff & Boukili (1984); [18] Annersten & Hålenius (1976); [19] Richardson (1975); [20] Kleim & Lehmann (1979); [21] Richardson (1976); [22] Finch, Gainsford & Tennant (1982); [23] Bish (1977); [24] Sherman & Vergo (1988a); [25] Karickhoff & Bailey (1973); [26] Sherman & Vergo (1988b); [27] Manceau & Calas (1985); [28] Faye (1974); [29] Cervelle (1991); [30] Burns, Clark & Stone (1966); [31] Belsky, Rossman, Prewitt & Gasparik (1984); [32] Ford & Hitchman (1979); [33] Clark & Burns (1967); [34] Hofmeister & Rossman (1983); [35] Faye(1969); [36] Lehmann & Bambauer (1973) .

spectra) and 4,650 to 4,700 cm^{-1} (γ spectra) were assigned to Fe^{2+} ions in the channels. Yet another interpretation offered of the 10,600 cm^{-1} band of cordierite is that it originates from Fe^{2+} ions replacing tetrahedrally coordinated Al^{3+} ions in the cordierite structure (Pollak, 1976; Vance and Price, 1984). Thus, while the origin of the band at 10,800 cm^{-1} remains unclear, the two absorption bands at 10,200 and 8,700 cm^{-1} may be assigned to Fe^{2+} ions in octahedral sites of the cordierite structure, yielding estimated Δ_o and CFSE values of about 9,150 cm^{-1} and 3,900 cm^{-1}, respectively.

5.7.4 Tourmalines

The optical spectra of blue tourmalines have attracted considerable attention focused mainly on assignments of $Fe^{2+} \rightarrow Fe^{3+}$ IVCT bands, positions of crystal field bands for Fe^{2+} ions expected to be located in two different octahedral sites, and intensification mechanisms of these crystal field bands (e.g., Faye *et al.*, 1968., 1974; Wilkins *et al.*, 1969; Burns, 1972a; Smith, 1977, 1978a; Mattson and Rossman, 1984, 1987b). Curve-resolved spectra yielded two sets of paired bands (Faye *et al.*, 1974). One set at 14,500 cm^{-1} and 9,500 cm^{-1} assigned to Fe^{2+} in the Al or *c*-site (point group C_1; mean Al–O = 192.9 pm) yielded Δ_o and CFSE values of about 11,000 cm^{-1} and 4,900 cm^{-1}, respectively. The second set of bands at 13,200 cm^{-1} and 7,900 cm^{-1} attributed to Fe^{2+} in the Mg or *b*-site (point group C_m; mean Fe–O = 202.5 pm) provided Δ_o and CFSE values of approximately 10,000 cm^{-1} and 4,500 cm^{-1}, respectively.

Crystal field spectra for a Cr^{3+}–bearing tourmaline (Manning, 1968b) contain bands at 17,000 cm^{-1} and 24,000 cm^{-1}, while a V^{3+}-bearing tourmaline has absorption bands at 16,400–16,700 cm^{-1} and 22,700–23,900 cm^{-1} (Schmetzer, 1982). Correlations with crystal structure refinements of Cr^{3+}- and V^{3+}-bearing tourmalines (Nuber and Schmetzer, 1979; Foit and Rosenberg, 1979) indicate that these absorption bands originate from transition metal ions located mainly in the Mg sites. In some black tourmalines, broad, intense bands located at 18,400 cm^{-1} and 22,700 cm^{-1} and at 19,200 cm^{-1} in pink 'watermellon' tourmalines (Manning, 1968b) have been assigned to Mn^{3+} ions (Manning, 1973).

5.7.5 Eudialyte

Another silicate mineral besides gillespite that contains planar four-coordinated Fe^{2+} ions is eudialyte, ideally $Na_{16}(REE,Ca)_6Fe_3Zr_3(Si_3O_9)_2(Si_9O_{27})_2(OH,Cl)_4$, a complex cyclosilicate (Rastsvetaeva and Borutsky, 1988) also containing significant amounts of lanthanide or rare earth elements (REE). The

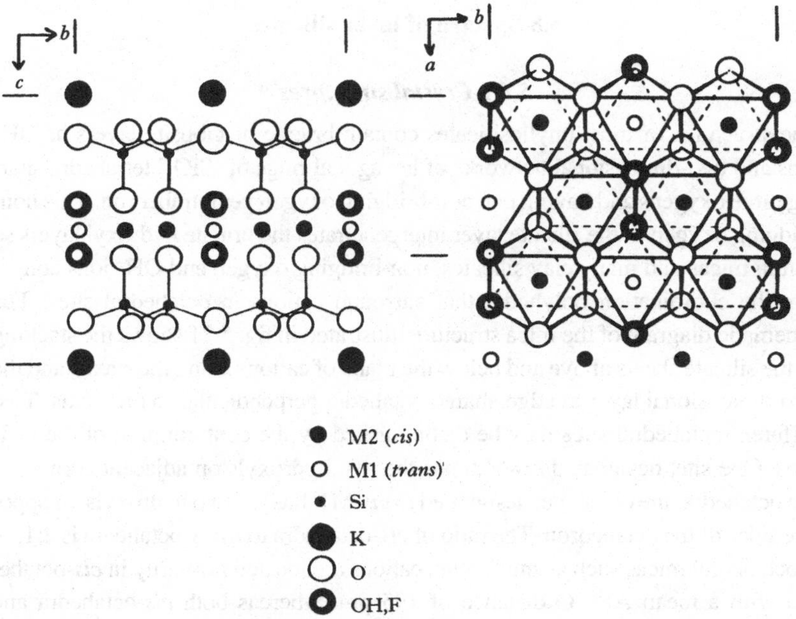

Figure 5.21 Schematic diagram of the mica crystal structure. Left: projected onto (100) showing the locations of the cation sites and OH^- anions; right: projected onto (001), showing the configurations of the *cis*-M2 and *trans*-M1 coordination polyhedra (from Dyar & Burns, 1986).

$[FeO_4]$ group in eudialyte forms a rectangle of four oxygens with the Fe^{2+} ions shifted slightly beyond the plane, as in gillespite. The Fe–O distances in eudialyte, 203–206 pm, are slightly larger than those in gillespite (200 pm; see fig. 3.19). Crimson and pink optically positive eudialytes are characterized by a weakly pleochroic absorption band around 18,900 cm^{-1} with less intense bands occurring at 10,900 cm^{-1}, 7,150 cm^{-1} and 4,000 cm^{-1} (Pol'shin *et al.*, 1991). At low temperatures, the intensity of the 18,900 cm^{-1} band decreases whereas that of the 10,900 cm^{-1} band increases, indicating different origins. In spectra of brown and yellow-brown optically negative eudialytes, bands at 10,900 cm^{-1} and 4,000 cm^{-1} predominate. The two bands at 18,900 cm^{-1} and 7,150 cm^{-1} originate from Fe^{2+} ions in the planar rectangular site, and are analogous to those of gillespite. The other two bands at 10,900 cm^{-1} and 4,000 cm^{-1} have been assigned to five-coordinated Fe^{2+} ions located within a pyramidal site consisting of the four rectangular oxygen atoms (Fe–O = 209 pm) and an axial OH^- ion located 221 pm above the Fe^{2+} ion.

5.8 Spectra of layer silicates

5.8.1 Crystal structures

The structures of most phyllosilicates contain brucite or gibbsite layers of OH^- ions and two-dimensional networks of hexagonal rings of $[SiO_4]$ tetrahedra sharing three oxygens and having one non-bridging oxygen per tetrahedron. This non-bridging oxygen of the silicate layer interpenetrates the brucite hydroxyl layers so that in micas and many clay silicates, non-bridging oxygen and OH^- ions constitute the coordination polyhedra that surround cations in octahedral sites. The schematic diagram of the mica structure illustrated in fig. 5.21 shows the stacking of the silicate sheets above and below the plane of cations along the *c* axis, and the two-dimensional layer of edge-shared octahedra perpendicular to the *c* axis. Two different octahedral sites may be distinguished by the configuration of the OH^- ions. One site, designated *cis*-M2, has the two hydroxyls on adjacent corners of the octahedra; the other site, designated *trans*-M1, has the two hydroxyls on opposite sides of the octahedron. The ratio of *cis*-octahedra to *trans*-octahedra is 2:1. In dioctahedral micas such as muscovite, cations are located primarily in *cis*-octahedra with a mean Al^{3+}–O distance of 195 pm, whereas both *cis*-octahedra and *trans*-octahedra are occupied in trioctahedral micas of the phlogopite–annite (biotite) series with mean Mg^{2+}–O distances in phlogopite of 206.4 pm and 207.6 pm, respectively. In the chlorite crystal structure, brucite layers containing Mg^{2+} (and some Al^{3+} ions) octahedrally coordinated to OH^- ions alone alternate with the double silicate sheets along the *c* axis. Crystal field spectral data for layer silicates contained in table 5.8 are described below.

5.8.2 Muscovites

Optical spectra of transition metal-bearing muscovites (Rossman, 1984) include Cr^{3+}-bearing micas (Faye, 1968a; Neuhaus, 1960; Calas *et al.*, 1984) with absorption bands near 16,200 cm^{-1} and 23,800 cm^{-1}, V^{3+}-bearing micas (Schmetzer, 1982) consisting of bands at 16,950 cm^{-1} and 24,000 cm^{-1} ($2M_1$ muscovite) or 16,200 cm^{-1} and 23,500 cm^{-1} (roscoelite), and Mn^{3+}-bearing muscovites containing absorption bands at 13,200 cm^{-1}, 18,250 cm^{-1} and 21,850 cm^{-1} (Richardson, 1975, 1976; Annersten and Hålenius, 1976).

5.8.3 Biotites

The numerous optical spectral measurements of biotites (e.g., Faye, 1968b; Robbins and Strens, 1972; Kleim and Lehmann, 1979; Smith and Strens, 1976;

Smith, 1977; Smith *et al.*, 1980, 1983) reviewed by Rossman (1984) have centred on origins of the diagnostic basal-plane pleochroism attributed to $Fe^{2+} \rightarrow Fe^{3+}$ and $Fe^{2+} \rightarrow Ti^{4+}$ IVCT transitions (table 4.2). Extractable from these spectra, however, are absorption bands around 11,900 cm^{-1} and 8,900 cm^{-1} attributed to octahedrally coordinated Fe^{2+} ions. These bands lead to estimated values for Δ_o and CFSE of about 10,000 cm^{-1} and 4,400 cm^{-1}, respectively, representing averaged values for Fe^{2+} ions in the *trans*-M1 and two *cis*-M2 octahedra.

Absorption spectra of manganophyllite measured with light polarized in the cleavage plane of the mica contain bands at 14,500 cm^{-1}, 19,050 cm^{-1}, 21,400 cm^{-1} and 23,800 cm^{-1} attributable to Jahn–Teller splitting of the 3*d* orbitals of Mn^{3+} ions. An alternative assignment of these bands is that they represent pair excitations (cf. §3.7.3) involving tetrahedral Fe^{3+} and octahedral Mn^{2+} ions in the phlogopite structure (Smith *et al.*, 1983). Fitzroyite, a titanian phlogopite, has an asymmetric absorption band centred near 22,000 cm^{-1} attributable to Ti^{3+} ions.

5.8.4 Miscellaneous layer silicates

5.8.4.1 *Chlorites* have been studied spectroscopically mainly on account of $Fe^{2+} \rightarrow Fe^{3+}$ IVCT bands near 14,300 cm^{-1} that contribute to their optical spectra (e.g., White and Keester, 1966; Faye, 1968b; Smith and Strens, 1976; Smith, 1977). Two other bands centred near 11,500 cm^{-1} and 9,500 cm^{-1} provide estimates for the crystal field parameters of Fe^{2+} ions in chlorite of Δ_o = 11,200 cm^{-1} and CFSE = 4,300 cm^{-1}. Crystal spectra of Cr^{3+}-bearing chlorite, kämmererite, yield absorption bands at 18,450 cm^{-1} and 25,000 cm^{-1}, giving Δ_o = 18,450 cm^{-1} and a CFSE of 22,140 cm^{-1} for octahedrally coordinated Cr^{3+} ions surrounded by OH^- ions in the brucite sheets. The spectra of other Cr^{3+}-bearing clay silicates have been described (Calas *et al.*, 1984), including clinochlore and stichtite.

5.8.4.2 *Chloritoid.* Optical spectra of chloritoids, again studied mainly on account of the $Fe^{2+} \rightarrow Fe^{3+}$ IVCT band at 16,300 cm^{-1} (Faye *et al.*, 1968; Hålenius *et al.*, 1981), also contain features assignable to CF transitions in Fe^{2+} ions. These cations are located in the M1B positions in brucite layers which are surrounded by four OH^- ions and two *trans-* non-bridging oxygens belonging to isolated $[SiO_4]$ tetrahedra in silicate layers in the chloritoid structure. The two absorption bands at 10,900 cm^{-1} and 8,000 cm^{-1} yield approximate values of Δ_o = 9,000 cm^{-1} and CFSE = 4,050 cm^{-1}, respectively, for the Fe^{2+} ions.

5.8.4.3 '*Garnierite.*' A number of early investigations of garnierites yielded crystal field bands characteristic of octahedrally coordinated Ni^{2+} ions (Lakshman and Reddy, 1973b; Tejedor-Tejedor *et al.*, 1983; Faye, 1974; Cervelle, 1991). Garnierite, in fact, is a complex mixture of several phyllosilicates, including nepouite (Ni lizardite) and pimelite (Ni kerolite or talc). More recent diffuse reflectance spectral measurements of hand-picked samples provided spectral data for Ni^{2+} in Mg–Fe–Ni lizardite (Cervelle and Maquet, 1982; Manceau and Calas, 1985), nepouite, pimelite and Mg–Ni kerolite (Manceau and Calas, 1985). These data are included in table 5.19 discussed later (§5.10.11).

5.8.4.4. *Gillespite*, $BaFeSi_4O_{10}$, a sheet silicate containing rings of four linked $[SiO_4]$ tetrahedra and accommodating Fe^{2+} ions in square planar coordination at ambient pressures (see fig. 3.19), has been the focus of numerous spectroscopic and crystallographic studies, including the polarized spectral measurements described in §3.8.2 (Burns *et al.*, 1966; Strens, 1966b; Mackay *et al.*, 1979). Above 2.6 GPa, a phase change occurs leading to puckering of the silicate sheets and a change of symmetry of the four-coordination site from square-planar to distorted tetrahedral (Hazen and Burnham, 1974). High-pressure spectral measurements of gillespite (Abu-Eid *et al.*, 1973) made above the phase transition show new absorption bands occurring at 17,200 cm^{-1} and 7,150 cm^{-1} which are displaced from positions near 20,000 cm^{-1} and 8,000 cm^{-1} in the 1 atmosphere spectra (fig. 3.3). Other optical spectral measurements of phases isostructural with gillespite include Cu^{2+} in cuprorivaite, $CaCuSi_4O_{10}$ (Egyptian blue) (Clark and Burns, 1967; Ford and Hitchman, 1979), and Cr^{2+} in synthetic $CaCrSi_4O_{10}$ (Belsky *et al.*, 1984).

5.9 Spectra of framework silicates

In framework silicates, including quartz and feldspars, all oxygens of the $[SiO_4]$ and $[(Si,Al)O_4]$ tetrahedra are shared with other tetrahedra, producing cavities which generally accommodate cations larger than the transition metal ions. As a result, concentrations of transition metals in framework silicates are generally low. In some cases, notably Fe^{3+}, trivalent cations may replace Al in the $[(Si,Al)O_4]$ tetrahedra. Thus, gem quality yellow sanidine, orthoclase and labradorite produce optical spectra with an assemblage of peaks at 20,700 cm^{-1}, 22,656 cm^{-1}, 24,000 cm^{-1} and 26,500 cm^{-1} originating from tetrahedrally coordinated Fe^{3+} ions (Faye, 1969; Hofmeister and Rossman, 1983). The spectra of plagioclase feldspars found in rocks from the Moon, as well as some terrestrial sources, such as those illustrated in fig. 5.22, give broad

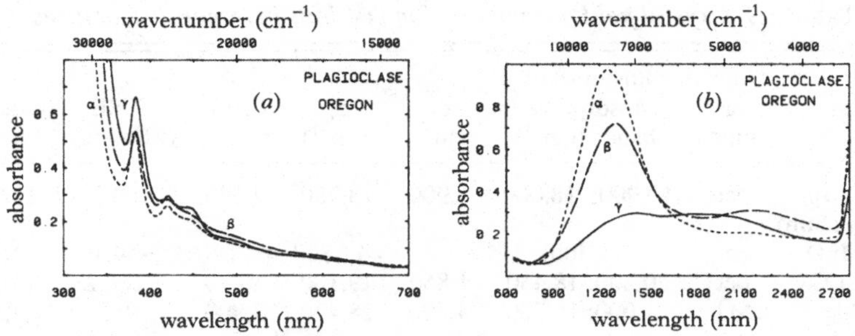

Figure 5.22 Polarized absorption spectra of iron in plagioclase feldspar (from Hofmeister & Rossman, 1983). (*a*) Visible region spectra with peaks originating from Fe^{3+} ions in tetrahedral coordination; (*b*) near-infrared spectra attributed to Fe^{2+} ions in the Ca sites. The labradorite, from Rabbit Hills, Oregon, contains about 0.1 wt per cent Fe_2O_3 and 0.3 wt per cent FeO. (Crystal thickness 1.0 cm.)

absorption bands centred around 8,500 cm^{-1} and 4,500 cm^{-1} (Bell and Mao, 1972c, 1973a,b; Mao and Bell, 1973a,b; Hofmeister and Rossman, 1983). The intensity of these bands correlate with iron content of the feldspar, which rarely exceeds 1.0 wt per cent Fe. The 8,500 cm^{-1} and 4,500 cm^{-1} bands appear to originate from crystal field transitions in Fe^{2+} replacing Ca^{2+} ions in seven-coordinated sites with average metal–oxygen distances around 250 pm. The blue or green colours of amazonite and other feldspars do not originate from the presence of transition metal ions. Instead, broad polarization-dependent bands occuring at 15,500–16,000 cm^{-1} have been attributed to colour centres involving coexisting structurally bound H_2O and Pb, perhaps present as Pb^{3+} ions (Hofmeister and Rossman, 1985, 1986).

Some varieties of coloured quartz contain trace amounts of transition metal ions (10^{-2} to 10^{-3} weight per cent), including rose quartz, amethyst and citrine, often as interstitial cations. In rose quartz, an absorption band near 19,500 cm^{-1} may result from an IVCT transition between structural Ti^{4+} and either Ti^{3+} or Fe^{2+} located in interstitial octahedral sites (Cohen and Makar, 1985). The optical spectra of amethyst show absorption bands at 28,570 cm^{-1} and 18,520 cm^{-1} with a region of minimum absorption around 22,100 cm^{-1} being responsible for the violet colour of amethyst (Cohen and Hassan, 1974; Lehmann, 1975; Cox, 1977). The colour and spectra of amethyst have been attributed to structural Fe^{4+} and interstitial Fe^{2+} ions induced in irradiated quartz crystals. The yellow colour of some citrines appears to originate from O $\rightarrow$ Fe^{3+} CT involving iron assimilated as non-structural impurities (Balitsky and Balitskaya, 1986).

Table 5.9. *Crystal field parameters for Ti^{3+} ions in mineral structures*

Mineral	Site or symmetry	Positions of absorption bands (cm^{-1})	δ_3 (cm^{-1})	Δ_o (cm^{-1})	CFSE (cm^{-1})	CFSE (kJ/g.ion)	Sources of data
$[Ti(H_2O)_6]^{3+}$ (Cs Ti alum)	oct	19,900; 18,000	1,900	18,950	7,580	−90.7	[1]
Ti^{3+} MgO	oct	(11,300)		≈11,300	≈4,520	−54.1	[2]
Ti^{3+} Al_2O_3	oct	20,300; 18,450	1,850	19,100	8,870	−106.2	[3]
diopside (Allende)	M1	21,000; 16,500	4,500	18,470	7,900	−91.5	[1,4,5]
pyroxene (Apollo 11)	M1	21,500; 16,000	5,500	18,500	7,650	−94.5	[6]
kosmochlor	M1	22,000		22,000	<8,800	≈−105.3	[7]
synthetic $NaTiSi_2O_6$	M1	≈22,000; 15,600	(6,300)	(18,770)	(8,000)	≈−95.7	[1,8]
zoisite	M3	22,500		22,500	<9,000	≈−107.7	[9]
mica	M1, M2	22,000 to 24,000		23,000	9,200	−110.0	[10]
basaltic glass	oct	18,200		18,200	7,280	−87.1	[11]
hibonite	M5 (trig.bipyr.)	13,990					[12,13]

Sources of data : [1] Burns & Huggins (1973); [2] Drickamer & Frank (1973); [3] McClure (1962); [4] Dowty & Clark (1973); [5] Mao & Bell (1974a); [6] Burns, Abu-Eid & Huggins (1972a); [7] Abs-Wurmbach, Langer & Oberhänsli (1985); [8] Prewitt, Shannon & White (1972); [9] Faye & Nickel (1970b); [10] Faye (1968a); [11] Nolet, Burns, Flamm & Besancon (1979); [12] Burns & Burns (1984b); [13] Ihinger & Stolper (1986).

5.10 Summary of crystal field spectra

5.10.1 Ti^{3+}

Information derived from the crystal field spectra of Ti^{3+}-bearing phases, including synthetic oxides and silicate minerals, is summarized in table 5.9. Parameters listed there include the energy separations of upper-level e_g orbitals (δ_3), mean octahedral crystal field splittings (Δ_o), and estimated crystal field stabilization energies (CFSE). Although only one absorption band, corresponding to the $^2T_{2g} \rightarrow {}^2E_g$ transition, is expected in octahedrally coordinated Ti^{3+} ions, $3d^1$ (cf. fig. 3.11), effects of dynamic Jahn–Teller splitting of the excited ^2E_g state, coupled with site occupancies of distorted coordination polyhedra in some of the structures (e.g., corundum and pyroxene M1 site), lead to broadened asymmetric bands or two separate peaks in the absorption spectra. The CFSE's estimated from the spectra are approximate values due to unknown splittings of lower-level t_{2g} orbital energy levels. The CFSE values are seen to decrease in the order

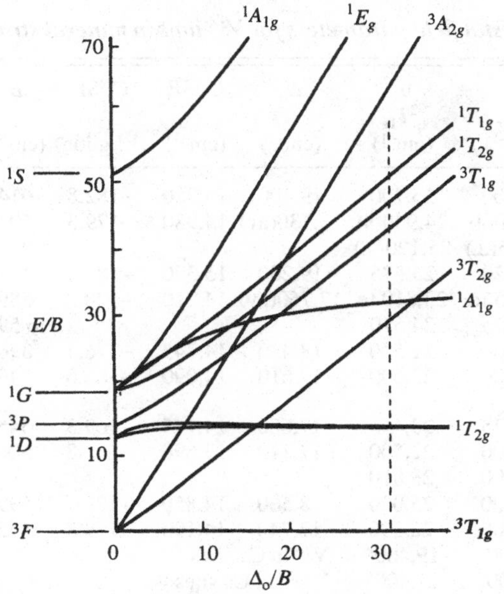

Figure 5.23 Tanabe–Sugano energy level diagram for V^{3+}, $3d^2$, in octahedral coordination calculated from spectral data for V^{3+}-doped corundum. Transitions in V^{3+} in Al_2O_3 are shown by the dashed line.

Al-sites in micas, zoisite, corundum > hydrated cation >
Mg sites in pyroxene M1 > basaltic glass >> periclase. (5.18)

5.10.2 V^{3+}

Spectral data for a variety of vanadium-bearing minerals, often occurring as gems, are summarized in table 5.10. Assignments of the absorption bands can be made with reference to the Tanabe–Sugano energy diagram for the V^{3+} ion, $3d^2$, shown in fig. 5.23. The spectroscopic ground term of the field-free V^{3+} ion, ^3F, yields three high spin-multiplicity (triplet) crystal field states, $^3T_{1g}$, $^3T_{2g}$ and $^3A_{2g}(F)$ with a fourth ($^3T_{1g}$) being derived from the ^3P term, when V^{3+} ions occur in octahedral coordination. The ground state of V^{3+} is $^3T_{1g}(^3F)$, representing the electronic configuration $(t_{2g})^2(e_g)^0$, with one of the three t_{2g} orbitals unoccupied. Three spin-allowed transitions are expected to triplet excited states, corresponding to the $^3T_{1g} \rightarrow {}^3T_{2g}(^3F)$ (band υ_1), $^3T_{1g} \rightarrow {}^3T_{1g}(^3P)$ (band υ_2) and $^3T_{1g} \rightarrow {}^3A_{2g}(F)$ (band υ_3) transitions. However, only the first two bands are measurable in optical spectra of V^{3+}-bearing minerals, the third band, located in the ultraviolet, generally being obscured by OMCT bands. In non-

Table 5.10. *Crystal field parameters for V^{3+} ions in mineral structures*

Mineral	υ_1 $^3T_{2g}$ (cm^{-1})	υ_2 $^3T_{1g}$ (cm^{-1})	Δ_o (cm^{-1})	CFSE (cm^{-1})	CFSE (kJ/g.ion)	*B* (cm^{-1})	Sources of data
$[V(H_2O)_6]^{3+}$	17,800	25,700	19,100	15,280	−182.8	614	[1]
corundum	17,510(||)	24,930(||)	18,730(av)	14,980	−179.3	594	[2,3]
	17,465(⊥)	25,120(⊥)					[4]
spinel	17,920	25,575	19,200	15,360	−183.8	595	[5]
chrysoberyl	16,285 to 16,835	23,810 to 24,510	17,760(av)	14,210	−170.0	589 to 593	[6]
pyrope	17,500	22,500	18,430	14,745	−176.4	396	[6,7]
grossular	16,500	23,600	17,610	14,090	−168.6	549	[6,7]
(tsavorite)							[8]
grossular	16,475	23,475	17,680	14,145	−169.3	545	[9]
goldmanite	16,000	22,500	17,110	13,690	−163.8	506	[6,7]
andalusite	17,700	23,810					[10]
kyanite	17,000	25,000	18,560	14,850	−177.7	599	[6]
zoisite	16,700	22,500	17,740	14,190	−169.8	455	[6,11,12]
(unheated)	13,400*	19,400*	V^{2+} in Ca sites				[6,11,12]
tanzanite	13,700*	17,400*	VO^{2+} in Ca sites				[6,11,12]
(heated zoisite)	18,900*	27,000*	V^{4+} in octahedral sites				[6,11,12]
diopside	15,700						[6]
spodumene	16,000	23,500	17,260	13,810	−165.2	578	[6]
tremolite	15,500	22,500	16,650	13,320	−159.4	543	[6]
beryl	16,000	23,800	17,240	13,790	−165.0	603	[13]
beryl	16,450	23,260	17,600	14,080	−168.5	530	[6,14]
	15,300	25,000	16,500	13,200	−158.0		
tourmaline	16,700(|)	22,700(||)	17,370(av)	13,900	−166.3	515	[6]
	16,400(⊥)	23,900(⊥)					[6]
axinite	17,400	25,000	18,660	14,930	−178.6	591	[6]
kornuperine	15,200	23,000	16,420	13.140	−157.2	602	[6]
muscovite	16,950	24,000	18,130	14,505	−173.6	549	[6]
roscoelite	16,200	23,500	17,400	13,920	−166.6	567	[6]
diopside–albite glass	15,700	21,700	16,700	13,360	−159.9	469	[15]
hibonite	25,000*	V^{3+} in five-fold sites (trigonal bipyramid)					[16,17]

* These transitions do not originate from the $^3T_{1g}$ ground state of octahedrally coordinated V^{3+} ions

Sources of data : [1] Lever (1984) p. 402; [2] McClure (1962); [3] White, Roy & Crichton (1967); [4] Rahman & Runciman (1971); [5] Drifford & Charpin (1967); [6] Schmetzer (1982); [7] Schmetzer & Ottemann (1979); [8] Schmetzer, Berdesinski & Traub (1975); [9] Rossman (1988); [10] Carlson & Rossman (1988); [11] Faye & Nickel (1970b); [12] Tsang & Ghose (1971); [13] Beckwith & Troup (1973); [14] Ghera & Lucchesi (1987); [15] Keppler (1992) [16] Burns & Burns (1984b); [17] Ihinger & Stolper (1986).

cubic minerals, degeneracies of the ground and excited states are resolved so that energies of bands υ_1 and υ_2 differ slightly in each polarized spectrum when V^{3+} ions occur in distorted octahedral sites in the crystal structures. Hence, the band positions listed in table 5.10 are averages for the two (uniaxial) or three (biaxial) polarized spectra. The table lists the crystal field splitting (Δ_o) and Racah *B* parameters estimated from the Tanabe–Sugano energy level diagram shown in fig. 5.23 (see also Lever, 1984, p. 816), together with approximate crystal field stabilization energies estimated from CFSE = $0.8\Delta_o$, which is applicable to the $V^{3+}(3d^2)$ cation (table 2.2). The CFSE data show a general trend of decreasing CFSE in the order

$$\text{spinel} > \text{corundum} > \text{kyanite} > \text{pyrope} > \text{muscovite} > \text{zoisite} > \text{grossular} > \text{beryl} > \text{tourmaline} > \text{pyroxene} > \text{amphibole}. \quad (5.19)$$

The absorption spectra of vanadium-bearing zoisite before and after heating in air to produce the gem tanzanite illustrated earlier (fig. 4.11) are worthy of attention. In the unheated zoisite, the two bands at 22,500 cm^{-1} and 16,700 cm^{-1} are consistent with similar bands in the spectra of other V^{3+}-bearing minerals (table 5.10), and probably originate from V^{3+} ions in the zoisite M3 sites (cf. fig. 4.4), which are occupied by other trivalent transition metal ions in epidote. However, the bands at 13,400 cm^{-1} and 19,400 cm^{-1} in unheated zoisite are inconsistent with octahedral V^{3+} ions. One possibility is that they originate from V^{2+} ions occurring in the Ca^{2+} sites. This assignment was based on electron paramagnetic resonance measurements (Tsang and Ghose, 1971) and also by analogies with absorption spectra of the hydrated V^{2+} ion, $[V(H_2O)_6]^{2+}$,which produces crystal field bands at 12,350 cm^{-1} and 18,500 cm^{-1} (Lever, 1984, p. 415). In the heated zoisite (tanzanite), oxidation to tetravalent vanadium has probably occurred. If so, the absorption band at 27,000 cm^{-1} may be assigned to V^{4+} ($3d^1$) ions formed by oxidation of V^{3+} in the M3 site. The other three bands at 18,900, 17,400 and 13,700 cm^{-1} in tanzanite may be attributed to vanadyl complex ions, VO^{2+} (Hutton, 1971), formed by oxidation of V^{2+} ions in the Ca sites. Note that the VO^{2+} ion was identified in a green vanadium-bearing apophyllite, $KCa_4Si_8O_{20}(F,OH).8H_2O$, based on absorption bands at 20,235 cm^{-1}, 15,480 cm^{-1} and 12,060 cm^{-1} (Rossman, 1974). These bands correlate with those observed in vanadyl compounds at 25,300 cm^{-1}, 16,000 cm^{-1} and 13,100 cm^{-1}. The VO^{2+} ions in apophyllite are believed to occupy both the seven-fold capped trigonal prismatic calcium site and the eight-fold tetragonal prismatic potassium site having analogies to the Ca sites in zoisite.

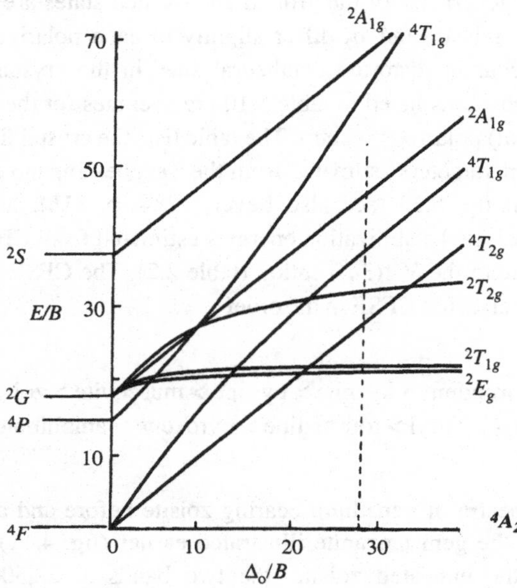

Figure 5.24 Energy level diagram for Cr^{3+}, $3d^3$, in octahedral coordination calculated from spectral data for ruby. Transitions in Cr^{3+} in Al_2O_3 are shown by the dashed line.

5.10.3 Cr^{3+}

The Tanabe–Sugano diagram illustrated in fig. 5.24 shows that the field-free Cr^{3+} ion, $3d^3$ again has two spectroscopic terms of high spin-multiplicity, ^4F, and ^4P. The lowest energy crystal field state for Cr^{3+} ions in octahedral coordination is $^4A_{2g}(^4F)$, representing the unique configuration $(t_{2g})^3(e_g)^0$. Three spin-allowed crystal field transitions are to be expected in optical spectra of octahe-

---

Sources to table 5.11

Sources of data : [1] Lever (1984) p. 419; [2] Low (1957); [3] Drickamer & Frank (1973) p. 74;[4] McClure (1962); [5] Neuhaus (1960); [6] Reinen (1969); [7] Sviridov, Sevastyanov, Orekhova, Sviridova & Veremeichik (1973); [8] Mao & Bell (1975a); [9] Farrell & Newnham (1965); [10] Schmetzer, Bank & Gübelin (1980); [11] Amthauer (1976); [12] Abu-Eid (1976); [13] Scheetz & White (1972); [14] Ikeda, Schneider, Akasaka & Rager (1992); [15] Rossman, Grew & Dollase (1982); [16] Carlson & Rossman (1988); [17] Langer (1988); [18] Petrov, Schmetzer & Eysel (1977); [19] Burns & Strens (1967); [20] Schmetzer & Berdesinski (1978); [21] Rossman (1988); [22] Mao, Bell & Dickey (1972); [23] Cohen & Janezic (1983); [24] Abs-Wurmbach, Langer & Oberhänsli (1985); [25] Parkin & Burns (1980); [26] Manning (1968a); [27] Schmetzer (1982); [28] Calas, Manceau, Novikoff & Boukili (1984); [29] Bish (1977); [30] Keppler (1992).

Table 5.11. *Crystal field parameters for Cr^{3+} ions in mineral structures*

Mineral	Site or symmetry	Mean M–O distance (pm)	υ_1 (= Δ_o) $^4T_{2g}$ (cm^{-1})	υ_2 $^4T_{1g}$ (cm^{-1})	CFSE (cm^{-1})	CFSE (kJ/g.ion)	B (cm^{-1})	Sources of data
$[Cr(H_2O_6)]^{3+}$	oct		17,400	24,600	20,880	–249.9	728.6	[1]
periclase	oct	210.9	16,200	22,700	19,440	–232.6	650.6	[2,3]
ruby (corundum)	oct	191.3	18,150	24,700	21,780	–260.6	635.6	[4]
eskolaite	oct	199.0	16,670	21,750	20,000	–239.3	475.4	[5]
$YCrO_3$ (perovskite)	oct	198.0	16,450	22,100	19,740	–236.2	541.7	[6]
spinel	oct	192.6	18,520	24,900	22,225	–265.9	612.1	[7]
chromite	oct	199.0	17,700	22,400	21,240	–254.2	430.6	[8]
alexandrite (chrysoberyl)	M2	193.8	17,700	23,550	21,240	–254.2	556.1	[9,10]
pyrope	oct	188.7	17,790	24,150	21,350	–255.5	615.6	[11]
grossular	oct	192.4	16,500	22,990	19,800	–236.9	645.4	[11]
uvarovite	oct	198.5	16,667	22,727	20,000	–239.3	589.1	[12]
forsterite	M1	209.4	16,900	23,500	20,280	–242.7	654.9	[13]
mullite	M1	191.1	18,850	25,300	22,620	–270.7	618.0	[14]
sillimanite	M1	191.2	16,130	23,640	19,360	–231.6	800.1	[15]
andalusite	M1	193.5	14,286	22,222	17,140	–205.1	961.9	[16]
kyanite	oct	189.6 to 191.8	16,500	24,000	19,800	–236.9	789.5	[17]
topaz	oct	187.1	18,150	25,300	21,780	–260.6	711.4	[18]
tawmawite (epidote)	M3	203.6	15,500	24,000	18,600	–222.6	1017.1	[19]
zoisite	M3	196.7	15,100	21,800	18,120	–216.8	697.4	[20]
enstatite	M1	207.8	15,480	22,625	18,580	–222.3	717.7	[21]
diopside	M1	207.7	15,500	22,000	18,600	–222.6	661.0	[22]
spodumene	M1	192.1	16,250	23,510	19,500	–233.3	758.0	[23]
kosmochlor (ureyite)	M1	200.1	15,458	21,742	18,550	–222.0	631.9	[24]
tremolite	M1,2,3	206.4 to 208.4	16,310	23,530	19,570	–234.2	750.7	[5]
emerald (beryl)	oct	190.6	16,130	23,530	19,360	–232.0	782.5	[5,25]
tourmaline	Mg Al	202.5 192.6	17,000	24,000	20,400	–244.1	707.1	[26]
kornuperine	M4	197.6	17,300	23,000	20,760	–248.4	541.5	[27]
fuchsite (mica)	M2		15,820	23,580	18,980	–227.2	851.7	[5]
muscovite	M2	194.0	16,200	23,800	19,440	–232.6	812.9	[28]
clinochlore	oct		18,200	25,300	21,840	–261.3	704.3	[28]
kammererite	oct		18,450	25,000	22,140	–264.9	632.9	[5]
chlorite	oct		18,020	25,640	21,625	–258.8	777.4	[29]
smectite	oct		16,800	24,000	20,160	–241.2	738.5	[28]
stichtite	oct		18,500	25,300	22,200	–265.6	663.0	[28]
diopside–albite glass	oct		15,400	23,000	18,480	–221.1	837.0	[30]

drally coordinated Cr^{3+}, corresponding to the ${}^4A_{2g} \rightarrow {}^4T_{2g}({}^4F)$ (band υ_1), ${}^4A_{2g} \rightarrow {}^4T_{1g}({}^4F)$ (band υ_2), and ${}^4A_{2g} \rightarrow {}^4T_{1g}({}^4P)$ (band υ_3) transitions. However, only the first two bands are generally measurable in optical spectra of Cr^{3+}-bearing minerals. In addition, the spin-forbidden ${}^4A_{2g} \rightarrow {}^2E_g({}^2G)$ transition is often observed as sharp peaks near band υ_1. In low-symmetry environments, the ${}^4T_{2g}$ and ${}^4T_{1g}({}^4F)$ states of Cr^{3+} are resolved into additional levels, so that bands υ_1 and υ_2 are often polarization dependent, as in the case of ruby (fig. 4.10). The ${}^2E_g$ state may also be resolved into two levels so that two sharp peaks are often observed on the flanks of the broader band υ_1. As noted earlier (§5.3.2.1), these peaks constitute the ruby R_1 and R_2 lines in laser technology. The ${}^4A_{2g} \rightarrow {}^4T_{2g}$ transition (band υ_1) provides a direct measure of the octahedral crystal field splitting parameter, Δ_o, while the Racah B parameter may be calculated from bands υ_1 and υ_2 by the relationship

$$B = \frac{(2\upsilon_1 - \upsilon_2)(\upsilon_2 - \upsilon_1)}{(27\upsilon_1 - 15\upsilon_2)}. \qquad (5.20)$$

Assembled in table 5.11 are the crystal field parameters derived from the spectra of several Cr^{3+}-bearing minerals. For non-cubic minerals, the energies of bands υ_1 and υ_2 represent values averaged from the polarized absorption spectra. Crystal field stabilization energies calculated from band υ_1, and corresponding to 1.2 x Δ_o (table 2.2), decrease in the order

mullite > spinel > ruby, topaz > pyrope, chlorites > tourmaline > chromite >
forsterite > uvarovite, eskolaite > grossular, kyanite >
sillimanite > tremolite, emerald >muscovite > epidotes,
Mg pyroxenes > andalusite. (5.21)

There is a general correlation of decreasing Δ_o with increasing average metal–oxygen distances in the mineral structures hosting the Cr^{3+} ions (Schmetzer, 1982). Also, Cr^{3+} ions in coordination sites with higher polyhedral volumes (see Appendix 7) also tend to have lower Δ_o values.

5.10.4 Cr^{2+}

Although Cr^{2+} ions are rare and unstable in terrestrial minerals, their presence is suspected in olivines and pyroxenes from the Earth's Mantle and the Moon (Burns, 1975a; Smith, 1971). Crystal field spectra exist for these silicates, as well as other synthetic Cr^{2+}-bearing phases, and parameters are summarized in table 5.12. Just one spin-allowed transition, corresponding to ${}^5E_g \rightarrow {}^5T_{2g}$, might

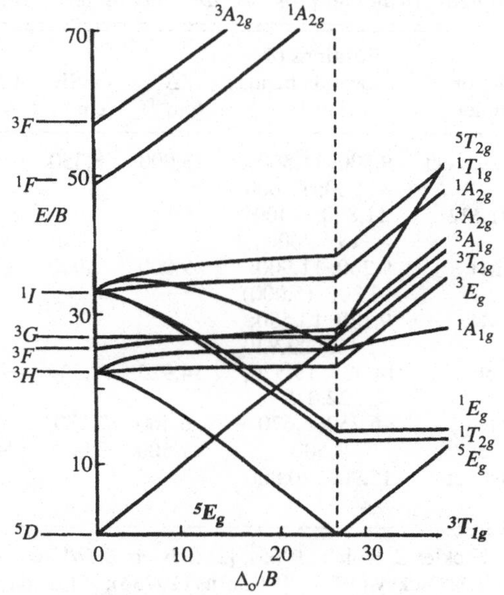

Figure 5.25 Energy level diagram for cations with $3d^4$ configurations, such as Cr^{2+} and Mn^{3+}, in octahedral coordination. The dashed line indicates the change from high-spin to low-spin configurations.

be expected in octahedrally coordinated $3d^4$ cations such as Cr^{2+} (fig. 5.25). However, effects of Jahn–Teller distortion of octahedral sites occupied by Cr^{2+} ions lead to resolution of the ^5E_g and $^5T_{2g}$, states to additional states, resulting in several absorption bands in optical spectra of Cr(II) compounds. Tetrahedral Cr^{2+} ions, which appear to exist in spinels, give rise to a broad absorption band located in the region 6,250 to 7,500 cm^{-1} representing the $^5T_2 \rightarrow {}^5E$ transition (fig. 3.11). As noted previously (§3.8.2), the spectra of Cr^{2+} ions in square-planar sites in synthetic $CaCrSiO_4$ provide a direct measure of the energy separation (2,505 cm^{-1}) between the d_{z^2} and d_{xz}, d_{yz} levels which is not attainable from polarized spectral measurements of Fe^{2+} in gillespite (table 3.7).

The cause of colour in natural and synthetic chromium-bearing blue diopsides has been widely debated and assignments of absorption bands in their polarized spectra remain controversial (Mao *et al.*, 1972; Burns, 1975a,b; Ikeda and Yagi 1977, 1982; Schreiber, 1977, 1978). One interpretation is that low-spin Cr^{3+} ions in tetrahearal sites in the pyroxene structure are responsible for the colour and spectra of blue diopsides (Ikeda and Yagi, 1977, 1982).

Table 5.12. *Crystal field parameters for Cr^{2+}-bearing minerals and compounds*

Mineral	Site or symmetry	Positions of absorption bands (cm^{-1})	Δ (cm^{-1})	CFSE (cm^{-1})	CFSE (kJ/g.ion)	Sources of data
$[Cr(H_2O)_6]^{2+}$	J–T dist. oct	9,500; 11,500; 13,900; 17,000	13,900	8,350	−99.8	[1]
forsterite	M1,M2	11,800; 6,400 to 6,700				[2,3]
diopside	M1	9,200; 11,900; 13,300; (5,000)	9,000	7,900	−94.5	[4,5]
diopside	M1	9,170; 13,550; 17,150; 22,830				[6]
$CaCrSi_4O_{10}$ (Cr^{2+} gillespite)	sq. planar	14,925; 19,570; 22,075	14,925	18,330	−219.3	[7]
spinel	tet	6,250; 6,670	6,460	2,580	−30.9	[8,9]
chromite	tet	7,500	7,500	3,000	−35.9	[10]
diopside–albite glass	dist.oct	15,900; 20,800				[11]

Sources of data : [1] Fackler & Holah (1965); [2] Scheetz & White (1972); [3] Burns (1975a); [4] Mao, Bell & Dickey (1972); [5] Burns (1975b); [6] Ikeda & Yagi (1982); [7] Belsky, Rossman, Prewitt & Gasparik (1984); [8] Ulmer & White (1966); [9] Greskovich & Stubican (1966); [10] Mao & Bell (1975a); [11] Keppler (1992).

Although high-spin and low-spin electronic configurations are theoretically possible for tetrahedrally coordinated $3d^3$ cations (table 2.3), it is extremely unlikely that the energy separation between the lower-level *e* orbitals and the higher energy t_2 orbitals (fig. 2.7) is sufficiently large to induce spin-pairing in Cr^{3+} if they were to occupy tetrahedral sites. Certainly, tetrahedral Cr^{3+} ions are unknown in any other Cr(III) chemical compound because the high octahedral CFSE attainable by Cr^{3+} ions induces them to occupy octahedral sites. The occurrence of tetravalent Cr^{4+} ions substituting for Si in the diopside structure has also been proposed (Schreiber, 1977). By possessing the $3d^2$ configuration, the Cr^{4+} ion not only acquires relatively high CFSE in tetrahedral coordination (table 2.3), but its tetrahedral ionic radius (41 pm) is comparable to that of Al^{3+} (39 pm) known to substitute for Si^{4+} in pyroxenes. A third possibility favoured here is that octahedral Cr^{2+} ions are present in blue diopsides (Mao *et al.*, 1972; Burns, 1975a,b) and that Jahn–Teller distortion predicted for this $3d^4$ cation is made possible by its occupancy of the deformed M1 site of the clinopyroxene structure. It is also significant that energies of several of the absorption bands measured for Cr-bearing blue diopsides are similar to those observed for aqueous Cr^{2+} ions (table 5.12).

5.10.5 Mn^{3+}

The Mn^{3+} ion, like its isoelectronic counterpart Cr^{2+} with the $3d^4$ configuration, is thermodynamically unstable, eqs (2.5) and (2.6), but occurs, nevertheless, in several minerals in distorted octahedral sites (table 6.1) where it is stabilized by the Jahn–Teller effect (§2.12). Crystal field spectral data are available for a variety of Mn^{3+}-bearing minerals, parameters for which are summarized in table 5.13. In a regular octahedral site, Mn^{3+} is expected to give rise to a single spin-allowed absorption band representing the $^5E_g \rightarrow {}^5T_{2g}$ transition (figs 5.25 and 3.11). However, effects of dynamic and static Jahn–Teller distortion result in at least three absorption bands in optical spectra of Mn^{3+}-bearing minerals, which often show marked polarization dependencies. As a result, all $3d$ orbital energy separations of Mn^{3+} are usually attainable from the crystal field spectra, including the splittings of upper-level e_g orbitals, δ_3, listed in table 5.13. The large values of δ_3 correlate with large static distortions of octahedral sites in many of the crystal structures accommodating the Mn^{3+} ions (table 6.1), particularly in the andalusite (M1), epidote (M3), mica (M2), clinopyroxene (M1) and alkali amphibole (M2) structures. In some cases, notably garnet where there is negligible static distortion of the octahedral site, values of δ_3 can only be a manifestation of the dynamic Jahn–Teller effect (Arni *et al.*, 1985).

Estimates of crystal field splitting (Δ_o or 10 Dq) and stabilization energy (CFSE) parameters for Mn^{3+} in distorted octahedral obtained by two different methods are listed in table 5.13. The first method is based on the orbital baricentre illustrated in fig. 4.5 which was used to derive $3d$ orbital energy levels of Mn^{3+} in piemontite. Here, CFSE's for the Mn^{3+} $3d^4$ configuration are calculated from $(0.6\Delta_o + \frac{1}{2}\delta_3)$. The second method, yielding values in parenthesis in table 5.13, which is based on eqs (3.14) to (3.17) derived from ligand field parameters for tetragonally distorted octahedra, provides mean 10 Dq values averaged for the equatorial and axial splittings. The CFSE's for Mn^{3+} in sites approximating tetragonally distorted octahedra are then calculated as $\{[0.6 \times 10\,Dq_{(av)}] + \frac{1}{2}\delta_3\}$. The data summarized in table 5.13 show that there is reasonable agreement between energy splittings and stabilization energies derived by the two methods. Trends of decreasing CFSE's for Mn^{3+}-bearing minerals are

andalusites > micas > epidotes > tourmaline >
pyroxene, amphiboles > corundum, garnets (5.22)

5.10.6 Mn^{2+}

Divalent manganese occurring as high-spin cations in all minerals and possessing the unique electronic configuration $(t_{2g})^3(e_g)^2$ with single occupancy of all

Table 5.13. *Crystal field parameters for Mn^{3+} ions in mineral structures*

Mineral	Site or symmetry	Positions of absorption bands (cm^{-1})	δ_3 (cm^{-1})	Δ_o* (cm^{-1})	CFSE‡ (cm^{-1})	CFSE‡ (kJ/g.ion)	Sources of data
Mn^{3+} Al_2O_3 (corundum)	oct	18,750; 20,600	1,850	19,370	11,620	−139.1	[1]
grossular (garnet)	oct	17,000; 20,400	3,400	18,150	10,890	−130.3	[2]
calderite (garnet)	oct	19,200; 21,000	1,800	19,800	11,880	−142.2	[3]
viridine (andalusite)	M1 dist.oct	14,100; 21,500; 23,300	14,100	15,050 (15,650)	16,080 (16,440)	−192.4 (−196.7)	[4–6]
kanonaite (andalusite)	M1 dist.oct	16,000; 22,000; 22,600	15,900	14,400 (12,630)	16,640 (15,530)	−199.1 (−185.8)	[5,7,8]
yoderite	M2/M3 trig.bipyr.	16,900; 20,600		21,040	14,900	−178.3	[9,10]
piemontite (epidote)	M3 dist.oct	12,000; 18,170; 22,000	12,000	13,450 (12,726)	14,070 (13,640)	−168.4 (−163.2)	[5,11,13]
blanfordite (omphacite)	M1 dist.oct	7,700 to 8,500; 16,800; 17,300; 19,000	8,100	13,650 (14,300)	12,240 (12,630)	−146.5 (−151.1)	[14]
rhodonite	M1,M2 or M3	6,530; 18,520	6,530				[15]
juddite (riebeckite)	M2 dist.oct	5,800 to 6,800; 16,500; 18,500	6,100	14,820 (16,180)	12,040 (12,760)	−144.1 (−152.7)	[14]
morganite (beryl)	oct	(18,020); 18,520; 20,200; (28,170)					[16]
tourmaline		9,000; 13,500; 18,400; 22,700	9,000	13,900	12,840	−153.6	[17]
manganophyllite (phlogopite)¶	M2	14,850; 19,050; 21,400; (23,800)	14,850	12,800 (14,760)	15,105 (14,060)	−180.7 (−168.2)	[18,19]
muscovite	M2	13,200; 18,250; 21,850	13,200	13,450 (16,370)	14,670 (16,420)	−175.5 −196.5	[20]

apophyllite		20,400					[21]
montmorillonite	M1,M2	10,400; 18,800; 20,600; 22,400	10,400	15,400 (16,301)	14,440 (14,980)	−172.8 −179.3	[22]
diopside–albite glass	dist.oct	13,700; 21,400	13,700	14,550	15,580	−186.4	[23]

* Values of Δ_o calculated by the baricentric method (e.g. fig. 4.4). Values in parenthesis are 10 $Dq_{(av)}$ parameters calculated by eqs (3.13) to (3.15).

‡ CFSE calculated as $(0.6\,\Delta_o + \frac{1}{2}\delta_3)$. Values in parenthesis calculated as $(6\,Dq_{(av)} + \frac{1}{2}\delta_3)$.

¶ Alternatively, Fe^{3+}(tet) – Mn^{2+}(oct) pair absorption [19].

Sources of data : [1] McClure (1962); [2] Frendrup & Langer (1981); [3] Langer & Lattard (1984); [4] Abs-Wurmbach, Langer & Tillmans (1977); [5] Smith, Halenius & Langer (1982); [6] Rossman (1988); [7] Abs-Wurmbach, Langer, Seifert & Tillmanns (1981); [8] Langer (1988); [9] Abu-Eid, Langer & Seifert (1978); [10] Langer, Smith & Halenius (1982); [11] Burns & Strens (1987); [12] Xu, Zheng & Peng (1982); [13] Kersten, Langer, Almen & Tillmanns (1987); [14] Ghose, Kersten, Langer, Rossi & Ungretti (1986); [15] Gibbons, Ahrens & Rossman (1974); [16] Wood & Nassau (1968); [17] Manning (1973); [18] Burns (1970), p. 58; [19] Smith, Halenius, Annersten & Ackermann (1983); [20] Annersten & Hålenius (1976); [21] Rossman (1974); [22] Sherman & Vergo (1988); [23] Keppler (1992).

Table 5.14. *Crystal field parameters for Mn^{2+} ions in mineral structures*

Mineral	υ_1 $^4T_1(^4G)$ (cm^{-1})	υ_2 $^4T_2(^4G)$ (cm^{-1})	υ_3 $^4E,^4A_1(^4G)$ (cm^{-1})	υ_4 $^4T_2(^4D)$ (cm^{-1})	υ_5 $^4E(^4D)$ (cm^{-1})	Δ (cm^{-1})	B (cm^{-1})	Sources of data
$[Mn(H_2O)_6]^{2+}$	18,700	23,120	24,960	27,980	29,750	8,480	671	[1]
ilesite $MnSO_4.4H_2O$	19,800	23,470	25,000	28,330	29,850	7,500	693	[2]
manganosite MnO	16,530	20,830	23,920			9,790	786	[2]
rhodochrosite $MnCO_3$	18,300	22,700	24,750	27,850	29,200	7,500	636	[2]
lithiophilite $LiMnPO_4$	20,370	23,310	24,870	28,170	29,580	6,800	673	[2]
tephroite Mn_2SiO_4	17,390	22,680	24,630	27,700	28,570	8,750	577	[2]
glaucochroite $CaMnSiO_4$	17,640	22,220	24,450	27,700	28,570	8,750	588	[2]
spessartine $Mn_3Al_2Si_3O_{12}$	20,800	23,500	24,500	27,200		7,000	664	[3–6]
leucophoenicite $Mn_7(SiO_4)_3(OH)_2$	18,590	22,930	24,510	28,090	28,820	6,160	616	[2]
bustamite $(Mn,Ca)_3Si_3O_9$	20,200	23,600	24,500	28,100	29,100		664	[3,4]
rhodonite $CaMn_4(SiO_3)_5$	18,600	22,200	24,240			8,250	590	[2–4,7–9]
pyroxmangite $CaMn_6(SiO_3)_7$	18,200	23,000	24,450	27,700	29,100		671	[3,4]
serandite	19,250	23,100	24,500	27,700	29,350		679	[3,4]

inesite $Ca_2Mn_7Si_{10}O_{28}(OH)_2.5H_2O$	18,970	22,880	24,570	27,780	29,150	7,500	693	[2]
friedelite $Mn_8Si_6O_{15}(OH)_{10}$			24,650		29,200		650	[3,4]
ganophyllite $NaMn_5Si_6O_{15}(OH)_5.2H_2O$			24,450		29,050		657	[3,4]

Sources of data : [1] Lever, (1984), p. 449; [2] Keester & White (1968); [3] Manning (1969a); [4] Manning (1970); [5] Moore & White (1972); [6] Parkin & Burns (1980); [7] Lakshman & Reddy (1973a); [8] Gibbons, Ahrens & Rossman (1974); [9] Marshall & Runciman (1975).

five $3d$ orbitals, acquires zero CFSE in the crystal structures. The non-degenerate sextet ground state of Mn^{2+} is $^6A_{1g}$ in octahedral coordination (or $^6A_1$ in tetrahedral sites) and remains non-degenerate in low-symmetry environments. Tanabe–Sugano energy level diagrams for $3d^5$ cations such as Mn^{2+} illustrated earlier (figs 3.10 and 3.16) show that all excited states are either quartets or doublets (table 3.3), so that every crystal field transition in Mn^{2+} is spin-forbidden. However, the characteristic pink colours of Mn(II) minerals indicate that relaxation of this selection rule occurs, often as a result of exchange interactions with next-nearest neighbour cations, so that weak absorption bands are measurable in optical spectra.

Crystal field spectral data for Mn^{2+}-bearing minerals are summarized in table 5.14. The two lowest energy crystal field transitions, $^6A_{1g} \rightarrow {}^4T_{1g}(^4G)$ (band υ_1) and $^6A_{1g} \rightarrow {}^4T_{2g}(^4G)$ (band υ_2) often occur as weak, very broad bands in the visible region. The $^6A_{1g} \rightarrow {}^4A_{1g}, {}^4E_g(^4G)$ transition (band υ_3), on the other hand, is more conspicuous as a very sharp peak centred around 24,500 to 25,000 cm^{-1}. Transitions to higher energy quartet states, including $^6A_{1g} \rightarrow {}^4T_{2g}(^4D)$ (band υ_4) and $^6A_{1g} \rightarrow {}^4E_g(^4D)$ (band υ_5), are often obscured by absorption edges of oxygen $\rightarrow$ metal CT bands particularly when Fe is also present in the mineral. The splitting parameters Δ_o or $10\,Dq$, are generally evaluated from Tanabe–Sugano diagrams such as fig. 3.10 or 3.16. The Racah B parameter, however, may be determined directly from the energies of bands υ_3 and υ_5, since $\upsilon_5 - \upsilon_3 = 7B$.

5.10.7 Fe^{3+}

The Fe^{3+} ion, $3d^5$, which is isoelectronic with Mn^{2+}, also has a sextet $^6A_{1g}$ (or $^6A_1$) ground state, so that all crystal field transitions to excited quartet states are again spin-forbidden. Nevertheless, the $^6A_{1g} \rightarrow {}^4T_{1g}(^4G)$ (band υ_1), $^6A_{1g} \rightarrow {}^4T_{2g}(^4G)$ (band υ_2) and, particularly, $^6A_{1g} \rightarrow {}^4A_{1g}, {}^4E_g(^4G)$ (band υ_3) transitions are usually observed in optical spectra of most Fe^{3+}-bearing minerals, energies of which are summarized in table 5.15. The presence of Fe^{2+} ions, which contribute more intense spin-allowed Fe^{2+} crystal field and $Fe^{2+} \rightarrow Fe^{3+}$ intervalence charge transfer transitions (table 3.6), usually obscure bands υ_1 and υ_2 originating from Fe^{3+} ions, but not band υ_3. Thus, the sharp peak located near 22,000 cm^{-1} (band υ_3) is particularly diagnostic of Fe^{3+} ions in minerals. Band υ_4 [$^6A_{1g} \rightarrow {}^4T_{2g}(^4D)$] and band υ_5 [$^6A_{1g} \rightarrow {}^4E_g(^4D)$] located towards the ultraviolet are often obscured by OMCT transitions. Note also that due to the negatively sloping $^4T_1(^4G)$ and $^4T_2(^4G)$ levels relative to the $^6A_1(^6S)$ ground state in the Tanabe–Sugano diagram for $3d^5$ cations (figs 3.10 and 3.16), bands υ_1 and

υ_2 in spectra of tetrahedrally coordinated Fe^{3+} ions occur at *higher* energies than do the corresponding bands for octahedrally coordinated Fe^{3+} ions, contrary to expectations based on the $\Delta t = -\frac{4}{9}\Delta_o$ relationship, eq. (2.7). Thus, in octahedral [$Fe^{3+}O_6$] clusters, the ${}^6A_{1g} \rightarrow {}^4T_{1g}({}^4G)$ transition occurs around 11,000 cm^{-1}, whereas in tetrahedral [$Fe^{3+}O_4$] clusters, the ${}^6A_1 \rightarrow {}^4T_1({}^4G)$ transition is near 22,000 cm^{-1}. In dense oxide structures containing Fe^{3+} ions in adjacent edge- or face-shared octahedra, paired transitions may lead to additional absorption bands in optical spectra. For example, the ${}^6A_1 + {}^6A_1 \rightarrow {}^4T_1 + {}^4T_1$ paired transition between Fe^{3+} located in face-shared octahedra in the corundum structure is responsible for the the absorption band observed at 18,690 cm^{-1} in yellow sapphire (figs 3.16 and 4.16) and at 18,900 cm^{-1} in hematite (fig. 10.2), the energies of which are approximately double those of corresponding ${}^6A_{1g} \rightarrow {}^4T_{1g}(G)$ (band υ_1) transitions in the single isolated Fe^{3+} ions (cf. tables 5.15 and 10.2).

5.10.8 Fe^{2+}

In high-spin Fe^{2+} ions occurring in oxides and silicates at ambient pressures, only one spin-allowed crystal field transition is expected in optical spectra. This corresponds to the ${}^5T_{2g} \rightarrow {}^5E_g$ (octahedral site) or ${}^5E \rightarrow {}^5T_2$ (tetrahedral site), respectively (see figs 3.8, 3.11 and 5.26). However, due to site occupancies of Fe^{2+} ions in distorted octahedra in many structures and dynamic Jahn–Teller splitting of the excited ${}^5E_g$ state, two or more separate peaks are generally observed in crystal field spectra of most Fe^{2+}-bearing minerals, due to resolution of the ${}^5T_{2g}$ and ${}^5E_g$ crystal field states into additional levels. In most cases, splittings of upper-level e_g orbitals may be obtained directly from the absorption spectra. However, lower-level splittings of t_{2g} orbitals are poorly constrained, affecting estimates of the crystal field splittings and stabilization energies of Fe^{2+} ions in oxide and silicate minerals, which are summarized in table 5.16. The data plotted in fig. 5.27 confirm the general trend of decreasing crystal field splitting with increasing metal–oxygen distances in many oxide and silicate minerals (Faye, 1972).

The values of Fe^{2+} CFSE's that are derived throughout this chapter by assuming baricentric splittings of t_{2g} and e_g orbitals in low-symmetry or distorted sites are listed in table 5.16 (see also fig. 7.6). For minerals hosted by Al^{3+} ions, the CFSE values are seen to decrease in the order

$$\text{tourmaline 'Al site'} > \text{sillimanite} > \text{kyanite} > \text{beryl} > \text{corundum} > \text{euclase}. \quad (5.23)$$

Table 5.15. *Crystal field parameters for Fe^{3+} ions in mineral structures*

Mineral	Site or symmetry	υ_1 $^4T_1(^4G)$ (cm^{-1})	υ_2 $^4T_2(^4G)$ (cm^{-1})	υ_3 $^4E,^4A_1(^4G)$ (cm^{-1})	υ_4 $^4T_2(^4D)$ (cm^{-1})	υ_5 $^4E(^4D)$ (cm^{-1})	Δ (cm^{-1})	B (cm^{-1})	Sources of data
$[Fe(H_2O)]_6^{3+}$	O_h	12,600	18,500	24,300			13,700		[1]
periclase MgO	O_h	10,000	13,500	21,740	25,120	27,500	13,400	650	[2]
corundum (yellow sapphire) Al_2O_3	C_3	9,700 9,570	18,700* 14,350	22,200 22,120	25,600 25,860	26,700 26,570	15,270	480	[3–5]
hematite α-Fe_2O_3	C_3	11,300 11,600	15,410	22,520 23,800	24,690	26,320 26,700	 15,900	543 410	[6] [7]
maghemite γ-Fe_2O_3	oct + tet	10,700	≈15,000	≈23,000		27,000	15,410	580	[6]
synthetic γ-$LiAlO_2$	tet	14,500 16,300	18,500 19,600	21,600	22,300	25,600	8,350	570	[8,9]
goethite α-FeOOH	oct	10,900	15,400	≈23,000			15,300	590	[6,10]
lepidocrocite γ-FeOOH	oct	10,400	15,400	≈23,000			15,950	610	[6,10]
grossular $Ca_3Al_2Si_3O_{12}$	C_{3i}	13,111		22,865 23,121	23,592	27,040	12,800	557	[11,12]
andradite $Ca_3Fe_2Si_3O_{12}$	C_{3i}	12,453	16,650	22,701 22,999	24,000	27,000	12,600	593	[11–13]
		11,920	16,370 17,640	22,700 22,960	26,040	26,950	13,000	607	[14]
		11,700	17,000	22,670 22,900					[15]

vesuvianite	oct	14,500	17,500	19,800	23,500	27,200	≈11,000	630	[16]
$Ca_{10}Mg_2Al_4Si_9O_{34}(OH)_4$			to 21,600						
kyanite	M1 to M4	16,000		22,400	26,500	27,000		585	[17]
Al_2SiO_5	C_1	to 17,000		23,400					
sillimanite	tet or oct	(16,130)		21,740	24270	27,700			[18]
Al_2SiO_5									
calcic pyroxene	M1	14,000	16,700	22,200			≈11,500		[19,20]
$Ca(Mg,Fe)(Si,Al)_2O_6$									
beryl	D_3	12,300	17,500	23,600	26,500		≈12,000		[21]
$Be_3Al_2Si_6O_{18}$		14,200	20,000						
epidote	M3	9,470	16,480	21,200	24,800		≈12,500		[22]
$Ca_2FeAl_2Si_3O_{12}(OH)$		to 12,300	to 17,900	22,000					
muscovite	M2	11,600	14,700	21,500	26,600	27,400	≈14,500	670	[23–27]
$KAl_2(Si_3Al)\,O_{10}(OH)_2$	C_1	17,200	19,650	22,600					
phlogopite	tet	19,200	20,300	22,700	25,000		≈5,800		[28,29]
nontronite	oct	10,692	15,775	22,400	26,509	27,067	15,050	614	[30,31]
quartz	tet	18,700	20,200	22,500	24,800	27,300	≈6,000	690	[32]
SiO_2									
orthoclase	tet		20,700	22,650	24,000	26,500	≈5,700	550	[33–35]
coquimbite	three oct	12,850	17,830	23,090	24,100				[36]
$Fe_2(SO_4)_3.9H_2O$				23,420					
botryogen	oct	10,650	20,080	23,150	24,270				[36,37]
$MgFe(SO_4)_2(OH).7H_2O$									
magnesiocopiapite	oct	11,570	18,020	23,260					[36]
$MgFe_4(SO_4)_6(OH)_2.20H_2O$									
butlerite	oct	10,870	20,500	23,580					[37]
$FeSO_4(OH).2H_2O$									
jarosite	oct	10,720	21,190	23,040					[37]
$KFe_3(SO_4)_2(OH)_6$									

Table 5.15 *continued*

Mineral	Site or symmetry	υ_1 $^4T_1(^4G)$ (cm^{-1})	υ_2 $^4T_2(^4G)$ (cm^{-1})	υ_3 $^4E,^4A_1(^4G)$ (cm^{-1})	υ_4 $^4T_2(^4D)$ (cm^{-1})	υ_5 $^4E(^4D)$ (cm^{-1})	Δ (cm^{-1})	B (cm^{-1})	Sources of data
fibroferrite $FeSO_4(OH).5H_2O$	oct	11,900	18,000	23,640					[37]
stewartite $MnFe_2(PO_4)_2(OH)_2.8H_2O$	oct	11,360		23,360					[37]
guildite $CuFe(SO_4)_2(OH).4H_2O$	oct	11,440		23,260					[38]

* Paired transitions $^6A_1 + {}^6A_1 \rightarrow {}^4T_1 + {}^4T_1$ also observed (or calculated) as follows: sapphire, 18,690 cm^{-1}; hematite, 18,900 cm^{-1}; maghemite, 19,600 cm^{-1}; goethite and lepidocrocite, 20,800 cm^{-1}; nontronite, 19,200 cm^{-1}.

Sources of data : [1] Lever (1984), p. 449; [2] Blazey (1977); [3] Lehmann & Harder (1970); [4] Ferguson & Fielding (1972); [5] Krebs & Maisch (1971); [6] Sherman & Waite (1985); [7] Marusak, Messier & White (1980); [8] Waychunas & Rossman (1983); [9] Sherman (1985a); [10] Mao & Bell (1974b); [11] Manning (1969c); [12] Moore & White (1972); [13] Manning (1967a); [14] Lin (1981); [15] Parkin & Burns (1980); [16] Manning (1968b); [17] Faye & Nickel (1969); [18] Rossman, Grew & Dollase (1982); [19] Bell & Mao (1972c); [20] Burns, Parkin, Loeffler, Leung & Abu-Eid (1976); [21] Low & Dvir (1960); [22] Burns & Strens (1967); [23] Faye (1968a); [24] Richardson (1975); [25] Annersten & Hålenius (1976); [26] Richardson (1976); [27] Finch, Gainsford & Tennant (1982); [28] Faye & Hogarth (1969) ; [29] Hogarth, Brown & Pritchard (1970); [30] Karickhoff & Bailey (1973); [31] Sherman & Vergo (1988b); [32] Lehmann & Bambauer (1973); [33] Faye (1969); [34] Manning, (1970); [35] Hofmeister & Rossman (1983); [36] Rossman (1975); [37] Rossman (1976); [38] Wan, Ghose & Rossman (1978).

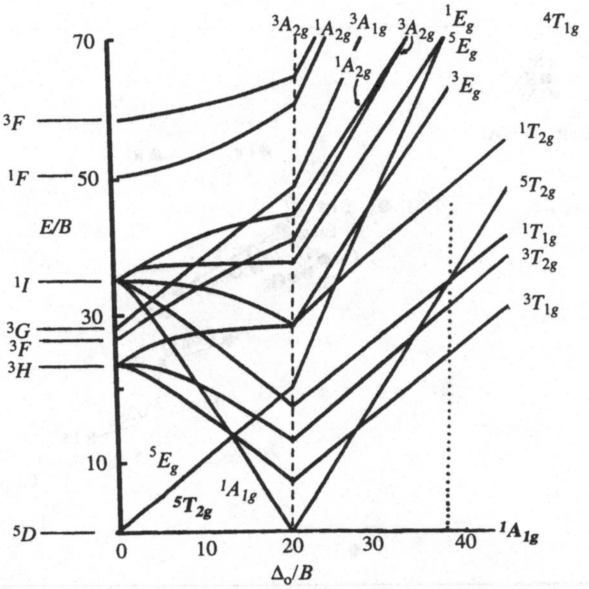

Figure 5.26 Energy level diagram for cations with $3d^6$ configurations, such as Fe^{2+} and Co^{3+}, in octahedral coordination. Transitions in the $[Co(H_2O)_6]^{3+}$ ion are shown by the dotted line. The dashed line indicates the change from high-spin to low-spin configurations.

In ferromagnesian silicates, the order of decreasing CFSE values is

$$\begin{gathered}\text{ringwoodite} > \text{olivine M1} > \text{tourmaline 'Mg site'} > \text{olivine M2},\\ \text{biotite M1+M2, periclase, chlorite} > \text{pyroxene M1},\\ \text{chloritoid} > \text{amphibole M1+M2+M3, babingtonite} >\\ \text{pyroxene M2} > \text{garnet, staurolite} > \text{plagioclase},\\ \text{amphibole M4} \gg \text{spinel}. \end{gathered} \tag{5.24}$$

5.10.9 Co^{3+}

The Co^{3+} ion, $3d^6$, is isoelectronic with Fe^{2+}, so that its energy level diagram corresponds to that illustrated in figure 5.26. High-spin Co^{3+} ions occur in K_3CoF_6, the spectra of which contain absorption bands at 11,400 cm^{-1} and 14,500 cm^{-1} (Lever, 1984, p. 460), indicating dynamic Jahn–Teller splitting of the $^5E_g(^5D)$ excited state involved in the $^5T_{2g} \rightarrow {}^5E_g$ transition in the paramagnetic Co^{3+} ion (Cotton and Meyers, 1960).

With the exception of some fluorides, however, most Co(III) compounds are diamagnetic indicating the presence of low-spin Co^{3+} ions with the $(t_{2g})^6$ configuration. Accordingly, the ground state is $^1A_{1g}(^1I)$ and the two lowest energy excited singlet states are $^1T_{1g}(^1I)$ and $^1T_{2g}(^1I)$ (fig. 5.26). Two spin-allowed

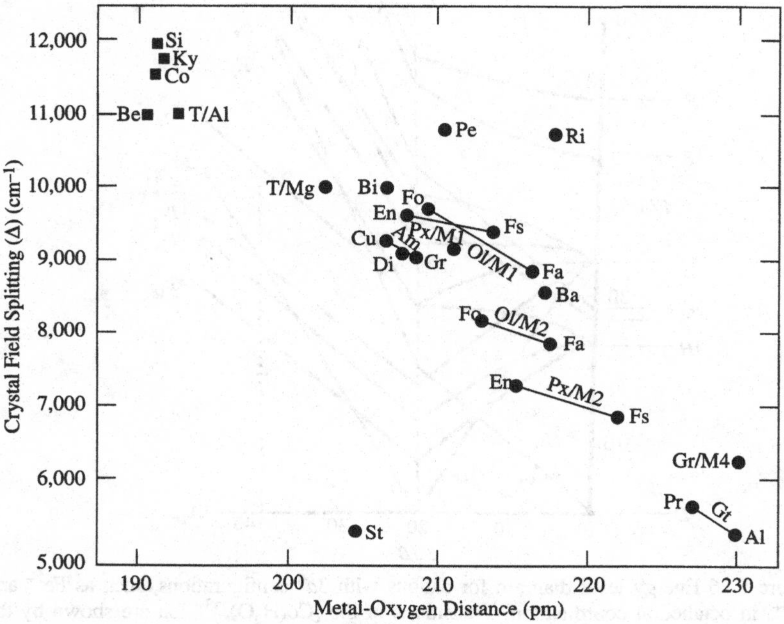

Figure 5.27 Crystal field splitting correlated with average metal–oxygen distances in Fe^{2+}-bearing silicate and oxide minerals. Squares: Al^{3+} sites; circles: Mg^{2+}–Fe^{2+} sites. Legend to mineral symbols is given in table 5.16.

transitions, ${}^1A_{1g} \rightarrow {}^1T_{1g}({}^1I)$ (band υ_1) and ${}^1A_{1g} \rightarrow {}^1T_{2g}({}^1I)$ (band υ_2) are observed in the spectra of many Co(III) compounds (Lever, 1984, p. 463), examples of which are listed in table 5.17. In $[Co(H_2O)_6]^{3+}$, for example, these transitions are responsible for absorption bands at 16,500 cm^{-1} and 23,150 cm^{-1}. Two weaker spin-forbidden peaks also occur at 8,000 cm^{-1} [${}^1A_{1g} \rightarrow {}^3T_{1g}({}^3H)$] and 12,500 cm^{-1} [${}^1A_{1g} \rightarrow {}^3T_{2g}({}^3H)$]. The energy separation between the ${}^1T_{1g}(I)$ and ${}^1T_{2g}(I)$ states is approximately 16 B while the energy of the ${}^1T_{1g}$ state is about 10 $Dq - C$. (Lever, 1984, p. 462). By assuming $C \approx 4\,B$ for these Racah parameters (table 11.1), estimates of 10 Dq and B for the $[Co(H_2O)_6]^{3+}$ ion provide values of 18,550 cm^{-1} and 512 cm^{-1}, respectively. In Co^{3+}–doped corundum, the spin-allowed transitions in low-spin Co^{3+} are located near 15,550 cm^{-1} and 23,000 cm^{-1}, yielding a value of 17,410 cm^{-1} for 10 Dq or Δ_o (table 5.2).

Cyanide complexes enable direct comparisons to be made between crystal field parameters of low-spin Fe^{2+} and Co^{3+} ions. In $[Fe(CN)_6]^{4-}$, the two spin-allowed bands occur at 31,000 cm^{-1} and 37,040 cm^{-1}, while in $[Co(CN)_6]^{3-}$ they are located at 32,400 cm^{-1} and 39,000 cm^{-1}. Thus, for low-spin Co^{3+} in $[Co(CN)_6]^{3-}$, 10 Dq is about 34,500 cm^{-1}, whereas for low-spin Fe^{2+} in

Table 5.16. *Crystal field parameters for Fe^{2+} ions in mineral structures*

Mineral	Site or symmetry	Peaks (cm^{-1}) ($^5T_{2g} \rightarrow {}^5E_g$)	δ_3 (cm^{-1})	Δ (cm^{-1})	CFSE (cm^{-1})	CFSE (kJ/g.ion)	Sources of data	Legend to Figs 5.27 and/or 7.6
$[Fe(H_2O)_6]^{2+}$	oct	10,400; 8,300	2,100	9,020	3,640	–43.6	§3.3	Fe^{2+}
periclase ($Wü_{26}$)	oct (O_h)	10,000; 11,600	1,600	10,800	4,320	–51.7	§5.3.1	Pe
corundum (blue sapphire)	oct (C_3)	≈11,500		≈11,500	≈4,600	–55.0	§4.7.3.3	Co
spinel	tet (T_d)	4,830	0	4,830*	2,900*	–34.7	§5.3.3	Sp
garnet (Alm_{1-100})	cube (D_{2h})	4,400; 5,800; 7,600	(1,100)	5,710 to 5,300†	3,700 to 3,460†	–44.3 to –41.4	§5.4.1	Gt (Pr–Al)
olivine (Fa_{1-100})	M1 (C_i)	8,000; 11,000	3,000	9,760 to 8,830	4,700 to 4,250	–56.3 to –50.9	§5.4.2.3	Ol (Fo–Fa)
	M2 (C_s)	9,500 to 9,100	~400	8,250 to 7,930	4,420 to 4,280	–52.9 to –51.2		
ringwoodite (γ-Fe_2SiO_4)	oct (D_{3d})	11,430		10,760	4,970	–59.5	§5.4.3	Ri
kyanite	oct	11,760		≈11,760	4,700	–56.3	§5.4.4.2	Ky
sillimanite	oct	11,960		≈11,960	4,780	–57.2	§5.4.4.3	Si
euclase	oct	11,630; 8,000	3,600	9,050	4,350	–52.1	§4.7.2.8	Eu
staurolite	tet	4,550 to 7,000	~2,500	5,300*	3,700*	–44.3	§5.4.5	St
diopside–hedenbergite	M1	10,200; 8,475	1,725	9,100	3,890	–46.6	§5.5.3	Di
enstatite–ferrosilite (Fs_{14}–Fs_{86})	M1	10,700; 8,550	2,150	9,650 to 9,400	4,200 to 4,010	–50.3 to –47.9	§5.5.4	Px (En–Fs)
	M2	11,000 to 10,400; 5,400 to 4,900	5,600	7,300 to 6,900	3,820 to 3,660	–45.7 to –43.8		

Table 5.16 *continued*

Mineral	Site or symmetry	Peaks (cm^{-1}) ($^5T_{2g} \rightarrow {}^5E_g$)	δ_3 (cm^{-1})	Δ (cm^{-1})	CFSE (cm^{-1})	CFSE (kJ/g.ion)	Sources of data	Legend to Figs 5.27 and/or 7.6
pigeonite	M1	10,750 to 10,400	2,200	9,400	4,000	–47.9	§5.5.5	Pg
(Fs_{57})	M2	8,550 to 4,880		6,650	3,560	–42.6		
babingtonite	Fe(1)	10,580; 7,720	2,860	8,600	3,900	–46.7	§4.7.2.8	Ba
rhodonite	M4‡	4,250; 6,700						
	(5–CN)	8,040; 9,750						
cummingtonite	M4	~10,000; 4,200	~5,800	6,300	3,520	–42.1	§5.6.3	Am
–grunerite				to 6,000	to 3,400	to –40.7		(Cu–Gr)
(35–95%	M1,M2,M3	10,100; 8,500	1,600	9,250	3,950	–47.3		
$Fe_7Si_8O_{22}(OH)_2$)				to 9,050	to 3,850	to –46.1		
calcic	M4	9,710; 4,050	5,660	5,890	3,350	–40.1	§5.6.	At
amphiboles	M1,M2,M3	(10,200; 8,470)		(9,100)	(3,890)	(–46.6)		(Am)
alkali	M1,M3	10,000; 8,600	1,400	9,100	3,870	–46.3	§5.6.5	(Am)
amphiboles								
beryl	Al/oct	12,200; 10,300	1,900	11,000	4,650	–55.7	§5.7.2	Be
	channel	4,760; 12,200						
cordierite	Mg/oct	10,200; 8,700	1,500	9,150	3,900	–46.7	§5.7.3	Co
tourmaline	Al/oct	14,500; 9,500	5,000	11,000	4,900	–58.6	§5.7.4	T/Mg
	Mg/oct	13,200; 7,900	5,300	10,000	4,500	–53.9		T/Al
biotite	M1,M2	11,900; 8,900	3,000	10,000	4,400	–52.7	§5.8.3	Bi
chlorite	Mg/oct	11,500; 9,500	2,000	10,200	4,300	–51.5	§5.8.4	Cl
chloritoid	Mg/oct	10,900; 8,000	2,900	9,000	4,050	–48.5	§5.8.4	Ch
gillespite	(D_{4h})	20,000; 8,300		11,600	6,125	–77.3	§3.8.2	Gi
plagioclase	Ca	8,500; 4,500	4,000	5,750	3,550	–42.5	§5.9	Pl
diopside–albite glass	oct	9,260; 5,400	3,860	6,580	3,380	–40.5	§8.5.2	gl

* tetrahedral CF splitting; † cubic CF splitting; ‡ five-fold coordination

Table 5.17. *Crystal field parameters for octahedrally coordinated low-spin Co^{3+} and Fe^{2+} ions*

Complex ion or chemical compound	υ_1 ${}^1T_{1g}({}^1I)$ (cm^{-1})	υ_2 ${}^1T_{2g}({}^1I)$ (cm^{-1})	B (cm^{-1})	10 *Dq* or Δ_o (cm^{-1})	CFSE* (cm^{-1})	CFSE (kJ/g.ion)	Sources of data
$[Co(H_2O)_6]^{3+}$	16,500	24,700	512	18,550	20,895	−250.0	[1]
Co^{3+} Al_2O_3	15,550	23,000	466	17,410	18,165	−217.4	[2,3]
$[Co(NO_3)_6]^{3-}$	15,200	22,500	456	17,025	17,235	−206.2	[1]
$[Co(NH_3)_6]^{3+}$	21,200	29,550	522	23,290	32,265	−386.1	[1]
$[Co(CN)_6]^{3-}$	32,400	39,000	412	34,050	58,095	−695.2	[1]
$[Fe(CN)_6]^{4-}$	31,000	37,040	377	32,510	58,875	−704.5	[1]

* Corrected for pairing energies (table 2.5): Co^{3+} = 23,625 cm^{-1}; and Fe^{2+} = 19,150 cm^{-1}.

Sources of data : [1] Lever (1984), p. 463; [2] McClure (1962); [3] Muller & Gunthard (1966).

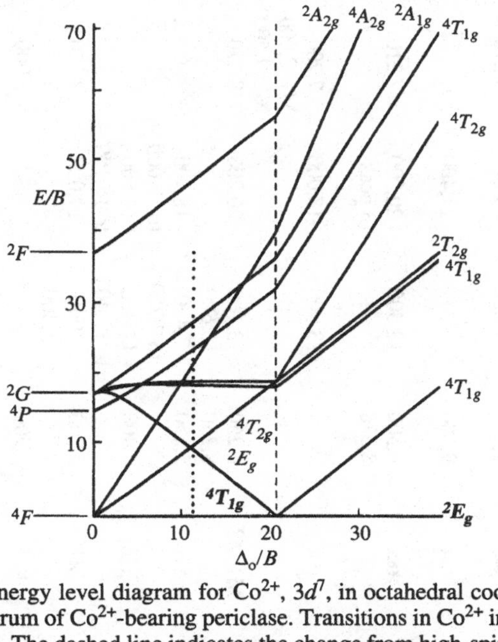

Figure 5.28 Energy level diagram for Co^{2+}, $3d^7$, in octahedral coordination calculated from the spectrum of Co^{2+}-bearing periclase. Transitions in Co^{2+} in MgO are shown by the dotted line. The dashed line indicates the change from high-spin to low-spin configuration.

$[Fe(CN)_6]^{4-}$, 10 *Dq* is approximately 32,500 cm^{-1}. Comparisons between the spectra of cyanide complexes of Fe(II) and Co(III) also enable assessments to be made of the locations of absorption bands for minerals in the Earth's Mantle that might contain low-spin Fe^{2+} ions. This topic is discussed in §9.7.6.

Table 5.18. *Crystal field parameters for Co^{2+}-bearing minerals and compounds*

Mineral or synthetic phase	Site or symmetry	υ_1 $^4T_{2g}$ (cm^{-1})	υ_2 $^4A_{2g}$ (cm^{-1})	υ_3 $^4T_{1g}(^4P)$ (cm^{-1})	Δ_o (cm^{-1})	CFSE (cm^{-1})	CFSE (kJ/g.ion)	B (cm^{-1})	Sources of data
$[Co(H_2O)]_6^{2+}$	oct	8,100	(16,000)	19,400	9,054	7,240	−86.6	589	[1]
periclase MgO	oct	8,500	17,200	19,600	9,270	7,420	−88.8	845	[2–4]
corundum Al_2O_3	oct	9,100		21,400					[5,6]
Co_2SiO_4 olivine (α spectrum)	oct	6,500 8,330	12,500	20,000 (3 peaks)	7,210	6,270	−80.4	451	[7]
Co_2SiO_4 (reflectance)	oct		13,000	18,000	7,200 M1: 8,000 M2: 7,500	6,250 6,400 6,000	−74.8 −76.6 −71.8		[3,8]
Co^{2+} enstatite $(Co_{0.15}Mg_{0.85})SiO_3$	M1,M2	6,250 8,250	13,100 15,600	20,200	8,370	6,700	−80.2	936	[9]
Co^{2+} diopside	M1	5,250	7,400 to 7,700	18,900 to 19,000					[9]
$CaCoSi_2O_6$	M2	4,900	11,300 to 11,400	14,500 to 15,100					
Co^{2+} diopside $CaMg_{0.2}Co_{0.8}Si_2O_6$	M1	5,200	7,900 15,400	19,000	8,020	6,420	−76.8	864	[9]
zincite ZnO	tet*	3,900			3,900	4,680	−56.0	700	[2,4]
spinel $MgAl_2O_4$	tet*	4,100	7,200	17,200	Δ_t = 4,100	4,920	−58.9	830	[4]

lusakite Co staurolite	tet*	(4,000)	6,150 7,050 7,850	15,650 17,120 18,780	$\Delta_t \approx 4{,}000$	4,800	–57.5		fig.4.7
diopside–albite glass	tet*	(4,040)	7,000	15,500	$\Delta_t = 4{,}040$	4,850	–58.0	787	[10]
$Sr(Co,Zn,W)O_3$ perovskite	oct				7,700				[4]
$CoCdTiO_3$ ilmenite	oct				6,900				[4]
ZrO_2 cubic zirconia	cube†	7,940; 9,800; 16,810(s); 17,920; 19,420 (s)						[11]	

* Transitions in tetrahedral Co^{2+}; † Transitions in eight-fold coordinated Co^{2+}.

Sources of data : [1] Lever (1984), p.481; [2] Low (1958b); [3] Pappalardo, Wood & Linares (1961b); [4] Reinen (1970); [5] McClure (1962); [6] Muller & Gunthard (1966); [7] Rossman (1988); [8] Goodgame & Cotton (1961); [9] White, McCarthy & Scheetz (1971); [10] Keppler (1992); [11] R. G. Burns, unpubl. [cubic ZrO_2 with 12 mole % Y_2O_3 and 0.3 wt. % CoO.]

5.10.10 Co^{2+}

The energy level diagram for octahedrally coordinated high-spin Co^{2+}, $3d^7$, is illustrated in fig. 5.28 and shows the ground state to be $^4T_{1g}(^4F)$. Three spin-allowed transitions are expected to triplet excited states, corresponding to the $^4T_{1g} \rightarrow {}^4T_{2g}(^4F)$ (band υ_1), $^4T_{1g} \rightarrow {}^4A_{2g}(^4F)$ (band υ_2) and $^4T_{1g} \rightarrow {}^4T_{1g}(^4P)$ (band υ_3) transitions. When Co^{2+} ions occur in distorted octahedral sites, the $^4T_{1g}(^4F)$, $T_{2g}(^4F)$ and $^4T_{1g}(^4P)$ states are resolved into additional levels, so that several peaks on absorption bands are expected when Co^{2+} ions are in low-symmetry environments. Moreover, spin-forbidden transitions to the $^2T_{1g}(^2G)$ and $^2T_{2g}(^2G)$ states also occur, often appearing as sharp peaks or prominent shoulders in the spectra of Co^{2+}-bearing silicates. As a result, positive assignments of Co(II) spectra are often very difficult. The few available crystal field spectral data for Co^{2+}-bearing minerals are summarized in table 5.17. Table 5.17 lists approximate crystal field splittings and Racah *B* parameters estimated from the Tanabe–Sugano energy level diagram shown in fig. 5.27 (see also Lever, 1984, p. 816).

When Co^{2+} ions occur in tetrahedral coordination, the ground-state becomes $^4A_2(^4F)$, and spin-allowed transitions are to the $^4T_2(^4F)$, $^4T_1(^4F)$ and $^4T_1(^4P)$ states (fig. 3.12). Note that the $^4A_2 \rightarrow {}^4T_2(^4F)$ transition provides a direct measure of Δ_t. When Co^{2+} ions are present in distorted tetrahedral sites, as in the cobaltian staurolite lusakite (§4.4.3), each of the excited triplet crystal field states is resolved into several levels, so that numerous polarization-dependent absorption bands occur in the visible to near infra-red spectra (cf. fig. 4.7).

5.10.11 Ni^{2+}

For Ni^{2+} ions, $3d^8$, in octahedral coordination, the ground state shown in fig. 5.29 is $^3A_{2g}$ (^3F), corresponding to the unique electronic configuration $(t_{2g})^6(e_g)^2$. Three spin-allowed crystal field transitions are expected in the crystal field spectra of Ni(II) compounds, corresponding to the $^5A_{2g} \rightarrow {}^3T_{2g}(^3F)$ (band υ_1), $^3A_{2g} \rightarrow {}^3T_{1g}(^3F)$ (band υ_2), and $^3A_{2g} \rightarrow {}^3T_{1g}(^3P)$ (band υ_3) transitions. In addition, spin-forbidden transitions to the $^1E_g(^1D)$ (band υ_4) and $^1A_{1g}(^1G)$ states are also observed in the visible and near-ultraviolet regions. In fact, the $^1E_g(^1D)$ level lies so close to the $^3T_{1g}(^3F)$ level that bands υ_1 and υ_4 often overlap in spectra of spectra of Ni(II) compounds. In addition, in low-symmetry environments, the $^3T_{1g}(^3F)$ and $^1E_g(^1D)$ states, as well as $^3T_{2g}(^3F)$ and $^3T_{1g}(^3P)$, are resolved into additional energy levels, so that several polarization dependent spin-allowed and spin-forbidden bands may contributed to the crystal field spectra of Ni^{2+}-bearing silicate minerals.

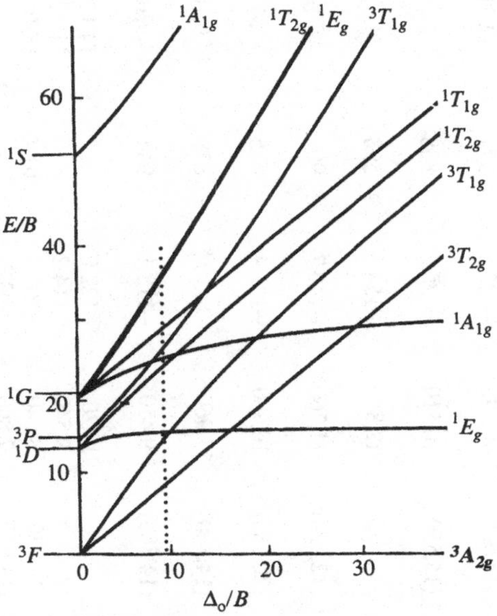

Figure 5.29 Energy level diagram for Ni^{2+}, $3d^8$, in octahedral coordination calculated from the spectrum of Ni^{2+}-bearing periclase. Transitions in Ni^{2+} in MgO are shown by the dotted line.

The crystal field parameters of minerals containing Ni^{2+} ions are summarized in table 5.19. Note that the energy of the first transition, band υ_1, for Ni^{2+} in octahedral coordination provides a direct measure of the crystal field splitting parameter Δ_o. Crystal field stabilization energies for Ni^{2+} derived from band υ_1 decrease in the order

$$\text{corundum} > \text{spinel} > \text{periclase} > \text{forsterite M1} > \text{diopside} > \text{forsterite M2} > \text{phyllosilicates} > \text{enstatite M1,M2}. \quad (5.25)$$

5.10.12 Cu^{2+}

The electronic configuration of octahedrally coordinated Cu^{2+}, $(t_{2g})^6(e_g)^3$, corresponding to the ^2E_g (^2D), ground state (figs 3.5 and 3.11), makes this $3d^9$ cation highly vulnerable to Jahn–Teller distortion (§2.12). As a result, all Cu(II) compounds contain Cu^{2+} ions in low-symmetry environments, ranging from tetragonally distorted (elongated) octahedra to square-planar coordination sites (table 6.1). Because of such coordination environments, crystal field spectra of Cu(II) compounds contain asymmetric or multiple absorption bands.

Table 5.19. *Crystal field parameters for Ni^{2+} ions in mineral structures*

Mineral or synthetic phase	Site or symmetry	$\upsilon_1 (= \Delta_o)$ $^3T_2(^3F)$ (cm^{-1})	υ_2 $^3T_1(^3F)$ (cm^{-1})	υ_3 $^3T_1(^3P)$ (cm^{-1})	υ_4 $^1E(^1G)$ (cm^{-1})	CFSE (cm^{-1})	CFSE (kJ/g.ion)	B (cm^{-1})	Sources of data
$[Ni(H_2O)_6]^{2+}$	oct	8,510	13,868	25,460	(15,200)	10,210	−122.2	930	[1]
periclase MgO	oct	8,600	13,500	24,600	(14,900)	10,320	−123.5	971	[2,3]
bunsenite NiO	oct	8,760	14,085	24,100	(15,550)	10,510	−125.8	808	[4]
corundum Al_2O_3	oct	(10,700)	18,000	25,000		12,840	−153.6	598	[5,6]
spinel $MgAl_2O_4$	oct	9,800	16,000	27,000	(12,920)	11,760	−140.7	907	[7]
Ni_2SiO_4 spinel	oct	9,150	14,780	22,550	(20,180)	10,980	−131.4	625	[8]
liebenbergite Ni_2SiO_4 olivine	M1	9,000 to 9,960	15,900	26,200	(14,300)	11,960*	−143.1*	878	[4,9,10]
	M2	7,000 to 8,000	13,000	24,000	(14,300)	9,450	−113.1	944	[4,9,10]
forsterite Mg_2SiO_4	M1	9,000	14,000	25,000		10,200	−122.0		[4,10, 11,12]
	M2	6,600	12,800	23,400		9,540	−114.2		
enstatite $MgSiO_3$	M1	7,200	12,970	24,510	(14,930)	8,640	−103.4	1046	[3,10]
	M2	6,500	12,440	21,550	(14,930)	7,800	−93.3		
Ni^{2+} enstatite $(Mg_{95}Ni_5)SiO_3$	M1,M2	6,900	12,700	24,500		8,280	−99.1	1039	[13]
Ni^{2+} diopside $CaNiSi_2O_6$	M1	8,400	13,800	24,500		10,080	−120.6	881	[13]
garnierite Ni clay silicates	oct	9,100	15,200	26,300	(13,000)	10,920	−130.7	936	[1,14–17]

kerolite (talc) $(Mg,Ni)_3Si_4O_{10}(OH)_2$	oct	8,950	17,325	25,800	(13,868)	8,950	−107.1		[1]
lizardite $(Mg,Ni)_3Si_2O_5(OH)_4$	oct	8,845				8,845	−106.1		[1]
nepouite $Ni_3Si_4O_{10}(OH)_2$	oct	9,120				9,120	−109.1		[1]
diopside–albite glass	oct	5,260	10,200	22,420	(15,870, 18,520)	6,310	−75.5	845	[18]
Ni^{2+} diopside glass	tet. & trig.bipyr.	15,700; 18,000; (5,000); (8,300) 10,700; 22,000							[19]
zincite ZnO	tet	4,050†				3,240	−38.8	795	[3]
$NiCr_2O_4$ spinel	tet	4,150†				3,320	−39.7		[7]
cubic zirconia ZrO_2	cubic‡	11,765; 17,180; (18,180); 22,680; 24,100							[20]

* Value probably too high: see §5.4.2.4; † $^3T_1 \rightarrow {}^3T_2$ transition in tetrahedral Ni^{2+}; ‡ transitions in eight-fold coordinated Ni^{2+}

Sources of data : [1] Manceau & Calas (1985); [2] Low (1958a); [3] Pappalardo, Wood & Linares (1961a); [4] Rossman, Shannon & Waring (1981); [5] McClure (1962); [6] Muller & Gunthard (1966); [7] Reinen (1970); [8] Yagi & Mao (1977); [9] Hu, Langer & Boström (1990); [10] Rossman (1988); [11] Rager, Hosoya & Weiser (1988); [12] Wood (1974); [13] White, McCarthy & Scheetz (1971); [14] Lakshman & Reddy (1973b); [15] Faye (1974); [16] Tejedor-Tejedor, Anderson & Herbillon (1983); [17] Cervelle (1991); [18] Keppler (1992); [19] Galoisy & Calas (1991); [20] R. G. Burns, unpubl. [cubic ZrO_2 with 12 mole % Y_2O_3 and 0.3 wt. % NiO]

Table 5.20. *Crystal field parameters for Cu^{2+}-bearing minerals and compounds*

Mineral or compound	Site symmetry	Positions of absorption bands (cm^{-1})	Δ (cm^{-1})	Sources of data
$[Cu(H_2O)_6]^{2+}$	Jahn-Teller dist. oct.	10,500; 13,000; 14,500	13,000	[1]
$(NH_4)_2Cu(H_2O)_6(SO_4)_2$		6,400; 11,500; 10,650; 12,350		[2]
chalcanthite	tet.bipyr	12,000; 20,166	12,500	[3]
$CuSO_4.5H_2O$	(4x) H_2O; (2x) SO_4^{2-}	(16,524; 18,177)		
guildite	tet. bipyr	7,780; 12,820; (11,440 = Fe^{3+})		[4,5]
$CuFe_2(SO_4)_2$	(4x) H_2O; (2x) SO_4^{2-}			
$(OH).4H_2O$				
malachite	two very dist. oct.	9,200; 13,000; 22,000		[6]
$Cu_2CO_3(OH)_2$	CO_3^{2-} and H_2O	11,919; 16,482; 18,160; 20,166	12,500	[3,7]
azurite	sq.planar + pyramid	13,000; 16,100	13,000	[6]
$Cu_3(CO_3)_2(OH)_2$	CO_3^{2-} and H_2O			
chrysocolla		11,800; 14,000	11,800	[6]
$(Cu.Al)_2H_2Si_2O_5$				
$(OH)_4.nH_2O$				
syn.Cu gillespite	sq. planar	12,900; 15,800; 18,800	12,900	[8]
$BaCuSi_4O_{10}$				
cuprorivaite	sq. planar	12,740; 16,130; 18,520	12,740	[9]
$CaCuSi_4O_{10}$				
orthopyroxene	M2	7,280; 12,200		[10]
$MgSiO_3/Cu^{2+}$				
diopside-	dist.oct.	12,820	12,820	[11]
albite glass				

Sources of data : [1] Holmes & McClure (1957); [2] Hitchman & Waite (1976); [3] Lakshman & Reddy (1973c); [4] Wan, Ghose & Rossman (1978); [5] Rossman (1988); [6] Andersen (1978); [7] Hunt & Salisbury (1971); [8] Clark & Burns (1967); [9] Ford & Hitchman (1979); [10] Rossman (1980); [11] Keppler (1992).

Summarized in table 5.20 are the few crystal field spectral data for Cu^{2+}-bearing minerals.

5.11 Summary

Chapter 5 summarizes the crystal field spectra of transition metal ions in common rock-forming minerals and important structure-types that may occur in the Earth's interior. Peak positions and crystal field parameters for the cations in several mineral groups are tabulated. The spectra of ferromagnesian silicates are described in detail and correlated with the symmetries and distortions of the Fe^{2+} coordination environments in the crystal structures. Estimates are made of the CFSE's provided by each coordination site accommodating the Fe^{2+} ions. Crystal field splitting parameters and stabilization energies for each of the transition metal ions, which are derived from visible to near-infrared spectra of oxides and silicates, are also tabulated. The CFSE data are used in later chapters to explain the crystal chemistry, thermodynamic properties and geochemical distributions of the first-series transition elements.

5.11 Background reading

Burns, R. G. (1968) Optical absorption in silicates. In *The Application of Modern Physics to the Earth and Planetary Interiors*. (S. K. Runcorn, ed.; J. Wiley & Sons, New York), pp. 191–211.

Burns, R. G. (1985b) Electronic spectra of minerals. In *Chemical Bonding and Spectroscopy in Mineral Chemistry*. (F. J. Berry & D. J. Vaughan eds; Chapman & Hall, London), pp. 63–101.

Hawthorne, F. C. (1981b) Amphibole spectroscopy. In *Amphiboles and Other Hydrous Pyriboles – Mineralogy*. (D. R. Veblan, ed.; Mineral. Soc. Amer. Publ.), *Rev. Mineral.*, 9A, 103–39

Hofmeister, A. M. & Rossman, G. R. (1985) Color in feldspars. In *Feldspar Mineralogy, 2nd edn.* (P. H. Ribbe, ed.; Mineral Soc. Amer. Publ.), *Rev. Mineral.*, 2, 281–96.

Marfunin, A. S. (1979) *Physics of Minerals and Inorganic Materials.* (Springer Verlag, New York), ch. 6.

Rossman, G. R. (1980) Pyroxene spectroscopy. In *Pyroxenes*. (C. T. Prewitt, ed.; Mineral Soc. Amer Publ., *Rev. Mineral.*, 7, 93–115.

Rossman, G. R. (1984) Spectroscopy of micas. In *Micas*. (S. W. Bailey, ed.; Mineral Soc. Amer. Publ.), *Rev. Mineral.*, 13, 145–81.

Rossman, G. R. (1988) Optical spectroscopy. In *Spectroscopic Methods in Mineralogy and Geology*. (F. C. Hawthorne, ed.; Mineral. Soc. Amer. Publ.), *Rev. Mineral.*, 18, 207–54.

Vaughan, D. J. (1990) Some contributions of spectral studies in the visible (and near visible) light region to mineralogy. In *Absorption Spectroscopy in Mineralogy*. (A. Mottana & F. Burragato, eds; Elsevier Science Publ., Amsterdam), pp. 1–38.

6

Crystal chemistry of transition metal-bearing minerals

> *A great deal has been written about the crystal-field model for first-row transition metal cations in the distorted octahedra of olivines. – – Its predictions are useful for – – rationalizing the intra- and inter-crystalline cation partitioning.*
>
> G. E. Brown Jr, *Rev. Mineral.*, *5* (2nd edn), 333 (1982)

6.1 Introduction

The crystal chemistry of many transition metal compounds, including several minerals, display unusual periodic features which can be elegantly explained by crystal field theory. These features relate to the sizes of cations, distortions of coordination sites and distributions of transition elements within the crystal structures. This chapter discusses interatomic distances in transition metal-bearing minerals, origins and consequences of distortions of cation coordination sites, and factors influencing site occupancies and cation ordering of transition metals in oxide and silicate structures, which include crystal field stabilization energies

6.2 Interatomic distances in transition metal compounds

One property of a transition metal ion that is particularly sensitive to crystal field interactions is the ionic radius and its influence on interatomic distances in a crystal structure. Within a row of elements in the periodic table in which cations possess completely filled or efficiently screened inner orbitals, there should be a decrease of interatomic distances with increasing atomic number for cations possessing the same valence. The ionic radii of trivalent cations of the lanthanide series for example, plotted in fig. 6.1, show a relatively smooth contraction from lanthanum to lutecium.* Such a trend is determined by the

* There is a small break in slope after gadolinium in some plotted ionic radii, notably for octahedrally coordinated cations, attributed to crystal field effects on electrons in the 4*f* orbitals. This is discussed in Phillips & Williams (1966, vol. 2, p. 106).

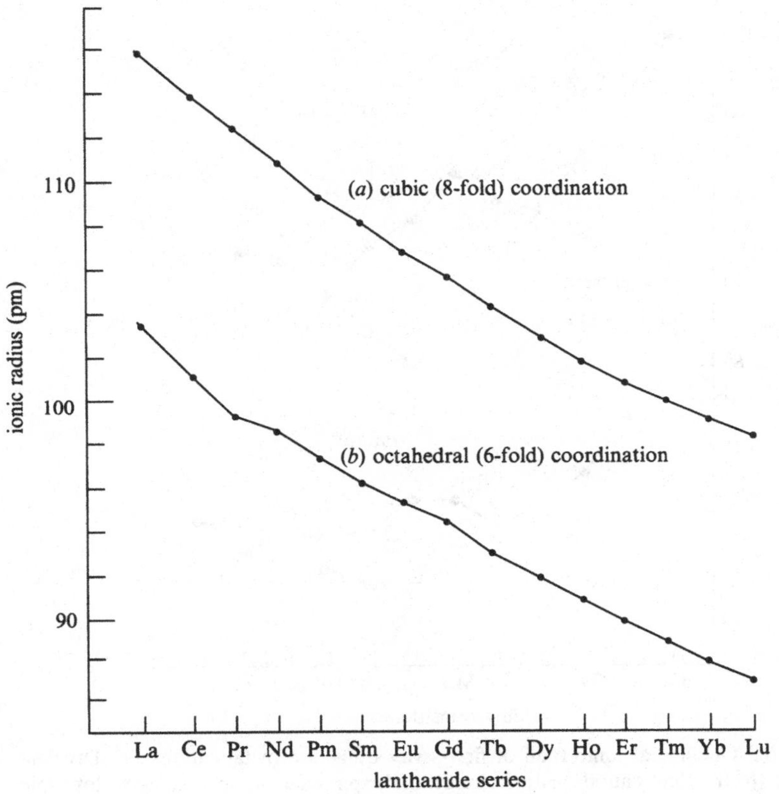

Figure 6.1 Ionic radii of the lanthanide cations in oxides. (*a*) Eight-fold coordination; (*b*) octahedral coordination (data from Shannon, 1976).

regular increase of nuclear charge along the lanthanide series which outweighs the opposing effect of increasing interelectronic repulsion. Because electrons do not screen entirely the positive nuclear charge, successive electrons added to the 4*f* orbitals along the series between La^{3+} and Lu^{3+} are drawn in more closely to the nucleus so that surrounding anions may approach the cation more closely.

Cations of the first transition series do not conform to the smooth pattern for the lanthanide elements shown in fig. 6.1. This is illustrated in fig. 6.2*a* by the radii of divalent cations in oxides containing transition metal ions in high-spin states. There is an overall decrease of octahedral ionic radius from Ca^{2+} to Zn^{2+}, but values first decrease to V^{2+}, then rise to Mn^{2+}, decrease to Ni^{2+}, and rise again to Zn^{2+}. The characteristic double-humped curve shown in fig. 6.2*a* has

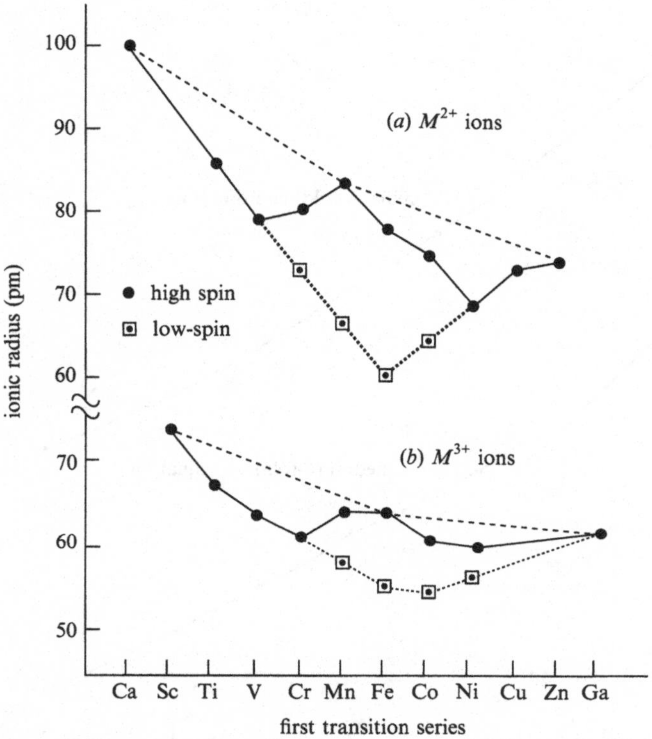

Figure 6.2 Octahedral ionic radii of first-series transition metal cations. (*a*) Divalent cations; (*b*) trivalent cations. Filled circles: high-spin cations; open squares: low-spin cations (data from Shannon, 1976).

been explained by crystal field theory (van Santen and van Wieringen, 1952; Hush and Pryce, 1957, 1958; Hush, 1958).

In cations located in octahedral sites in oxide structures, electrons in t_{2g} orbitals interact with and repel oxygen ions less strongly than do electrons in e_g orbitals. In passing from one cation to the next along the series, with the positive nuclear charge increasing by one, a greater decrease in metal–oxygen distance is expected when an electron is added to a t_{2g} orbital than upon entering an e_g orbital. Large decreases in interatomic distances are expected as the first three electrons are added to t_{2g} orbitals, with smaller decreases and even increases occurring as the fourth and fifth electrons are added to e_g orbitals. A similar pattern applies to cations with $3d^6$ to $3d^{10}$ configurations. This explanation accounts for the observed minimum interatomic distances in compounds of V^{2+} and Ni^{2+} relative to the other divalent transition metal ions. It appears, for example, when comparisons are made of metal–oxygen distances in olivine, pyroxene and silicate spinel structures containing divalent Mn, Fe, Co and Ni (see Appendix 7).

A similar explanation applies to ionic radii of trivalent transition metal ions in octahedral sites in oxides, such as compounds of $LaM^{3+}O_3$ with the perovskite structure, where $M^{3+} = Sc^{3+}$, Ti^{3+}, etc. The trend is illustrated in fig. 6.2*b*. The correlation is not so perfect as it is for divalent cations in oxides due to the absence of data for compounds containing Cu^{3+} and Zn^{3+} ions, and because of the change from high-spin to low-spin configuration between Fe^{3+} and Co^{3+}. However, the comparatively small ionic radius of the Co^{3+} ion, $(t_{2g})^6$, in oxides may be explained by the absence of electrons in e_g orbitals. The unusually small cobalt–oxygen distances found in other rare earth–cobalt perovskites, Co_2O_3 corundum, Co_3O_4 spinel and CoOOH polymorphs, as well as iron–sulphur distances in pyrite (FeS_2) and violarite ($FeNi_2S_4$), are indicative of transition metal ions in low-spin states. In some sulphides and related minerals, such low-spin configurations may result from π-bond formation (§11.5).

6.3 Jahn–Teller distortions in crystal structures

The interpretation of cation radii trends in §6.2 is applicable to transition metal ions in regular octahedral coordination. Deviations are shown by Mn^{3+} and Cu^{2+}, which, as noted in §2.12, are cations with $3d^4$ and $3d^9$ configurations and are susceptible to Jahn–Teller distortions in six-coordinated sites. Such distortions have important effects on metal–oxygen distances and crystal structures containing these cations, which are often unique or deviate from common structure types. For example, whereas fluorides of most divalent cations have rutile structures, the compounds CrF_2 and CuF_2 have distorted rutile structures (Wells, 1984). Indeed, the crystal structures of most Cu(II) minerals do not resemble those of other divalent transition metal oxygen-bearing minerals because of the Jahn–Teller effect in Cu^{2+} ions. Examples listed in table 6.1 include tenorite, malachite, azurite and dioptase which contain Cu^{2+} ions in sites so distorted from regular octahedral symmetry that they approach square planar coordination. A similar situation applies to Mn(III) minerals and compounds (Shannon *et al.*, 1975). Thus, MnF_3 has a structure distorted from the VF_3 structure to which the fluorides of most trivalent transition metal ions conform. Hausmannite, groutite and manganite also contain Mn^{3+} ions in distorted octahedral sites, deviating from spinel, ramsdellite and pyrolusite (rutile) structures, respectively. In the mixed-valence Mn oxide minerals romanechite (psilomelane) and todorokite, Mn^{3+} ions are accommodated in the most distorted octahedral sites (§8.7.3.1). In the Mn^{3+} clinopyroxene blanfordite, the M1 polyhedron becomes increasingly distorted from octahedral symmetry by

Table 6.1. *Cation sites in minerals showing Jahn–Teller distortions*

Mineral (formula)	Cation site	Configuration of coordination site	Metal–oxygen distances: Range (pm)	Metal–oxygen distances: Average (pm)	Sources of data
		Mn^{3+} minerals			
hausmannite Mn_3O_4	oct	tetrag. dist. oct.	(4x) 193.0; (2x) 228.2	193.0 (av.4) 204.7 (av.6)	[1]
Mn_3O_4 (high P)	oct	elong. oct.	191.4; 191.6; 192.5; 194.7; 230.4; 232.2	192.6 (av.4) 205.5 (av.6)	[2]
bixbyite Mn_2O_3	M1	oct.	(6x) 200.9	200.9 (av.6)	[3]
	M2	dist. oct.	(2x) 193.0; (2x) 203.3; (2x) 214.7	203.3 (av.6)	
marokite $CaMn_2O_4$	oct	elong. oct.	190.2; 191.8; 192.6; 196.3; 237.1; 247.8	192.7 (av.4) 209.3 (av.6)	[4]
groutite α-MnOOH	oct	elong. oct.	(av.4) 193.2; (av.2) 225.9	193.2 (av.4) 204.1 (av.6)	[5]
manganite γ-MnOOH	oct	elong. oct.	(av.4) 192.3; (av.2) 226.7	192.3 (av.4) 203.8 (av.6)	[6]
romanechite $(Ba,H_2O)_2Mn_5O_{10}$	M2	elong. oct.	(av.4) 191.6; (av.2) 205.6	191.6 (av.4) 196.2 (av.6)	[7]
todorokite $(Na,Ca)(Mn,Mg)_6O_{12}.xH_2O$	M2	dist. oct.	(2x) 194.0; (2x) 195.0; 189.0; 198.0	194.0 (av.6)	[8]
	M4	dist. oct.	(2x) 192.0; (2x) 198.0; 190.0; 205.0	196.0 (av.6)	
braunite $Mn^{2+}Mn^{3+}{}_6O_8SiO_4$	M2	dist. oct.	(2x) 186.6; (2x) 202.8; (2x) 221.2	203.5 (av.6)	[9]
	M3	severely dist. oct.	(2x) 190.6; (2x) 197.1; (2x) 226.7	204.8 (av.6)	
	M4	dist. oct.	(2x) 192.1; (2x) 194.6; (2x) 224.0	203.6 (av.6)	

kanonaite $(Mn,Al)_2SiO_5$	M1	elong. oct.	(2x) 185.0; (2x) 191.6; (2x) 224.2	200.3 (av.6)	[10,11]
	M2	dist. trig. bipyr.	186.2; 184.7; 192.1; (2x) 180.6	184.8 (av.5)	
piemontite $Ca_2(Mn,Fe)Al_2$	M3	very dist. oct.	(2x) 227.4; (2x) 203.1 190.0; 186.1	206.2 (av.6)	[12]
$Si_3O_{12}(OH)$	M1	dist. oct.	(2x) 187.2; (2x) 194.1 (2x) 198.5	193.3 (av.6)	
norrishite $KLiMn_2Si_4O_{12}$	M2	very dist. oct.	(2x) 185.5; (2x) 202.3 (2x) 223.3	203.7 (av.6)	[13]
			Cu^{2+} minerals		
tenorite CuO	Cu	sq. planar	(2x) 195.0; (2x) 196.0	195.5 (av.4)	[14]
			(2x) 278.0	223.0 (av.6)	
malachite $Cu_2CO_3(OH)_2$	M1	very dist. oct.	OH^-: 189.9; 191.5 CO_3^{2-}: 198.7; 207.1;	196.8 (av.4)	[15]
			CO_3^{2-}: 252.0; 263.0	217.0 (av.6)	
	M2	very dist. oct.	CO_3^{2-}: 204.2; 211.5 OH^-: 192.2; 192.4	200.1 (av.4)	
			OH^-: 235.3; 236.2	212.0 (av.6)	
azurite $Cu_3(CO_2)_2(OH)_2$	M1	sq. planar	OH^-: (2x) 188.0 CO_3^{2-}: (2x) 198.0	193.0 (av.4)	[16]
	M2	sq. pyr.	OH^-: 199.0; 204.0	199.0 (av.4)	
			CO_3^{2-}: 192.0; 201.0; 238.0	206.8 (av.5)	
turquoise $CuAl_6(PO_4)_4(OH)_8$	Cu	dist. oct.	OH^-: (2x) 191.5; (2x) 210.9	201.2 (av.4)	[17,18]
			H_2O: (2x) 242.2	214.9 (av.6)	
chalcanthite $CuSO_4.5H_2O$	Cu	tetrag. bipyr.	H_2O: (2x) 198.0; (2x) 199.0	198.5 (av.4)	[19]
			SO_4^{2-}: (2x) 238.0	205.7 (av.6)	
guildite $CuFe(SO_4)_2(OH).4H_2O$	Cu	tetrag. bipyr.	H_2O: (2x) 195.1; 201.6; 202.1	198.5 (av.4)	[20]
			SO_4^{2-}: 220.0; 232.8	207.8 (av.6)	
ransomite $CuFe_2(SO_4)_4.6H_2O$	Cu	tetrag. bipyr.	H_2O: (2x) 201.3; (2x) 195.6	198.5 (av.4)	
			SO_4^{2-}: (2x) 243.7	213.5 (av.6)	[21]

Table 6.1 *continued*

Mineral (formula)	Cation site	Configuration of coordination site	Metal–oxygen distances: Range (pm)	Metal–oxygen distances: Average (pm)	Sources of data
dioptase $Cu_6Si_6O_{18}.6H_2O$	Cu	sq. planar	(4x) 196.0	196.0 (av.4)	[22]
kinnoite $Cu_2Ca_2Si_3O_{10}.2H_2O$	Cu(1)	sq. pyr.	O: (2x) 194.0; (1x) 245.0 H_2O: (2x) 194.0	194.0 (av.4) 204.2 (av.5)	[23]
	Cu(2)	tetrag. bipyr.	O: (2x) 197.0; (2x) 224.0 H_2O: (2x) 196.0	196.5 (av.4) 205.7 (av.6)	
synth. $BaCuSi_2O_6$	Cu	dist. sq. planar	(4x) 192.6	192.6 (av.4)	[24]
cuprorivaite $BaCuSi_4O_{10}$	Cu	sq. planar	(4x) 191.0	191.0 (av.4)	[25]
Cu buserite Cu Mn(IV) oxide	Cu	tetrag. dist. oct.	(4x) 193.0; (2x) 283.0	193.0 (av.4) 223.0 (av.6)	[26]
		Cr^{2+} phases			
synth. $CaCrSi_4O_{10}$	Cr	sq. planar	(4x) 200.0	200.0 (av.4)	[27]
synth. pyroxene $(Mg_{1.4}Cr_{0.6})Si_2O_6$	M2	very dist. oct.	200.9; 203.5; 209.5; 218.2; 228.2; 264.9	220.9 (av.6)	[28]

Sources of data : [1] Jarosch (1987); [2] Ross, Rubie & Paris (1990); [3] Geller (1971); [4] Lepicard & Protas (1966); [5] Dent-Glasser & Ingram (1968); [6] Dachs (1963); [7] Turner & Post (1988); [8] Post & Bish (1988); [9] Moore & Araki (1976); [10] Weiss, Bailey & Rieder (1981); [11] Abs-Wurmbach, Langer, Siefert & Timanns (1981); [12] Dollase (1969); [13] Tyrna & Guggenheim (1991); [14] Åsbrink & Norrby (1970); [15] Süsse (1967); [16] Gattow & Zemann (1958); [17] Cid-Dresdner (1965); [18] Guthrie & Bish (1991); [19] Bacon & Curry (1962); [20] Wan, Ghose & Rossman (1978); [21] Wood (1970); [22] Heide & Boll-Dornberger (1955); [23] Laughon (1971); [24] Finger, Hazen & Hemley (1989); [25] Pabst (1959); [26] Stouff & Boulégue (1988); [27] Belky, Rossman, Prewitt & Gasparik (1984); [28] Angel, Gasparik & Finger (1989).

the Jahn–Teller effect, as do the M2 sites of the Mn^{3+}-bearing alkali amphiboles winchite and juddite (Ghose *et al.*, 1986). The epidote M3 and M1 sites also become increasingly distorted when occupied by Mn^{3+} ions in piemontite (Dollase, 1969). Most of these Mn(III) minerals contain Mn^{3+} ions in tetragonally elongated octahedra. The bipyramidal distortion of the andalusite M1 octahedral site in kanonaite induced by the Jahn–Teller effect is surpassed by very few Mn^{3+} octahedra whose structures have been refined (Weiss *et al.*, 1981). In fact, the increased distortion of the andalusite M1 octahedral sites may be correlated with shifts of absorption bands in crystal field spectra of the $(Al_{1-x}Mn_x)AlSiO_5$ series (Abs-Wurmbach *et al.*, 1981). In the trioctahedral mica norrishite, on the other hand, the M2 site accommodating the Mn^{3+} ions is compressed along one axis (Tyrna and Guggenheim, 1991). Finally, it may be noted that the structure of $NaNiO_2$ is distorted from the sodium chloride superstructure shown by other $NaM^{3+}O_2$ phases as a result of Jahn–Teller distortion by low-spin Ni^{3+} ions with the $3d^7$ or $(t_{2g})^6(e_g)^1$ configuration.

The examples of minerals affected by Jahn–Teller distortions that are listed in table 6.1 demonstrate that the concept of ionic radius is not a rigorous atomic property when applied to crystal structures containing the Cr^{2+}, Mn^{3+} and Cu^{2+} ions. Other consequences of Jahn–Teller distortions in mineral structures are discussed in §6.8.3.2 and elsewhere (Strens, 1966a; Walsh *et al.*, 1974).

6.4 Crystal chemistry of spinels

The crystal chemistry of spinels exhibits several unusual features that were elegantly explained in one of the earliest applications of crystal field theory (McClure, 1957; Dunitz and Orgel, 1957). Most oxide spinels contain trivalent and divalent cations and have the general formula $X^{2+}Y^{3+}{}_2O_4$. The structure-type is that of the mineral, $MgAl_2O_4$, which is cubic and contains eight formula units. The spinel structure illustrated earlier (fig. 4.19) shows the cubic close-packed arrangement of oxygen ions and some of the octahedral and tetrahedral sites occupied by cations. In each unit cell eight of the total sixty-four tetrahedral sites and sixteen of the thirty-two octahedral sites are occupied by cations. When the tetrahedral sites are occupied only by divalent cations and the octahedral sites by trivalent ions alone, the structure is termed a *normal* (*N*) spinel, $X[Y_2]O_4$.* When trivalent ions fill the eight tetrahedral sites and the remaining eight trivalent and eight divalent cations occupy the octahedral sites, the spinel

* In fact, the site occupancy of a *normal* spinel is abnormal. According to the radius ratio concept, larger divalent cations should occupy octahedral sites and smaller trivalent cations might be expected to favour the tetrahedral sites. The mineral spinel, $MgAl_2O_4$, which is predominantly *normal*, does not conform with radius ratio criteria.

Table 6.2. *Experimental and theoretical cation distribution in spinels*

$Y \Rightarrow$	Al^{3+}		V^{3+}		Cr^{3+}		Mn^{3+}		Fe^{3+}		Co^{3+}		Ga^{3+}	
X ⇓	Exp	Th	Exp	Th	Exp	Th	Exp	Th	Exp	Th	Exp	Th	Exp	Th
Mg^{2+}	$\frac{7}{8}I$	0	*N*	*N*	*N*	*N*	–	*N**	*I*	0	*N*	*N*	*I*	0
Zn^{2+}	*N*	0	*N*	*N*	*N*	*N*	*N**	*N**	*N*	0	*N*	*N*	*N*	0
Cd^{2+}	*N*	0	*N*	*N*	*N*	*N*	*N*	*N**	*N*	0	*N*	*N*	*N*	0
Mn^{2+}	*N*	0	*N*	*N*	*N*	*N*	*N**	*N**	*I*	0	*N*	*N*	*N*	0
Fe^{2+}	*N*	*I*	*N*	*N+I*	*N*	*N*	–	*N**	*I*	*I*	*N*	*N*	*N*	*I*
Co^{2+}	*N*	*I*	*N*	*N+I*	*N*	*N*	*I+N*	*N**	*I*	*I*	*N*	*N*	*N*	*I*
Ni^{2+}	$\frac{3}{4}I$	*I*	–	*I*	*N*	*N*	$\frac{7}{8}I$	*I+N**	*I*	*I*	$\frac{7}{8}I$	*N*	*I*	*I*
Cu^{2+}	*I*	*I**	–	*I**	*N*	*N*	$\frac{3}{4}I$	*N**	$\frac{7}{8}I$*	*I**	*N*	*N*	*N*	*I**

N = *normal* spinel; *I* = *inverse* spinel; 0 = no prediction; – = no data; * = tetragonally distorted.
Sources of data : [1] Dunn, McClure & Pearson (1965), p. 86; [2] Hill, Craig & Gibbs (1979); [3] Lenglet, Guillamet, D'Huysser, Durr & Jørgensen (1986); [4] Navrotsky & Kleppa (1967).

is termed an *inverse* (*I*) spinel, $Y[XY]O_4$. A *random* or statistical cation distribution would be formulated $(X_{1/3}Y_{2/3})[X_{2/3}Y_{4/3}]O_4$. Intermediate spinels also exist and are formulated as $(X_{1-n}Y_n)[X_nY_{2-n}]O_4$, so that $n = 0$, $\frac{2}{3}$ and 1 correspond to *normal, random* and *inverse* spinels, respectively (Navrotsky and Kleppa, 1967). For example, if seven divalent and one trivalent cations occupy the tetrahedral sites, the intermediate spinel is termed one-eighth *inverse*, with $n = \frac{1}{8}$, which corresponds to the cation distribution in the mineral spinel ($MgAl_2O_4$).

One of the unusual features of spinel crystal chemistry is that some transition metals form *normal* spinels and others *inverse*. The spinel-types are summarized in table 6.2. The site occupancy patterns were considered to be anomalous until they were explained by crystal field theory (McClure, 1957; Dunitz and Orgel, 1957).

Measurements of absorption spectra of oxides, glasses and hydrates of transition metal ions have enabled crystal field stabilization energies (CFSE's) in tetrahedral and octahedral coordinations to be estimated in oxide structures (see table 2.5). The difference between the octahedral and tetrahedral CFSE is called the octahedral site preference energy (OSPE), and values are summarized in table 6.3. The OSPE's may be regarded as a measure of the affinity of a transition metal ion for an octahedral coordination site in an oxide structure such as spinel. Trivalent cations with high OSPE's are predicted to occupy octahedral sites in spinels and to form *normal* spinels. Thus, Cr^{3+}, Mn^{3+}, V^{3+}

Table 6.3. *Octahedral site preference energies of transition metal ions in oxide structures*

Number of 3*d* electrons	Cation			Octahedral CFSE (E_o) (kJ/mole)	Tetrahedral CFSE (E_t) (kJ/mole)	Octahedral Site Preference Energy (OSPE) (kJ/mole)	n_o/n_t *
0	Ca^{2+}	Sc^{3+}	Ti^{4+}	0	0	0	0
1		Ti^{3+}		−87.4	−58.6	−28.8	15
2		V^{3+}		−160.2	−106.7	−53.5	158
3		Cr^{3+}		−224.7	−66.9	−157.8	2.9×10^6
4	Cr^{2+}			−100.4	−29.3	−71.1	829
4		Mn^{3+}		−135.6	−40.2	−95.4	8,208
5	Mn^{2+}	Fe^{3+}		0	0	0	0
6	Fe^{2+}			−49.8	−33.1	−16.7	5
6		Co^{3+}		−188.3	−108.8	−79.5	1,827
7	Co^{2+}			−92.9	−61.9	−31.0	19
8	Ni^{2+}			−122.2	−36.0	−86.2	3,440
9	Cu^{2+}			−90.4	−26.8	−63.7	407
10	Zn^{2+}	Ga^{3+}	Ge^{4+}	0	0	0	0

* $n_o/n_t = \exp-[(E_o-E_t)/RT]$ at $T = 1{,}000$ °C.
Sources of data: [1] McClure (1957); [2] Dunitz & Orgel (1957).

and Co^{3+} (low-spin) form exclusively *normal* spinels (table 6.2). Similarly, divalent cations with large OSPE's are expected to form *inverse* spinels by favouring the octahedral sites. Thus, Ni^{2+} and Cu^{2+} have a strong tendency to form *inverse* spinels. Cations such as Fe^{3+} and Mn^{2+}, which have zero OSPE's, form both *normal* and *inverse* spinels depending on the site preferences of the other cations in the structures.

The extent of agreement between the observed and predicted spinel-type is remarkably good, considering that the crystal field stabilization energy contributes less than ten per cent to the lattice energy. This factor, discussed later (§7.2), is illustrated by the data plotted in fig. 7.1. It cannot be assumed that the CFSE is the sole determiner of cation distribution in the spinel structure (Navrotsky and Kleppa, 1967; Glidewell, 1976), but it evidently cannot be ignored. Such an analysis of site occupancies in spinels could not be deduced, for example, from conventional crystal chemical concepts of ionic radius and radius ratio. More quantitative treatments of the thermodynamics of cation distributions in simple spinels have been given (Navrotsky and Kleppa, 1967), including an assessment of the role of electronic entropy in transition metal spinels (O'Neill and Navrotsky, 1983) which is described in §7.4. Thus, electronic entropy reinforces the ability of V^{3+} to form *normal* spinels and for Ni^{2+} to tolerate tetrahedal sites in some spinels (table 6.2).

The octahedral site preference energy parameter listed in table 6.3, applied originally to spinel crystal chemistry, has had a profound influence in transition metal geochemistry following its introduction into earth science literature in 1964 (Burns and Fyfe, 1964; Curtis, 1964). The use of such site preference energies to explain distribution coefficients of transition metal ions in coexisting minerals and phenocryst/melt systems are described in §7.6, §7.8 and §8.5.3.

6.5 Site occupancies in silicate structures

The site preferences shown by cations in the spinel structure demonstrate that transition metal ions prefer coordination sites that bestow on them greatest electronic stability. In addition, certain cations deform their surrounding in order to attain enhanced stability by the Jahn–Teller effect. These two features suggest that similar factors may operate and cause enrichments of cations in specific sites in silicate structures, leading to cation ordering or intersite (intracrystalline) partitioning within individual minerals which, in turn, may influence distribution coefficients of cations between coexisting phases.

The concept of cation ordering, or relative enrichment of ions in specific coordination sites, may be explained with reference to the crystal structures of ferromagnesian silicate minerals. The olivine, pyroxene, amphibole and mica groups may be regarded as solid-solutions of magnesium silicate and iron silicate components, such as the forsterite–fayalite, enstatite–ferrosilite, tremolite–actinolite, and phlogopite–annite series, respectively. The magnesium silicate, iron silicate and all intermediate Mg^{2+}–Fe^{2+} silicate compositions in a solid-solution series have the same basic crystal structure. However, each of these ferromagnesian silicates contains two or more crystallographic positions in the crystal structure which can accommodate Mg^{2+} and Fe^{2+}, as well as other transition metal ions. The Mg^{2+} ions must occupy all of the coordination sites in the magnesium silicate end-member, while in the iron silicate end-member Fe^{2+} ions will fill the same sites. Cation ordering occurs when Mg^{2+} or Fe^{2+} ions show relative enrichments in a particular site in the crystal structure of a ferromagnesian silicate of intermediate composition. When other transition metal ions such as Ni^{2+}, Co^{2+} and Mn^{2+} are present, often in minor amounts and substituting for Mg^{2+} and Fe^{2+} ions in the ferromagnesian silicates, they may preferentially occupy or be enriched in one of the Mg^{2+} or Fe^{2+} sites.

In aluminosilicates, too, including epidote and the Al_2SiO_5 polymorphs andalusite, kyanite and sillimanite, Al^{3+} ions occupy several coordination sites. Again, trivalent cations such as Cr^{3+}, V^{3+}, Mn^{3+} and Fe^{3+} may show cation ordering and be relatively enriched in one specific Al^{3+} site in a crystal structure.

The mineral structures listed in Appendix 7, together with the crystal structures of common ferromagnesian silicate and aluminosiliate minerals illustrated throughout chapters 3, 4 and 5, indicate that there are wide ranges of type, symmetry and distortion of oxygen atoms constituting the coordination sites. Consequently, the relative energies of transition metal 3*d* orbitals vary between the sites with the result that cations with specific electronic configurations may be stabilized in certain sites relative to others. This effect may induce cation ordering within a particular structure and also produce metal enrichment in one crystal structure relative to a coexisting phase in mineral assemblages in rocks.

In the following sections the observed and predicted enrichments of transition metal ions in silicate structures are discussed. The partitioning of the cations between coexisting minerals is considered in chapter 7.

6.6 Measurements of site populations in crystal structures

The conventional method for determining cation ordering and site populations within a crystal structure is by diffraction techniques using X-ray, electron and neutron sources. For determining site occupancies of transition metal ions, these methods have been supplemented by a variety of spectroscopic techniques involving measurements of Mössbauer, electron paramagnetic resonance (EPR or ESR), X-ray absorption (EXAFS and XANES), X-ray photoelectron (XPS), infrared and optical absorption spectra.

6.6.1 X-ray diffraction techniques

Analyses of electron density distributions have enabled the positions of major elements of high atomic weights such as iron and other transition elements to be located relative to lighter elements such as magnesium and aluminium in mineral crystal structures. The widespread availability of automated X-ray diffractometry and least squares refinement programs have increased the availability of site occupancy data for transition metal ions in most contemporary crystal structure refinements.

Site populations of less abundant cations estimated from X-ray diffraction measurements have limited accuracies particularly when iron is present, since diffraction phenomena involve the cooperative scattering effects of many atoms in the unit cell. It is very difficult, and sometimes impossible in X-ray structure refinements, to distinguish between cations of different valences, such as Fe^{2+} and Fe^{3+}, Mn^{2+} and Mn^{3+} or Ti^{3+} and Ti^{4+}, and between neighbouring elements in the periodic table with similar scattering factors, such as iron and manganese. Neutron diffraction techniques, however, have been used to distinguish Fe^{2+} and Mn^{2+} in silicate structures (e.g., Ballet *et al.*, 1987). When significant differences of ionic radii exist, cation site occupancies may be estimated from interatomic distances. Notable examples where X-ray structure refinements have yielded cation site occupancies of silicates include a variety of natural and synthetic transition metal-bearing Mg^{2+} olivines and pyroxenes (e.g., Ghose *et al.*, 1975, Rajamani *et al.*, 1975, Francis and Ribbe, 1980; Bish, 1981, Byström, 1987; Ottonello *et al.*, 1989).

6.6.2 Channel enhanced X-ray emission spectroscopy

The technique of channeling-enhanced X-ray emission (CHEXE) has enabled cation site occupancies to be determined in various minerals, including transition metal ions in spinels and ferromagnesian silicates (Taftø, 1982; Taftø and Spence, 1982; Smyth and Taftø, 1982; McCormick *et al.*, 1987). The method, which is based on relative intensities of X-ray peaks measured on crystals with diameters as small as 50 nm under the electron microscope, is particularly useful for determining site occupancies of minor elements with concentrations as low as 0.05 atom per cent in a structure. The most important criterion for the determination of element distribution in a mineral by this technique is that the cation sites should lie on alternating crystallographic planes. In order to make quantitative site population estimates, additional information is required, particularly the occupancy of at least one element in one of the sites or in another site that lines up with one of the sites of interest (McCormick *et al.*, 1987). For example, cation site occupancies by CHEXE measurements have been determined from X-ray peak intensity ratios of Si to Ni, Mn, Cr and Fe in forsterite, as well as thermal disordering of these cations in heated olivines (Smyth and Taftø, 1982).

6.6.3 Mössbauer spectroscopy

The Mössbauer effect, or recoilless emission and resonant absorption of nuclear gamma rays by specific isotopes in solids, is observed in two transition elements, ^{57}Fe and ^{61}Ni, as well as in ^{67}Zn. Whereas measurements involving nickel and zinc are extremely difficult, iron-bearing minerals with a natural isotopic abundance of 2.2% ^{57}Fe (Appendix 2) are conveniently studied by Mössbauer spectroscopy and this technique has yielded considerable information about the crystal chemistry of iron in numerous rock-forming silicates (Coey, 1985; Hawthorne, 1988) and planetary materials (Burns, 1992b). Each distinct iron species in a mineral may give rise to separate peaks in a

Mössbauer spectrum, the positions and separations of which depend on the oxidation state, electronic configuration, magnetic ordering, coordination number and symmetry of the iron atoms in a mineral structure. Computed peak areas enable Fe^{3+}/Fe^{2+} ratios and site populations of Fe^{2+} ions in crystal structures to be readily determined for several minerals without interference from coexisting Mn, Cr and other transition elements.

6.6.4 Electron paramagnetic resonance

Investigations of minerals by electron paramagnetic resonance (EPR) spectroscopy are mainly centred on transition metal ions with odd numbers of unpaired electrons. In particular, Mn^{2+}, Fe^{3+}, Cr^{3+}, V^{4+}, Cu^{2+} and Ti^{3+}, are conducive to EPR measurements when they are present in low concentrations in diamagnetic host crystals such as Mg and Al oxides and silicates containing negligible ferrous iron contents. Valence states of paramagnetic ions, as well as sites occupied by the cations and relative proportions in each site, may be determined by EPR (synonymous with electron spin resonance, ESR, because paramagnetism is due to electron spin), particularly when the cation sites have different symmetries. Mineralogical applications of EPR are described elsewhere (Vassilikou-Dova and Lehmann, 1987; Calas, 1988).

6.6.5 Hydroxyl stretching frequencies in infrared spectra

Site populations in hydroxyl-bearing minerals may be estimated from measurements of O–H stretching frequencies of the fundamental band located at 3,600 to 3,700 cm^{-1} or from peaks in the first overtone region at 7,050 to 7,200 cm^{-1} (Burns and Strens, 1966; Farmer, 1974). The O–H vibration spectra of end-member amphiboles such as tremolite, $Ca_2Mg_5Si_8O_{22}(OH)_2$, or grunerite, $Fe_7Si_8O_{22}(OH)_2$, consist of single peaks. However, in infrared spectra of intermediate Mg–Fe^{2+} amphiboles and biotites, multiple sets of peaks are present, the number and relative intensity of which depend on the distributions of cations in crystallographic positions that are coordinated to the hydroxyl group (Burns and Strens, 1966; Wilkins, 1967). The technique has been widely used to obtain directly cation site populations of the amphibole M1 and M3 positions (see fig. 5.18) and the mica M1 and M2 positions (see fig. 5.21). In combination with Mössbauer spectroscopy, the amphibole M2 site populations may also be obtained indirectly (e.g., Burns and Greaves, 1971). Cation distributions derived from hydroxyl stretching frequences of amphiboles have been critically reviewed (Strens, 1974; Hawthorne 1981b, 1983), and data for micas are summarized elsewhere (Farmer, 1974; Rossman, 1984).

6.6.6 Crystal field spectra

Variations of extinction coefficients and spectrum profiles with changes in chemical composition of a mineral provide information on cation ordering in the structure. Examples involving Al^{3+}–Mn^{3+} ordering in epidotes and andalusites are discussed in §4.4.2 and §4.5, and Mn^{2+}–Fe^{2+} ordering in olivine is illustrated in fig. 4.8. Compositional variations of intensities of absorption bands in polarized spectra of orthopyroxenes described in §5.5.4. (fig. 5.15) have yielded Fe^{2+}/M2 site populations (Goldman and Rossman, 1979), while similar trends in the crystal field spectra of synthetic Mg–Ni olivines described in §5.4.2.4 (fig. 5.12) have yielded site occupancy ratios of Ni^{2+} ions in the olivine M1 and M2 sites (Hu *et al.*, 1990).

6.6.7 X-ray absorption spectra (XANES and EXAFS)

X-ray absorption spectroscopy is a versatile element-specific structural probe for elucidating the crystal chemistry of several elements, including transition metal ions, even when present in low concentrations (Brown *et al.*, 1988). Fine structure near the immediate vicinity of the X-ray absorption edge, the XANES region, provides information about cation valence, ligand type, coordination number and site geometry (Waychunas *et al.*, 1983). Beyond the XANES region, above the edge absorption maximum, are the fine structure oscillations, or EXAFS region, which contain information about the number, type and distances of neighbouring atoms about the absorbing element under investigation (Waychunas *et al.*, 1986). The two techniques have been used, for example, to elucidate valences, electronic structures, and site occupancies of Fe and Ti in a number of oxide and silicate minerals (Waychunas *et al.*, 1983; 1986; Waychunas, 1987), the distrbutions of Co, Ni, Cu and Zn in Mn(IV) oxides (Manceau *et al.*, 1987; Stouff and Boulégue, 1988), and the site occupancies of Ni in phyllosilicates (Manceau and Calas, 1985, 1986).

6.7 Site occupancies in silicate minerals

6.7.1 Olivines

6.7.1.1 Fe^{2+} ions.

A number of X-ray crystal structure refinements of lunar and terrestrial Mg^{2+}–Fe^{2+} olivines spanning wide composition ranges have revealed that Fe^{2+} ions are often slightly enriched in the smaller M1 sites (e.g., Finger, 1970; Brown and Prewitt, 1973; Smyth and Hazen, 1973; Wenk and Raymond, 1973; Will and Nover, 1979; Basso *et al.*, 1979; Nover and Will, 1981; Princivalle and Secco, 1985). However, some olivines in xenoliths from the Earth's Mantle show small Fe^{2+} enrichments in the olivine M2 sites (Princivalle and Secco, 1985; Princivalle, 1990). These trends for Mg^{2+}–Fe^{2+} olivines were confirmed in Mössbauer spectra measured above 400 °C at which temperature the two closely overlapping ferrous doublets originating from Fe^{2+} ions in the M1 and M2 sites become resolvable (e.g., Bush *et al.*, 1970; Finger and Virgo, 1971; Virgo and Hafner, 1972; Shinno, 1981).

The partitioning of Mg^{2+} and Fe^{2+} ions between the M1 and M2 sites of olivine is represented by an exchange reaction

$$Fe^{2+}{}_{M2} + Mg^{2+}{}_{M1} = Fe^{2+}{}_{M1} + Mg^{2+}{}_{M2} \tag{6.1}$$

and is expressed by the intracrystalline or intersite partition coefficient, K_d, which may be written as either

$$K_d = [(Fe^{2+})_{M1}/(Fe^{2+})_{M2}]/[(Mg^{2+})_{M1}/(Mg^{2+})_{M2}] \tag{6.2}$$

or

$$K_d = [(Fe^{2+}/Mg^{2+})_{M1}]/[(Fe^{2+}/Mg^{2+})_{M2}]. \tag{6.3}$$

The condition for random distribution of the cations on the two sites corresponds to $K_d = 1$. For Mg^{2+}–Fe^{2+} exchange in olivines, K_d often exceeds 1.0 when plotted against mole fraction of fayalite (Rajamani *et al.*, 1975; Princivalle, 1990), indicating a slight ordering of Fe^{2+} ions in M1 sites.

Crystal structure refinements of olivines at elevated temperatures (Brown and Prewitt, 1973; Smyth and Hazen, 1973; Aikawa *et al.*, 1985) indicate that distortions of the M1 and M2 coordination polyhedra increase with rising temperature, the distortion increasing more rapidly for the M1 site. These patterns for natural olivines correlate with X-ray structure refinements of olivine single crystals equilibrated at high temperatures under controlled oxygen fugacities (Princivalle, 1990; Ottonello *et al.*, 1990). The K_d values increase regularly from <1.0 at temperatures below 400 to 600 °C, depending on composition, to >1.0 at higher temperatures, indicating slight increases of Fe^{2+} occupancies of the olivine M1 sites as they become more distorted at elevated temperatures. At constant P and T, values of K_d increase with increasing fayalite content of the olivine. There is ambiguity over the influence of oxygen partial pressure, f_{O2}. Early observations of two volcanic forsteritic olivines indicated that K_d values increase with decreasing f_{O2} (Will and Nover, 1979; Nover and Will, 1981), whereas more recent experiments suggest that varying f_{O2} has a negligi-

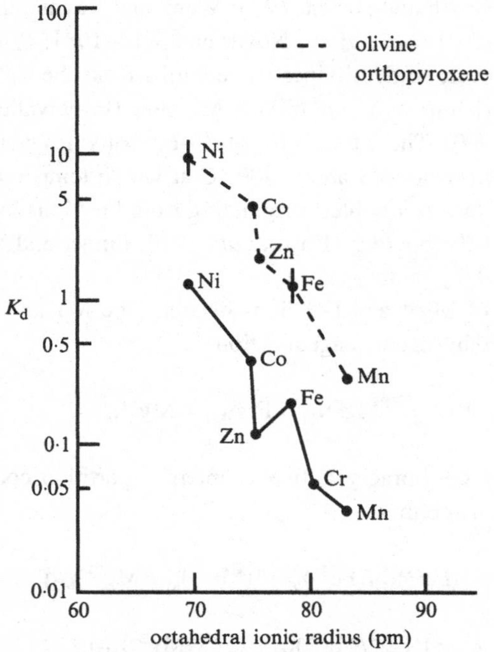

Figure 6.3 Log–linear plot of the intersite partition coefficient, K_d, versus ionic radii of divalent cations in olivines and orthopyroxenes (modified from Rajamani *et al.*, 1975).

ble influence on intracrystalline order–disorder in Mg^{2+}–Fe^{2+} olivines (Ottonello *et al.*, 1990).

6.7.1.2 *Ni^{2+} ions.*

The strong enrichment of Ni^{2+} ions in the smaller M1 sites in liebenbergite (Bish, 1981) has been confirmed for a variety of synthetic Mg^{2+}–Ni^{2+} and Fe^{2+}–Ni^{2+} olivines by X-ray structure refinements (Rajamani *et al.*, 1975; Bish, 1981; Boström, 1987; Ottonello *et al.*, 1989), Mössbauer spectroscopy (Annersten *et al.*, 1982; Nord *et al.*, 1982; Ribbe and Lumpkin, 1984), CHEXE (Taftø and Spence, 1982; Smyth and Taftø, 1982) and crystal field spectral measurements (Hu *et al.*, 1990) described in §5.4.2.4.

The strong preference of Ni^{2+} ions for olivine M1 sites is graphically demonstrated in plots of K_d versus ionic radii of divalent cations illustrated in fig. 6.3. For exchange reactions of the type formulated in eq. (6.1), K_d values lie in the range 9 to 17 for Ni^{2+}–Mg^{2+} partitioning in olivine (Rajamani *et al.*, 1975; Bish, 1981; Boström, 1987), a minimum value being observed for olivine compositions with about 65 mole per cent Ni_2SiO_4 (Boström, 1987). The K_d values decrease slightly at elevated temperatures (Smyth and Taftø, 1982; Ottonello *et al.*, 1989), correlating with slightly reduced distortion of the olivine M2 site as Ni^{2+} ions are disordered into this position.

6.7.1.3 Other transition metal ions

The X-ray structure refinement of a synthetic Mg–Co^{2+} olivine (Ghose and Wan, 1974; Ghose *et al.*, 1975) show that Co^{2+} ions also favour the M1 sites. The K_d data plotted in fig. 6.3 indicate that cobalt is less strongly enriched than Ni^{2+} ions in the M1 sites. The X-ray structure refinements (Ghose *et al.*, 1975) and K_d data plotted in fig. 6.3 show that Zn^{2+} ions also favour the olivine M1 sites.

In manganiferous olivines, the evidence from crystal field spectral measurements for strong ordering of Mn^{2+} ions in the larger M2 sites of the olivine structure (Burns, 1970) described in §4.5 was subsequently confirmed by X-ray structure refinements (Ghose *et al.*, 1975; Francis and Ribbe, 1980), neutron diffraction studies (Ballet *et al.*, 1987), CHEXE (Taftø and Spence, 1982), vibrational spectra in the mid-infared region (Huggins, 1973) and Mössbauer spectral measurements (Annersten *et al.*, 1984) of natural and synthetic Mg^{2+}–Mn^{2+} and Fe^{2+}–Mn^{2+} olivines. In Mössbauer spectral studies of Fe^{2+}–Zn^{2+} orthosilicates, Fe^{2+} were shown to be relatively enriched in the olivine M2 sites as Zn^{2+} ions preferentially occupy the M1 sites of fayalite (Ericsson and Filippidis, 1986). In willemite, Fe^{2+} ions are relatively enriched in one of the two tetrahedral Zn^{2+} sites (Ericsson and Filippidis, 1986).

In a synthetic forsterite doped with 2 x 10^{-4} per cent chromium, Cr^{3+} ions were shown by EPR measurements (Rager, 1977) to be slightly enriched in the M1 sites resulting in an occupancy ratio M1:M2 = 60:40, corresponding to 1.2 x10^{-4} per cent Cr^{3+} in the M1 site. In ferric-bearing forsterites, EPR measurements indicated that Fe^{3+} ions have a preference for the M2 sites (Weeks *et al.*, 1974).

6.7.2 Orthopyroxenes and pigeonites

6.7.2.1 Fe^{2+} ions

In a metamorphic hyperthene, $(Mg_{0.93}Fe_{1.07})Si_2O_6$, an early X-ray crystal structure determination showed that ninety per cent of the M2 positions and only fifteen per cent of the M1 positions are occupied by Fe^{2+} ions (Ghose, 1965). This strong preference of ferrous iron for the larger M2 sites was confirmed for a wide range of orthopyroxene compositions by more recent structure refinements (e.g., Burnham *et al.*, 1971; Ghose and Wan, 1973; Ghose *et al.*, 1975) and several Mössbauer spectral studies (e.g., Bancroft *et al.*, 1967; Virgo and Hafner, 1969; Saxena and Ghose, 1971; Besancon, 1981; Anovitz *et al.*, 1988).

The crystal field spectral measurements of orthopyroxenes described in §5.5.4 also demonstrate the relative enrichments of Fe^{2+} ions in the pyroxene M2 sites (Goldman and Rossman, 1979).

In pigeonites, X-ray diffraction (e.g., Clark *et al.*, 1971; Brown *et al.*, 1972; Takeda *et al.*, 1974) and Mössbauer spectral measurements (Bancroft and Burns, 1967b) of samples from lunar and volcanic rocks demonstrate that there is a strong enrichment of Fe^{2+} ions in the M2 sites, but the enrichments are slightly smaller than those in metamorphic orthopyroxenes with similar iron contents due, in part, to the higher Ca contents of pigeonites.

6.7.2.2 Other transition metal ions

The strong preference of Mn^{2+} ions for the orthopyroxene M2 sites, first demonstrated indirectly by Mössbauer spectroscopy (Bancroft *et al.*, 1967), was subsequently confirmed by X-ray structure refinements of synthetic Mg^{2+}–Mn^{2+} (Ghose *et al.*, 1975) and Mg^{2+}–Mn^{2+}–Co^{2+} (Hawthorne and Ito, 1977) orthopyroxenes. These X-ray measurements, as well as those of synthetic Mg^{2+}–Co^{2+} and Mg^{2+}–Zn^{2+} orthopyroxenes (Ghose *et al.*, 1975), also showed that Co^{2+} and Zn^{2+} ions are both relatively enriched in the orthopyroxene M2 site. In a synthetic Mg^{2+}–Cr^{2+} pyroxene, the Cr^{2+} ions are strongly enriched in the M2 site (Angel *et al.*, 1989), causing increased distortion of this site as a result of the Jahn–Teller effect (§6.3).

The relative enrichments of divalent cations in the orthopyroxene M1 or M2 sites are demonstrated by the intersite K_d data plotted in fig. 6.3. Most of the divalent transition elements have K_d values less than one, indicating that they favour the larger, more distorted M2 sites. In contrast, Ni^{2+} ions are enriched in the smaller M1 sites of synthetic Mg^{2+}–Ni^{2+} orthopyroxene (Ghose *et al.*, 1975), having a K_d value of about two. Note that the relative enrichment of Ni^{2+} ions in the orthopyroxene M1 sites is not as strong as it is in the olivine M1 sites.

6.7.3 Amphiboles

Cation distributions in amphibole structures, the site occupancy data for which were obtained mainly from X-ray crystal structure refinements, Mössbauer spectroscopy and hydroxyl stretching frequency measurements in the infrared region, have been critically reviewed (Strens, 1974; Hawthorne, 1981a,b; 1983).

In Mg^{2+}–Fe^{2+} amphiboles of the anthophyllite and cummingtonite–grunerite series, Fe^{2+} ions are strongly enriched in the M4 positions, depleted in the M2 positions, and generally randomly distributed over the M1 and M3 positions. In manganiferous amphiboles of the tirodite–dannemorite series, Mn^{2+} ions are more strongly ordered than Fe^{2+} in the M4 positions, displacing iron into M2 positions so these positions now have higher Fe^{2+} site occupancies than the M1 and M3 positions.

Calcic amphiboles of the tremolite–actinolite series show complex variations of Fe^{2+} ordering. A common Fe^{2+} site enrichment pattern is M1 > M3 > M2 >> M4. In F^--bearing amphiboles, however, Fe^{2+} are relatively enriched in the M2 site due to F^- ion avoidance described later (§6.7.7).

In alkali amphiboles, in which sodium fills the M4 positions, trivalent cations (Fe^{3+}, Al^{3+}) are largely confined to the smaller distorted octahedral M2 sites, so that the ordering pattern for Fe^{2+} ions becomes M3 > M1 >> M2. In the manganese-bearing alkali amphiboles winchite and juddite, Mn^{3+} ions are ordered in the M2 sites (Ghose *et al.*, 1986), in which they are stabilized by the Jahn–Teller effect (§6.3).

6.7.4 Biotites

Numerous structure refinements of biotites reviewed elsewhere (Bailey, 1984), together with Mössbauer spectral measurements of suites of specimens from igneous and metamorphic rocks (Bancroft and Brown, 1975; Dyar and Burns, 1986; Dyar, 1987), have revealed that Fe^{2+} ions are slightly enriched in the *cis*-M2 sites which are marginally smaller in size than the centrosymmetric *trans*-M1 sites. In metamorphic biotites, the Fe^{2+} ions are more highly enriched in the *cis*-M2 sites than in biotites from igneous rocks.

6.7.5 Epidotes

Crystal structure refinements show that Fe^{3+} and Mn^{3+} ions are strongly enriched in the very distorted M3 site of the epidote structure (Dollase, 1968; Gabe *et al.*, 1973; Stergiou and Rentzeperis, 1987), but these cations also appear to occupy M1 sites, too, in manganiferous epidotes (Dollase, 1969). These site occupancies correlate with Mössbauer spectral data for Fe^{3+} (Bancroft *et al.*, 1967; Dollase, 1971) and crystal field spectra of Mn^{3+} in piemontites (Burns and Strens, 1967; Smith *et al.*, 1982) described in §4.4.2 and §5.4.4. EPR measurements of clinozoisites (Vassilikou-Dova and Lehmann, 1987) indicate that Cr^{3+} and Fe^{3+} ions occupy the M3 sites.

6.7.6 Other mineral structures

Chrysoberyl, Al_2BeO_4, which is iso-structural with olivine, exists as the gems alexandrite, $(Al,Cr)_2BeO_4$, and sinhalite, $(Al,Fe)_2BeO_4$. Structure refinements of these gem minerals (Farrell *et al.*, 1963), as well as EPR measurements

(Forbes, 1983), show that Cr^{3+} and Fe^{3+} ions preferentially populate the slightly larger Al2 sites corresponding to the mirror plane acentric M2 sites of the olivine structure. The EPR measurements indicated that 78 per cent of the Cr^{3+} ions in a synthetic alexandrite occur in the Al2 positions. This contrasts with the slight enrichment of Cr^{3+} in the M1 sites of forsterite (Rager, 1977).

In tourmalines, transition metal ions, including Cr^{3+}, V^{3+} and iron cations, are relatively enriched in the larger 'Mg sites' relative to the smaller 'Al sites' (Foit and Rosenburg, 1979; Nuber and Schmetzer, 1979).

6.7.7 Fluorine avoidance in hydroxysilicates

A common observation of hydroxyl-bearing ferromagnesian silicates is that minerals with high Mg/Fe ratios show more extensive substitution of F^- ions for hydroxyl groups than iron-rich compositions. This feature, referred to as 'Fe–F avoidance', is commonly observed in amphiboles, micas and tourmalines (Rosenberg and Foit, 1977). Spectroscopic studies of F^--bearing biotites demonstrated that OH groups are preferentially coordinated to Fe^{2+} ions (Sanz and Stone, 1979), implying that F^- ions specifically avoid those sites which require coordination with Fe^{2+} ions. In F^--bearing amphiboles, however, Fe^{2+} ions are relatively enriched in M2 positions where the coordinating ligands are all non-bridging oxygen ions, and avoid the M1 and M3 sites which are coordinated to F^- ions that substitute for OH^- ions. Such Fe–F avoidance is a direct consequence of the spectrochemical series (§2.9.2), in which Δ for F^- is much lower than that for oxygen ligands, indicating a higher CFSE for Fe^{2+} coordinated to OH^- than to F^- ions.

6.8 Explanations of cation ordering

The X-ray structure refinements and spectroscopic measurements described in §6.7 demonstrate that transition metal ions are ordered to varying degrees in the crystal structures of numerous ferromagnesian and aluminosilicate minerals. The enrichment of a transition metal ion relative to Mg^{2+} or Al^{3+} in a specific coordination site results from the interplay of several crystal chemical and bond energy factors, including the crystal field stabilization energy.

6.8.1 Cation size

Since site enrichments in a crystal structure are expressed relative to a host cation, the primary factor affecting the cation distributions is the comparative ionic radius of Mg^{2+} or Al^{3+} to a transition metal ion of similar charge. Among divalent cations, octahedral ionic radii tabulated in Appendix 3 and plotted in fig. 6.2*a* increase in the order

$$Ni^{2+} \text{ (69 pm)} < Mg^{2+} \text{ (72 pm)} < Zn^{2+} \text{ (74 pm)} < Co^{2+} \text{ (74.5 pm)} <$$
$$Fe^{2+} \text{ (78 pm)} < Cr^{2+} \text{ (80 pm)} < Mn^{2+} \text{ (83 pm)} < Ca^{2+} \text{ (100 pm)}, \quad (6.4)$$

while for trivalent cations plotted in fig. 6.2*b* the order is

$$Al^{3+} \text{ (53.5 pm)} < \text{low-spin } Co^{3+} \text{ (54.5 pm)} <$$
$$\text{low-spin } Ni^{3+} \text{ (56 pm)} < Cr^{3+} \text{ (61.5 pm)} <$$
$$Ga^{3+} \text{ (62 pm)} < V^{3+} \text{ (64 pm)} < Fe^{3+} = Mn^{3+} \text{ (64.5 pm)} <$$
$$Ti^{3+} \text{ (67 pm)} < Sc^{3+} \text{ (74.5 pm)}. \quad (6.5)$$

In ferromagnesian silicates, therefore, Ni^{2+} ions are expected to be enriched over Mg^{2+} in smallest octahedral sites, with the other divalent transition metal ions favouring larger sites in the crystal structures. Thus, based on the ionic radius criterion alone, the olivine M1 and pyroxene M1 sites would be expected be enriched in Ni^{2+}, with the other divalent cations showing preferences for the larger olivine M2 and pyroxene M2 sites. Similarly, in aluminosilicates, all trivalent transition metal ions are predicted to show preferences for the largest $[AlO_6]$ octahedron.

Confusing the cation size criterion, however, is the fact that ranges of individual metal–oxygen distances exist in most coordination sites in ferromagnesian silicates (e.g., table 5.6; Appendix 7). Moreover, ranges of metal–oxygen distances within sites are wider than differences between the ionic radii of Mg^{2+} and individual transition metal ions competing with one another for cation sites in the crystal structures. In forsterite, Fa_{10}, for example, metal–oxygen distances for the two sites illustrated in fig. 5.8 are (Birle *et al.*, 1968)

$$\text{olivine M1 site: average M1–O} = 210.1 \text{ pm; range} = 207.5 \text{ to } 214.1 \text{ pm;}$$
$$\text{olivine M2 site: average M2–O} = 213.5 \text{ pm; range} = 205.7 \text{ to } 222.1 \text{ pm.} \quad (6.6)$$

Thus, although the average metal–oxygen distance of the forsterite M2 site is 3.4 pm larger than the M1 site, there are two individual M2–O distances that are actually shorter than the smallest M1–O distance (Appendix 7). These differences are comparable to ionic radii differences between Mg^{2+}–Ni^{2+} or Mg^{2+}–Co^{2+} pairs. The situation is even worse for the orthopyroxene structure in which there is a large spread of metal–oxygen distances particularly for the M2 site (table 5.6; Appendix 7).

Therefore, predictions based on ionic radii values indicating that Ni^{2+} ions will prefer the smaller M1 sites of olivine and orthopyroxene and that Co^{2+} ions will favour the larger M2 sites in both minerals are rather subjective. On the other hand, the data summarized in Appendix 7 show that all Ni^{2+}–O and Co^{2+}–O distances in Ni_2SiO_4 and Co_2SiO_4 relative to corresponding Mg^{2+}–O distances in forsterite, including individual, ranges and average metal–oxygen distances for both olivine M1 and M2 sites, do conform with the relative order of ionic radii of the cations, eq. (6.4).

6.8.2 Crystal field stabilization energy

Zinc and the divalent transition metal ions Ni^{2+}, Co^{2+}, Fe^{2+}, Cr^{2+} and Mn^{2+} all have ionic radii comparable to Mg^{2+}, eq. (6.4), making it possible for atomic substitution to occur in ferromagnesian silicates. However, electronic configurations differ significantly between these cations resulting in a wide range of CFSE (table 2.5) attained by the cations in a crystal structure. The data for the cations in regular octahedral sites in oxides summarized in table 2.5 show that CFSE's (kJ/g.ion) decrease (i.e. become less negative) in the order

$$Ni^{2+}(-29.6) > Cr^{2+}(-23.8) > Co^{2+}(-21.3) > Fe^{2+}(-11.9) > Mn^{2+}, Zn^{2+}, Mg^{2+} \text{ (zero)}. \qquad (6.7)$$

Furthermore, the inverse fifth-power dependence of crystal field splitting on metal–oxygen distance, eq. (2.17), indicates that an increased CFSE may be acquired by most transition metal ions when they occupy smaller coordination sites. As a result of this enhanced stability, Ni^{2+}, Cr^{2+}, Co^{2+} and Fe^{2+} are predicted to have preferences for the olivine M1 and pyroxene M1 sites having smaller average metal–oxygen distances than the corresponding M2 sites (eq. (6.7); Appendix 7) and to display relative enrichments for the M1 sites in the order: $Ni^{2+} > Cr^{2+} > Co^{2+} > Fe^{2+} > Mn^{2+}, Zn^{2+}$. This order closely resembles the measured site occupancies (Ghose *et al.*, 1975; Rajamani *et al.*, 1975; Walsh *et al.*, 1976) shown in fig. 6.3.

In aluminosilicates each high-spin trivalent transition metal ion has a larger ionic radius than the host Al^{3+} ion. However, cations such as Cr^{3+} (−249.9 kJ/mole), V^{3+} (−182.8 kJ/mole) and Mn^{3+} (−150.8 kJ/mole), which acquire particularly large CFSE's in octahedral coordination (table 2.5), are induced to enter $[AlO_6]$ octahedra, and not five-coordinated $[AlO_5]$ (andalusite, yoderite) or tetrahedral $[AlO_4]$ (sillimanite) sites, by the enhanced stabilization bestowed in octahedral crystal fields. Although Co^{3+} ions have not been positively identified in silicate minerals, the strong enrichment of cobalt in natural and syn-

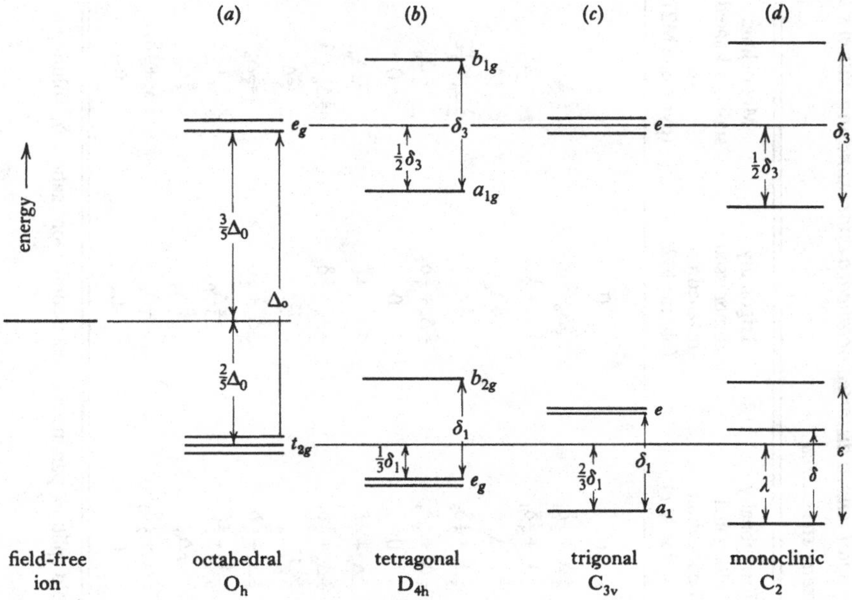

Figure 6.4 Energy level diagram for 3*d* orbitals of transition metal ions located in six-coordinated sites of different symmetries. (*a*) Octahedral; (*b*) tetragonal (elongated along the tetrad axis); (*c*) trigonal (compressed along the triad axis); (*d*) monoclinic (cf. fig. 6.4). The average metal–oxygen distances are assumed to be the same for each site.

thetic Mn(IV) oxides discussed in §8.7.4 results from the very high CFSE attained by low-spin Co^{3+} ions (54.5 pm) substituting for octahedrally coordinated Mn^{4+} ions (53 pm) with comparable ionic radii.

6.8.3 Site distortion

In some cases, the CFSE attained by a transition metal ion in a regular octahedral site may be enhanced if the coordination polyhedron is distorted. This effect is potentially very important in most silicate minerals since their crystal structures typically contain six-coordinated sites that are distorted from octahedral symmetry. Such distortions are partly responsible for the ranges of metal–oxygen distances alluded to earlier, eq. (6.6). Note, however, that the displacement of a cation from the centre of a regular octahedron, such as the comparatively undistorted orthopyroxene M1 coordination polyhedron (fig. 5.16), also causes inequalities of metal–oxygen distances.

In low-symmetry, distorted coordination sites, the 3*d* orbitals of a transition metal ion are resolved into additional energy levels. As a result, the energy of

Table 6.4. *Theoretical crystal field stabilization energies of transition metal ions in distorted octahedra approximating cation sites in olivine and orthopyroxene structures*

Cation	Electronic configuration	Octahedron (regular) (pyroxene M1)	Tetragonally elongated octahedron (olivine M1)	Trigonally compressed octahedron (olivine M2)	Monoclinic six-coordinated site (pyroxene M2)
Ca^{2+} Sc^{3+} Ti^{4+}	[Ar]	0	0	0	0
Ti^{3+} V^{4+}	$(t_{2g})^1$	$\frac{2}{5}\Delta_o$	$\frac{2}{5}\Delta_o+\frac{1}{3}\delta_1$	$\frac{2}{5}\Delta_o+\frac{2}{3}\delta_1$	$\frac{2}{5}\Delta_o+\lambda$
V^{3+} Cr^{4+}	$(t_{2g})^2$	$\frac{4}{5}\Delta_o$	$\frac{4}{5}\Delta_o+\frac{2}{3}\delta_1$	$\frac{4}{5}\Delta_o+\frac{1}{3}\delta_1$	$\approx\frac{4}{5}\Delta_o$
V^{2+} Cr^{3+} Mn^{4+}	$(t_{2g})^3$	$\frac{6}{5}\Delta_o$	$\frac{6}{5}\Delta_o$	$\frac{6}{5}\Delta_o$	$\frac{6}{5}\Delta_o$
hs Cr^{2+} Mn^{3+}	$(t_{2g})^3(e_g)^1$	$\frac{9}{5}\Delta_o$	$\frac{3}{5}\Delta_o+\frac{1}{2}\delta_3$	$\frac{3}{5}\Delta_o$	$\frac{3}{5}\Delta_o+\frac{1}{2}\delta_3$
ls Cr^{2+} Mn^{3+}	$(t_{2g})^4$	$\frac{8}{5}\Delta_o$	$\frac{8}{5}\Delta_o+\frac{1}{3}\delta_1$	$\frac{8}{5}\Delta_o+\frac{2}{3}\delta_1$	$\approx\frac{4}{5}\Delta_o$
hs Mn^{2+} Fe^{3+}	$(t_{2g})^3(e_g)^2$	0	0	0	0
ls Mn^{2+} Fe^{3+}	$(t_{2g})^5$	$2\Delta_o$	$2\Delta_o+\frac{2}{3}\delta_1$	$2\Delta_o+\frac{1}{3}\delta_1$	$\approx 2\Delta_o$
hs Fe^{2+} Co^{3+}	$(t_{2g})^4(e_g)^2$	$\frac{2}{5}\Delta_o$	$\frac{2}{5}\Delta_o+\frac{1}{3}\delta_1$	$\frac{2}{5}\Delta_o+\frac{2}{5}\delta_1$	$\frac{2}{5}\Delta_o+\lambda$
ls Fe^{2+} Co^{3+}	$(t_{2g})^6$	$\frac{12}{5}\Delta_o$	$\frac{12}{5}\Delta_o$	$\frac{12}{5}\Delta_o$	$\frac{12}{5}\Delta_o$
hs Co^{2+} Ni^{3+}	$(t_{2g})^5(e_g)^2$	$\frac{4}{5}\Delta_o$	$\frac{4}{5}\Delta_o+\frac{2}{3}\delta_1$	$\frac{4}{5}\Delta_o+\frac{1}{3}\delta_1$	$\approx\frac{4}{5}\Delta_o$
ls Co^{2+} Ni^{3+}	$(t_{2g})^6(e_g)^1$	$\frac{9}{5}\Delta_o$	$\frac{9}{5}\Delta_o+\frac{1}{2}\delta_3$	$\frac{9}{5}\Delta_o$	$\frac{9}{5}\Delta_o+\frac{1}{2}\delta_3$
Ni^{2+}	$(t_{2g})^6(e_g)^2$	$\frac{6}{5}\Delta_o$	$\frac{6}{5}\Delta_o$	$\frac{6}{5}\Delta_o$	$\frac{6}{5}\Delta_o$
Cu^{2+}	$(t_{2g})^6(e_g)^3$	$\frac{3}{5}\Delta_o$	$\frac{3}{5}\Delta_o+\frac{1}{2}\delta_3$	$\frac{3}{5}\Delta_o$	$\frac{3}{5}\Delta_o+\frac{1}{2}\delta_3$
*Zn^{2+} Ga^{3+} Ge^{4+}	$(t_{2g})^6(e_g)^4$	0	0	0	0

hs and ls are high-spin and low-spin configurations, respectively; Δ_o, crystal field splitting parameter in octahedral coordination; δ_1, splitting of t_{2g} orbital group; δ_3, splitting of e_g orbital group.
* Zn^{2+} occurs in tetrahedral coordination in most oxide structures.

at least one of the lower-level t_{2g} orbitals is further stabilized in a distorted octahedral site. The upper-level e_g orbitals, too, may be split into separate energy levels, except in a trigonally distorted octahedron. Such splittings of the 3*d* orbital energy levels are depicted in fig. 6.4 for transition metal ions in sites having octahedral, tetragonal, trigonal and lower (orthorhombic, monoclinic or triclinic) symmetries. The average metal–oxygen distances are assumed to be identical in each site. Table 6.4 lists the CFSE's predicted for each transition metal ion in a variety of distorted octahedral sites based on the 3*d* orbital energy separations shown in fig. 6.4.

6.8.3.1 Fe^{2+} ions

The electronic configuration of the Fe^{2+} ion, $3d^6$, is such that five electrons occupy singly each of the 3*d* orbitals and the sixth electron fills the most stable orbital in a distorted coordination site. The Fe^{2+} ion thus attains an increment to its octahedral CFSE, normally amounting to $0.4\Delta_o$ in a regular octahedral site, as a result of its sixth 3*d* electron entering the lowest energy t_{2g} orbital relative to regular octahedral coordination. Provided the average metal–oxygen distances remain equal, Fe^{2+} ions are predicted to be stablized in a distorted six-coordinated site and to be enriched in that environment relative to a more regular octahedral site.

If average metal–oxygen distances were identical in each coordination polyhedron, the following enrichments of Fe^{2+} relative to Mg^{2+} ions might be expected within the structures of individual ferromagnesian silicates, based on the approximate site configurations summarized earlier in table 3.5 and the energy level diagrams depicted in fig. 6.4,

$$
\begin{aligned}
\text{orthopyroxene and pigeonite: } & \text{M2} > \text{M1} \\
Mg^{2+}\text{–}Fe^{2+} \text{ amphiboles: } & \text{M4} > \text{M2} > \text{M1} \approx \text{M3} \\
\text{calcic amphiboles: } & \text{M2} > \text{M1} = \text{M3} \\
\text{biotite: } & \text{M2} \geq \text{M1} \\
\text{olivine: } & \text{M1} = \text{M2}.
\end{aligned}
\tag{6.8}
$$

This order is based on the assumptions that average metal–oxygen distances are identical and that all oxygen ligands have the same bond-type in the coordination sites compared. However, in many ferromagnesian silicate, the most distorted coordination site generally has the largest average metal–oxygen distance (Appendix 7), and oxygen ligands differ between sites. As a result, the average octahedral crystal field splitting parameter, Δ_o, will be smaller for relatively larger distorted sites. However, the overall CFSE attained in a distorted site may be higher than that acquired in a regular octahedral site as a result of the additional stabilization induced by the sixth $3d$ electron of Fe^{2+} occupying the lowest energy t_{2g} orbital. This effect applies to olivines in which the pseudo-trigonally distorted M2 site has the smaller Δ_o value (table 5.5) but the CFSE acquired by Fe^{2+} ions in this site is comparable to that obtained in the pseudo-tetragonally elongated M1 octahedral site. As a result, only minor Fe^{2+} cation ordering occurs in the olivine structure. A similar situation applies to the amphibole M1 and M3 sites and the biotite M1 and M2 sites.

The estimated CFSE's of Fe^{2+} ions in different sites of individual ferromagnesian silicates show that Fe^{2+} ions, although receiving higher CFSE's in the pyroxene M1 site and the amphibole M1+M2+M3 sites, are strongly enriched in the very distorted pyroxene M2 and amphibole M4 sites. Other criteria discussed later (§6.8.4) have had to be invoked to explain these apparent anomalies.

6.8.3.2 Other transition metal ions

Several qualitative predictions and interpretations of the relative enrichments of transition metal cations in different coordination sites having octahedral, tetragonal, trigonal and lower symmetries may be made based on the theoretical stabilization energies summarized in table 6.4. The predicted and observed enrichments of transition metal ions in ferromagnesian silicate crystal structures are summarized in table 6.5.

The Ni^{2+}, Cr^{3+} and low-spin Co^{3+} ions do not acquire additional stabilization in distorted octahedral sites. They are expected to favour smaller sites that more closely approximate octahedral symmetry than other available sites in the crystal structures. As noted in §6.8.2, the high octahedral CFSE's acquired by these three cations in small octahedral sites in silicate and oxide structures accounts for the observed relative enrichments of Ni^{2+} in the olivine M1 and orthopyroxene M1 sites, the sole occupancy by Cr^{3+} of pyroxene M1 sites, and the occurrence and stability of low-spin Co^{3+} in Mn(IV) oxides.

In the case of Cu^{2+} and Cr^{2+}, compounds of which are susceptible to Jahn–Teller distortions (§6.3; table 6.1), these cations are predicted to show strong preferences for the most distorted orthopyroxene M2 and amphibole M4 sites. A similar explanation accounts for the observed enrichments of Mn^{3+} ions in the distorted andalusite M1, alkali amphibole M2, epidote M3 and, perhaps, epidote M1 sites (table 6.1). The presence of significant amounts of chromium in olivines from the Moon and as inclusions in diamond may be due to the presence of Cr^{2+} ions, and not Cr^{3+}, in the distorted M1 and M2 sites of the olivine structure (Burns, 1975b), in which Jahn–Teller stability may be attained. A similar factor accounts for the stability and site occupancy of the Cr^{2+} ion in the orthopyroxene M2 site (table 6.1).

Table 6.5. *Predicted and observed cation ordering in olivines and pyroxenes*

Cation	Olivine M1 site preferred Predicted			Olivine Observed	Orthopyroxene M1 site preferred Predicted			Orthopyroxene Observed
	ionic radius	CFSE	distortion		ionic radius	CFSE	distortion	
Fe^{2+}	no	yes	yes	M1 = M2	no	yes	no	M1 < M2
Ni^{2+}	yes	yes	–	M1 > M2	yes	yes	–	M1 > M2
Co^{2+}	no	yes	yes	M1 > M2	no	yes	no	M1 < M2
Mn^{2+}	no	–	–	M1 < M2	no	–	–	M1 < M2
Cr^{2+}	no	yes	yes	?	no	yes	no	M1 < M2
Zn^{2+}	no	–	–	M1 > M2	no	–	–	M1 < M2
Ti^{3+}	yes	yes	yes	?	yes	yes	yes	M1 > M2
V^{3+}	yes	yes	yes	?	yes	yes	yes	M1 > M2
Cr^{3+}	yes	yes	–	M1 > M2	yes	yes	–	M1 > M2
Mn^{3+}	yes	yes	yes	?	yes	yes	yes	M1 > M2
Fe^{3+}	yes	–	–	M1 < M2	yes	–	–	M1 > M2

The Co^{2+}, Ti^{3+} and V^{3+} ions are expected to prefer either distorted or small octahedral sites. Thus, Co^{2+} should be slightly enriched in the orthopyroxene M2 and cummingtonite M4 sites, favour the pseudo-tetragonally distorted olivine M1 site, and be randomly distributed over the amphibole M1, M2 and M3 sites. The V^{3+} and Ti^{3+} ions are expected to occupy the orthopyroxene M1 and alkali amphibole M2 sites, and to be enriched in distorted epidote M3 sites. As noted earlier, the occurrence and stability of Ti^{3+} ions in lunar and meteoritic clinopyroxenes (§4.4.1) may be explained by the availability of the distorted octahedal M1 site in the calcic clinopyroxene structure.

The site occupancies of Mn^{2+} and Fe^{3+} ions with zero CFSE's should not be influenced by crystal field effects. Instead the distributions of these cations are controlled largely by geometric and electrostatic considerations. Thus, Mn^{2+} ions which are larger than Mg^{2+} and Fe^{2+}, show relative enrichments in the larger orthopyroxene M2, cummingtonite M4 and olivine M2 sites. On the other hand, the predominance of Fe^{3+} ions in alkali amphibole M2 sites may be explained by size factors and also by charge balance considerations. The M2 site not only has the smallest average metal–oxygen distance in the alkali amphibole structure, but it is in close proximity to the M4 positions occupied by Na^{+} ions (fig. 4.14). Similar factors could lead to the enrichment of other trivalent ions in the M2 positions of amphiboles, in general, in which there are substantial amounts of alkali metal ions in the neighbouring M4 positions. Such site occupancies are not governed solely by crystal field effects, although crystal field stabilization energies are probably higher in the smaller M2 sites of alkali amphiboles than in the larger M1 and M3 sites. The fact that the alkali amphibole M2 sites are also significantly distorted from octahedral symmetry explains the occurrence and stability of Mn^{3+} ions in juddite and winchite (Ghose *et al.*, 1986).

6.8.4 Covalent bonding

The three criteria for predicting cation ordering, based on ionic radii, crystal field stabilization energy and site distortion, while being useful for qualitatively explaining intersite distribution patterns of transition metals and related elements, have been critically evaluated (e.g., Brown, 1982). These assessments have highlighted additional criteria that need to be taken into consideration in more general interpretations of cation order–disorder behaviour in minerals. These criteria are based on ionicity–covalency relationships.

The intersite partition coefficients for divalent transition metal ions plotted in fig. 6.3 show similar trends for synthetic olivines and orthopyroxenes. The sequence of decreasing K_d values, $Ni^{2+} > Co^{2+} > Fe^{2+} > Mn^{2+}$, is also the order of decreasing CFSE, eq. (6.7). However, despite the large differences between the two cation environments in the orthopyroxene structure, Ni^{2+} ions are only slightly ordered in the smaller, more regular orthopyroxene M1 sites having comparable dimensions and oxygen bond-type to the olivine M1 sites that are so strongly preferred by Ni^{2+} ions. In the olivine structure, non-bridging oxygen ligands, alone, constitute the two coordination polyhedra and average metal–oxygen distances in the M1 and M2 sites are comparable (eq. (6.6); Appendix 7). If cation size, site configuration and CFSE were solely responsible for the site preferences, then Ni^{2+} ions should exhibit a greater enrichment than it does in the orthopyroxene M1 site relative to the M2 site (Rajamani *et al.*, 1975). Furthermore, Fe^{2+} ions show very strong preferences for pyroxene M2 sites even though CFSE's estimated from absorption spectra of Mg–Fe^{2+} orthopyroxenes are higher for the pyroxene M1 site (eqs (5.12) and (5.13); table 5.7). Similar discrepencies apply to Co^{2+} ions (Ghose and Wan, 1974) which, although significantly enriched in the smaller distorted olivine M1 sites, still favour the larger distorted M2 site in the orthopyroxene structure. Covalent bonding has been invoked to explain these discrepancies.

From charge balance considerations, the oxygen ligands surrounding the orthopyroxene M2 site are neutral, on the average, whereas the M1 site carries a nett negative charge (Brown, 1982). This has led to the suggestion that the more ionic cations should favour the orthopyroxene M1 site whereas cations exhibiting covalent characteristics should enter the orthopyroxene M2 sites (Burnham *et al.*, 1971; O'Nions and Smith, 1973; Ghose and Wan, 1974). That the orthopyroxene M2 site has a higher degree of covalency than the M1 site is implied by the smaller isomer shift parameter for Fe^{2+}/M2 site cations in Mössbauer spectra of orthopyroxenes (Burnham *et al.*, 1971). Therefore, transition metal ions such as Fe^{2+}, Co^{2+} and Ni^{2+} which are considered to be more covalent than Mg^{2+}, are predicted to prefer the orthopyroxene M2 site. For Ni^{2+} alone, this effect is offset by the high octahedral CFSE that this cation acquires in the smaller orthopyroxene M1 site, leading to the relative enrichment of Ni^{2+} ions alone in the latter site (fig. 6.3). Note that the Racah *B* parameter data for Mg–Ni olivines also indicate higher covalency in the M1 site (§5.4.2.4).

These qualitative explanations are supported by structure energy calculations. Electrostatic calculations and charge density studies of olivines (Alberti and Vezzalini, 1978; Bish, 1981; Fujino *et al.*, 1981; Tamada *et al.*, 1983; Bish and Burnham, 1984) suggest that the olivine M2 site is more ionic than the M1 site, so that Mg^{2+} ions with higher ionic character should favour the M2 sites and the more covalent transition metal ions should favour the olivine M1 sites, thereby contributing to the strong enrichments of Ni^{2+} and Co^{2+}, as well as Zn^{2+}, in the olivine M1 sites. This covalency effect is offset in the case of larger cations such as Mn^{2+}, and to some extent Fe^{2+}, which prefer to occupy the larger olivine M2 sites. Similar electrostatic energy calculations for Mg, Fe and Co orthopyroxenes also indicate that the orthopyroxene M1 sites carry higher nett positive charges than the M2 sites (Sasaki *et al.*, 1982), suggesting that the more covalent transition metal ions should favour the orthopyroxene M2 sites. Again, the tendency for Ni^{2+} ions, alone, to be enriched in the pyroxene M1 sites indicates that CFSE factors dominate its site occupancy.

Electrostatic factors have also been invoked for chrysoberyl (iso-structural with olivine) to explain the reluctance of trivalent cations, such as Cr^{3+} in alexandrite and Fe^{3+} in sinhalite, to replace Al^{3+} ions in the smaller M1 sites (Ganguli, 1977). Each [$M1O_6$] octahedron shares more edges with adjacent octahedra than do the [$M2O_6$] octahedra in the serrated chains in the olivine structure (fig. 5.8) leading to greater cation–cation repulsions in the M1 site and a reluctance for trivalent transition metal ions larger than host Al^{3+} ions to enter this site. Highly charged cations thus favour the [$M2O_6$] octahedra. A similar factor may also contribute to the Fe^{3+}/M2 site occupancy of ferrifayalites and laihunite (Schaefer, 1985).

6.9 Summary

Chapter 6 describes how crystal field stabilization energies influence the crystal chemistry of minerals containing the transition elements. Site occupancies of the cations in oxide and silicate structures are also discussed.

Cation radii and interatomic distances. Trends in the ionic radii of the transition elements reflect the distribution of electrons in their 3*d* orbitals. Octahedral ionic radii are larger for cations containing electrons in e_g orbitals, which project towards the surrounding oxygen ions. Unusually short interatomic distances may indicate transition metal ions in low-spin electronic states. Coordination sites containing ions subject to the Jahn–Teller effect are usually distorted, resulting in a range of metal–oxygen and oxygen–oxygen distances in the site.

Crystal chemistry of spinels. A classic example showing that transition metal ions display distinct site preferences in oxides stems from studies of spinel crystal chemistry. The spinel structure contains tetrahedral and octahedral sites; *normal* and *inverse* forms exist in which divalent and trivalent ions, respectively, fill the tetrahedral sites. The type of spinel formed by a cation is related to its octahedral site preference energy (OSPE), or difference between crystal field stabilization energies in octahedral and tetrahedral coordinations in an oxide structure. Trivalent and divalent cations with large site preference energies (e.g., Cr^{3+} and Ni^{2+}) tend to form *normal* and *inverse* spinels, respectively. The type of spinel adopted by cations with zero CFSE (e.g., Fe^{3+} and Mn^{2+}) is controlled by the preferences of the second cation in the structure.

Cation distributions in silicate minerals. Measurements of site populations by conventional diffraction (X-ray, electron and neutron) techniques and a variety of spectroscopic methods (e.g., crystal field, Mössbauer, infrared, EPR, XANES and EXAFS) show that cation ordering is prevalent in transition metal-bearing minerals. The enrichments of transition metal ions relative to Mg^{2+} or Al^{3+} in specific sites results from the interplay of several crystal chemical and bonding factors, including cation size, site distortion, crystal field stabilization energy and covalency factors. In ferromagnesian silicates, appreciable ordering of Fe^{2+} is observed in the orthopyroxene, pigeonite, cummingtonite and anthophyllite structures, while olivine, calcic amphibole, and biotite structures show less pronounced Fe^{2+}–Mg^{2+} ordering. The near-random Fe^{2+}-occupancy of the olivine M1 and M2 sites and the strong enrichments of Fe^{2+} in pyroxene M2 and amphibole M4 sites may be explained by site distortion which stabilizes the sixth 3*d* electron of ferrous iron. The strong preference of Mn^{2+}, which acquires zero CFSE in silicate minerals, for the larger pyroxene M2, amphibole M4 and olivine M2 sites correlates with the relatively large ionic radius of Mn^{2+}. The high octahedral CFSE of Ni^{2+} overrides other bond energy factors, contributing to strong enrichments of nickel in the M1 sites of olivines and pyroxenes. Cations showing the Jahn–Teller effect, such as Mn^{3+}, Cr^{2+} and Cu^{2+}, are predicted and observed to be stabilized in the most distorted sites in mineral crystal structures. The Cr^{3+} and Co^{3+} ions which, like Ni^{2+}, receive very high CFSE in octahedral coordination, are predicted to favour the smallest sites, whereas the Ti^{3+}, V^{3+} and Co^{2+} ions should be relatively enriched in small, distorted octahedral sites.

6.10 Background reading

Bailey, S. W. (ed.) (1984) *Micas.* (Mineral. Soc. Amer. Publ.), chs 1 & 2.
Brown, G. E. (1982) Olivines and silicate spinels. In *Orthosilicates, 2nd edn.* (P. H. Ribbe, ed.; Mineral. Soc.Amer. Publ.), *Rev. Mineral.,* 5, 275–381.
Burns, R. G. (1968) Enrichments of transition metal ions in silicate structures. In *Symposium on the Origin and Distribution of Elements. Section V. Terrestrial Abundances.* (L. H. Ahrens, ed.; Pergamon Press, Oxford), pp. 1151–64.
Hawthorne, F. C. (1983) The crystal chemistry of the amphiboles. *Canad. Mineral.,* 21, 173-480.
Orgel, L. E. (1966) *An Introduction to Transition-Metal Chemistry: Ligand-Field Theory, 2nd edn.* (Methuen, London), chs 4 & 5.
Prewitt, C. T. (ed.) (1980) *Pyroxenes.* (Mineral. Soc. Amer. Publ.), ch. 2.
Ribbe, P. H. (ed.) (1982) *Orthosilicates, 2nd edn.* (Mineral. Soc. Amer. Publ.), chs 2, 6–11.

7

Thermodynamic properties influenced by crystal field energies

These observations (non-linear heats of hydration) suggest the following hypothesis: In the absence of crystal-field effects the thermodynamic properties – – would evolve steadily along the transition series.

L. E. Orgel, *Journ. Chem. Soc.*, p. 4756 (1952).

7.1 Introduction

One of the most successful applications of crystal field theory to transition metal chemistry, and the one that heralded the re-discovery of the theory by Orgel in 1952, has been the rationalization of observed thermodynamic properties of transition metal ions. Examples include explanations of trends in heats of hydration and lattice energies of transition metal compounds. These and other thermodynamic properties which are influenced by crystal field stabilization energies, including ideal solid-solution behaviour and distribution coefficients of transition metals between coexisting phases, are described in this chapter.

7.2 Influence of CFSE on thermodynamic data

7.2.1 Graphical correlations

Crystal field stabilization energies derived spectroscopically from absorption bands in the visible to near-infrared region, including the crystal field spectral measurements of minerals described in chapter 5, are enthalpy terms and, as such, might be expected to contribute to bulk properties such as lattice energies and solvation energies of transition metal compounds. If cations were spherically symmetrical and no preferential filling of $3d$ orbitals occurred, a given thermodynamic quantity would be expected to display smooth periodic variation in a series of transition metal compounds as a result of contraction of the cations. However, experimentally determined thermodynamic quantities of first series transition metal ions frequently fall on characteristic double-humped curves with maximum values for cations with three and eight 3*d* electrons. For example, the lattice energies of divalent oxides, sulfides, orthosilicates and trivalent fluorides plotted in fig. 7.1 show values for most transition metal compounds to be offset from smooth curves with linear trends passing through the values for cations possessing $3d^0$, $3d^5$ and $3d^{10}$ electronic configurations and zero CFSE's. Moreover, trends in the double-humped curves are remarkably uniform in view of the fact that more than one structure-type is represented in some series of transition metal compounds. This is particularly well illustrated by the lattice energies of divalent sulfides, which include data for four distinct structure-types. The lattice energy data for binary orthosilicates plotted in fig. 7.1 demonstrate that values documented for the olivines Ni_2SiO_4, Co_2SiO_4 and Fe_2SiO_4 (Ottonello, 1987) lie well above the Mn_2SiO_4–Zn_2SiO_4 tie-line.

The heats of hydration illustrated in fig. 7.2. display patterns that are similar to the lattice energy data plotted in fig. 7.1. Again, the values of ΔH_{hyd} across the transition metal series fall on double-humped curves. By interpreting these particular trends, Orgel established the connection between spectroscopy and thermodynamics. Orgel (1952) showed that when crystal field stabilization energies (CFSE's) estimated from visible-region absorption spectra of hydrated cations (table 2.5) are deducted from the measured heats of hydration, the 'corrected' values lie near the smooth curves passing through experimental values for cations with zero, five and ten 3*d* electrons that acquire zero CFSE (fig. 7.2). Such an analysis, together with the correlation between predicted and observed minima for Mn^{2+} and Fe^{3+} compounds, provides an elegant confirmation of the 3*d* orbital energy separations proposed by crystal field theory.

Two features arise from closer inspection of the plots of thermodynamic data in figs 7.1 and 7.2. First, crystal field stabilization energies are small by comparison with the absolute heats of hydration or lattice energies. Thus, crystal field stabilization energies, important as they are in explaining *differences* between transition metal ions, make up only a small fraction of the total energies of transition metal compounds. For example, the CFSE of Cr^{3+} contributes less than ten per cent to its heat of hydration, while in liebenbergite (Ni_2SiO_4) less than two per cent of the lattice energy can be attributed to the CFSE of Ni^{2+} ions located in the olivine M1 and M2 sites. Second, in the absence of spectral data, CFSE's could be estimated from curves such as those shown in figs 7.1 and 7.2, provided thermodynamic data were available for a suite of transition metal compounds. This method is of particular importance for opaque compounds which do not give absorption spectra suitable for evaluating 3*d* orbital energy separations. For example, the crystal field splittings and stabilization energies of divalent transition metal ions in sulphide, oxide and orthosilicate

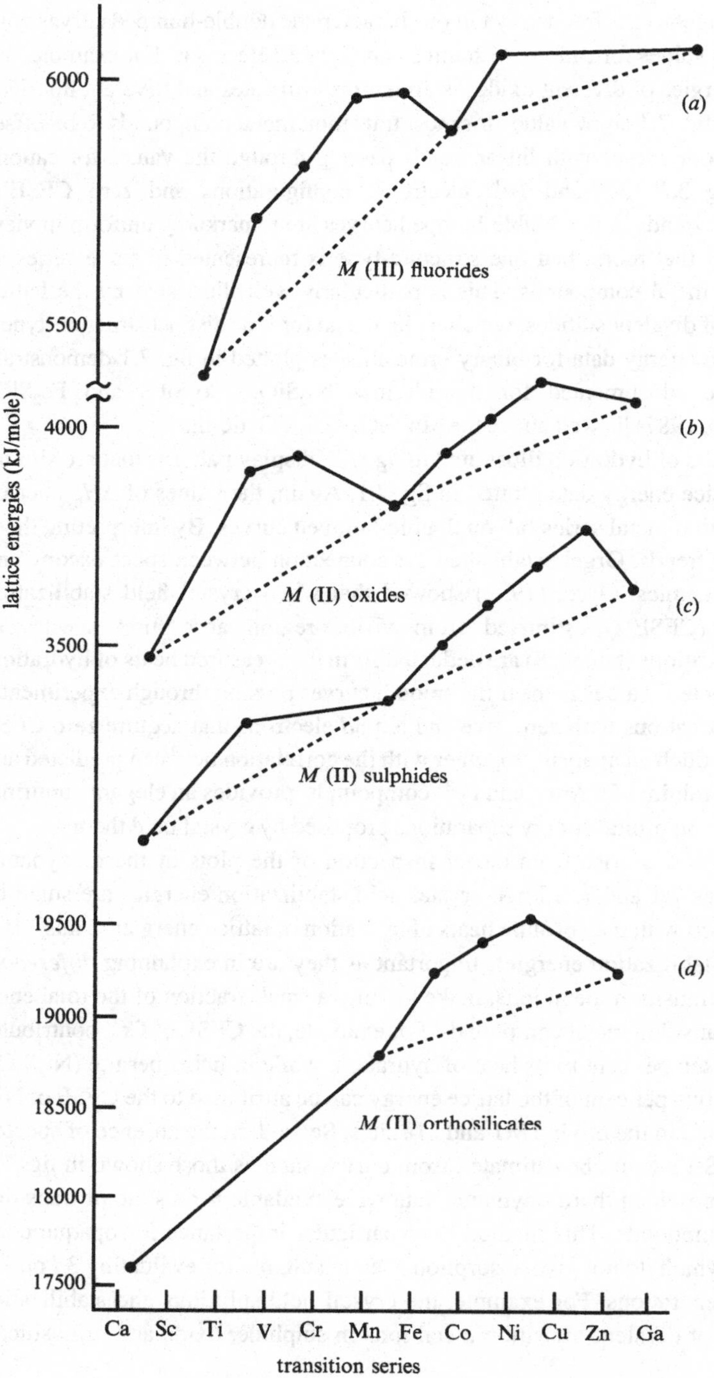
lattice energies (kJ/mole)
6000
5500
4000
3500
19500
19000
18500
18000
17500
(a)
M (III) fluorides
(b)
M (II) oxides
(c)
M (II) sulphides
(d)
M (II) orthosilicates
Ca
Sc
Ti
V
Cr
Mn
Fe
Co
Ni
Cu
Zn
Ga
transition series

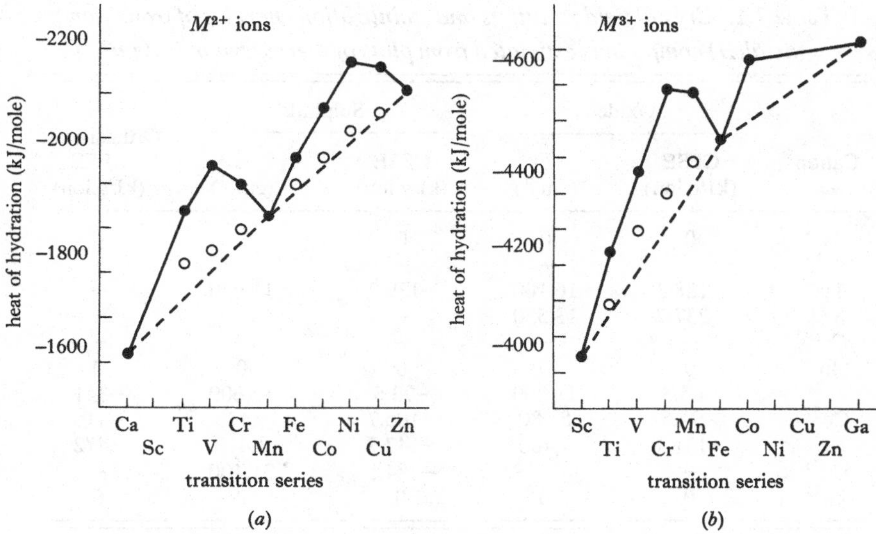

Figure 7.2 Heats of hydration of transition metal ions (*a*) M^{2+} ions; (*b*) M^{3+} ions. Filled circles: experimental; open circles: CFSE deducted. Note that experimental values lie on double-humped curves; when CFSE's (table 2.5) are deducted for each cation, the corrected values lie on smooth curves through the values for $3d^0$, $3d^5$ and $3d^{10}$ cations. [*Sources of data*: George & McClure, 1959; table 2.5.]

minerals that have been estimated from the data in fig. 7.1, are summarized in table 7.1. Comparison of the data for oxides in table 7.1 with those in tables 2.5 and 6.3 show that there is reasonable agreement between the Δ and CFSE data obtained from measurements of absorption spectra and plots of thermodynamic data. The CFSE's of Fe^{2+} and Ni^{2+} in the respective olivines determined from fig. 7.1 are higher than those estimated from absorption spectral data of fayalite, eqs. (5.2) and (5.3), and liebenbergite, eqs (5.4a) and (5.5a), perhaps indicating that the tie-line between Mn_2SiO_4 and Zn_2SiO_4 in fig. 7.1 should be curved rather than a straight line.

Caption to figure 7.1

Figure 7.1 Lattice energies for transition metal compounds and minerals. (*a*) Metal(III) fluorides (TiF_3, VF_3, CrF_3 and FeF_3 are isostructural; ScF_3 and GaF_3 have the ReO_3 structure; MnF_3 has a distorted VF_3 structure); (*b*) metal(II) oxides (ZnO has the wurtzite structure; the remainder have the NaCl structure); (*c*) metal(II) sulphides (CaS has the NaCl structure; ZnS has either the sphalerite, wurtzite or NaCl structure; the remainder have the NiAs structure); (*d*) metal(II) orthosilicates (Zn_2SiO_4 has the willemite structure; the remainder have the olivine structure). [*Sources of data*: George & McClure, 1959; Waddington, 1959; Ottonello, 1987.]

Table 7.1. *Crystal field splittings and stabilization energies of transition metal(II) compounds estimated from plots of thermodynamic data**

	Oxide		Sulphide		
Cation	CFSE (kJ/g.ion)	Δ (cm^{-1})	CFSE (kJ/g.ion)	Δ (cm^{-1})	Orthosilicate CFSE (kJ/g.ion)
Ca^{2+}	0	0	0	0	0
(Sc^{2+})	–	–	–	–	–
Ti^{2+}	−188.3	16,100	−139.2	11,900	–
V^{2+}	−237.7	13,550	–	–	–
Cr^{2+}	–	–	–	–	–
Mn^{2+}	0	0	0	0	0
Fe^{2+}	−65.3	11,200	−73.6	12,600	−341
Co^{2+}	−94.6	8,080	−106.7	9,120	−416
Ni^{2+}	−131.4	7,480	−147.7	8,410	−472
Cu^{2+}	–	–	−184.5	21,100	–
Zn^{2+}	0	0	0	0	0

* See fig. 7.1

7.2.2 Correlation with interatomic distances

The profiles of thermodynamic data plotted in figs 7.1 and 7.2 bear striking resemblances to the plots of ionic radii shown in fig. 6.2. In both situations, there are non-linear trends of data across the transition metal series with maximum deviations occurring for cations with $3d^3$ and $3d^8$ configurations. Such a similarity suggests that there is a connection between the thermodynamic and crystallographic data.

In the ionic model, binding energies represent the nett effects of attraction between charges on the anions and central cation, repulsion between the anions and all electrons on the cation, and repulsion between the nuclei. The lattice energy, U_0, of a binary ionic solid, such as periclase, may be expressed by the Born equation, one form of which is

$$U_0 \frac{ANZ_AZ_Ce^2}{R}\left[1-\frac{1}{n}\right], \qquad (7.1),$$

where Z_A, Z_C are charges on the anion and cation separated by the equilibrium internuclear distance R; e is the electronic charge; N is Avogadro's number; A is the Madelung constant (1.747 6 for MgO); and n is an integer between 4 and 12 determined empirically from compressibility experiments. The value of n is about 4 for oxides. Equation 7.1 shows that there is an inverse relationship between lattice energy and cation–anion distance. Such a relationship bears analogies to the inverse fifth power dependence of crystal field splitting on cation–anion distances, eq. (2.17). Thus, the values of the lattice energy and the crystal field splitting parameter both increase with decreasing cation–anion distance. The contraction of cation–oxygen interatomic distances described in §6.2, which is largest for cations such as Cr^{3+} and Ni^{2+}, also correlates with the relatively higher CFSE's of these ions. It is apparent that these larger CFSE's contribute significantly to the thermodynamic quantities plotted in figs 7.1 and 7.2.

7.3 Ideal solution behaviour in silicate minerals

Many ferromagnesian silicate series, such as Mg^{2+}–Fe^{2+} olivines, pyroxenes and garnets, are generally assumed to be ideal solid-solutions of magnesium silicate and iron silicate components. This is due to atomic substitution of Mg^{2+} and Fe^{2+} ions for one another in the crystal structures which is facilitated by the cations having comparable sizes and charges. Similar factors assist the atomic substitution of Ni^{2+} and Co^{2+} in ferromagnesian silicates which are assumed to exhibit ideal solid-solution behaviour between the Mg or Fe silicate and corresponding Ni and Co silicate components. The validity of this assumption may be assessed by examining the criteria for ideal solution behaviour in the light of evidence for cation ordering (§6.7 and table 6.5) and compositional variations of CFSE's of transition metal ions in structural sites (e.g., eqs (5.2) to (5.5)).

7.3.1 Criteria for ideal solution behaviour

The criteria for ideal solution behaviour are discussed in most chemical thermodynamic texts (e.g., Lewis and Randall, 1961, p. 130; Nordstrom and Munoz, 1986, p. 401). The conditions may be summarized as follows.

First, Raoult's law must be obeyed, so that

$$a_i = x_i a_i^0 \text{, for all values of } x_i \ (0 < x_i < 1), \qquad (7.2a)$$

where a_i, a_i^0 are the activities of component i (e.g., Ni^{2+}) in solution and in the pure substance, respectively, and x_i is the mole fraction of component i in solution. Raoult's law may be a good approximation for non-ideal solutions when x_i is large. In highly dilute solutions, however, Henry's law may be applicable, in which

$$a = h_i x_i , \qquad (7.2b)$$

where h_i is a proportionality constant.

Second,

$$\mu_i = \mu_i^0 + RT \ln x_i \,, \qquad (7.3)$$

where μ_i, μ_i^0 are the chemical potentials of component i in the solution and pure phase, respectively.

Third, the free energy of mixing, ΔG_m, is given by

$$\Delta G_m = - RT \,\Sigma(x_i \ln x_i). \qquad (7.4)$$

Finally, the entropy of mixing or configurational entropy, $S_{conf,}$ must have a maximum value. The configurational entropy is given by (Nordstrom and Munoz, 1986)

$$S_{conf} = - R \,\Sigma_j(m_j \Sigma_i [x_{ij} \ln x_{ij}]), \qquad (7.5)$$

where m_j is the total number of atoms occupying the jth crystallographic site (expressed as atoms per formula unit) and x_{ij} is the mole fraction of the ith atom on the jth site. The S_{conf} is a maximum only if the atoms are randomly distributed throughout the solution.

The relationships expressed in eqs (7.2) to (7.5) lead to the following necessary conditions for ideal solid-solution behaviour. First, the heat of mixing, ΔH_m, must be zero

$$\Delta H_m = 0 \qquad (7.6)$$

and second, the volume of mixing, ΔV_m, must be zero

$$\Delta V_m = 0. \qquad (7.7)$$

7.3.2 Configurational entropy

The formation of an ideal solution between two or more components requires that the configurational entropy be the maximum value, eq. (7.4). This implies that ions must be randomly distributed over coordination sites in the crystal structure. Whenever cation ordering occurs in a structure, the configurational entropy is not the ideal maximum value. The evidence for cation ordering summarized in §6.7 indicates that few silicate minerals are ideal solutions.

For example, consider the orthopyroxene $(Fe_{0.395}Mg_{0.605})_2Si_2O_6$, site populations of which were estimated to be (Ghose *et al*, 1975)

$$\begin{aligned} &\text{M1 positions: } 19\%\ Fe^{2+},\ 81\%\ Mg^{2+}; \\ &\text{M2 positions: } 60\%\ Fe^{2+},\ 40\%\ Mg^{2+}. \end{aligned} \qquad (7.8)$$

If ions were randomly distributed among the M1 and M2 positions, the ideal maximum configurational entropy summed over the two positions would be

$$\Delta S^0_{conf} = -\,2R\,[(0.395 \ln 0.395) + (0.605 \ln 0.605)] = 11.16\ \mathrm{J/(deg.\ mole)}. \quad (7.9)$$

The actual entropy of mixing, which takes into account the measured site populations, is

$$\begin{aligned}\Delta S_{conf} &= -R[(0.19 \times 0.395) \ln (0.19 \times 0.395) + (0.81 \times 0.605) \ln (0.81 \times 0.605) \\ &\quad + (0.60 \times 0.395) \ln (0.60 \times 0.395) + (0.40 \times 0.605) \ln (0.40 \times 0.605)] \\ &= 10.21\ \mathrm{J/(deg.\ mole)} \quad (7.10)\end{aligned}$$

Therefore, the true configurational entropy is 0.95 J/(deg. mole) lower than the maximum value as a result of Fe^{2+}–Mg^{2+} ordering in the orthopyroxene structure. The cation ordering found in other members of the enstatite–ferrosilite series, as well as the synthetic Mg^{2+}–Ni^{2+}, Mg^{2+}–Co^{2+} and Mg^{2+}–Mn^{2+} pyroxenes (Ghose *et al.*, 1975; Hawthorne and Ito, 1977), shows that most transition metal-bearing orthopyroxenes are not ideal solid-solutions.

Similarly, the synthetic liebenbergite $(Ni_{0.51}Mg_{0.49})_2SiO_4$, in which site populations in the olivine structure were estimated to be (Boström, 1987; Ottonello *et al.*, 1989)

$$\begin{aligned}&\text{M1 positions: } 78.4\%\ Ni^{2+},\ 21.6\%\ Mg^{2+}; \\ &\text{M2 positions: } 23.6\%\ Ni^{2+},\ 76.4\%\ Mg^{2+}, \quad (7.11)\end{aligned}$$

has a configurational entropy of 10.20 J/(deg. mole). This value is 1.33 J/(deg. mole) lower than the maximum, ideal value of 11.53 J/(deg. mole). Similar discrepencies between the maximum and observed configurational entropies for other Mg^{2+}–Ni^{2+} olivines are summarized in table 7.2.

The cation ordering measured in other Mg^{2+}–Ni^{2+} olivines (Rajamani *et al.*, 1975; Bish, 1981; Smyth and Taftø, 1982; Boström, 1987; Ottonello *et al.*, 1989; Hu *et al.*, 1990), as well as in Mg^{2+}–Co^{2+} (Ghose *et al.*, 1975), Mg^{2+}–Mn^{2+} (Ghose *et al.*, 1975; Francis and Ribbe, 1980) and Mg^{2+}–Zn^{2+} (Ghose *et al.*, 1975) olivines, indicates that these binary orthosilicates are not ideal solid-solutions of the Mg silicate and transition metal silicate components. The forsterite–fayalite series, however, approaches a more random distribution of Mg^{2+} and Fe^{2+} ions in the crystal structures. For example, in the lunar olivine, $(Fe^{2+}{}_{0.503}Mg^{2+}{}_{0.497})_2SiO_4$ displaying a small degree of cation ordering (Ghose and Wan, 1975),

Table 7.2. *Crystal chemical and thermodynamic data for Mg^{2+}–Ni^{2+} olivines*

Olivine composition: $(Mg_{1-x}Ni_x)_2SiO_4$:	x_{Ni} =	0.02	0.30	0.36	0.51	0.69	0.75	1.00
[1] M1 site occupancy:	Ni =	>0.03	0.536	0.620	0.784	0.898	0.934	1.00
	Mg =	~0.97	0.464	0.376	0.216	0.102	0.066	0.00
[1] M2 site occupancy:	Ni =	<0.01	0.064	0.100	0.236	0.482	0.434	1.00
	Mg =	~0.99	0.936	0.900	0.764	0.518	0.566	0.00
[2] Intersite partition coefficient	K_d =	–	16.89	14.94	11.75	9.46	10.85	–
[3] Intersite free energy, kJ/mole	ΔG =	–	–27.62	–26.4	–23.82	–21.92	–23.45	–
[4] Configurational entropy, ideal, J/(deg. mole)	S^0_{conf} =	–	10.16	10.86	11.52	10.30	9.35	–
[5] Configurational entropy, actual, J/(deg. mole)	S_{conf} =	–	7.10	9.27	10.20	8.86	7.66	–
[6] Entropy of mixing, J/(deg. mole)	ΔS_m =	–	–3.06	–1.59	–1.32	–1.44	–1.69	–
[7] Crystal field stabilization energy, kJ/mole: M1 site	CFSE =	–136.4	–139.9	–140.5	–141.6	–142.4	–142.6	–143.1
[7] Crystal field stabilization energy, kJ/mole: M2 site	CFSE =	–105.8	–106.1	–106.4	–107.4	–109.3	–109.9	–113.2
[8] CFSE of mixing, kJ/mole	$CFSE_m$ =	–	–4.89	–6.04	–5.65	–3.73	–3.17	–

[1] Boström (1987)

[2] $K_d = [(Ni^{2+})_{M1}/(Ni^{2+})_{M2}]/[(Mg^{2+})_{M1}/(Mg^{2+})_{M2}$
or $K_d = [(Ni^{2+}/Mg^{2+})_{M1}]/[(Ni^{2+}/Mg^{2+})_{M2}]$

[3] $\Delta G = -2.303\boldsymbol{RT}\log K_d$, using synthesis temperatures listed by Boström (1987)

[4] Calculated by eq. (7.4): $S^0_{conf} = -2.303\boldsymbol{R}\,\Sigma_j[m_j\Sigma(x_{ij}\log x_{ij})]$, using bulk chemical data [see eq.(7.9)]

[5] Calculated by eq. (7.4): $S_{conf} = -2.303\boldsymbol{R}\,\Sigma_j[m_j\Sigma(x_{ij}\log x_{ij})]$, using site occupancy data in reference 1 [see eq. (7.10)]

[6] $\Delta S_m = S_{conf} - S^0_{conf}$

[7] CFSE from Hu, Langer & Boström (1990)

[8] Calculated by eq. (7.14)

$$\text{M1 positions: } 52.5\%\ Fe^{2+},\ 47.5\%\ Mg^{2+};$$
$$\text{M2 positions: } 48.0\%\ Fe^{2+},\ 52.0\%\ Mg^{2+}, \quad (7.12)$$

the actual configurational entropy (11.52 J/(deg. mole)) is very close to the ideal value (11.53 J/(deg. mole)).

Since cation ordering exists in other multisite ferromagnesian silicates (e.g., pigeonite, anthophyllite, cummingtonite–grunerite and actinolite series) and aluminosilicates (e.g., epidote, andalusite) discussed earlier (§6.7), such minerals, strictly speaking, cannot be ideal solid-solutions. Only in minerals where single-site atomic substitution is possible, such as the M1 site of calcic clinopyroxenes, the octahedral site of garnet and the deformed cube site of pyrope–almandine garnets, can configurational entropies attain maximum values. Other factors, however, may annul ideal solid-solution behaviour in such minerals when atomic substitution of Fe^{2+} and other transition metal ions occurs in the structures. One of these factors is the heat of mixing.

7.3.3 Enthalpy of mixing

A necessary condition for an ideal solid-solution behaviour is that there be zero heat of mixing in forming the solution from its components, eq. (7.6). This condition cannot be fulfilled when differences exist between CFSE's of cations in the end-member components and in the solid-solutions.

This differential CFSE factor is demonstrated by the formation of hortonolite, $(Mg_{0.5}Fe_{0.5})_2SiO_4$, by the mixing of forsterite and fayalite components. To a close approximation, Mg^{2+} and Fe^{2+} ions may be assumed to be randomly distributed in the olivine structure eq., (7.12), so that 0.5 Fe^{2+} ions per formula unit occupy each of the M1 and M2 positions. The CFSE's of the Fe^{2+} ion in the M1 and M2 sites of hortonolite are approximately −53.2 and −52.0 kJ/mole, respectively, and in fayalite the corresponding CFSE are −50.9 kJ/mole (M1 site) and −51.2 kJ/mole (M2 site) (eqs (5.2) and (5.3); table 5.16). The formation of hortonolite may be represented as follows:

$0.5\,Mg_2SiO_4$ +	$0.5\,Fe_2SiO_4$	=	$(Mg_{0.5}Fe_{0.5})_2SiO_4$	
forsterite	fayalite		hortonolite	
CFSE = 0	CFSE (M1)	= 0.5 (−50.9)	CFSE (M1)	= 0.5 (−53.2)
		= −25.45 kJ		= −26.65 kJ
	CFSE (M2)	= 0.5 (−51.2)	CFSE (M2)	= 0.5 (−52.0)
		= −25.60 kJ		= −26.00 kJ
	total CFSE	= −51.05 kJ	total CFSE	= −52.65 kJ .

(7.13)

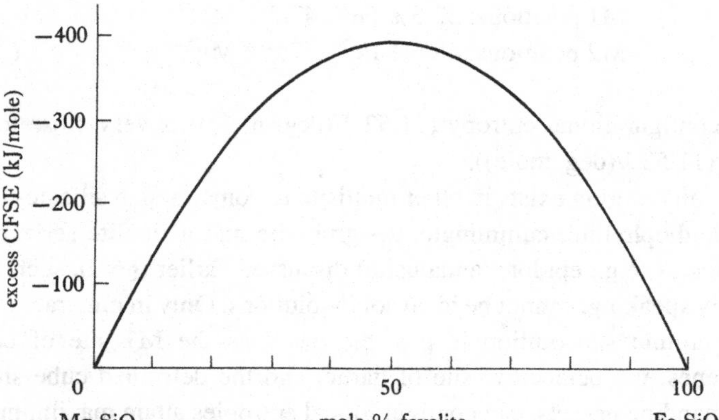

Figure 7.3 Excess crystal field stabilization energy of mixing for Mg–Fe^{2+} olivines of the forsterite–fayalite series.

Therefore, according to this calculation, there is an excess CFSE (enthalpy) of mixing of about −1.6 kJ/(mole of olivine). Similarly, the formation of all intermediate olivines by mixing of Mg_2SiO_4 and Fe_2SiO_4 components results in an excess CFSE of mixing. This is illustrated in fig. 7.3. These results imply that there is a heat of mixing term and that the olivine series is not an ideal solid-solution of forsterite and fayalite.

When similar calculations are carried out for olivines showing strong cation ordering, the different site populations in the structure must be taken into account when evaluating the excess CFSE of mixing. This may be illustrated by the formation of a liebenbergite, $(Mg_{0.49}Ni_{0.51})_2SiO_4$, by the mixing of Mg_2SiO_4 and Ni_2SiO_4 components. The cation distribution discussed previously, eq. (7.11), indicates that the percentages of Ni^{2+} ions in the two olivine sites are 78.4 per cent (M1) and 23.6 per cent (M2). Using the estimates of the CFSE's of Ni^{2+} in the two sites originally derived from crystal field spectral measurements of Mg^{2+}-Ni^{2+} olivines (eqs (5.4) and (5.5); Hu *et al.*, 1990), the stabilization energies listed in table 7.2 correspond to −141.6 kJ/mole (M1 site) and −107.4 kJ/mole (M2 site) for the intermediate liebenbergite and −143.1 and −113.2 kJ/mole for corresponding sites in the Ni_2SiO_4 end-member. The formation of intermediate liebenbergite may be represented by the equation:

$0.49\ Mg_2SiO_4$ +	$0.51\ Ni_2SiO_4$=	$(Mg_{0.216}Ni_{0.784})^{M1}(Mg_{0.764}Ni_{0.236})^{M2}SiO_4$
forsterite	liebenbergite	Mg liebenbergite
CFSE = 0	CFSE (M1) = 0.51 (−143.1) = −73.0 kJ	CFSE (M1) = 0.784 (−141.6) = −111.0 kJ

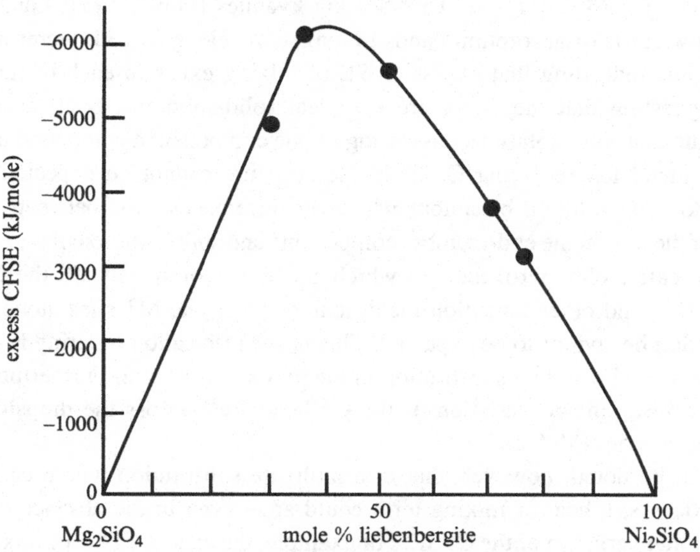

Figure 7.4 Excess CFSE of mixing for Mg^{2+}–Ni^{2+} olivines of the forsterite–liebenbergite series (see table 7.2).

$$\begin{aligned} \text{CFSE (M2)} &= 0.51\,(-113.2) & \text{CFSE (M2)} &= 0.236\,(-107.4) \\ &= -57.7\text{ kJ} & &= -25.35\text{ kJ} \\ \text{total CFSE} &= -130.7\text{ kJ} & \text{total CFSE} &= -136.35\text{ kJ.} \end{aligned} \tag{7.14}$$

This calculation shows that the formation of intermediate liebenbergite by mixing of Mg_2SiO_4 and Ni_2SiO_4 components is accompanied by an excess CFSE of mixing of −5.65 kJ/mole. These results suggest that Mg^{2+}–Ni^{2+} olivines depart considerably from ideal solution behaviour (Bish, 1981). This is further demonstrated in fig. 7.4 by the compositional variation of excess CFSE of mixing for the suite of synthetic Mg^{2+}–Ni^{2+} olivines for which site occupancy and CFSE data are available (table 7.2).

Solid-solutions involving other transition metal ions also display non-ideality by having excess CFSE of mixing. For example the compositional variations of Δ and the CFSE of Mn^{3+} in epidotes (Burns and Strens, 1967; Langer, 1988) and andalusites (Abs-Wurmbach *et al.*, 1981; Langer, 1988) show that these aluminosilicates are not ideal solid-solutions of Al^{3+} and Mn^{3+} end-member silicates. Such evidence for non-ideality is further supported by the existence of Mn^{3+} ordering in their crystal structures (§6.5). Similarly, spectral measurements of Cr^{3+} ions in several solid-solution series, such as Al_2O_3–

Cr_2O_3, $MgAl_2O_4$–$MgCr_2O_4$, and Cr^{3+}-bearing kyanites (Poole, 1964; Langer, 1988), show shifts of absorption bands to longer wavelengths with increasing Cr^{3+} contents, indicating that excess CFSE of mixing exists in each of these series, suggesting that they, too, are not ideal solid-solutions. Such results demonstrate that solid-solutions possessing as one component a compound of a transition metal ion receiving CFSE in the structure cannot be expected to conform to ideal solution behaviour whenever differences exist between the CFSE's of the ions in the end-member components and solid-solutions.

Only in calcic clinopyroxenes, in which Ca^{2+} ions completely fill the M2 sites and Fe^{2+} and other transition metal ions occur in the M1 sites alone, is ideal solution behaviour to be expected. This is because cation ordering is not possible in one-site atomic substitution in the pyroxene M1 site. Furthermore, there is an insignificant variation of the CFSE of Fe^{2+} across the diopside–hedenbergite series (§5.5.3).

It should be noted, however, that in a multisite substitution where cation ordering occurs, a heat of mixing term could arise even in the absence of a compositional variation of the CFSE. For example, the excess CFSE of mixing would increase to about −8.20 kJ/mole in solid-solution formation of the liebenbergite $(Mg_{0.49}Ni_{0.51})_2SiO_4$, eq. (7.14), if the CFSE of Ni^{2+} ions in the M1 and M2 sites were to remain constant at −143.1 and −113.2 kJ/mole, respectively.

7.4 Contributions of electronic entropy

Variations of electronic entropies, defined in §2.14 by eq. (2.27)

$$S_{el} = -R\,\Sigma(P_i \ln P_i) \qquad (2.27)$$

and resulting from electron occupancies of degenerate energy levels, have been invoked in a number of crystal chemical and petrological applications (Wood and Strens, 1972; Burns and Sung, 1978; Wood, 1981; O'Neill and Navrotsky, 1983; Sherman, 1988). Values of S_{el} for transition metal ions in regular octahedral and tetrahedral sites are summarized in table 7.3. It is evident from this table that the high-spin Fe^{2+} ($3d^6$) ion, for example, has electronic entropies of 9.13 J/(deg. g.ion) and 5.76 J/(deg. g.ion) in octahedral and tetrahedral sites, respectively, resulting from occupancies of its sixth $3d$ electron in either the three-fold degenerate t_{2g} orbitals (octahedral) or the two-fold degenerate e orbitals (tetrahedral) at low temperature. Note that the value of S_{el} is zero for low-spin Fe^{2+} in octahedral sites because no degeneracy exists when the six $3d$ electrons fill the t_{2g} orbitals.

When high-spin Fe^{2+} is present in a distorted (low-symmetry) site, however,

Table 7.3. *Electronic entropies for transition metal ions in octahedral and tetrahedral coordinations and high-spin (hs) and low-spin (ls) states*

Electronic structure		Cation			Electronic configuration	S_{el} octahedral	S_{el} tetrahedral	ΔS_{el} (oct – tet) J/(deg.mole)	ΔS_{el} (ls–hs) J/(deg.mole)
	$3d^0$	Ca^{2+}	Sc^{3+}	Ti^{4+}	[Ar]	0	0	0	–
	$3d^1$		Ti^{3+}	V^{4+}	$(t_{2g})^1$	$R\ln3$	$R\ln2$	3.37	–
	$3d^2$		V^{3+}	Cr^{4+}	$(t_{2g})^2$	$R\ln3$	0	9.13	–
	$3d^3$	V^{2+}	Cr^{3+}	Mn^{4+}	$(t_{2g})^3$	0	$R\ln3$	–9.13	–
hs	$3d^4$	Cr^{2+}	Mn^{3+}		$(t_{2g})^3(e_g)^1$	$R\ln2$	$R\ln3$	–3.37	–
ls	$3d^4$	Cr^{2+}	Mn^{3+}		$(t_{2g})^4$	$R\ln3$	–	–	–3.37
hs	$3d^5$	Mn^{2+}	Fe^{3+}		$(t_{2g})^3(e_g)^2$	0	0	0	–
ls	$3d^5$	Mn^{2+}	Fe^{3+}		$(t_{2g})^5$	$R\ln3$	–	–	9.13
hs	$3d^6$	Fe^{2+}	Co^{3+}		$(t_{2g})^4(e_g)^2$	$R\ln3$	$R\ln2$	3.37	–
ls	$3d^6$	Fe^{2+}	Co^{3+}		$(t_{2g})^6$	0	–	–	–9.13
hs	$3d^7$	Co^{2+}	Ni^{3+}		$(t_{2g})^5(e_g)^2$	$R\ln3$	0	9.13	–
ls	$3d^7$	Co^{2+}	Ni^{3+}		$(t_{2g})^6(e_g)^1$	$R\ln2$	–	–	3.37
	$3d^8$	Ni^{2+}			$(t_{2g})^6(e_g)^2$	0	$R\ln3$	–9.13	–
	$3d^9$	Cu^{2+}			$(t_{2g})^6(e_g)^3$	$R\ln2$	$R\ln3$	–3.37	–
	$3d^{10}$	Zn^{2+}	Ga^{3+}	Ge^{4+}	$(t_{2g})^6(e_g)^4$	0	0	0	–

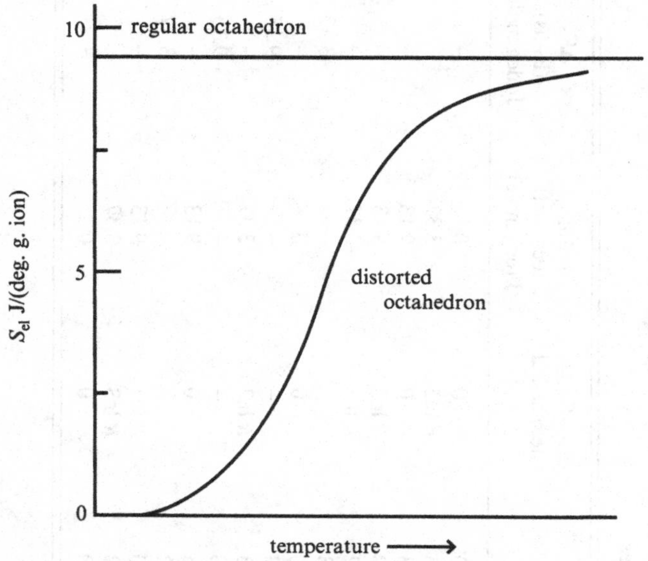

Figure 7.5 Comparison of the electronic entropy of Fe^{2+} in regular and distorted octahedral sites (modified from Wood, 1981).

its sixth 3*d* electron becomes localized in the lowest-lying 3*d* orbital, resulting in zero electronic entropy. Thus, although site distortions may increase the CFSE (or electronic enthalpy) of high-spin Fe^{2+} ions (§6.8.3), these effects are counterbalanced by a marked decrease of electronic entropy. Such effects influence intracrystalline (intersite) partitioning (or cation ordering) and intercrystalline (interphase) partitioning of Fe^{2+} ions, as well as thermodynamic properties of ferromagnesian silicate minerals (Wood, 1981).

At elevated temperatures, the entropy difference between regular and distorted sites is reduced due to thermal population of higher energy orbitals by the sixth 3*d* electron of Fe^{2+}. The probabilities, P_0 and P_1, of this sixth electron occupying the lowest and next-lowest orbitals separated by an energy δ_1 at temperature T is given by the Maxwell–Boltzmann equation,

$$P_1/P_0 = \exp(-\delta_1/RT). \qquad (7.15)$$

As the temperature rises, the probability of the sixth electron occupying orbital energy levels above the lowest level increases, and so too does the S_{el} of Fe^{2+} located in a distorted site, eq. (2.27). However, when Fe^{2+} is present in a regular octahedron, its sixth 3*d* electron remains delocalized over the three equivalent t_{2g} orbitals so that S_{el} remains constant at 9.13 J/(deg. g.ion). Therefore, entropy differences between regular and distorted sites decrease with rising temperature (Wood and Strens, 1972), which is illustrated in fig. 7.5.

Since disordering of the sixth $3d$ electron of Fe^{2+} contributes to the electronic entropy as well as the enthalpy (CFSE) of this cation, these electronic terms should be incorporated in molar entropies and heat capacities estimated for ferromagnesian silicates by oxide-sum methods (Wood, 1981). For example, the observation that the molar entropy calculated for fayalite at 298 K, 162.8 J/(deg. mole), exceeds the measured value, 148.6 J /(deg. mole), by 14.2 J/(deg. mole) may be attributed to electronic entropy differences for Fe^{2+} between regular octahedral sites in FeO and the distorted octahedral sites of fayalite. The entropy effect arises from the fact that the sixth $3d$ electron of Fe^{2+} is disordered over three t_{2g} orbitals in FeO but becomes localized in low-lying orbital levels in fayalite due to additional crystal field splitting of the t_{2g} orbitals in the distorted M1 and M2 sites of fayalite (cf. fig. 5.17). A similar effect cannot occur in Mn^{2+} since its $3d^5$ configuration has all orbitals singly occupied. As a result, the molar entropy of Mn_2SiO_4 calculated by the oxide-sum method matches the measured value. Electronic entropies also influence transition pressures for the olivine $\rightarrow$ spinel transformation in the Earth's Mantle (Burns and Sung, 1978) discussed in chapter 9 (§9.9.1).

Differences of electronic entropies between tetrahedral and octahedral sites, ΔS_{el}, have been applied to spinel crystal chemistry (O'Neill and Navrotsky, 1983). Thus, the large $\Delta S_{el}^{(oct-tet)}$ value for V^{3+} reinforces its relatively small OSPE (table 6.3), thereby enabling V^{3+}-spinels to be exclusively *normal* (table 6.2). Randomization of site occupancies in $NiAl_2O_4$, on the other hand, may be attributed to the large negative $\Delta S_{el}^{(oct-tet)}$ value for Ni^{2+} which offsets its large OSPE.

Differences of electronic entropy also exist between high-spin and low-spin configurations (table 7.3). An alternative equation, which is regarded to be more valid than eq. (2.27) (Sherman, 1988), for evaluating the change of electronic entropy during a high-spin to low-spin transition, is

$$\Delta S_{el}^{(ls-hs)} = R \ln [n_d^{ls}(2S^{ls} + 1)] - R \ln [n_d^{hs}(2S^{hs} + 1)] \qquad (7.16a)$$

or

$$\Delta S_{el}^{(ls-hs)} = R \ln [n_d^{ls}(2 \Sigma m_s^{ls} + 1)] - R \ln [n_d^{hs}(2 \Sigma m_s^{hs} + 1)] \,, \qquad (7.16b)$$

where n_d^{ls} and n_d^{hs} are the orbital degeneracies of the low-spin (ls) and high-spin (hs) states, respectively, and the $(2S + 1)$ terms correspond to the spin-multiplicities (§3.4.1). For an electronic configuration with spin S there are $(2S + 1)$ degenerate electronic states. When the ground state of Fe^{2+} changes from $^5T_{2g}$ (with $S = 2$ and $n_d^{hs} = 3$) to $^1A_{1g}$ (with $S = 0$ and $n_d^{ls} = 1$), the change of elec-

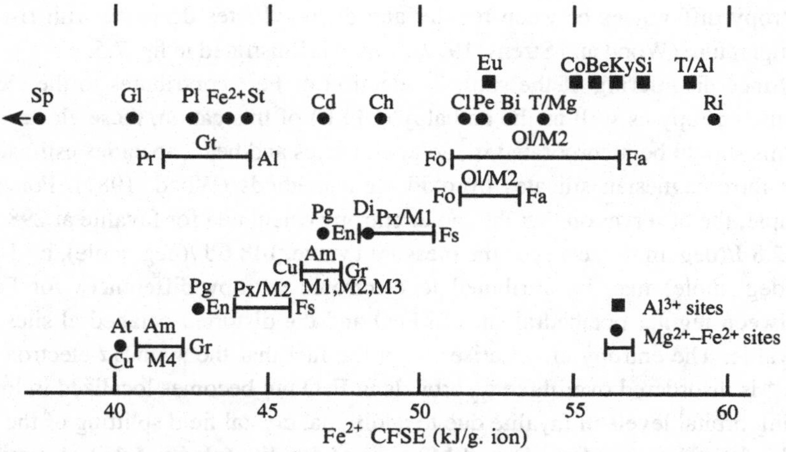

Figure 7.6 Ranges of CFSE of Fe^{2+} ions in ferromagnesian silicates. Legend to mineral symbols is given in table 5.16.

tronic entropy is −22.5 J/(deg. g.ion) compared to −9.13 J/(deg. g.ion) calculated from eq. (2.27). Such a decrease of electronic entropy would again offset the large increase of CFSE of Fe^{2+} acquired in the low-spin configuration.

7.5 Iron:magnesium ratios in coexisting ferromagnesian silicates

The observed relative enrichments of Fe^{2+} ions in coordination sites within individual silicate minerals were discussed in §6.7 and cation ordering trends shown by olivines and orthopyroxenes were summarized in table 6.5. These intersite partitioning patterns are partially explained by the relative CFSE's attained by Fe^{2+} ions in each coordination site of the mineral structures (§6.8.3.1).

The ranges of CFSE for Fe^{2+} ions in mineral structures derived in chapter 5 (table 5.16) are plotted in fig. 7.6 and may used to assess the Fe/Mg ratios in coexisting ferromagnesian silicates. If it is assumed that Fe^{2+} ions are enriched in crystallographic sites bestowing highest CFSE and that equilibrium distribution of iron took place during mineral formation, the following order of decreasing relative iron enrichment is predicted in igneous mineral structures

$$\text{olivine M1} > \text{olivine M2} > \text{orthopyroxene M1} > \text{pigeonite M1} > \text{calcic clinopyroxene M1} \approx \text{amphibole M1,M2,M3} > \text{orthopyroxene M2} \approx \text{pigeonite M2} > \text{amphibole M4}. \quad (7.17)$$

Based on bulk chemical data, which provide metal concentrations averaged over all crystallographic sites in a mineral, the Fe^{2+}/Mg^{2+} ratios of coexisting minerals in igneous rocks might be expected to decrease in the order olivine > pyroxenes > amphiboles. Taking into account the strong enrichment of Fe^{2+} ions in the pyroxene M2 sites, however, the order of decreasing Fe^{2+}/Mg^{2+} ratios is anticipated to be:

olivine ≈ orthopyroxene > pigeonite > calcic clinopyroxene >
calcic amphibole > plagioclase > silicate glass. (7.18)

Similarly, in metamorphic assemblages the predicted order of decreasing Fe^{2+}/Mg^{2+} ratios based on bulk chemical analyses is:

aluminosilicates > olivine > biotite > chlorite > chloritoid >
pyroxenes > cordierite > staurolite > amphiboles > garnet ≈ plagioclase. (7.19)

In general, there is reasonably close agreement between the relative iron enrichments predicted by crystal field theory and those observed in coexisting mineral assemblages (e.g., Ramberg and DeVore, 1951; Burns, 1968a; Grover and Orville, 1969; Obata *et al.*, 1974; Matsui and Banno, 1969; Matsui and Nishizawa, 1974; Harley, 1984). The most serious exceptions to the predicted orders are pyrope–almandine garnets which strongly fractionate Fe^{2+} over coexisting ferromagnesian silicate phases (O'Neill and Wood, 1979; Harley, 1984), despite the comparatively low CFSE attained by Fe^{2+} ions in the eight-coordinated site in the garnet structure.

The predicted orders of relative Fe^{2+} enrichments expressed in eqs (7.18) and (7.19) are strictly valid only for mineral formation at 25 °C and atmospheric pressure, the conditions under which absorption spectral measurements producing the CFSE data shown in fig. 7.6 were made. In order to apply the CFSE data summarized in table 5.16 and fig. 7.6 to igneous and metamorphic assemblages, and in the absence of spectral data for each ferromagnesian silicate at elevated temperatures and pressures, it must be assumed that all phases show similar variations of CFSE with rising temperatures and pressures.

In addition, the predicted orders in eqs (7.18) and (7.19) do not take into account phase equilibrium relationships involving coexisting minerals. For example, olivine should always be enriched in Fe^{2+} ions relative to orthopyroxene, according the the CFSE data plotted in fig. 7.6. Moreover, studies of the system: $MgO–FeO–SiO_2$ at atmospheric pressure (Bowen and Schairer, 1935) showed that magnesium olivine crystallizes first and becomes increasingly enriched in iron before pyroxene commences to crystallize. The olivine maintains its early lead of higher Fe/Mg ratio than the coexisting pyroxene. Thus, phase relationships support predictions from crystal field theory that Fe^{2+} ions will be enriched in olivine relative to pyroxene at all stages of magmatic crystallization, providing equilibrium crystallization prevailed. If early formed olivine crystals were removed, or complete resorption of olivine did not take place, during magmatic crystallization the apparent coexisting pyroxene phase might possess the higher Fe/Mg ratio.

In pro-grade metamorphic reactions, on the other hand, effects of intersite cation ordering at elevated temperatures could influence Fe/Mg ratios of orthopyroxene–olivine assemblages in granulite facies rocks. For the olivine–orthopyroxene exchange reaction

$$Mg_2SiO_4 + Fe_2Si_2O_6 = Fe_2SiO_4 + Mg_2Si_2O_6, \quad (7.20)$$

the interphase partition coefficient, K_D, given by

$$K_D{}^{ol/py} = [(Fe/Mg)_{ol}/(Fe/Mg)_{py}] \quad (7.21)$$

varies at 700 °C from values of $K_D{}^{ol/py} < 1$ for the most magnesian compositions to $K_D{}^{ol/py} \approx 3$ for the most iron-rich bulk compositions (Sack, 1980). At elevated pressures and temperatures, $K_D{}^{ol/py}$ also exceeds 1 (Matsui and Nishizawa, 1974). The explanation for these trends again lies in the strong cation ordering of Fe^{2+} in M2 sites in enstatites which is offset in iron-rich assemblages by the higher CFSE attained by Fe^{2+} ions in both M1 and M2 sites of olivine relative to the pyroxene M1 site.

7.6 Distributions of divalent transition metal ions between coexisting ferromagnesian silicates

The interpretation of relative enrichments of iron in coexisting silicates presented in §7.5 hinges on the availability of CFSE data for Fe^{2+} ions in coordination sites for a wide variety of rock-forming minerals. The CFSE data are more sporadic for other transition metal ions in coexisting minerals, particularly the divalent cations.

The CFSE values of Ni^{2+}-bearing minerals summarized earlier in table 5.19 decrease in the order

corundum > spinel > ringwoodite > olivine > clay silicates >
periclase > pyroxenes. (7.22)

Thus, nickel should be enriched relative to pyroxenes in dense Upper Mantle minerals (e.g., silicate spinel and olivine) and their low-temperature alteration products (e.g., garnierite).

In the absence of complete CFSE data for divalent cations in all ferromagnesian silicates, qualitative arguments may be used to deduce relative enrichments of these ions in coexisting minerals using crystal chemical information summarized in chapter 6 (§6.7 and §6.8).

The various factors discussed in §6.8 which contribute to cation ordering or intersite partitioning of transition metal ions in a crystal structure, may be arranged in sequences as follows.

First, the sizes of coordination sites, as gauged by mean metal–oxygen distances or polyhedral volumes (Appendix 7), decrease in the order

amphibole M4 > pyroxene M2 > olivine M2 > olivine M1 >
amphibole M2 > pyroxene M1 ≈
amphibole M1 > amphibole M3 = biotite M1,M2. (7.23)

Second, the octahedral crystal field splitting parameters, values of which are higher for smaller sites, are expected to decrease in the same order as eq. (7.23).

Third, the extent of distortion of each site from regular octahedral symmetry based on the quadratic elongation (Appendix 7) decreases in the order:

amphibole M4 > pyroxene M2 > olivine M2 = olivine M1 >
amphibole M1,M3 = biotite M1,M2 > pyroxene M1 = amphibole M2.
(7.24)

Finally, other factors, including the increased covalent bonding with bridging oxygen atoms in the amphibole M4 and pyroxene M2 sites and the slightly larger ionic character of the olivine M1 site, also influence distributions of the cations in coexisting minerals.

On the basis of the sequences expressed in eqs (7.23) and (7.24), the following site enrichments might be expected for each divalent transition metal in the olivine, pyroxene, amphibole and biotite structures

Mn^{2+}: amphibole M4 > pyroxene M2 > olivine M2 > olivine M1 > amphibole M2 > pyroxene M1 > amphibole M1,M3 > biotite M1,M2 ;
(7.25)

Co^{2+}: pyroxene M2 > olivine M1 > olivine M2 > amphibole M4 > pyroxene M1 = amphibole M1,M2,M3 = biotite M1,M2 ; (7.26)

Ni^{2+}: olivine M1 > olivine M2 > pyroxene M1 = biotite M1,M2 ≈ amphibole M1,M2,M3 > pyroxene M2 > amphibole M4 . (7.27)

If these site enrichment factors apply to bulk mineral assemblages, the transition metal/magnesium ratios might be expected to follow the trends

Mn/Mg: olivine ≥ Mg amphibole ≥ orthopyroxene > Ca pyroxene = Ca amphibole > biotite ; (7.28)

Co/Mg: olivine > orthopyroxene > Mg amphibole > Ca pyroxene > Ca amphibole ; (7.29)

Ni/Mg: olivine > orthopyroxene > Ca pyroxene ≥ Ca amphibole = Mg amphibole . (7.30)

In general, there is good agreement between predicted and observed enrichments of these divalent cations in coexisting ferromagnaesian silicates (e.g., Mercy and O'Hara, 1967; Hakli and Wright, 1967; Matsui and Banno, 1969; Glassley and Piper, 1978; Dupuy *et al.*, 1980).

7.7 Distributions of trivalent transition metal ions in mineral assemblages

Trends in CFSE data derived from the absorption spectral measurements described in chapter 5 may be used to predict or interpret element partitioning involving M^{3+} cations between coexisting aluminosilicate minerals. In many cases, however, the CFSE data represent 'average' values for cations in several Al^{3+} sites in individual crystal structures. Nevertheless, the stabilization energies are applicable to interphase partitioning trends based on bulk chemical data for coexisting minerals.

7.7.1 Ti^{3+} partitioning

The CFSE data for Ti^{3+}-bearing minerals summarized in table 5.9 decrease in the order

mica > zoisite > corundum > hydrated cation > pyroxene M1 >> periclase. (7.31)

Thus, Ti^{3+} ions attain highest CFSE in minerals hosted by Al relative to Mg phases.

On Earth, the Ti(III) oxidation state is unstable. However, Ti^{3+}-bearing minerals are well-characterized in some meteorites and Moon rocks, generally coexisting with Ti^{4+} in such phases as calcic pyroxene, ulvöspinel, hibonite and ilmenite. In hibonite, $CaAl_{12}O_{19}$, a refractory phase in carbonaceous chondrites, EPR and optical spectral data indicate that Ti^{3+} ions are present (Ihinger and Stolper, 1986; Live *et al.*, 1986). The trivalent Ti ions may be stabilized in the five-coordinated trigonal bipyramidal M5 site of the hibonite structure

(Burns and Burns, 1984b). In Moon rocks, Ti^{3+}/Ti^{4+} ratios might be expected to decrease in the order: ilmenite > ulvöspinel > pyroxene.

7.7.2. V^{3+} partitioning

The crystal field spectral data for V^{3+}-bearing minerals summarized in table 5.10 indicate that CFSE's decrease in the order

spinel > corundum > kyanite = pyrope > muscovite >
grossular > beryl > tourmaline > pyroxene > tremolite. (7.32)

As a minor element in rock-forming minerals, vanadium might be expected to be fractionated in micas relative to amphiboles and pyroxenes, and to be strongly enriched in high-grade metamorphic assemblages. The relative enrichment of vanadium in hibonite in refractory inclusions of carbonaceous chondrites again may be attributed to the occurrence and stability of V^{3+} ions in the five-coordinated M5 sites (Burns and Burns, 1984b).

7.7.3 Cr^{3+} partitioning

The CFSE data for Cr^{3+}-bearing minerals summarized in table 5.11 suggest the following relative enrichments of Cr^{3+} ions

spinel > corundum, topaz > pyrope, chlorites > tourmalines >
chromite > forsterite > kyanite, grossular > sillimanite > tremolite,
beryl > muscovite > epidote, pyroxenes > andalusite. (7.33)

Thus, Cr^{3+} ions are predicted to be relatively enriched in magmatic spinels, high-grade metamorphic mineral assemblages containing spinel, corundum and pyrope, and chlorite-bearing low-grade metamorphic rocks. In mantle xenoliths, the higher Cr contents of kyanites compared to coexisting pyroxene and garnet assemblages correlate with the high CFSE of Cr^{3+} attained in kyanite (Langer, 1988).

7.7.4 Mn^{3+} partitioning

For Mn^{3+}-bearing minerals, the CFSE data summarized in table 5.13 suggest the relative enrichments to be

andalusites (kanonaite) > micas > epidotes (thulite, piemontite) >
tourmalines >clinopyroxenes (blanfordite), amphiboles (juddite) >
garnets, corundum. (7.34)

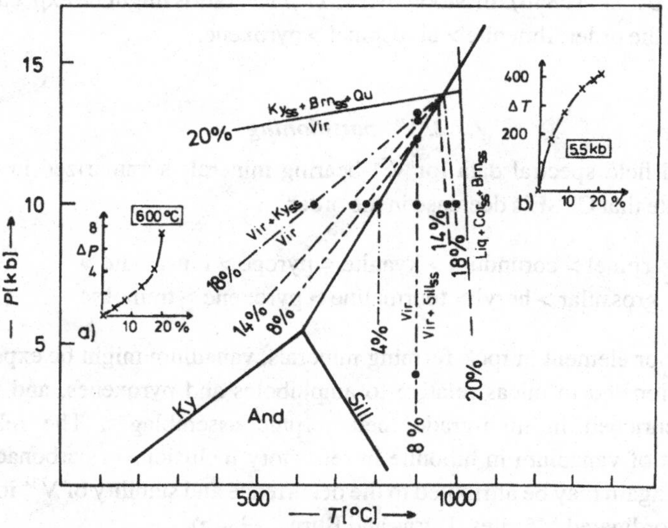

Figure 7.7 Phase relationships in the system: Al_2SiO_5–Mn_2SiO_5 in pressure–temperature–composition space projected onto the *P*,*T* plane (from Abs-Wurmbach *et al.*, 1983; Langer, 1988). Note the large increase of the andalusite stability field (And) with increasing Mn^{3+} contents plotted as mole per cent theoretical Mn_2SiO_5 end-member. Insets (a) and (b) show how the triple point at 500 °C and 3.8 kb for pure Al_2SiO_5 increases with rising Mn^{3+} content of andalusite. [Legend to Mn–Al solid-solutions (ss): Ky = kyanite; Sill = sillimanite; Brn = braunite; Cor = corundum; Vir = viridine; Qu = quartz.]

Thus, as a consequence of the Jahn–Teller effect, which causes Mn^{3+} ions to be strongly enriched in mineral structures providing very distorted coordination sites (§6.4; table 6.1), trivalent manganese is predicted and observed to be enriched in low-grade metamorphic assemblages containing andalusite or epidote, thereby affecting the stability fields of these minerals (Strens 1966a, 1968; Abs-Wurmbach *et al.*, 1983; Langer, 1988).

7.7.4.1 Influence of CFSE on Al_2SiO_5 phase relationships

Besides requiring a Jahn–Teller distorted site to stabilize Mn^{3+} ions in a mineral structure, the formation of trivalent manganese necessitates rather high oxygen fugacities provided by the Mn_2O_3/MnO_2 buffer. Equilibrium experiments in the system Al_2SiO_5–'Mn_2SiO_5' not only show Mn^{3+} ions to be strongly enriched in andalusite relative to kyanite and sillimanite, but also that the stability field of Mn^{3+}-bearing andalusites is increased (Abs-Wurmbach *et al.*, 1983; Langer, 1988). This effect is illustrated in fig. 7.7 where breakdown curves of various Mn^{3+}–substituted andalusites are shown. In andalusite solid-solutions containing up to 20 mole per cent of the theoretical Mn_2SiO_5 end-member component, the triple point in the pure Al_2SiO_5 system at 500 °C and 3.8 kbar is shifted beyond 500 °C (P = 3.8 kbar) and 8 kbar (T = 500 °C) with increasing Mn^{3+} content. This enormous enlargement of the stability field of manganiferous andalusites is a manifestation of the exceptionally large CFSE acquired by Mn^{3+} in the andalusite M1 site.

7.8 Distributions of transition metals between crystals and melts

If two phases are in equilibrium with one another at some temperature and pressure, a transition metal ion will be distributed between the phases in such a way as to minimize the free energy of the two-phase assemblage. This should generally result in the transition element being concentrated in the phase giving largest crystal field stabilization energy.

Consider two phases α and β containing octahedral sites available for cation occupancy. The distribution of cations between these sites will depend on the relative values of the CFSE for the two sites. If $\mathrm{CFSE}^{\alpha} > \mathrm{CFSE}^{\beta}$, then the Nernst distribution coefficient, D, expressing the concentration ratio of transition element, M, between the two phases

$$D_{\mathrm{M}}^{\alpha/\beta} = [\mathrm{M}]^{\alpha}/[\mathrm{M}]^{\beta}, \tag{7.35}$$

should increase with increasing CFSE for the cation. In an exchange reaction involving Mg in a ferromagnesian silicate, the interphase (or intercrystalline) partition coefficient normalized to Mg, which is given by

$$K_{\mathrm{D}}^{\mathrm{M/Mg}} = [(\mathrm{M/Mg})^{\alpha}]/[(\mathrm{M/Mg})^{\beta}]\ , \tag{7.36}$$

should also increase with increasing CFSE.

The data in table 2.5 show that CFSE's for divalent cations increase in the order Mn < Fe < Co < Ni. From this the following orders are predicted for the distribution coefficients and the interphase (intercrystalline) partition coefficients:

$$D_{\mathrm{Ni}}^{\alpha/\beta} > D_{\mathrm{Co}}^{\alpha/\beta} > D_{\mathrm{Fe}}^{\alpha/\beta} > D_{\mathrm{Mn}}^{\alpha/\beta} \tag{7.37a}$$

$$K_{\mathrm{D}}^{\mathrm{Ni/Mg}} > K_{\mathrm{D}}^{\mathrm{Co/Mg}} > K_{\mathrm{D}}^{\mathrm{Fe/Mg}} > K_{\mathrm{D}}^{\mathrm{Mn/Mg}}\ . \tag{7.37b}$$

Similarly, the distribution coefficients for trivalent cations in octahedral sites are predicted to show the order:

$$D_{\mathrm{Cr}}^{\alpha/\beta} > D_{\mathrm{V}}^{\alpha/\beta} > D_{\mathrm{Ti}}^{\alpha/\beta} > D_{\mathrm{Fe}}^{\alpha/\beta} > D_{\mathrm{Al}}^{\alpha/\beta}\ . \tag{7.38}$$

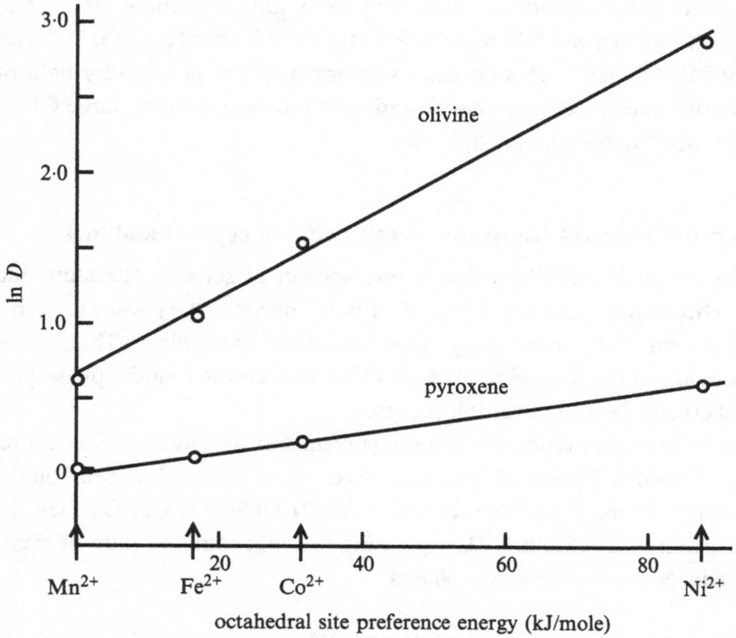

Figure 7.8 Relationship between the octahedral site preference energy and distribution coefficient of divalent transition metal ions partitioned between olivine or pyroxene crystals and the basaltic groundmass (modified from Henderson & Dale, 1969; Henderson, 1982, p. 147).

These sequences have been commonly observed in numerous investigations of the partitioning of transition metal ions between silicate melts and olivine or pyroxene crystals (e.g., Banno and Matsui, 1973; Roeder, 1974; Duke, 1976; Irving, 1978; Takahashi, 1978; Lindstrom and Weill, 1978; Colsen *et al.*, 1988)

A relationship between octahedral site preference energies (table 6.3) and distribution coefficients has been demonstrated for transition metal ions partitioned between olivine or pyroxene crystals and the groundmass of oceanic basalts, which is assumed to represent the composition of the magma from which the ferromagnesian silicates crystallized (Henderson and Dale, 1969; Dale and Henderson, 1972). Plots of ln D against OSPE, such as those illustrated in fig. 7.8, show linear trends between the two parameters.

An explanation for the linear trends observed in fig. 7.8 is as follows (Henderson, 1982, p. 146). At constant pressure and composition, the dependence of the distribution coefficient on temperature can be expressed by

$$\ln D^{\alpha/\beta} = -\Delta H/RT + b\,, \tag{7.39}$$

where ΔH is the difference between heats of solution of a transition element in phase α (crystal) and phase β (melt) and b is an integration constant. During crystallization of magma, transition elements are partitioned between octahedral sites plus environments with lower coordination numbers in the melt and the six-coordinated sites in the olivine or pyroxene structures (§8.5). Therefore, the octahedral site preference energy (table 6.3), or a comparable CFSE difference parameter (table 8.1), appears to be a major contributor to the ΔH term in eq. (7.39), influencing the partitioning of Ni^{2+}, Co^{2+}, Fe^{2+} and Mn^{2+} ions between ferromagnesian silicates and coexisting melts (or magma). The data plotted in fig. 7.8 indicate that in magma at constant temperature, crystal field effects contribute significantly to fractionation patterns of divalent transition metal ions between groundmass and phenocrysts.

7.8.1 Thermodynamic behaviour of Ni

Of all of the transition elements, nickel is particularly important as a petrogenetic indicator because it partitions so strongly into olivine, the major constituent of the upper mantle and most stony meteorites. As a result, nickel has provided a powerful tool for constraining the compositions of mantles of terrestrial planets and for investigating the petrogenesis of basaltic and andesitic magmas (e.g., Hart and Davis, 1978). However, considerable debate has centred on the thermodynamic properties of Ni, particularly with regard to its obedience to Henry's law (e.g., Mysen, 1978; Nabelek, 1980; Drake and Holloway, 1981). The configurational entropy and CFSE of mixing data discussed earlier (§7.3.2 and §7.3.3) indicate that Ni^{2+} partitioning in melt/olivine systems will become increasingly non-ideal when Ni concentration levels exceed a few mole per cent Ni_2SiO_4 in olivine phenocrysts.

In a classic study of Hawaiian lava samples quenched at known temperatures (Hakli and Wright, 1967), the distribution of nickel between coexisting olivine, pyroxene and glass was determined. It was suggested that such Ni partitioning formed the basis of a geothermometer based on eq. (7.39). Subsequent investigations suggested that temperature exerts the dominant influence on the distribution coefficients of Ni, with changes of melt composition apparently playing a minor role (Roeder, 1974; Leeman and Lindstrom, 1978; Takahashi, 1978). However, other studies (e.g., Watson, 1977; Hart and Davis, 1978; Colson *et al.*, 1988; Kinzler *et al.*, 1990) have demonstrated that the partitioning of Ni and other transition metals between olivine and melt is not the strong function of temperature originally proposed (Hakli and Wright, 1967). Instead, distribution coefficients depend also on the bulk composition of a melt which, in turn, influences the polymerization of silicate liquids. Such factors which result in decreased CFSE with increased polymerization of a silicate melt, are examined further in chapter 8 (§8.5.2).

7.9 Summary

Chapter 7 discusses some of the thermodynamic properties of transition metal compounds and minerals that are influenced by crystal field effects. The characteristic double-humped curves in plots of thermodynamic data for suites of transition metal-bearing phases originate from contributions from the crystal field stabilization energy. However, these CFSE's, important as they are for explaining differences between individual cations, make up only a small fraction of the total energy of a transition metal compound. In the absence of spectroscopic data, CFSE's could be evaluated from the double-humped curves of thermodynamic data for isochemical compounds of the first transition series.

Ideal solution behaviour. Two properties of an ideal solid-solution that are influenced by the crystal field are: first, the entropy of mixing which should be a maximum; and second, the enthapy of mixing which should be zero when the solution is formed from its components. Compositional variations of the CFSE of Fe^{2+} in most ferromagnesian silicates and of Ni^{2+} in Mg–Ni olivines, together with observations of cation ordering (e.g., Fe^{2+} in orthopyroxene M2 sites; Ni^{2+} in olivine M1 sites), indicate that few silicates are ideal solid-solutions of Mg, Fe and Ni silicate components. Compositional variations of CFSE imply a heat of mixing, and cation ordering reduces the configurational entropy from the ideal maximum value.

Electronic entropy. The S_{el} resulting from electron occupancies of degenerate energy levels of some transition metal ions in regular octahedral sites (e.g., Fe^{2+}, Co^{2+}, Ti^{3+}, V^{3+}) is greatly reduced when cations occur in low-symmetry coordination environments. However, S_{el} differences between regular and distorted sites decrease with rising temperature as a result of thermal disordering of electrons in close-spaced orbital energy levels. Since the S_{el} contributes to the total entropy of a ferromagnesian silicate, discrapencies of molar entropies and heat capacities from values estimated from component oxides are observed in fayalite, for example. S_{el} may reinforce site occupancies already favoured by high CFSE.

Fe/Mg ratios in coexisting silicates. The CFSE's acquired by Fe^{2+} ions in ferromagnesian silicates obtained from spectral measurements may be used to explain Fe/Mg ratios in coexisting silicate minerals. The orders of decreasing CFSE between igneous mineral assemblages olivine ~ orthopyroxene > pigeonite > Ca pyroxene > amphiboles

and between metamorphic minerals,

olivine > biotite > chlorite > chloritoid > pyroxenes > cordierite > staurolite > amphiboles > garnet

correlate with a general decreasing Fe^{2+}/Mg^{2+} ratio, except for pyrope–almandine garnets which are anomalously enriched in Fe.

Partitioning of transition metals between coexisting phases. CFSE data for transition metal ions derived from absorption spectra of minerals hosting them (chapter 5), together with observed and predicted site occupancy data (chapter 6), provide explanations of distribution coefficients of divalent and trivalent cations in mineral assemblages. Ni^{2+}, Cr^{3+} and V^{3+} with high octahedral site preference energies are concentrated in octahedral sites in dense anhydrous silicates. The Jahn–Teller effect in Mn^{3+} not only induces this cation to be strongly enriched in distorted coordination sites, but also increases the pressure–temperature stability field of minerals (e.g., andalusite) hosting this cation. Partition coefficients of cations (e.g., Ni^{2+}) between silicate minerals and coexisting melts are also strongly influenced by larger CFSE's attained in octahedral sites in crystal structures.

7.10 Background reading

Burns. R. G. (1985) Thermodynamic data from crystal field spectra. In *Microscopic to Macroscopic: Atomic Environments to Mineral Thermodynamics.* (S. W. Kieffer & A. Navrotsky, eds; Mineral. Soc. Amer. Publ), *Rev. Mineral.* 14, 277–316.

Langer, K. (1988) UV to NIR spectra of silicate minerals obtained by microscope spectrometry and their use in mineral thermodynamics and kinetics. In *Physical Properties and Thermodynamic Behaviour of Minerals.* (E. K. H. Salje, ed.; D. Reidel Publ. Co.), pp. 639–85.

Lewis, G. N. & Randall, M (1961) *Thermodynamics., 2nd edn.* (Revised by K. S. Pitzer & L. Brewer; McGraw–Hill Book Co., New York), 723 pp.

Nordstrom, D. K. & Munoz, J. L. (1986) *Geochemical Thermodynamics.* (Benjamin/Cummings Publ., Menlo Park), 477 pp.

Orgel, L. E. (1952) The effects of crystal fields on the properties of transition metal ions. *J. Chem. Soc.*, pp. 4756–61.

Ottonello, G. (1987) Energies and interactions in binary (*Pbnm*) orthosilicates: A Born parameterization. *Geochim. Cosmochim. Acta,* 51, 3119–35.

Wood, B. J. (1981) Crystal field electronic effects on the thermodynamic properties of Fe^{2+} minerals. In *Thermodynamics of Minerals and Melts.* (R. C. Newton, A. Navrotsky & B. J. Wood, eds; Springer–Verlag, New York), *Adv. Phys. Geochem.*, 1, 63–84.

8

Trace element geochemistry: distribution of transition metals in the Earth's crust

– – although ionic radius and charge are important factors in determining (trace) element distributions, there are other energy factors that need to be considered.

P. Henderson, *Inorganic Geochemistry*
(Pergamon Press, 1982), p. 134

8.1 Introduction

One outcome of interpreting transition metal geochemistry by crystal field theory is that the theory has enabled some of the basic concepts of geochemistry to be critically evaluated and defined more rigorously. In earlier chapters, crystal field theory was used to explain why some transition elements deviate from periodic crystal chemical and thermodynamic trends shown by other cations with similar charges and ionic radii. In this chapter, criteria for interpreting trace element geochemistry are examined. Examples are highlighted where fractionation patterns applicable to many elements sometimes deviate for transition metal ions. Crystal field effects are shown to be dominant factors influencing the distributions of several of the transition elements in crustal processes during the petrogenesis of igneous, sedimentary and metamorphic rocks.

8.2 Trace elements

The classification of chemical elements into major and minor or trace element categories is somewhat arbitrary. Thermodynamically, a minor element may be defined as one that is partitioned between coexisting phases in compliance with laws of dilute solutions, such as Henry's law, eq. (7.2b). In geochemical parlance, however, trace elements are usually categorized on the basis of abundance data. In this context, the mineral, rock or environment containing the chemical elements must be defined as well as the concentration boundary separating a major and trace element.

Eight elements dominate the continental and oceanic crustal abundances summarized in Appendix 1. They are oxygen, silicon, aluminium, iron, calcium, magnesium, sodium and potassium, which together constitute almost 96 wt per cent of the oceanic crust and more than 98.5 wt per cent of the continental crust (Appendix 1). Four other elements, titanium, hydrogen, phosphorus and manganese, have crustal abundances exceeding about 0.1 wt per cent. If an arbitrary boundary between major and trace elements is drawn at 0.1 wt per cent or approximately 1,000 parts per million (ppm), only three of the transition elements, iron, titanium and manganese, would be classed as major elements of the oceanic crust. The remaining elements, including scandium (38 ppm), vanadium (250 ppm), chromium (270 ppm), cobalt (47 ppm), nickel (135 ppm) and copper (86 ppm) have crustal abundances below 1,000 ppm and would be classified as trace elements.

If the elements were classified on the basis of their abundances in the Earth as a whole and the boundary between major and trace elements was still maintained at 1,000 ppm, then nickel, chromium and cobalt, together with iron and manganese would rank as major elements, with titanium being relegated to a trace element. In extreme situations of metal concentrations in ore bodies, such as stratiform chromite deposits in layered intrusions or porphyry copper deposits in calc–alkaline igneous rocks, chromium and copper might be regarded as major elements. Since the Earth's continental crust is the most accessible source of samples for measuring and interpreting element distribution patterns during mineral-forming processes, a classification of the elements into major and trace categories is usually based on their relative crustal abundances. Therefore, the majority of the transition elements must be regarded as trace elements in the crust.

8.3 Trace element distribution rules

8.3.1 Background

The historical development of trace element geochemistry has been founded on chemical analyses of minerals and rocks and on theoretical interpretations of the abundance data. In the pioneering stages of geochemistry, considerable analytical data were obtained for the occurrences, distributions and abundances of most of the elements in rocks, minerals and other geologic media principally by F. W. Clarke and co-workers at the US Geological Survey. These data were first published in 1908 as a US Geological Survey bulletin under the title *The Data of Geochemistry* which was revised five times before 1924 (Clarke, 1924). The acquisition of geochemical data was continued by Goldschmidt and co-workers during the 1920's and 1930's, and compiled by numerous geochemists thereafter (e.g., Goldschmidt, 1954; Rankama and Sahama, 1950). Such analytical studies have culminated in recent systematic compilations for the individual elements in the *Handbook of Geochemistry* (Wedepohl, 1969–78) and for different geochemical systems in revised bulletins of the *Data of Geochemistry* series (Fleisher, 1962 addenda). As a result, the geochemical distributions of individual elements are well-known.

Contemporary with these classical developments of analytical geochemistry were the advances made of X-ray and neutron diffraction techniques for determinating the crystal structures of minerals, beginning in the 1920's and continuing to the present day. More recently, crystal structure refinements have been complemented by a variety of spectroscopic techniques which have provided information on cation valences, site occupancies and nearest-neighbour environments in the mineral structures. Several examples were described in chapter 6, including crystal chemical data summarized in tables 6.2 and 6.5.

During the Goldschmidt era in the mid-1930's, therefore, considerable information became available on the sizes of ions and the distributions of elements in minerals. The stage was set for Goldschmidt to formulate certain principles for the distribution of elements in rocks and minerals (Goldschmidt, 1937). Goldschmidt postulated that during the evolution of the Earth there have been various stages of fractionation of the elements. First, a primary distribution had occurred between the metallic Core and the silicate Crust and Mantle containing massive amounts of sulphides. Second, a secondary distribution has been taking place during recycling processes in the Upper Mantle and Crust as the result of plate tectonic activity along subduction zones and spreading centres. Third, a redistribution is occurring on the Earth's surface through sedimentary and biological processes.

Goldschmidt's ideas on the primary distribution of the elements in the Earth have not been seriously challenged (see, however, Burns and Fyfe, 1966a). From studies of minerals in meteorites and phases from blast furnaces, Golschmidt classified the elements as *siderophilic* if they are inert (relative to iron) and enter the metallic phase, *chalcophilic* if they are concentrated in sulphides, *lithophilic* if they are concentrated in silicates and *atmophilic* if they are gaseous and are present in the atmosphere. Those elements enriched in organisms were also classed as *biophilic*.

In the secondary distribution of the elements in the Upper Mantle and Crust, Goldschmidt believed that during the formation of a crystalline phase the principal factor controlling the behaviour of an element is the size of its ions. From tables of ionic radii one could predict the crystal structures in which a given ion was most likely to occur. This concept has immense qualitative value in crystal chemistry, and has been remarkably successful in view of the assumptions made about ionic radii in complex silicate structures discussed in §6.8.1. This aspect is examined further in §8.4.1.

8.3.2 The Goldschmidt Rules

On the basis of ion size and charge, Goldschmidt deduced certain principles governing trace element distribution. The principles have become known as the Goldschmidt Rules. They are:

(1) If two ions have the same radius and charge, they will enter into solid-solution in a given mineral in amounts proportional to their abundances. The trace element is 'camouflaged' by the major element.
(2) If two ions have similar radii and identical charges, the smaller ion will be preferentially concentrated in early fractions of a crystallizing mineral. The bond was assumed to be weakened for the larger ion leading to a lower melting point.
(3) If two ions have similar radii but different charges, the ion with the higher charge will enter a crystal structure preferentially. When the trace element has a higher charge, it is 'captured' by the major element and enters early fractions. If the trace element has a lower charge, it is 'admitted' by the major element and enters late fractions.

At first, the Goldschmidt Rules appeared to be a useful guide for explaining distributions of trace elements in rocks and minerals. However, in the two decades following their enunciation (Goldschmidt, 1937), several exceptions to the rules were found. One implication of the Goldschmidt Rules, for example, is that during the formation of a mineral, a trace element is incorporated into a crystal structure if it is of suitable size to replace the host element in a certain site. When a difference in size exists for two ions with the same charge, the larger ion should be concentrated in the lower temperature fraction. This was referred to as the enrichment principle (Shaw, 1953). Shaw pointed out that, although the enrichment principle may apply to an ideal binary solid-solution series, it cannot apply to a system with a maximum or minimum melting point. The alkali feldspars illustrate this criticism. At a water vapour pressure of 5 kb in granitic magma, for example, the binary system: $NaAlSi_3O_8$–$KAlSi_3O_8$ has a minimum melting point at about 30 wt per cent $KAlSi_3O_8$, Since the octahedral ionic radius of K^+, 133 pm, is larger than that of Na^+, 95 pm, the Goldschmidt Rules apply to only 30 per cent of the phase diagram.

The failure of the Goldschmidt Rules in other cases, such as accounting for the geochemical behaviour of zinc, was attributed to effects of covalent bonding (Fyfe, 1951, 1954). The rules are stated in terms of ionic radius and charges, implying that the inverse square law of electrostatic interaction embodied in the Born equation, eq. (7.1), accounts for bonding energies of trace elements in minerals. The Zn^{2+} ion, 74 pm, has a radius intermediate between Mg^{2+} (72 pm) and Fe^{2+} (77 pm). Zinc, therefore, would be expected to occur in ferromagnesian silicates by analogy with other divalent trace elements with similar ionic radii and charges, such as Ni^{2+} (69 pm) and Co^{2+} (74.5 pm). However, zinc appears to discriminate against octahedral sites in ferromagnesian silicates and to form discrete silicate minerals such as willemite (Zn_2SiO_4) and hemimorphite ($Zn_4Si_2O_7(OH)_2.H_2O$), in which the Zn^{2+} ion occurs in tetrahedral coordination. The non-adherence of zinc and other B-subgroup elements to the Goldschmidt Rules was attributed to effects of covalent bonding.

8.3.3 Ringwood's modifications of the Goldschmidt Rules

Several attempts were made in the 1950's to modify the Goldschmidt Rules in order to extend their range of applicability to ionic and covalent elements by including additional parameters such as ionization potentials, electronegativity and bond energies (Ahrens, 1953; Ringwood, 1955; Nockolds, 1966). Ringwood noted that most of the deviations from the Goldschmidt Rules arise when comparisons are made between major elements of A-subgroups and trace elements of B-subgroups of the periodic table. He noted that melting points of compounds of A-subgroup elements are generally higher than those of compounds of B-subgroup elements with the same crystal structure. He deduced from this that melting points reflect relative bond energies of ions in crystal structures. Ringwood believed that the relative bond energies of the ions were also reflected in the atomic property of electronegativity. This led him to suggest the following additional criterion governing trace element distribution:

'For two ions with similar valences and ionic radii the one with the lower electronegativity will be preferentially incorporated because it forms a stronger and more ionic bond than the other'.

Even when modified by electronegativity, bond energy and ionization potential criteria the Goldschmidt Rules still lacked generality. In particular, they were unable to explain the geochemical behaviour of some transition elements, most notably nickel and chromium. Prior to the more recent and comprehensive compilations of ionic radii (Shannon and Prewitt, 1969; Whittaker and Muntus, 1970; Shannon, 1976), values tabulated by Pauling (1960, p. 514), Goldschmidt (1954, p. 88) and Ahrens (1952) were used to interpret trace element geochemistry. The then accepted ionic radius of Ni^{2+} (e.g., 69 pm in Ahrens, 1952) was larger than that of Mg^{2+} (66 pm), leading to the expectation based on the Goldschmidt Rules that Mg^{2+} would enter crystallizing minerals ahead of Ni^{2+} ions. Nickel, however, was found to be invariably enriched in magnesian olivines appearing at early stages during fractional crystallization of basaltic magma, contrary to the Goldschmidt Rules. Furthermore, the electronegativity of Ni^{2+}, $\chi_{Ni} = 1.8$, is higher than that of magnesium, $\chi_{Mg} = 1.2$, suggesting that Ni^{2+} should be enriched in late crystal fractions relative to Mg^{2+} according to Ringwood's modification of the Goldschmidt Rules. Moreover, the behaviour of nickel could not be explained by phase relationships in the simple binary system: Mg_2SiO_4–Ni_2SiO_4 (Ringwood, 1956), in which early crystals are impoverished and late liquids enriched in Ni on account of the higher melting point of forsterite. Ringwood attributed the observed nickel enrichment in natural olivines to the preferential replacement of the larger Fe^{2+} ions by the smaller Ni^{2+} ions in the olivine structure. In retrospect, there is some validity to this interpretation, as indicated by fig. 6.3, which shows that both Ni^{2+} and Fe^{2+} are enriched relative to Mg^{2+} in the olivine M1 sites. The apparent anomalous behaviour of nickel subsequently sparked numerous investigations of element partitioning in crystal/melt systems, examples of which are discussed later (§8.5.5 and §8.5.6).

Apart from a fundamental flaw in correlating melting points with bond energies (Burns and Fyfe, 1967b), the modified Goldschmidt Rules of trace element geochemistry were found to lack generality particularly with regard to the transition elements, with the result that dissatisfaction has been expressed over the utility of the rules for predicting trace element distribution.

8.3.4 Ionic radii and distribution coefficients

The ionic radius criterion for interpreting geochemical distributions of trace elements was given a boost in the early 1970's when correlations were shown to exist between ionic radii and partition coefficients of some trace elements (Onuma *et al.*, 1968; Higuchi and Nagasawa, 1969; Jensen, 1973). The influence of cation radius and charge on trace element distribution patterns was demonstrated by measurements of the distribution coefficient, *D*, defined by

$$D = [M]_{\text{crystal}}/[M]_{\text{melt}} \tag{8.1}$$

and expressing relative concentrations of trace element *M* in a crystal and the coexisting melt. By analysing phenocrysts and groundmass from a number of igneous intrusives, distribution coefficients for several trace elements in different minerals could be evaluated. Graphical plots of such distribution coefficients versus ionic radius for different cations appeared to give relatively smooth curves for ions with similar charges (Onuma *et al.*, 1968; Higuchi and Nagasawa, 1969; Jensen, 1973). For calcic pyroxene phenocrysts, distribution coefficients for suites of trace element cations having identical charges lie on smooth curves with peaks at two distinct ionic radius values near those for the host Mg^{2+} and Ca^{2+} ions in the pyroxene M1 and M2 coordination sites. These patterns suggested that ionic radius and crystal structure strongly influence the partitioning of trace elements (Philpotts, 1978). Such effects are simply manifestations, though, of basic crystal chemical principles enunciated by Goldschmidt (1954) and Pauling (1960).

Notably absent in the plots of phenocryst/melt distribution coefficients, however, are data for transition elements such as Cr, V and Ni. Subsequent measurements of distribution coefficients for trivalent transition elements between calcic pyroxene phenocrysts and basalt groundmass revealed significant deviations of Cr^{3+}, and to a lesser extent V^{3+}, from smooth curves defined by other trivalent cations (Henderson, 1979). As stated in the preface of this chapter, Henderson concluded that although ionic radius and charge are important factors determining element distribution, there appear to be other factors affecting the partitioning of trace elements between minerals and magma. These include crystal field stabilization energies of certain transition metal ions discussed later (§8.5).

8.3.5 Discussion

The various 'rules' that were enunciated to explain element distributions and partitioning in crystal/melt systems have had a profound influence in crystal chemistry and on interpretations of trace element geochemistry. They have served as useful guiding principles for predicting mineral occurrences and explaining the locations of trace elements in crystal structures. Thus, given the ionic radius and valence of a trace element, assessments can be made of the crystal structures most likely to accommodate that element. Well-known examples include

(1) Ni^{2+}, Co^{2+}, Li^{+}, and Sc^{3+} readily occupying Mg^{2+} and Fe^{2+} sites in the olivine, orthopyroxene, and calcic clinopyroxene structures;
(2) trivalent lanthanide, Mn^{2+} and Na^{+} ions replacing calcium in the diopside, anorthite, eudialyte and apatite structures;
(3) the substitution of Rb^{+}, Tl^{+}, Ba^{2+} and Sr^{2+} ions for potassium in potash felspars, zeolites and certain Mn(IV) oxides such as cryptomelane and romanechite; and
(4) atomic substitution of OH^{-} by F^{-} anions in micas, amphiboles, tourmalines and topaz.

However, the lack of generality of the trace element distribution rules, particularly with regard to the transition elements, has caused dissatisfaction over the precision of the 'rules' for predicting element fractionation patterns. This has led to critical evaluations of the empirical rules, which have highlighted fundamental flaws in their applications to geological media.

8.4 Evaluation of principles governing trace element distribution

Because the Goldschmidt Rules and their various modifications lacked generality, they were critically evaluated (Shaw, 1953; Burns and Fyfe, 1966a, 1967b; Whittaker, 1967). Fundamental misconceptions were demonstrated concerning the electronegativity of an element, correlations between melting points and bond energies, and effective ionic radius of a cation in distorted coordination sites where there are wide ranges of metal–oxygen distances (Burns and Fyfe, 1967b). The major criticisms of the 'rules' are thermodynamically based, however.

8.4.1 Ionic radius criterion

Although the ionic radius criterion of Goldschmidt continues to serve as a useful principle of crystal chemistry, attention has been drawn to limitations of it (Burns and Fyfe, 1967b; Burns, 1973). As noted earlier, the magnitude of the ionic radius and the concept of radius ratio (i.e. cation radius/anion radius) has proven to be a valuable guide for determining whether an ion may occupy a specific coordination site in a crystal structure. However, subtle differences between ionic radii are often appealed to in interpretations of trace element distributions during mineral formation.

Theoretically, the radius of an ion extends from the nucleus to the outermost orbital occupied by electrons. The very nature of the angular wave function of an electron, which approaches zero asymptotically with increasing distance from the nucleus, indicates that an atom or ion has no definite size. Electron density maps compiled in X-ray determinations of crystal structures rarely show zero contours along a metal–anion bond.

Ionic radii estimated independently by Goldschmidt and Pauling were obtained from interatomic distances in simple ionic crystals such as halides and oxides of metals at laboratory temperatures. These tables of ionic radii (Goldschmidt, 1954; p. 88; Pauling, 1960, p. 514), as well as the more recent compilations (Ahrens, 1952; Shannon and Prewitt, 1969; Whittaker and Muntus, 1970; Shannon, 1976), are based on *average* interatomic distances within a particular coordination site. The values are strictly valid only for ions in regular octahedral, cubic, or tetrahedral coordinations in these simple crystal structures and might not be expected to apply rigorously to complex structures such as silicate minerals. The crystal structures of ferromagnesian silicates described in earlier chapters show that cations usually occur in distorted coordination sites and that there is a large range of bond distances within each cation site (see table 5.6; Appendix 7). For example, in the forsterite structure the metal–oxygen distances vary from about 205 pm to almost 221 pm. If oxygen is assigned an ionic radius of 140 pm, the ionic radius of Mg^{2+} varies from 65 pm to 81 pm in the M1 and M2 sites of this olivine. The extremes are greater in the orthopyroxene structure where distances in enstatite range from about 199 pm to 245 pm (table 5.6). A trace element substituting for a major element such as magnesium in olivine or orthopyroxene will 'see' coordination polyhedra in the structures in which metal–oxygen distances may vary within individual and between different Mg^{2+} sites by almost 50 pm. Therefore, predictions and interpretations of relative enrichment based on small differences in ionic radii, including those based on distribution coefficient data (Jensen, 1973), should be evaluated critically in view of the large variability of interatomic distances within common rock-forming silicate minerals.

It should also be noted that during the 1950–60 era, the major element/trace element associations in minerals based on ionic radius or crystallographic criteria often degenerated into an analytical relationship (Burns, 1972b; 1973). At that time, geochemists tended to interpret their analytical data for bulk rock samples in terms of relationships between tabulated ionic radii values for major and trace element constituents without specific reference to a particular mineral structure. Fortunately, this fallacious trend is no longer followed because microbeam analytical techniques enable compositions of host phenocrysts and glasses to be measured directly. However, since the geometries and dimensions of coordination sites of an individual major cation may differ from one mineral to another and also within a particular mineral structure, the ionic radius criterion is too simple for explaining element partitioning in mineral structures, let alone trends in bulk rock analyses.

8.4.2 Melting point criterion

Melting points have been used as criteria of bond strength and bond type. In some series of chemical compounds, there is a correlation between melting point and lattice energy of ionic crystals; in alkali metal halides, for example, there is a steady fall in melting point and a progressive decrease in lattice energy with rising atomic number of the alkali metal or halide ion (Burns and Fyfe, 1967b). This correlation has led to the widely held view that if two compounds have similar crystal structures and cell dimensions, a melting point difference is indicative of different bond energies in the crystals.

The concept of melting point may be put on a more quantitative basis. The process of melting is a thermal one and is characterized by the collapse of molecular units in a crystal to a disordered array. Energy is required to rupture the crystal lattice so that the heat of fusion, ΔH_{fus}, is positive. In addition, the solid–liquid transition involves an increase of randomness and the entropy of fusion, ΔS_{fus}, is also positive. Melting point is the temperature of fusion (T_{fus}) at which solid and liquid phases are in equilibrium. Therefore,

$$G_l = G_s, \text{ or } \Delta G_{fus} = 0 \tag{8.2}$$

and

$$\Delta H_{fus} = T_{fus}\,\Delta S_{fus}. \tag{8.3}$$

Thus,

$$T_{fus} = \Delta H_{fus}/\Delta S_{fus} \tag{8.4}$$

or

$$T_{fus} = (H_l - H_s)/(S_l - S_s), \tag{8.5}$$

where G_l, G_s, H_l, H_s, and S_l, S_s are the free energies, bond energies and entropies of the liquid (l) and solid (s) phases, respectively.

Equation 8.5 shows that factors which determine the melting point are the *differences* in bond energies and entropies between liquid and solid phases, and not the *absolute* energy of either phase. In general, heats of fusion are small (5 to 130 kJ/mole) by comparison with lattice energies (400 to 20,000 kJ/mole), indicating that bond energies are comparable in the solid and liquid phases. Any property of an atom which leads to large lattice energies for its crystalline compounds may still be effective in the liquid state. The lattice energy of fayalite, for example, is about 19,270 kJ/mole (fig. 7.1*d*) and the heat of fusion is about 70 kJ/mole, suggesting that the lattice energy of molten Fe_2SiO_4 (T_{fus} = 1,205 °C) is about 19,200 kJ/mole, or only 0.35 per cent smaller than that of crystalline fayalite.

8.4.3 Thermodynamic criteria

Most of the discussions of trace element distribution during mineral formation have considered bonding forces of ions in crystalline phases only. Thus, bonding of ions in media from which minerals have crystallized (e.g., magma or aqueous phases) have been largely ignored even though the magnitude of the lattice energies of solid and liquid phases may be comparable. This neglect has been criticized on the grounds that stabilities of ions in both the mineral and the medium from which the mineral crystallized have to be considered (Burns and Fyfe, 1966a, 1967b).

Element fractionation may take place through four different processes. These may be represented by the general reactions

$$X_m + YZ_s \rightarrow Y_m + XZ_s, \qquad (8.6)$$

where the two species X and Y in the melt (m) compete for a site in a solid (s) crystallizing from the melt;

$$XZ_s + YZ'_s \rightarrow XZ'_s + YZ_s, \qquad (8.7)$$

where the two species are distributed between the two solid phases;

$$X_{aq} + YZ_s \rightarrow Y_{aq} + XZ_s, \qquad (8.8)$$

where the two species are fractionated between an aqueous phase (aq) and a solid; and

$$X_g + YZ_s \rightarrow Y_g + XZ_s, \qquad (8.9)$$

where the two species are present in a low-pressure, low-density gas phase (g).

Equation (8.6) represents magmatic crystallization, eq. (8.7) metamorphic recrystallization, eq. (8.8) sedimentary processes and eq. (8.9) gas-phase deposition.

In order to predict the direction and extent of the equilibria represented in these equations, stabilities of ions in each phase must be assessed. In §8.3.2 it was noted that underlying the Goldschmidt Rules, is a consideration of the lattice energies of crystalline phases. These rules are most valid for predicting trace element distribution in gas-phase deposition, eq. (8.9), which, however, is now the least significant process in mineral formation in the Earth, though the process may have influenced element partitioning during condensation of solar nebular gas in Earth's earliest history. For example, enrichments of vanadium and titanium in hibonites, $CaAl_{12}O_{19}$, occurring in refractory inclusions of carbonaceous chondrites may be the result of high crystal field stabilization energies acquired by the Ti^{3+} and V^{3+} ions in the trigonal bipyramidal M5 sites of the hibonite crystal structure that were not attainable in the solar nebular gas phase (Burns and Burns, 1984b).

Whenever mineral formation takes place in a medium more complex than the gas phase, such as a magma or an aqueous solution, it is necessary to evaluate thermodynamic data for all steps of a cyclic reaction such as eq. (8.10) before the relative enrichment of an ion can be assessed.

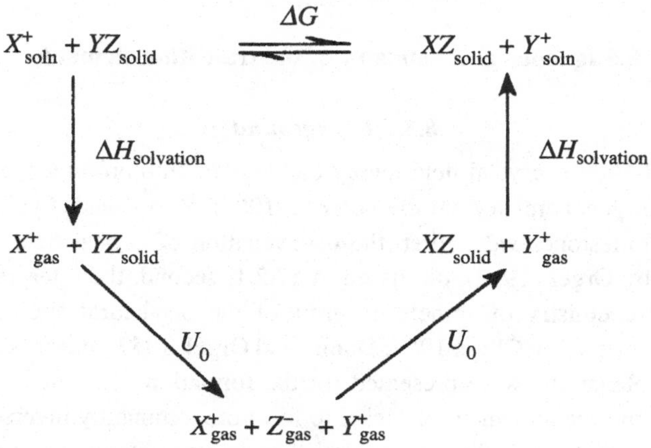

(8.10)

The direction and extent of eq. (8.10) depends on:

(i) the relative lattice energies, U_0, of the solids;
(ii) the solvation energies, $\Delta H_{\text{solvation}}$, of the ions in a silicate melt or aqueous solution; and
(iii) the overall change in free energy, ΔG, which is small by comparison with U_0 and $\Delta H_{\text{solvation}}$.

Values of free energies of reactions such as those occurring in magma are not available and other approximations must be made. Furthermore, values of ΔG are small by comparison with lattice energies of solids and heats of solvation of the ions (Burns and Fyfe, 1966a). The balance of the lattice energies and heats of solvation between reacting species controls the direction of element distribution reactions. In order to understand and make valid predictions for distributions of elements in mineral-forming processes, it is necessary to have thermodynamic data such as heats of solvation of ions in magma or pegmatitic fluids, lattice energies of silicate minerals and values of the free energies of reactions in melts at temperatures around 1,000 °C. Such information is not yet available but may be attainable by electrostatic energy calculations (Ottonello, 1987).

There are some cases, however, where with limited data, valid *comparisons* can be made between transition metals ions and other cations of similar charge and size. It is possible to interpret qualitatively trends in the geochemical distribution of transition metal ions by crystal field theory.

8.5 Igneous geochemistry of the transition elements

8.5.1 Background

By the mid-1960's, crystal field theory had entered into prominence in transition metal geochemistry (Burns *et al.*, 1964; Burns and Fyfe, 1967a). Important milestones were: first, the re-juvenation of crystal field theory in chemistry by Orgel (1952), discussed in §7.2.1; second, the interpretation of the crystal chemistry of spinels in terms of the octahedral site preference energy parameter (McClure, 1957; Dunitz and Orgel, 1957), described in §6.4. Elegant explanations were presented for the formation of normal spinels by chromium and the tendency for nickel to form predominantly inverse spinels. This work demonstrated for the first time that strong fractionation of transition metal ions occurs between different coordination sites in a crystal structure.

The third milestone was the explanation of the fractionation patterns of transition elements during crystallization of the Skaergaard magma by R. J. P. Williams. In one of the most thoroughly documented accounts of trace element distribution during magmatic crystallization, Wager and Mitchell (1951) had measured the concentrations of several elements in minerals that had separated from the magma at different stages of crystallization. Williams (1959; see also Phillips and Williams, 1966, vol. 2, p. 611) examined the analytical data of Wager and Mitchell and showed that a correlation exists between the order of uptake of transition metal ions from the magma and the relative crystal field stabilization energies of the cations in octahedral coordination. According to Williams the process of removal of an ion from a liquid into a crystal is characterized by an increase in average coordination number of the ion in a solid and a decrease in interatomic distances in the solid phase. Ions which are particularly stabilized in octahedral coordination gain stability by passing from a melt of irregular coordination to a mineral providing more regular octahedral coordination sites. Williams' plots of the data from the Skaergaard intrusion are shown in fig. 8.1. The enrichments of Cr^{3+} and Ni^{2+} in early crystals were attributed to the high crystal field stabilization energies of these cations in octahedral coordination.

The views of Williams were extended by subsequent writers. Curtis (1964) considered the stability of ions in tetrahedral coordination in minerals as well as in octahedral sites. While this approach is applicable to spinels crystallizing from the magma, it does not apply to silicate minerals in which the transition metal ions occupy only six-coordinated sites. Burns and Fyfe (1964) presented crystal field spectral data indicating that cations such as Ni^{2+} are present in both octahedral and tetrahedral sites in silicate glasses assumed to approximate

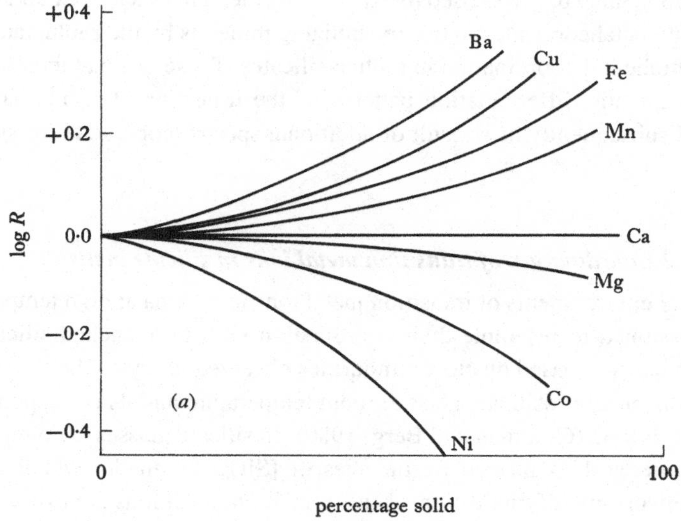

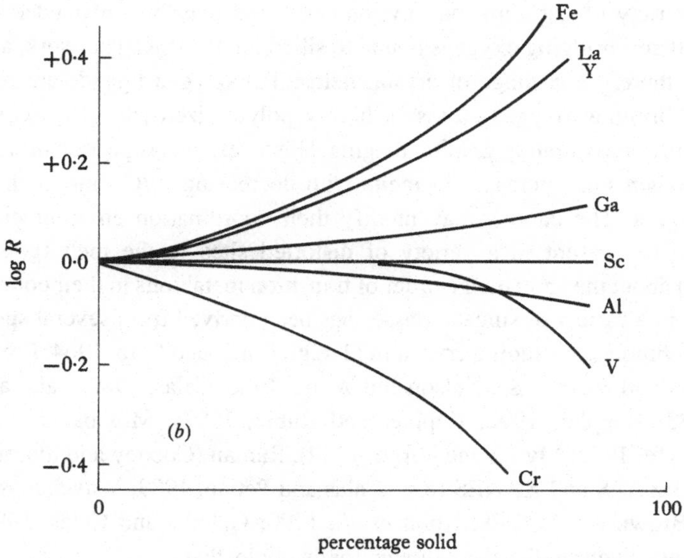

Figure 8.1 The uptake of transition metal ions into silicate minerals crystallizing from basaltic magma. (a) Divalent cations; (b) trivalent cations. The enrichment factor, *R*, is the ratio of the concentration of the element in the magma after *X* per cent solidification to the concentration of the element in the initial liquid (after Williams, 1959).

magma. They stressed that relative bond energies of cations in magma and crystal structures have to be assessed. The octahedral site preference energy parameter (§6.4; table 6.3) was used to explain why elements such as Cr and Ni are driven into octahedral sites in the crystallizing minerals by their reluctance to exist in tetrahedral coordination in molten silicates. These original interpretations of magmatic differentiation patterns of the transition elements have been refined subsequently as a result of additional spectroscopic data for silicate glasses.

8.5.2 Coordination of transition metal ions in silicate melts

The structural environments of transition metal ions in magma at high temperatures are assumed to resemble their coordination sites in quenched silicate glasses, a conclusion based on close similarities observed in crystal field spectra of Fe^{2+} ions in a borosilicate glass at room temperature and its corresponding melt at 1,260 °C (Goldman and Berg, 1980). In silicate glasses, a continuous random network is formed by linkages of [SiO_4] tetrahedra which are bound by network-modifying cations. Most transition metal ions are expected to substitute for network-modifying cations, but some may replace silicon in the network-forming [SiO_4] tetrahedra. Thus, in the glass structure, the cations exist in a variety of coordination environments and may be surrounded by bridging and non-bridging oxygens bound to silicon in the [SiO_4] network, and experience, therefore, a range of crystal fields. The oxygen ligands are predominantly bridging oxygen atoms in highly polymerized [SiO_4] networks such as melts approximating granitic magma. However, the proportion of non-bridging oxygen ions increases in melts with decreasing Si/O ratio such as basaltic magma. The cations may modify their coordination environments, though, and be present in a variety of distorted sites in the melt (glass). Information about the short-range order of transition metal ions in their coordination sites in a variety of silicate glasses has been derived from several spectroscopic techniques, including crystal field (e.g., Burns and Fyfe, 1964; Fox *et al.*, 1982; Nelson *et al.*, 1983; Nelson and White, 1986; Calas, 1982; Calas and Petiau, 1983; Keppler, 1992; Keppler and Rubie, 1992), Mössbauer (e.g., Boon and Fyfe, 1972; Mysen and Virgo, 1989), Raman (Cooney and Sharma, 1990), and EXAFS and XANES (e.g., Calas and Petiau, 1983; Waychunas *et al.*, 1988; Brown *et al.*, 1988; Kohn *et al.*, 1990; Galoisy and Calas, 1991) measurements. Such studies have shown, for example, that

(1) Co^{2+} ions predominate in network-modifying tetrahedral sites in most glass compositions. In albite melts at elevated pressures, there is a change of

coordination from tetrahedral to octahedral, indicating an increase of CFSE for the Co^{2+} ions;

(2) Mn^{2+} ions occupy either a distribution of octahedral and tetrahedral sites or distorted non-centrosymmetric sites. The fraction of non-centrosymmetric, tetrahedral or distorted sites increases with decreasing number of non-bridging oxygens so that tetrahedral $[MnO_4]$ clusters occur in orthosilicate glasses;

(3) Fe^{2+} ions occur in a variety of distorted octahedral sites in basaltic glasses, with a small proportion of Fe^{2+} ions possibly occurring in tetrahedral sites. Tetrahedral Fe^{2+} ions predominate, however, in Fe_2SiO_4-bearing orthosilicate glasses;

(4) The site occupancy of nickel is controversial. Octahedrally coordinated Ni^{2+} ions in distorted sites appear to predominate. Some Ni^{2+} ions may also occur in network-modifying tetrahedral and trigonal bipyramidal (five-fold) sites, such as Ni_2SiO_4- and $CaNiSi_2O_6$-bearing glasses. In albite glasses at high pressures, the coordination of Ni^{2+} ions changes from tetrahedral to regular or distorted octahedral symmetry, with an increase of CFSE for Ni^{2+} in the melt. Thus, Ni^{2+} as well as Co^{2+} could become less compatible in igneous fractionation processes under high pressure (Keppler and Rubie, 1992); and

(5) Octahedral Cr^{3+}, V^{3+} and Mn^{3+} ions predominate in silicate glasses, while Fe^{3+} ions are mainly in network-forming (tetrahedral) sites.

Crystal field spectral measurements of transition metal ions doped in a variety of silicate glass compositions (e.g., Fox *et al.*, 1982; Nelson *et al.*, 1983; Nelson and White, 1986; Calas and Petiau, 1983; Keppler, 1992) have produced estimates of the crystal field splitting and stabilization energy parameters for several of the transition metal ions, examples of which are summarized in table 8.1. Comparisons with CFSE data for each transition metal ion in octahedral sites in periclase, MgO (divalent cations) and corundum , Al_2O_3 (trivalent cations) and hydrated complexes show that CFSE differences between crystal and glass (e.g., basaltic melt) structures,

$$\text{CFSE}_{\text{crystal}} - \text{CFSE}_{\text{glass}} , \tag{8.11}$$

generally decrease (i.e. become less negative) in the orders

$$Ni^{2+} > Co^{2+} > Fe^{2+} > (Mn^{2+}) \tag{8.12}$$

and

$$Ti^{3+} > Cr^{3+} > V^{3+} > Mn^{3+} > (Fe^{3+}) . \tag{8.13}$$

Table 8.1. *Crystal field stabilization energies of transition metals in silicate glasses*

Cation	Albite–diopside glass (50/50 wt %) Δ (cm^{-1})	Albite–diopside glass (50/50 wt %) CFSE (kJ/g.ion)	Silicate glasses CFSE (kJ/g.ion) G	Silicate glasses CFSE (kJ/g.ion) B	Difference CFSE MgO – glass (kJ/g.ion) G	Difference CFSE MgO – glass (kJ/g.ion) B	Difference CFSE Al_2O_3 – glass (kJ/g.ion) G	Difference CFSE Al_2O_3 – glass (kJ/g.ion) B	Difference CFSE hydrate – glass (kJ/g.ion) G	Difference CFSE hydrate – glass (kJ/g.ion) B
Ti^{3+}	–	–	–46.0	–56.5	–8.4	2.1	–60.2	–49.7	–54.7	–34.2
V^{3+}	16,700	–159.9	–174.4	–174.4	–	–	–4.8	–4.8	–8.4	–8.4
Cr^{3+}	15,400	–221.1	–215.9	–225.9	–16.7	– 6.7	–44.7	–34.7	–34.0	–24.0
Mn^{3+}	17,550	–126.0	–111.7	–151.5	–	–	–27.4	12.4	–39.1	0.7
Cr^{2+}	18,350	–131.7	–110.3	–110.3	–	–	–	–	10.5	10.5
Fe^{2+}	7,330	–35.1	–43.1	–48.5	–8.6	–3.2	–	–	–1.9	3.3
Co^{2+}	4,040*	–29.0*	–73.6	–73.6	–7.8	–7.8	–41.2	–41.2	–15.5	–15.5
Ni^{2+}	5,260	–75.5	–77.0	–79.5	–48.6	–46.1	–76.6	–74.1	–45.1	–42.6
Cu^{2+}	12,820	–92.0	–92.9	–97.9	–	–	–	–	–0.4	4.6
Sources of data	[1]	[2]	[3]	[3]	[4]	[4]	[5]	[5]	[6]	[6]

* Tetrahedral coordination.

Sources of data : [1] From Keppler (1992); Δ for Ni^{2+} selected to be 5,260 cm^{-1}

[2] Calculated from reference [1] for octahedral cations, except tetrahedral Co^{2+}

[3] From Calas & Petiau (1983): B = basaltic glass; G = granitic glass

[4] Difference: doped periclase, MgO (table 5.1), and reference [3] glasses (basalt B and granite G)

[5] Difference: doped corundum, Al_2O_3 (table 5.2), and reference [3] glasses (basalt B and granite G)

[6] Difference: hexahydrated cations (table 2.5) and reference [3] glasses (basalt B and granite G)

8.5.3 Explanation of magmatic differentiation patterns of transition metals

The structures of liquids, molten salts and quenched glasses have been shown by X-ray diffraction, thermal neutron studies, infrared and Raman spectroscopy to have some degree of long-range order provided by the network-forming cations. Other spectroscopic techniques (e.g., X-ray absorption spectra) providing information about the short-range scale indicate that, in general, metal–anion interatomic distances are similar or slightly shorter, next-nearest-neighbour distances are longer, and the average coordination numbers of cations are smaller in the liquid (glass) than the solid at its melting point.

When cations are dissolved in silicate liquids in which a variety of coordination sites of different symmetries are present, a distribution of ions between the sites will occur if thermal energies are comparable with site energy differences. In silicate melts of granitic and basaltic compositions, for example, network-forming tetrahedral and network-modifying six-fold, five-fold and four-fold coordination sites predominate. Transition metal ions may occur in each of these coordination environments in the silicate melt. The cations, however, are rarely found in tetrahedral sites in silicate minerals which crystallize from a magma, but they do occur in six-coordinated, approximately octahedral sites in these minerals Therefore, during magmatic crystallization, partitioning of cations takes place between six-fold, five-fold and four-fold coordination sites in a magma and the octahedral sites in a crystal. The process may be represented by the equation

$$(M^{n+}{}_{4\text{–fold}} + M^{n+}{}_{5\text{–fold}} + M^{n+}{}_{6\text{–fold}})_{\text{magma}} \rightarrow (M^{n+}{}_{\text{octahedral}})_{\text{crystal}} \, . \qquad (8.14)$$

The enthalpy change of this reaction is partly related to the change in CFSE of a cation as it passes out of a magma into the crystalline phase. The magnitude of the octahedral site preference energy parameter, (OSPE = CFSE_{oct} − CFSE_{tet}), discussed in §6.4 (table 6.3), as well as the ($\text{CFSE}_{\text{crystal}} - \text{CFSE}_{\text{glass}}$) difference data summarized in table 8.1, eqs (8.11) to (8.13), provide measures of the relative affinities of cations in a magma for octahedral sites in silicate minerals. Thus, based on the OSPE parameter, the relative orders of uptake of cations into a crystallizing silicate mineral are expected to be

$$\text{Ni} > (\text{Cr}) > (\text{Cu}) > \text{Co} > \text{Fe} > \text{Mn, Ca, Zn} \quad \text{for } M^{2+} \text{ ions} \qquad (8.15)$$

and

$$\text{Cr} > (\text{Mn}) > \text{Co} > \text{V} > \text{Ti} > \text{Fe, Sc, Ga} \quad \text{for } M^{3+} \text{ ions} \, . \qquad (8.16)$$

These predicted orders are in good agreement with enrichment patterns observed during crystallization of basaltic magma (fig. 8.1). The agreement is better when cations producing Jahn–Teller distortions of their coordination polyhedra are excluded from the predicted orders. These cations, which are shown in parenthesis in (8.15) and (8.16), prefer the presumably more deformable sites in silicate melts and crystallize in late-stage minerals. Such an argument may be applicable to copper mineralization and, perhaps, Cr^{2+} uptake if oxygen fugacities are sufficiently low to stabilize divalent chromium in the magma (Schreiber and Haskin, 1976). Equation (8.16) may not apply to Mn^{3+} either, since this cation is inherently unstable with respect to Mn^{2+} plus Mn(IV) oxides, eqs. (2.5) and (2.6). The uptake of Mn^{3+} ions requires a mineral that provides a very distorted coordination site to bestow Jahn–Teller stability on it (§6.2). For Ti^{3+}, its high $CFSE_{(crystal-glass)}$ value may account for the prevalence of trivalent Ti in igneous minerals in basalts on the Moon which crystallized at lower oxygen fugacities than those existing in terrestrial basaltic magma.

Certain qualifications should be noted in the use of the octahedral site preference energy parameter alone to explain patterns of magmatic crystallization. First, the OPSE parameter applies to element partitioning only between octahedral and tetrahedral sites. If other coordination symmetries such as trigonal bipyramidal (five-fold) sites exist in a melt (Galoisy and Calas, 1991) then other CFSE-difference parameters should be employed. The $3d$ orbital energy level data for alternative coordination symmetries listed in table 2.4 indicate that site preference energies estimated as differences between six-fold (octahedral) and five-fold (trigonal bipyramidal) coordinations would result in very high preferences of cations with $3d^1$, $3d^2$, $3d^3$, $3d^6$, $3d^7$ and $3d^8$ configurations (e.g., Ti^{3+}, V^{3+}, Cr^{3+}, Fe^{2+}, Co^{2+}, and Ni^{2+}, respectively) for octahedral sites and strong enrichments of $3d^4$ and $3d^9$ cations (e.g., Mn^{3+}, Cr^{2+} and Cu^{2+}) in trigonal bipyramidal sites in a melt.

A second qualification is that the OSPE parameter, as well as the $CFSE_{(crystal-glass)}$ difference parameters, eq. (8.11), are enthalpy terms and as such cannot be used to determine the affinity of a reaction. Free energy changes must be considered. However, since entropy changes are expected to be similar for all transition metal ions of the same valency, the use of an enthalpy term such as CFSE to interpret element fractionation may be a valid approximation when comparisons are made between transition metal ions of similar radius and charge.

Thirdly, cations are not likely to be equally distributed between six-fold, five-fold and four-fold coordination sites in the magma. According to the Maxwell–Boltmann distribution law, the ratio of cations in octahedral and tetrahedral sites, n_o/n_t, the crystal field stabilization energies of which are E_o and E_t, respectively, above a reference level, U_0 is

$$n_o/n_t = \exp[-(E_o - E_t)/RT] \,. \tag{8.17}$$

Since lattice energies are comparable for structures containing similar cations in octahedral and tetrahedral sites (cf. NiO and ZnO in fig. 7.1*b*), the n_o/n_t ratios in table 6.3 indicate that there are higher proportions of transition metal ions in octahedral sites than in tetrahedral sites in magmas despite the fact that lattice energies of silicate crystals and melts are almost two orders of magnitude higher than those of the CFSE's, E_o and E_t (cf. fig. 7.1*b* and *d*).

A fourth qualification is that octahedral site preference energies (or related CFSE-difference parameters), which are calculated from spectral data measured at ambient temperatures, are assumed to apply to magmatic temperatures. Few spectral measurements have been made of transition metals in oxide and silicate melts at elevated temperatures (Goldman and Berg, 1980), with the result that CFSE data are not available for transition metal ions in magma. When extending the room temperature spectroscopic data to magmatic temperatures, the assumption made is that the relative orders of octahedral site preference energies remain approximately constant over the temperature range 25 °C to 1,200 °C.

Finally, the use of CFSE's for evaluating relative stabilities of ions in octahedral and tetrahedral coordinations has been criticised (Katzin, 1961; Glidewell, 1976). It was pointed out that CFSE's refer to small electronic differences relative to the mean potential or baricentre of the ion in its environment. Katzin believed that the position of the baricentre in the scale of potential energy is not necessarily the same in tetrahedral and octahedral coordinations, and that relative stabilities of ions cannot be deduced from absorption spectra alone. It should be noted, however, that the lattice energies of transition metal compounds shown in fig. 7.1 display relatively smooth variations across a series of ions, irrespective of the crystal structure of the compound or mineral. This feature is particularly well illustrated by the lattice energies of divalent sulphides where at least four structure-types with cations in different coordination symmetries, including octahedral, tetrahedral and square planar sites, are represented. This suggests that the baricentre of 3*d* orbital energy levels is largely independent of coordination number of a transition metal ion in a compound. This is supported by computed molecular orbital energy level diagrams for octahedral $[FeO_6]$ and tetrahedral $[FeO_4]$ clusters described in §11.6.2 and §11.6.3 in which electrons in core orbitals (1*s*, 2*s*, 2*p*, 3*s*, 3*p*) and non-bonding oxygen 2*p* molecular orbitals appear to have comparable energy levels for iron in different coordination sites.

A further conclusion may be drawn from the equilibrium represented by eq. (8.14) and the influence that changes of silicate melt compositions have on the

OSPE or any of the $CFSE_{(crystal-melt)}$ difference parameters. This concerns the distribution coefficients of transition metal ions between igneous minerals and magma of granitic or basaltic compositions. The original spectral measurements of Ni^{2+}-bearing glasses (Burns and Fyfe, 1964) indicated that the proportion of tetrahedral sites in silicate melts increases with increasing amounts of alkali metal ions and silica, and decreasing amounts of alumina in the glass. More recent EXAFS, XANES and crystal field spectral measurements suggest that Ni^{2+} ions in five-coordinated sites may also exist in some silicate glasses (Galoisy and Calas, 1991). The crystal field spectral data also indicate a decrease of crystal field splitting of Ni^{2+} and other transition metal ions with increasing polymerization of the silicate melt (Calas and Petiau, 1983; Keppler, 1992). The more polymerized the melt, the more are transition metals expected to be partitioned preferentially into crystallizing minerals. As a result, the equilibrium in eq. (8.14) should be driven further to the right in granitic melts than in basaltic melts. This accounts for the distribution coefficients of transition metal ions possessing large octahedral site preference energies being greater for granitic melts than for magma of basaltic composition. Such an effect would also account for the paucity of certain transition metals in granitic and residual melts.

The prediction that site symmetry in magmas influences cation partitioning (Burns and Fyfe, 1964) was borne out by subsequent experimental investigations involving Ni^{2+} (e.g., Irvine and Kushiro, 1976) and Cr^{3+} (Irvine, 1974, 1975; Barnes, 1986) which are described later (§8.5.5 and §8.5.6).

8.5.4 The dilemma over nickel

The enrichment of nickel in early fractions of minerals crystallizing from magma is an enigma (Burns and Fyfe, 1966b). It was noted earlier (§8.3.3) that the behaviour of nickel defies explanation not only by the empirical rules enunciated by Goldschmidt and by Ringwood, but also because melting relations in the two-component system: Mg_2SiO_4–Ni_2SiO_4 show that nickel–bearing olivines have lower melting points than forsterite. Thus, in the binary system cation ratios and distribution coefficients (D) are

$$(\mathrm{Ni/Mg})_{\mathrm{ol}} < (\mathrm{Ni/Mg})_{\mathrm{melt}} \text{ or } D_{\mathrm{Ni}}^{\mathrm{ol/melt}} < D_{\mathrm{Mg}}^{\mathrm{ol/melt}} \qquad (8.18)$$

whereas in the basaltic magma they appear to be

$$(\mathrm{Ni/Mg})_{\mathrm{ol}} > (\mathrm{Ni/Mg})_{\mathrm{melt}} \text{ or } D_{\mathrm{Ni}}^{\mathrm{ol/melt}} > D_{\mathrm{Ni}}^{\mathrm{ol/melt}}\,. \qquad (8.19)$$

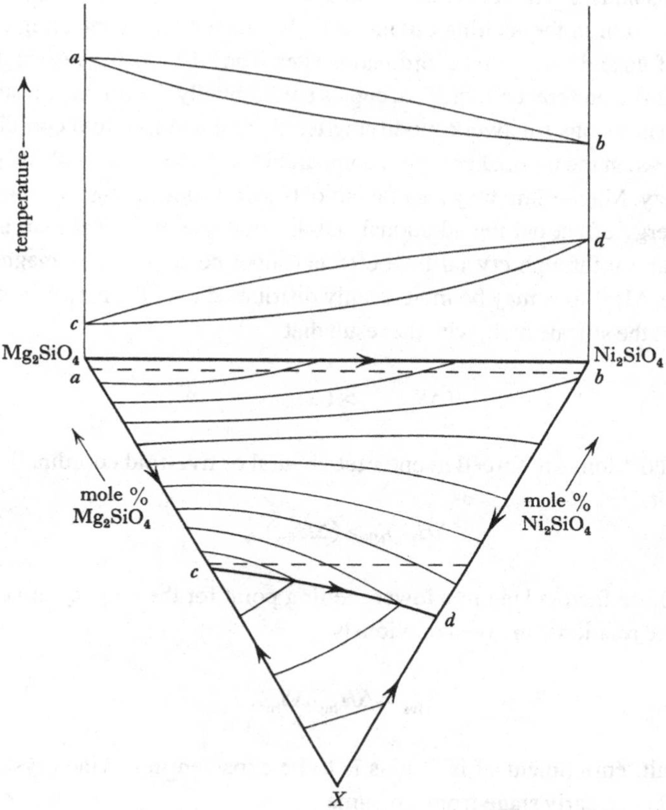

Figure 8.2 Phase diagram for the system: Ni_2SiO_4–Mg_2SiO_4–X. Component X is a phase which gives a liquid containing a high proportion of tetrahedral coordination sites. Curve *a–b* shows the liquidus–solidus relations in the binary system: Ni_2SiO_4–Mg_2SiO_4. Curve *c-d* shows how the relative melting points of Ni_2SiO_4 and Mg_2SiO_4 might change in melts rich in component X with Ni_2SiO_4 having the higher melting point.

In order to account for the relative enrichment of nickel over magnesium during the fractionation of olivine from magma it was postulated (Burns and Fyfe, 1966b) that solidus–liquidus relations in the binary solid-solution series: Mg_2SiO_4–Ni_2SiO_4 are inverted in melts containing large proportions of tetrahedral (Burns and Fyfe, 1964; Cooney and Sharma, 1990; Kohn *et al.*, 1990) and trigonal bipyramidal (Galoisy and Calas, 1991) coordination sites. Such an inversion, which is illustrated in fig. 8.2, has been demonstrated in the ternary system: $Na_2Si_2O_5$–Mg_2SiO_4–Ni_2SiO_4, as well as in other systems containing the components K_2O, SiO_2 and $(Mg,Ni)_2SiO_4$ (Irvine and Kushiro, 1976; Takahashi,

1978). This unusual effect may be explained by comparing interactions of nickel and magnesium in the melting olivine with the silicate melt containing octahedral, five-fold and tetrahedral coordination sites. The Ni^{2+} ion, because of its high octahedral site preference energy, occupies preferentially the octahedral sites in a melt and enters other network-modifying tetrahedral and five-fold coordination sites only when the thermal energy is comparable with the octahedral site preference energy. Magnesium may also favour octahedral sites through a comparable lattice energy effect but the additional stabilization energy available to a transition metal ion through crystal field effects cannot be acquired by magnesium. Therefore, Mg^{2+} ions may be more evenly distributed over the network-modifying sites in the silicate melt, with the result that

$$(\Delta S_{\text{fus}})_{\text{Mg}} > (\Delta S_{\text{fus}})_{\text{Ni}} \,. \tag{8.20}$$

Because Ni^{2+} ions are forced to enter tetrahedral or five-fold coordination sites in the melt,

$$(\Delta H_{\text{fus}})_{\text{Ni}} > (\Delta H_{\text{fus}})_{\text{Mg}} \,. \tag{8.21}$$

Both of these factors lead to a lower melting point for the magnesium component by the relationship given previously

$$T_{\text{fus}} = \Delta H_{\text{fus}} / \Delta S_{\text{fus}} \,. \tag{8.4}$$

As a result, enrichment of Ni^{2+} ions is to be expected in olivine crystals that separated at an early stage from a magma.

8.5.5 Experimental studies of nickel partitioning

The mechanism of nickel enrichment suggested in §8.5.4 was subsequently confirmed in studies of the partitioning of Ni and Mg between olivine crystals and silicate liquids in the system: $(Mg,Ni)_2SiO_4$–K_2O–$6SiO_2$, initially containing 4 wt per cent NiO (Irvine and Kushiro, 1976). The plots of partition coefficients versus inverse temperature are illustrated in fig. 8.3. Values of K_D increase as the melt becomes enriched in silica and K_2O in accord with prediction (Burns and Fyfe, 1964), but the variation also bears a strong inverse relationship to temperature. Figure 8.3 shows a curvilinear relationship in the lower part of the temperature range and a marked inflection near 1,600 °C. A second curvilinear relationship is indicated at temperatures above the inflection point. Thus, above 1,650 °C the K_D values correlate with Ringwood's (1956) observation of nickel enrichment in the melt. It was inferred from the

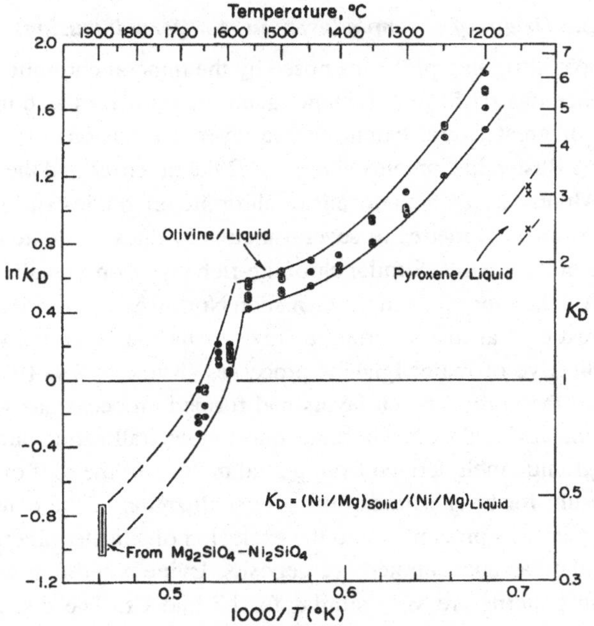

Figure 8.3 Temperature variation of the partition coefficient of Ni between Mg^{2+}–Ni^{2+} olivines and melts in the system: $(Mg,Ni)_2SiO_4$–$K_2O.6SiO_2$ (from Irvine & Kushiro, 1976). Below 1575 °C, Ni^{2+} ions are strongly partitioned into the olivine crystals as the melt becomes more siliceous.

inflection in the plotted K_D values in fig. 8.3 that a change in the structure of the melt occurs below 1,575 °C (Irvine and Kushiro, 1976).

Irvine (1974) proposed a slight modification of the equilibrium represented by eq. (8.14) as follows

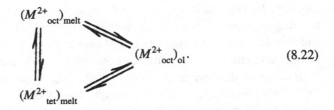

$$\begin{array}{ccc} (M^{2+}{}_{oct})_{melt} & \rightleftharpoons & \\ \updownarrow & & (M^{2+}{}_{oct})_{ol}. \\ (M^{2+}{}_{tet})_{melt} & \rightleftharpoons & \end{array} \qquad (8.22)$$

As the melt becomes more siliceous, the equilibria for Ni^{2+} shift strongly in favour of the crystallizing olivine phase due to decreased CFSE's attained in more highly polymerized silicate liquids. Apparently, the proportion of tetrahedral sites in the melt rises with increasing $K_2O + SiO_2$ components.

8.5.6 Origin of chromite layers in stratiform intrusions

One of the most intriguing problems posed by the mineral chromite, $FeCr_2O_4$, is its occurrence in stratiform ultramafic–gabbroic intrusions as thin, concentrated layers of great lateral extent. These layered intrusions are especially common in the Bushveld Complex and Great Dyke in Africa and the Stillwater Complex in Montana, in which dozens of chromite ore bodies variously measuring from a few centimetres to several metres in thickness extend for distances of tens of kilometres. Similar chromite-rich layers on a smaller scale are found in the Muskox intrusion in the Canadian Northwest Territories. The layers hold interest both as major chromite ore deposits and as a remarkable phenomenon indicative of major igneous processes (Irvine, 1974. 1975). Irvine postulated that the chromite-rich layers had formed on occasions when their parental magma deviated from a normal course of crystallization due to contamination by granitic melt derived from partial melting of the roof of the intrusion. As a result, for brief periods during crystallization of the contaminated magma, chromite was precipitated to the exclusion of silicate minerals and so could accumulate as concentrated ore deposits. Irvine noted that whole-rock differentiation patterns are very similar for Ni and Cr. The distribution of chromite shows a cyclic variation similar to that of Ni, supporting the suggestion that disseminated chromite was fractionated with the olivine in dunite sequences in the intrusion. The similar petrological behaviour of Ni and Cr in some way reflects their free energy characteristics in the magma, an inference supported by the fact that both elements have exceptionally high octahedral site preference energies and therefore probably came from the same type of site in the silicate melt. It was suggested that increased polymerization of the magma caused by the joint enrichment of silica and alkalis and the decrease of liquidus temperature caused by fractional crystallization of the olivine and chromite acted to reduce the number of octahedral sites in the melt that were energetically favourable for occupancy. Thus, the remaining Cr^{3+} and Ni^{2+} in the magma, along with other octahedrally coordinated cations such as Mg^{2+} and Fe^{2+} were forced to enter tetrahedral sites in the melt or to transfer to octahedral sites in the crystals of olivine or chromite according to eq. (8.22).

Because of the very large octahedral site preference energies of Ni^{2+} and Cr^{3+}, their partitioning would be strongly biased in favour of the crystals. Hence, equilibria in eq. (8.22) were increasingly shifted to the right for Ni^{2+} and Cr^{3+}, with the result that these cations were preferentially expelled from the magma at similar rates. The Ni^{2+} ions entered the olivine structure whereas the Cr^{3+} effectively controlled the precipitation of chromite. Irvine (1974, 1975) demonstrated experimentally that the fractionation path of a silicate melt initially crystallizing olivine and chromite can be altered through contamination by siliceous material so that the melt subsequently precipitates first chromite and then orthopyroxene, giving the rock sequence peridotite–chromitite–orthopyroxenite observed in lower parts of of the cyclic units containing the chromite-rich layers.

In chromite-bearing layered intrusions such as the Stillwater, Muskox, Great Dyke, and Bushveld complexes, the crystallizing minerals accumulate entirely on their floors and the roof contact is continuously exposed to the high temperatures of the basic magma. Consequently, any sialic rocks along the roof, such as pelitic schists or granite in country rock, are subject to melting, yielding a granitic liquid that tends to float on top of the basic magma and to remain separate because of its low density and high viscosity. The basic magma is contaminated by mixing with large amounts of salic melts. Subsequent events could be very complicated but in general the contamination should act to lower the actual temperature of the magma, reduce its liquidus temperature, and cause chromite to crystallize alone. The contamination mechanism proposed by Irvine (1974, 1975) is potentially a very powerful one as it is a possible means for producing not only chromitite layers, but also magnetite, sulfide and other magmatic ore deposits.

8.6 Transition element metallogenesis and plate tectonics

Several geochemical and geophysical properties of the Earth's interior have been elegantly explained by the theory of plate tectonics. For example, the concept of lithospheric plate subduction has been invoked to account for the strong space–time relationships between metallogenic provinces and ore genesis (e.g., Sillitoe, 1972, 1973; Sawkins, 1990;). The metallogenesis patterns of certain transition elements, notably Cu, Mn, Co, Ni and Cr, at convergent plate margins reflect fractionation processes during partial fusion of the Mantle and of subducted oceanic lithosphere, which may be explained by crystal field theory (Burns, 1977). Applications to the enrichments of Ni and Cr in certain lherzolites, the mineralizations of Cu and Mn in porphry copper deposits, the occurrence of Cr in olivines included in diamonds from kimberlites and the nickel concentrations in alpine peridotites and laterite deposits are described below.

8.6.1 Enrichments of Ni and Cr in lherzolites

The observation that Cr and Ni are enriched in earliest minerals crystallizing from magma such as chromite and forsteritic olivines discussed in earlier sections (§8.5) has led to the widespread use of the abundances of these particular elements to establish differentiated rock sequences (e.g., Hart and Davis, 1978; Shimizu and Allegre, 1978). Conversely, Ni and Cr abundances may serve as an index of partial melting. For example, Carter (1970) showed that the concentrations of Ni and Cr in certain lherzolites are consistent with the hypothesis that these xenoliths represent the refractory residuum resulting from partial fusion of the Mantle (Burns, 1973).

A variety of field, petrographic and chemical data indicate that certain lherzolites are residuals of partial fusion, while some peridotite xenoliths represent the resultant liquid. Carter (1970) demonstrated that the Mg/(Mg + Fe) ratio of the olivines in lherzolites is an index of the degree of partial melting, with rising magnesium content being indicative of increased partial melting of the lherzolite. Nickel and chromium are reluctant to enter the molten phase because they are stabilized in the mineral structures. During partial melting of the minerals in a lherzolite, the cations Ni^{2+} and Cr^{3+} encounter the equilibrium

$$(M^{n+}{}_{\mathrm{oct}})_{\mathrm{crystal}} \underset{\text{crystallization}}{\overset{\text{fusion}}{\rightleftharpoons}} (M^{n+}{}_{\text{6-fold}} + M^{n+}{}_{\text{5-fold}} + M^{n+}{}_{\text{4-fold}})_{\mathrm{melt}} \qquad (8.23)$$

or

$$(M^{2+}{}_{\mathrm{oct}})_{\mathrm{olivine}} \rightleftharpoons (M^{2+}{}_{\mathrm{oct}})_{\mathrm{melt}}; \quad (M^{2+}{}_{\mathrm{oct}})_{\mathrm{olivine}} \rightleftharpoons (M^{2+}{}_{\mathrm{tet}})_{\mathrm{melt}}; \quad (M^{2+}{}_{\mathrm{oct}})_{\mathrm{melt}} \rightleftharpoons (M^{2+}{}_{\mathrm{tet}})_{\mathrm{melt}} \qquad (8.24)$$

which correspond to eqs (8.14) and (8.22), respectively. Since Ni^{2+} and Cr^{3+} have particularly high octahedral site preference energies, the equilibria in eqs (8.23) and (8.24) lie heavily to the left. Nickel and chromium, therefore, remain behind in the refractory residual minerals of lherzolite. Transition elements such as Mn^{2+}, Fe^{3+} and Ti^{4+}, which acquire no CFSE and have zero octahedral site preference energies, are more amenable to the tetrahedral and octahedral sites of a melt. These ions are therefore depleted in the residual lherzolite minerals. Note that the equilibria represented by eqs (8.23) and (8.24) apply to cations coordinated to oxygen. Different fractionation patterns may occur for some elements (e.g., Cu and Ni) in magmas containing a molten sulphide phase (Fleet *et al.*, 1981). In addition, the simple model does not take into account the effects of variable oxygen fugacities. During magmatic crystallization, high oxygen fugacities promote the formation of magnetite, resulting in an impoverishment of such transition metals as Fe, Ti and V. Partial fusion under low oxygen fugacities would accentuate the enrichment of Fe and Ti into the early melts.

8.6.2 Cu and Mn mineralization in porphyry copper systems

One of the most spectacular metallogenic relationships is that involving porphyry copper deposits, consisting of chalcopyrite, covellite and metallic Cu disseminated in calc–alkaline igneous rocks, which occur consistently over active and recently active subduction zones throughout the circum-Pacific, Caribbean and Alpine orogenic belts (Sillitoe, 1972; Sawkins, 1990). Sillitoe (1973) has convincingly depicted the development and configuration of a typical porphyry copper system by piecing together various field examples which expose different levels of such systems. Most interesting is the recognition that porphyry copper deposits span the boundary between plutonic and volcanic environments by being situated in and about the cupolas and stocks of calc–alkaline plutons underlying stratovolcanoes. A significant feature is that Cu is not the only mineralization common to a porphyry copper system. Sillitoe (1973) clearly demonstrated that the porphyry copper system is typically marked by Cu, Mo, Pb, Zn and precious metal veins in the basement of a stratovolcano, and by Cu, Fe, gypsum, and native S in the sublimates at high temperature fumeroles. In view of the global distributions of Cu and Mn ore deposits, a close space–time relationship between porphyry copper deposits, manganese ore deposits, and subduction zones has been demonstrated (Thonis and Burns, 1975), leading to the suggestion that volcanogenic processes generating porphyry copper systems also produce manganese ore deposits. Copper, itself, need not be present, however, for the porphyry copper system to exist and bear manganese.

Explanations for the correlations between porphyry copper deposits, calc–alkaline magmatism, and subduction zones stem from the belief that subducted oceanic lithosphere undergoes frictional heating along a Benioff Zone (Ringwood, 1969), leading to the partial fusion of oceanic crust and consequent rise of calc–alkaline or andesitic magmas. Other processes have occurred, including partial melting of the oceanic crust with subsequent modification by the Upper Mantle as well as partial melting of the Upper Mantle triggered by volatiles rising from the subducted lithosphere. All such processes involve the oceanic lithosphere, an important observation because the oceanic crust and Upper Mantle are enriched in transition elements relative to the continental crust (Appendix 1). High abundances of Mn, Cu, Ni, Fe, and Co also occur in the pelagic sediments (§8.7.4) that overlie pillow basalts containing significant concentrations of the minor elements Mn, Cu and Ni.

During partial fusion of the oceanic lithosphere, Mn and Cu are among the first transition elements to enter the calc–alkaline magma either because of their negligible CFSE's or due to intolerance for silicate mineral structures. Subsequently, a partitioning of the Cu and Mn from magma to a fluid phase containing water, HCl, HF, CO_2 and other volatiles may occur, and the metals are probably transported hydrothermally through the porphyry copper system as Mn(II) and Cu(I) or Cu(II) halide or bisulfide complexes. At lower temperatures in aqueous environments, hydrolysis occurs as a response to cooling, reaction with wall rock, and interactions with the lower-temperature, less saline groundwater system. This leads to the formation of extremely insoluble sulphide, oxide and metallic Cu minerals found in porphyry copper deposits which subsequently undergo supergene alteration (Blain and Andrew, 1977). Manganese is transported to higher levels in the system and may escape out of the volcanic pile into the submarine environment to be deposited as Mn^{2+}-bearing carbonates and silicates or oxidized to very insoluble Mn(IV) oxide minerals in the eugeosynclinal environment.

8.6.3 Nickel in alpine peridotites

As indicated in the previous section, oceanic crust and overlying pelagic sediments are enriched in several transition elements, notably Mn, Cu, Ni, Co and Fe. At active oceanic ridges, hydrothermal fluids rich in Mn and Fe, and to a lesser extent Cu and Ni, pour into seawater, forming chimney sulfide deposits and metalliferous sediments at submarine spreading centres. During seafloor spreading, these metalliferous deposits and underlying pillow basalts become progressively buried by pelagic sediments. Another prime environment for the accumulation of transition metals, notably Mn, Fe, Ni, Cu and Co, is ferromanganese oxide encrustations and nodules that form at the seawater–sediment and seawater–basalt interfaces (§8.7.3). The sources of these transition elements may be biological and volcanogenic. It is this assemblage – nodules, pelagic sediments, metalliferous deposits and oceanic basalts enriched in Mn, Fe, Cu and Ni – that undergoes partial fusion along subduction zones at convergent plate margins producing calc–alkaline magmatism described earlier (§8.6.2). A question that arises is: if the Mn and Cu emerge through a porphyry copper system into the crust, what happens to the nickel?

Crystal field theory predicts that Ni^{2+}, because of its high CFSE in octahedral coordination in minerals, should remain behind in refractory phases resulting from partial fusion of the oceanic lithosphere and become enriched in the Upper Mantle. It is possible, however, that the nickel reappears in alpine peridotites or obducted ophiolite suites.

Significant features of the ultramafic rocks in alpine peridotites and ophiolites are that they are associated with orogenic regions near plate margins, and that the forsteritic olivines contain appreciable nickel concentrations, typically 0.2 wt per cent Ni. Frequently, the peridotites are serpentinized and in equatorial regions, the ultramafic rocks are weathered to laterite deposits. Many of these laterite deposits (e.g., New Caledonia, Cuba, Solomon Islands, Philippine Islands, Indonesia) are important nickel ore deposits containing as much as 3 wt per cent Ni in clay silicate assemblages such as garnierite which replace decomposed olivine in weathered ultramafic rocks (Sawkins, 1990). The enrichment of nickel in these garnierite deposits is a direct consequence of the high CFSE of Ni^{2+}; it is very difficult for permeating aqueous solutions to leach Ni^{2+} ions from octahedral sites in the oxide mineral structures (§8.7.1).

8.6.4 Chromium in olivine inclusions in diamond

Diamonds from kimberlites contain minute inclusions of a number of minerals, of which forsteritic olivines Fo_{92-95}, are the most abundant (Meyer and Boyd, 1972). Significant amounts of chromium (up to 0.4 oxide total, expressed as Cr_2O_3) occur in the olivine inclusions. Several crystal chemical arguments suggest that divalent Cr, and not Cr^{3+} ions, replace Mg in the olivine structure (Burns, 1975a,b). Certain lunar olivines, which crystallized under reducing conditions on the Moon, also contain significantly high Cr contents (Bell, 1970; Smith, 1971). This has led to the widespread belief that the presence of Cr^{2+} ions in olivines included in diamonds are indicative of low oxygen fugacities in the Earth's Mantle. An alternative explanation is that low oxidation states of certain transition metal ions, including Cr^{2+}, are formed and may be stabilized at very high pressures and temperatures in the Mantle (Burns, 1976a; §9.8.4). The occurrence of Cr^{2+} ions is further enhanced by the Jahn–Teller stabilization they may acquire in distorted sites in Mantle minerals, including the olivine and pyroxene M1 and M2 sites (Burns, 1975a,b).

Chromium(II)-bearing minerals are expected to be highly susceptible to pressure-released oxidation of Cr^{2+} to Cr^{3+} ions under the relatively lower pressures and more oxidizing conditions existing in the Earth's Crust. However, observations that high confining pressures are maintained in crystals included in diamonds may account for the retention of the Cr(II) oxidation state in the olivine inclusions brought to the Earth's surface.

The source of diamond-bearing kimberlites is uncertain but the stability field of diamond implies a deep-seated origin. Kimberlites are generally believed to have resulted from metamorphism or melting at the deep end of a subducted oceanic plate, the source material being the oceanic lithosphere with entrained hydrous and carbonaceous material (Gurney, 1990). Emplacement probably took place under continents parallel to fold mountain ranges, as inferred from the African kimberlite fields and their relationship to the Cape Fold Range. Such an origin conforms with the very high CFSE of octahedrally coordinated Cr^{3+} ions, which renders this cation one of the last to be remobilized during partial fusion of the oceanic lithosphere. However, at the very high pressure and temperature in the deep Upper Mantle, the formation and stability of divalent Cr^{2+} in the olivine structure relative to Cr^{3+} may become a dominant factor governing the geochemistry of chromium in the Upper Mantle. On the other hand, in the Lower Mantle, Cr^{2+} ions may be stabilized in distorted Mg sites of the $MgSiO_3$ perovskite phase (§9.8.1).

8.7 Sedimentary geochemistry of the transition elements

Several interrelated chemical processes are involved in sedimentary geochemistry. During weathering, aqueous solutions act on minerals in igneous and metamorphic rocks. Soluble ions are leached, and silicate minerals are hydrolysed to insoluble clay silicates and hydrated oxides of Al, Fe and Mn. Leached ions are transported and precipitated in environments when solubility products are exceeded. Throughout the weathering, leaching and transportation stages, oxidation states of ions are susceptible to change, especially for those transition metals which show a range of valences. The aqueous phase dominates sedimentary geochemical processes, and in order to account for the behaviour of transition metal ions, relative stabilities of hydrated and complex ions in solution (enthaphies of hydration) and crystal structures (lattice energies) need to be assessed, eq. (8.9). Some trends have been noted from thermodynamic data, which suggest that oxide phases with highest standard enthalpies of formation (ΔH_f^0) are the least vulnerable to dissolution (Brimhall, 1987). Thus, cations existing in high oxidation states have large ΔH_f^0 values, including TiO_2, MnO_2 and Fe_2O_3 (relative to FeO), and are least soluble. In addition to such thermodynamic considerations, factors involving the mechanism and rates of dissolution and aqueous oxidation reactions have to be considered.

8.7.1 Leaching of ions and break-down of silicates

During leaching of a mineral an ion is removed from a site in a crystal structure to an aqueous phase. Most transition metal ions are present in six-coordinated sites in silicate minerals and exist in aqueous solutions as hexahydrated ions, $[M(H_2O)_6]^{n+}$. The crystal field splittings summarized in table 2.5 indicate that CFSE's of ions in aqueous solutions and oxide structures are comparable,

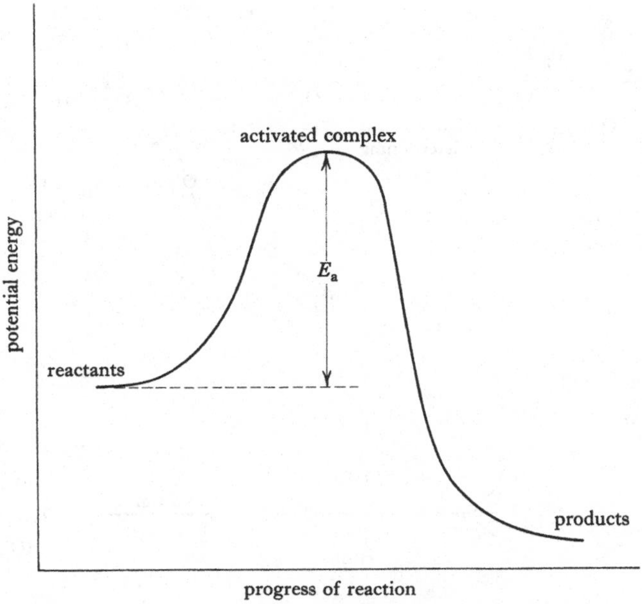

Figure 8.4 Activation energy barrier to be surmounted by a system in a substitution reaction.

unless a cation attains increased or decreased stability in a distorted coordination site, which may be forced by the crystal structure. As a result the extent of leaching depends on the ease with which ions are removed from a crystal structure. Kinetic and mechanistic factors must be considered when deciding whether an ion is labile to leaching. Crystal field theory can provide insights into substitution reactions involving transition metal ions (Basolo and Pearson, 1967; Huheey, 1983).

Ions on surfaces of crystals, and along fractures and cleavage planes, show the greatest susceptibility to substitution reactions and removal in solution due to the large number of broken bonds existing at the surface. Leaching of ions from a crystal structure depend on whether water molecules or other reacting species can penetrate into the environment about the ion. According to the transition state theory of reaction rates (Glasstone *et al.*, 1941; Murphy and Helgeson, 1987), the rate at which a substitution reaction proceeds depends on the properties of the transition state or activated complex. This activated species is the configuration of the system at its highest potential energy when the reaction is passing over the activation energy barrier depicted in fig. 8.4. The rate at which the reaction takes place depends on the energy difference or

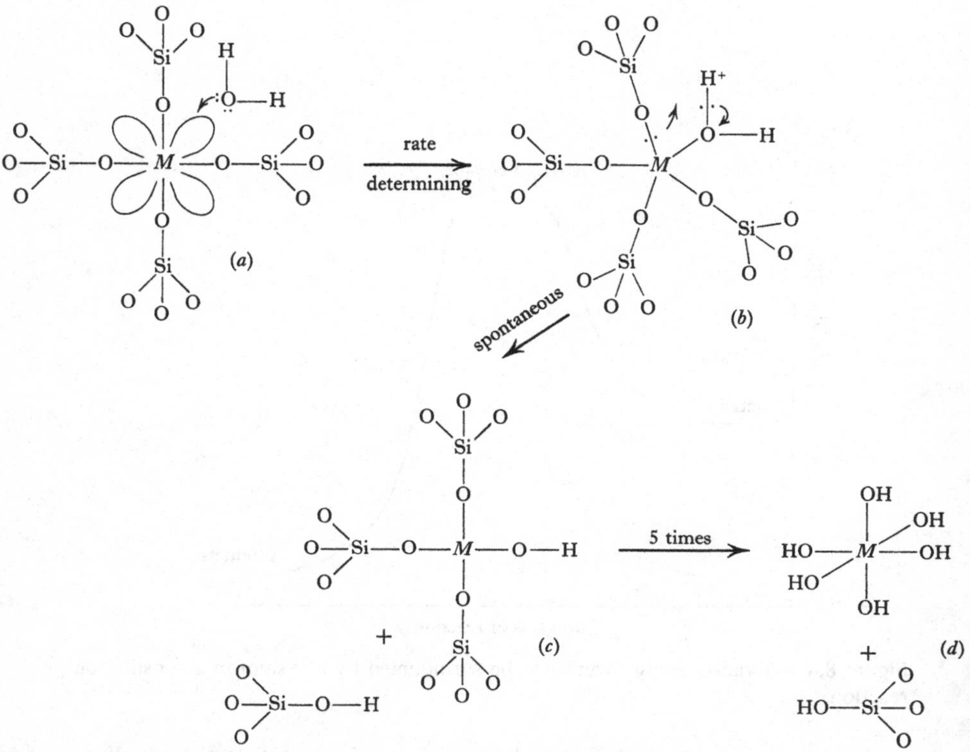

Figure 8.5 A mechanism for the dissolution and hydrolysis of a ferromagnesian silicate. The figure is a two-dimensional representation of the reaction described in the text (§8.7.1). Oxygens of two additional silicate groups lie above and below the plane.

activation energy, E_a, between the reacting species and transition state. The smaller the activation energy, the faster the reaction proceeds, neglecting entropy of activation terms.

Substitution reactions involving transition metal ions in silicate minerals probably bear resemblances to substitution nucleophilic reactions of octahedral complexes (Basolo and Pearson, 1967), which may be either unimolecular (S_N1) or bimolecular (S_N2). In S_N1 reactions, one of the ligands coordinated to the central cation removes itself along one of the cartesian axes leaving an activated complex with the configuration of a five-coordinated square pyramid onto which another ligand may become attached to restore six-fold coordination to the octahedrally coordinated ion. Such S_N1 reactions probably take place at the outermost surface of a crystal face where broken bonds occur and five-coordinated cation species may already exist that are receptive to incoming ligands (i.e. H_2O) molecules.

Reactions via the S_N2 mechanism, on the other hand, may influence substitution reactions along dislocations and stacking faults in crystal interiors . The reactions proceed through the transient formation of a seven-coordinated activated complex possessing the configuration of a pentagonal bipyramid. Such a transition state is formed most readily by the substituting ligand entering the octahedral transition metal cluster along a plane in which one of the low-energy t_{2g} orbitals is empty. An analogous reaction involving hydrolysis of a transition metal ion in a silicate may be represented in two dimensions by fig. 8.5. A water molecule approaches the central transition metal *M* along a vacant t_{2g} orbital (fig. 8.5*a*), forming a seven-coordinated transition state (fig. 8.5*b*). The H_2O molecule forms the seventh group and is bonded through the lone–pair electrons on the oxygen atom. Formation of the transition state is the rate-determining step in the reaction. The transition state disproportionates spontaneously forming a metal hydroxysilicate and hydroxysilicate residue (fig. 8.5*c*). Repetition of the process ultimately yields a metal hydroxide or hydrated oxide plus a hydroxysilicate or hydrated silica residue (fig. 8.5*d*).

During S_N2 reactions, transition metal ions having $3d^1$ and $3d^2$ electronic configurations, and possessing at least one empty t_{2g} orbital, might be expected to react more rapidly than a cation possessing a $3d^3$ configuration, in which all t_{2g} orbitals are singly occupied. In order for substitution reactions to take place in ions with more than three 3*d* electrons, energy is required to pair electrons in t_{2g} orbitals so as to leave one t_{2g} orbital vacant to initiate substitution. Therefore, ions with high-spin $3d^4$, $3d^5$ and $3d^6$ configurations are predicted to show intermediate reaction rates and those with $3d^7$, $3d^8$ and $3d^9$ configurations to display very slow reaction rates, since no t_{2g} orbital can be vacated entirely in these ions. Ions with low-spin $3d^6$ configurations would be expected to show very low rates of substitution.

These arguments may be formulated more rigorously by comparing CFSE's for ions with different $3d^n$ configurations in octahedral and square pyramidal or pentagonal bipyramidal coordination sites, corresponding to S_N1 or S_N2 reactions, respectively. The CFSE data, derived from the relative energy levels of 3*d* orbitals in different coordination symmetries (table 2.4), are summarized in table 8.2. The difference, ΔE_a, between the CFSE of an ion in a six-coordinated site prior to a substitution reaction and the value for a five- or seven-coordinated site in a transient activated complex formed during the reaction may be regarded as a contribution to the total activation energy, E_a. A large positive value of ΔE_a results in a slow reaction, since the rate constant for a bimolecular substitution reaction, k_2, is related to the activation energy by

$$k_2 = A \exp(-E_a/RT) . \qquad (8.25)$$

Table 8.2. *Crystal field activation energies* for substitution reactions*

Number of 3d electrons		Cation			Octahedral CFSE (6-fold)	Square pyramidal CFSE (5-fold)	Pentagonal bipyramidal CFSE (7-fold)	Activation energies, ΔE_a: S_N1 reaction (6 → 5)	Activation energies, ΔE_a: S_N2 reaction (6 → 7)
	0	Ca^{2+}	Sc^{3+}	Ti^{4+}	0	0	0	0	0
	1		Ti^{3+}	V^{4+}	0.4	0.457	0.528	−0.057	−0.128
	2		V^{3+}		0.8	0.914	1.056	−0.114	−0.256
	3		Cr^{3+}	Mn^{4+}	1.2	1.000	0.774	+0.200	+0.426
	4	Cr^{2+}	Mn^{3+}		0.6	0.914	0.492	−0.314	+0.108
hs	5	Mn^{2+}	Fe^{3+}		0	0	0	0	0
ls	5	Mn^{2+}	Fe^{3+}		2.0	1.914	1.830	+0.086	+0.170
hs	6	Fe^{2+}			0.4	0.457	0.528	−0.057	−0.128
ls	6		Co^{3+}		2.4	2.000	1.548	+0.400	+0.852
hs	7	Co^{2+}	Ni^{3+}		0.8	0.914	1.056	−0.114	−0.256
ls	7	Co^{2+}	Ni^{3+}		1.8	2.000	1.548	−0.200	+0.252
	8	Ni^{2+}			1.2	1.000	0.774	+0.200	+0.426
	9	Cu^{2+}			0.6	0.914	0.492	−0.314	+0.108
	10	Zn^{2+}	Ga^{3+}	Ge^{4+}	0	0	0	0	0

* Expressed as fractions of the octahedral crystal field splitting parameter, Δ_o.
hs and ls are high-spin and low-spin configurations, respectively.

The immediate conclusion from inspection of table 8.2 is that transition metal ions with $3d^3$, $3d^8$ and low-spin $3d^6$ configurations are the ones most adversely affected, so far as CFSE is concerned, by forming a seven-coordinated or five-coordinated transition state. Activation energies are predicted to be high in reactions involving the Cr^{3+}, Ni^{2+} and Co^{3+} ions, with the result that rates of substitution reactions involving these cations will be slow.

A large negative value of ΔE_a implies that a cation gains stability by forming an activated complex with pentagonal bipyramidal or square pyramidal symmetry. On this basis, cations such as Fe^{2+}, Co^{2+}, Ti^{3+}, and V^{3+}, as well as the $3d^5$ cations with zero CFSE, are predicted to show rapid reaction rates of substitution or leaching from a crystal structure.

It is assumed in the calculations of the ΔE_a values in table 8.2 that cations initially occupy regular octahedral sites and that electrons remain in the same orbitals that they occupied in the octahedral site while passing over to the seven-coordinated or five-coordinated transition state.

Analogous arguments may be applied to substitution reactions of transition metal ions in silicates. The values of ΔE_a are probably influenced by the distorted coordination sites. Thus, some cations stabilized in low-symmetry environments, particularly Fe^{2+}, Mn^{3+} and Cu^{2+}, may show slower reaction rates during dissolution processes due to the reluctance of the cations to leave silicates accommodating them in distorted coordination sites (e.g., Fe^{2+} in pyroxene M2 sites; Mn^{3+} in epidote M3 or andalusite M1 sites).

Few experimental studies have been made of the leaching behaviour of transition metal ions in silicates. The data that are available (Vinogradov, 1959; Hawkins and Roy, 1963; Schorin, 1983; Casey and Westrich, 1991) are in excellent agreement with predictions of crystal field theory; nickel, chromium and cobalt are among the group of elements that are most resistant to leaching, whereas manganese, copper, zinc and gallium are readily removed during weathering of basaltic rocks. The enrichments of nickel, cobalt and chromium in weathered igneous and ultramafic rocks and derived laterite deposits may be due, in part, to the resistance of these cations to substitution reactions and subsequent leaching, and partly to their high stabilities in octahedral sites in crystal structures of residual minerals discussed in §8.6.1 and §8.6.3. Note, however, that it is impossible to discriminate between kinetic and thermodynamic factors when examining final assemblages.

8.7.2 Oxidation of transition metal ions in sedimentary processes

Under the relatively oxidizing environments existing near the Earth's surface provided by aerated aqueous solutions in contact with the atmosphere, several

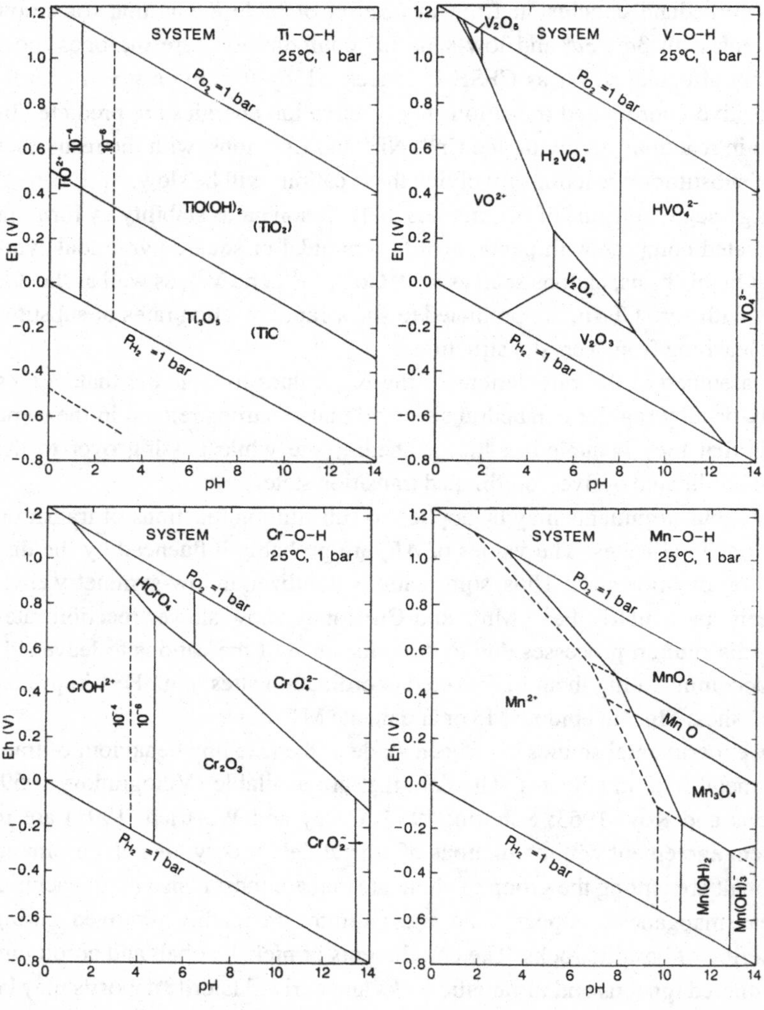

Figure 8.6 Eh–pH diagrams for elements of the first transition series (from Brookins, 1990). The stability fields are calculated assuming that each dissolved transition metal has an activity of 10^{-6} (*continued on facing page*).

of the transition elements can occur in higher oxidation states than the simple cations originally present in igneous mineral assemblages. These oxidation states are influenced by the pH, dissolved CO_3^{2-} concentrations and sulphur species present in the hydrosphere. The stability fields for different oxidation states of each transition element are portrayed in Eh–pH diagrams (Brookins, 1990), selected examples of which are illustrated in fig. 8.6. General features

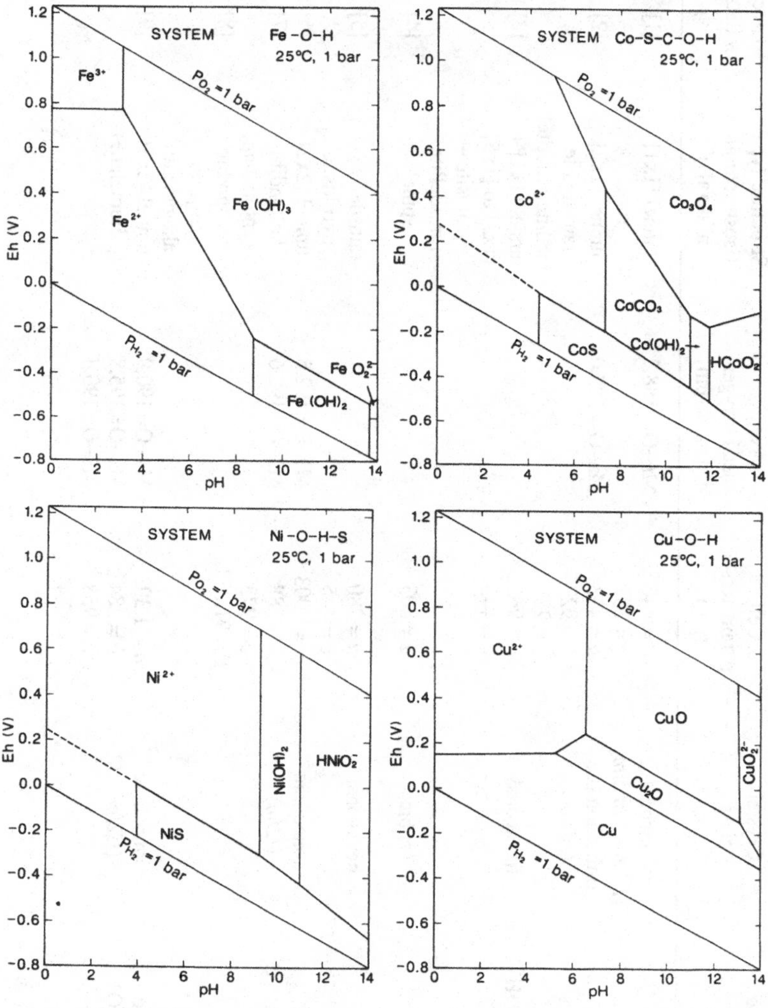

of these Eh–pH diagrams for oxidizing aqueous environments at 298 K and normal atmospheric pressure include

(1) The Ni(II) and Cu(II) oxidation states are stable throughout the entire pH range;

(2) dissolved Co(II) and Mn(II) can exist only in acidic and near-neutral pH solutions, while dissolved Cr(III) is restricted to acidic solutions;

(3) the Co(III), Mn(IV) and Cr(VI) oxidation states become stabilized in more alkaline solutions;

(4) the Fe(II) state is thermodynamically stable only in acidic solutions, with either dissolved Fe(III) ions becoming stabilized under very strongly acidic

Table 8.3. *Crystal chemical data for Mn(IV) oxides*

Mineral (phase)	Approximate formula	Crystal class	Cell parameters (pm)	Average metal–oxygen distances (pm)	Structure-type (isostructural minerals)	Sources of data
pyrolusite	β-MnO_2	tetragonal ($P4_2/m\ 2_1/n\ 2/m$)	$a = 440$ $c = 287$	Mn–O = 188.7	tunnel [1x1] rutile	[1]
ramsdellite	MnO_2	orthorhombic (*Pbnm*)	$a = 453$ $b = 927$ $c = 287$	Mn–O = 189.0	tunnel [1x2] ramsdellite (goethite,groutite)	[2]
nsutite (γ-MnO_2)	$(Mn^{4+}Mn^{3+})(O,OH)_2$	hexagonal	$a = 965$ $c = 443$		tunnel [1x1] + [1x2] to [1x5] pyrolusite + ramsdellite	[3]
MnO_2 spinel (synthetic)	λ-MnO_2	cubic	$a = 803$		spinel	[4]
ε-MnO_2 (synthetic)	ε-MnO_2	hexagonal	$a = 280$ $c = 445$		ramsdellite + pyrolusite	[5]
hollandite (α-MnO_2)	$(Ba,K,Pb,Na)_{1-2}$ $Mn_8(O,OH)_{16}\cdot xH_2O$	monoclinic (*I2/m*)	$a = 1,003$ $b = 288$ $c = 973$ $\beta = 91.03°$	M1–O= 192.5 M2–O= 192.6	tunnel [2x2] hollandite (cryptomelane coronodite manjiroite akaganéite)	[6] [7] [8]
romanechite	(Ba,K,Mn,Co) $Mn_5O_{10}\cdot xH_2O$	monoclinic (*C2/m*)	$a = 1,393$ $b = 284$ $c = 968$ $\beta = 92.4°$	M1–O= 190.9 M2–O= 195.8 M3–O= 190.4	tunnel [2x3] (psilomelane)	[9]

todorokite	$(Ca,Na,K,Ba)_{<1.0}$ $(Mn,Mg)_6O_{12}\cdot xH_2O$	monoclinic (*P2/m*)	$a = 976$ $b = 284.1$ $c = 968$ $\beta = 94.1°$	M1–O= 180.0 M2–O= 194.0 M3–O= 187.0 M4 –O= 196.0	tunnel [3x3] todorokite	[10]
chalcophanite	$ZnMn_3O_7.3H_2O$	trigonal ($R\bar{3}$)	$a = 753$ $c = 2{,}079$	Mn–O= 190.6 Zn–O= 210.4	layered [2x*n*] chalcophanite	[11]
birnessite	(Na,K,Ca,Mg,Mn^{2+}) $Mn_7O_{14}\cdot xH_2O$	monoclinic (*C2/m*)	$a = 505$ $b = 285$ $c = 705$ $\beta = 96.65°$	Mn–O= 194.0 Mg–O= 213.0	layered [2x*n*] chalcophanite	[12] [13]
vernadite (δ-MnO_2)	(Mn,Co,Ca) $(OH,O)_2\cdot xH_2O$	hexagonal		Mn–O= 194.0	layered [2x*n*]	[13]
lithiophorite	$Al_{14}Li_6(OH)_{42}$ $Mn^{2+}{}_3Mn^{4+}{}_{18}O_{42}$	trigonal ($P3_1$)	$a = 1{,}337$ $c = 2{,}820$		layered	[14]
buserite	Na Mn oxide hydrate	hexagonal	$a = 841$ $c = 1{,}001$	Mn–O= 194.0 M^{2+}–O= 203.0	layered [3x*n*]	[13]
asbolane	Co–Ni Mn oxide	hexagonal	$a = 282;304$ $c = 960$		mixed layer	[15 [16,17]

Sources of data : [1] Baur (1976); [2] Byström (1949); [3] Turner & Buseck (1983); [4] Hunter (1981); [5] de Wolff, Visser, Giovanoli & Brütsch (1978); [6] Turner & Buseck (1979); [7] Post, von Dreele & Buseck (1982); [8] Post & Bish (1989); [9] Turner & Post (1988); [10] Post & Bish (1988); [11] Post & Appleman (1988); [12] Post & Veblen (1990); [13] Stouff & Boulégue (1988); [14] Pauling & Kamb (1982); [15] Manceau, Llorca & Calas, (1987); [16] Chukhrov, Gorshkov, Vitovskaya, Drits, Sivtsov & Rudnitskaya (1980); [17] Chukhrov, Gorshkov, Drits, Shterenberg & Sakharov (1983).

conditions or insoluble Fe(III) oxide phases being precipitated under less acidic and alkaline conditions; and

(5) titanium occurs only in the Ti(IV) state, while the V(IV) state is stable up to pH 3.5 with V(V) anionic species becoming thermodynamically stable in higher pH solutions.

Other noteworthy features of the Eh–pH diagrams (fig. 8.6) are the preponderance of transition metal oxidation states with [Ar]$3d^0$ inert gas configurations [Ti(IV), V(V), Cr(VI)], and the stabilities of the $3d^3$ [Mn(IV)], $3d^5$ [Fe(III)] and low-spin $3d^6$ [Co(III)] electronic configurations in aerated aqueous environments possessing high Eh and/or neutral and alkaline pH values. Particularly important in the sedimentary environment are the thermodynamic stabilities and low solubilities of high valence oxide minerals consisting of polymorphs of TiO_2, MnO_2 and Fe_2O_3 or FeOOH (Brimhall, 1987). These phases host many minor elements, either through atomic substitution in their crystal structures or by adsorption on surfaces of microcrystalline particles. These factors lead to the strong enrichments of Co, Ni, Cu and Zn in Mn(IV) and Fe(III) oxides phases occurring in marine sediments.

8.7.3 Authigenic manganese oxides

A remarkable aspect of transition metal geochemistry is the enrichment of manganese during weathering. Here, the status of manganese changes from a trace element in igneous rocks to host element in Mn(IV) oxides that are often associated with Fe(III) oxide minerals. A large number of secondary oxide, hydrated oxide, and oxide hydroxide or oxyhydroxide phases of Mn and Fe are known, and many of them are listed in table 8.3. Most of these oxides have been positively identified or suggested to occur in weathered continental rocks and in marine sediments (Burns and Burns, 1981). Several of the Mn(IV) and Fe(III) oxides possess similar crystal structures, as indicated in table 8.3.

8.7.3.1 Manganese oxides with tunnel structures

The simplest, relatively stoichiometric polymorphs of MnO_2 have crystal structures based on close-packed lattices of O^{2-} ions. In the β-MnO_2 (pyrolusite), ramsdellite and γ-MnO_2 (nsutite) structures, the oxygens are approximately hexagonally close-packed. They are in cubic close-packing, however, in the synthetic λ-MnO_2 polymorph possessing the spinel structure. The different arrangements of Mn^{4+} cations in the octahedral sites distinguish the pyrolusite, ramsdellite and nsutite crystal structures.

The fundamental structural unit in Mn(IV) oxides is the $[MnO_6]$ octahedron, in which Mn^{4+} ions with their $3d^3$ electronic configuration acquire exceptionally high CFSE, perhaps exceeding 300 kJ/g ion (cf table 2.5). The $[MnO_6]$

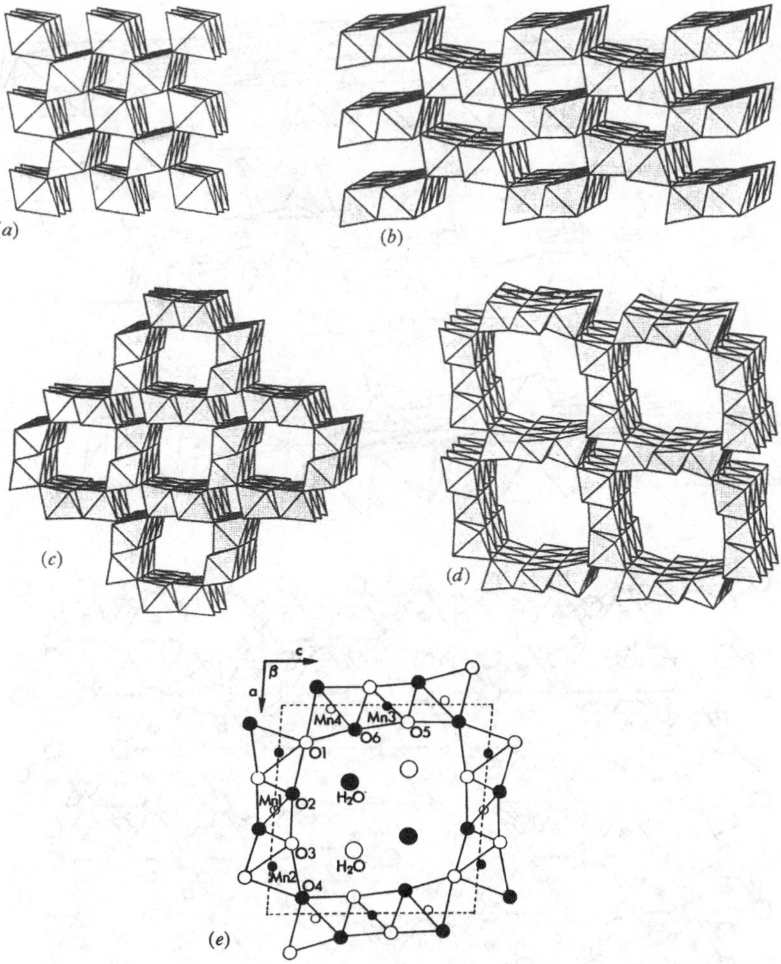

Figure 8.7 Structures of Mn(IV) oxides with tunnel structures formed by different linkages of $[MnO_6]$ octahedra. (a) Pyrolusite with single chains of edge-shared octahedra forming [1x1] tunnels; (b) ramsdellite with double chains of edge-shared octahedra forming [1x2] tunnels; (c) hollandite with [2x2] tunnels formed by a framework of double chains of edge-shared octahedra; (d) todorokite with [3x3] tunnels formed by a framework of treble chains of edge-shared octaheda; (e) a single [3x3] tunnel in the todorokite structure showing.the central M1 and M3 octahedra occupied by Mn^{4+} ions, while lower valency cations (e.g., Mn^{3+}, Mg^{2+}, Ni^{2+}, Cu^{2+}, etc.) occupy larger M2 and M4 sites. Note that H_2O molecules and large cations such as Ba^{2+}, K^+, Pb^{2+}, Ag^+ and hydrated Ca^{2+} and Na^+ are located in the [2x2] and [3x3].

octahedra are linked by corner-sharing and edge-sharing, resulting in the variety of tunnel and layer structure-types listed in table 8.3 and illustrated in figs 8.7 and 8.8. In the pyrolusite structure (isostructural with rutile, TiO_2), the

(a)

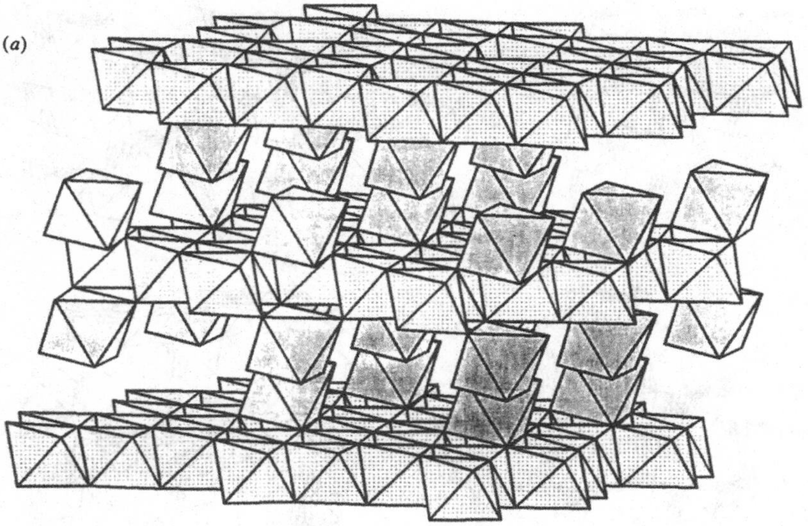

(b)

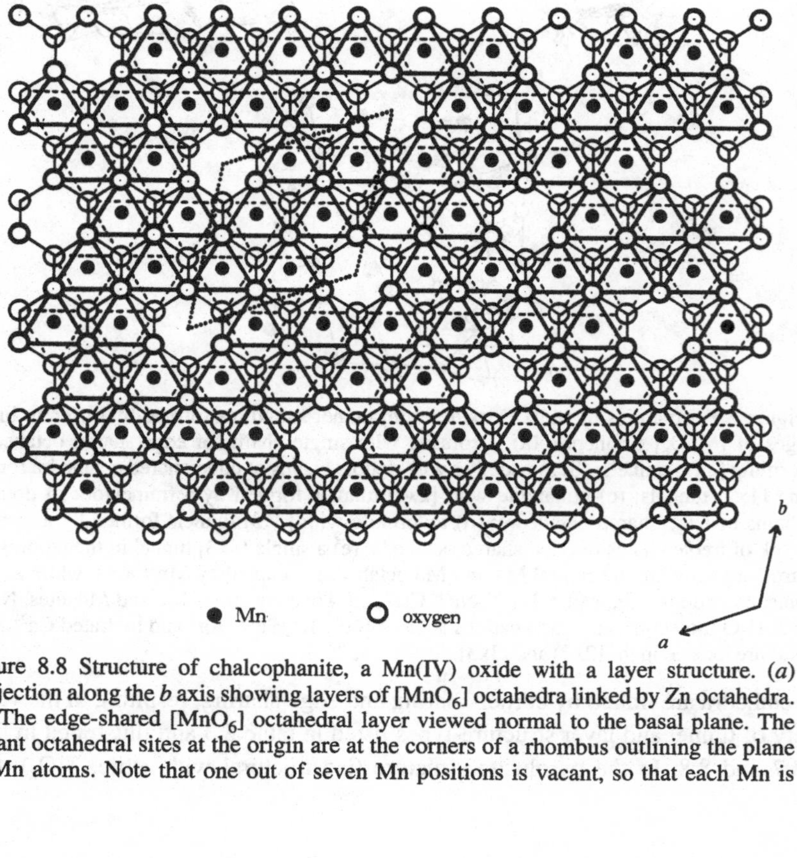

Figure 8.8 Structure of chalcophanite, a Mn(IV) oxide with a layer structure. (*a*) Projection along the *b* axis showing layers of [MnO_6] octahedra linked by Zn octahedra. (*b*) The edge-shared [MnO_6] octahedral layer viewed normal to the basal plane. The vacant octahedral sites at the origin are at the corners of a rhombus outlining the plane of Mn atoms. Note that one out of seven Mn positions is vacant, so that each Mn is

[MnO_6] octahedra share edges to form single chains running parallel to the *c* axis (fig. 8.7*a*). All [MnO_6] octahedra are equivalent and the average Mn–O distance is 188.7 pm. The unit cell dimension, $c = 287.2$ pm, represents the Mn^{4+}–Mn^{4+} interatomic distance. This distance, or multiples of it, is a common cell parameter in structures of tetravalent manganese oxides. The single chains of [MnO_6] octahedra in pyrolusite are crosslinked with neighbouring single chains by corner-sharing of oxygen atoms of adjacent octahedra to give tetragonal symmetry. In the β-MnO_2 (rutile) structure, the unoccupied octahedral sites may be considered to form single lines of vacancies parallel to the *c* axis, and these define 'tunnels' with dimensions corresponding to the width of one [MnO_6] octahedron, which are designated as [1x1] tunnels. Note, however, that in the pyrolusite structure no gaping holes exist *per se* in the close-packed oxygen lattice.

Ramsdellite contains double chains of linked [MnO_6] octahedra. The octahedra are again linked together by edge-sharing, and the double chains run parallel to the *c* axis (fig. 8.7*b*). The double chains of linked octahedra are further crosslinked to adjacent double chains to give orthorhombic symmetry in ramsdellite. The unoccupied octahedral sites form double lines of vacancies parallel to the *c* axis defining [1x2] dimensional tunnels, although once again there are no gaping holes in the hexagonally close-packed lattice of oxygen ions defining the ramsdellite structure. The similarity of widths of the tunnels in one dimension in the pyrolusite and ramsdellite structures permit domain intergrowths of these phases found in nsutite (γ-MnO_2). Thus, γ-MnO_2 contains irregular intergrowths of pyrolusite [1x1] and ramsdellite [1x2] units. Some nsutites contain up to [1x5] dimensional tunnels.

In α-MnO_2 possessing the hollandite structure, a close-packed oxygen lattice no longer exists, but pairs of ramsdellite-like double-chains of edge-shared [MnO_6] octahedra are linked at right angles by corner-sharing to produce [2x2] dimensional tunnels (fig. 8.7*c*). These [2x2] tunnels provide genuine cavities capable of accommodating large cations such as K^+ (cryptomelane), Ba^{2+} (hollandite), Na^+ (manjiroite), Pb^{2+} (coronadite), etc., as well as NH_4^+, H_2O, NH_3, etc., groups found in synthetic α-MnO_2 phases. Larger [2x3] tunnels occur in the romanechite (psilomelane) structure-type (table 8.3), while [3x3] tunnels commonly occur in todorokite (fig. 8.7*d*).

Lattice imaging by high resolution transmission electron microscopy (HRTEM) of fibrous manganese(IV) oxide minerals demonstrated the exis-

---

Caption for Fig. 8.8 (contd.).

adjacent to a vacancy. Zn^{2+} ions are located above and below the vacancies and each cation is bonded to three oxygens in the [MnO_6] layers and to three H_2O molecules in the water layer.

tence of complex intergrowths of cryptomelane [2x2] and romanechite [2x3] tunnels (Turner and Buseck, 1979). In some specimens, isolated structures with tunnel widths greater than the [2x3] chains were observed, in some cases up to [2x7] octahedra wide. Subsequent HRTEM studies of todorokites revealed that this Mn(IV) oxide also exhibited a variety of linkages (Turner and Buseck, 1981; Turner *et al.*, 1982). Thus, in addition to the predominant [3x3] tunnels, tunnel widths ranging from [3x2] (cf. romanechite) to [3x7] were found in todorokites from continental ore deposits and seafloor manganese nodules.

A nomenclature scheme for families of Mn(IV) oxides was proposed in which the symbol T(*m,n*), where T denotes a tunnel structure and *m,n* are the widths of infinite chains of edge-shared [MnO_6] octahedra forming the walls (*m*) and floors or ceilings (*n*) of the tunnels. Thus, $m = 1$ defines the nsutite family in which T(1,1) and T(1,2) symbolize pyrolusite and ramsdellite, respectively. Intergrowths of these fundamental units characterize synthetic γ-MnO_2 and naturally–occurring nsutites. Similarly, T(2,*n*) includes both the α-MnO_2 – hollandite (cryptomelane) group, T(2,2), and the romanechite (psilomelane) group, T(2,3), together with the observed coherent intergrowths of T(2,4) to T(2,7) tunnel structures found in fibrous Mn(IV) oxide minerals (Turner and Buseck, 1979). The todorokite family is represented by T(3,*n*) with T(3,3) the most common structure-type (Turner and Buseck, 1981).

Lower valence cations also exist in romanechite and todorokite with the 'walls' and 'floors' of the [2x3] and [3x3] tunnels of romanechite and todorokite, respectively, holding the key to the site occupancies. The double chains of edge-shared [MnO_6] in romanechite contain only Mn^{4+} ions which are present in the M3 sites, the mean M3–O distance of which is 190.4 pm (table 8.3). The 'walls', however, consisting of triple chains of edge-shared octahedra (defined by the romanechite unit cell parameter $c = 968$ pm), contain Mn^{4+} ions in the central M1 site (average M1–O distance = 191 pm) which are flanked by Mn^{3+} ions in the M2 sites. The six-coordinated M2 site is significantly distorted from octahedral symmetry, as indicated by the large range of metal–oxygen distances (190–212 pm, average 196 pm), attributable to the Jahn–Teller effect in Mn^{3+} ions. In the todorokite structure-type portrayed in fig. 8.7*d*, the 'walls' and 'floors' are similar to the 'walls' in romanechite with low-valence cations occurring in outer M2 and M4 sites in the triple chains of edge-shared octahedra and with only Mn^{4+} ions located in the inner M1 and M3 sites, the latter having smaller average metal–oxygen distances than the M2 and M4 sites (table 8.3). The M4 site is also the most distorted octahedron, suggesting that Jahn–Teller stabilized cations such as Mn^{3+} and Cu^{2+} are more favourably accommodated in that site.

The cation occupancies of todorokite may also account for the relative stability of this mineral towards oxidation. Todorokite is destabilized by the presence of substantial Mn^{2+} and Mn^{3+} ions in the structures, because these cations are vulnerable to oxidation. Thus, Mn^{2+}-bearing todorokites show oxidation to manganite and vernadite. However, replacement of Mn^{2+} or Mn^{3+} by Mg^{2+}, Ni^{2+}, Zn^{2+} and Cu^{2+}, which are not susceptible to oxidation, stabilizes todorokite, particularly in manganese nodule deposits.

8.7.3.2 Manganese oxides with layer structures

The structural model for Mn(IV) oxides with layer structures is chalcophanite, ideally $ZnMn_3O_7.3H_2O$, illustrated in fig. 8.8. The chalcophanite structure (Post and Appleman, 1988) consists of single sheets of water molecules between layers of edge-shared $[MnO_6]$ octahedra, with Zn^{2+} ions located between the water layer and oxygens of the $[MnO_6]$ layer. The stacking sequence along the c axis is thus –O–Mn–O–Zn–O–Mn–O– (fig. 8.8a), and the perpendicular distance between two consecutive $[MnO_6]$ layers is about 750 pm. Vacancies exist in the layers of linked $[MnO_6]$ octahedra so that one out of every seven octahedral sites is unoccupied by Mn. Each $[MnO_6]$ octahedron thus shares edges with five neighbouring octahedra and is adjacent to a vacancy (fig. 8.8b). A particularly significant feature of chalcophanite is that the Zn atoms are located above and below the vacancies in the Mn layers and are coordinated to three oxygens of the $[MnO_6]$ layer (fig. 8.8a). Each Zn^{2+} ion completes its coordination with three water molecules so as to form an irregular coordination polyhedron. The chemical compositions of natural chalcophanites often differ significantly from the ideal formula, with variable water contents and a deficiency of Mn^{4+} ions but with additional cations such as Mg^{2+}, Ni^{2+}, Ag^{+} and Mn^{2+} so as to exceed four metal atoms per $ZnMn_3O_7.3H_2O$ formula unit.

The existence of ordered vacancies in the $[MnO_6]$ octahedral layers of chalcophanite and the proximity of divalent cations to these vacancies are important features in structures proposed for the birnessite group, $(Na,K,Ca,Mg,Mn^{2+})Mn_7O_{14}.\ xH_2O$, which show some variability (Post and Veblen, 1990). Layers of edge-shared $[MnO_6]$ octahedra separated by about 705 pm again enclose sheets of H_2O molecules and exchangeable Na^+, Ca^{2+}, etc., cations. Fewer vacancies apparently occur in the layer of $[MnO_6]$ octahedra of birnessite. However, the vacancies may be less ordered than in chalcophanite, contributing to different superstructures observed in natural and synthetic birnessite phases. Vernadite, $(Mn,Ca,Co)(O,OH)_2.\ xH_2O$, analogous to synthetic δ-MnO_2, appears to be related to birnessite with disordering along the layer-stacking axis. The synthetic buserite phase, formed as a precursor to synthetic birnessite, appears to have additional layers of water molecules between the [MnO_6] layers which are separated by about 1,000 pm, so that it is commonly referred to as the '10 Å phase'. Cations such as Ni^{2+}, Co^{2+}, Mg^{2+} and Ca^{2+} appear to stabilize synthetic buserite from dehydration (Giovanoli *et al.*, 1975) and collapse to birnessite, known as the the '7 Å phase'. Lithiophorite also has a layer structure in which edge-shared [MnO_6] octahedra alternate with layers of [(Al,Li)$(OH)_2$] octahedra separated by about 960 pm. Substitution of Li by Ni and Co occurs in lithiophorite formed by lateritic weathering of ultramafic rocks in New Caledonia (Manceau *et al.*, 1987). Asbolane, another mixed layer Mn(IV) oxide associated with lithiophorite, contains two incommensurable sublattices with different *a* parameters but identical *c* dimensions of about 950 pm (Chukrov *et al.*, 1980, 1983). One sublattice consists of continuous layers of edge-shared [MnO_6] octahedra which are interleaved with a second sublattice containing discontinuous layers of [(Ni,Co)$(OH,H_2O)_6$] octahedra, Ca^{2+} ions and H_2O molecules.

Most of the Mn(IV) oxide minerals listed in table 8.3 occur in weathered continental rocks, and often constitute important manganese ore deposits. However, several of the minerals, notably todorokite, birnessite, vernadite and, perhaps, buserite and asbolane, are major constituents of seafloor hydrothermal crusts near spreading centres and in manganese nodule deposits.

8.7.4 Crystal chemistry of manganese nodules

Iron and manganese are major constituents of seafloor ferromanganese oxide deposits, their presence being made conspicuous by the colours that their hydrated oxides impart to the sediments. Thus, orange, brown and red hues indicate the predominance of Fe(III) oxides and silicates, while dark brown, chocolate, and black colourations signify the presence of Mn(IV) oxides. Pale green or blue colours, on the other hand, may be indicative of Fe(III) and mixed-valence Fe(III)–Fe(II) oxides or clay silicates, including glauconite. Not only do Fe and Mn predominate over other heavy metals in pelagic sediments, but the crystallinity and structures of their hydrated oxide phases contribute to the uptake and removal of other transition elements (e.g., Cu, Ni, Co, Ti) and trace metals (e.g., Zn, Mo, Pb, Ce) from seawater and pore water into the ferromanganese oxide phases in the sediments. Another characteristic feature of Fe and Mn is that their mineralogy and geochemistry in the marine environment are strongly influenced by oxidation–reduction reactions.

Manganese nodules occur on the floors of all oceans of the world (Mero, 1965; Cronan, 1980, ch. 5; Halbach *et al.*, 1988). They are typical rock and consist of mixtures of intimately intergrown crystallites of various minerals, including several authigenic manganese oxides (e.g., todorokite, vernadite, birnessite, buserite), iron oxides (e.g., goethite, feroxyhyte, ferrihydrite), zeolite and silica phases, and a variety of detrital minerals, organic matter and colloidal materials. The nodules usually have nucleated around pumice, altered volcanic glass or igneous rock and pebbles. Manganese, iron, cobalt, copper, nickel and zinc are enriched in marine manganese nodules relative to seawater by factors exceeding a million, and the concentrations of these elements in the nodules are more than a hundred times greater than their respective crustal abundances. As a result, manganese nodules are an important potential resource for these transition elements (Burns and Burns, 1977; Halbach *et al.*, 1988).

Interelement relationships have been deduced from bulk chemical analyses of the agglomeration of material constituting manganese nodules. The distributions of individual elements within microcrystalline aggregates have been measured directly by microbeam analytical techniques (Burns and Fuerstenau, 1966; Burns and Burns, 1978; Moore *et al.*, 1981). These measurements show that there is a pronounced element association between manganese, nickel, copper, zinc and, frequently, cobalt, particularly in manganese nodules containing todorokite. There is a negative correlation between manganese and iron. Sometimes cobalt appears to be enriched in regions of high iron concentrations in nodules containing intimate intergrowths of vernadite–feroxyhyte assemblages (Burns and Fuerstenau, 1966).

The enrichment of transition metal ions in todorokite-bearing nodules may be explained by referring to fig. 8.7 which shows Mn(IV) oxides with tunnel structures. As noted earlier (§8.7.3.1), the 'walls' and 'floors' of the [3x3] tunnels of todorokite are flanked by the relatively larger M2 and M4 octahedral sites (fig. 8.7*e*) which can accommodate divalent cations of Ni, Cu and Co, with Cu^{2+} ions probably favouring the more distorted M4 octahedra. The observations of intergrowths of variable tunnel widths in todorokite, particularly in phases occurring in manganese nodules (Turner *et al.*, 1982; Siegel and Turner, 1983), point to other important factors contributing to metal enrichments in these seafloor deposits. In the 'floors' and 'ceilings' of the todorokite tunnels, cation vacancies within the bands of edge-shared [MnO_6] octahedra probably become more prevalent the wider the 'floor' dimensions. Such Mn^{4+} vacancies are essential features of a number of divalent cation-bearing Mn(IV) oxide phases such as chalcophanite (fig. 8.8) and possibly birnessite. In these layer structures, the Mn^{4+} vacancies in the sheets of edge-shared octahedra dictate their crystal chemistries because divalent cations are bonded to positions above and below the vacancies (cf. Stouff and Boulégue, 1988; Manceau and Combes, 1988). Similar Mn^{4+} vacancies may also be present in the (001) planes of todorokite where multiple-width edge-shared [MnO_6] domains exist. These vacancies not only may nucleate faults, kinks, and twinning observed in HRTEM micrographs of todorokite fibres, but they also could influence the crystal chemistry and site occupancies of the tunnel interiors.

Therefore, four types of atomic substitution may contribute to the crystal chemistry of todorokite- and vernadite-bearing manganese nodules.

(1) Substitution of Mn^{2+} and Mn^{3+} ions in M2 and M4 sites by Mg^{2+}, Ni^{2+}, Cu^{2+}, Co^{2+}, Zn^{2+} and other cations having ionic radii in the range 65–80 pm;
(2) inclusion in the tunnels, in which a variety of large cations (K^+, Ba^{2+}, Ag^+, Na^+, Ca^{2+}, Pb^{2+}) and H_2O molecules may be accommodated;
(3) replacement of Mn^{4+} ions in the M1 and M3 sites by cations of similar ionic radii (53–55 pm), such as low-spin Co^{3+} ions discussed later (§8.7.5); and
(4) incorporation as hydrated cations adsorbed adjacent to Mn^{4+} vacancies in the 'floors' and 'ceilings', particularly in todorokites with large tunnel widths.

Such a diversity of structural positions enables seafloor manganese nodule deposits to enrich a variety of strategic metals.

8.7.5 Oxidation state of cobalt in Mn(IV) oxides

The marked dissimilarity between cobalt and nickel, particularly in manganese nodules from seamounts, is noteworthy. These elements as divalent cations exhibit similar geochemical behaviours in igneous rocks and crustal processes, where they coexist in silicate and oxide crystallizing from magma. In seawater, too, cobalt and nickel probably occur as oxidation state (II) species. The fractionation of cobalt and nickel in manganese nodules is due to a change of valence of cobalt ions (Burns, 1965b, 1976b; McKenzie, 1970; Glasby, 1975; Dillard *et al.*, 1982).

It is noteworthy that Co^{3+} ions have a low-spin configuration in many oxide structures, including corundum Co_2O_3, spinel Co_3O_4 and perovskite $LaCoO_3$ phases (§9.7.1), and that the ionic radius of low-spin Co^{3+} (54.5 pm) is remarkably similar to that of Mn^{4+} (53 pm) (Appendix 3). Low-spin Co^{3+} ions have been shown to exist in several natural and synthetic Mn(IV) oxide phases (e.g., Dillard *et al.*, 1982; Manceau *et al.*, 1987). The strong fractionation of cobalt into vernadite-bearing manganese nodules, as well as synthetic and terrestrial manganese oxides, is the result of low-spin Co^{3+} substituting for Mn^{4+} ions in the [MnO_6] octahedra, leading to a very high CFSE for Co^{3+} (see table 2.5). Another consequence of replacement of Mn^{4+} by Co^{3+} is that charge compensation may be acquired by other cations being adsorbed onto the layers of edge-sharing $[(Mn,Co)O_6]$ octahedra, contributing to the diversity of metals enriched in marine manganese oxide deposits.

8.8 Partitioning of transition metal ions during metamorphic processes

Metamorphic processes involve the recrystallization of minerals under various degrees of isochemical conditions. The reaction

$$XZ_s + YZ'_s \rightarrow XZ'_s + YZ_s \qquad (8.7)$$

gives no indication of the mechanism by which a metamorphic process takes place. The extremely slow rates of diffusion of ions through crystal structures, often leading to chemically zoned porphyroblasts, suggest that a reaction medium must be present to facilitate metamorphic reactions.

At the onset of diagenesis of wet sediments and metamorphism of lithified rocks there are substantial amounts of intergranular aqueous and saline solutions. As the grade of metamorphism increases, the porosity and permeability of a rock decrease and intergranular fluids become less continuous and more localized. As a result, the availability of ions for metamorphic reactions becomes increasingly dependent on the composition of the rock *in situ*.

The mechanism of a metamorphic reaction might be visualized as follows

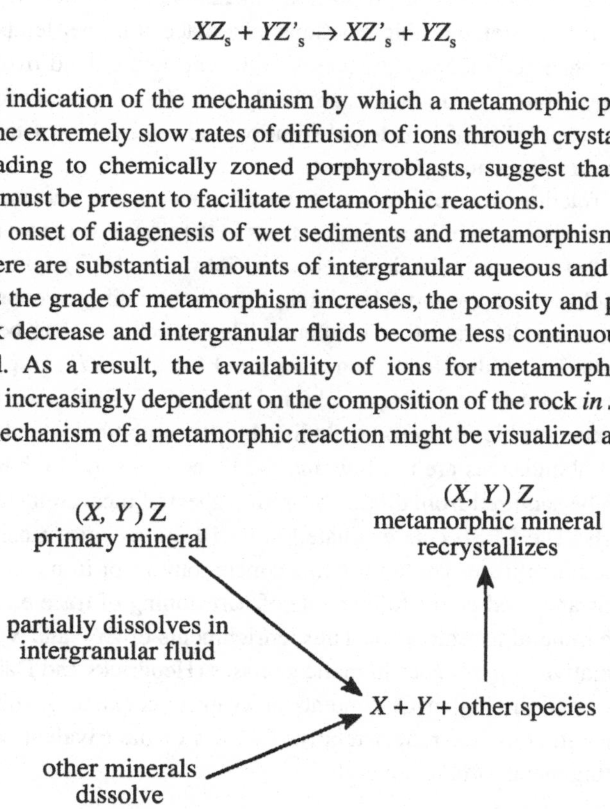

(8.26)

Thus, existing minerals partially dissolve in the intergranular fluid with precipitation and growth of the new mineral taking place at the expense, both compositionally and spatially, of the existing minerals.

Factors similar to those in sedimentary processes are involved in metamorphic reactions. The susceptibility of a cation in a mineral to dissolution and recrystallization in a new phase depends on the relative stability of the ion in each crystal structure and the ease of removal of the ion from the structure. Thus, kinetic and thermodynamic factors again determine the fractionation of elements in metamorphic reactions. By analogy with sedimentary processes, the Mn^{4+}, Cr^{3+}, Ni^{2+} and Co^{3+} ions, because of their high CFSE's and low vulnerabilities to substitution reactions, are predicted to show the slowest rates of solution and might be expected to be relatively enriched in remnant precursor minerals. The distributions of major and trace elements in gneisses of the Adirondack Mountains, for example, show that there are enrichments of Mg, Ti, V, Cr, Co, Ni and Cu and depletions of Al, Sc, Mn, Fe^{2+} and Fe^{3+} in the remnant biotite with increasing proportions of almandine garnet (Engel and Engel, 1960, 1962). Metamorphic reactions take place at higher temperatures and pressures than sedimentary processes, however, and extend over longer periods of time. Thus, kinetic factors are likely to be less important for determining the susceptibility of ions to substitution and removal in metamorphic reactions than in sedimentary processes.

Chemical fractionation of transition metal ions during metamorphism depends on the relative stabilities of the cations in the crystal structures of minerals involved in the crystallization processes. Schwartz (1967) demonstrated that crystal field effects are more important than ionic radii in accounting for distributions of certain transition elements. This conclusion has been confirmed by numerous studies (e.g., Annersten and Ekstrom, 1971; Dupuy *et al.*, 1980; Rosler and Bouge, 1983; Hendricks and Dahl, 1987; Dahl *et al.*, 1993).

In most metamorphic minerals, including biotite, amphiboles and garnets, trace element abundances are too low and the Fe contents too high to enable CFSE data to be acquired from direct absorption spectral measurements on the porphyroblasts. The CFSE data evaluated in §5.10 for specific transition elements enriched in silicates containing low concentrations of iron enable some interpretations and predictions to be made of partitioning of trace elements in metamorphic mineral assemblages. Thus, enrichments of Cr^{3+} and V^{3+} in calcic garnets relative to pyroxenes in mafic gneisses (Hendricks and Dahl, 1987; Langer, 1988) and the higher Cr contents of kyanites coexisting with garnets and pyroxenes in eclogites reflect relative CFSE's of the trivalent cations in these coexisting metamorphic minerals.

8.9 Summary

Chapter 8 discusses aspects of the geochemistry of the first series transition elements, most of which are trace elements in the Earth's Crust (that is, their crustal abundances are below 1,000 ppm).

Trace element distribution rules. Attempts to explain the distributions of transition metal ions in crustal rocks and minerals by empirical rules based on cation size and charge have been generally unsuccessful. This has led to critical appraisals of the rules and criteria on which they are based.

Ionic radius. The wide variation of metal–oxygen distances within individual coordination sites and between different sites in crystal structures of silicate minerals warns against too literal use of the radius of a cation, derived from interatomic distances in simple structures. Relationships between cation radius and phenocryst/glass distribution coefficients for trace elements are often anomalous for transition metal ions (Cr^{3+}, V^{3+}, Ni^{2+}), which may be attributed to the influence of crystal field stabilization energies.

Thermodynamic considerations. A rigorous thermodynamic analysis shows that empirical rules which consider bonding forces of ions in crystalline phases alone are invalid. It is necessary to compare binding forces of ions in a mineral and the medium from which that mineral crystallized. For transition elements, this requires information about relative CFSE's of the cations in coexisting minerals, silicate melts, aqueous solutions and hydrothermal fluids.

Igneous geochemistry. Although bond energies of ions in magma are not known, comparisons can be made between the behaviours of transition elements themselves and other cations with similar radii and charges during magmatic crystallization. Spectroscopic measurements of transition metal ions in silicate glasses (assumed to represent magma) indicate that the cations occur in six-fold, five-fold and four-fold coordination sites in silicate melts, but occupy octahedral sites in igneous minerals (olivine, pyroxenes, amphiboles and micas). The octahedral site preference energy and other $CFSE_{(crystal - glass)}$ difference parameters provide measures of relative affinities of transition metal ions for a silicate mineral. There is close agreement between the predicted and observed orders of uptake of transition metal ions during magmatic differentiation of basaltic magma. Crystal field stabilization also accounts for the metallogenesis patterns of several transition elements during plate tectonic processes, including porphyry copper systems, nickel in alpine peridotites, and chromite deposits in ultramafic intrusions.

Sedimentary geochemistry. The aqueous phase dominates weathering, leaching, transportation, and precipitation processes in the sedimentary cycle. The behaviour of transition metal ions to chemical attack depends on the relative stabilities of hydrated and complex ions in solution and bonded cations in crystal structures. The break-down of minerals and leaching of ions takes place through substitution reactions, which depend on kinetic and mechanistic factors. Arguments based on crystal field theory and the transition state theory of reaction rates show that the cations most resistant to substitution reactions are those with $3d^3$, $3d^8$, and low-spin $3d^6$ configurations, such as Cr^{3+}, Ni^{2+} and Co^{3+}, which also acquire high CFSE's in octahedral coordination. These factors contribute to the enrichments of Cr, Ni and Co in laterite deposits. Oxidation of manganese to the tetravalent state results in a substantial increase of CFSE and a great structural diversity of Mn(IV) oxide minerals. These minerals are often enriched in other transition metal ions, notably Ni, Cu and Co in marine manganese nodule deposits, with low-spin Co^{3+} substituting for Mn^{4+} ions of comparable ionic radius.

Metamorphic geochemistry. Since element fractionation during metamorphism probably involves intergranular aqueous and saline solutions, factors similar to those in sedimentary processes operate during metamorphic recrystallization. Cations with highest CFSE that are least susceptible to substitution reactions (e.g., Cr^{3+}, Ni^{2+}) are the ones showing relative enrichments in remnant precursor minerals.

8.10 Background reading

Basolo, F. D. & Pearson, R. G. (1967) *Mechanisms of Inorganic Reactions: A Study of Metal Complexes in Solution, 2nd edn.* (J. Wiley & Sons, New York).

Burns, R. G. (1973) The partitioning of trace transition elements in crystal structures: A provocative review with applications to Mantle geochemistry. *Geochim. Cosmochim. Acta*, 37, 2395–403.

Burns, R. G. & Fyfe, W. S. (1967) Trace element distribution rules and their significance. *Chem. Geol.*, 2, 89–104.

Henderson, P. (1982) *Inorganic Geochemistry.* (Pergamon Press), ch. 6.

Huheey, J. E. (1983) *Inorganic Chemistry, 3rd edn.* (Harper & Row, New York), ch. 11.

Schwartz, H. P. (1967) The effect of crystal field stabilization on the distribution of transition metals between metamorphic minerals. *Geochim. Cosmochim. Acta*, 31, 503–17.

9

Mantle geochemistry of the transition elements: optical spectra at elevated temperatures and pressures

It is embarassing that the pressure experiments show the dependence of Δ upon the distance R between the metal and the ligand to be fairly close to that given by the point charge model.

S. Sugano & S. Oshnishi, in
Material Science of the Earth's Interior.
(I. Sunagawa, ed., Terra Scientific Publ. Co.,
Tokyo, 1984), p. 174

9.1 Introduction

Considerable interest centres on the Mantle constituting, as it does, more than half of the Earth by volume and by weight. Attention has been focussed on several problems, including the chemical composition, mineralogy, phase transitions and element partitioning in the Mantle, and the geophysical properties of seismicity, heat transfer by radiation, electrical conductivity and magnetism in the Earth. Many of these properties of the Earth's interior are influenced by the electronic structures of transition metal ions in Mantle minerals at elevated temperatures and pressures. Such effects are amenable to interpretation by crystal field theory based on optical spectral data for minerals measured at elevated temperatures and pressures.

In the Mantle, temperatures range up to several thousands of degrees Kelvin and pressures may exceed 100 GPa in the deep interior, attaining 136 GPa at the Core–Mantle boundary. In the past two decades, the optical spectra of several minerals and synthetic analogues have been measured at high pressures and elevated temperatures simulating conditions in the interior of the Earth. The results of many of these high *P* and *T* spectral measurements are reviewed in this chapter and applications of the spectral data are described to transition metal-bearing Mantle minerals. Of particular interest are the thermodynamic stabilities and properties of ferromagnesian silicates (olivine, pyroxenes, gar-

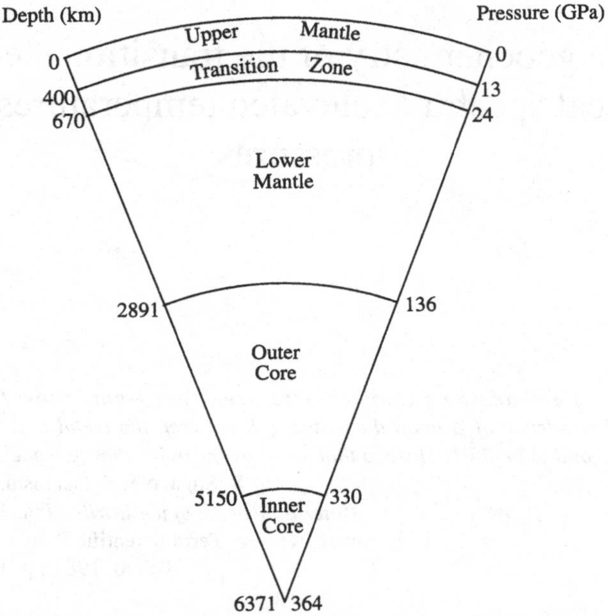

Figure 9.1 Major subdivisions of the Earth (from Liu & Bassett, 1986). The figure shows the pressures and depths of the major boundaries deduced from seismic data.

net) in the Upper Mantle relative to spinel polymorphs in the Transition Zone, and post-spinel phases such as perovskite and periclase in the Lower Mantle.

9.2 Chemical composition of the Mantle

Seismic velocity data indicate that the Earth's interior has a layered structure, being broadly divided into Crust, Mantle and Core. As indicated in fig. 9.1, the Mantle is further subdivided into three distinct regions: the Upper Mantle to a depth of about 400 km is separated from the Lower Mantle by the Transition Zone between 400 km and about 670 km. The chemical composition of the Mantle has been estimated from element abundances of carbonaceous chondrites, compositions of Mantle xenoliths, studies of ophiolite sequences and basalt petrogenesis (Ringwood, 1975; Basaltic Volcanism Study Project, 1981; Liu and Bassett, 1986; Jeanloz and Knittle, 1989). The Upper Mantle corresponds to garnet peridotite with a modal mineralogy dominated by olivine, coexisting with pyroxenes plus garnet, in which the ratio Fe/(Mg + Fe) is about 0.10. The major elements in the Mantle are oxygen, magnesium and iron, with smaller amounts of Al, Ca and Na. Element abundance data summarized in Appendix 1 indicate that the concentrations of Fe, Cr, Ni, Ti and Mn, collectively amounting to almost 7 wt per cent or 2.5 atom per cent of the Upper Mantle, are relatively higher than the remaining transition elements. However, there is an appreciable depletion of Fe, Co, Ni, Cu and the noble metals in the Mantle relative to their abundances in carbonaceous chondrites, indicating that these siderophilic elements have been fractionated into the Core. The geochemical data indicate that the more lithophilic elements Cr, V and Mn are also depleted in the Earth's Mantle as well as on the Moon relative to their Mg-normalized primordial abundances (e.g., Drake *et al.*, 1989; Ringwood *et al.*, 1991). This may be attributed to siderophilic behaviour of Cr, V and Mn at high pressures and temperatures during Core formation, contrasting with the lithophilic properties of these elements at the pressures, temperatures and oxygen fugacities in the Earth's Upper Mantle and the Moon. Such a similarity underlies the 'giant impact' hypothesis that the Moon was derived from the Earth's Mantle after Core formation had occurred.

The chemical composition of the Lower Mantle below 670 km is essentially unknown. It has often been assumed to be the same as the Upper Mantle with the seismic discontinuity at 670 km representing a phase change to denser polymorphs rather than a chemical boundary (Liu and Bassett, 1986). However, some models of the Earth's interior suggest that the Mantle is stratified with the Upper Mantle and Lower Mantle convecting separately, leading to compositional density differences between these two regions. There is a commonly held view that the Lower Mantle has a higher Fe/(Mg+Fe) ratio than the Upper Mantle (Liu and Bassett, 1986; Jeanloz and Knittle, 1989).

9.3 Mineralogy of the Mantle

Numerous experimental investigations have demonstrated that common rock-forming minerals such as olivine, pyroxene and garnet assemblages occurring in the Earth's Crust and Upper Mantle are transformed to denser phases at high pressures and temperatures existing in the Lower Mantle. The Transition Zone represents the region in the Earth's interior where these phase changes occur. Examples of such phase transitions are summarized in fig. 9.2. The onset of the Transition Zone at 400 km represents the depth at which the olivine structure (termed the α-phase) transforms to wadsleyite, more commonly called the β-phase. At high pressures, the β-phase transforms to ringwoodite with the spinel structure (γ-phase) around 570 km. Note that the β-phase is sometimes referred to as the 'modified spinel' phase. This is a misnomer because wadsleyite is a sorosilicate, better formulated as $(Mg,Fe)_4OSi_2O_7$, containing pairs of linked $[SiO_4]$ tetrahedra, in contrast to olivine and silicate spinel, the crystal structures

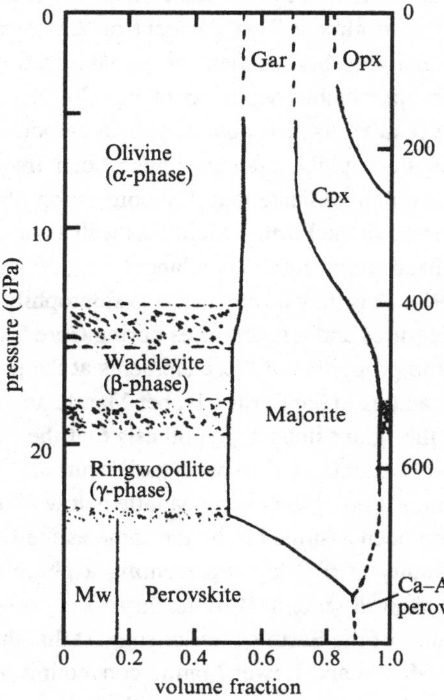

Figure 9.2 Minerals predicted to occur in the Earth's Mantle (based on Ito & Takahashi, 1989). The volumetric fractions represent the modal mineralogy of a peridotitic Mantle down to the Lower Mantle. Olivine, wadsleyite and ringwoodite correspond to the α-, β- and γ-phases of $(Mg,Fe)_2SiO_4$; Gar = garnet compositions spanning pyrope $[(Mg,Fe)_3Al_2(SiO_4)_3]$ – majorite $[(Mg,Fe)_3(Al,Cr,Si)_2(SiO_4)_3]$; Cpx = clinopyroxene; Opx = orthopyroxene; Mw = magnesiowüstite [(Mg,Fe)O]; perovskite = $[(Mg,Fe)SiO_3]$; Ca–Al perovskite = $[(Ca,Na)(Al,Si)O_3]$.

of which contain isolated $[SiO_4]$ tetrahedra (figs 4.19 and 5.8). Wadsleyite also contains three different oxygen ligands, in contrast to just one oxygen bond-type occurring in the olivine and silicate spinel phases. Thus, in addition to the non-bridging oxygens bonded to one Si atom (present also in the olivine and silicate spinel structures), oxygens bridging two Si atoms as well as isolated O^{2-} anions exist in the wadsleyite crystal structure. The β-phase (wadsleyite) also contains three six-coordinated cation sites, compared to two such sites in olivine and only one octahedral site in spinel, with indications of some Fe^{2+} ordering in the crystal structure of wadsleyite (Sawamoto and Horiuchi, 1990).

The $\alpha \rightarrow \beta \rightarrow \gamma$ phase transitions involving $(Mg,Fe)_2SiO_4$ are isochemical. However, below 670 km where the Lower Mantle begins, disproportionation

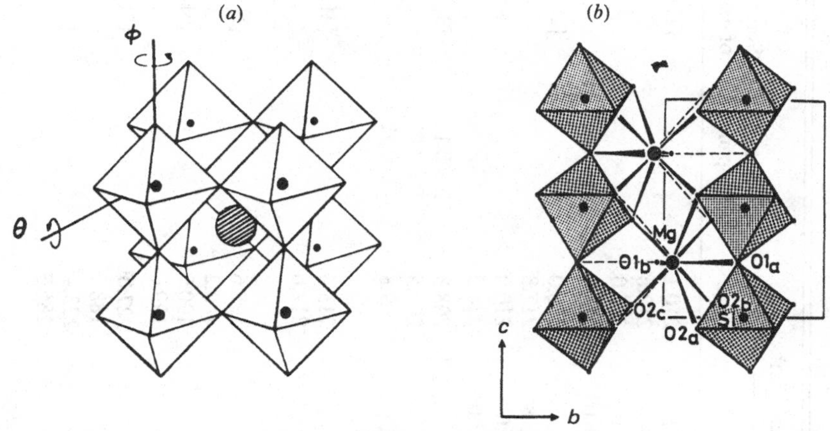

Figure 9.3 The perovskite crystal structure. (*a*) Ideal cubic phase showing corner-shared octahedra surrounding the twelve-coordinated cuboctahedral site; (*b*) projection of the orthorhombic perovskite $MgSiO_3$ structure at 6.7 GPa along the *c* axis (from Kudoh *et al.*, 1987). Rotation and tilting of $[SiO_6]$ octahedra relative to the cubic structure are shown. The eight shorter bonds (Mg–O distances = 199 to 244 pm) are indicated by solid lines and the four longer bonds (Mg–O = 277 to 315 pm) by dashed lines.

reactions occur and the spinel structure breaks down to a mixture of silicate perovskite $[(Mg,Fe)SiO_3]$ plus periclase $[(Mg,Fe)O]$.

The pyroxene + garnet phases of the Upper Mantle also transform to denser phases initially involving majorite garnet, then the ilmenite structure and eventually the perovskite phase. Each of these phase transformations entails a progressive change of coordination number of Si from tetrahedral to octahedral. Thus, the common ferromagnesian silicates of the Upper Mantle, all containing Fe^{2+} ions in distorted six-coordinated sites and tetrahedrally coordinated Si, transform to dense oxide structures in which cations acquire higher coordination numbers or occupy more regular octahedral sites.

Structural information for selected high pressure mineral phases are summarized in table 9.1. Note that metal–oxygen distances are shortened in the crystal structures at elevated pressures. In the orthorhombic perovskite structure illustrated in fig. 9.3, corner-sharing octahedra (the B sites) provide the framework with the three oxygens of each triangular face of the octahedra forming a distorted cuboctahedron surrounding the A site. Cations in the A site thus have twelve near-neighbour oxygen atoms but with metal–oxygen distances exhibiting a wide range of values. Note also that cations in the A and B sites are only 279 pm apart. Rotation and tilting of the B-site octahedra produce distortion of the cuboctahedron forming the A site. This is the cause of the wide range of metal–oxygen distances and ill-defined coordination number for the A site (table 9.1).

Table 9.1. *Coordination sites in high-pressure phases in the mantle*

Mineral	Pressure*	Site	Point symmetry	Metal–Oxygen distances (pm) Range	Mean†	Sitte occupancy	Source of data
forsterite	1 atm	M1	C_i	(2x) 207.4; (2x) 208.3; (2x) 214.5	210.1		[1]
(α-phase)		M2	C_s	(2x) 206.4; (2x) 220.8; 204.5; 216.6	212.6		
Mg_2SiO_4	5.0 GPa	M1	C_i	(2x) 205.0; (2x) 209.0; (2x) 203.0	206.0		
		M2	C_s	(2x) 206.0; (2x) 227.0; 192.0; 206.0	211.0		
fayalite	1 atm	M1	C_i	(2x) 210.8; (2x) 212.5; (2x) 221.8	215.0		[2]
(α-phase)		M2	C_s	(2x) 206.6; (2x) 229.1; 211.5; 223.8	217.8		
Fe_2SiO_4	4.2GPa	M1	C_i	(2x) 202.0; (2x) 211.0; (2x) 217.0	210.0		
		M2	C_s	(2x) 209.0; (2x) 226.0; 217.0; 233.0	228.0		
wadsleyite	1 atm	M1	C_i	(4x) 205.6; (2x) 213.6	208.3	Fe=0.11	[3]
(β-phase)		M2	C_s	(4x) 209.4; 203.5; 210.0	208.5	Fe=0.05	
$(Mg_{0.9}Fe_{0.1})$		M3	C_2	(2x) 202.9; (2x) 213.0; (2x) 213.6	209.8	Fe=0.12	
$O.Si_2O_7$							
ringwoodite	1 atm	oct	D_{3d}	(6x) 213.6	213.6		[4]
(γ-phase)	4 GPa			(6x) 212.6	212.6		
Fe_2SiO_4 spinel							
ruby	1 atm	Al	C_3	(3x) 185.6; (3x) 197.1	198.6		[5]
Al_2O_3/Cr^{3+}	8 GPa	Al	C_3	(3x) 184.3; (3x) 194.1	189.2		
Co_2O_3	1 atm	ls Co	C_3	(3x) 188.0; (3x) 192.0	190.0		[6]
(corundum)		hs Co		(3x) 186.0; (3x) 212.0	199.0		
pyrope	1 atm	Mg	D_2	(4x) 219.7; (4x) 234.3	227.0‡		[7]
		Al	C_{3i}	(6x) 188.7	188.7		
	5.6 GPa	Mg		(4x) 215.6; (4x) 230.7	223.4‡		[7]
		Al		(6x) 186.3	186.3		

diopside	1 atm	M1	C_2	(2x) 205.1; (2x) 206.0; (2x) 211.9	207.6	[8]
	5.3 GPa	M1		(2x) 201.1; (2x) 203.6; (2x) 207.5	204.1	
gillespite I	1 atm	Fe	$\approx D_{4h}$	(4x) 198.4	198.4	[9]
	0.9 GPa			(4x) 196.2	196.2	
gillespite II	2.1 GPa	Fe	C_{2v}	(2x) 195.3; (2x) 203.1	199.2	
	4.5 GPa			(2x) 192.1; (2x) 203.1	197.6	
periclase	1 atm	oct	O_h	(6x) 210.6	210.6	[10]
	1,042 °C			(6x) 213.2	213.2	
	2.4 GPa			(6x) 209.6	209.6	
$MgSiO_3$	1atm	Mg	C_s	(2x) 205.2; 201.4; 209.6;		[11,12]
(perovskite)				(2x) 227.8; (2x) 242.7;	220.3‡	
				(2x) 312.0; 276.3; 298.8	247.3¶	
		Si	C_i	(2x) 178.3; (2x) 179.6; (2x) 180.1	179.3	
$MgSiO_3$	9.6 GPa	Mg	C_s	(2x) 198.6; 203.8; (205.6;		[11,13]
(perovskite)				(2x) 221.7; (2x) 242.0;	216.8‡	
				(2x) 313.9; 276.3; 298.8	244.7¶	
		Si	C_i	(2x) 176.2; (2x) 179.2; (2x) 178.0	177.8	

* 1 atm = 1.013×10^{-4} GPa;

† Average of six distances; ‡ Average of eight distances; ¶ Average of twelve distances.

Sources of data : [1] Hazen (1976b); [2] Hazen (1977); [3] Sawamoto & Horiuchi (1990); [4] Finger, Hazen & Yagi (1979); [5] Finger & Hazen (1980); [6] Chenavas, Joubert & Marezio (1971); [7] Hazen & Finger (1978); [8] Levien & Prewitt (1981); [9] Hazen & Finger (1983); [10] Hazen (1976a); [11] Kudoh, Ito & Takeda (1987); [12] Kudoh, Prewitt, Finger, Darovskikh & Ito (1990); [13] Mao, Hemley, Fei, Shu, Chen, Jephcoat & Wu (1991).

9.4 Effects of temperature and pressure on optical spectra

9.4.1 Background

Before describing optical spectra of Mantle minerals in detail, it is perhaps useful to summarize predicted effects of temperature and pressure on absorption bands of transition metal-bearing minerals. These factors may be gleaned from earlier chapters describing concepts of crystal field theory (chapter 2) and types of electron excitation processes generating optical spectra (chapters 3 and 4). Different effects of P and T are anticipated for crystal field (CF), metal–metal intervalence charge transfer (IVCT) and oxygen–metal charge transfer (OMCT) transitions.

9.4.2 Crystal field spectra

According to the crystal field model, the crystalline electrostatic field of ligands in a crystal structure influences the $3d$ orbital energy levels of a central transition metal ion in a coordination site. There is a two-fold influence of *temperature* on a crystal structure. First, increased thermal motion results in increased amplitudes of atoms vibrating about their crystallographic positions. Second, thermal expansion causes small increases in interatomic distances, so that according to eq. (2.25)

$$\frac{\Delta_T}{\Delta_0} = \left(\frac{V_0}{V_T}\right)^{5/3} = [1 + \alpha(T - T_0)]^{-5/3} \qquad (2.25)$$

values of Δ are expected to decrease at elevated temperatures. Since the width of an absorption band in a crystal field spectrum is related to fluctuating metal–ligand distances in an oscillating system (§3.8), absorption bands are expected to broaden and band centres move to slightly longer wavelengths due to thermal expansion. Such displacements of absorption bands to lower energies (longer wavelengths) are termed 'red-shifts'.

Increased *pressure*, on the other hand, results in compression of a crystal structure. The compressibilities of oxides and silicates show that the molar volumes of these phases decrease with rising pressure, indicating that interatomic distances within coordination sites become shorter. This is demonstrated by the data for Mantle minerals at elevated pressures summarized in table 9.1. Equation (2.20)

$$\frac{\Delta_P}{\Delta_0} = \left(\frac{R_0}{R_P}\right)^5 \qquad (2.20)$$

relating the crystal field splitting to the inverse fifth power of the interatomic distance, together with eq. (2.24)

$$\frac{d\Delta}{dP} = \frac{5\Delta}{3\kappa} \qquad (2.24)$$

showing that the pressure variation of crystal field splitting depends on the site incompressibility, both indicate that a small contraction of metal–oxygen distance produces a large increase in Δ. As a result, energy separations between 3*d* orbitals increase with rising pressure, and absorption bands are expected to move to shorter wavelengths. These migrations of absorption bands to higher energies (shorter wavelengths) are termed 'blue-shifts'. Thus, high pressures and elevated temperatures are expected to have compensatory effects on band maxima of absorption bands in crystal field spectra. However, high pressures and elevated temperatures are both expected to intensify absorption bands due, respectively, to effects of increased covalency and increased vibronic coupling.

9.4.3 Intervalence charge transfer

Metal–metal intervalence charge transfer transitions taking place between adjacent cations in edge- or face-shared coordination polyhedra are strongly polarization dependent and are facilitated by short metal–metal interatomic distances (§4.7.2). The probability of such transitions might be expected to be enhanced and intensities increased at high pressures. Rising temperature decreases the intensity of IVCT transitions, in contrast to most crystal field transitions which generally intensify at elevated temperatures.

9.4.4 Oxygen → metal charge transfer

High-energy ultraviolet light generally induces electron transfer between cations and nearest-neighbour oxygens forming the coordination polyhedron about the metal. However, absorption bands extend into the visible region (§4.7.3). The energies of OMCT transitions depend on the cation and the symmetry of its coordination site; for octahedrally coordinated cations, OMCT energies have been calculated to decrease in the order $Cr^{3+} > Ti^{3+} > Fe^{2+} > Fe^{3+} > Ti^{4+}$, eq. (4.5). With rising pressures and temperatures, absorption edges of OMCT bands show red-shifts, with increased absorption in the visible region in high *P* and high *T* spectra.

9.5 Optical spectral data at high *P* and *T*

9.5.1 General trends

The electronic spectra of a variety of transition metal-bearing oxide and silicate minerals have been measured at high pressures and/or elevated temperatures. Trends for absorption bands originating from crystal field (CF), metal–metal intervalence charge transfer (IVCT) and oxygen → metal charge transfer (OMCT) transitions are summarized in tables 9.2, 9.3 and 9.4, respectively.

Table 9.2. *Crystal field spectra of silicate and oxide minerals at high pressures and temperatures*

Mineral or phase	Composition or cation site	*P* or *T* range	Observations	Sources of data
			Oxides:	
periclase	MgO Fe^{2+}/oct	5 GPa	Fe^{2+} spin-allowed bands show blue–shifts: 983 cm^{-1} by 55 cm^{-1}/GPa; and 11,810 cm^{-1} by 72 cm^{-1}/GPa	[1–3]
magnesio–wüstite		30 GPa	Fe^{2+} spin-allowed bands show blue-shifts, but obscure by oxygen → Fe charge transfer	[4]
ruby	Al_2O_3 Cr^{3+}/oct	static &	Cr^{3+} spin-allowed bands show blue-shifts at 10 GPa (static);	[5]
		shock	at 46 GPa (shock): blue-shift 18,020 to 19,880 cm^{-1}; at 32 GPa (shock): blue-shift 24,690 to 25,910 cm^{-1}; Racah *B* parameter decreases from 630 to 590 cm^{-1}; red-shift of spin-forbidden peak at 14,405 cm^{-1}	[6,7]
		400 °C	Cr^{3+} spin-allowed bands broaden, intensify; show red-shifts: 18,315 to 17,890 cm^{-1}; and 24,815 to 24,390 cm^{-1}	[8]
spinel	$MgAl_2O_4$ Fe^{2+}/tet	5 GPa	Fe^{2+} spin-allowed band at 4,830 cm^{-1} shows blue-shift of 30 cm^{-1}/GPa	[2]
magnesio–chromite	$MgCr_2O_4$ Cr^{3+}/oct	2 GPa	Cr^{3+} spin-allowed band shifts: 17,700 to 18,100 cm^{-1}	[9]
ringwoodite	γ–Fe_2SiO_4 Fe^{2+}/oct	30 GPa	Fe^{2+} spin-allowed band at 11,000 cm^{-1} shows blue-shift	[10,11]
γ–Ni_2SiO_4	Ni_2SiO_4 Ni^{2+}/oct	12 GPa	Ni^{2+} spin-allowed bands intensify considerably; blue-shifts: 9,150 to 10,100 cm^{-1}; 14,781 to 15,920 cm^{-1}; and 22,550 to 23,750 cm^{-1}; spin-forbidden peak shifts: 20,180 to 20,830 cm^{-1}; Racah *B* parameter decreases: 671 to 638 cm^{-1}	[12]

Table 9.2 *continued*

Mineral or phase	Composition or cation site	*P* or *T* range	Observations	Sources of data
			Olivines:	
fayalite	α–Fe_2SiO_4	4 to 20 GPa	α spectrum: Fe^{2+}/M1 site spin-allowed bands shift: 7,350 to 8,333 cm^{-1}; and 11,100 to 10,100 cm^{-1}	[10,11, 13–15]
		2.5 to 20 GPa	β spectrum: Fe^{2+}/M1 site spin-allowed bands shift: 7,962 to 8,299 cm^{-1}; and 9,010 to 9,000 cm^{-1}	
		2.5 to 20 GPa	Fe^{2+} spin-forbidden peaks shift: 16,242 to 15,974 cm^{-1}; and 22,075 to 22,039 cm^{-1}	
		400 °C	γ spectrum: Fe^{2+}/M2 site spin-allowed band shifts: 9,260 to 9,435 cm^{-1}; intensified	[16]
			β spectrum: Fe^{2+}/M1 site spin-allowed bands shift: 8,200 to 8,000 cm^{-1}; and 10,750 to 10,420 cm^{-1}; intensified	
forsterite	Fa_{8-10}	2.5 GPa	Fe^{2+}/M2 site spin-allowed band at 9,300–9,400 cm^{-1} shows 130 cm^{-1}/GPa shift to 9,900–10,000 cm^{-1}	[2]
	Fa_{12}	400 °C	γ spectrum: Fe^{2+}/M2 site spin-allowed band shifts: 9,525 to 9,800 cm^{-1}; intensified β spectrum: Fe^{2+}/M1 site spin-allowed bands shift: 9,175 to 8,850 cm^{-1}; and 12,195 to 11,905 cm^{-1}; intensified	[16]
	Fa_{8-16}	1,400 °C	β and γ spectra: Fe^{2+}/M2 site spin-allowed band broadened and intensified appreciably	[17]
			Pyroxenes:	
ortho-pyroxene	Fs_{23}	5 GPa	Fe^{2+}/M2 site spin-allowed band at 11,068 cm^{-1} shows 143 cm^{-1}/GPa shift to 12,050 cm^{-1}	[2]
enstatite	Fs_{14}	400 °C	γ spectrum: Fe^{2+}/M2 site spin-allowed bands: shift of 11,110 to 10,870 cm^{-1} with intensification; negligible shift of 5,555 cm^{-1} band, but loss of intensity	[16]
ferrosilite	$FeSiO_3$	2.5 GPa	α spectrum: Fe^{2+}/M2 site spin-allowed bands shift: 4,878 to 5,411 cm^{-1}; and 10,627 to 10,923 cm^{-1}	[18]
			γ spectrum: Fe^{2+}/M1 site spin-allowed bands shift: 8,333 to 9,049 cm^{-1}; and 10,549 to 10,846 cm^{-1}	
pigeonite (lunar)	$Wo_6En_{65}Fs_{29}$	400 °C	Fe^{2+}/M2 site spin-allowed band shifts: 10,640 to 10,310 cm^{-1}, intensified	[16]

Table 9.2 *continued*

Mineral or phase	Composition or cation site	*P* or *T* range	Observations	Sources of data
augite (lunar)	Fe^{2+}/M2 Ti^{3+}/M1	5 GPa	spin-allowed Fe^{2+} band at 9,805 cm^{-1} and Ti^{3+} band at 20,920 cm^{-1} show blue-shifts and intensify; negligible shifts of Fe^{2+} spin-forbidden peaks at 18,180 and 19,760 cm^{-1}	[13]
omphacite	Fe^{2+}/oct	4 GPa	Fe^{2+}/M1 site spin-allowed bands at 10,750 and 8,695 cm^{-1} show blue-shifts; Fe^{3+} spin-forbidden band at 22,730 cm^{-1} shows negligible shift	[13]
fassaite (Allende)	Ti^{3+}/oct	4 & 30 GPa	Ti^{3+}/M1 site spin-allowed bands at 21,280 and 16,670 cm^{-1} show blue-shifts	[19]
			Garnets:	
almandine	Fe^{2+}/ dist.cube	5, 10 & 17.3 GPa	Fe^{2+}/8–fold site spin-allowed bands at 8,130 and 5,880 cm^{-1} show blue–shifts of 50 to 60 cm^{-1}/GPa	[2,20]
		12 GPa	Fe^{2+}/8-fold site spin-allowed bands initially at 7,800 and 6,100 cm^{-1} show blue-shifts of 60 cm^{-1}/GPa and 90 cm^{-1}/GPa, respectively; red-shifts occur for spin-forbidden peaks at 17,600 cm^{-1}, 19,100 cm^{-1}, 19,900 cm^{-1}, and 21,100 cm^{-1}	[21]
		400 °C	All Fe^{2+} spin-allowed bands show red shifts: 7,690 to 7,620 cm^{-1}; 5,850 to 5,695 cm^{-1}; and 4,175 to 4,150 cm^{-1}; they broaden and intensify appreciably	[8]
spessartine	Mn^{2+}/ dist.cube	12 GPa	Mn^{2+}/8–fold site spin-forbidden peaks initially at 24,500, 24,300, 23,750, 23,200, 21,600 and 20,800 cm^{-1} all show red-shifts	[21]
andradite	Fe^{3+}/oct	10 GPa	Fe^{3+} spin-forbidden peak at 22,730 cm^{-1} shows no significant shift	[13]
		300 °C	Fe^{3+} spin-forbidden peaks show small red-shifts: 24,510 to 22,520 cm^{-1}; 18,620 to 17,330 cm^{-1}; and 11,710 to 11,750 cm^{-1}; all peaks intensify	[8]
uvarovite	Cr^{3+}/oct	19.7 GPa	Cr^{3+} spin-allowed bands show blue–shifts: 16,667 to 17,668 cm^{-1}; and 22,727 to 23,687 cm^{-1}; Racah *B* parameter decreases: 589 to 577 cm^{-1}	[13]
spessartine	Mn^{2+}/ dist.cube	400 °C	Mn^{2+} spin-forbidden bands show small red-shifts: 24,510 to 24,330 cm^{-1}; and 20,000 to 19,420 cm^{-1}; slight intensification.	[8]

Table 9.2 *continued*

Mineral or phase	Composition or cation site	*P* or *T* range	Observations	Sources of data
			Epidotes:	
epidote	Fe^{3+}/M3 site	300 °C	Fe^{3+} spin-forbidden bands show small shifts: 21,880 to 21,370 cm^{-1}; 21,010 to 20,530 cm^{-1}; 16,210 to 16,390 cm^{-1}; and 10,190 to 9,460 cm^{-1}	[8]
piemontite	Mn^{3+}/M3 site	19.7 GPa	All Mn^{3+} spin-allowed bands intensify; blue-shifts: 12,000 to 12,250 cm^{-1}; 18,170 to 20,200 cm^{-1}; and 22,000 to 23,250 cm^{-1}	[13]
			Fe(III) oxyhydroxides:	
goethite	α–FeOOH Fe^{3+}/oct	2 &10 GPa	General decrease of intensity; red-shift of Fe^{3+} spin-forbidden bands; e.g., β spectrum: 15,504 to 15,279 cm^{-1}; and 10,929 to 10,730 cm^{-1}; powdered sample: 11,099 to 10,204 cm^{-1} at 10 GPa.	[22]
lepidocrocite	γ–FeOOH Fe^{3+}/oct	2 &10 GPa	General decrease of intensity; red-shift of Fe^{3+} spin-forbidden bands; e.g., β spectrum: 13,699 to 13,530 cm^{-1}; and 10,142 to 9,950 cm^{-1}; powdered sample: 10,917 to 9,775 cm^{-1} at 10 GPa.	[22]
akaganeite	β–FeOOH Fe^{3+}/oct	7 GPa	In powdered sample, poorly defined band at 11,765 cm^{-1} becomes obscured by absorption edge; shoulder near 10,000 cm^{-1} intensifies.	[23]
gillespite	$BaFeSi_4O_{10}$ Fe^{2+}/sq. planar	2.6 GPa	Below 2.6 GPa, blue-shift of Fe^{2+} spin-allowed bands at 20,000 and 8,300 cm^{-1}; above 2.6 GPa, phase change leads to Fe^{2+} in distorted tetrahedral site and red-shifts to about 16,700 and 7,140 cm^{-1}	[24]

Sources of data : [1] Shankland (1968); [2] Shankland, Duba & Moronow (1974); [3] Goto, Ahrens, Rossman & Syono (1980); [4] Mao (1973); [5] Stephens & Drickamer (1961); [6] Gaffney & Ahrens (1973); [7] Goto, Ahrens & Rossman (1979); [8] Parkin & Burns (1980); [9] Mao & Bell (1975a); [10] Mao & Bell (1972a); [11] Mao & Bell (1972b); [12] Yagi & Mao (1977); [13] Abu-Eid (1976); [14] Smith & Langer (1982a); [15] Smith & Langer (1982b); [16] Sung, Singer, Parkin & Burns (1977); [17] Shankland, Nitsan & Duba (1979); [18] Mao & Bell (1971); [19] Mao & Bell (1974a); [20] Balchan and Drickamer (1959); [21] Smith & Langer (1983); [22] Mao & Bell (1974b); [23] Taylor, Mao & Bell (1974); [24] Abu-Eid, Mao & Burns (1973).

Table 9.3. *Intervalence charge transfer transitions in minerals at high pressures*

Mineral	Cations involved	*P* or *T* range	Observations	Sources of data
vivianite	$Fe^{2+} \rightarrow Fe^{3+}$	30 GPa	Band at approx. 700 nm intensifies considerably, shows only minor red-shift	[1,2]
pyroxenes, amphiboles,	$Fe^{2+} \rightarrow Fe^{3+}$	30 GPa	Strong intensification for absorption bands near 700 nm	[1,2]
tourmalines		100 – 300 K	Decreased intensity of IVCT bands at elevated *T*	[3]
omphacite	$Fe^{2+} \rightarrow Fe^{3+}$	4 GPa	Band at 665 nm intensifies, shifts to 690 nm	[4]
pyroxene (Allende)	$Ti^{3+} \rightarrow Ti^{3+}$	30 GPa	Band at 660 nm intensifies considerably, negligible energy shift	[5]
pyroxene (Angra dos Reis)	$Fe^{2+} \rightarrow Ti^{4+}$	5.2 GPa	Band at 485 nm intensifies and shifts to 522 nm.	[6]
kyanite	$Fe^{2+} \rightarrow Fe^{3+}$	84 – 493 K	Band at 615 nm decreases in intensity at elevated *T*	[3]
sapphire	$Fe^{2+} \rightarrow Ti^{4+}$	84 – 493 K	Bands at 590 nm and 775 nm lose intensity at high *T*	[3]
biotite, cordierite	$Fe^{2+} \rightarrow Fe^{3+}$	43 – 587 K	IVCT bands decrease in intensity at elevated *T*	[3]

Sources of data : [1] Bell & Mao (1974); [2] Mao (1976); [3] Smith & Strens (1976); [4] Abu–Eid (1976); [5] Mao & Bell (1974a); [6] Hazen, Bell & Mao (1977).

In general, blue-shifts of CF absorption bands to higher energies are observed with rising pressures, but pressure-induced shifts of band maxima often deviate from the inverse fifth power dependence on metal–oxygen distance expressed in eq. (2.20). Discrepencies are attributed to variations of local compressibility for different cations in host mineral structures, changes of site distortion, and increased covalent bonding. Specific examples are discussed later. Rising temperature usually shifts CF bands to slightly lower energies, in accord with eq. (2.25). The most important effect of temperature, however, is to broaden and significantly increase intensities of CF bands to a greater extent than rising pressure (table 9.2).

On the other hand, the effect of pressure on IVCT transitions dominates the temperature variations (table 9.3). Intensities increase considerably at high pressures but decrease with rising temperature. In general, energies of homonuclear IVCT transitions (e.g., $Fe^{2+} \rightarrow Fe^{3+}$; $Ti^{3+} \rightarrow Ti^{4+}$) are relatively

Table 9.4. *Oxygen $\rightarrow$ metal charge transfer transitions in minerals at high pressures and temperatures*

Mineral or phase	Formula or composition	Cation	*P* or *T* range	Observations	Sources of data
magnesio-wüstite	(Mg,Fe)O	Fe^{2+} (?Fe^{3+})	30 GPa	Strong absorption of visible region at ≈ 2 GPa ($Wü_{56}$), 8 GPa ($Wü_{39}$), and 15 GPa ($Wü_{22}$) which obscures Fe^{2+} crystal field bands near 10,000 cm^{-1}	[1,2]
forsterite	α-Mg_2SiO_4	Fe^{2+}	30 GPa	Strong red-shifts of UV absorption edge into visible region for all Fe^{2+}-bearing olivines	[1,2]
			14 GPa	Point on absorption edge at 31,000 cm^{-1} shows red-shift of 290 cm^{-1}/GPa	[3]
fayalite	α-Fe_2SiO_4	Fe^{2+}	3.4 GPa	O $\rightarrow$ Fe^{2+} CT band at 34,500 cm^{-1} intensifies, undergoes red-shift: 190 cm^{-1}/GPa	[4]
			30 GPa	Above 5 GPa, absorption edge extends well into visible region; above 30 GPa, obscures Fe^{2+} CF bands near 10,000 cm^{-1}	[2,5]
			400 to 1,400 °C	Red-shift of absorption edge towards visible region	[6,7]
ringwoodite	γ-Fe_2SiO_4	Fe^{2+}	30 GPa	Strong red-shift of absorption edge across visible region; above 20 GPa obscures Fe^{2+} CF band at 11,000 cm^{-1}	[5,8]
goethite & lepidocrocite	α- and γ-FeOOH	Fe^{3+}	10 GPa	Absorption edges of O $\rightarrow$ Fe^{3+} CT starting at 16,700 cm^{-1} shift beyond 12,500 cm^{-1}, obscuring some Fe^{3+} spin-forbidden CF bands	[9]
crocoite	$PbCrO_4$	Cr^{6+}	5 to 10 GPa	O $\rightarrow$ Cr^{6+} CT band at 25,000 to 20,000 cm^{-1} intensifies and broadens considerably; absorption edge at 18,180 cm^{-1} shifts to 11,000 cm^{-1} by 5 to 6 GPa	[2,10]
vanadinite	$Pb_5(VO_4)_5Cl$	V^{5+}	5 to 10 GPa	O $\rightarrow$ V^{5+} CT band at 28,500 to 22,250 cm^{-1} intensifies and broadens considerably	[10]
wulfenite	$PbMoO_4$	Mo^{6+}	10 GPa	Absorption edge of O $\rightarrow$ Mo^{6+} CT shows strong red-shift	[2,11, 12]

Sources of data : [1] Mao (1973); [2] Mao (1976); [3] Balchan & Drickamer (1959); [4] Abu–Eid & Langer (1978); [5] Mao & Bell (1972a); [6] Sung, Singer, Parkin & Burns (1977); [7] Shankland, Nitsan & Duba (1979); [8] Mao & Bell (1972b); [9] Mao & Bell (1974b); [10] Abu-Eid (1976); [11] Bell & Mao (1974); [12] Mao & Bell (1975b).

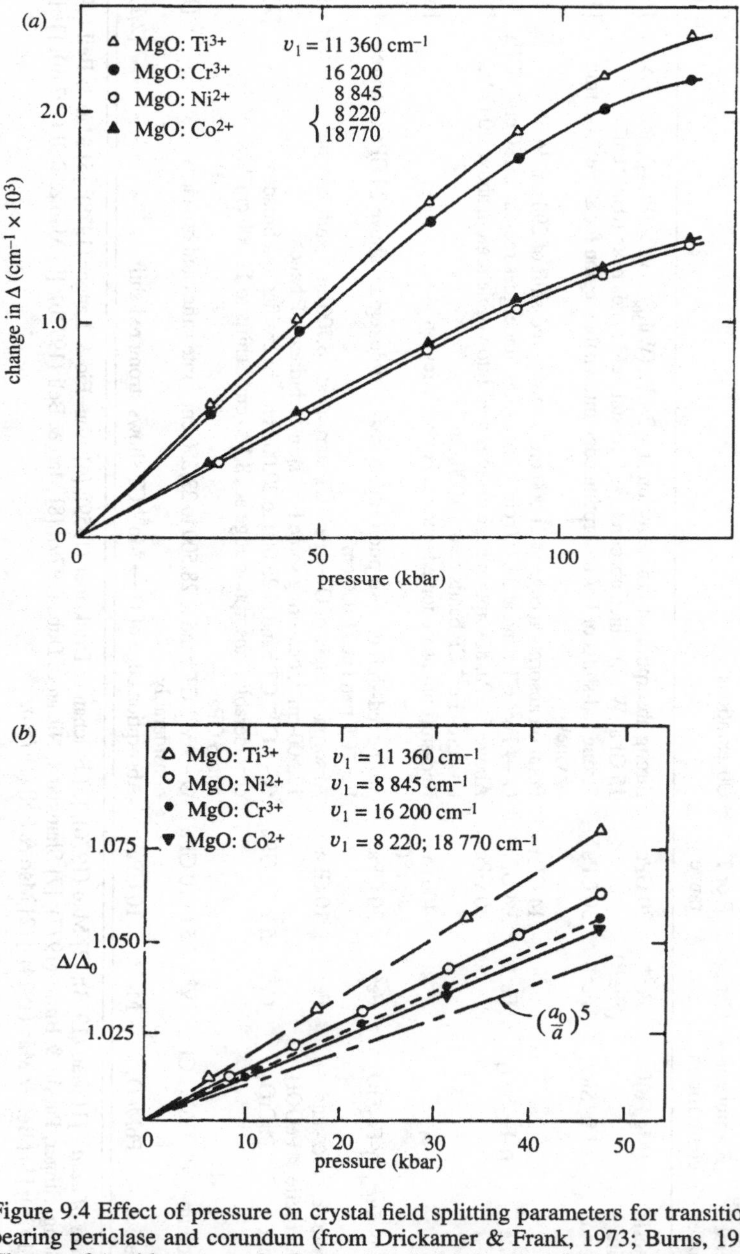

Figure 9.4 Effect of pressure on crystal field splitting parameters for transition metal-bearing periclase and corundum (from Drickamer & Frank, 1973; Burns, 1985a). (*a*) Change of Δ with pressure for four cations in MgO; (*b*) and (*c*) (*on facing page*) pressure variations of Δ with changes of the unit cell a_0 dimension of MgO and Al_2O_3.

(*c*)

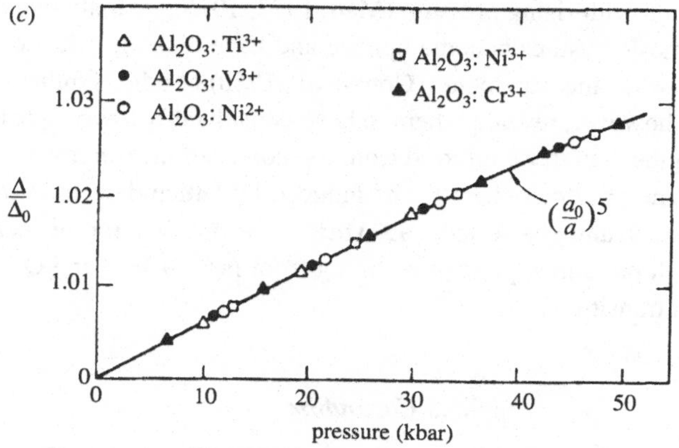

insensitive to pressure variations, whereas heterogeneous IVCT transitions (e.g., $Fe^{2+} \rightarrow Ti^{4+}$) show red-shifts to longer wavelengths at high pressures. In both cases, pressure-induced variations of IVCT bands are distinct from CF bands. For oxygen → metal charge transfer transitions, increased pressure and temperature both cause red-shifts of absorption edges into the visible region (table 9.4).

In the following sections, results are described for the more important minerals and structure-types occurring in the Mantle.

9.5.2 Periclase

Effects of pressure on crystal field spectra of transition metal-doped periclases (Drickamer and Frank, 1973) are shown in fig. 9.4*a.* Crystal field bands move to higher energies so that Δ values rise at elevated pressures for MgO containing Ti^{3+}, Cr^{3+}, Ni^{2+} and Co^{2+}, indicating an increase of CFSE for each of these cations at high pressures. The optical spectra of Fe^{2+}-doped periclase (cf. fig. 5.2) and magnesiowüstites show similar pressure-induced shifts of the Fe^{2+} CF bands (Shankland *et al.*, 1974; Mao, 1973). However, pressure variations of Δ (and hence CFSE) deviate from the Δ versus R^{-5} dependence, eq. (2.17). Thus, as shown in fig. 9.4*b*, the fractional changes in Δ_p at elevated pressures over the one atmosphere value (Δ_0) plotted against the change in lattice parameter (where a_0 = 421.2 pm for MgO at one atmosphere) are somewhat higher than the predicted values. This suggests that the compressibility of a transition metal ion centred in its octahedral site in the periclase structure is significantly higher than that of the [MgO_6] octahedron.

In high-pressure spectra of some iron-rich magnesiowüstites, absorption edges attributed to oxygen → Fe charge transfer transitions shift rapidly into the visible region with rising pressure (Mao, 1973, 1976), so that at very high pressures magnesiowüstites become opaque and have high electrical conductivities. Shock-wave measurements (Goto *et al.*, 1980) of other synthetic magnesiowüstites, however, revealed them to have considerably lower opacities at short wavelengths than those inferred from diamond cell measurements (Mao, 1973, 1976). The presence of Fe^{3+} ions induced by lattice defects in magnesiowüstites (Hirsch and Shankland, 1991) may be responsible for the increased opacities at high pressures, perhaps resulting from $Fe^{2+} \rightarrow Fe^{3+}$ and $O \rightarrow Fe^{3+}$ charge transfer transitions.

9.5.3 Corundum

High-pressure spectral measurements of transition metal-doped Al_2O_3 (Drickamer and Frank, 1973), including several studies of ruby (e.g., Stephens and Drickamer, 1961; Goto *et al.*, 1979), show pressure-induced shifts of the crystal field bands to higher energies. Results summarized in fig. 9.4*c* indicate increased Δ values for several cations hosted by Al_2O_3 that are in remarkably close agreement with trends predicted from the Δ versus R^{-5} relationship, eq. (2.20). This suggests that, in contrast to periclase, the compressibilities of $[AlO_6]$ and transition metal-centred octahedra are comparable in the corundum structure.

9.5.3.1 Ruby

Measurements of ruby spectra at elevated temperatures (McClure, 1962; Parkin and Burns, 1980) have revealed that there is a general broadening, intensification (integrated areas), and shift of band maxima to longer wavelengths between room temperature and 900 °C, the effect being most pronounced for the 18,000 cm^{-1} band, particularly in the E||c spectrum. Above 500 °C, the colour of a typical ruby becomes green (Poole, 1964).

The effect of increased pressure on ruby optical spectra (cf. fig. 4.10) is to cause intensification and a blue-shift of the spin-allowed transitions to shorter wavelengths at high pressures (Drickamer and Frank, 1973; Gaffney and Ahrens, 1973; Goto *et al.*, 1979). For example, in unpolarized light, the $^4A_2 \rightarrow {}^4T_2$ transition moves from 18,020 cm^{-1} to 19,880 cm^{-1} at 46 GPa, and the $^4A_2 \rightarrow {}^4T_1$ transition shifts from 24,690 cm^{-1} to 25,910 cm^{-1} at 32 GPa. On the other hand, the ruby R_1 fluorescence line, representing one peak of the spin-forbidden $^4A_2 \rightarrow {}^2E(G)$ transition doublet, shows a red-shift, decreasing from about 14,405 cm^{-1} (694.2 nm) at 1 atm (1.013 x 10^5 Pa) to about 13,600 cm^{-1} at 112 GPa. This pressure-induced shift of the ruby fluorescence spectrum is utilized as an *in situ* pressure gauge in experiments using the diamond anvil cell

(Mao and Bell, 1978; Mao *et al.*, 1978). When small grains of ruby are added to the diamond cell, the pressure P may be calculated from the equation

$$P[\text{GPa}] = 0.380\,8\left\{\left[1+\frac{(\Delta\lambda)}{694.2}\right]^5 - 1\right\} \tag{9.1}$$

where $\Delta\lambda$ is the pressure-induced wavelength shift of the ruby R_1 fluorescence line initially at 694.2 nm at atmospheric pressure. Equation (9.1) is applicable to static pressures up to 172 GPa. Under quasi-hydrostatic conditions, the equation becomes (Mao *et al*, 1986)

$$P[\text{GPa}] = 0.248\,4\left\{\left[1+\frac{(\Delta\lambda)}{694.2}\right]^{7.665} - 1\right\}, \tag{9.2}$$

which is applicable up to 80 GPa.

9.5.4 Spinels

Spectral measurements of Fe^{2+}-bearing oxide spinels ($MgAl_2O_4$) indicate a blue-shift of the tetrahedral Fe^{2+} CF band (Shankland *et al.*, 1974). Similar shifts of the octahedral Cr^{3+} CF bands occur in magnesiochromites (Mao and Bell, 1975a). Of greater relevance so far as the Earth's interior is concerned are the high-pressure spectral measurements of octahedrally coordinated cations occurring in silicate spinels (γ-phase or ringwoodite). Studies of γ-Fe_2SiO_4 indicate that increased pressure induces a blue-shift of the Fe^{2+} CF band, initially located near 11,430 cm^{-1}, towards the visible region (Bell and Mao, 1969; Mao and Bell, 1972a). Similar trends are shown by the octahedral Ni^{2+} CF bands in γ-Ni_2SiO_4 (Yagi and Mao, 1977). A strong pressure-induced red-shift of the oxygen $\rightarrow$ Fe absorption edge causes the Fe^{2+} CF band in γ-Fe_2SiO_4 to become obscured above 20 GPa. This effect, occurring also in magnesiowüstites and olivines, may influence geophysical properties of the interior, including radiative heat transfer and electrical conductivity discussed in §9.10.

9.5.5 Garnets

As noted in §5.4.1 and illustrated in fig. 5.5, Fe^{2+} ions located in the eight-coordinated distorted cube site in almandine garnets produce CF bands located near 4,400 cm^{-1}, 5,800 cm^{-1} and 7,600 cm^{-1}. However, only the latter two bands have been investigated at high pressures due to experimental difficulties below 5,000 cm^{-1} (Shankland *et al.*, 1974; Bell and Mao, 1969; Balchan and Drickamer,

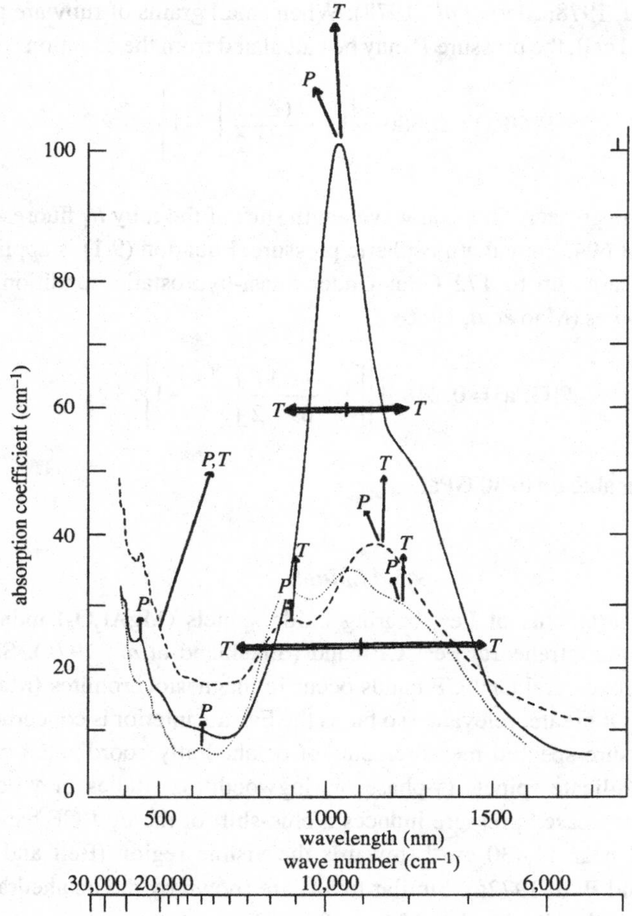

Figure 9.5 Characteristic absorption spectral profiles of olivine (cf. fig. 5.9). Superimposed on the spectra are arrows showing pressure- and temperature-induced variations of positions, widths and intensities of Fe^{2+} crystal field peaks, and red-shift of the oxygen → Fe charge transfer absorption edge (from Burns, 1982).

1959; Smith and Langer, 1983). Strong blue-shifts of these two Fe^{2+} CF bands are observed such that by 10.1 GPa they have moved to 6,700 cm^{-1} and 8,200 cm^{-1}, respectively (Smith and Langer, 1983). In uvarovite, the Cr^{3+} CF bands also show pronounced blue-shifts (Abu-Eid, 1976). In andradite, the weak spin-forbidden Fe^{3+} CF bands are unaffected by pressure (Abu-Eid, 1976).

The spectra of different garnets measured at 400 °C (Parkin and Burns, 1980) show temperature-induced red-shifts in almandine of all three Fe^{2+} CF bands to 4,150 cm^{-1}, 5,850 cm^{-1} and 7,620 cm^{-1}. Increased temperature also induces small red-shifts and intensification of spin-forbidden CF peaks located in the visible region originating from Mn^{2+} and Fe^{3+} ions in spessartine and andradite, respectively.

9.5.6 Olivines

Several spectral measurements at high pressures and temperatures have been made of Mg–Fe^{2+} olivines (Bell and Mao, 1969; Mao and Bell, 1971, 1972a,b; Shankland *et al.*, 1974; Abu-Eid, 1976; Smith and Langer, 1982a,b; Langer, 1988), and general trends are summarized in fig. 9.5 (Burns, 1982). The dominant absorption band at 9,600 to 9,260 cm^{-1} (1,040 to 1,080 nm), originating from Fe^{2+}/M2 site cations and most conspicuous in γ-polarized (E‖*a*) spectra, shifts by about 130 cm^{-1}/GPa to shorter wavelengths at high pressures (Shankland *et al.*, 1974). A similar blue-shift is observed for one of the Fe^{2+}/M1 site CF bands located near 8,000 cm^{-1} (1,200 nm), while the other band situated near 11,000 cm^{-1} (900 nm) shows minor variation or a small red-shift to longer wavelengths (Bell and Mao, 1969; Mao and Bell, 1972a,b; Abu-Eid, 1976). These trends have been interpreted as indicating that the olivine M1 site becomes less distorted at high pressures (Abu-Eid, 1976; Huggins, 1976; Hazen, 1977). All Fe^{2+} CF bands in olivine intensify at elevated pressures. The conspicuous weak spin-forbidden Fe^{2+} peaks near 16,260 cm^{-1} (615 nm) and 22,075 cm^{-1} (453 nm) also intensify slightly and show small pressure-induced displacements to longer wavelengths (Mao and Bell, 1972a).

Numerous high-temperature spectral measurements of olivines, often complicated by oxidation of structural Fe^{2+} ions to produce segregations of nanocrystalline Fe_2O_3, show negligible or small red-shifts of the Fe^{2+} CF bands (Fukao *et al.*, 1968; Aronson *et al.*, 1970; Sung *et al.*, 1977; Burns and Sung, 1978; Shankland *et al.*, 1979; Singer and Roush, 1985). However, as shown schematically in fig. 9.5, significant thermal broadening and intensification of these absorption bands occurs.

The ultraviolet absorption edges of several olivine compositions have been studied extensively (Bell and Mao, 1969; Mao and Bell, 1972a; Abu-Eid and Langer, 1978; Burns and Sung, 1978; Shankland *et al.*, 1979; Smith and Langer, 1982a,b, 1983) following predictions (Drickamer, 1965) that at sufficiently high pressure (14 GPa) and temperature (1,000 °C) in the Upper Mantle the absorption edge has moved well into the infrared. Such pressure-induced red-shifts of this absorption edge are accompanied by opacity in fayalite above 15 GPa and a rapid increase of its electrical conductivity. Opacity in forsterite Fa_{18} is barely discernible by 30 GPa (Mao, 1973). However, above this pressure, even that forsterite showed increased electrical conductivity with rising pressure. The significance of these observations in connection with electrical conductivity and heat transfer by radiation through the Mantle is discussed in §9.10.

9.5.7 Pyroxenes

In orthopyroxenes, the very intense Fe^{2+}/M2 site bands near 11,000 cm^{-1} and 5,000 cm^{-1} in spectra measured at atmospheric pressure (cf. fig. 5.15), as well as the Fe^{2+}/M1 site band located at 8,333 cm^{-1}, show pressure-induced blue-shifts (Shankland *et al.*, 1974; Mao and Bell, 1971). At elevated temperatures, the Fe^{2+}/M2 site '1 micron' (11,000 cm^{-1}) band shows negligible thermal shifts (Sung *et al.*, 1977; Singer and Roush, 1985). However, the '2 micron' (5,000 cm^{-1}) band shows a significant red-shift in orthopyroxenes and a blue-shift in clinopyroxenes (Singer and Roush, 1985). These effects, which have important applications in remote-sensed spectral measurements of hot planetary surfaces, are described in chapter 10 (§10.7).

Pressure-induced intensification and blue-shifts of absorption bands around 470 to 478 nm (21,280 to 20,920 cm^{-1}) in extraterrestrial titanian pyroxenes enabled Ti^{3+} ions to be identified in meteorites (Mao and Bell, 1974a) and specimens from the Moon (Abu-Eid, 1976). These effects are described in §4.4.1 and §4.7.2.5.

9.5.8 Silicate perovskite

Optical spectral data are not currently available for the (Mg,Fe)SiO_3 perovskite phase. However, approximate locations of absorption bands may be predicted based on data for other ferromagnesian silicates. In the perovskite A site, average metal–oxygen distances to the eight nearest-neighbour oxygens are comparable to those in pyrope–almandine garnets (table 9.1), so that the Fe^{2+} CF bands in silicate perovskites should be located in the region 4,500–8,000 cm^{-1}. If Fe^{3+} ions are present in octahedral B sites as a result of lattice defects in the perovskite phase (Hirsch and Shankland, 1991), IVCT transitions between Fe^{2+} (A site) and Fe^{3+} (B site) cations separated by only 279 pm might produce a broad intense band in the 14,000 to 16,000 cm^{-1} (710 to 625 nm) region (cf. $Fe^{2+} \rightarrow Fe^{3+}$ IVCT bands in tables 4.2 and 9.3). Such Fe^{3+} ions might also cause the absorption edge of $O \rightarrow Fe^{3+}$ charge transfer bands to extend into the visible region in optical spectra of perovskites.

9.6 Polyhedral bulk moduli from high pressure spectra

Elastic constants of minerals are the key to understanding geophysical properties of the Earth's interior. Bulk modulus and rigidity parameters, for example, influence the velocities of seismic waves through the Earth. Numerous experimental and semi-empirical approaches have been developed to evaluate bulk moduli or incompressibilities of rock-forming minerals with varying degrees of self-consistency between the various methods (Anderson and Anderson, 1970; Anderson, 1972; Hazen and Finger, 1982). Contributions to the crystal bulk modulus by component polyhedral bulk moduli of individual coordination sites have also been determined, using X-ray data and crystal structure refinements at elevated pressures (Hazen and Finger, 1979, 1984). Polyhedral bulk moduli of oxide and silicate minerals may also be derived from measurements of pressure-induced variations of crystal field spectra of transition metal-bearing phases (Shankland *et al.*, 1974; Abu-Eid, 1976; Burns, 1985a, 1987a). The spectrally derived polyhedral bulk modulus parameter, κ^s, resembles the X-ray determined polyhedral bulk modulus, κ^X (Hazen and Finger, 1982).

The bulk modulus of a mineral, κ, represents the pressure-dependence of its molar volume, V, defined by

$$\kappa = -V\left(\frac{dP}{dV}\right) = \frac{1}{\beta}, \tag{9.3}$$

where β is the volume compressibility of the crystal. In general κ increases as the crystal is compressed at elevated pressures. By convention, crystal bulk moduli of minerals are usually expressed as zero pressure (1 bar or $\approx 10^{-4}$ GPa) values, denoted as κ_0, and their pressure derivatives κ_0' (where $\kappa_0' = d\kappa/dP$) at near-zero pressure. For many oxide and silicate minerals, κ_0' is approximately 4.0.

Several empirical equations of state (EOS), representing correlations between pressure and molar volume data have been defined, one of which is the Birch–Murnaghan EOS,

$$P = \frac{3}{2}\kappa_0\left[\left(\frac{V_0}{V}\right)^{7/3} - \left(\frac{V_0}{V}\right)^{5/3}\right]\left\{1 + \frac{3}{4}(\kappa_0' - 4)\left[\left(\frac{V_0}{V}\right)^{2/3} - 1\right]\right\}, \tag{9.4}$$

where V and V_0 are the molar volumes of the crystal at elevated pressure and 1 bar, respectively. Since κ_0' approximates 4 for many solids, the first-order Birch–Murnaghan EOS becomes

$$P = \frac{3}{2}\kappa_0\left[\left(\frac{V_0}{V}\right)^{7/3} - \left(\frac{V_0}{V}\right)^{5/3}\right]. \tag{9.5}$$

Equations (9.4) and (9.5) indicate that an experimental technique providing pressure variations of a volume–related parameter may be utilized to estimate

Table 9.5. *Polyhedral bulk moduli of transition metal-bearing minerals**

Mineral (structure)	Cation	Site	κ_0^s (spectral) (GPa)	Spectral data sources	κ_0^X (X-ray) (GPa)	X–ray data sources	κ_0 (bulk crystal) (GPa)	Bulk moduli sources
periclase	Mg^{2+}	oct			161	[15]	162.1	[20]
magnesiowüstite	Fe^{2+}	oct	148	[1]	153 (FeO)	[15]	$162.7+17x_{Fe}$	[21]
(periclase)	Ni^{2+}	oct	185	[2]	196 (NiO)	[15]		
(periclase)	Co^{2+}	oct	139	[2]	185 (CoO)	[15]		
(periclase)	Cr^{3+}	oct	133	[2]				
corundum	Al^{3+}	oct			240	[15]	257	[20]
ruby	Cr^{3+}	oct	303	[2]	230 (Cr_2O_3)	[15]		
ruby	Cr^{3+}	oct	693	[3]				
(corundum)	V^{3+}	oct	303	[2]	180	[15]		
spinel ($MgAl_2O_4$)	Fe^{2+}	tet	263	[1]			196.9	[20]
chromite	Cr^{3+}	oct	145	[4]				
γ-Ni_2SiO_4	Ni^{2+}	oct	148	[5]	150	[15,16]		
γ-Fe_2SiO_4	Fe^{2+}	oct	179	[6]			197	[20]
							$213–16x_{Fe}$	[21]
pyrope	Mg^{2+}	8-fold			130	[15,17]	173	[20]
almandine	Fe^{2+}	8-fold	143 to183	[7]				
grossular	Al^{3+}	oct			220	[15,17]		
uvarovite	Cr^{3+}	oct	453	[8]			165.2	[22]
diopside	Mg^{2+}	M1			105	[18]	113	[20]
fassaite	Fe^{2+}	M1	145	[9,10]	140	[19]		

fassaite	Ti^{3+}	M1	195	[11]		
ferrosilite	Fe^{2+}	M1	≈71	[8]	$108-10x_{Fe}$	[21]
ferrosilite	Fe^{2+}	M2	≈72	[12]		
fayaliite	Fe^{2+}	M1	≈140	[12]	137.9	[20]
fayalite	Fe^{2+}	M2	≈97	[13]	$129.1+8.4x_{Fe}$	[21]
piemontite	Mn^{3+}	M3	266	[8]		
perovskite	Fe^{2+}	A	(70 to 140)	[14]	261	[23,24]

* Based on Burns (1985a, 1987a)

Sources of data : [1] Shankland, Duba & Woronow (1974); [2] Drickamer & Frank (1973); [3] Goto, Ahrens & Rossman (1979); [4] Mao & Bell (1975a); [5] Yagi & Mao (1977); [6] T. Yagi & H.-K. Mao, unpubl; [7] Smith & Langer (1983); [8] Abu-Eid (1976); [9] Hazen, Bell & Mao (1977); [10] Mao, Bell & Virgo (1977); [11] Mao & Bell (1974a); [12] Mao & Bell (1971); [13] Smith & Langer (1982a,b); [14] estimated, based on values for almandine and pyroxene M2 sites; [15] Hazen & Finger (1979); [16] Finger, Hazen & Yagi (1979); [17] Hazen & Finger (1978); [18] Levien & Prewitt (1981); [19] Hazen & Finger (1977); [20] Watanabe (1982); [21] Jeanloz & Thompson (1983); [22] Bass (1986); [23} Kudoh, Ito & Takeda (1987); [24] Mao, Hemley, Fei, Shu, Chen, Jephcoat & Wu (1991).

zero-pressure bulk moduli. One such parameter is the crystal field splitting, Δ, which is related to the spectrally derived bulk modulus, κ^s, by

$$\kappa^s = \frac{5}{3}\Delta\left(\frac{dP}{d\Delta}\right). \tag{9.6}$$

Substitution into the first-order Birch–Murnaghan EOS, eq. (9.5), enables $\kappa_0{}^s$ to be calculated from

$$\kappa_0{}^s = \frac{2}{3}P\left[\left(\frac{\Delta_P}{\Delta_0}\right)^{7/5} - \left(\frac{\Delta_P}{\Delta_0}\right)\right]^{-1}, \tag{9.7}$$

where the spectrally determined zero-pressure polyhedral bulk modulus, $\kappa_0{}^s$, may be determined from the crystal field splitting parameters, Δ_P and Δ_0, measured at elevated pressure and 1 bar, respectively.

Bulk moduli data of transition metal-bearing minerals derived from the high-pressure crystal field spectral data cited in table 9.2 are summarized in table 9.5. There is remarkably good agreement between spectrally determined polyhedral bulk moduli ($\kappa_0{}^s$) and those obtained from high-pressure X-ray data ($\kappa_0{}^X$). The data for Fe^{2+} in structures providing six-fold coordination sites (periclase, γ-Fe_2SiO_4, olivine, pyroxenes) and in the spinel (tetrahedral sites) and garnet (eight-coordinated sites) structures conform with the pattern of higher incompressibilities for sites having low coordination numbers (Hazen and Finger, 1982). The $[CrO_6]$ octahedron is significantly incompressible in the corundum and garnet structures, perhaps reflecting the very large CFSE of Cr^{3+} ions in these structures.

Equation (9.7), which embodies the Δ versus R^{-5} relationship, is more applicable to mineral structures in which transition metal ions occupy regular, or only slightly distorted, octahedral, tetrahedral or cubic coordination polyhedra. When the coordination polyhedra are very distorted, it becomes increasingly difficult to evaluate splittings of lower-level t_{2g} orbitals and, hence, to accurately estimate the Δ parameter from crystal field spectra. The spectral method for evaluating polyhedral bulk moduli is best suited, therefore, to transition metal-bearing periclase, garnet, spinel and perovskite structure-types which, fortuitously, are major phases of the Lower Mantle. Furthermore, because these phases are cubic (isotropic), crystal field spectra do not show polarization dependencies, thereby facilitating evaluations of the Δ parameter. On the other hand, estimated values of $\kappa_0{}^s$ for Upper Mantle minerals such as olivine and pyroxenes are less accurate due to their anisotropies, the occurrence of transition metal ions in two different sites, effects of cation ordering in very distorted coordination polyhedra and the consequent pleochroic spectra of these minerals in polarized light.

9.7 Spin-pairing transitions

9.7.1 Background

The various phase transformations occurring in the Mantle, including those described in §9.3, have one feature in common: they are characterized by decreased molar volumes. As a result, each phase change is favoured by rising pressure accompanying increased depth in the Mantle. Since the change of electronic configuration in a transition metal ion from high-spin to low-spin leads to a contraction of metal–oxygen distances, such as those observed for Co(III) in oxide phases (§6.2; fig. 6.2.*b*), spin-pairing should be induced by increased pressure. Indeed, spin-pairing transitions are well-documented in Co(III) oxides possessing structure-types relevant to the Mantle (Burns, 1989b), including Co_2O_3 corundum (Chenavas *et al.*, 1971), Co_3O_4 spinel (O'Neill, 1985; Mocala *et al.*, 1992) and $LaCoO_3$ and other rare earth cobalt perovskite phases (Burns, 1989b; Liu and Prewitt, 1991). The spin-pairing transition of cobalt in Co_2O_3 is accompanied by a five per cent decrease of average Co–O distances (table 9.1). In Co_3O_4, $LaCoO_3$ and other rare earth cobalt perovskites, low-spin and high-spin Co^{3+} ions coexist over a range of temperatures (Burns, 1989b). Since Fe^{2+} is isoelectronic with Co^{3+}, similar high-spin to low-spin transitions are possible in dense Fe^{2+}-bearing phases. Some sulphide minerals such as pyrite (FeS_2) and violarite ($FeNi_2S_4$) already contain low-spin Fe^{2+} at normal pressures, while pressure-induced spin-pairing transitions have been induced in Fe^{2+} ions present in MnS_2 (Bargeron *et al.*, 1971; Drickamer and Frank, 1973). Such observations have led to the suggestion that spin-pairing transitions may occur in Fe^{2+} in dense oxide phases of the Mantle (Fyfe, 1960).

9.7.2 Features of high-spin and low-spin states

Two alternative electron distributions between t_{2g} and e_g orbitals corresponding to high-spin and low-spin states are possible for octahedrally coordinated transition metal ions with $3d^4$ to $3d^7$ configurations (c.f. table 2.2), including Fe^{2+}, $3d^6$, the most abundant cation. Its ground-state configurations, high-spin $(t_{2g})^4(e_g)^2$ or low-spin $(t_{2g})^6$, have either four (quintet state) or zero (singlet state) unpaired electrons, respectively. An intermediate triplet spin-state, $(t_{2g})^5(e_g)^1$, with two unpaired electrons is also possible, by analogy with Co^{3+} in $LaCoO_3$ at elevated temperatures (Burns, 1989b). Properties of Fe^{2+} ions with

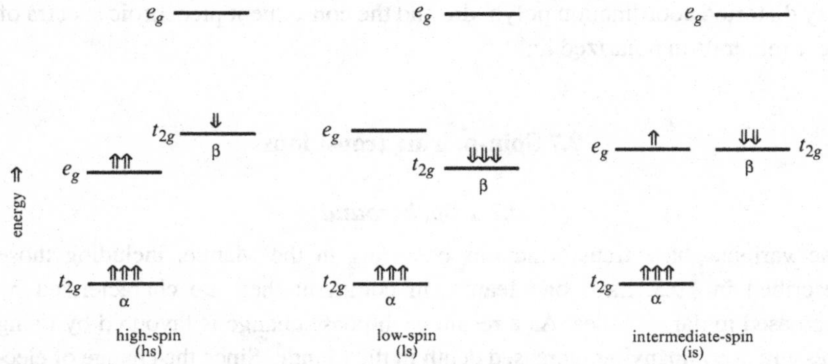

Figure 9.6 Electronic configuration of $3d^6$ transition metal ions, Fe^{2+} and Co^{3+}, in octahedral coordination. The schematic energy levels for high-spin (hs), low-spin (ls) and intermediate-spin (is) states are shown. Clockwise and anticlockwise spins are designated by α and β (from Burns, 1989b).

each of these electronic configurations are summarized in fig. 9.6. Large differences of magnetic susceptibility, ionic radius and CFSE exist between the high-spin and low-spin states of Fe^{2+}. Among the cations that exist in high-spin and low-spin states, Fe^{2+} has the lowest spin-pairing energy, amounting to less than 19,150 cm^{-1} (table 2.5), as estimated from spectra of gaseous transition metal ions (Huheey, 1983, p. 380). Factors that influence the adoption of either of these spin-states are discussed in §2.5. Thus, Hund's rule which states that ground states with maximum spin-multiplicities minimize interelectronic repulsion, favours the adoption of the high-spin configuration. On the other hand, an increased crystal field stabilization energy, achieved when all six 3*d* electrons fill the t_{2g} orbitals, favours the low-spin configuration. However, this gain of CFSE is offset by having to pair two more electrons in the t_{2g} orbitals of ferrous iron.

9.7.3 Evidence for spin-pairing transitions in iron minerals

Positive identification of high-spin to low-spin transitions in minerals containing Fe^{2+} ions coordinated to oxygen has been elusive. Low-spin Fe^{2+} ions do exist in violarite ($FeNi_2S_4$) and pyrite (FeS_2) in which Fe–S distances are about 226 pm. This distance is approximately 15 per cent shorter than that in hauerite (MnS_2) containing high-spin Mn^{2+} (§11.5). Moreover, reversible spin-pairing of Fe^{2+} has been induced in iron-bearing MnS_2, commencing at pressures of 4 GPa and becoming complete by about 14 GPa (Bargeron *et al.*, 1971).

The existence of low-spin Fe^{2+} in pyrite formed the basis of an early estimate that spin-pairing of Fe^{2+} would occur in olivine at a depth in the Mantle of about 1,400 km (Fyfe, 1960). Subsequently, deerite, $Fe^{2+}{}_6Fe^{3+}{}_3Si_6O_{20}(OH)_5$, occurring in high-pressure blueschist metamorphic rocks, was shown to have an anomalous temperature dependency of its magnetic susceptibility (Carmichael *et al.*, 1966), leading to the suggestion that low-spin iron cations occur in this silicate mineral. However, Mössbauer spectral measurements of deerite (Amthauer *et al.*, 1980) revealed that such anomalies originate from thermally activated electron delocalization in this opaque silicate mineral (§4.8).

In gillespite, $BaFeSi_4O_{10}$, the crystal field spectra illustrated in fig. 3.3 indicate a maximum energy separation between the 3*d* orbital energy levels of about 20,000 cm^{-1} (table 3.7) which exceeds the spin-pairing energy (19,150 cm^{-1}) for the field-free gaseous Fe^{2+} ion (table 2.5). An instantaneous and reversible colour change in gillespite from red to colourless, which was observed when crystals mounted on (001) cleavage planes were viewed in a diamond cell at pressures of about 2.6 GPa, led to the suggestion that a pressure-induced spin-pairing transition had occurred in Fe^{2+} ions located in the square planar sites (Strens, 1966b). However, X-ray structure refinements of gillespite (Hazen and Burnham, 1974; Hazen and Finger, 1983), optical spectroscopy (Abu-Eid *et al.*, 1973), and Mössbauer spectral measurements (Huggins *et al.*, 1976), all made in diamond anvil cells at high pressures, demonstrated that the pressure-induced red to blue colour change is caused by the Fe^{2+} square planar coordination site becoming a flattened tetrahedron as a result of a phase change in gillespite. No pressure-induced spin-state change had been induced in Fe^{2+} in gillespite after all.

Shock-wave experiments causing phase changes in wüstite, $Fe_{0.94}O$, and Fe_2O_3 have also produced controversial evidence for spin-pairing transitions in iron cations (Jeanloz and Ahrens, 1980; Jackson and Ringwood, 1981; Goto *et al.*, 1982; Syono *et al.*, 1984).

9.7.4 Calculated spin-pairing transition pressures in the Mantle

There have been several semi-empirical calculations of transition pressures and metal–oxygen distances needed to induce spin-pairing transitions in Fe^{2+} in the Mantle (Fyfe, 1960; Strens, 1966b, 1969, 1976; Gaffney, 1972; Gaffney and Anderson, 1973; Oshnishi, 1978; Oshnishi and Sugano, 1981; Sugano and Oshnishi, 1984; Sherman, 1988). Most of these approaches using concepts of crystal field theory indicate that low-spin Fe^{2+} ions are likely to exist in the Lower Mantle. For example, using the Δ versus R^{-5} relationship, eq. (2.20), the Birch–Murnaghan EOS, eq. (9.5), and pressure variations of the Racah *B* parameter described in chapter 11 (§11.2.4), spin-pairing in Fe^{2+} in FeO was predicted to occur at about 25 to 40 GPa, and at 70 to 130 GPa for other transition metal ions in MnO, CoO and Fe_2O_3 phases (Oshinishi, 1978; Goto *et al.*, 1982; Ohnishi and Sugano, 1981). Molecular orbital calculations, which demonstrated that pressure-induced spin-pairing of Fe^{2+} is unlikely to occur in gillespite (Tossell, 1976), indicated that the onset of a high-spin to low-spin transition could take place in Fe^{2+} in FeO in the deepest half of the Lower Mantle (Sherman, 1988). The calculations also suggested that at depths exceeding 1,700 km, the majority of Fe^{2+} in magnesiowüstite could exist in the low-spin state (Sherman, 1991). Evidence for the partitioning of Fe^{2+} ions into the octahedral sites of synthetic $(Mg,Fe)SiO_3$ perovskites (Jackson *et al.*, 1987; Williams *et al.*, 1989) raises the possibility that low-spin Fe^{2+} ions might also exist in perovskite phases in the Lower Mantle.

9.7.5 Consequences of spin-pairing transitions in the Mantle

The presence of low-spin Fe^{2+} ions in magnesiowüstite and perovskite phases at depth in the Lower Mantle would have interesting ramifications. First, since there is a decrease in the number of unpaired electrons and, hence, magnetic susceptibility when spin-pairing takes place in iron and other transition metal ions (fig. 9.6), the change to a low-spin configuration in Fe^{2+} at high pressures will affect magnetic properties of the Lower Mantle. Second, pressure-induced contraction of interatomic distances could lead to enrichments of low-spin Fe^{2+} ions and increased density of host phases at depth within the Earth (Strens, 1969; Gaffey and Anderson, 1973). If two compounds contain cations of the same charge, the one with the larger cation generally has the lower melting point. At low pressures, the Fe end-members of ferromagnesian silicate solid-solution series have lower melting points than the Mg end-members, with the result that crystallization leads to enrichment of Mg in the mineral and Fe in the liquid. At pressures above the spin-pairing transition point, however, low-spin Fe^{2+} may have a smaller ionic radius than Mg^{2+} and possibly lead to a reversal of melting point relationships (cf. fig. 8.2). Phase heterogeneity could also result, leading to a Mg-rich silicate perovskite containing Mg^{2+} ions in the A sites coexisting with a Fe-rich perovskite with low-spin Fe^{2+} in the B sites. These two factors could have produced enrichment of iron the the Lower Mantle and the production of a relatively magnesium-rich Upper Mantle from residual silicate melts during the thermal evolution of the Earth. Such hypotheses, which have an important bearing on the mechanism of differentiation of the Earth, the scale of Mantle convection and whether convective decoupling occurs between the Upper Mantle and Lower Mantle, may ultimately depend on pressure-induced spin-pairing transitions in iron.

9.7.6 Crystal field spectra of low-spin Fe^{2+} in Mantle minerals

If low-spin Fe^{2+} ions are present in periclase and perovskite phases in the Lower Mantle, the question arises where absorption bands might be located in crystal field spectra of these low-spin Fe^{2+}-bearing oxide structures. The spectra of isoelectronic Co(III) compounds provide a clue. In octahedrally coordinated low-spin Co^{3+} ions, two absorption bands corresponding to the spin-allowed transitions $^1A_{1g} \rightarrow {}^1T_{1g}$ and $^1A_{1g} \rightarrow {}^1T_{2g}$ (see figs 3.8 and 5.26) occur at 16,500 cm^{-1} and 24,700 cm^{-1} in the $[Co(H_2O)_6]^{3+}$ complex, and at 32,400 cm^{-1} and 39,000 cm^{-1} in $K_3Co(CN)_6$ (table 5.17). Low-spin Fe^{2+} ions occur in $K_4Fe(CN)_6$, the crystal field spectrum of which contains absorption bands at 31,000 cm^{-1} and 37,040 cm^{-1}. These energies for low-spin Fe^{2+}, which is octahedrally coordinated by CN^- anions, are thus displaced by 1,400 cm^{-1} and 1,960 cm^{-1} to lower energies relative to corresponding bands for low-spin Co^{3+} surrounded by six CN^- anions. Assuming similar displacements relative to low-spin Co^{3+} ions octahedrally coordinated to oxygen in the hexahydrated ion, absorption bands at 15,100 cm^{-1} (662 nm) and 22,660 cm^{-1} (441 nm) might be expected for low-spin Fe^{2+} present in octahedral sites in oxides. Since blue-shifts and intensification of these crystal field bands are expected at elevated pressures (table 9.2), periclase and perovskite phases containing low-spin Fe^{2+} ions would absorb strongly in the visible region. This contrasts with absorption bands for high-spin Fe^{2+} ions which are located towards the infrared region in Lower Mantle minerals.

9.8 Distributions of transition metal ions in the Mantle

Inversion of phase relationships induced by spin-pairing in Fe^{2+} ions provides one mechanism for possibly enriching this transition element in the Lower Mantle. Other, more general mechanisms influencing element fractionations, are the effects of pressure on relative sizes, crystal field stabilization energies, bond-types and oxidation states of the cations.

9.8.1 Changes of coordination number

A guiding principle of crystal chemistry is that the coordination number of a cation depends on the radius ratio, R_c/R_a, where R_c and R_a are the ionic radii of the cation and anion, respectively. Octahedrally coordinated cations are predicted when $0.414 < R_c/R_a < 0.732$, while four-fold (tetrahedral) and eight- to twelve-fold (cubic to dodecahedral) coordinations are favoured for radius ratios below 0.414 and above 0.732, respectively. The ionic radii summarized in Appendix 3 indicate that in octahedral coordination with oxygen (ionic radius 140 pm), divalent transition metal ions have radius ratios in the range 0.52 to 0.59 lying towards the upper limit (0.732) predicted for octahedral coordination, whereas the trivalent cations with $R_c/R_a = 0.38$ to 0.48 straddle the lower limit (0.414).

A dominant theme pervading the phase transformations predicted in the Mantle (fig. 9.2) is that the coordination numbers of host cations, notably Si and Mg, increase in the denser polymorphs, particularly in silicate perovskites. This is a direct consequence of the greater compressibility of the highly polarizable oxygen anions compared to the cations, resulting in an increase of radius ratio for most cations at high pressures. Since divalent cations lie nearer to the upper limit of radius ratio for octahedral coordination, they are more susceptible to pressure-induced increases of coordination number than are the trivalent cations, except in the case of low-spin Fe^{2+}. Thus, high-spin ions such as Cr^{2+}, Mn^{2+}, Fe^{2+} and Co^{2+} are expected to acquire coordination numbers greater than six at high pressures and to occur in perovskite A sites in the Lower Mantle. However, cations such as Ti^{4+}, Ti^{3+}, V^{3+}, Cr^{3+} and Fe^{3+} are expected to be remain in octahedral coordination throughout the Mantle and to occur in magnesiowüstite and perovskite B sites.

9.8.2 Changes of CFSE

According to the Δ versus R^{-5} relationship, eq. (2.17), crystal field splittings and stabilization energies should increase with decreasing metal–oxygen distances. Rising pressure produces shorter interatomic distances (table 9.1) so that Δ and CFSE values are expected to increase at high pressures provided the coordination symmetry remains unchanged. The crystal field spectral data for minerals summarized in table 9.2, which generally show pressure-induced blue-shifts of the crystal field bands, confirm that Δ's, and hence CFSE's, increase with rising pressure. As a result, strong preferences for octahedral sites are expected for cations acquiring large CFSE's, such as Cr^{3+}, V^{3+}, Ti^{3+} and Ni^{2+}, thereby aiding their enrichments in magnesiowüstite and the perovskite B sites. The reduced CFSE's attained in sites with eight- to twelve-coordinations may also induce high-spin Fe^{2+} ions, as well as other large divalent cations acquiring CFSE's such as Cr^{2+} and Co^{2+}, to occupy octahedral sites provided by the Lower Mantle phases. This is suggested by experimental studies of Fe^{2+}–Mg^{2+} partitioning in post-spinel phases (Yagi *et al.*, 1979; Bell *et al.*, 1979; Mao *et al.*, 1982; Fei *et al.*, 1991) discussed in §9.9.2.

9.8.3 Changes of bond-type

Interpretations of the crystal chemistry of transition metal ions by crystal field theory are based on a predominatly ionic model of the chemical bond. As interatomic distances decrease at high pressures, the covalent character of metal–oxygen bonds are expected to increase significantly. One measure of the change of the ionic–covalent character of bonds in transition metal-bearing minerals is the Racah *B* parameter discussed in chapter 11 (§11.2). High-pressure optical spectral measurements of Cr^{3+} in ruby (Drickamer and Frank, 1973) and garnet (Abu-Eid, 1976; Abu-Eid and Burns, 1976) show that the Racah *B* parameter decreases by less than three per cent per 10 GPa, suggesting that the degree of covalent character of chromium–oxygen bonds increases by less than 20 per cent towards the Lower Mantle. Similar estimates may be obtained for iron–oxygen bonds from pressure variations of the isomer shift parameter in high-pressure Mössbauer spectra of silicate minerals (Huggins, 1976). Molecular orbital theory and energy level calculations described in chapter 11 indicate that while significant covalent bonding occurs in transition metal-bearing oxides and increases somewhat at elevated pressures, there is remarkably good agreement for calculated crystal field energies with the R^{-5} law (Tossell, 1976; Sherman, 1991).

9.8.4 Changes of oxidation state

A variety of spectroscopic evidence has demonstrated that transition metal ions appear to undergo reversible pressure-induced reduction to lower oxidation states at very high pressures (Burns, 1976a). Originally, Mössbauer spectroscopy demonstrated that in most synthetic iron compounds and in some minerals there is a reduction of Fe^{3+} to Fe^{2+} ions at high pressures (Drickamer *et al.*, 1970; Burns *et al.*, 1972c; Huggins, 1976), leading to one viewpoint that negligible amounts of ferric iron occur in the Lower Mantle (Tossell *et al.*, 1972). The processes are reversible once effects of locked–in strain are removed by pulverizing samples returned to atmospheric pressure. Optical spectral measurements have demonstrated that pressure-induced reduction of Mn^{3+} to Mn^{2+} and Cu^{2+} to Cu^{+} ions also occur at high pressures (Wang and Drickamer, 1973; Ahsbahs *et al.*, 1974; Gibbons *et al.*, 1974).

The available data (Burns, 1976a) show a trend towards the stabilization of progressively lower oxidation states at high pressures across the first transition series. Such observations indicate that higher oxidation states characteristic of the Earth's surface (Ti^{4+}, Cr^{3+}, Fe^{3+}, Ni^{2+}) may become unstable under the high *P*,*T* conditions of the Lower Mantle. Exotic oxidation states such as Ti(III), Cr(II), Fe(I), and Ni(I) could be prevalent towards the Core–Mantle boundary, particularly if hosted by sulphide phases.

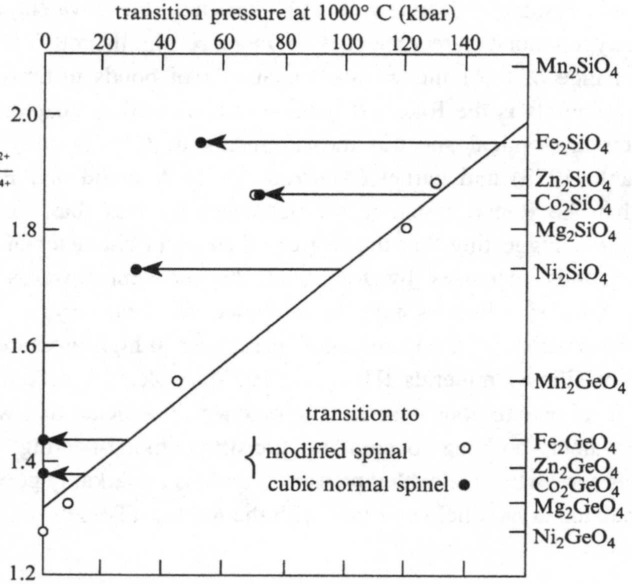

Figure 9.7 Transition pressures at 1,000 °C for various olivines transforming to the β-phase ('modified' spinel or wadsleyite) or γ-phase (spinel or ringwoodite) as a function of the ionic radius ratio: divalent cation (R^{2+}) to Si^{4+} or Ge^{4+} (M^{4+}) (from Syono *et al.*, 1971). Note that the cations acquiring excess CFSE in spinel over olivine (e.g., Fe^{2+}, Co^{2+}, Ni^{2+}) deviate from a linear trend.

9.9 Influence of CFSE on phase equilibria in the Mantle

The presence of transition metal ions in mineral structures may significantly modify phase equilibria at high pressures as a result of increased CFSE acquired by certain cations in the dense oxide phases believed to constitute the Lower Mantle. The additional electronic stabilization can influence both the depth in the Mantle at which a phase transition occurs and the distribution coefficients of transition metals in coexisting dense phases in the Lower Mantle.

9.9.1 The olivine → spinel phase transition

Experimental phase equilibrium studies have confirmed deductions from seismic velocity data that below 400 km, olivine and pyroxene, the major constituents of Upper Mantle rocks, are transformed to denser polymorphs with the garnet, γ-phase (spinel) and β-phase (wadsleyite) structures (fig. 9.2). In transformations involving olivine to the β- or γ-phases, transition pressures appear to be anomalously low for several silicates and germanates containing specific transition metal ions (Syono *et al.*, 1971). Thus, data plotted in fig. 9.7 correlating transition pressures, P_t, with the ratio of the ionic radius of a divalent cation to that of Si^{4+} or Ge^{4+}, indicate that phases containing Fe^{2+}, Co^{2+} and Ni^{2+} deviate significantly from a linear trend. For example, the transition pressures at 1000 °C for Fe_2SiO_4, Co_2SiO_4 and Ni_2SiO_4 are expected to be about 15, 13 and 10 GPa, respectively, from the P_t versus r_R^{2+}/r_M^{4+} plot in fig. 9.7. The experimental P_t values are, in fact, much lower, corresponding to about 5, 7 and 3 GPa for Fe_2SiO_4, Co_2SiO_4 and Ni_2SiO_4, respectively. The apparent discrepencies were attributed to excess CFSE's for Fe^{2+}, Co^{2+} and Ni^{2+} ions in the spinel structure compared to those provided in the olivine structure (Syono *et al.*, 1971).

Attempts to quantify the effect of CFSE on the transition pressure, P_t, of olivine → spinel phase changes have utilized crystal field parameters estimated from high-pressure absorption spectra (Mao and Bell, 1972b; Yagi and Mao, 1977). Thus, CFSE was estimated to lower the P_t of Fe_2SiO_4 by about 9.8 GPa, conforming more closely with the linear trend of cation radius ratio versus transition pressure shown in fig. 9.7. In the case of Ni_2SiO_4, however, the CFSE difference for Ni^{2+} between the olivine and spinel phases indicated that P_t should be lowered by 14 GPa which, although apparently the correct direction according to fig. 9.7, is too large in comparison with other olivine → spinel transitions. In order to define more accurately the role of CFSE on phase transitions, it was suggested (Yagi and Mao, 1977) that information is needed on pressure dependencies of CFSE, compressibilities of the phases, and energy levels of cations in distorted octahedral sites.

In a more rigorous evaluation of the effect of crystal field stabilization (CFS) on olivine → spinel transformations in the system: Mg_2SiO_4–Fe_2SiO_4, free energy changes, ΔG_{CFS}, due to different crystal field splittings in γ-Fe_2SiO_4 and fayalite were estimated from the relationship

$$\Delta G_{CFS} = \Delta H_{CFS} - T\Delta S_{CFS} \tag{9.8}$$

as functions of P and T using crystal field spectral data for the two phases at high pressures and elevated temperatures (Burns and Sung, 1978). In eq. (9.8), ΔH_{CFS} and ΔS_{CFS} are differences of crystal field stabilization enthalpies and electronic configurational entropies (§7.4) between Fe_2SiO_4 spinel and fayalite at the transition temperature and pressure. The calculations indicated that ΔG_{CFS} is always negative, showing that CFS always promotes the olivine → spinel transition in Fe_2SiO_4 and expands the stability field of spinel at the expense of olivine in the system: Mg_2SiO_4–Fe_2SiO_4. Because of crystal field effects, the transition pressure for the olivine → spinel phase change in Fe_2SiO_4 is lowered by about 5 GPa at 1,000 °C. Since olivine in the Upper Mantle contains approximately ten mole per cent Fe_2SiO_4, the transition pressures may be decreased by as much as 0.5 GPa due to the presence of Fe^{2+} ions in the crystal structure. This means that the depth of the olivine → spinel transformation in a typical Upper Mantle forsteritic olivine is 15 km shallower than it would be if iron were absent from this silicate.

9.9.2 Partitioning of iron in post-spinel phases in the Lower Mantle

The transformation of olivine to spinel which delineates the onset of the Transition Zone in the Mantle is isochemical. However, in the Lower Mantle the $(Mg,Fe)_2SiO_4$ stoichiometry appears to be unstable with respect to denser oxide structure-types. One post-spinel transformation that has been the subject of numerous investigations is the spinel to periclase plus perovskite transition

$$\underset{\text{spinel}}{(Mg,Fe)_2SiO_4} \rightarrow \underset{\text{perovskite}}{(Mg,Fe)SiO_3} + \underset{\text{magnesiowüstite}}{(Fe,Mg)O} \,, \tag{9.9}$$

which is believed to occur below 670 km (fig. 9.2). Such a disproportionation reaction raises the possibility that breakdown products will have different Fe/Mg ratios. Semi-quantitative estimates indicate that the CFSE attained by Fe^{2+} in octahedral sites in the periclase structure ($0.4\Delta_o$) will be higher than that acquired in the sites with eight- to twelve-fold coordination in the perovskite structure, amounting to approximately $0.6\Delta_d$ or $0.3\Delta_o$, assuming identical metal–oxygen distances in the sites and $\Delta_d = 0.5\Delta_o$ [cf. eq. (2.9)].

In investigations of phase relations in the system: $MgO–FeO–SiO_2$ at high pressures and temperatures (Yagi *et al.*, 1979; Bell *et al.*, 1979; Mao *et al.*, 1982; Fei *et al.*, 1991), compositions of $(Mg_{1-x}Fe_x)_2SiO_4$ yielded magnesiowüstites with higher Mg/Fe ratios than coexisting perovskites; for example,

$$(Mg_{0.85}Fe_{0.15})_2SiO_4 \rightarrow (Mg_{0.96}Fe_{0.04})SiO_3 + (Mg_{0.74}Fe_{0.26})O \,. \tag{9.10}$$

The strong partitioning of iron into magnesiowüstites is the result of higher CFSE of Fe^{2+} ions in the periclase structure.

Various estimates have been made of the CFSE of Fe^{2+} in dense oxide structures modelled as potential Mantle mineral phases (Gaffney, 1972; Burns, 1976a). All estimates indicate that octahedrally coordinated Fe^{2+} (in periclase, for example) has a considerably higher CFSE than Fe^{2+} ions in the eight- to twelve-coordination sites in the perovskite structure. Thus, the CFSE of Fe^{2+} in magnesiowüstite was estimated to exceed that acquired in perovskite by more than 53 kJ/mole. This large CFSE factor is predicted to cause iron enrichment in magnesiowüstite and a depletion of iron in the coexisting perovskite phase in the Lower Mantle.

There are two factors that may complicate this simple model of iron fractionation in the Lower Mantle. The first is a possible phase change of MgO from the NaCl to CsCl structure-type, which would produce a change of coordination symmetry from octahedral (six-fold) to cubic (eight-fold). Assuming identical metal–oxygen distances in the two sites and $\Delta_c = -\frac{8}{9}\Delta_o$ [eq. (2.8)], the CFSE of high-spin Fe^{2+} in the CsCl structure-type could be *higher* ($0.6\Delta_c = 0.6 \times \frac{8}{9}\Delta_o = 0.533\Delta_o$) than that in octahedral coordination in magnesiowüstite ($0.4\Delta_o$), depending on actual metal–oxygen distances in the two structures. The second complication is the possibility of a spin-pairing transition in Fe^{2+} ions discussed in §9.8. The smaller ionic radius of low-spin Fe^{2+} could induce iron to enter octahedral sites in the perovskite structure, particularly in aluminous perovskites (Burns, 1989b), leading to iron enrichment in the $(Mg,Fe)SiO_3$ phase of the deep Lower Mantle.

9.10 Geophysical properties of the Earth's interior

9.10.1 Radiative heat transport in the Mantle

Transfer of heat by radiation is an important contributor to heat flow at elevated temperatures in the Mantle. However, it has long been known that minerals absorbing radiation in the near-infrared and visible regions control the radiative heat transport mechanism in the Mantle (Clark, 1957). As a result, numerous attempts have been made to assess the temperature- and pressure-induced variations of absorption bands in relevant Fe^{2+}-bearing Mantle minerals, including garnet, pyroxenes, olivine, spinel, and magnesiowüstite, on the effective radiative conductivity of the Earth's interior (Pitt and Tozer, 1970; Shankland, 1970; Aronson *et al.*, 1970; Schatz and Simmons, 1972; Mao and Bell, 1972a; Nitsan and Shankland, 1976; Shankland *et al.*, 1979).

The energy transfer of photons through a grey body (i.e. one in which absorption by photons is finite, non-zero and independent of wavelength) is given by (Stacey, 1969, p. 248)

$$K_r = \frac{16\, n^2 S\, T^3}{3\, \alpha}. \qquad (9.11)$$

Here K_r is the effective radiative conductivity, n is the mean refractive index, T

is temperature, α is the mean absorption coefficient (table 3.2), and Stefan's constant, S, has the value

$$S = \frac{2\pi^5 k^4}{15c^2 h^3} = 5.67 \times 10^{-8}\,\mathrm{J\,m^{-2}K^{-4}s^{-1}}, \qquad (9.12)$$

where k is Boltzmann's constant, h is Planck's constant and c is the velocity of light. Stefan's constant appears in the equation

$$\frac{\mathrm{d}E}{\mathrm{d}t} = S\,T^4, \qquad (9.13)$$

which relates the power (i.e. rate of energy emission per unit area) of an ideal black body to absolute temperature T.

Because K_r increases proportionally to T^3, it was regarded to be a dominant mechanism of heat transfer under the high-temperature conditions of the Mantle. Here, temperatures range from 1,500 to 2,000 °C within the Transition Zone to about 3,500 °C at the Core–Mantle boundary (Ringwood, 1975; Liu and Bassett, 1986). At these temperatures, the peak maximum of radiation energy flux for a black body corresponds to about 1,400 nm (1,500 to 2,000 °C) and 800 nm (3,500 °C), respectively. Therefore, it is important to assess how much radiation in the 800 to 1,400 nm wavelength region is absorbed by electronic excitations in transition metal-bearing Mantle minerals.

The high P and T spectral data summarized in tables 9.2, 9.3 and 9.4 indicate that so far as CF transitions are concerned, there is a general pressure-induced intensification and blue-shift of absorption bands towards the visible region. However, temperature-induced intensification appears to predominate over pressure variations. In iron-bearing olivine, silicate spinel, and magnesiowüstite, however, the red-shift of the absorption edge due to oxygen → iron CT transitions and, perhaps, intensification of $Fe^{2+} \rightarrow Fe^{3+}$ IVCT transitions are likely to absorb radiation in the visible region at very high temperatures.

Radiative heat transport through olivine has been discussed extensively (e.g., Fukao *et al.*, 1968; Shankland, 1970; Schatz and Simmons, 1972; Schärmeli, 1979; Shankland *et al.*, 1979). The radiative thermal conductivity, K_r of forsteritic olivine increases with rising temperature and would contribute to heat flow in the Upper Mantle (Shankland *et al.*, 1979). However, values of K_r for olivine are considered to be rather low to satisfactorily explain the dissipation of the Earth's internal heat by radiation and lattice conduction alone. Note, however, that Fe^{2+} CF transitions in almandine, pyroxenes (M2 site) and, perhaps, silicate perovskites absorb strongly in the wavelength range 1,250 to

2,200 nm (see figs 5.6 and 5.19; §9.5.8), the region where the radiation energy flux for a black body is a maximum in the temperature range 1,000 to 1,500 °C. Thus, Fe^{2+} ions in pyroxene or garnet phases in the Upper Mantle and Transition Zone, and perhaps silicate perovskite in the Lower Mantle, would reduce radiative heat transfer in these regions of the Earth's interior.

There is greater uncertainty about radiative heat transfer through silicate spinels in the Transition Zone and magnesiowüstite plus perovskite assemblages in the Lower Mantle. The strong pressure-induced visible-region absorption in γ-Fe_2SiO_4 and magnesiowüstite at 30 GPa suggest that these phases would effectively block black-body radiation at wavelengths shorter than 1,500 nm (Mao, 1973). This conclusion is at variance with other results obtained from shock-compressed absorption spectral measurements of two magnesiowüstites, $Wü_{14}$ and $Wü_{26}$, (Goto *et al.*, 1980). Such measurements demonstrated that absorption of visible-region radiation was considerably lower in these specimens than in magnesiowüstites used in static measurements performed in the diamond cell, which were inferred to contain some ferric iron. Under highly reducing conditions (i.e. absence of Fe^{3+} ions) in the Lower Mantle, radiative conductivity through magnesiowüstite would be comparable with phonon lattice conductivity at depths in the Earth of the order of 1,000 km.

In the Lower Mantle, if high-spin Fe^{2+} ions are present in magnesiowüstite, they would effectively block black-body radiation in the red and near-infrared regions. The presence of coexisting Fe^{3+} ions resulting from lattice defects (Hirsch and Shankland, 1991) would absorb black-body radiation in the visible region as a result of absorption by $Fe^{2+} \rightarrow Fe^{3+}$ IVCT and oxygen → Fe charge transfer transitions which are strongly intensified by pressure. Similar effects would result from transitions involving Fe^{2+} (A site) → Fe^{3+} (B site) IVCT involving these cations in close proximity (279 pm) in the perovskite structure. The presence of low-spin Fe^{2+} ions in the periclase and perovskite phases, however, would totally confine the crystal field transitions to the visible region (§9.7.6), so that radiative heat transfer could occur at near-infrared wavelengths in the Lower Mantle.

9.10.2 Electrical conduction in the Mantle

The magnitude and mechanism of electrical conductivity within the Earth have also been studied extensively. The radial distribution of electrical conductance in the Earth's interior may be estimated from studies of secular variations of the Earth's magnetic field. These measurements indicate a steep increase of electrical conductivity between depths of 400 and 1,000 km in the Mantle

(Stacey, 1969; Banks, 1972; Mao, 1973), apparently correlating with the phase changes occurring in the Transition Zone. As a result, electrical conductivities of the Lower Mantle are some four to five orders of magnitude higher than those in the Upper Mantle. Several explanations have been proposed for the relatively higher electrical conductivity of the Lower Mantle, including charge transfer transitions observed in high-pressure optical spectra of oxide and silicate minerals (tables 9.3 and 9.4).

At a given pressure, the electrical conductivity σ is related to absolute temperature by

$$\sigma = {}_i\Sigma\, \sigma_{0i} \exp\left(-E_a/kT\right), \tag{9.14}$$

where E_a is the activation energy, k is the Boltzmann constant, and σ_{0i} is the electrical conductivity originating from a specific conducting mechanism i. One such mechanism involves extrinsic conduction of the type oxygen → metal CT, information about which may be obtained from high-pressure optical spectral and electrical conductivity measurements of Mantle mineral assemblages. Another mechanism for high electrical conductivity is $Fe^{2+} \rightarrow Fe^{3+}$ IVCT involving clusters of iron cations in the crystal structures in mixed-valence minerals(§4.8).

The numerous studies of olivines at ambient pressures have shown that several factors affect electrical conductivities, including increasing iron contents, oxygen fugacity, and temperature (Duba and Nicholls, 1973; Shankland, 1975). Pressure also significantly influences electrical conductivities. Several studies have shown that the electrical conductivity of fayalite increases by several orders of magnitude at very high pressures (Mao and Bell, 1972a, 1973c; Mao, 1973; Mashimo *et al.*, 1980). Even in forsteritic olivines, electrical conductivities increase significantly at high pressures (Mao, 1973; Schulien *et al.*, 1978). Similar pressure-induced increases are observed also for the spinel γ-Fe_2SiO_4 and various magnesiowüstites (Mao and Bell, 1972a; Mao, 1973). These data suggested that at 3 GPa the electrical conductivity of magnesiowüstite in the Lower Mantle is six orders of magnitude higher than forsteritic olivine in the Upper Mantle, correlating with the rapid variations of electrical conductivity suggested in the Transition Zone.

Temperature variations of electrical conductivities of olivines and magnesiowüstites at high pressures indicate that activation energies are small, and show a systematic decrease with increasing pressure (Mao and Bell, 1972a; Mao, 1973; Mashimo *et al.*, 1980). These observations correlate with the pressure- and temperature-enhanced opacities and red-shifts of the oxygen → metal CT absorption edges into the visible region spectra of iron-bearing olivine, silicate spinel and periclase phases described earlier. Such spectral variations, together with the rapid increases and large values of electrical conductivities, indicate that an extrinsic conduction mechanism of the type oxygen → iron CT is important in the Lower Mantle. Thus, the very electronic transitions that could inhibit radiative heat transfer in the Earth's interior may enhance its electrical conductivity.

Similar mechanisms appear to facilitate electrical conductivity in perovskites. Laboratory studies of electrical conductivity of anhydrous $(Mg,Fe)SiO_3$ perovskites and perovskite–magnesiowüstite assemblages at high pressures and temperatures in laser-heated diamond anvil cells have yielded values several orders of magnitude lower than estimates of the electrical conductivity of the Lower Mantle (Li and Jeanloz, 1990). The electrical conductivity increased considerably in similar experiments conducted with more iron-rich assemblages (Li and Jeanloz, 1991a), matching results obtained for more magnesian compositions using externally-heated diamond anvil cells (Peyronneau and Poirier, 1989). In other electrical conductivity measurements of hydrous silicate assemblages formed from enstatite, Fs_{12}, at Lower Mantle conditions, high electrical conductivities were attributed to a mixed-valence hydrous 'phase D' approximating the composition $(Mg,Fe)SiH_2O_4$ and believed to contain coexisting Fe^{2+} and Fe^{3+} ions (Li and Jeanloz, 1991b).

To account for these results for $(Mg,Fe)SiO_3$ perovskites, various charge transfer mechanisms have been proposed (Li and Jeanloz, 1990; Hirsch and Shankland, 1991; Sherman, 1991). Lattice defects permitting Fe^{3+} ions to exist in the perovskite structure give rise to oxygen → Fe and $Fe^{2+} \rightarrow Fe^{3+}$ charge transfer transitions, the latter being facilitated by the close proximity (279 pm) of the A sites (Fe^{2+}) to the B sites (Fe^{3+}) in the perovskite structure. The opacity of the hydrous 'phase D' indicates that extensive electron delocalization may occur in its crystal structure.

9.11 Summary

Chapter 9 describes how crystal field energy data obtained from measurements of electronic spectra of minerals at elevated pressures and temperatures may be applied to geophysical and geochemical features of the Mantle.

Composition and mineralogy of the Mantle. The Earth's Mantle consists of Upper and Lower regions separated by the Transition Zone at depths between about 350 km and 650 km. Several phase changes occur in the Transition Zone in which common ferromagnesian silicates of the Upper Mantle, all containing Fe^{2+} ions in distorted six-coordinated sites and tetrahedrally coordinated Si, transform to dense oxide structures with cations occupying regular octahedral or higher coordination sites in periclase and perovskite phases. Although concentrations of Fe, Cr and Ni in the Mantle are significantly higher than their crustal abundances, all of the transition elements are depleted relative to their chondritic abundances due to siderophilic behaviour at high pressures during formation of the Core.

Effects of temperature and pressure on absorption spectra. Energies of crystal field transitions undergo significant blue-shifts to shorter wavelengths at high pressures but show minor shifts at elevated temperatures. Accordingly, CFSE's of Fe^{2+}, Cr^{3+} and Ni^{2+} increase at elevated pressures. Rising temperatures intensify CF bands moreso than do increasing pressures. Energies of metal → metal intervalence charge transfer bands show relatively minor variations with increasing *P* and *T*. However, the IVCT bands intensify considerably at high pressures. Absorption edges of oxygen → metal charge transfer transitions located in the ultraviolet show prominent red-shifts into the visible region at high *P* and *T*. The pressure-induced variations of crystal field splittings provide polyhedral bulk moduli data for transition metal-bearing coordination sites in garnet, spinel, periclase and perovskite structure-types in the Mantle.

Spin-pairing in transition metal ions. High-spin to low-spin transitions are likely in Fe^{2+} ions occurring in the dense oxide phases of the Lower Mantle, by analogy with isoelectronic Co^{3+}-bearing corundum, spinel and perovskite phases. Such spin-pairing transitions in Fe^{2+}-minerals producing reduced magnetic susceptibility would influence magnetic properties of the Lower Mantle. Contraction of interatomic distances and decreased molar volumes of low-spin Fe^{2+}-bearing phases could invert phase relationships in crystal–melt systems, leading to enrichment of iron in Lower Mantle minerals.

Influence of CFSE on phase equilibria and element partitioning in Mantle minerals. Pressure-induced changes of CFSE, coordination number, bond-type and oxidation state affect the distributions and stabilities of transition metal ions in Mantle minerals. The higher CFSE's of Fe^{2+} and Ni^{2+} attained in silicate spinels relative to olivines depress transition pressures of the olivine → spinel transformation. As a result, the onset of the Transition Zone at about 350 km, which is delineated by the olivine → spinel transition in $(Mg_{0.9}Fe_{0.1})_2SiO_4$, is about 15 km shallower than it would be if Fe^{2+} were absent. Strong partitioning of iron into magnesiowüstite at the onset of the Lower Mantle near 650 km, where silicate spinel disproportionates to (Mg,Fe)O plus $(Mg,Fe)SiO_3$, may result from the higher CFSE attained by Fe^{2+} ions in the periclase structure relative to perovskite.

Radiative heat and electrical conductivity of the Mantle. The pressure-induced blue-shifts of Fe^{2+} CF bands, coupled with red-shifts of oxygen → Fe charge transfer absorption in Fe^{2+}-bearing silicate and oxide phases, result in high absorption of visible-region radiation at high *P* and *T*. As a result, although the radiative thermal conductivity of olivine contributes to heat flow in the Upper Mantle, the magnesiowüstite and silicate perovskite phases in the Lower Mantle may adversely affect radiative heat transport through the Earth's interior. On the other hand, intensification of oxygen → Fe CT transitions in the latter phases may enhance the extrinsic electrical conductivity of the Lower Mantle.

9.12 Background reading

Burns, R. G. (1982) Electronic spectra of minerals at high pressures: How the Mantle excites electrons. In *High-Pressure Researches in Geoscience*. (W. Schreyer, ed.; E. Schweizerhart'sche Verlagsbuchhandlung, Stuttgart), pp. 223–46.

Drickamer, H. G. & Frank, C. W. (1973) *Electronic Transitions and the High Pressure Chemistry and Physics of Solids*. (Chapman & Hall, London), 220 pp.

Jeanloz, R. & Knittle, E. (1989) Density and composition of the Lower Mantle. *Phil. Trans. Royal Soc. London*, A328, 377–89.

Langer, K. (1990) High pressure spectroscopy. In *Absorption Spectroscopy in Mineralogy* (A. Mottana & F. Burragato, eds; Elsevier Science Publ., Amsterdam), pp. 227–84.

Liu, L.-G. & Bassett, W. A. (1986) Chemical and mineral composition of the Earth's interior. In *Elements, Oxides, and Silicates*. (Oxford Univ. Press), pp. 234–44.

Sherman, D. M. (1988) High-spin to low-spin transition of iron(II) oxides at high pressures: possible effects on the physics and chemistry of the Lower Mantle. In *Structural and Magnetic Phase Transitions in Minerals*. (S. Ghose, J. M. D. Coey & E. Salje, eds; Springer–Verlag, New York), *Adv. Geochem.*, 7, 113–28.

Sherman, D. M. (1991) The high pressure electronic structure of magnesiowüstite (Mg,Fe)O: Applications to the physics and chemistry of the Lower Mantle. *J. Geophys. Res.*, 96, 14299–312.

10

Remote-sensing compositions of planetary surfaces: applications of reflectance spectra

> *Absorption bands in the visible and near-infrared spectra of Moon and Mars – – correlate well with a narrow choice of minerals. – – they provide a perhaps unique means of remote analysis of some of the abundant mineral phases on the surfaces of the bodies.*
>
> J. B. Adams, *Science*, *159*, 1453 (1968)

10.1 Introduction

Earlier chapters have demonstrated that spectral features of most rock-forming minerals in the visible to near-infrared region originate from the presence of transition elements in their crystal structures. Iron and titanium have higher crustal abundances on terrestrial planets relative to other transition elements and, consequently, are expected to contribute significantly to the reflectance spectra of planetary surfaces. Spectral profiles of sunlight reflected from planetary surfaces, when correlated with measured optical spectra of rock-forming minerals, may be used to detect the presence of individual transition metal ions, to identify constituent minerals, and to determine modal mineralogies of regoliths on terrestrial planets. The origin and applications of such remote-sensed reflectance spectra measured through Earth-based telescopes are described in this chapter.

10.2 Chemical composition of the terrestrial planets

Properties of the terrestrial planets that are central to this chapter are summarized in table 10.1 and information about element abundances is contained in Appendix 1. The crustal abundance data for the Earth indicate the presence of relatively high concentrations of Fe, and to a lesser extent Ti, compared to other first-series transition elements. However, the terrestrial abundance of Fe is smaller than its cosmic abundance due to chemical fractionation during major Earth-forming processes involving Core formation, Mantle evolution and chemical weathering of igneous minerals exposed to the atmosphere. Other first-series transition elements, notably Cu in porphyry calc–alkaline rocks (§8.6.2), Ni in ultramafic rocks (§8.6.3), and Mn in sub-aqueous fissures (§8.7.3), may be concentrated in local areas where deeply eroded igneous rocks, ore deposits, or hydrothermal veins outcrop at the Earth's surface. On Earth, the most common and stable oxidation states of transition metal-bearing minerals occurring in near-surface environments are Fe(II), Fe(III), Mn(IV), Mn(II), Ti(IV), Ni(II), Cu(II), Cr(III) and Cr(VI). Electronic transitions involving these cations dominate the spectra of terrestrial minerals described in earlier chapters. Earth's atmosphere contains N_2 and O_2, which are spectrally inactive in the visible to near-infrared region. However, the small amounts of CO_2, water vapour and ozone have important telluric effects on light reaching spectrometers attached to Earth-based telescopes observing planetary bodies.

Different evolutionary histories of other terrestrial planets have influenced the relative concentrations of the transition elements compared to their cosmic abundances, as suggested by geochemical data for surface rocks on the Moon, Mars and Venus (Appendix 1). Chemical analyses of lunar samples returned from the Apollo and Luna missions show that minerals and glasses occurring on the Moon contain high concentrations of Fe and Ti existing as oxidation states Fe(II), Ti(III) and Ti(IV). Some lunar minerals, notably olivine and opaque oxides, also contain significant amounts of Cr(II), Cr(III) and Mn(II). The lack of an atmosphere on the Moon simplifies interpretation of remote-sensed reflectance spectra of its surface.

On Mars, the *in situ* X-ray fluorescence analyses of its surface during the Viking Lander experiments indicate an iron-rich regolith in which Fe(III) appears to predominate. Although the atmospheric pressure of Mars is much lower than that on Earth, its constituent CO_2 influences reflectance spectral profiles of the Martian surface. Chemical analyses of the surface of Venus performed during the Venera and Vega missions indicate that the composition of the Venusian surface resembles that of Earth by having similar crustal abundances of Fe and Ti. However, the high atmospheric pressure on Venus with its constituent CO_2, SO_2 and other spectrally-active gases prevents light reflected from the Venusian surface reaching Earth. Asteroids such as Vesta and Ceres which may be sources of basaltic achondrites such as eucrites and diogenites, also appear to be characterized by Fe(II), Ti(IV) and perhaps Ti(III)-bearing phases. Although no geochemical data exist for Mercury, this planet may contain relatively low concentrations of transition metal ions, perhaps existing in low oxidation states.

Table 10.1. *Surface features of terrestrial planets measured by remote-sensing techniques*

Planet	Distance from Sun (AU)*	Distance from Earth (10^6 km)	Equatorial diameter (km)	Temperature (K)†	Atmospheric pressure (atmos)	¶Surface mineralogy and petrology; and atmospheric constituents
Earth	1	–	12,756	315	1	Igneous, sedimentary and metamorphic rocks; atmosphere contains N_2, O_2, Ar, CO_2 and H_2O vapour.
Mercury	0.25	91.7	4,878	650	–	Lunar-like surface but with lower Fe^{2+} content; traces of atmospheric Na implies sputtering from Na-bearing minerals.
Venus	0.75	41.4	12,104	740	96	Oxidized basaltic surface containing massive sulphides and Fe^{2+}-Fe^{3+} minerals (e.g., magnetite, ilvaite); CO_2-dominated atmosphere with some CO, HCl, HF and SO_2; H_2SO_4 in clouds.
Mars	1.5	78.3	6,796	280	0.01	Mafic silicates in dark regions, pyroxenes detected locally, olivine suspected; Fe^{3+} phases in bright regions probably include nanophase hematite and sulphate phases; these and clay silicates in global dust are oxidative weathering products; H_2O (frost) condensates; ice at north pole; 'dry ice' (CO_2) at south pole.
Moon	1	0.384	3,476	400	–	Mare basalts contain olivine, Fe^{2+}-bearing plagioclase and basaltic glasses, Fe^{2+}-Ti^{3+}-bearing pyroxenes and glasses, and abundant Fe–Ti–Cr–opaque oxides; highland anorthosites contain shocked and unshocked plagioclases; vitrification (agglutinates) by impact events; solar wind darkening; existence of many lithic units unsampled by Apollo and Luna missions; olivine-rich rocks (troctolites) in central peaks of craters.

Asteroids:						Fe^{2+}-bearing pyroxenes, olivine and plagioclase; some asteroids have fine-grained regolith; impact-produced glasses are rare; at least 14 asteroid spectral classes exist; close resemblances between some asteroid and meteorite spectra suggest parent asteroid bodies; Vesta and basaltic achondrites indicate a differentiated parent body.
Vesta	2.5–3	204	540	180	–	
Pallas	2.5–3	264	610	175	–	
Ceres	2.5–3	264	1,020	175	–	

* Astronomical units: 1 AU = 1.496 x 10^8 km.; † Maximum day-time high temperature at the equator;
¶ Based on McCord, T. M., ed. (1988) *Reflectance Spectroscopy in Planetary Science: Review and Strategy for the Future*. (NASA Spec. Rept.), **493**, 37 pp.

Plate tectonic activity, which is responsible on Earth for subduction zones, spreading centres and obducted ophiolites, as well as associated ore deposits of Cu, Cr and Ni described in §8.6, appears to have been less significant on other terrestrial planets. As a result, local enrichments of these and other transition elements (apart from Fe and Ti) are probably absent on the Moon, Mercury, Venus, Mars and the asteroids. Since Fe and Ti minerals are predominant on terrestrial planets, electronic spectra of Fe^{2+} and Fe^{3+} in silicates and oxides influenced by Ti^{4+} and Ti^{3+} are expected to dominate remote-sensed spectra of their surfaces.

The primary silicate minerals containing Fe and Ti cations on terrestrial planets are olivine, pyroxenes and, perhaps, amphiboles and micas, together with glass and opaque oxide phases. During weathering processes on planetary surfaces, a variety of secondary minerals dominated by Fe^{3+} ions, including oxides, oxyhydroxides, clay silicates and sulphates, are formed by interactions with volatiles in the atmosphere or by hydrothermal activity associated with impact events. How well the electronic spectra of primary ferromagnesian silicate phases can be distinguished from one another and from those of secondary ferric-bearing assemblages is discussed in this chapter.

Each terrestrial planet has spawned its own unique set of problems that have had to be solved in order to better understand the mineral assemblages contributing to remote-sensed reflectance spectral profiles of light scattered from its surface. Selected examples are described throughout the chapter. First, origins of reflectance spectra of minerals and their generation by sunlight penetrating the surface of a planet are discussed. Reflectance spectra from the Moon are then described together with laboratory investigations of mineral mixtures. Effects of temperature on reflectance spectra are considered next, because they refine interpretations of lunar spectra and set constraints on mineral constituents of the surface of Mercury. The need to understand spectral features arising from oxidative weathering products of mafic igneous rocks leads to a description of reflectance spectra of ferric oxide minerals because their spectral characteristics underlie interpretations of remote-sensed spectra of Mars, as well as Venus, which are also complicated by atmospheric absorption at critical wavelengths. Finally, the information that spectral measurements of meteorites brings to bear on the mineralogy of asteroids is discussed. Emerging from such examples is the central theme that information about the surface mineralogy of terrestrial planets, complemented by measurements of samples from the Earth, Moon, Mars and meteorites, has resulted from reflectance spectroscopy used as a remote-sensing tool with Earth-based telescopes. The focus is on optical spectra of pyroxenes, olivine and iron oxide phases, because these minerals appear to dominate the surface mineralogy and visible to near-infrared spectrum profiles, and are particularly diagnostic of the evolution of terrestrial planets.

10.3 Origin of reflectance spectra

Regoliths or soils on surfaces of planets consist of an intimate mixture of different minerals, including primary igneous rocks and alteration products formed by chemical and mechanical weathering and impact events. Such assemblages are present in a range of particle sizes which scatter and absorb sunlight incident on them. The total reflectance, R_T, from such a multicomponent polycrystalline assemblage is the sum of the surface or specular reflectance, R_S, and volume or diffuse reflectance, R_V. The surface reflectance is the portion of the reflected radiation that has not penetrated inside any particle, whereas the volume reflectance is that portion which has been transmitted through some part of one or more particles. There are limiting conditions under which either R_S or R_V is predominant. When the absorption coefficient and mean size of the scattering particles are large, then $R_T \approx R_S$. However, when absorption coefficients and particle sizes are small, which applies to most regolith materials over wide wavelength ranges, then $R_T \approx R_V$ and the Kebulka–Munk theory of diffuse reflectance becomes valid (Wendlandt and Hecht, 1966; Kortum, 1969; Morris *et al.*, 1982).

The Kebulka–Munk remission function, $f(R_\infty)$, is defined as

$$f(R_\infty) = \frac{(1-R_\infty)^2}{2R_\infty} = \frac{\alpha}{S}, \tag{10.1}$$

where R_∞ is the reflectance from an infinitely thick sample which, for most powdered silicate and oxide minerals, occurs within depths of just a few millimeters. The parameters α and S are the absorption coefficient (table 3.2) and scattering coefficient, respectively. Equation 10.1 assumes that the incident radiation is monochromatic and that the scattering processes show no wavelength dependence. A wavelength dependence may occur when dimensions of scattering particles are comparable to or less than the wavelength, λ, of the incident radiation. In visible to near-infrared spectra, this corresponds to dimensions of 0.4 to 2.5 microns (table 3.1). Equation (10.2) is then expressed as

$$f(R_\infty) = \frac{(1-R_\infty)^2}{2R_\infty} = C\left(\frac{\alpha}{S}\right)\lambda^n, \tag{10.2}$$

where C is a constant and exponent n lies within the limits $1 \leq n \leq 4$, the value $n = 4$ being characteristic of Rayleigh scattering.

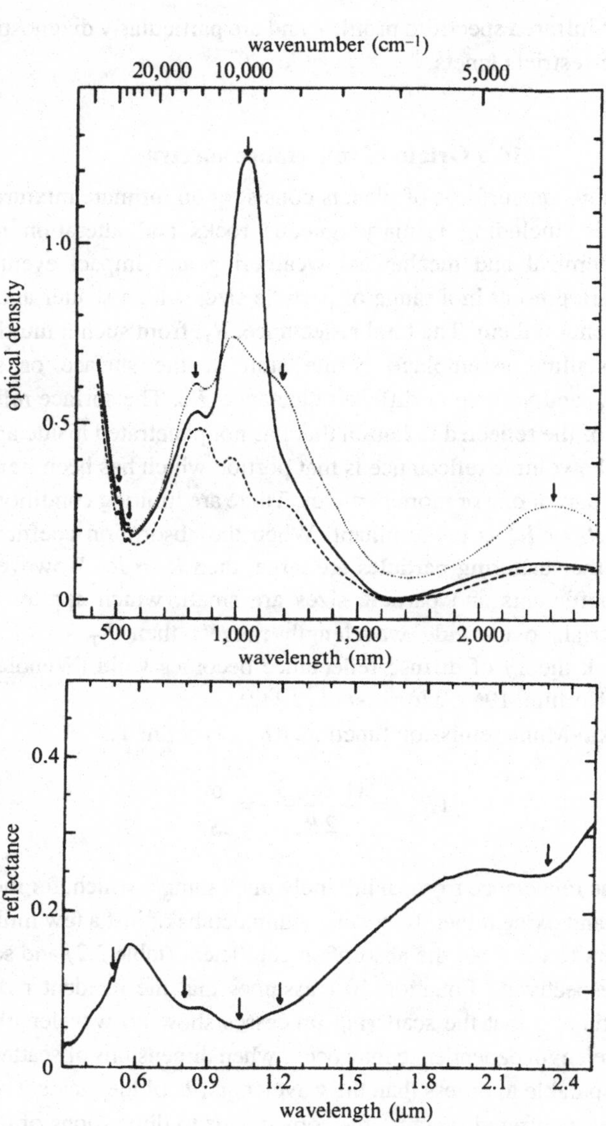

Figure 10.1 Comparisons of visible to near-infrared spectra of calcic pyroxene in transmitted and reflected light. Polarized absorption spectra of single crystals are correlated with the reflectance spectrum of a powdered sample of the same mineral (cf. fig. 5.14).

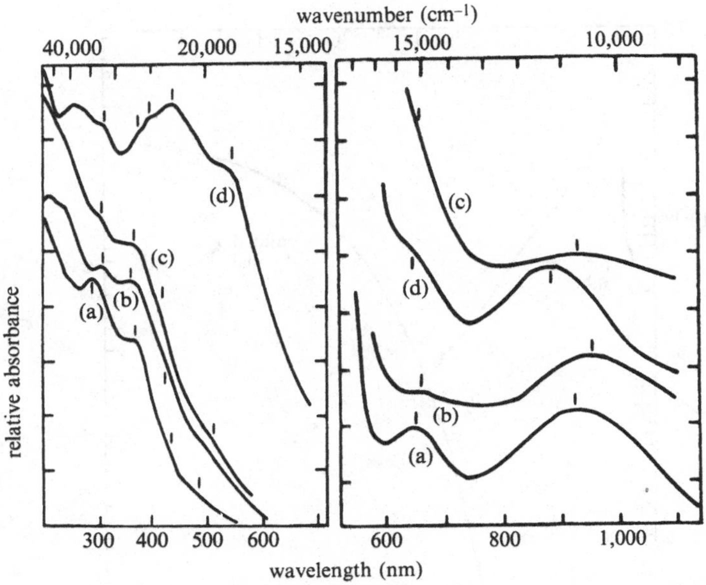

Figure 10.2 Absorption spectra of Fe(III) oxides in the ultraviolet–visible region (left) and visible–near-infrared region (right) (from Sherman & Waite, 1985). (*a*) Goethite; (*b*) lepidocrocite; (*c*) maghemite; and (*d*) hematite. Measured reflectance spectra were converted into absorption spectra by applications of the Kebulka–Munk function. The vertical bars indicate band positions (listed in table 10.2).

Reflectance spectra are usually measured using a diffuse reflectance accessory with an integrating sphere attached to a spectrophotometer. Spectra are referenced against a reflectance standard, such as smoked MgO, barite or Halon powder. The latter is a commercial fluorocarbon that does not absorb water or suffer radiation damage as does MgO. Each of these standards is virtually free of spectral features in the wavelength range 0.3 to 2.5 μm.

In reflectance spectral profiles, regions of minimum reflectance correspond to peak maxima in absorption spectra measured in transmitted light. Comparisons for a pyroxene are shown in fig. 10.1, in which polarized spectra of single crystals are correlated with the diffuse reflectance spectrum of the powdered mineral. Much resolution is lost in reflectance spectral profiles of mineral powders. Thus, bands are broadened and contrasts between band intensities are diminished in reflectance spectra compared to single-crystal polarized spectra. Nevertheless, significant differences of spectrum profiles and peak positions do exist between different minerals, and this forms the basis for identifying mineral asemblages by remote-sensing measurements of planetary surfaces. By using the Kebulka–Munk functions, eqs (10.1) or

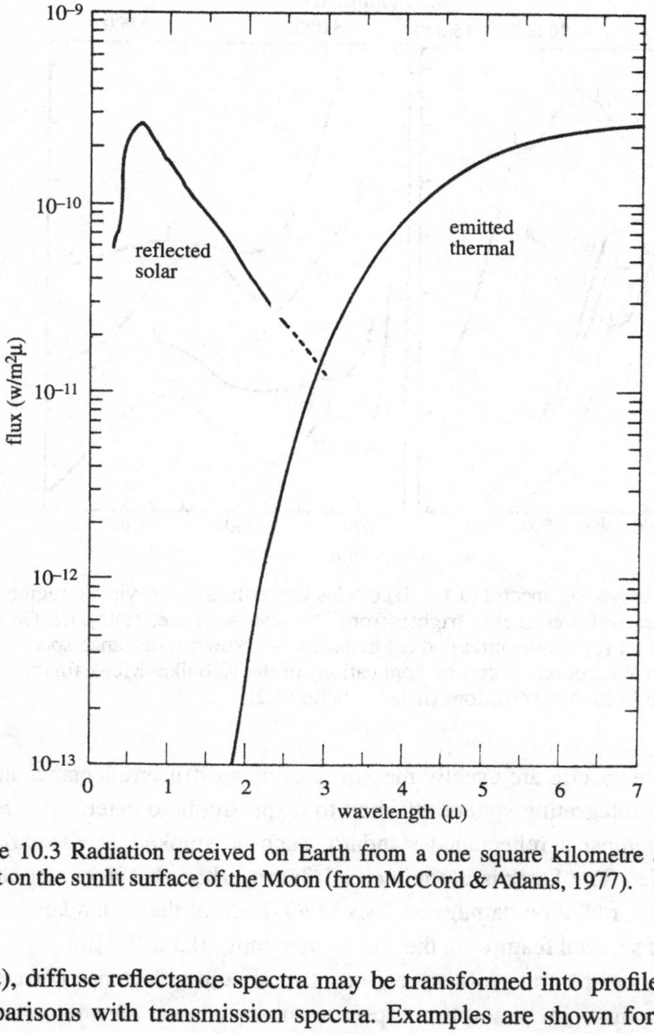

Figure 10.3 Radiation received on Earth from a one square kilometre area of mare basalt on the sunlit surface of the Moon (from McCord & Adams, 1977).

(10.2), diffuse reflectance spectra may be transformed into profiles for direct comparisons with transmission spectra. Examples are shown for Fe^{3+} oxide minerals in fig. 10.2. Here, reflectance spectra measured for goethite, lepidocrocite, maghemite and hematite have been converted to absorption spectra by application of the Kebulka–Munk function (Sherman and Waite, 1985).

10.4 Measurement of telescopic reflectance spectra

In remote-sensed reflectance spectra of planetary surfaces measured through Earth-based telescopes, the Sun is the illuminating source. Light reaching

Earth from a reflecting planetary object such as the Moon or Mars has travelled a complex path from the Sun. If a planet has an atmosphere, the sunlight is first scattered and absorbed by constituents in the atmosphere before undergoing surface reflection and diffuse scattering by heated particles in the regolith. The Sun as a quasi-black-body radiator at 6,000 °C is an efficient radiator of energy in the visible to near-infrared region. However, the planetary body itself becomes a radiator somewhere in the spectral region between about 1.6 μm and 5 μm, depending on the distance of a terrestrial planet from the Sun, and emits back into space a greater proportion of thermal energy from its surface than the energy of scattered light received from the Sun. For example, the spectral flux received on Earth from a square kilometre of sunlit mare basalt on the Moon's surface is plotted in fig. 10.3. Here, the solar radiation flux consists of two parts: a specular reflectance component that has been scattered by particles on the Moon' surface; and a diffuse relectance component that has been absorbed by the regolith, converted to heat and emitted as thermal radiation. Other shining planetary objects have similar energy flux profiles to that of the Moon. However, the relative contribution of the emitted thermal radiation flux increases from hotter planetary surfaces such as Mercury which has a higher daytime surface temperature ($\approx$650 K) than that on the Moon ($\approx$400 K). When the emitted thermal radiation becomes the major fraction of the reflected energy at a given wavelength, interpretations of measured reflectance spectra become very difficult. Figure 10.3 also shows that in the ultraviolet region (wavelengths shorter than about 0.4 μm), the reflected solar flux decreases rapidly. Therefore, the region of most interest in remote-sensed spectroscopic studies of planetary objects is confined to the wavelength range 0.35 to 5 μm where reflected incident solar light is usually the primary source flux from surfaces and atmospheres of sufficient energy to be measured accurately.

Upon entering the Earth's atmosphere, some absorption by atmospheric gases occurs. Below 0.3 μm, absorption by ozone results in the atmosphere being effectively opaque to ultraviolet radiation. Absorption by CO_2 occurs at 1.33, 1.44, 1.60, 1.95, 2.02 and 2.08 μm, while traces of H_2O in the atmosphere absorb very weakly at 0.82 and 0.95 μm and strongly at 1.4, 1.9 and 2.85 to 3.1 μm. These telluric effects may be eliminated, however, by ratioing the energy flux from the reflecting planet to that from a nearby standard star measured in the heavens at the same time. The spectral reflectance is usually scaled to unity at a selected wavength (e.g., 0.56 or 1.02 μm) to standardize albedo effects (i.e. overall reflectance) so that attention can be focussed on absorption spectral features rather than background.

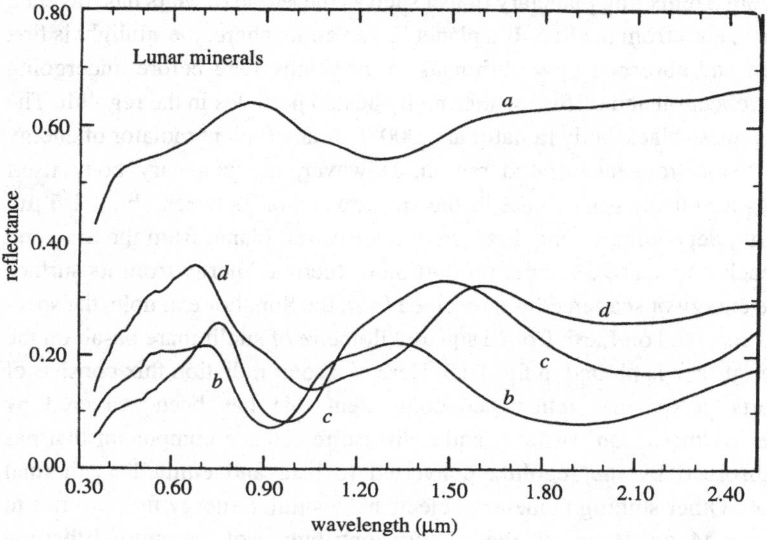

Figure 10.4 Diffuse reflectance spectra of powdered lunar minerals containing Fe^{2+} ions. (*a*) Plagioclase feldspar; (*b*) orthopyroxene; (*c*) calcic clinopyroxene; and (*d*) olivine (courtesy of C. Pieters).

10.5 Reflectance spectra of ferrous silicates in Moon rocks

Ferrous iron in silicate minerals typically gives rise to crystal field bands in the 1 μm region, as demonstrated by the spectra illustrated throughout chapter 5. The positions and intensities of these Fe^{2+} CF bands serve as calibration standards for identifying individual minerals in diffuse reflectance spectra of powdered rocks and surfaces of planets (e.g., Burns *et al.*, 1972a,b; Adams, 1974, 1975; Burns and Vaughan, 1975; Bell *et al.*, 1975; Vaughan and Burns, 1977; Burns, 1989a). Examples are illustrated in fig. 10.4. Lunar calcic plagioclase feldspars containing small amounts of Fe^{2+} replacing Ca^{2+} ions in large coordination sites, produce a weak crystal field band centred around 1.25 μm (§5.9). The 1.05 μm band in olivine spectra originates from Fe^{2+} ions in M2 sites, while Fe^{2+} ions in M1 sites produce the features at about 0.85 and 1.25 μm (§5.4.3.2). Pyroxenes, alone, produce crystal field bands in the 2 μm region originating from Fe^{2+}/M2-site cations (§5.5), the positions of which are more diagnostic of the pyroxene structure-type than the band near 1 μm. Note, however, that tetrahedrally coordinated Fe^{2+} ions in oxide structures and silicate glasses also absorb in the 1.8 to 1.9 μm region (Bell *et al.*, 1975; Nolet *et al.*, 1979), which can complicate interpretations of pyroxene crystal chemistry (White and Keester, 1966; Bancroft and Burns, 1967a) and remote-sensed

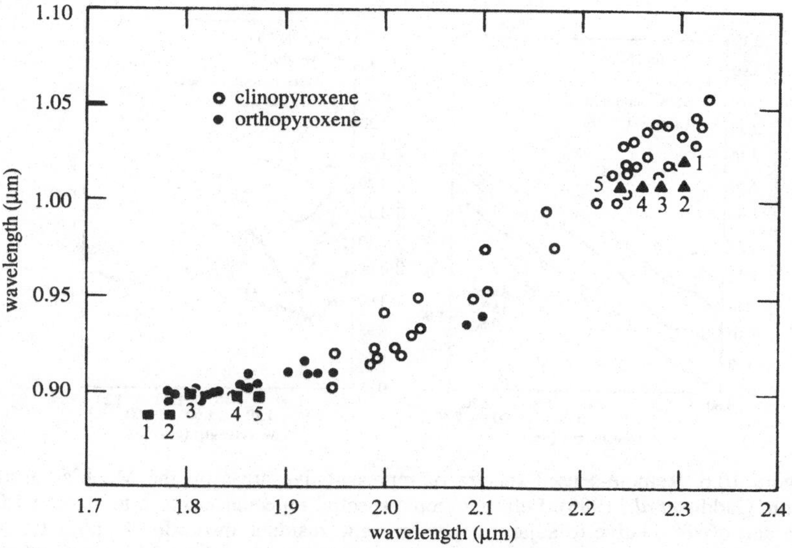

Figure 10.5. The '1 μm' versus '2 μm' pyroxene spectral determinative curve widely used to identify compositions and structure-types of pyroxenes on planetary surfaces (from Adams, 1974). Circles refer to room-temperature data. Numbered squares (orthopyroxene $En_{86}Fs_{14}$) and triangles (clinopyroxene $Wo_{42}En_{51}Fs_7$) represent spectral data obtained at the temperatures (1) 80 K; (2) 173 K; (3) 273 K; (4) 373 K; and (5) 448 K (modified from Singer & Roush, 1985).

reflectance spectra of basaltic glass-bearing assemblages (Farr *et al.*, 1980; Dyar and Burns, 1981).

The spectra illustrated in figs 10.1 and 10.4*b,c* indicate that there are structural and compositional dependencies of the pyroxene absorption bands, which are demonstrated in plots of the 'one micron' band versus the 'two micron' band such as that illustrated in fig. 10.5 (Adams, 1974; Singer and Roush, 1985). The occurrence of Fe^{2+} ions in the very distorted, asymmetric pyroxene M2 sites is responsible for these two bands. As noted in chapter 5, the positions of the intense pyroxene Fe^{2+}/M2-site CF bands in the 0.9 to 1.03 μm and 1.8 to 2.3 μm regions serve to distinguish orthorhombic and monoclinic pyroxenes, to indicate pyroxene structure-type, and to estimate Fe^{2+} compositions (§5.5). These spectral characteristics are embodied in a variety of pyroxene determinative curves (e.g., Burns *et al.*, 1972a; Adams, 1974; Hazen *et al.*, 1978; Singer and Roush, 1985; Cloutis and Gaffey, 1991). The one illustrated in fig. 10.5 is particularly applicable to powdered samples and has been widely used to interpret remote-sensed reflectance spectra of planetary surfaces. Note, however, that fig. 10.5 is applicable only to iron-bearing pyroxenes in which

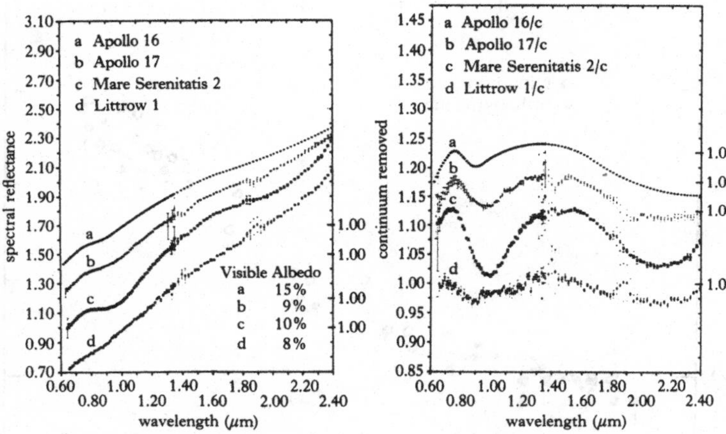

Figure 10.6. Remote-sensed spectra of representative areas on the Moon's surface (from Gaddis *et al.*, 1985). Left: telescopic spectral reflectance scaled to unity at 1.02 μm and offset relative to adjacent spectra; right: residual absorption features for the same measurements after a straight line continuum extending from 0.73 μm to 1.6 μm has been removed. (a) Highland soil sampled at the Apollo 16 landing site; (b) high-Ti mare basalt at the Apollo 17 landing site; (c) low-Ti mare basalt at Mare Serenitatis; and (d) pyroclastic deposits at Taurus–Littrow.

Fe^{2+} ions are present in M2 sites. Pure enstatite ($MgSiO_3$), stoichiometric calcium pyroxenes along the diopside–hedenbergite series [$Ca(Mg,Fe^{2+})Si_2O_6$], and Ti-bearing pyroxenes would remain unidentified on the basis of '1 micron' band versus '2 micron' band correlations. Note also that the intensities of the '1 μm' band are affected by overlapping bands from M1-site Fe^{2+} ions, particularly in orthopyroxenes, contributions from which might be indicated, though, by the other pyroxene M1-site Fe^{2+} CF transitions around 1.15 μm (cf. figs 5.14 and 5.15). This relatively weak pyroxene band, in turn, overlaps olivine and plagioclase Fe^{2+} CF spectral features (fig. 10.4). Also, as noted earlier, tetrahedrally coordinated Fe^{2+} ions in basaltic glasses and spinels which also absorb near 1.8 to 2.0 μm may interfere with the pyroxene '2 μm' band. Despite these problems and effects of temperature described later, the '1 μm' versus '2 μm' pyroxene determinative curve shown in fig. 10.5 has been widely used to map basaltic rocks on terrestrial planets by remote-sensing reflectance spectroscopy.

10.6 The Moon: Problems of mineral mixtures

The close proximity of the Moon to Earth and its lack of an atmosphere have resulted in the compilation of a large number of telescopic reflectance spectra from diverse areas of the lunar surface (McCord *et al.*, 1981), including mare basalts (Pieters, 1978), pyroclastic mantling deposits (Gaddis *et al.*, 1985) and lunar highlands (Pieters, 1986). Representative spectra are illustrated in fig. 10.6. A prominent feature of fig. 10.6 (left), and, indeed, all reflectance spectra of the Moon, is the positive continuum slope (i.e. an overall increase of reflectivity towards longer wavelengths). Superimposed on this positive continuum are absorption features in the 1 and 2 μm regions, suggesting that pyroxenes are the dominant minerals at each site. These pyroxene features are accentuated after a straightline continuum extending from 0.73 to 1.6 μm has been removed, as in fig. 10.6 (right). Correlations with the '1 μm' versus '2 μm' determinative curve in fig. 10.5 indicate the presence of orthopyroxene in highland soil sampled at the Apollo 16 mission landing site (fig. 10.6*a*), pigeonite and subcalcic augite in high-Ti mare basalt at the Apollo 17 site (fig. 10.6*b*), and calcic augite in low-Ti basalt at Mare Serenitatis (fig. 10.6*c*) (Gaddis *et al.*, 1985). However, since mono-minerallic assemblages are extremely unlikely on the Moon, the pyroxene-dominated spectra shown in fig.10.6 must have contributions from Fe^{2+} in other phases in the regolith, particularly in the low-albedo reflectance spectrum from pyroclastic mantled mare at the Littrow region (fig. 10.6*d*). Mineralogical and optical spectral studies of samples returned from the Moon indicate that pyroxenes coexisting with olivine, plagioclase feldspar, volcanic and impact glasses, and opaque ilmenite and spinels are the most likely phases to modify the pyroxene 1 and 2 μm bands. The problem of how one mineral interferes with the spectral features of another has been addressed by laboratory reflectance spectral measurements of various mineral mixtures (e.g., Nash and Conel, 1974; Singer, 1981; Cloutis *et al.*, 1986; Mustard and Pieters, 1987; Sunshine *et al.*, 1990).

Examples of experimental measurements of mixed-mineral assemblages are illustrated by the spectral data for suites of orthopyroxene–clinopyroxene and orthopyroxene–olivine mixtures in fig. 10.7. The gradual shifts of the 1 and 2 μm bands are clearly seen for orthopyroxene–clinopyroxene mixtures (fig. 10.7; left). While deleterious broadening of the 1 μm band occurs, the 2 μm band appears to be resolvable into component bands in pyroxene mixtures. The orthopyroxene–olivine mixtures (fig. 10.7; right) demonstrate that the diagnostic pyroxene 2 μm band is conspicuous even when pyroxene is a minor constituent. Thus, in the 75 per cent olivine/25 per cent orthopyroxene mixture, the pyroxene 2 μm band is already prominent, while the pyroxene 1 μm band has almost obscured the characteristic features for olivine at 0.9, 1.05 and 1.25 μm. Figure 10.8 showing reflectance spectra of mixtures of orthopyroxene with plagioclase feldspar and magnetite vividly demonstrates how the presence of an opaque oxide phase drastically lowers the albedo (i.e. overall reflectance)

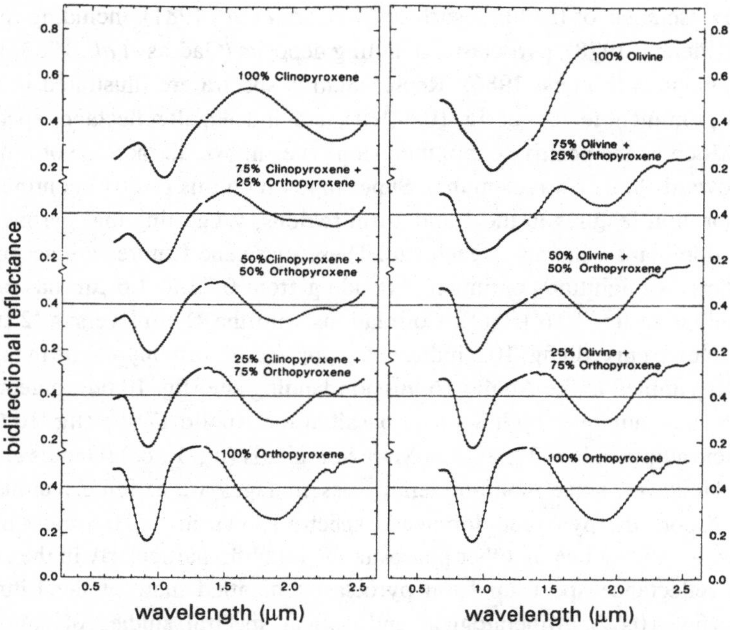

Figure 10.7 Reflectance spectra of mixed-mineral assemblages (modified from Singer, 1981). Left: orthopyroxene ($En_{86}Fs_{14}$) – clinopyroxene ($Wo_{41}En_{51}Fs_{7}$) mixtures; right: orthopyroxene – olivine ($Fo_{85}Fs_{15}$) mixtures. Mineral proportions are expressed as wt per cent. Note how pyroxene dominates the mineral-mixture spectra. Olivine causes broadening of the pyroxene 1 μm band but another olivine feature persists near 1.25 μm.

and intensities of the diagnostic pyroxene absorption features. Metallic iron in chondritic meteorites is also detrimental to pyroxene spectra (Gaffey, 1976).

Such results for mixed-mineral assemblages enable more elaborate interpretations to be made of the remote reflectance spectra of the Moon shown in fig. 10.6 (Gaddis *et al.*, 1985). Olivine and mixed-pyroxene assemblages are responsible for the overall profiles obtained from mare basalts and lunar highlands (figs 10.6*a*–*c*). However, band-broadening and low-albedo spectra of dark mantling pyroclastic deposits (fig. 10.6*d*) indicate the presence of volcanic glass which has been partially devitrified to opaque ilmenite-bearing assemblages, such as that those found in Fe–Ti orange glass spherules from Shorty Crater at the Apollo 17 site (Vaughan and Burns, 1973).

The ultimate objective of remote reflectance spectral measurements is to obtain quantitative estimates of the modal mineralogy of unexplored surfaces of the Moon and, indeed, other terrestrial planets. This necessitates difficult and elaborate spectrum-curve fitting procedures, which has been the focus of detailed research (e.g., Roush and Singer, 1986; Cloutis *et al.*, 1986; Mustard

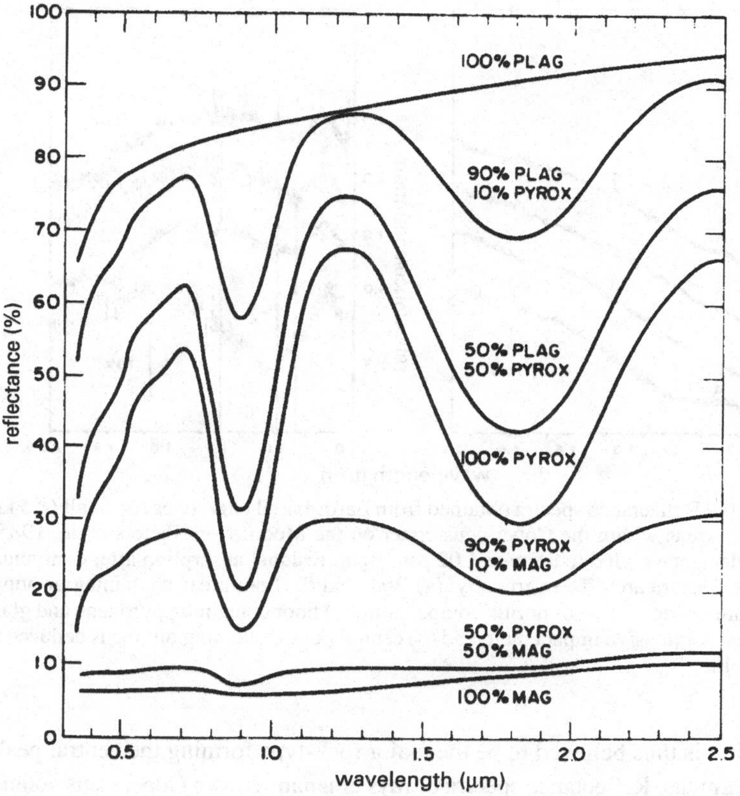

Figure 10.8 Diffuse reflectance of orthopyroxene (PYROX) with plagioclase feldspar (PLAG) and magnetite (MAG) (from Adams, 1974, attributed to C. M. Pieters). Mineral proportions are expressed as wt per cent. Note how the opaque oxide phase swamps the diagnostic pyroxene bands at 1 and 2 μm.

and Pieters, 1987; Sunshine *et al.*, 1990). Nevertheless, valuable petrological information may be deduced from the reflectance spectral profiles alone. For example, telescopic reflectance spectra for ≈5 km diameter areas within the 95 km diameter lunar crater Copernicus shown in fig. 10.9 indicate that wall (fig. 10.9*a*) and floor (fig. 10.9*b*) areas with the weak band centred near 0.92 μm diagnostic of orthopyroxenes are typical highland soils of noritic composition and predominate in Copernicus (Pieters *et al.*, 1985). Other floor areas (fig. 10.9*c*) contain high proportions of glass-bearing impact melt. However, central peaks of Copernicus (fig. 10.9*d*) are quite different from the walls and floors. They exhibit a broad multiple band centred near 1.05 μm indicating that olivine is the principal ferromagnesian silicate mineral (Pieters, 1982).

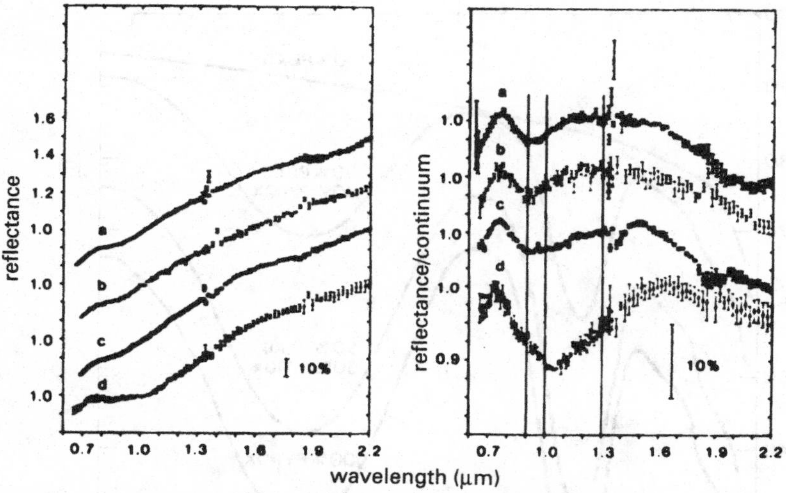

Figure 10.9 Reflectance spectra obtained from Earth-based telescopes for small (< 5 km diameter) areas within the Copernicus crater on the Moon (from Pieters *et al.*, 1985). Left: reflectance scaled to unity at 1.02 μm; right: residual absorption after continuum removal. Spectra are offset vertically. (a) Wall and (b) floor areas containing orthopyroxene are deduced to be of noritic composition; (c) floor containing pyroxene and glass is an area of extensive impact melt; and (d) central peak containing olivine is deduced to be troctolite.

Troctolite is thus believed to be the major rock-type forming the central peaks of Copernicus. Reflectance spectra of rays emanating from Copernicus contain more calcic-rich pyroxenes indicative of pigeonite–augite assemblages (Pieters *et al.*, 1985).

10.7 Mercury and the Moon: problems of high temperatures

Since reflectance spectra measured through telescopes are produced by sunlight impinging upon the surface of a planet, the effects of temperature on spectral reflectivities of minerals need to be carefully assessed, particularly when large differences exist between planetary surfaces and ambient conditions on Earth under which calibration measurements have been made. Such situations arise for Mercury (≈650 K), Venus (≈740 K), and the Moon (≈400 K), day–time high temperatures of which at their equators are higher than those on Earth (table 10.1). Corresponding sunlit surfaces of Mars (≈280 K) and asteroids (≈175 K), on the other hand, have much lower temperatures. Numerous laboratory investigations of temperature-induced changes of mineral spectra have been undertaken, therefore, in order to interpret modal miner-

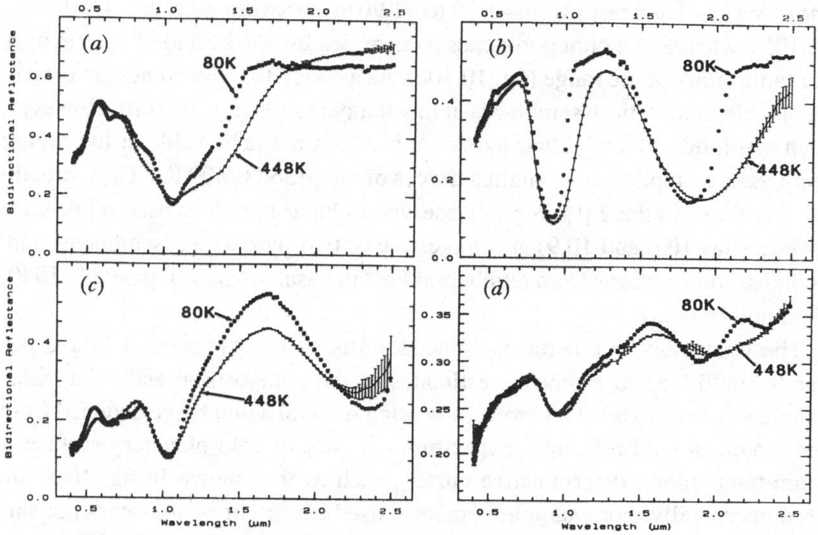

Figure 10.10 Temperature-induced variations of reflectance spectra of Fe^{2+}-bearing silicate mineral and basaltic assemblages (adapted from Singer & Roush, 1985). (*a*) Olivine, $Fo_{89}Fa_{11}$; (*b*) orthopyroxene, $En_{86}Fs_{14}$; (*c*) clinopyroxene, $Wo_{42}En_{51}Fs_{7}$; and (*d*) basaltic assemblage, comprising orthopyroxene, pigeonite–augite intergrowth, olivine and plagioclase feldspar. Note that the temperature variations of the '1 μm' and '2 μm' bands for the two pyroxenes are plotted in fig. 10.5.

alogies of hot and cold planetary surfaces (Sung *et al.*, 1977; Osborne *et al.*, 1978; Nolet *et al.*, 1979; Parkin and Burns, 1980; Singer and Roush, 1985).

Two important consequences of elevated temperatures on visible to near-infrared spectra discussed in chapter 9 are: first, positions and intensities of crystal field bands within a transition metal-bearing mineral are affected; and, second, thermal emission by the host mineral, itself, occurs. The effects of temperature on reflectance spectral profiles are demonstrated by the data for Fe^{2+} in olivine, pyroxenes, and basaltic assemblages shown in fig. 10.10. The three components of the olivine spectra centred around 1 μm become better resolved at lower temperatures but show insignificant shifts of band minima (fig. 10.10*a*). However, reflectance spectra of pyroxenes (figs 10.10*b* and *c*) show dramatic changes of band shape with rising temperature, particularly at the longer wavelength edges, which may result from thermal population of vibrational levels of the crystal field states (cf. §3.9.5; fig. 3.14). In common with olivines, the pyroxene 1 μm bands also show only minor wavelength shifts. However, in marked contrast to the 1 μm bands, the pyroxene 2 μm bands show major differences of temperature-induced shifts; for orthopyroxenes, this band increases from ≈1.80 to ≈1.90 μm between 80 K and 448 K (fig. 10.10*b*), whereas for clinopyroxenes it decreases from ≈ 2.35 to ≈2.25 μm over the same temperature range (fig. 10.10*c*). As a result, two pyroxenes are clearly resolvable in basaltic assemblages at low temperatures (fig. 10.10*d*). However, such resolution of individual pyroxene bands is not achievable at the higher temperatures applicable to sunlit surfaces of the Moon (≈400 K). Thus, broad–band features in the 2 μm region observed in lunar remote-sensed reflectance spectra (figs 10.6 and 10.9) are suggestive of two-pyroxene assemblages and, perhaps, contributions from tetrahedral Fe^{2+} in basaltic glass (Farr *et al.*, 1980; Dyar and Burns, 1981).

The contrasting temperature-induced shifts of the pyroxene 1 and 2 μm bands could lead to erroneous estimates of the composition and, to a lesser extent, structure-type of a pyroxene-bearing mineral assemblage deduced from the remote-sensed reflectance spectrum of a hot or cold planetary surface if room-temperature determinative curves, such as that shown in fig. 10.5, are used uncritically. For example, remote-sensed spectra of planets with hot surfaces, such as Mercury and the Moon, would lead to overestimates of Fe^{2+} contents of the orthopyroxenes and underestimated Fe^{2+} contents of the clinopyroxenes (Singer and Roush, 1985). Planets with cold surfaces, such as Mars and the asteroids, could produce opposite results. On the other hand, the room-temperature data underlying the pyroxene determinative curve shown in fig. 10.5 may impose constraints on the compositions of pyroxenes deduced from telescopic spectra of a planet with very high surface temperatures, such as Mercury.

The hot surface of Mercury, as well as reflecting sunlight, also behaves as a quasi-black-body radiator and emits thermal energy back into space. As a result, the spectral reflectance of Mercury rises sharply above ≈1.5 μm due to its thermal emissivity obscuring any possible contribution from a Fe^{2+}-pyroxene 2 μm band. The close proximity of Mercury to the Sun also makes telescopic reflectance spectral measurements of its surface very difficult, so that attempts to identify pyroxenes on Mercury and to estimate their compositions from the 1 μm band alone have produced ambiguous results (McCord and Clark, 1979; Vilas, 1985). For example, a weak broad band resolved at 0.89 μm (McCord and Clark, 1979) and corresponding to the lowest limit for an orthopyroxene in the room-temperature pyroxene determinative curve shown in fig. 10.5 might imply the presence of enstatite in low-iron basalts on the surface of Mercury. However, temperature-induced variations of the orthopyroxene 1 μm band, such as those portrayed in fig. 10.10*b* showing that Fe^{2+} contents of orthopyroxenes are overestimated from high-temperature spectra, suggest that the wavelegth of the 0.89 μm feature is impossibly low to be assigned to Fe^{2+} in orthopyroxene on Mercury. Pure end-member enstatite or diopside could occur on Mercury, but these pyroxenes would not be identified by spectral reflectance. The 0.89 μm band, if it exists, could be indicative of ferric-bearing augites (Burns *et al.*, 1976; Straub *et al.*, 1991) in the regolith of Mercury since Fe^{3+} crystal field transitions intensify considerably but do not shift much at elevated temperatures (table 9.2; Parkin and Burns, 1980).

10.8 Venus and Mars: problems of atmospheres and surface weathering products

The surface of the Moon and Mercury are conducive to telescopic spectral measurements, albeit complicated by effects of high temperatures, because they lack atmospheres. The presence of atmospheric gases on Venus and Mars, however, impose severe problems on the measurement and interpretation of Earth-based remote-sensed reflectance spectra obtained from these planets. The problem is most acute for Venus because its hot surface is masked by the dense atmosphere which strongly absorbs and scatters visible to near-infrared radiation. However, Soviet Venera missions to the Venusian surface have yielded spectrophotometric data in the form of multispectral images at three wavelengths in the visible region (0.44, 0.54 and 0.63 μm) from which surface mineralogy has been deduced (Pieters *et al.*, 1986). After correcting for effects of orange colouration due to the atmosphere, the surface of Venus appears to be dark without significant colour. Correlations with high-temperature laboratory reflectance spectra of oxidized basaltic materials suggested that the basaltic surface of Venus contains ferric-bearing minerals possibly formed from oxidation of pyroxenes (Straub *et al.*, 1991) and olivine (Pieters *et al.*, 1986).

The surface of Mars, on the other hand, is visible and has been accessible to several Earth-based telescopic reflectance spectral measurements in the visible and infrared regions (McCord *et al.*, 1982; Singer, 1982; Bell et al., 1990a,b; Pinet and Chevrel, 1990), as well as *in situ* multispectral images of the surface taken during the 1977 Viking orbiter and lander experiments (Singer, 1985) and the 1989 Phobus II spacecraft mission to Mars (e.g., Erard *et al.*, 1991). However, while thermal emissivity of the relatively cold surface of Mars becomes dominant only beyond the 5 μm wavelength region, the presence of atmospheric CO_2 and traces of H_2O mask critical regions in the near-infrared spectra needed for positive identification of ferromagnesian silicates in the Martian regolith. To a close approximation, Mars' surface is composed of bimodal high- and low-albedo regions (Singer, 1985; Bell *et al.*, 1990a,b) which give rise to the typical 'bright-region' and 'dark-region' spectra illus-

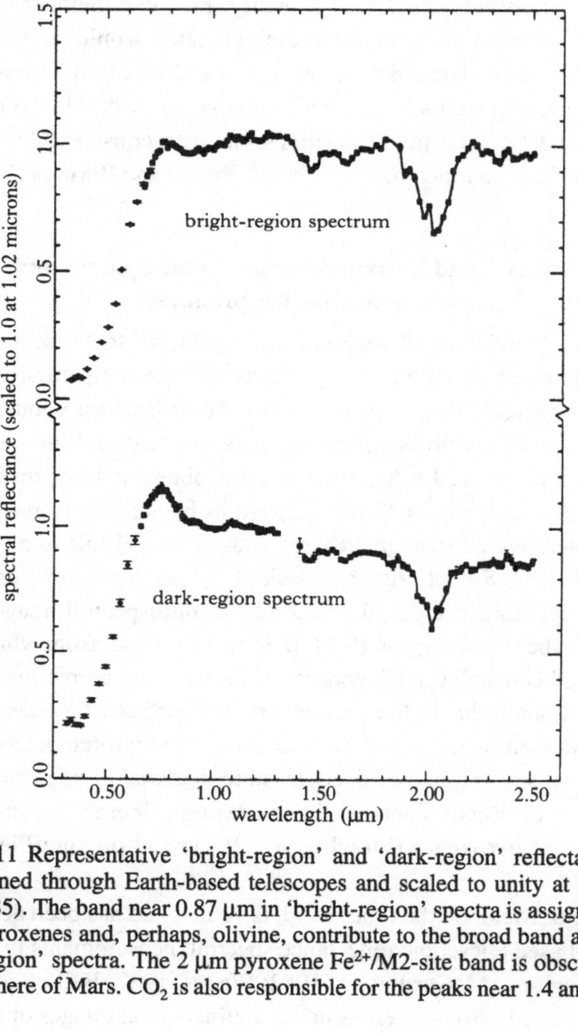

Figure 10.11 Representative 'bright-region' and 'dark-region' reflectance spectra of Mars obtained through Earth-based telescopes and scaled to unity at 1.02 μm (from Singer, 1985). The band near 0.87 μm in 'bright-region' spectra is assigned to Fe^{3+} ($^6A_1 \rightarrow {}^4T_1$). Pyroxenes and, perhaps, olivine, contribute to the broad band at 0.9 to 1.1 μm in 'dark-region' spectra. The 2 μm pyroxene Fe^{2+}/M2-site band is obscured by CO_2 in the atmosphere of Mars. CO_2 is also responsible for the peaks near 1.4 and 1.62 μm.

trated in fig. 10.11. 'Bright-region' spectra are dominated by spectral features at ≈0.87 μm and ≈0.62 μm, attributed to crystal field transitions in Fe^{3+} ions (table 5.15); atmospheric CO_2 is responsible for the sharper features at 1.45, 1.62 and 1.9 to 2.1 μm. The Fe^{3+} spectral features lack specificity and a variety of ferric-bearing phases have been suggested as oxidative weathering products on the Martian surface, including a variety of poorly crystalline oxides, clay silicate and sulphate phases (Sherman *et al.*, 1982; Singer, 1985; Burns, 1986, 1988; Morris *et al.*, 1985, 1989, 1990). In the 'dark-region' spectra, although features due to Fe^{3+} ions may be present, they do not obscure contributions from Fe^{2+} ions in the 1 μm region (Singer, 1985; Pinet and Chevrel, 1990). These spectral features indicate the presence of subcalcic pyroxenes and, perhaps, olivine in iron-rich basalts believed to occur on Mars' surface. Unfortunately, interference by atmospheric CO_2 at 1.9 to 2.1 μm makes it extremely difficult to identify the pyroxene structure-type and composition from its diagnostic 2 μm band. However, correlations of telescopic spectra of dark regions of Mars with reflectance spectra of shergottites have been made (McFadden, 1989; Singer and McSween, 1992). These meteorites, which are believed to have originated from the surface of Mars (§10.9), contain pigeonite–augite (± olivine) assemblages giving broad spectral features at 1 and 2.1 μm, indicative of basaltic rocks on the Martian surface.

10.8.1 Spectra of ferric oxides

The vulnerability of igneous ferromagnesian silicate and iron sulfide minerals to atmospheric oxidation on terrestrial planets results in the formation of numerous Fe^{3+}-bearing phases, including a variety of oxide, oxyhydroxide, clay silicate and sulphate minerals found in soils on Earth, gossans capping sulphide ore deposits, and oceanic sediments. Because some of these minerals could have formed in Martian regolith and on the surface of Venus, numerous visible to near-infrared spectral measurements have been made on candidate Fe(III) phases (Sherman *et al.*, 1982; Sherman and Waite, 1985; Morris *et al.*, 1985, 1989, 1990; Morris and Lauer, 1990; Townsend, 1987; Straub *et al.*, 1991). Representative absorption spectral profiles of four such minerals, goethite, lepidocrocite, maghemite and hematite, are illustrated in fig. 10.2. Table 10.2 summarizes structural data and peak assignments of these and other Fe^{3+}-bearing minerals that have figured in discussions of Mars surface mineralogy.

The visible and near-infrared spectra of goethite, lepidocrocite, nontronite and jarosite resemble one another and reflect similarities between the coordination environments in each mineral. Thus, Fe^{3+} ions occur in edge-shared octahedral sites formed by oxygen and hydroxyl ligands. The broad bands centred around 0.90 μm (≈11,100 cm^{-1}) and 0.64 μm (≈15,100 cm^{-1}) are at typical energies for the $^6A_1 \rightarrow {}^4T_1(^4G)$ and $^6A_1 \rightarrow {}^4T_2(^4G)$ spin-forbidden CF transitions within Fe^{3+} ions octahedrally coordinated to oxygen ligands (table 5.15). The spectrum profile of hematite, however, differs from those of the other Fe(III) minerals, particularly in the visible region, where bands centred near 0.55 and 0.43 μm are conspicuous. Their intensification results from the unique structural environment of Fe^{3+} ions in hematite.

Table 10.2. *Crystal chemical and spectral data for ferric minerals relevant to Mars*

Property	Hematite	Maghemite	Goethite	Akaganeite	Lepido-crocite	Nontronite	Jarosite
Formula:	α-Fe_2O_3	γ-Fe_2O_3	α-FeOOH	β-FeOOH	γ-FeOOH	$Na_{0.3}Fe_2(Si,Al)_4O_{10}(OH)_2.xH_2O$	$(K,Na,H_3O)Fe_3(SO_4)_2(OH)_6$
Structure-type:	corundum	spinel	ramsdellite	hollandite	lepidocrocite	smectite	jarosite
Fe^{3+} site:	trig.dist. octahedron	octahedron+ tetrahedron	octahedron O^{2-}; OH^-	Fe1 & Fe2 both oct.	octahedron O^{2-}; OH^-	octahedra O^{2-}; OH^- *cis* ; *trans*	octahedron SO_4^{2-}; OH^-
Fe–O distances: (pm)	(3x) 211.5 (3x) 194.5	(6x) 205.9 (oct) (4x) 188.7 (tet)	(2x) 195.7 (1x) 192.6 (2x) 209.5 (1x) 209.5	192 to 213 and 180 to 232	(2x) 200.0 (2x) 200.0 (2x) 201.0		(2x) 205.8 (SO_4^{2-}) (4x) 197.5 (OH^-)
Mean (pm):	203.0	205.9(oct) 188.7(tet)	202.1	204 and 204	200.0		200.3
Magnetic ordering temperature (K):	955	> 850	400	299	77	<4.2	60
CF peaks (nm):							
$^6A_1 \rightarrow {}^4T_1(^4G)$	885	935	917	850	960	935	933
$^6A_1 \rightarrow {}^4T_2(^4G)$	649	510	649		485	634	≈600
$^6A_1 \rightarrow {}^4A_1, {}^4E(^4G)$	444	435	435		435	455	434
$^6A_1 \rightarrow {}^4T_2(^4D)$	405					377	
$^6A_1 \rightarrow {}^4E(^4D)$	380	370	365		360	369	
$^6A_1 \rightarrow {}^4T_1(^4P)$	320	315	285		305		
Paired transition (nm):							
$2[(^6A_1 \rightarrow {}^4T_1(^4G)]$	530	510	480		485	448	

OMCT peaks (nm):							
$6t_{1u} \rightarrow 2t_{2g}$	270	250	250		240		
$1t_{2u} \rightarrow 2t_{2g}$			250		210		
CF parameters (cm^{-1}):							
10 *Dq* or Δ	14,000	15,410	15,320		15,950	15,050	
Racah *B*	540	560	590		610	614	
Racah *C*	3,410	3,510	3,490		3,470	3,268	
Sources of data:							
(spectra)*	[1–5]	[2–4]	[2–4,6–8]	[9]	[2–4,6]	[10,11]	[7,12]
(structure)†	[A]	[B]	[C]	[D]	[E]	[F]	[G]

Other candidate Fe(III) oxides not listed include: magnetite (Fe_3O_4); feroxyhyte, δ'-FeOOH; ferrihydrite, $Fe_5O_7(OH).4H_2O$.

* *Sources of spectral data*:

[1] Marusak, Messier & White (1980); [2] Sherman, Burns & Burns (1982); [3] Sherman & Waite (1985); [4] Morris, Lauer, Lawson, Gibson, Nace & Stewart (1985); [5] Morris, Agresti, Laauer, Newcomb, Shelfer & Murali (1989); [6] Mao & Bell (1974); [7] Townsend (1987); [8] Cerville (1991); [9] Taylor, Mao & Bell (1974) (1975); [10] Karickhoff & Bailey (1973); [11] Sherman & Vergo (1988); [12] Rossman (1976).

† *Sources of structural data*:

[A] Blake, Hessevick, Zoltai & Finger (1966); [B] Greaves (1983); [C] Szytula, Burewicz, Dimitrijevic, Krasnicki, Rzany, Todorovic, Wanic & Wolski (1968); [D] Post & Buchwald (1991); [E] Oles, Szytula & Wanic (1970); [F] Güven (1988); [G] Menchetti & Sabelli (1976).

In the hematite crystal structure which is isostructural with corundum (fig. 5.4), planes of hexagonal close-packed O^{2-} ions are stacked along the *c* axis and accommodate Fe^{3+} ions in two-thirds of the available octahedral sites. Pairs of Fe^{3+} ions are located along the *c* axis above and below vacant octahedral sites and the adjacent [FeO_6] octahedra share three oxygens across a common triangular face linking the two octahedra. The Fe^{3+}–O distances to the three face-shared oxygens are shorter (194.5 pm) than those to the other three oxygens (211.5 pm), so that the Fe^{3+} ions are not centrally located in their coordination sites. The Fe^{3+}–Fe^{3+} distances between pairs of face-shared octahedra are 289 pm parallel to the *c* axis. Each [FeO_6] octahedron also shares edges with three neighbouring [FeO_6] octahedra perpendicular to the *c* axis, and Fe^{3+}–Fe^{3+} distances across edge-shared octahedra are 297 pm. Although all Fe^{3+} sites in the hematite structure are crystallographically equivalent, there are two magnetically inequivalent sites. All cations in a given plane of edge-shared octahedra perpendicular to the *c* axis are equivalent but magnetically inequivalent to cations in adjacent planes separating the face-shared octahedra. Below the Curie temperature of crystalline hematite (955 K), Fe^{3+} ions are strongly antiferromagnetically coupled.

The presence of Fe^{3+} ions in asymmetric trigonally distorted octahedral sites contributes to the overall intensification of spin-forbidden bands in crystal field spectra of hematite. However, magnetic coupling between the antiferromagnetically ordered Fe^{3+} ions in crystalline hematite also intensifies these bands and results in an additional band at 0.53 μm (18,700 cm^{-1}). This band represents a paired transition between cations in the face-shared [FeO_6] octahedra, by analogy with that seen in spectra of isostructural yellow sapphire (fig. 3.16). Small particle sizes of Fe_2O_3 and atomic substitution of Al^{3+} for Fe^{3+} ions diminish magnetic coupling interactions, thereby lowering intensities of visible-region spectra of nanocrystalline and aluminous hematites (Morris *et al.*, 1985, 1989; Morris and Lauer, 1990).

Although magnetic coupling between Fe^{3+} ions in edge-shared [FeO_6] octahedra also occurs in goethite, lepidocrocite, nontronite and jarosite, leading to intensification of ferric spin-forbidden Fe^{3+} CF bands relative to magnetically-dilute minerals such as garnets and coquimbite, intensities of paired transitions in the 0.485 to 0.550 μm region are significantly reduced relative to hematite (fig. 10.2). Molecular orbital calculations (Sherman, 1985b) support experimental evidence (Rossman, 1975, 1976) showing that intensification of Fe^{3+} CF transitions is smaller when Fe^{3+} are bridged by OH^- rather than O^{2-} anions.

10.8.2 Ferric-bearing assemblages on Mars

The overall colour of Mars, together with telescopic reflectance spectral resolution of the 0.87 and 0.62 μm bands attributable to the $^6A_1 \rightarrow {}^4T_1(^4G)$ and $^6A_1 \rightarrow {}^4T_2(^4G)$ crystal field transitions in octahedral Fe^{3+} ions, suggests that ferric-bearing phases are present in the rusty 'bright' regions of Mars. 'Dark-region' reflectance spectra, on the other hand, show broad bands near 1.0 and 2.1 μm, attributable to Fe^{2+} CF bands in relatively unoxidized pyroxenes and olivine. The high iron contents of Martian regolith were demonstrated during the 1977 Viking Lander XRF experiments (Toulmin *et al.*, 1977), in which high sulphur concentrations (assumed to be sulphate) were also analysed (see Appendix 1). Although a magnetic phase inferred to be maghemite (γ-Fe_2O_3) (Hargraves *et al.*, 1979), feroxyhyte (δ'-FeOOH) (Burns, 1980) or nanophase hematite (Morris *et al.*, 1985, 1989) was demonstrated by the Viking magnetic properties experiment, the overall mineralogy of Martian regolith has not been positively identified.

The more obvious crystalline Fe(III) minerals most commonly found on oxidized surfaces of the Earth, such as hematite and goethite, appear to be eliminated on Mars on account of mismatches of relative intensities of the crystal field bands in their visible-region spectra and the rapid decrease of reflectance below 0.7 μm. Magnetic ordering in crystalline hematite adversely enhances the intensities of the $^6A_1 \rightarrow {}^4T_1$ (≈0.60 μm) and $^6A_1 \rightarrow {}^4E,{}^4A_1$ (≈0.44 μm) transitions relative to remote-sensed reflectance spectral profiles of Mars' surface in the 0.40 to 0.70 μm region (cf. figs. 10.2 and 10.11). Alternative materials whose visible to near-infrared spectra compare more favourably with Martian 'bright-region' spectra, include nanocrystalline Fe_2O_3 phases (Sherman *et al.*, 1982; Morris *et al.*, 1985, 1989), Fe^{3+}-exchanged clays (Banin *et al.*, 1985) containing ferrihydrite (Bishop *et al.*, 1992), palagonite (Gooding, 1978; Morris *et al.*, 1990), and gossaniferrous ferrihydrite–jarosite–opal–clay silicate assemblages (Hunt and Ashley, 1979; Burns, 1987b, 1988). Each of these candidates comprises poorly crystalline or X-ray amorphous Fe(III) phases, an essential property for reducing magnetic ordering of constituent Fe^{3+} ions and decreasing the intensity of crystal field bands in the visible region. Inferences can be drawn from each of these terrestrial analogues about weathering processes on Mars. Thus, nanophase Fe(III) oxides could be produced by ablation of surface basalt flows during global dust-storms and photochemical oxidation of ferromagnesian silicates exposed to the Martian atmosphere. Palagonites suggest the intrusion of basaltic magma into permafrost on the frozen Martian surface. Clay silicates indicate chemical weathering of igneous minerals in an aqueous environment. And, gossans imply the presence of sulphide mineralization and perhaps ore deposits associated with the iron-rich basalts on Mars. Oxidative weathering of such sulphide mineralization produces acid groundwater which facilitates chemical weathering of the silicate minerals (Burns, 1987b, 1988; Burns and Fisher, 1990). Thus, near-surface ablation, dissolution, oxidation and hydrolysis of basaltic materials would produce poorly crystalline and colourful ferric sulphate (jarosite, botryogen, copiapite, coquimbite), oxide (ferrihydrite, hematite, goethite), and clay silicate (montmorillonite, nontronite, hisingerite) minerals all of which could contribute to the red colour and ferric-like spectra of the Martian surface.

10.8.3 Ferric-bearing assemblages on Venus

Evidence for ferric-bearing minerals on the surface of Venus, which was derived from spectral reflectance data measured during Soviet Venera missions (Pieters *et al.*, 1986), poses a dilemma. Hematite is thermodynamically unstable (Fegley *et al.*, 1991) and is rapidly converted to magnetite (Straub and Burns, 1992) under temperature and oxygen fugacity conditions believed to exist on the surface of Venus. Ferrifayalites and laihunite (§4.8.3.4), too, are rapidly reduced to Fe^{2+}–olivines under Venusian T–f_{O2} surface conditions. However, other mixed-valence Fe^{2+}–Fe^{3+} silicates could exist on the surface of Venus, including ilvaite (§4.8.2) and dehydroxylated silicates such as oxyhornblendes and oxybiotites (Burns and Straub, 1992). The high electrical conductivities of Fe^{2+}–Fe^{3+} silicates such as ilvaite at Venusian surface temperatures render these mixed-valence minerals viable candidates, in addition to magnetite (Fe_3O_4), pyrite (FeS_2) and perovskite ($CaTiO_3$) (Fegley et al., 1992; Klose *et al.*, 1992), that could be responsible for the high radar-reflectivity surfaces observed on most mountainous terranes on Venus (Pettengill *et al.*, 1992).

10.9 Reflectance spectra of meteorites and asteroids

Before the advent of the Apollo and Luna missions which retrieved samples from the Moon's surface, meteorites provided the only source of extraterrestrial materials and raised questions about their sources from parent bodies such as asteroids. Visible to near-infrared reflectance spectroscopy, therefore, has been applied extensively to laboratory investigations of meteorites and to remote-sensed measurements of many asteroids (Gaffey, 1976; Gaffey and McCord, 1978; McFadden *et al.*, 1982, 1984; Bell and Keil, 1988).

Spectral reflectance curves for the range of meteorite types are illustrated in fig. 10.12. These spectra demonstrate the diagnostic features of the various

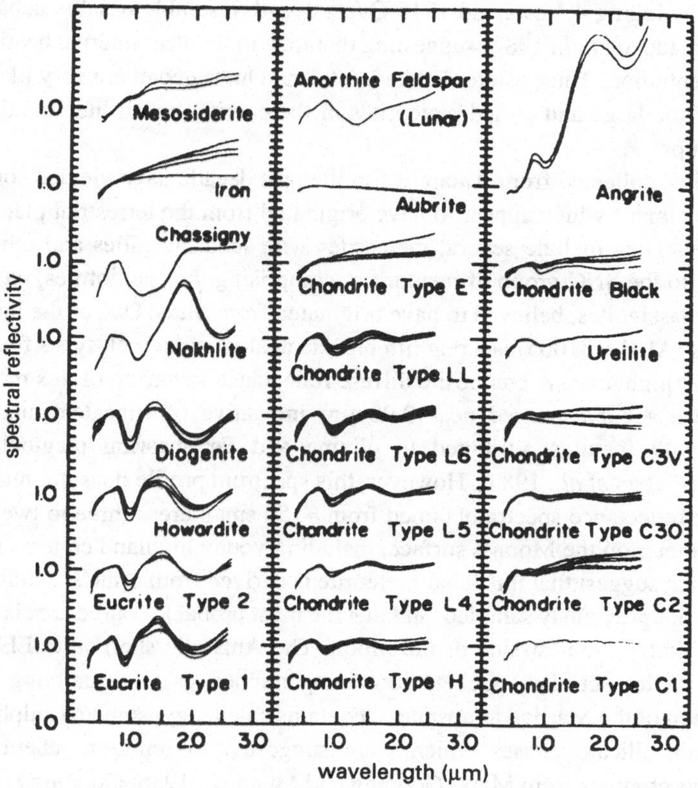

Figure 10.12 Normalized spectral reflectance curves for the range of meteorite types (from Gaffey, 1976).

meteorite types, including the presence or absence of absorption bands, their positions and relative intensities, their symmetries and widths, and other properties such as continuum slope, curvature, and inflection points (Gaffey, 1976). Again, noteworthy features of the meteorite spectra are the prominent pyroxene 1 and 2 μm bands, the relative intensities and asymmetries of which have been used to characterize the pyroxenes and to estimate olivine/pyroxene ratios in meteorites. Relating these meteorite spectra to those of asteroids has been difficult due to the faintness of these objects in space. Nevertheless, telescopic spectral measurements have led to discoveries of several large asteroids containing olivine-rich dunite-like rocks (Cruikshank and Hartmann, 1984), as well as other asteroids with wide ranges of olivine/pyroxene ratios, which are inconsistent with an ordinary chondritic composition but suggestive of affinities with stony-iron meteorites (Bell and Keil, 1988). The spectra of some asteroids, including 4 Vesta and 1915 Quetzalcoatl, resemble basaltic achondrites (McFadden *et al.*, 1982) suggesting that they underwent internal heating and differentiation. Thus, asteroids like Vesta and Quetzalcoatl are very likely sources of the large and varied meteorites of the eucrite, howardite, and diogenite groups.

Meteorites collected from Antarctica in the past decade have yielded some unique specimens which appear to have originated from the terrestrial planets themselves. They include several meteorites with lunar affinities and others belonging to the SNC group of meteorites, comprising the *s*hergottites, *n*ahklites and *c*hassignites, believed to have originated from Mars. One of the lunar meteorites (ALHA 81005) is a regolith breccia analogous to rock-types found in the lunar highlands. A composite diffuse reflectance spectrum of this meteorite shows a band centred near 0.98 μm indicative of subcalcic augite, together with features attributed to olivine and Fe^{2+}-bearing plagioclase feldspars (Pieters *et al.*, 1983). However, this spectrum profile does not match telescopic reflectance spectra obtained from ≈150 small areas three to twenty km in diameter on the Moon's surface, including young highland craters. The spectral data suggest that this lunar meteorite is derived from a surface unit on the Moon not previously sampled, and that the most probable source area is the near-side limb or the far-side of the Moon. The Antarctic shergottite EETA 79001, with impact-glass pockets containing trapped gases resembling the composition of the Martian atmosphere, contains calcite, gypsum and sulphur-rich aluminosilicate phases which were suggested to represent chemical weathering products from Mars (Gooding and Muenow, 1986; Gooding *et al.*, 1988). However, negligible ferric iron was resolved in Mössbauer spectral measurements of the impact glasses, while the pyroxene-dominated matrix of EETA 79001 contains less than 2 wt per cent Fe^{3+} (Solberg and Burns, 1989).

10.10 Future measurements and missions

Reflectance spectroscopy has proven to be the most powerful and versatile remote-sensing technique for determining surface mineralogy, chemical compositions and lithologies of planetary objects, as well as constituents of their atmospheres. Table 10.1 summarizes information that has been deduced for the terrestrial planets based on spectral properties of light in the visible and near-infrared regions reflected from their surfaces.

The planetary reflectance spectral data constituting Table 10.1 traditionally have been obtained with Earth-based telescopes. Such remote-sensed measurements are limited by telescope availability, favourable observational conditions, and optimum viewing alignments of the planetary objects. As a result, comparatively few high quality telescopic spectra (perhaps numbering several hundreds) are available for the solar system planets and their satellites. However, this situation could change dramatically.

Future spacecraft missions to solar system objects are primarily being oriented towards remote-sensing experiments, in contrast to the soft-landed *in situ* experiments and sample-return initiatives during the 1970's and 1980's. Because reflectance spectroscopy has become one of the most important investigative techniques in the planetary sciences, current and planned space missions for the 1990's and 21st century should include visible and near-infrared spectrometers in their instrument payloads. Reflectance spectral measurements from space would provide more favourable viewing geometries, eliminate problems due to telluric water and CO_2, and improve the resolution of areas scanned on a nearby planetary surface.

Studies of a qualitative nature described earlier in the chapter have demonstrated that spectral reflectance profiles of planetary surface materials are influenced primarily by the chemical compositions and abundances of constituent minerals. There is now an increasing awareness of second-order effects, such as temperature, viewing geometry, grain size and particle packing, on positions and intensities of diagnostic mineral spectral features. Research is needed to quantify these effects and to more closely simulate physical properties of planetary regoliths when they reflect sunlight. Such projects include spectral measurements for mineral assemblages having different temperatures, particle sizes, grain-size packings, and modal mineral proportions, over a range of angles of incident and reflected light and in confining atmospheres having a variety of pressures and compositions. Other effects such as atmospheric weathering and radiation sputtering processes under different intensities and exposure times also need to be assessed. The ultimate goal of such laboratory investigations is to develop a theoretical model for deducing the modal mineralogy of an area on a planetary surface from its reflectance spectrum measured under known lighting conditions.

10.11 Summary

Chapter 10 describes how spectral measurements of sunlight reflected from surfaces of planets, when correlated with experimental visible to near-infrared spectra of rock-forming minerals, have been used to detect transition metal ions, to identify constituent minerals, and to determine modal mineralogies of regoliths on terrestrial planets.

Compositions of terrestrial planets. Geochemical data derived from lunar samples returned by the Apollo and Luna missions to the Moon, *in situ* chemical analyses by instruments sent to Mars and Venus, and meteorites collected on Earth show variations of transition element abundances. These may be attributable to different evolutionary histories of the terrestrial planets. The surface of Venus resembles Earth by having similar crustal abundances of Fe and Ti, whereas Mars' surface is more iron-rich. Some mare basalts are Ti-rich. Oxidation states that predominate in the regoliths of these terrestrial planets are Fe(II), together with Ti(III) on the Moon and Fe(III) on Mars and, perhaps, Venus.

Origin and measurement of reflectance spectra. For mineral identification purposes, reflected light that has penetrated inside regolith materials (volume or diffuse reflectance), rather than surface-scattered light (or specular reflectance), is measured spectrally. The diffuse reflectance spectra give profiles in which minima correspond to peak maxima in absorption spectra but with much loss of spectral resolution.

Minerals contributing to remote-sensed spectra. Reflectance spectra measured through Earth-based telescopes may contain absorption bands in the one micron and two micron wavelength regions originating from crystal field transitions within Fe^{2+} ions. Pyroxenes with Fe^{2+} ions in M2 sites dominate the spectra, and the resulting '1 μm' versus '2 μm' spectral determinative curve is used to identify compositions and structure-types of pyroxenes on surfaces of the Moon, Mercury, Mars and the asteroids, after correcting for experimentally determined temperature-shifts of peak positions. Olivines and Fe^{2+}-bearing plagioclase feldspars also give diagnostic peaks just beyond 1μm, while tetrahedral Fe^{2+} ions in glasses absorb in the 2 μm region as well. Opaque ilmenite, spinel and metallic Fe phases mask all of these Fe^{2+} spectral features. Laboratory studies of mixed-mineral assemblages have enabled coexisting Fe^{2+} minerals to be identified in remote-sensed spectra of regoliths.

Typical results from telescopic spectral measurements. Noritic rocks (orthopyroxene) in the lunar highlands, troctolites (olivine) in central peaks of impact craters such as Copernicus, and high-Ti and low-Ti mare basalts (Ti^{3+}-pyroxenes) have been mapped on the Moon's surface. The Venusian atmosphere prevents remote-sensed spectral measurements of its surface mineralogy, while atmospheric CO_2 and ferric-bearing materials in the regolith of Mars interfere with pyroxene characterization in 'bright-region' and 'dark-region' spectra. Correlations with asteroid reflectance spectra indicated that Vesta was the source of basaltic achondrite meteorites.

10.12 Background reading

Adams, J. B. (1975) Interpretations of visible and near-infrared diffuse reflectance spectra of pyroxenes and other rock-forming minerals. In *Infrared and Raman Spectroscopy of Lunar and Terrestrial Minerals.* (C. Karr Jr., ed.; Academic Press, New York), pp. 91–116.

Bell, P. M., Mao, H. K. & Rossman, G. R. (1975) Absorption spectroscopy of ionic and molecular units in crystals and glasses. In *Infrared and Raman Spectroscopy of Lunar and Terrestrial Minerals.* (C. Karr Jr, ed.; Academic Press, New York), pp. 1–38.

Burns, R. G. (1989) Spectral mineralogy of terrestrial planets: Scanning their surfaces remotely. *Mineral. Mag.*, 53, 135–51.

Burns, R. G. (1992) Origin of electronic spectra of minerals in the visible to near-infared region. In *Remote Geochemical Analysis: Elemental and Mineralogical Composition.* (C. Pieters & P. A. J. Englert, eds; Camb. Univ. Press), ch. 1.

Burns, R. G. & Vaughan, D. J. (1975) Polarized electronic spectra. In *Infrared and Raman Spectroscopy of Lunar and Terrestrial Minerals.* (C. Karr Jr., ed.; Academic Press, New York), pp. 39–76.

Clark, R. N., King, T. V. V., Klejwa, M. & Swayze, G. A. (1990) High resolution relectance spectroscopy of minerals. *J. Geophys. Res.*, 95, 12653–80.

Gaffey, M. J. & McCord, T. B. (1978) Asteroid surface materials: Mineralogical characterization from reflectance spectroscopy. *Space Sci. Rev.*, 21, 555–628.

Hunt, G. R. (1982) Spectroscopic properties of rocks and minerals. In *Handbook of Physical Properties of Rocks, vol. 1.* (R. S. Carmichael, ed., CRC Press, Boca Raton, Florida), pp. 295–385.

McCord, T. M., editor (1988) *Reflectance Spectroscopy in Planetary Science: Review and Strategy for the Future.* NASA Spec. Rept. 493 (Planet. Geol. Geophys. Progr.), 37pp.

Pieters, C. M. (1978) Mare basalt types on the front side of the Moon: A summary of spectral reflectance data. *Proc. 9th Lunar Planet. Sci. Conf., Geochim. Cosmochim. Acta, Suppl. 9* (Pergamon Press), pp. 2, 825–50.

Pieters, C. M. (1986) Composition of the lunar highland crust from near-infrared spectroscopy. *Rev. Geophys.*, 24, 557–78.

Singer, R. B. (1985) Spectroscopic observations of Mars. *Adv. Space Res.*, 5, 59–68.

11

Covalent bonding of the transition elements

Theories of chemical bonding – – fall into one of two categories: those which are too good to be true and those which are too true to be good.

F. A. Cotton, *J. Chem. Educ.*, 41, 475 (1964).

11.1 Introduction

In earlier chapters, allusions were made to the effects of covalent bonding. For example, covalent interactions were invoked to account for the intensification of absorption bands in crystal field spectra when transition metal ions occupy tetrahedral sites (§3.7.1); patterns of cation ordering for some transition metal ions in silicate crystal structures imply that covalency influences the intracrystalline (or intersite) partitioning of these cations (§6.8.4); and, the apparent failure of the Goldschmidt Rules to accurately predict the fractionation of transition elements during magmatic crystallization was attributed to covalent bonding characteristics of these cations (§8.3.2).

A fundamental assumption underlying the crystal field model of chemical bonding is that ligands may be treated as point negative charges with no overlap of metal and ligand orbitals. Thus, $3d$ electrons are assumed to remain entirely on the transition metal ion with no delocalization into ligand orbitals. This situation is never realized, even in ionic structures such as periclase (MgO) and forsterite (Mg_2SiO_4), let alone bunsenite (NiO), liebenbergite (Ni_2SiO_4) or fayalite (Fe_2SiO_4), in which metal–oxygen bonds have some degree of covalent character and electrons in metal orbitals participate in the bonding. Some of the fundamental features of crystal field theory are contrary to expectation or are impossible to derive using the point charge model (Cotton, 1964). For example, it is impossible to understand the spectrochemical series, eq. (2.19), in any ionic model, to explain why ligands such as H_2O

and NH_3 give larger crystal field splittings than do the O^{2-} and OH^- anions, and to account for why the Δ_o value for CN^- is so much larger than that for F^-. Moreover, if instead of using the point charge model, allowances are made for the actual sizes of anions, one obtains the incorrect ordering of cation t_{2g} and e_g orbitals (Cotton, 1964).

Considerable evidence derived from optical spectra, electron spin resonance, nuclear magnetic resonance, Mössbauer spectroscopy, magnetic properties and quantitative molecular orbital energy level calculations, indicates that orbital overlap or covalent bonding does occur between metal and ligands in transition metal compounds including minerals. Such evidence shows that ligands play a more active role in determining the 3*d* orbital energy levels of cations than acknowledged by crystal field theory. Nevertheless, as the preceding chapters have demonstrated, simple concepts of crystal field theory do provide a useful basis for understanding several properties of transition metal ions in ionic structures such as silicate and oxide minerals, particularly when comparisons are made between them individually and with other elements in the periodic table.

It turns out that qualitative results of crystal field theory and molecular orbital (MO) theory are rather similar even though their basic premises, namely purely electrostatic interactions *versus* orbital mixing, appear to be so fundamentally different. Crystal field theory is more easily understood and readily applied than the molecular orbital model, but its physical reality is poor. Molecular orbital theory by requiring estimates of accurate wave functions and appropriate overlap integrals presents acute computational problems, particularly in multi-electron systems involving a transition metal and its nearest-neighbour ligands. These factors underlie the cynical quotation at the preface to this chapter. Despite the enormity of computing energy levels of every electron in a transition metal located in a crystal structure, rapid advances are being made to model molecular complexes of transition metal ions, including coordination clusters relevant to minerals. Such molecular orbital calculations are producing quantitative estimates of energy levels which are in surprisingly good agreement with spectroscopically determined energy separations. Some results of these molecular orbital calculations are described in this chapter.

Another manifestation of covalent bonding relates to the sulphide mineralogy of the transition elements. Although earlier chapters have stressed properties of transition metal ions in oxides and silicates, an important feature of these elements is the frequency of their geochemical association with B-subgroup non-metal and 'metalloid' elements such as sulphur, selenium, tellurium, phophorus, arsenic and antimony. The chalcophilic properties of iron, cobalt, nickel and copper in the crust are well known and are important eco-

Table 11.1. *Racah parameters for field-free (gaseous) transition metal ions*†

Number of 3*d* electrons	Cation			Racah *B* parameter	Racah *C* parameter	Ratio *C/B*
$3d^0$	Ca^{2+}	Sc^{3+}	Ti^{4+}	-	-	-
$3d^1$		Ti^{3+}	V^{4+}	-	-	-
$3d^2$	Ti^{2+}			718	2,629	3.7
		V^{3+}		861	4,165	4.8
			Cr^{4+}	1,039	4,238	4.1
$3d^3$	V^{2+}			766	2,855	3.7
		Cr^{3+}		918	3,850	4.2
			Mn^{4+}	1,064	-	-
$3d^4$	Cr^{2+}			830	3,430	4.1
		Mn^{3+}		965*	3,675	3.8
$3d^5$	Mn^{2+}			960	3,325	3.5
		Fe^{3+}		1,015*	-	-
$3d^6$	Fe^{2+}			1,058	3,901	3.7
		Co^{3+}		1,100	-	-
$3d^7$	Co^{2+}			971	4,366	4.5
		Ni^{3+}		1,115*	-	-
$3d^8$	Ni^{2+}			1,041	4,831	4.6
$3d^9$	Cu^{2+}			-	-	-
$3d^{10}$	Zn^{2+}	Ga^{3+}	Ge^{4+}	-	-	-

† From Lever (1984), p. 115; * From Huheey (1983), p. 447.

nomically in the formation of ore deposits. Such associations suggest that there are factors inherent in the electronic configurations of some transition metals which bestow increased stability on them when they are in combination with the B-subgroup elements of the periodic table. In order to interpret such behaviour, types of covalent bond formation need to be considered. A discussion of some of the qualitative features of molecular orbital theory forms the basis of the present chapter. First, however, spectroscopic indicators of covalency in mineral structures are described, followed by examples of qualitative and computed molecular orbital energy level diagrams.

11.2 Covalency parameters from optical spectra

11.2.1 Racah parameters

Visible-region spectra, which are the principal source of CFSE data for transition metal ions, also provide a measure of relative covalent bonding interactions of cations in host structures and enable estimates to be made of changes

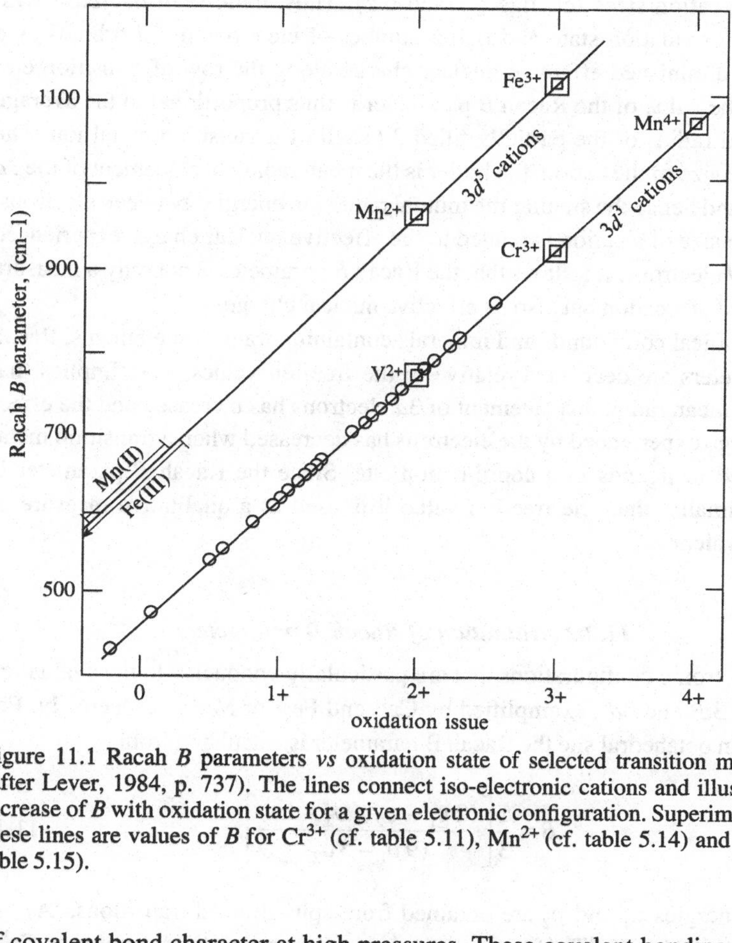

Figure 11.1 Racah B parameters *vs* oxidation state of selected transition metal ions (after Lever, 1984, p. 737). The lines connect iso-electronic cations and illustrate the increase of B with oxidation state for a given electronic configuration. Superimposed on these lines are values of B for Cr^{3+} (cf. table 5.11), Mn^{2+} (cf. table 5.14) and Fe^{3+} (cf. table 5.15).

of covalent bond character at high pressures. These covalent bonding interactions are expressed semi-empirically in the ligand field model of transition metal spectroscopy in terms of the Racah B parameter, which in a field-free (gaseous) cation is one of the parameters expressing electron interaction energies resulting from coulombic and exchange forces. The other is the Racah C parameter. Free-ion values of the Racah B and C parameters are summarized in table 11.1. Note that the C/B ratio approximates four in many cases.

The Racah parameters listed in table 11.1 indicate that the values of B *increase* with increasing oxidation state and number of $3d$ electrons in the first transition metal series. These features are also demonstrated by the free-ion Racah B values for cations with the $3d^3$ and $3d^5$ configurations plotted in fig. 11.1. Similar factors, namely high valences and large number of $3d$ electrons, influence cation sizes (cf. figs 6.1 and 6.2). Thus, ionic radii *decrease* with increasing oxidation state and rising number of electrons in 3*d* orbitals as a result of diminished effective nuclear charge along the row of transition elements. The value of the Racah *B* parameter is thus proportional to the average reciprocal radius of the partially-filled 3*d* shell of a transition metal ion. The larger the size of the cation, the larger is the mean radial displacement of the 3*d* orbitals and hence the smaller the mutual repulsion energies between electrons. Since the size of a cation is related to the effective nuclear charge experienced by the 3*d* electrons, it follows that the Racah *B* parameter is not only a measure of the size of a cation but also its effective nuclear charge.

In chemical compounds and minerals containing transition elements, Racah *B* parameters are decreased relative to the free-ion values. This implies that both the mean radial displacement of 3*d* electrons has increased and the effective charge experienced by the electrons has decreased when a transition metal is bonded to ligands in a coordination site. Since the Racah *B* parameter is always smaller than the free-ion value it is used as a qualitative measure of bond covalency.

11.2.2 Evaluation of Racah B parameters

Two electronic configurations that are particularly conducive to the evaluation of *B* are $3d^3$ and $3d^5$, exemplified by Cr^{3+} and Fe^{3+} or Mn^{2+}, respectively. For Cr^{3+} in an octahedral site the Racah B parameter is calculated from

$$B = \frac{1}{3}\left[\frac{(2\upsilon_1 - \upsilon_2)(\upsilon_2 - \upsilon_1)}{(9\upsilon_1 - 5\upsilon_2}\right], \tag{11.1}$$

where energies υ_1 and υ_2 are obtained from spin-allowed transitions, ${}^4A_{2g} \rightarrow {}^4T_{2g}(F)$ and ${}^4A_{2g} \rightarrow {}^4T_{1g}(F)$, respectively, in the crystal field spectra of Cr^{3+}-bearing minerals. Such spectral data are summarized in table 5.11.

Many of the Cr^{3+} Racah *B* parameters listed in table 5.11 are superimposed on the free-ion data for divalent (V^{2+}), trivalent (Cr^{3+}) and tetravalent (Mn^{4+}) cations with $3d^3$ configurations plotted in fig. 11.1. This graphical representation of the Racah *B* parameter indicates a lowering of formal charge on chromium in Cr(III)-bearing minerals as a result of covalent bonding. The effect is particularly noticable in end-member uvarovite ($Ca_3Cr_2Si_3O_{12}$) and eskolaite (Cr_2O_3) compositions, as well as dense structure-types such as ruby, spinel and pyrope.

Racah *B* parameters have also been calculated for Mn^{2+} and Fe^{3+} in several minerals (Manning, 1970; Keester and White, 1968). In these $3d^5$ cations, two of the energy separations between crystal field states that are relatively insensitive to crystal field strength are expressed by

$$^6A_{1g}(^6S) - {}^4A_{1g},{}^4E_g(^4G)\text{: } \upsilon_3 = 10B + 5C \tag{11.2}$$

and

$$^6A_{1g}(^6S) - {}^4E_g(^4D)\text{: } \upsilon_5 = 17B + 5C\,, \tag{11.3}$$

so that the Racah B parameter may be obtained from

$$\upsilon_5 - \upsilon_3 = 7B\,. \tag{11.4}$$

Values of B derived from crystal field spectra of Mn^{2+}- and Fe^{3+}-bearing minerals, which are listed in tables 5.14 and 5.15, are plotted in fig. 11.1. As for trivalent Cr, the formal charges on divalent Mn and trivalent Fe in oxide and silicate minerals are much lower than the field-free cation values as a result of covalent bonding in the crystal structures.

11.2.3 The nephelauxetic ratio

The results for Cr^{3+} and the $3d^5$ cations Fe^{3+} and Mn^{2+} show that it is possible to derive values of the Racah B parameter for transition metal compounds from absorption bands in their crystal field spectra, enabling comparisons to be made with field-free ion values. In all cases, there is a decrease of the Racah B parameter for the bonded cations relative to the gaseous ions, which is indicative of diminished repulsion between $3d$ electrons in chemical compounds of the transition metals. This reduction is attributable to electron delocalization or covalent bonding in the compounds. Such decreases of Racah B parameters are expressed as the *nephelauxetic* (*Greek*: cloud expanding) ratio, β, given by

$$\beta = \frac{B}{B_0}, \tag{11.5}$$

where B_0 and B are values for the field-free cation and chemically bonded transition element, respectively. β is always less than one.

Two mechanisms contribute to the decrease of Racah B parameters. First, lone pairs of electrons from the ligand may penetrate the $3d$ shell of the transition metal and screen its $3d$ electrons from the nucleus, thereby decreasing the effective nuclear charge experienced by the electrons and expanding the $3d$ shell. This mechanism is termed *central field covalency*. In the second mechanism, referred to as *symmetry restricted covalency*, delocalization of the transition metal 3*d* electrons occurs onto the ligands. Thus, electrons in e_g orbitals become σ^*-antibonding (defined later in §11.3) while t_{2g} electrons may become π-bonding, each spending time on the ligand. As indicated later (§11.3), central field covalency predominates in oxides and silicates with symmetry restricted covalency becoming more prominent in sulphide minerals. Both mechanisms, because they induce electron delocalization onto the ligands, reduce overall repulsion of the cation's electrons and result in lowering of the Racah *B* parameter.

Ligands can be arranged in order according to the extent to which they reduce interelectronic repulsion, as gauged by the Racah *B* parameter or nephelauxetic β ratio. This order constitutes the nephelauxetic (or cloud-expanding) series, a limited version of which is

$$\text{free cation} > F^- > H_2O > OH^- \geq \text{Si–O} > O^{2-} > NH_3 > CO_3^{2-} > Cl^- \\ CN^- > Br^- > S^{2-} > I^- > Se^{2-}. \quad (11.6)$$

Such an order is intuitively one of increasing covalent bonding characteristics of the ligands based on their polarizabilities. The nephelauxetic series departs significantly from the spectrochemical series described in §2.9.2, which is based on relative values of the crystal field splitting parameter, Δ [see eq. (2.19)].

By maintaining the ligand constant a nephelauxetic series for cations may be derived. For example, from absorption spectra of metal fluoride compounds, the order of decreasing nephelauxetic ratio is

$$Mn^{2+} > V^{2+} > Ni^{2+} > Co^{2+} > Fe^{2+} > Cr^{3+} > V^{3+} > Fe^{3+} > \\ Co^{3+} > Mn^{4+}, \quad (11.7)$$

which also departs from the spectrochemical series for these cations [see eq. (2.18)]. The nephelauxetic series suggests that Fe^{2+} and Fe^{3+} have higher covalent bond characters than Ni^{2+} and Cr^{3+}, respectively, contributing to observed enrichments of Fe^{2+} ions in pyroxene M2 and olivine M2 sites (§6.8.4).

11.2.4 Pressure variations of the Racah B parameter

Although the nephelauxetic ratio is only a qualitative indicator of relative covalent bond character of transition metal ions, interesting trends have been established, nevertheless, for minerals containing the cations Cr^{3+}, Fe^{3+} and Mn^{2+}, for which Racah *B* parameters are readily obtained using eqs (11.1) and (11.4) (Keester and White, 1968; Manning, 1970; Abu-Eid and Burns, 1976).

For phases containing low concentrations of Cr^{3+}, there is a trend towards smaller Racah *B* parameters and lower nephelauxetic ratios with decreasing metal–oxygen distances in the host structures (Abu-Eid and Burns, 1976; Amthauer, 1976; Schmetzer, 1982). This trend is also consistent with the results from crystal field spectra of uvarovite at high pressures which indicate a 2.5 per cent decrease of Racah *B* parameter over a 20 GPa pressure range, consistent with compression of the Cr^{3+}-bond. A similar decrease occurs for ruby over the pressure range 10 GPa (Drickamer and Frank, 1973). High-pressure spectral measurements of Mn^{2+}-bearing garnets (Smith and Langer, 1983) again show a decrease in ionicity of Mn^{2+}–O bonds of less than one per cent at 11.2 GPa. These relatively small decreases of Racah *B* parameters of Mn^{2+}- and Cr^{3+}-bearing minerals suggest that their covalent bonding characters do not change dramatically over pressure ranges applicable to the Earth's Mantle.

11.3 Qualitative molecular orbital diagrams

Covalency of ligands clearly influences the positions and intensities of absorption bands in crystal field spectra of oxides and silicates, so that it is pertinent to discuss the types of covalent bonds that exist when transition elements are present in mineral structures. In this section, the more qualitative aspects of molecular orbital theory are described.

The fundamental premise of molecular orbital theory is that the overlap of orbitals depends on the spatial and symmetry properties of metal and ligand orbitals. Principles of group theory are used to ascertain which orbitals may or may not overlap, based on symmetry and directional requirements. The results of these considerations may be summarized as follows.

11.3.1 Formation of σ molecular orbitals

The five $3d$, single $4s$ and three $4p$ orbitals of a free transition metal collectively have comparable energies (compared to the $3s$, $3p$ and core orbitals with lower principal quantum numbers). Six of these orbitals, the d_{z^2} and $d_{x^2-y^2}$ (e_g group), single s (a_{1g}) and p_x, p_y and p_z (t_{1u}) orbitals, have components of their wave functions that are directed along the cartesian axes. When these orbitals are empty they are able to overlap with filled orbitals belonging to six ligands in octahedral coordination to form six bonding σ molecular orbitals and six antibonding σ* molecular orbitals. The bonding molecular orbitals represent the maximum positive overlap between symmetric wave functions and are more stable than the individual atomic orbitals of uncombined metal and ligand atoms. The corresponding antibonding molecular orbitals are less stable

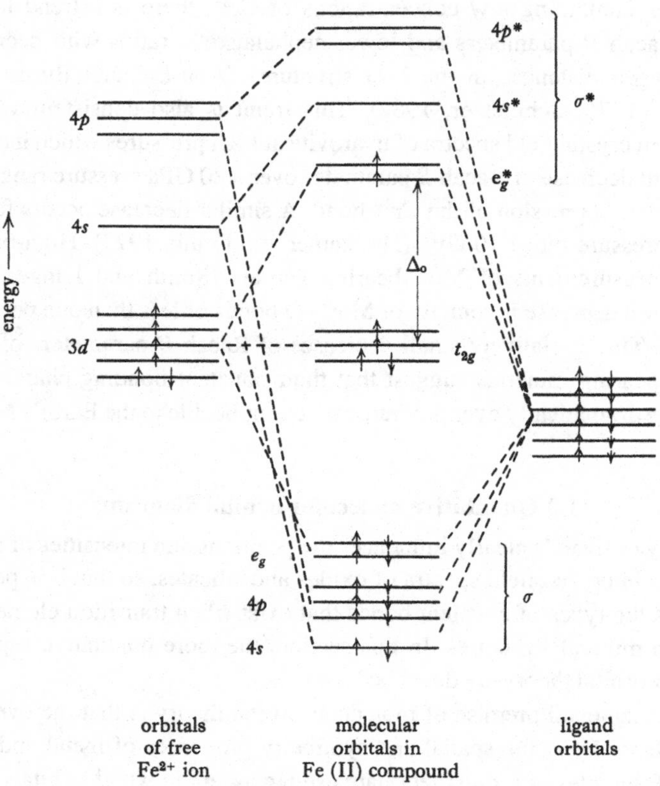

Figure 11.2 Qualitative molecular orbital energy level diagram for the Fe^{2+} ion in octahedral coordination. The diagram refers to σ-bond formation only.

than the component atomic orbitals. The d_{xy}, d_{yz} and d_{zx} (t_{2g} group) orbitals projecting between the ligand atoms may either remain non-bonding or form π-bonding molecular orbitals with the ligands. The formation of π molecular orbitals is considered later (§11.3.2).

The energy relationships for an octahedral cluster, $[FeL_6]$, consisting of a transition element (Fe^{2+}) and six ligands (L) that do not form π-bonds are shown in fig 11.2. There are certain important implications of this qualitative energy level diagram. If a molecular orbital is nearer in energy to one of the atomic orbitals used to construct it than to the other one, it has much more of the character of the first atomic orbital than the second one. Thus, the relationships in fig. 11.2 imply that the six bonding σ molecular orbitals have more of the character of ligand orbitals than of metal orbitals. Electrons occupying these orbitals will be mainly ligand electrons rather than metal electrons.

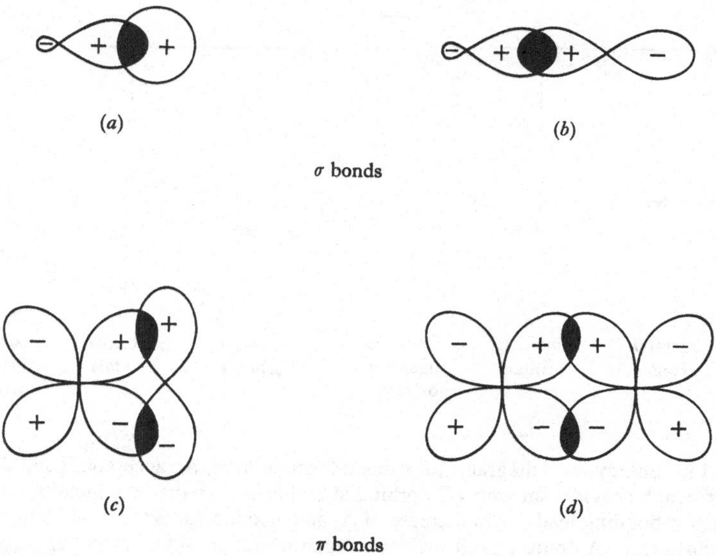

Figure 11.3 Diagrammatic representation of metal–ligand bonds. (*a*) σ-bond formed by the overlap of a ligand *s* orbital and a hybrid metal orbital composed of the 4*s*, three 4*p* and two e_g orbitals; (*b*) σ-bond formed by the overlap of a ligand *p* orbital and the hybrid metal orbital; (*c*) π-bond formed by the overlap of the ligand p_z orbital and a metal t_{2g} orbital; (*d*) π-bond formed by the overlap of a ligand 3*d* orbital and a metal t_{2g} orbital. Signs of the wave functions are indicated; shaded regions represent areas of overlap.

Conversely, electrons that occupy any of the antibonding molecular orbitals are predominantly metal electrons. Any electrons in the t_{2g} orbitals will be purely metal electrons when there are no π molecular orbitals. Thus, the t_{2g} and antibonding e_g* orbitals are predominantly of metal 3*d* orbital character.

The energy separation between the non-bonding t_{2g} and the antibonding e_g* orbitals is Δ_o or 10 *Dq*, and the 3*d* electrons are distributed between the two types of orbitals, being governed by factors similar to those which determine high-spin and low-spin configurations in crystal field theory (§2.5). Therefore, the parameter Δ_o bears a similar relationship in the two theories. In crystal field theory, however, Δ_o is interpreted as the *difference* in repulsion energy between electrons in pure orbitals of the t_{2g} and e_g groups located on the cation by the spherical negative charges of surrounding ligands; in the molecular orbital theory, on the other hand, Δ_o depends on the *strength* of metal–ligand bonds and sharing of electrons takes place between ligand orbitals and the metal 3*d* orbitals.

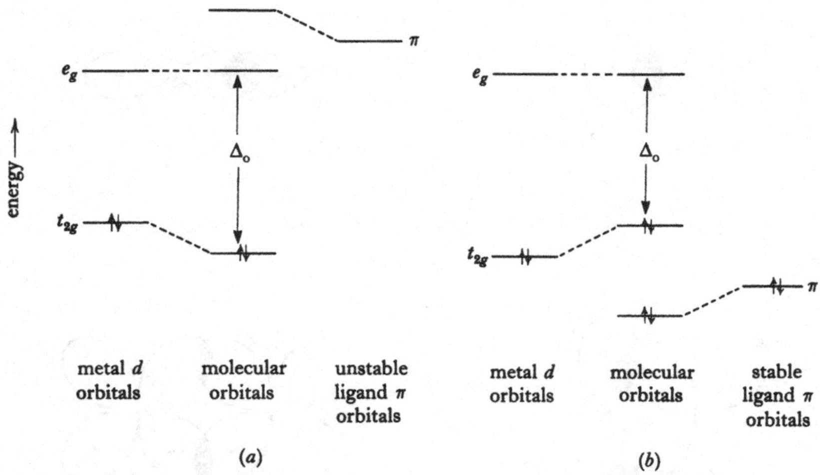

Figure 11.4 Energy level diagrams for π-bond formation. (*a*) An acceptor ligand such as sulphur which provides an empty 3*d* orbital at higher energy than the metal t_{2g} orbital. Note that π-bonding leads to an increase of Δ_o and also increases the total bond energy of the cluster. (*b*) A donor ligand such as oxygen which provides a filled $2p_z$ orbital at lower energy than the metal t_{2g} orbital. Note that Δ_o for oxygen.ligands is diminished by π-bonding and there is little gain in bonding energy.

11.3.2 Formation of π molecular orbitals

Overlap of the t_{2g} orbitals of the transition metal ion with certain *p* and *d* orbitals of the surrounding ligands leads to the formation of π molecular orbitals. Examples are shown diagrammatically in fig. 11.3. Formation of π molecular orbitals of the type shown in fig. 11.3*c*,*d* modify the energy level diagram shown in fig. 11.2. The effects depend on the energy of the ligand π orbitals relative to the metal t_{2g} orbitals, and whether the ligand π orbitals are filled or empty. Three different situations exist.

(1) In the first case, ligands possess filled π orbitals of lower energy than the metal t_{2g} orbitals. Interaction between ligand π and metal t_{2g} orbitals leads to destabilization of the metal t_{2g} orbitals relative to the antibonding e_g orbitals and decreases the value of Δ_o. This is shown by the energy level diagram in fig. 11.4*b*. Ligands of this type include oxygen and fluoride anions. The $2p_z$ orbital of oxygen, for example, lies at a lower energy than the metal 3*d* orbitals. In forming a π bond by overlap of the oxygen $2p_z$ with a metal t_{2g} orbital (fig. 11.3*c*), the bonding π molecular orbital would more resemble the oxygen orbital than the metal orbital, and conversely the antibonding π* molecular orbital will more closely resemble the metal orbital. Since all three 2*p* orbitals of oxygen are already filled by forming σ-bonds, these elec-

trons would occupy the resultant t_{2g} π molecular orbitals. Electrons from the metal 3*d* (t_{2g}) orbitals would be forced to occupy π* molecular orbitals at a higher energy than would be the case if π-bonding had not taken place. Since the e_g σ^*-antibonding molecular orbitals are unaffected by the π-interaction, Δ_o is reduced as a result of π-bonding (fig. 11.4*b*).

(2) In the second case, ligand π orbitals are less stable (at a higher energy) than the metal t_{2g} orbitals and are unoccupied. This situation occurs in compounds where sulphur, selenium, tellurium, phosphorus, arsenic and antimony are the ligands, including numerous ore minerals. Electrons in the metal t_{2g} orbitals extend into the empty ligand π orbitals, and the result of the interaction is to stabilize the metal t_{2g} orbitals. This is shown by the energy level diagram in fig. 11.4*a*. Formation of π-bonds causes the value of Δ_o to be greater than it would be if there were only σ-bonds between metal and ligands. The explanation for an increase of Δ_o is as follows. Although the bonding t_{2g} molecular orbital is lowered and the antibonding t_{2g}* molecular orbital is raised in a manner similar to the first case described above, the fact that the ligand t_{2g} orbitals are empty allows their t_{2g}* molecular orbitals to rise with no cost of energy while the bonding t_{2g} molecular orbitals are stabilized. π-bonding of this type can thus stabilize the cluster by increasing the bond energy. In addition, since the resulting bonding t_{2g} π molecular orbital is localized over both the metal and all of the ligand atoms (as opposed to the localized metal t_{2g} orbitals in which the electrons would have been without π-bonding), electron density is removed from the metal. When cations have low-oxidation states, electron density that tends to pile up via σ-bonding can be dispersed through π-bonding. This is not possible when the metal has a high formal oxidation state since it already carries a partial positive charge. Thus, a synergistic effect can cause the two bonding situations to help each other. The more electron density that π-bonding can transfer from the metal to the ligand, the more the metal can accept via σ-bonding. In turn, the more electron density that σ-bonding removes from the ligand, the more readily the ligand can accept electron density through π-bonding. Up to a certain point, the σ- and π-bonding interactions can augment one another. Note that π-bonding also provides an explanation of the spectrochemical series, eq. (2.19), that cannot be explained purely by electrostatic interactions implicit in the crystal field model.

(3) The third case includes those ligands which possess both filled and empty π orbitals. Examples are the Br^-, Cl^- and CN^- anions. However, in order to understand the chalcophilic properties of transition elements, only the second case involving π-bonding with metal t_{2g} orbitals needs to be considered.

11.4 π-bond formation in minerals

11.4.1 Chalcophilic properties of Fe, Co, Ni and Cu

The occurrence of π-bonds is particularly relevant in sulphide minerals of the transition metals. The previous section described how π-bond formation between transition metals and acceptor ligands, such as sulphur, selenium, tellurium, phosphorus, arsenic and antimony, involves the interaction of paired electrons in metal t_{2g} with empty ligand π orbitals (figs 11.3*c,d*). This results in an increase of Δ_o (fig. 11.4*a*) and a larger contribution to the CFSE illustrated by the data for transition metal (II) sulphides and oxides shown in fig. 7.1 and table 7.1. By analogy with crystal field theory, a large value of Δ_o favours the formation of low-spin electronic configurations in transition metal ions, in which electrons fill low-energy π-bonding t_{2g} orbitals instead of the high energy antibonding e_g* orbitals. Conversely, the number and extent of π-bonds that a transition element can form are increased when electrons fill the t_{2g} orbitals. Those transition metal ions which possess the necessary electronic configurations to form π-bonds are listed below

low-spin Cr^{2+} and Mn^{3+},	$(t_{2g})^4$,	forming one π orbital
low-spin Mn^{2+} and Fe^{3+},	$(t_{2g})^5$,	forming two π orbitals
high-spin Fe^{2+} and Co^{3+},	$(t_{2g})^4(e_g)^2$,	forming one π orbital
low-spin Fe^{2+} and Co^{3+},	$(t_{2g})^6$,	forming three π orbitals
high-spin Co^{2+} and Ni^{3+},	$(t_{2g})^5(e_g)^2$,	forming two π orbitals
low-spin Co^{2+} and Ni^{3+},	$(t_{2g})^6(e_g)^1$,	forming three π orbitals
Ni^{2+},	$(t_{2g})^6(e_g)^2$,	forming three π orbitals
Cu^{2+},	$(t_{2g})^6(e_g)^3$,	forming three π orbitals.

Since electronic configurations with filled t_{2g} orbitals are favourable to π-bond formation with acceptor ligands, the cations capable of forming most π-bonds in octahedral coordination are Ni^{2+}, Cu^{2+}, and low-spin Co^{2+}, Co^{3+} and Fe^{2+}. Thus, the fact that iron, cobalt, nickel and copper are chalcophilic and dominate sulphide, arsenide, etc., mineralogy is due to the capacity of these elements to form a large number of π-bonds with sulphur, arsenic, and other metalloid elements.

11.4.2 Interatomic distances in pyrites

In the crystal structure of pyrite, which is modelled on the periclase structure (fig. 5.1), Fe^{2+} ions occupy Mg^{2+} positions and the mid-points of $(S\text{–}S)^{2-}$ dimeric anions are located at the O^{2-} positions. Each Fe^{2+} ion is in octahedral coordination with one sulphur atom belonging to six different $(S\text{–}S)^{2-}$ dimers, and the

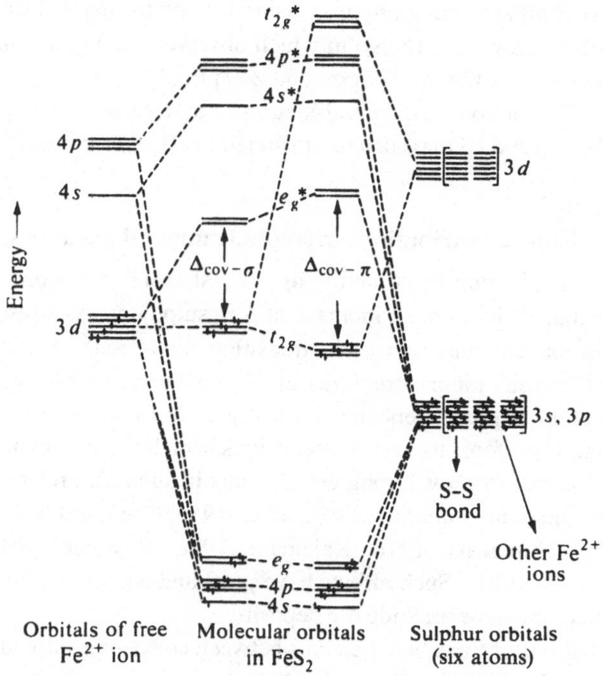

Figure 11.5 Molecular orbital energy level diagram for pyrite, FeS_2 (from Burns & Vaughan, 1970). Note the increase of Δ as a result of π-bond formation facilitated in low-spin Fe^{2+}.

octahedral sites are compressed along a trigonal axis. Each sulphur atom uses hybridized sp^3 orbitals and is tetrahedrally coordinated to three Fe^{2+} ions and the other sulphur atom of the $(S–S)^{2-}$ group. Pyrite is diamagnetic, which is indicative of low-spin Fe^{2+} ions, with the $(t_{2g})^6$ configuration listed in §11.4.1.

A qualititative molecular orbital energy level diagram for pyrite (FeS_2) is shown in fig. 11.5 (Burns and Vaughan, 1970). One hybrid sp^3 orbital from each of the six sulphur atoms in octahedral coordination forms a σ-bond with the central Fe^{2+} ion. The six σ molecular orbitals formed by Fe^{2+} originate from the two e_g, single 4*s* and three 4*p* orbitals as described in §11.3.1. The two electrons in each sulphur sp^3 hybrid fill completely the σ molecular orbitals, while the paired electrons in the three t_{2g} orbitals of low-spin Fe^{2+} form π-bonds with vacant t_{2g}-type 3*d* orbitals of the sulphur atoms. Since no electrons occupy antibonding e_g* orbitals in low-spin Fe^{2+} to repel $(S–S)^{2-}$ ligands, π-bonding is exceptionally efficient in pyrite. This results in a Fe–S distance in pyrite (226 pm) that is significantly shorter than the sum of ionic radii for S^{2-} plus low-spin

Fe^{2+} (245 pm). Similar bond-shortening due to π-bond formation also occurs in cattierite (CoS_2) and vaesite (NiS_2) in which observed (and radii sum) metal–sulphur distances are 232 pm (249 pm) and 240 pm (253 pm), respectively (cf. Appendix 3). Similar contracted bond-lengths in diarsenide and sulpharsenide minerals of Fe, Co and Ni may also be attributed to efficient π-bond formation.

11.5 Element distributions in sulphide mineral assemblages

The energy level diagram illustrated in fig. 11.5 shows how π-bond formation with sulphur ligands leads to an increase of Δ in sulphides. As a result, crystal field stabilization energies of certain transition metal ions in sulphides are higher than those in 'ionic' structures involving bonds with oxygen atoms. This stabilization is greatly enhanced when a cation occurs in the low-spin configuration. Therefore, transition metal ions which form π-bonds are predicted and observed to show strong enrichments in sulphides relative to coexisting silicate and oxide minerals (e.g., Nickel, 1954; Clark and Naldrett, 1972; MacLean and Shimazaki, 1976; Rajamani, 1976; Rajamani and Naldrett, 1978; Fleet *et al.*, 1981). Such effects have profound economic importance in the formation of massive sulphide ore deposits.

The distributions of transition elements between coexisting sulphide minerals can also be qualitatively explained by MO theory. In coexisting pyrite and pyrrhotite assemblages, there is relative enrichment of cobalt over nickel in pyrite, whereas nickel is enriched relative to cobalt in pyrrhotite (e.g., Bjorlykke and Jarp, 1950; Fleisher, 1955). These relative enrichments may be explained by the higher stabilization energy of low-spin Co^{2+} in pyrite compared to Ni^{2+}, and the higher stabilization energy of Ni^{2+} in pyrrhotite compared to high-spin Co^{2+}.

Other examples where qualitative MO theory was used to interpret structural stabilities of sulphide and related minerals of the transition elements include disulphide, sulpharsenide, diarsenide, skutterudite and thiospinel phases (e.g., Nickel, 1968 a,b; Burns and Vaughan, 1970; Vaughan *et al.*, 1971). More rigorous interpretations of structures and bonding of transition metal sulphides have utilized data from MO energy level calculations (e.g., Vaughan and Tossell, 1980a,b, 1983; Vaughan and Craig, 1985; Tossell and Vaughan, 1992).

11.6 Computed molecular orbital energy level diagrams

11.6.1 The computation procedure

Chemical bonding of transition metals in minerals may be determined more quantitatively by computing energy levels of their coordination polyhedra

(Johnson and Smith, 1972). Such molecular orbital calculations have provided electronic structures of several coordination clusters that have been chosen to approximate cation sites in mineral structures, including simple oxides (e.g., Tossell *et al.*, 1973a,b; Vaughan *et al.*, 1974; Loeffler *et al.*, 1974, 1975; Vaughan and Tossell, 1978; Sherman, 1984, 1985a,b; Sherman and Waite, 1985) and sulphides (Vaughan and Tossell, 1980a,b, 1983; Vaughan and Craig, 1985), as well as larger atom clusters (Sherman, 1987a,b, 1990, 1991).

The procedure for calculating energy level diagrams by the *self-consistent field Xα scattered wave* (SCF–Xα–SW) method is as follows (Sherman, 1984, 1991). An octahedral cluster such as $[FeO_6]^{10-}$ is partitioned into a set of (overlapping) spheres centred about divalent iron and each oxygen atom, and these are surrounded by an outer sphere. Within each atomic sphere the one-electron Schrödinger equation

$$\left[\frac{1}{2}\nabla^2 + V_c + V_x\right]\Psi_i = \varepsilon_i \Psi_i \qquad (11.8)$$

is solved for the orbitals (Ψ_i) and their energies (ε_i) using expressions for the Laplacian operator (∇^2) and the coulomb (V_c) and exchange (V_x) potentials. The latter are expressed in terms of the electronic charge density ρ,

$$\rho = \Sigma_i\, n_i\ \Psi_i^* \Psi_i\,, \qquad (11.9)$$

where n_i is the occupancy of orbital *i*. The coulomb potential is then evaluated by solving the Poisson equation, while the exchange potential is evaluated using Slater's Xα approximation

$$V_x = -\ 6\alpha\left(\frac{3\rho}{4\pi}\right)^{1/3}, \qquad (11.10)$$

where α is an adjustable scaling parameter.

After the individual atomic potentials have been calculated to self-consistency, they are superimposed to give an initial molecular potential. Within each of the Fe and O atomic spheres, and beyond the outer sphere surrounding these atoms, the coulomb or exchange potential resulting from the partitioning is spherically averaged. Within the intersphere region the superimposed potential is volume averaged to give a constant value. Once the initial molecular potential is set up, molecular orbitals are obtained by again solving the Schrödinger equation within each region of the cluster and matching the solutions at the sphere boundaries using the multiple Xα scattered wave theory. The resulting molecular orbitals are then used to define the molecular potential

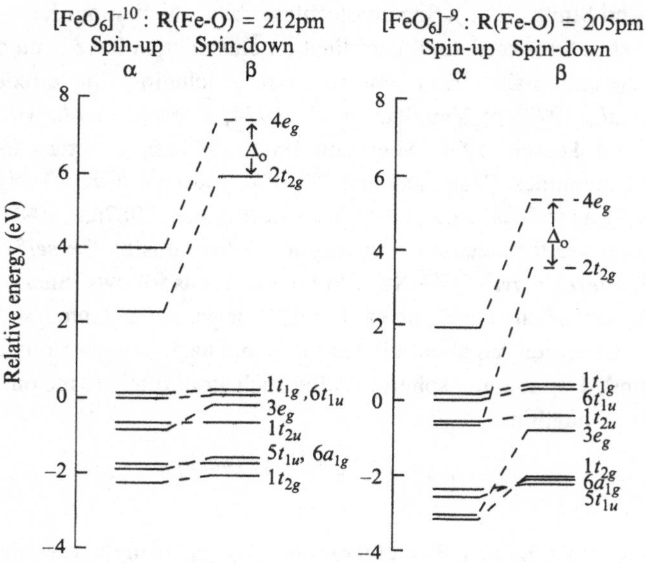

Figure 11.6 Molecular orbital energy level diagrams computed for iron octahedrally coordinated to oxygen. Left: divalent iron in the $[FeO_6]^{-10}$ cluster (based on Sherman, 1991); right: trivalent iron in the $[FeO_6]^{-9}$ cluster (from Sherman, 1985a). Orbital energies have been scaled relative to zero for the non-bonding $6t_{1u}$ level.

which, in turn, is used to define a new potential within each region of the cluster. The process is repeated until a self-consistent result is obtained.

In the Xα scattered wave approximation, the exchange potential for spin-up electrons may be different from that for spin-down electrons. In particular, when unpaired electrons are present, the exchange potentials, and hence the spin-up and spin-down orbitals and their energy levels, are different. Thus, MO calculations are performed using a 'spin-unrestricted' formalism so that separate orbital energy levels are given for spin-up (α) and spin-down (β) electrons.

It is important to note that energy differences between orbitals in the ground state electronic structure are not equal to energies of electronic transitions observed in optical spectra. This is because the orbital energies are a function of the interelectronic coulomb and exchange potentials which, in turn, are functions of the orbital occupancies. When an electronic transition occurs the orbital energies will relax about the new electronic configuration of the excited state. To compensate for this discrepancy between measured and calculated electronic transition energies, the orbital relaxation is taken into account by using a 'transition state' formalism in which the transition state is defined as having orbital occupancies midway between those found in the initial and final states.

11.6.2 MO diagrams for octahedral [FeO$_6$] clusters

Examples of computed molecular orbital energy level diagrams are illustrated in fig. 11.6 for the two clusters $[FeO_6]^{-9}$ and $[FeO_6]^{-10}$, representing Fe^{3+} and Fe^{2+} ions, respectively, bonded to oxygen atoms in regular octahedral sites (Sherman, 1985a, 1991). Energy differences between different spin alignments of electrons in the cation t_{2g} and e_g orbitals are distinguished in the diagrams by levels for spin-up (α) and spin-down (β) configurations. The energy levels shown in fig. 11.6 which are derived from the iron 3*d*, 4*s* and 4*p* atomic orbitals and the oxygen 2*p* orbitals fall into three groups. They are the valence, crystal field and low-energy conduction band molecular orbitals. Not shown on these energy level diagrams in fig. 11.6, are the core molecular orbitals comprising the iron 1*s*, 2*s*, 2*p*, 3*s* and 3*p* and the oxygen 1*s* and 2*s* atomic orbitals because they occur at much lower energies. These orbitals are assumed to be entirely localized on the iron or oxygen atoms and not involved in bonding. Each orbital energy level, nevertheless, is labelled according to group theory representations for octahedral (point group O_h) symmetry (cf. §2.3). The integers that prefix these representations increase sequentially in higher-energy molecular orbitals and designate the sequence of repetition of each symmetry notation (e.g., the O 1*s*, Fe 1*s*, O 2*s*, Fe 2*s*, etc., orbitals all have a_{1g} symmetries, but each is represented, respectively, by $1a_{1g}$, $2a_{1g}$, $3a_{1g}$, $4a_{1g}$, etc.). Reference zero on the energy scale is somewhat arbitrary in molecular orbital diagrams. In the MO diagrams shown in fig. 11.6 the reference zero energy is chosen to be that of the non-bonding $6t_{1u}$ orbitals in each [FeO$_6$] cluster.

The molecular orbital diagrams in fig. 11.6 depict the energy levels of non-bonding, bonding and antibonding orbitals in the octahedral clusters. The valence band orbitals include non-bonding oxygen orbitals labelled as the $1t_{1g}$, $6t_{1u}$, $1t_{2u}$ and $5t_{1u}$ levels, and orbitals with appreciable metal and oxygen characters represented by the $3e_g$, $6a_{1g}$ and $1t_{2g}$ levels. The atomic compositions of the molecular orbitals are given in table 11.2.

The set of orbitals responsible for the Fe–O bonds is similar to that described earlier in the qualitative MO diagram shown in fig. 11.2. The t_{1u}-type oxygen 2*p* orbitals overlap with the Fe 4*p* orbitals. The resulting $5t_{1u}$ molecular orbital is quite delocalized probably because the Fe 4*p* atomic orbital has a large radial extent. The a_{1g}-type oxygen 2*p* orbital combination overlaps with the Fe 4*s* atomic orbital to give the σ-bonding $6a_{1g}$ molecular orbital. The t_{2g}- and e_g-type oxygen 2*p* orbital combinations overlap with the Fe 3*d* orbitals to

Table 11.2. *Valence orbital energies and charge distributions in iron oxide coordination clusters*†

Spin-up (α) orbitals					Spin-down (β) orbitals				
Orbital†	Energy (eV)	% Fe	% O	% Inter-sphere	Orbital†	Energy (eV)	% Fe	% O	% Inter-sphere
divalent iron in octahedral $[FeO_6]^{-10}$ cluster									
Iron 4*s*, 4*p* conduction bands									
$7t_{1u}$* (0)	13.65	17	18	56	$7t_{1u}$* (0)	14.51	14	17	58
$7a_{1g}$* (0)	11.35	34	35	28	$7a_{1g}$* (0)	12.13	31	35	31
Iron 3*d* valence band									
$4e_g$* (2)	3.40	66	31	3	$4e_g$* (0)	7.20	85	13	2
$2t_{2g}$* (3)	0.80	85	9	6	$2t_{2g}$* (1)	5.69	83	7	10
Oxygen 2*p* valence band									
$1t_{1g}$ (3)	0.09	0	92	8	$1t_{1g}$ (3)	0.18	92	8	1
$6t_{1u}$ (3)	0	2	89	7	$6t_{1u}$ (3)	0.13	2	90	7
$1t_{2u}$ (3)	−0.64	0	85	14	$3e_g$ (2)	−0.09	11	86	1
$3e_g$ (2)	−0.90	33	66	0	$1t_{2u}$ (3)	−0.64	0	85	14
$5t_{1u}$ (3)	−1.75	2	75	22	$6a_{1g}$ (1)	−1.57	10	80	9
$6a_{1g}$ (1)	−1.85	12	79	9	$5t_{1u}$ (3)	−1.64	2	75	22
$1t_{2g}$ (3)	−2.18	9	66	24	$1t_{2g}$ (3)	−1.94	3	71	25
trivalent iron in octahedral $[FeO_6]^{-9}$ cluster									
Iron 4*s*, 4*p* conduction bands									
$7t_{1u}$* (0)	15.73	19	19	52	$7t_{1u}$* (0)	16.35	16	18	54
$7a_{1g}$* (0)	12.82	36	35	26	$7a_{1g}$* (0)	13.88	33	35	29
Iron 3*d* valence band									
$4e_g$* (2)	2.00	35	61	4	$4e_g$* (0)	5.43	75	24	1
$2t_{2g}$* (3)	−0.55	48	40	12	$2t_{2g}$* (0)	3.57	86	7	6

Oxygen 2p valence band											
$1t_{1g}$	(3)	0.19	0	92	8	$1t_{1g}$	(3)	0.43	0	91	8
$6t_{1u}$	(3)	0	2	89	8	$6t_{1u}$	(3)	0.28	2	89	8
$1t_{2u}$	(3)	−0.59	0	86	14	$1t_{2u}$	(3)	−0.35	0	85	14
$5t_{1u}$	(3)	−2.37	3	68	29	$3e_g$	(2)	−0.76	23	73	3
$6a_{1g}$	(1)	−2.51	11	75	14	$1t_{2g}$	(3)	−1.99	7	69	23
$3e_g$	(2)	−2.99	64	35	0	$6a_{1g}$	(1)	−2.08	9	75	15
$1t_{2g}$	(3)	−3.13	49	38	15	$5t_{1u}$	(3)	−2.13	2	68	29
trivalent iron in tetrahedral [FeO4]$^{-5}$ cluster											
Iron 4s conduction band											
$7a_1$*	(0)	14.91	22	21	38	$7a_1$*	(0)	15.64	16	18	39
Iron 3d valence band											
$6t_2$*	(3)	1.88	41	50	9	$6t_2$*	(0)	5.32	70	20	10
$2e$*	(2)	0.77	44	46	10	$2e$*	(0)	4.43	81	12	7
Oxygen 2p valence band											
$1t_1$	(3)	0	0	85	14	$1t_1$	(3)	0.48	0	86	13
$5t_2$	(3)	−1.41	3	64	33	$4t_2$	(3)	−0.62	21	68	10
$6a_1$	(1)	−1.82	11	71	18	$1e$	(2)	−0.79	11	68	20
$1e$	(2)	−2.33	52	35	13	$5t_2$	(3)	−1.01	9	70	21
$4t_2$	(3)	−2.64	55	40	4	$6a_1$	(1)	−1.19	9	70	21

† Orbitals with an asterisk are antibonding. The numbers in parentheses give the orbital occupancies. The reference zero energy is that of the non-bonding $6t_{1u}$ and $1t_1$ orbitals in the [FeO_6] and [FeO_4] clusters, respectively.
Sources of data : Sherman (1985a; 1991)

give the $1t_{2g}$ and $3e_g$ molecular orbitals. The most important bonding orbital is the $3e_g$ which corresponds to the Fe(3*d*) – O(2*p*) σ-bonding interaction. Of next importance is the $1t_{2g}$ orbital which corresponds to the Fe(3*d*) – O (2*p*) π-bonding interaction. The orbital composition data in table 11.2 suggest that there is a large degree of π-bonding in these octahedral $[FeO_6]$ clusters. The covalent bonding characters of these orbitals increase with increasing oxidation state of the iron atoms; higher electron density resides on trivalent Fe so that the $[FeO_6]^{-9}$ cluster becomes more stable than the $[FeO_6]^{-10}$ cluster for divalent Fe.

The $2t_{2g}$ and $4e_g$ molecular orbitals correspond to the 3*d* orbitals in the crystal field model. In the purely ionic bonding description of crystal field theory the $2t_{2g}$ and $4e_g$ molecular orbitals are the pure 3*d* orbitals of iron that have lost their degeneracy by electrostatic interactions with surrounding oxygen anions. However, in the molecular orbital description they are the antibonding equivalents of the $1t_{2g}$ and $3e_g$ bonding molecular orbitals. As the metal atom character of these bonding $1t_{2g}$ and $3e_g$ molecular orbitals increases, the ligand character of the antibonding $2t_{2g}$ and $4e_g$ molecular orbitals increases accordingly. Thus, as the $1t_{2g}$ and $3e_g$ molecular orbitals become more bonding in character, the $2t_{2g}$ and $4e_g$ crystal field orbitals become more antibonding. The average energy of the crystal field orbitals (relative to the oxygen 2*p* non-bonding orbitals such as $6t_{1u}$, $1t_{1g}$ and $1t_{2u}$) decreases with increasing oxidation state of the iron atom. Since the $2t_{2g}$ and $4e_g$ molecular orbitals are dominantly iron in character, their increasing stability is a consequence of the increased electronegativity of the ferric iron atom.

By analogy with the crystal field model, the octahedral crystal field splitting (Δ_o) is higher for ferric iron than for ferrous iron, due partly to a larger electrostatic interaction between the trivalent cation and the oxygen anions, but more importantly because of an increased degree of σ-bonding in the e_g orbitals relative to π-bonding character of the t_{2g} orbitals in the clusters. The $7a_{1g}$ and $7t_{1u}$ orbitals, which correspond to the Fe 4*p* and Fe 4*s* atomic orbitals, lie at energies in excess of 10 eV above the $2t_{2g}$ and $4e_g$ orbitals (table 11.2). They are the lowest-energy molecular orbitals of conduction band character and are the antibonding equivalents of the $6a_{1g}$, and $6t_{1u}$ bonding molecular orbitals. In a localized (ionic) description, they would correspond to the Fe 4*s* and 4*p* atomic orbitals. The molecular orbital calculations, however, show them to be extensively delocalized over the oxygen interatomic and extra-molecular regions. Note the energy differences between each of the two spin polarizations of the Fe^{2+} and Fe^{3+} ions; the calculated MO energy level of the spin-up configuration differs by as much as 4 eV from its equivalent spin-down configuration.

The orbital composition data summarized in table 11.2 indicate the extent of covalency of each orbital as measured by the donation of charge from the oxygen to the iron atoms in the octahedral clusters. The values are highly spin-polarization dependent, being of higher oxygen character for spin-down molecular orbitals than for spin-up molecular orbitals. In the $[FeO_6]^{-9}$ cluster, the α-spin e_g and t_{2g} antibonding molecular orbitals are occupied and cancel out much of the α-spin e_g and t_{2g} Fe–O bonding interactions. At the same time, the β-spin antibonding orbitals are unoccupied. Hence, most of the charge donated to the Fe^{3+} by the O^{2-} ions is of β-spin character. This contributes to antiferromagnetic ordering in ferric oxides (§10.8.1). The effective charges on the divalent and trivalent Fe atoms are reduced to about +0.94 and +1.3, respectively, as a result of covalent bonding (cf. fig. 11.1).

11.6.3 MO diagram for the tetrahedral [FeO$_4$] cluster

The molecular orbitals calculated for Fe^{3+} in the tetrahedral cluster $[FeO_4]^{-5}$ indicate that the antibonding e and t_2 molecular orbitals, corresponding to the iron $3d$ orbitals in a tetrahedral crystal field, are mostly localized on the iron atom (Sherman, 1985a). Furthermore, although allowed by symmetry, there appears to be little Fe $4p$ character in these orbitals, casting some doubt on the intensification mechanism of absorption bands in crystal field spectra of tetrahedrally coordinated cations (§3.7.1).

11.6.4 MO diagrams for low-symmetry coordination clusters

Molecular orbital cluster calculations have also been made for low-symmetry coordination polyhedra, including a trigonally distorted $[FeO_6]^{-9}$ octahedron modelling the Fe^{3+} site in hematite (Sherman, 1985a), and $[FeO_4(OH)_2]^{-7}$ clusters representing cation sites in goethite and lepidocrocite (Sherman, 1985b). In the trigonally distorted $[FeO_6]^{-9}$ cluster, there are covalency differences to the two types of oxygen bonded to iron. Surprisingly, the longer Fe–O(2) bonds (211.7 pm) are somewhat more covalent than the shorter Fe–O(1) bonds (195.0 pm). Apparently, strong repulsive interactions across shortest O(2)–O(2) bonds promote the Fe–O(2) bonding interaction. Results for the $[FeO_4(OH)_2]^{-7}$ polyhedron indicate that Fe–OH bonds are more ionic than Fe–O bonds, correlating with the relative positions of the OH^- anion and the H_2O ligand in the nephelauxetic series, eq. (11.6). Thus, nett charges on the oxygen atoms correspond to about −1.42 (Fe–OH) and −1.57 (Fe–O) in the $[FeO_4(OH)_2]^{-7}$ cluster and −1.72 (Fe–O) in the $[FeO_6]^{-9}$ cluster (Sherman, 1985b).

11.7 Molecular orbital assignments of electronic spectra

11.7.1 Crystal field spectra

The molecular orbital diagrams in fig. 11.6 may be used to interpret electronic transitions in optical spectra of Fe^{3+} and Fe^{2+}-bearing minerals, including crystal field, intervalence charge transfer and oxygen → metal charge transfer transitions (Sherman, 1985a,b, 1991; Sherman and Waite, 1985). Crystal field transitions in Fe^{2+} correspond to transferral of electrons between the antibonding $2t_{2g}$ and $4e_g$ molecular orbitals (fig. 11.6*a*). In the spin-unrestricted $X\alpha$ formalism, the crystal field splitting parameter Δ_o corresponds to the energy separation between the $2t_{2g}$ and $4e_g$ levels in the electronic configuration $(2t_{2g}{}^{\uparrow})^3(4e_g{}^{\uparrow})^2(2t_{2g}{}^{\downarrow})^{0.5}(4e_g{}^{\downarrow})^{0.5}$. Despite the strong Fe–O covalency, the electrons in these orbitals are mostly Fe 3*d* and are strongly localized on the Fe atom. The calculated energy separations between these levels, symbolized by Δ_o in fig. 11.6, correspond to 11,137 cm^{-1} for divalent Fe in the $[FeO_6]^{-10}$ octahedral cluster (Sherman, 1991) and 15,800 cm^{-1} for trivalent Fe in the $[FeO_6]^{-9}$ cluster (Sherman, 1985a). Similar calculations for the $[FeO_4]^{-5}$ cluster yielded Δ_t = 8,230 cm^{-1} for tetrahedrally coordinated Fe^{3+}. These values are in reasonable agreement with crystal field splitting parameters obtained from visible-region spectra of iron-bearing minerals (tables 5.15, 5.16 and 10.2).

11.7.2 Oxygen → metal charge transfer transitions

Charge transfer transitions between oxygen and iron cations in ultraviolet-region spectra may be assigned based on the molecular orbital diagrams shown in fig. 11.6. The lowest energy OMCT transitions in the octahedral clusters are from non-bonding molecular orbitals labelled $1t_{2u}$, $6t_{1u}$, and $1t_{1g}$ localized on the oxygen atoms to antibonding $2t_{2g}$ molecular orbitals localized on the iron cations. The energies of the $6t_{1u} \rightarrow 2t_{2g}$ and the $1t_{1u} \rightarrow 2t_{2g}$ transitions, each of which is both spin-allowed and parity (Laporte) allowed (§3.7), are calculated to be at 38,100 cm^{-1} and 43,600 cm^{-1}, respectively, in Fe^{3+} in the $[FeO_6]^{-9}$ cluster (Sherman, 1985a,b). In the $[FeO_6]^{-10}$ cluster representing octahedrally coordinated Fe^{2+} the $6t_{1u} \rightarrow 2t_{2g}$ OMCT transition is predicted to occur at 42,630 cm^{-1}. The lowest-energy OMCT transition, between non-bonding $1t_1$ and anti-bonding 2*e* molecular orbitals, in tetrahedrally coordinated Fe^{3+} is calculated to be centred at 40,400 cm^{-1}. Each of these OMCT bands is located well into the ultraviolet. However, since they are fully allowed and have intensities 10^3–10^4 times higher than those of crystal field transitions, absorption edges may extend into the visible region and overlap very weak spin-forbidden bands in Fe^{3+}- and Fe^{2+}-bearing minerals (§4.7.3).

11.7.3 Intervalence charge transfer (IVCT) transitions

11.7.3.1 Homonuclear IVCT transitions.

Insights into the mechanism of IVCT (or metal → metal charge transfer) transitions and electron delocalization processes in oxide minerals, including types of molecular orbitals involved, may be deduced from calculations of molecular orbital energy level diagrams for mixed-valence metal clusters (Sherman, 1987a, b, 1990). In these calculations, dimeric clusters such as $[Fe_2O_{10}]^{-15}$ corresponding to two edge-shared $[FeO_6]$ octahedra containing adjacent Fe^{2+} and Fe^{3+} ions are considered (Sherman, 1987a). Two different situations exist, corresponding to symmetric and antisymmetric mixed-valence systems. These are defined by the configurational coordinate q, expressed as $q = R_B - R_A$, where R_A and R_B are the mean Fe_A–O and Fe_B–O bond lengths in the linked octahedra. Symmetric clusters apply to Fe^{2+}–Fe^{3+} pairs in identical coordination sites such as the magnetite octahedral sites (cf. fig. 4.18), ilvaite Fe(A) sites (cf. fig. 4.19) or pyroxene M1 octahedra (cf. fig. 5.13) representatative of Class III mixed-valence compounds (table 4.3). The more general asymmetric clusters ($q = -\lambda$) correspond to Class II mixed-valence minerals in which Fe^{2+} and Fe^{3+} occur in different coordination polyhedra (e.g., amphibole M1, M2 and M3 sites shown in fig. 4.14) or are surrounded by different distributions of next-nearest-neighbour cations as a result of atomic substitution (e.g., Al and Mg dispersed among Fe^{3+} and Fe^{2+} cations in adjacent M2 and M1,M3 sites of amphiboles, fig. 4.14; and different cations occupying *cis*-M2 sites adjacent to Fe^{2+} or Fe^{3+} ions in *trans*-M1 sites of biotites, fig. 5.21).

Geometries of the two $[Fe_2O_{10}]^{-15}$ clusters used in the molecular orbital calculations of mixed-valence oxide and silicate minerals are illustrated in fig. 11.7*a,b*. Calculated molecular orbital diagrams for the two clusters are shown in fig. 11.7*c,d*. The MO diagram for the asymmetric ($q = -\lambda$) cluster (fig. 11.7*d*) more or less resembles the superposition of the MO diagrams for individual $[FeO_6]^{-9}$ and $[FeO_6]^{-10}$ clusters containing octahedrally coordinated Fe^{3+} and Fe^{2+} ions shown in fig. 11.6. However, there is a significant degree of metal–metal bonding across the shared octahedral edge resulting in partial delocalization of the sixth $3d$ (β-spin t_{2g}) electron of Fe^{2+} onto the Fe^{3+} site, resulting in a cation valence configuration of $Fe^{2.12+}Fe^{2.88+}$, each charge being lowered due to Fe–O bond covalency. In the symmetric $[Fe_2O_{10}]^{-15}$ cluster ($q = 0$), the Fe^{2+} β-spin electron is delocalized over both sites so that each Fe atom has a formal charge of $Fe^{2.5+}$. The delocalization of the Fe^{2+} β-spin electron can be seen in the wavefunction contours shown in fig. 11.7*e,f*. Complete delocalization results in the weak Fe–Fe bond evident in the wavefunction plot in fig.

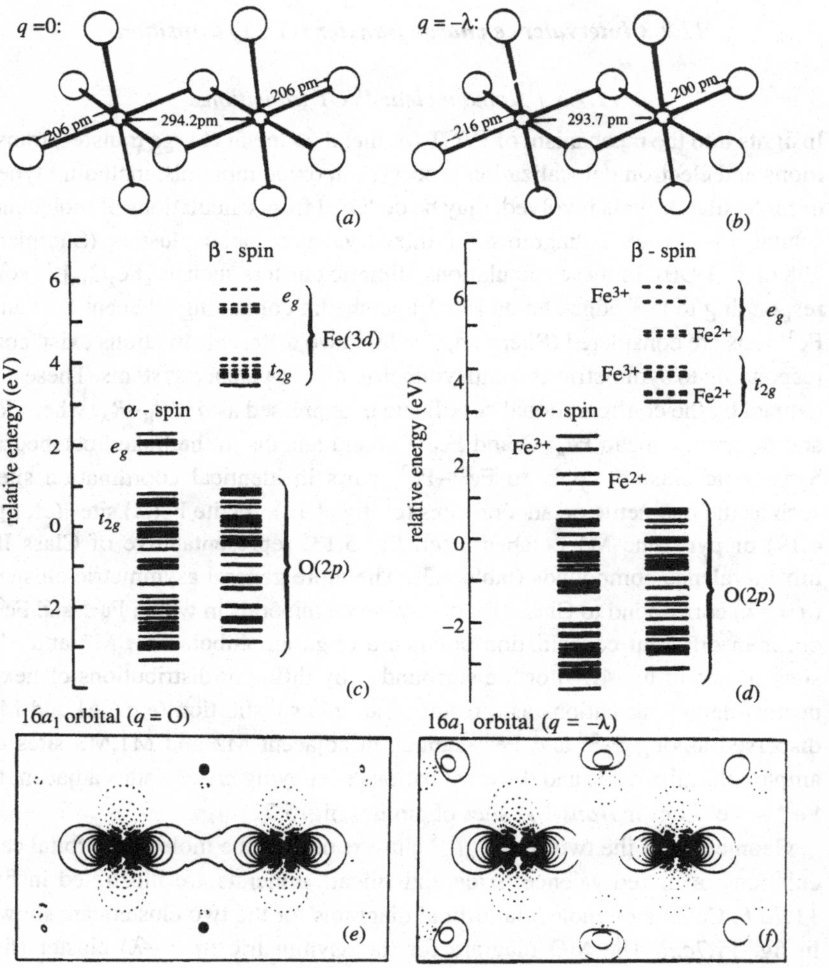

Figure 11.7 Molecular orbital descriptions for mixed-valence dimeric $[Fe_2O_{10}]^{-15}$ clusters corresponding to edge-sharing of Fe^{2+} and Fe^{3+} octahedra (from Sherman, 1987a). (*a,b*) Geometries of the clusters used to model the edge-shared octahedra. The $q = 0$ geometry corresponds to that found when the two sites are indistinguishable (cf. magnetite, ilvaite), whereas $q = -\lambda$ applies to Fe^{2+} and Fe^{3+} in different sites (e.g., amphiboles); (*c,d*) calculated molecular orbital energy level diagrams for each cluster geometry. All orbitals are singly degenerate. Dashed lines indicate empty orbitals; (*e,f*) wavelength contours for the Fe^{2+} β-spin electron in the ground state t_{2g} orbital labelled as the $16a_1$ representation.

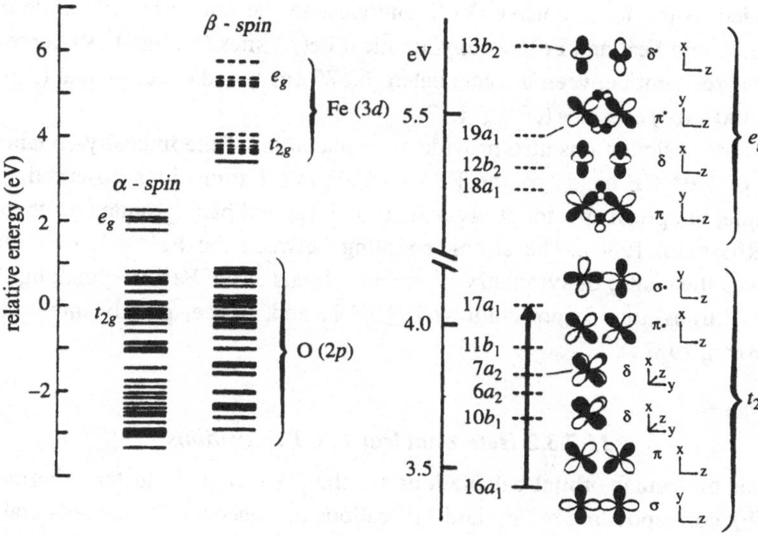

Figure 11.8 Details of the spin-down (β-spin) orbital energy levels at $q = 0$ showing the different types of Fe(3*d*)–Fe(3*d*) bonding interactions (from Sherman, 1987a). The only occupied orbital is the $16a_1$ level. Excitation of the β-spin electron between the $16a_1$ and $17a_1$ levels corresponds to a $Fe^{2+} \rightarrow Fe^{3+}$ IVCT transition observed in optical spectra of minerals (cf. table 4.2).

11.7*e*. Such Fe–Fe bonding provides the mechanism for electron delocalization in mixed-valence Fe^{2+}–Fe^{3+} minerals (Sherman, 1987a). Electron delocalization requires that the adjacent Fe atoms be ferromagnetically coupled, however.

The details of the β-spin Fe(3*d*) t_{2g} and e_g bands in the MO diagram for the symmetric $[Fe_2O_{10}]^{-15}$ cluster are expanded in fig. 11.8. The Fe^{2+} β-spin electron occupies the $16a_1$ molecular orbital which is Fe–Fe σ-bonding. Promotion of the β-spin electron from the $16a_1$ to the $17a_1$ molecular orbital (Fe–Fe σ-antibonding) corresponds to the optically induced IVCT transition (§4.7.2). The energy of this transition is calculated to be 6,775 cm^{-1} and 10,570 cm^{-1} for the symmetric and asymmetric $[Fe_2O_{10}]^{-15}$ clusters, respectively (Sherman, 1987a). These values appear to seriously underestimate observed values for $Fe^{2+} \rightarrow Fe^{3+}$ IVCT (table 4.2) plotted earlier in fig. 4.17. The discrepancies may be attributed to the fact that the cluster calculations describe symmetrical charge transfer (i.e. constant electrostatic potentials in both sites are forced to be the same), whereas in most silicates, the $Fe^{2+} \rightarrow Fe^{3+}$ IVCT transitions are asymmetric because Fe^{2+} and Fe^{3+} cations occupy crystallographically different sites and hence experience different electrostatic potentials. The difference in potential energies must be added to the finite cluster IVCT energies. In the one case, orthorhombic ilvaite, where Fe^{2+} and Fe^{3+} occupy identical Fe(A) sites (cf. fig. 4.19), there is better agreement between the calculated (6,675 cm^{-1}) and experimental (9,300 to 12,300 cm^{-1}) energies (cf. fig. 4.17).

The MO calculations also provide an explanation of the intensity-enhancement of Fe^{2+} CF bands by the $Fe^{2+} \rightarrow Fe^{3+}$ IVCT transitions observed, for example, in optical spectra of vivianite (fig. 4.12) and babingtonite (Amthauer and Rossman, 1984). The strong coupling between the Fe^{2+} and Fe^{3+} ions removes the centre of symmetry of the 3*d* orbitals at the Fe^{2+} site, making the Fe^{2+} CF transitions Laporte-allowed (§3.7.1) and, hence, greatly intensified (Sherman, 1987a).

11.7.3.2 Heteronuclear IVCT transitions

Similar molecular orbital calculations for the $[FeTiO_{10}]^{-14}$ cluster (Sherman, 1987b), corresponding to Fe^{2+} and Ti^{4+} cations in adjacent edge-shared octahedra, lead to an estimate of 18,040 cm^{-1} for the $Fe^{2+} \rightarrow Ti^{4+}$ IVCT energy, which agrees more favourably with values obtained from optical spectral measurements (table 4.2). Again, the calculations show that there is a weak chemical bond between the Fe and Ti atoms. The slight delocalization of the Fe^{2+} β-spin electron corresponds to a valence configuration $Fe^{2.18+}Ti^{3.82+}$, which is lowered again, however, by metal–oxygen bond covalency. More recent MO calculations extended to $[FeMnO_{10}]^{-n}$ clusters associated with Fe^{2+}–Mn^{3+}, Fe^{3+}–Mn^{2+} and Fe^{3+}–Mn^{3+} cation assemblages, have shown that the Fe^{2+}–Mn^{3+} configuration is unstable relative to the Fe^{3+}–Mn^{2+} configuration (Sherman, 1990). This indicates that Mn^{3+} cannot coexist with ferrous ions in silicates in which Fe^{2+} and Mn^{3+} occur in edge-shared octahedra. Optically induced $Mn^{2+} \rightarrow Fe^{3+}$ and $Mn^{3+} \rightarrow Fe^{3+}$ IVCT energies are estimated to be 14,900 cm^{-1} and 17,800 cm^{-1}, respectively, although Jahn–Teller distortion involving Mn^{3+} may increase the latter.

11.8 Structural stabilities of Mn(IV) oxides

One feature that distinguishes manganese from other transition elements is the large number of oxide minerals known to contain tetravalent cations. More than a dozen naturally occurring Mn(IV) oxides exist (Burns and Burns, 1979), many of which are listed in table 8.3, and the number is continuously being augmented by the synthesis of new MnO_2 phases (e.g., Hunter, 1981; Schumm *et al.*, 1985). These natural and synthetic polymorphs of 'MnO_2' display a variety of crystal structures with a diversity of linkages of $[MnO_6]$ octahedra illus-

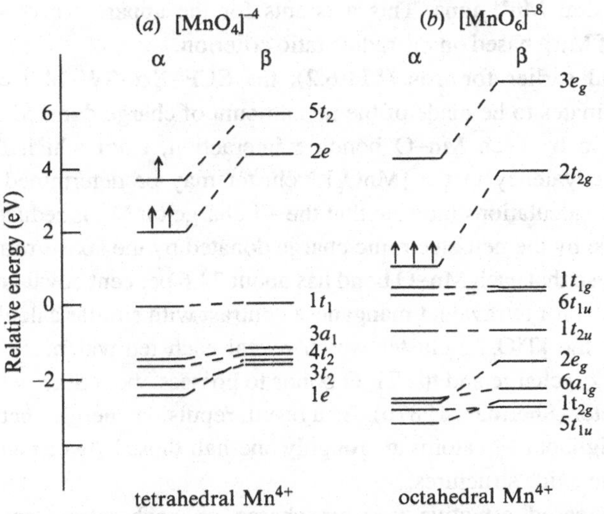

Figure 11.9 Molecular orbital energy level diagrams computed for tetravalent manganese. (*a*) Tetrahedral $[MnO_4]^{-4}$ cluster; (*b*) octahedral $[MnO_6]^{-8}$ cluster (based on Sherman, 1984).

trated earlier in figs 8.7 and 8.8. By comparison, tetravalent titanium forms fewer TiO_2 polymorphs and a more restricted structural array of linked $[TiO_6]$ octahedra (Waychunas, 1991). Octahedrally coordinated Ti^{4+} cations in oxides are expected in ionic structures based on radius ratio criteria. However, octahedral Mn^{4+} cations in oxide structures might not be expected because the radius ratio (Mn^{4+}/O^{2-} = 53pm/140pm = 0.378) falls in the range predicted for tetrahedral coordination (§9.8.1). One explanation for the apparent crystal chemical anomalies of tetravalent Mn is the high octahedral CFSE of Mn^{4+} ions (tables 2.5 and 5.2). Another is the high degree of covalent bonding character of Mn(IV) oxides compared to Ti(IV) oxides demonstrated by molecular orbital energy level calculations (Sherman, 1984; 1987b).

The molecular orbital energy level diagrams for two clusters, one corresponding to tetrahedral $[MnO_4]^{-4}$ with an interatomic distance Mn^{4+}–O^{2-} = 174 pm and the other to octahedral $[MnO_6]^{-8}$ with Mn^{4+}–O^{2-} = 188 pm, are shown in fig. 11.9. In the ground state of the $[MnO_6]^{-8}$ cluster, the three 3*d* electrons of Mn^{4+} occupy each of the three molecular orbitals labelled $5t_{2g}$, whereas in the $[MnO_4]^{-4}$ cluster, two of the 3*d* electrons occupy the two 2*e* molecular orbitals and the third enters one of the three $5t_2$ molecular orbitals. This tetrahedral $5t_2$ level lies 2.8 eV above the octahedral $2t_{2g}$ level, with the result that the $[MnO_6]^{-8}$ cluster is more stable than the $[MnO_4]^{-4}$ cluster. These calculations show that octahedrally coordinated Mn^{4+} cations are more stable in oxides than are tetrahedral Mn^{4+} ions. This accounts for the apparent crystal chemical anomaly of Mn^{4+} based on the radius ratio criterion.

As noted earlier for iron (§11.6.2), the SCF–Xα–SW MO calculations enable estimates to be made of the nett amount of charge donated to the manganese atom by each Mn–O bonding interaction, from which the relative degree of covalency in the $[MnO_6]^{-8}$ cluster may be determined (Sherman, 1984). The calculations indicate that the +4 charge on Mn is reduced to effectively +1.06 by the nett electronic charge donated by the six surrounding oxygen atoms, so that each Mn–O bond has about 73.6 per cent covalent character. These results for tetravalent manganese contrast with similar calculations performed on the $[TiO_6]^{-8}$ cluster which reveal each tetravalent Ti to have an effective +2.0 charge and the Ti–O bonds to possess about 50.2 per cent covalent character (Sherman, 1987b). As a result, repulsion energies between next-nearest-neighbour Mn atoms are roughly one-half those between tetravalent Ti atoms in the oxide structures.

The decreased effective positive charge on each tetravalent Mn atom bonded to oxygen provides an explanation for the diversity of linkages of $[MnO_6]$ octahedra in Mn(IV) oxides. As noted earlier (§8.7.3), the structures of all MnO_2 polymorphs contain $[MnO_6]$ octahedra linked to one another by edge- or corner-sharing to form infinite chains, layers, and three-dimensional tunnel networks (figs 8.7 and 8.8). The presence of shared edges allows for cation–cation interactions between Mn atoms separated by only 286 to 290 pm which would be extremely repulsive if each cation carried a +4 charge. However, the MO calculations described earlier indicate that each Mn atom has an effective charge of about +1 as a result of covalent bonding with oxygen. The reduced effective cationic charge enables each Mn to 'tolerate' several next-nearest-neighbour Mn atoms, accounting for chain (e.g., romanechite, todorokite, λ-MnO_2 spinel) and layer (e.g., chalcophanite, birnessite, vernadite, buserite) structures with as many as five or six next-nearest-neighbour Mn atoms in edge-shared $[MnO_6]$ octahedra. Titanium, on the other hand, appears to be far less tolerant towards repulsion by next-nearest-neighbour Ti^{4+} ions. Although double chains of edge-shared $[TiO_6]$ exist in anatase, brookite and priderite, the number of shared edges between adjacent $[TiO_6]$ does not exceed four. Indeed, the larger distortions of $[TiO_6]$ octahedra in anatase and brookite appear to reflect strong Ti^{4+}–Ti^{4+} repulsions despite indications from MO calculations that the formal charge on these tetravalent Ti atoms is reduced to +2.

Nevertheless, in the Mn(IV) oxides, some sort of 'safety valve' may be necessary to alleviate the cation–cation repulsive interactions between so many next-nearest-neighbour cations. This may be provided by cation vacancies and the presence of low-valence Mn cations in the linked octahedra. Thus, Mn^{3+} ions appear to coexist with tetravalent Mn in the double and treble chains of edge-shared octahedra forming the tunnels in romanechite, todorokite, hollandite and cryptomelane (fig. 8.7). The MO calculations for trivalent Mn in the $[MnO_6]^{-9}$ cluster (Sherman, 1984) reveal an effective charge of +0.92 on the trivalent Mn atom and about 69.3 per cent covalent character of each Mn–O bond, so that repulsive Mn^{3+}–Mn^{4+} interactions are again reduced considerably between Mn atoms in adjacent edge-shared octahedra. On the other hand, cation vacancies existing in the two-dimensional layers of edge-shared octahedra in chalcophanite and birnessite are such that each Mn atom is likely to be surrounded by five Mn atoms instead of six. This enables repulsive interactions to be alleviated by the proximity of a nearby cation vacancy (cf. fig. 8.8). However, uncompensated bonding by oxygen atoms surrounding these cation vacancies must be balanced by the occurrence of cations in sites above and below the vacancies. While such vacancies allow Zn^{2+}, Mn^{2+}, Mg^{2+}, etc., to occupy positions in the chalcophanite, birnessite, and buserite structures between the layers of edge-shared $[MnO_6]$ octahedra, they also provide the positioning of 'walls' in the tunnels of the hollandite and todorokite structure-types (Burns *et al.*, 1983, 1985).

11.9 Summary

In chapter 11, features of molecular orbital theory are described and applied to covalent bonding interactions in minerals. Qualitative MO diagrams are used to interpret the sulphide mineralogy and chalcophilic properties of the transition elements. Calculated MO energy level diagrams for metal–oxygen coordination clusters provide some quantitative insights into spectral features and crystal chemical properties of oxide minerals of Fe and Mn.

Racah parameters. The Racah *B* parameter, which is a measure of interelectronic repulsion and exchange interactions, provides a qualitative indication of bond covalency. Values of *B* derived from optical spectra are lower for transition metals bonded to ligands in a coordination site than for isolated gaseous cations. The nephelauxetic series represents the order of decreasing Racah *B* parameters and correlates with increasing covalent bonding characters of ligands coordinated to a transition metal.

Molecular orbitals. Overlap of certain atomic orbitals of the metal and ligand atoms in a coordination site leads to the formation of molecular orbitals. In octahedral coordination, σ-bonds are formed by the overlap of hybrid orbitals of the metal, comprising the 4*s*, three 4*p* and two e_g orbitals, with filled *s* or *p* orbitals of the ligand. Overlap of metal t_{2g} orbitals and certain *p* and *d* orbitals of the ligands leads to π-bond formation. As in crystal field theory, Δ_o or 10 *Dq* is the energy separation between levels of the t_{2g} and e_g orbital groups of the metal by the surrounding ligands. However, Δ_o in the molecular orbital theory depends more on the strength of metal–ligand bonds and not solely on differences of repulsion energy of electrons in t_{2g} orbital groups of the metal. π-bond formation between metals with filled t_{2g} orbitals and acceptor ligands (for example, sulphur, selenium, tellurium, phosphorus, arsenic) possessing empty *d* orbitals stabilizes the electrons which fill the metal t_{2g} orbitals, thereby increasing Δ_o and the CFSE.

π-bond formation in minerals. Cations with filled t_{2g} orbitals capable of forming most π-bonds in octahedral coordination are Ni^{2+}, Cu^{2+} and low-spin Co^{3+}, Co^{2+} and Fe^{2+}. The resulting higher Δ_o and increased CFSE accounts for the chalcophilic properties and predominance of these transition elements to form sulphide, arsenide, selenide, etc., minerals. Strong metal–ligand bonds induced by π-bonding also shortens interatomic distances, particularly for transition metal ions in low-spin states. As a result, observed metal–sulphur distances in some pyrite-group (FeS_2, CoS_2 and NiS_2) and thiospinel ($NiFe_2S_4$, Co_3S_4 and Ni_3S_4) minerals are shorter than those calculated from sums of ionic radii.

Element distributions in sulphide mineral assemblages. The increased Δ_o, and hence stabilization energies, resulting from π-bond formation in sulphides is also responsible for the enrichments of Ni, Cu, Co and Fe in sulphide minerals coexisting with silicates. Such strong chalcophilic properties of these transition elements has profound economic significance in the formation of massive sulphide ore deposits.

Molecular orbital diagrams. Calculations for $[FeO_6]^{-9}$ and $[FeO_6]^{-10}$ clusters representing Fe^{3+} and Fe^{2+} ions octahedrally coordinated to O^{2-} ions enable estimates to be made of bond covalences, effective charges on the atoms, electron occupancies and energy separations of molecular orbital energy levels. For example, in these $[FeO_6]$ octahedral clusters effective charges on the divalent and trivalent Fe atoms are about +0.94 and +1.3, respectively. In low-symmetry coordination clusters representing trigonally distorted $[FeO_6]^{-9}$ octahedra (e.g., hematite) or $[FeO_4(OH)_2]^{-7}$ octahedra (e.g., goethite), longer Fe–O bonds are more covalent than shorter Fe–O bonds, while Fe–OH bonds are more ionic than Fe–O bonds. Energy separations and covalent bond-characters derived from the MO calculations may be correlated with positions and intensities of CF and OMCT transitions observed in visible to near-infrared spectra of Fe oxide and silicate minerals. Similar calculations for dimeric $[Fe_2O_{10}]^{-15}$ clusters enable correlations to be made with spectrally determined $Fe^{2+} \rightarrow Fe^{3+}$ IVCT transitions.

Stabilities of Mn(IV) and Ti(IV) oxide structures. Covalent bonding in octahedral $[MnO_6]^{-8}$ clusters lowers the effective charge on the tetravalent manganese to about +1.06, compared to +2.0 for tetravalent Ti. Extensive edge-sharing of $[MnO_6]$ octahedra is thus facilitated in 'MnO_2' phases, accounting for the greater diversity of Mn(IV) oxides compared to Ti(IV) oxides.

11.10 Background reading

Gray, H. B. (1964) Molecular orbital theory for transition metal complexes. *J. Chem. Educ.*, 41, 1–12.

Huheey, J. E. (1983) *Inorganic Chemistry, 3rd edn.* (Harper & Row, New York,), ch 9.

Sherman, D. M. (1985) The electronic structures of Fe^{3+} coordination sites in iron oxides: Applications to spectra, bonding and magnetism. *Phys. Chem. Minerals*, 12, 161–75.

Sherman, D. M. (1987a) Molecular orbital (SCF–Xα–SW) theory of metal–metal charge transfer processes in minerals. I. Applications to $Fe^{2+} \rightarrow Fe^{3+}$ charge transfer and 'electron delocalization' in mixed-valence iron oxides and silicates. *Phys. Chem. Minerals,* 14, 355–64.

Tossell, J. A. (1985) Quantum mechanical models in mineralogy. In *Chemical Bonding and Spectroscopy in Mineral Chemistry.* (F. J. Berry & D. J. Vaughan, eds; Chapman & Hall, London), pp. 1–30.

Tossell, J. A. & Vaughan, D. J. (1992) *Theoretical Geochemistry: Applications of Quantum Mechanics in the Earth and Mineral Sciences.* (Oxford Univ. Press), 416 pp.

Appendix 1 Abundance data for the transition elements

Element	Solar System	Earth				Moon		Mars		Venus
	(Cosmic) (relative to O=466,000) (ppm)	Average Crust (ppm)	Oceanic Crust (ppm)	Upper Mantle (ppm)	Total Earth (ppm)	High Ti basalt (ppm)	Green glass (ppm)	Viking I lander (Chryse) (ppm)	SNC meteorite (ppm)	Vega 2 lander (ppm)
O	466,000	466,000	409,310	432,900	295,300	405,380	418,830	410,500	422,100	437,300
Si	34,370	277,200	231,000	213,900	15,200	176,400	210,900	205,300	238,000	212,800
Al	2,805	81,300	84,700	17,800	10,900	46,850	39,700	38,650	36,530	84,700
Mg	31,945	20,900	46,400	244,600	12,700	50,890	105,520	36,180	56,100	69,340
Na	1,615	28,200	20,800	2,500	5,700	2,670	965	≤15,800	9,870	14,840
S	20,205	260	-	-	19,300	-	-	26,800	-	18,800
Cl	225	130	-	-	-	-	-	8,000	-	≤4,000
K	180	25,900	1,250	200	700	415	249	<4,000	1,410	830
Ca	2,995	36,300	80,800	19,200	11,300	76,430	60,710	40,700	70,710	53,570
Sc	2	22	38	-	-	86	-	-	52	-
Ti	140	4,400	9000	900	500	77,940	2,280	3,720	5,400	1,200
V	18	135	250	-	-	50	165	-	312	-
Cr	860	100	270	3,100	2,600	3,030	-	-	1,575	-
Mn	640	950	1000	1,000	2,200	2,090	2,010	-	4,030	770
Fe	61,510	50,000	81,600	63,200	346,300	153,130	155,460	122,400	150,800	59,810
Co	160	25	47	-	1,300	-	75	-	38	-
Ni	3,540	75	135	700	15,800	2	153	-	36	-
Cu	40	55	86	-	23,900	6	-	-	54	≤3,000
Zn	100	70	85	-	-	-	-	-	62	≤2,000

Y-Cd	<3	220	<100	-	-	240	30	-	340	<1,000
La-Lu	<1	151	54	-	-	85	4	-	<16	-
Hf-Hg	<1	7	<10	-	-	<2	-	-	<2	-
Sources of data:	[1]	[2]	[2]	[3]	[2]	[4]	[4]	[5]	[6]	[7,8]

Sources of data : [1] Anders & Grevesse (1989); [2] Mason & Moore (1982) p. 46; [3] Liu & Bassett, W. A. (1986), p. 234; [4] Taylor (1982): Apollo 17 high-K basalt 74275; Apollo 15 green glass 15426; [5] Clark, Baird, Weldon, Tsusaki, Schnabel & Candelaria (1982): Viking 1, average Chryse; [6] Smith, Laul, Ma, Huston, Verkouteren, Lipschutz & Schmitt (1984): Shergotty; [7] Surkov, Barsukov, Moskalyeva & Kharyukova (1984): Venera 13 and 14; [8] Surkov, Moskalyeva, Kharyukova, Dudin, Smirnov & Zaitseva (1986): Vega 2.

Appendix 2 Isotopes of the transition elements

Element	Atomic number	Atomic weight	Isotopic composition: Stable isotopes (% abundance)						Isotopic composition: Natural radionuclide	Atom % of the Earth
potassium	19	39.0983	^{39}K 93.258	^{41}K 6.730					^{40}K* (0.0112%): $t_{1/2} = 5 \times 10^9$ y	0.05
calcium	20	40.08	^{40}Ca 96.941	^{42}Ca 0.647	^{43}Ca 0.135	^{44}Ca 2.086	^{46}Ca 0.004	^{48}Ca 0.187		0.75
scandium	21	44.9559	^{45}Sc 100							-
titanium	22	47.09	^{46}Ti 8.0	^{47}Ti 7.3	^{48}Ti 73.8	^{49}Ti 5.5	^{50}Ti 5.4			0.03
vanadium	23	50.9414	^{50}V 0.25	^{51}V 99.75						-
chromium	24	51.996	^{50}Cr 4.345	^{52}Cr 83.789	^{53}Cr 9.501	^{54}Cr 2.365			^{51}Cr**: $t_{1/2} = 27.8$ d	0.13
manganese	25	54.938	^{55}Mn 100						^{53}Mn: $t_{1/2} = 3.6 \times 10^6$ y	0.11
iron	26	55.847	^{55}Fe 5.8	^{56}Fe 91.72	^{57}Fe* 2.2	^{58}Fe 0.28				16.6
cobalt	27	58.9332	^{59}Co 100						^{56}Co**: $t_{1/2} = 77$ d; ^{60}Co**: $t_{1/2} = 5.26$ y	0.06
nickel	28	58.70	^{58}Ni 68.27	^{60}Ni 26.10	^{61}Ni* 1.13	^{62}Ni 3.59	^{64}Ni 0.91		^{59}Ni**: $t_{1/2} = 8 \times 10^4$ y	1.09

copper	29	63.546	^{63}Cu	^{65}Cu			
			69.17	30.83			
zinc	30	65.38	^{64}Zn	^{66}Zn	^{67}Zn*	^{68}Zn	^{70}Zn
			48.63	27.90	4.10	18.75	0.62

Source of data : *Handbook of Geochemistry* (K. H. Wedepohl, ed., Springer-Verlag, New York)
* Isotopes showing the Mössbauer effect; ** isotope occurs in lunar samples and meteorites.

Appendix 3 Ionic radii of transition metals and related cations (in pm*)

	4-fold coordination tetrahedral			5-fold coordination			6-fold[†] coordination octahedral					8-fold coordination cubic		
	M^{2+}	M^{3+}	M^{4+}	M^{2+}	M^{3+}	M^{4+}	M^{2+} ls	M^{2+} hs	M^{3+} ls	M^{3+} hs	M^{4+}	M^{2+}	M^{3+}	M^{4+}
Mg	57			66				72						
Al		39			48					53.5				
Ca								100				112		
Sc										74.5			87	
Ti			42					86		67	60.5			74
V								79		64	58			72
Cr			41				73	80		61.5	55			
Mn	66		39	75	58		67	83	58	64.5	53	96		
Fe	63	49			58		61	78	55	64.5	58.4	92	78	
Co	58		40	67			65	74.5	54.5	61	53			
Ni	55							69	56	60				
Cu	57			65				73						
Zn	60			68				74				90		
Ga		47			55					62				

Other cation radii:

V^{5+}	tet = 35.5;	oct = 54
Cr^{6+}	tet = 26	oct = 44
Mn^{7+}	tet = 25	oct = 46
Cu^{1+}	tet = 60	oct = 77
NH_4^+		oct = 151

Anion radii:

$BO_3^{3-} = 191$	$CO_3^{2-} = 164$	$HCO_3^{-} = 142$	$NO_3^{-} = 165$	$OH^{-} = 119$	$O^{2-} = 140$	$F^{-} = 133$
	$SiO_4^{4-} = 240$		$PO_4^{3-} = 238$	$SO_4^{2-} = 244$	$S^{2-} = 184$	$Cl^{-} = 181$
			$AsO_4^{3-} = 248$	$SeO_4^{2-} = 235$	$Se^{2-} = 198$	$Br^{-} = 196$
					$Te^{2-} = 221$	$I^{-} = 220$
			$VO_4^{3-} = 246$	$CrO_4^{2-} = 242$		$MnO_4^{-} = 215$

Sources of data : Shannon (1976); Huheey (1983), p. 78.
*100 pm = 1 Å; †Including values for cations with high-spin (hs) and low-spin (ls) electronic configurations.

Appendix 4 Nomenclature for atomic states and spectroscopic terms

A spectroscopic or Russell–Saunders state is designated ${}^{r}L$, where r is the electron spin multiplicity and L is the total atomic orbital angular momentum

$$L = \Sigma m_l$$

where m_l is the magnetic quantum number.

When $L = 0$, S spectroscopic state;
$L = 1$, P spectroscopic state;
$L = 2$, D spectroscopic state;
$L = 3$, F spectroscopic state;
$L = 4$, G spectroscopic state;
$L = 5$, H spectroscopic state;
$L = 6$, I spectroscopic state.

$$r = (2\Sigma m_s + 1),$$

where m_s is the spin quantum number and $m_s = \frac{1}{2}$ per unpaired electron.

For zero unpaired electrons, $\Sigma m_s = 0$ and $r = 1$: singlet state;
for one unpaired electron, $\Sigma m_s = \frac{1}{2}$ and $r = 2$: doublet state:
for two unpaired electrons, $\Sigma m_s = 1$ and $r = 3$: triplet state:
for three unpaired electrons, $\Sigma m_s = \frac{3}{2}$ and $r = 4$: quartet state:
for four unpaired electrons, $\Sigma m_s = 2$ and $r = 5$: quintet state:
for five unpaired electrons, $\Sigma m_s = \frac{5}{2}$ and $r = 6$: sextet state.

Appendix 5 Group theory nomenclature for crystal field states

States *a, b, e, t:* refer to one-electron descriptions according to symmetry properties.

States *A, B, E, T:* refer to mutlti-electron systems of corresponding symmetry.

Symbols *a, A:* non-degenerate; symmetric with respect to the principal axis of symmetry; no change of sign of the wave function when rotated through $2\pi/n$.
b, B: non-degenerate; antisymmetric with respect to the principal symmetry axis; change of sign of wave function when rotated through $2\pi/n$.
e, E: two-fold degenerate.
t, T: three-fold degenerate.

Subscripts *g:* symmetric under the operation of inversion through the centre of symmetry; no change of sign of the wave function upon inversion.
u: antisymmetric under the operation of inversion at the centre of symmetry; change in sign of wave function upon inversion.
1: refers to mirror planes parallel to a symmetry axis.
2: refers to diagonal mirror planes.

Superscripts $r = 1, 2, 3$, etc.: refer to spin multiplicities.

Examples *s* orbitals: a_{1g};
p orbitals: t_{1u};
d_{xy}, d_{xz}, d_{yz} orbitals: t_{2g};
d_{z^2}, $d_{x^2-y^2}$ orbitals: e_g.

Appendix 6 Correlations between the Schöenflies and Hermann–Mauguin symbols

	Point group or crystal class		
Crystal system	Schöenflies symbol	Hermann–Mauguin symbol	Examples in minerals*
cubic	O_h	$^4/_m\,\overline{3}\,^2/_m$ (or m3m)	periclase
	O	432	
	T_d	$\overline{4}3m$	spinel
	T_h	$^2/_m\,\overline{3}$ (or $m\,\overline{3}$)	
	T	23	
tetragonal	D_{4h}	$^4/_m\,^2/_m\,^2/_m$ (or $^4/_m mm$)	≈gillespite, ≈olivine M1
	D_4	$\overline{4}22$	
	D_{2d}	42m	
	C_{4v}	4mm	gillespite
	C_{4h}	$^4/_m$	
	S_4	$\overline{4}$	
	C_4	4	
hexagonal	D_{6h}	$^6/_m\,^2/_m\,^2/_m$ (or $^6/_m mm$)	
	D_6	$\overline{6}22$	
	D_{3h}	6m2	
	C_{6v}	6mm	
	C_{6h}	$^6/_m$	
	C_{3h}	6	
	C_6	6	
trigonal	D_{3d}	$\overline{3}\,^2/_m$ (or $\overline{3}m$)	brucite, spinel
	D_3	32	beryl
	C_{3v}	3m	≈corundum, ≈olivine M2
	C_{3i}	$\overline{3}$	garnet
	C_3	3	corundum

Crystal system	Point group or crystal class: Schöenflies symbol	Point group or crystal class: Hermann–Mauguin symbol	Examples in minerals*
orthorhombic	D_{2h}	$^2/_m{}^2/_m{}^2/_m$ (or mmm)	rutile, pyrolusite
	D_2	222	garnet
	C_{2v}	mm2 (or mm)	amphibole M4
monoclinic	C_{2h}	$^2/_m$	amphibole M3, biotite M2
	C_s	m	olivine M2; epidote M3;
	C_2	2	amphibole M1, M2; biotite M1
			clinopyroxene M1; andalusite M1
triclinic	C_i	$\bar{1}$	olivine M1; sillimanite M1
			epidote M1, M2;
	C_1	1	orthopyroxene M1, M2

* Symmetries of coordination sites accommodating transition metal ions.

Appendix 7 Coordination sites in host mineral structures accommodating transition metal ions

Mineral	Site	Point symmetry	Metal–oxygen distances[#] (pm)	Mean M–O (pm)	Polyhedral volume[◊]	Quadratic elongation[‡]
			oxides			
periclase (MgO)	$Mg^{[6]}$	O_h	(6x) 210.5	210.5	12.445	1.0000
corundum (Al_2O_3)	$Al^{[6]}$	C_3	(3x) 185.6; (3x) 196.9	191.3	9.066	1.0200
hematite (Fe_2O_3)	$Fe^{[6]}$	C_3	(3x) 194.5; (3x) 211.5	203.0	10.754	1.0264
eskolaite (Cr_2O_3)	$Cr^{[6]}$	C_3	(3x) 196.5; (3x) 201.6	199.0	10.312	1.0131
karelianite (V_2O_3)	$V^{[6]}$	C_3	(3x) 196.3; (3x) 206.2	201.2	10.719	1.0098
ilmenite	$Fe^{[6]}$	C_3	(3x) 207.8; (3x) 220.1	208.1	12.562	1.0271
($FeTiO_3$)	$Ti^{[6]}$	C_3	(3x) 187.4; (3x) 208.0	198.2	10.001	1.0277
spinel	$Mg^{[4]}$	T_d	(4x) 192.4	192.4	3.653	1.0000
($MgAl_2O_4$)	$Al^{[6]}$	D_{3d}	(6x) 192.6	192.6	9.371	1.0108
magnetite	$Fe^{[4]}$	T_d	(4x) 188.7	188.7	3.449	1.0000
(Fe_3O_4)	$Fe^{[6]}$	D_{3d}	(6x) 205.9	205.9	11.612	1.0015
chromite	$Fe^{[4]}$	T_d	(4x) 200.6	200.6	4.141	1.0000
($FeCr_2O_4$)	$Cr^{[6]}$	D_{3d}	(6x) 199.0	199.0	10.322	1.0123
			garnets			
pyrope	$Mg^{[8]}$	D_2	(4x) 219.7; (4x) 234.3	227.0	20.140	
($Mg_3Al_2Si_3O_{12}$)	$Al^{[6]}$	C_{3i}	(6x) 188.7	188.7	8.937	1.0014
almandine ($Fe_3Al_2Si_3O_{12}$)	$Fe^{[8]}$	D_2	(4x) 222.0; (4x) 237.8	229.9	20.930	
spessartine ($Mn_3Al_2Si_3O_{12}$)	$Mn^{[8]}$	D_2	(4x) 224.5; (4x) 240.6	232.6	21.650	
grossular ($Ca_3Al_2Si_3O_{12}$)	$Al^{[6]}$	C_{3i}	(6x) 192.4	192.4	9.491	1.0007
andradite ($Ca_3Fe_2Si_3O_{12}$)	$Fe^{[6]}$	C_{3i}	(6x) 202.4	202.4	11.046	1.0004
uvarovite ($Ca_3Cr_2Si_3O_{12}$)	$Cr^{[6]}$	C_{3i}	(6x) 198.5	198.5	10.413	1.0007

Mineral	Site	Point symmetry	Metal–oxygen distances# (pm)	Mean M–O (pm)	Polyhedral volume$^{\lozenge}$	Quadratic elongation‡
			aluminosilicates			
andalusite (Al_2SiO_5)	M1$^{[6]}$	C_2	(2x) 182.7*; (2x) 189.1; (2x) 208.6	193.5	9.539	1.0114
	M2$^{[5]}$	C_s	(2x) 181.4; 181.6*; 184.0; 189.9	183.6	5.153	
sillimanite (Al_2SiO_5)	M1$^{[6]}$	C_i	(2x) 186.8*; (2x) 191.4; (2x) 195.4	191.2	9.175	1.0109
	M2$^{[4]}$	C_s	(2x) 179.6; 171.1; 175.1	176.4	2.791	1.0062
yoderite [$Mg_2(Al,Fe)_6$	M1$^{[6]}$	C_1	192.9; 194.2; 194.7*; 197.2; 198.5†; 200.0	196.3		9.820
$Si_4O_{18}(OH)_2$]	M2$^{[5]}$	C_s	(2x) 192.7; 189.6*; 190.7; 200.8†	193.3		6.070
	M3$^{[5]}$	C_s	(2x) 186.8; 180.1; 184.3; 192.9*	186.2		5.250
kyanite (Al_2SiO_5)	M1$^{[6]}$	C_1	184.7; 184.8; 187.4*; 188.4*; 197.1; 198.7	190.2	8.977	1.0155
	M2$^{[6]}$	C_1	188.1; 188.9; 191.4*; 192.5; 193.0; 193.4*	191.3	9.136	1.0141
	M3$^{[6]}$	C_1	186.2; 188.3*; 188.5; 192.4; 196.8*; 198.6*	191.8	9.164	1.0180
	M4$^{[6]}$	C_1	181.6; 184.6*; 187.5; 191.1; 193.3; 199.8	189.6	8.921	1.0139
topaz [$Al_2SiO_4(OH.F)_2$]	M1$^{[6]}$	C_1	189.4; 190.2; 190.5; 191.1; 180.2^F; 180.8^F	187.1	8.654	1.0066
staurolite [$Fe_2Al_9Si_4O_{23}(OH)$]	Fe$^{[4]}$	C_s	(2x) 197.2; 204.3; 204.6	200.8	4.141	1.0026
zoisite [$Ca_2Al_3Si_3$	M1,2$^{[6]}$	C_1	184.3*; 184.9†; 185.0; 190.0; 192.6; 196.4	188.8	8.899	1.0066
$O_{12}(OH)$]	M3$^{[6]}$	C_s	(2x) 196.5; (2x) 213.3; 178.4; 182.2†	196.7	9.870	1.0237
epidote [Ca_2FeAl_2	M1$^{[6]}$	C_i	(2x) 184.3*; (2x) 193.9; (2x) 195.6	191.3	9.252	1.0065
$Si_3O_{12}(OH)$]	M2$^{[6]}$	C_i	(2x) 185.4; (2x) 187.0†; (2x) 192.7	188.3	8.853	1.0045
	M3$^{[6]}$	C_s	(2x) 198.5; (2x) 222.4; 186.0; 193.5*	203.6	10.864	1.0283
piemontite [$Ca_2(Mn,Fe)$	M1$^{[6]}$	C_i	(2x) 187.2*; (2x) 194.1; (2x) 198.5	193.3		
$Al_2Si_3O_{12}(OH)$]	M2$^{[6]}$	C_i	(2x) 185.7; (2x) 188.1†; (2x) 193.4	189.1		
	M3$^{[6]}$	C_s	(2x) 203.1; (2x) 227.4; 186.1; 190.0*	206.2		
ilvaite [$CaFe^{2+}{}_2Fe^{3+}$	M11$^{[6]}$	C_1	201.7†; 205.3; 208.6*; 211.4; 212.6; 222.2*	210.3	12.552	
$Si_2O_8(OH)$]	M12$^{[6]}$	C_1	197.6†; 201.7*; 203.6; 208.0; 210.2; 213.3*	205.7	11.262	
	M2$^{[6]}$	C_1	196.6†; 212.8*; 222.9; 224.6; 227.0; 228.7	218.8	13.592	

Mineral	Site	Point symmetry	Metal–oxygen distances[#] (pm)	Mean M–O (pm)	Polyhedral volume[◊]	Quadratic elongation[‡]
			olivines			
forsterite (Mg_2SiO_4)	M1[6]	C_i	(2x) 208.4; (2x) 206.8; (2x) 213.1	209.4	11.771	1.0260
	M2[6]	C_s	(2x) 206.6; (2x) 221.0; 204.5; 217.7	212.9	12.401	1.0260
fayalite (Fe_2SiO_4)	M1[6]	C_i	(2x) 212.1; (2x) 212.6; (2x) 223.6	216.1	12.737	1.0379
	M2[6]	C_s	(2x) 206.5; (2x) 229.5; 211.1; 223.3	217.7	13.072	1.0370
tephroite (Mn_2SiO_4)	M1[6]	C_i	(2x) 216.7; (2x) 220.0; (2x) 225.0	220.6	13.499	1.0398
	M2[6]	C_s	(2x) 215.5; (2x) 231.9; 213.7; 227.8	222.7	13.982	1.0367
Co olivine (Co_2SiO_4)	M1[6]	C_i	(2x) 209.1; (2x) 209.8; (2x) 216.7	211.9	12.144	1.0294
	M2[6]	C_s	(2x) 207.3; (2x) 222.3; 207.2; 218.7	214.2	12.606	1.0270
liebenbergite (Ni_2SiO_4)	M1[6]	C_i	(2x) 206.0; (2x) 206.4; (2x) 211.1	207.8	11.531	1.0254
	M2[6]	C_s	(2x) 205.3; (2x) 217.1; 204.3; 210.5	210.0	11.966	1.0215
			silicate spinels			
γ-Mg_2SiO_4	Mg[6]	C_{3v}	(6x) 207.0	207.0	11.780	1.0026
ringwoodite (γ-Fe_2SiO_4)	Fe[6]	C_{3v}	(6x) 213.7	213.7	12.912	1.0005
γ-Ni_2SiO_4	Ni[6]	C_{3v}	(6x) 206.3	206.3	11.663	1.0024
γ-Co_2SiO_4	Co[6]	C_{3v}	(6x) 210.3	210.3	12.332	1.0041
			pyroxenes			
enstatite ($MgSiO_3$)	M1[6]	C_1	200.7; 202.8; 204.5; 206.5; 215,1; 217.2	207.8	11.830	1.0088
	M2[6]	C_1	199.4; 203.2; 205.5; 208.8; 228.8[¶]; 244.7[¶]	215.1	12.460	1.0489
ferrosilite ($FeSiO_3$)	M1[6]	C_1	208.5; 208.6; 212.2; 212.8; 219.3; 219.5	213.5	12.810	1.0090
	M2[6]	C_1	198.7; 202.3; 212.3; 216.1; 245.3[¶]; 258.9[¶]	222.3	13.430	1.0700
Co-opx ($CoSiO_3$)	M1[6]	C_1	204.7; 206.0; 207.8; 208.9; 215.9; 218.5	210.3	12.250	1.0090
	M2[6]	C_1	198.2; 201.9; 207.2; 211.7; 238.8; 251.6	218.2	12.860	1.0600
diopside ($CaMgSi_2O_6$)	M1[6]	C_2	(2x) 205.0; (2x) 206.4; (2x) 211.5	207.7	11.850	1.0050
hedenbergite ($CaFeSi_2O_6$)	M1[6]	C_2	(2x) 206.8; (2x) 214.1; (2x) 218.4	213.1	12.810	1.0060
Co-cpx ($CaCoSi_2O_6$)	M1[6]	C_2	(2x) 207.1; (2x) 209.5; (2x) 213.6	210.1		

Mineral	Site	Point symmetry	Metal–oxygen distances# (pm)	Mean M–O (pm)	Polyhedral volume◊	Quadratic elongation‡
Ni-cpx ($CaNiSi_2O_6$)	$M1^{[6]}$	C_2	(2x) 205.9; (2x) 205.0; (2x) 210.1	207.0		
spodumene ($LiAlSi_2O_6$)	$M1^{[6]}$	C2	(2x) 182.0; (2x) 194.6; (2x) 199.7	192.1	9.260	1.0150
kosmochlor (ureyite) ($NaCrSi_2O_6$)	$M1^{[6]}$	C2	(2x) 195.0; (2x) 201.0; (2x) 204.2	200.1	10.550	1.0094
			amphiboles			
tremolite [$Ca_2Mg_5Si_8O_{22}(OH)_2$]	$M1^{[6]}$	C_2	(2x) 206.4; (2x) 207.8; (2x) 208.3	207.4	11.696	1.0108
	$M2^{[6]}$	C_2	(2x) 201.4; (2x) 208.3; (2x) 213.3	207.7	11.932	1.0073
	$M3^{[6]}$	C_{2h}	(2x) 205.7†; (4x) 207.0	206.6	11.498	1.0133
glaucophane [$Na_2Mg_3Al_2Si_8O_{22}(OH)_2$]	$M1^{[6]}$	C_2	(2x) 207.8; (2x) 208.2; (2x) 210.0	208.4	11.577	1.0237
	$M2^{[6]}$	C_2	(2x) 184.9; (2x) 194.3; (2x) 203.8	194.3	9.435	1.0121
	$M3^{[6]}$	C_{2h}	(2x) 207.7†; (4x) 210.3	209.4	11.783	1.0262
			cyclosilicates			
beryl ($Be_3Al_2Si_6O_{18}$)	$Al^{[6]}$	D_3	(6x) 190.6	190.6	8.939	1.0218
cordierite ($Mg_2Al_4Si_5O_{18}$)	$Mg^{[6]}$	C_2	(2x) 210.0; (2x) 211.4; (2x) 211.6	211.0	11.798	1.0406
	$T2^{[4]}$	C_s	(4x) 174.2	174.2	2.705	1.0025
tourmaline (dravite)	$Mg^{[6]}$	C_s	(2x) 201.8B; (2x) 213.7; 197.1†; 200.4†	202.5	10.715	1.0230
	$Al^{[6]}$	C_1	189.3; (2x) 190.0B; 193.2B; 195.5; 199.5†	192.9	9.379	1.0141
			micas			
phlogopite [$KMg_3AlSi_3O_{10}(OH)_2$]	$M1^{[6]}$	C_{2h}	(2x) 203.0†; (4x) 208.0	206.3	11.505	1.0120
	$M2^{[6]}$	C_2	(2x) 203.9†; (2x) 207.1; (2x) 208.3	206.4	11.447	1.0130
annite [$KFe_3AlSi_3O_{10}(OH)_2$]	M1[$^{[6]}$	C_{2h}	(2x) 211.9†; (4x) 212.3	212.1	12.530	1.0090
	$M2^{[6]}$	C_2	(2x) 208.3; (2x) 211.2; (2x) 211.9	211.1	12.210	1.0080
muscovite [$KAl_2Si_3AlO_{10}(OH)_2$]	$M1^{[6]}$	C_1	192.4†; 193.3†; 193.3; 194.0; 194.6; 196.2	194.0	9.355	1.0162
			miscellaneous structures			
wadsleyite ($Mg_4OSi_2O_7$ or β-Mg_2SiO_4)	$M1^{[6]}$	C_{2h}	(4x) 204.6; (2x) 211.5	206.9	11.731	1.0050
	$M2^{[6]}$	C_{2v}	(4x) 209.3; 203.5*; 209.5¶	208.4	11.966	1.0055
	$M3^{[6]}$	C_2	(2x) 201.6*; (2x) 212.3; (2x) 212.8	208.9	12.039	1.0072

Mineral	Site	Point symmetry	Metal–oxygen distances# (pm)	Mean M–O (pm)	Polyhedral volume◊	Quadratic elongation‡
$MgSiO_3$ ilmenite	Mg[6]	C_3	(3x) 199.0; (3x) 216.3	207.6	11.238	1.0429–
	Si[6]	C_3	(3x) 176.8; (3x) 183.0	179.9	7.592	1.0152
$MgSiO_3$ –perovskite	Mg[8]	C_s	201.4; 209.7; (2x) 205.2; (2x) 227.8; (2x) 242.7	220.3	20.100	
	Si[6]	C_i	(2x) 178.2; (2x) 179.6; (2x) 180.1	179.3	7.681	1.0005

In silicates, oxygen bond-types are predominantly non-bridging atoms, i.e. Si–O⁻ bound to one Si atom, except:
¶ denotes bridging oxygen atoms, i.e. Si–O–Si;
† denotes OH^- ions;
* denotes free O^{2-} ions (in silicates);
F denotes F^- ions;
B denotes oxygen bound to boron.
◊ Polyhedral volume, the volume of a coordination polyhedron (Hazen and Finger, 1982, Appendix III, p. 103).
‡ Quadratic elongation, <λ>, a measure of distortion of a coordination polyhedron (Robinson *et al.*, 1971),
defined as: $\langle\lambda\rangle = \Sigma\,[(l_i/l_o)^2/n]$, where l_i is the distance from the central atom to the *i*th coordinating atom, and l_o is the centre-to-vertex distance of a regular polyhedron of the same volume.
Sources of data : Hazen & Finger (1982), p. 103–111; Robinson, Gibbs & Ribbe (1971); Smyth & Bish (1988); Higgins, Ribbe & Nakajima (1982) for yoderite; Finger & Hazen (1987) for ilvaite

Appendix 8 Chemical and physical constants, units and conversion factors

International System (SI) base units

Physical Quantity	Unit	Symbol
length	metre	m
mass	kilogram	kg
time	second	s
electric current	ampere	A
temperature	kelvin	K
amount of substance	mole	mol
luminous intensity	candela	cd

Derived units

Physical Quantity	Unit	Symbol	Definition
frequency	hertz	Hz	s^{-1}
frequency	wavenumber	cm^{-1}	m^{-1}
energy	joule	J	$kg\ m^2\ s^{-2}$
force	newton	N	$J\ m^{-1}$
pressure	pascal	Pa	$N\ m^{-2}$
power	watt	W	$J\ s^{-1}$
electric charge	coulomb	C	A s
electric potential difference	volt	V	$J\ A^{-1}\ s^{-1}$
electric resistance	ohm	Ω	$V\ A^{-1}$
electric capacitance	farad	F	$A\ s\ V^{-1}$
magnetic flux	weber	Wb	V s
inductance	henry	H	$V\ s\ A^{-1}$
magnetic flux density	tesla	T	$V\ s\ m^{-2}$

Chemical and physical constants

Quality	Symbol	Value	
electronic charge	e	1.6021×10^{-19} C	= 4.8030×10^{-10} esu
electronic rest mass	m_e	9.1096×10^{-28} g	
proton mass	m_p	1.6726×10^{-24} g	
Bohr radius	a_o	52.9177 pm	= 0.5292 Å
Bohr magneton	μ_B	9.2741×10^{-24} A m^2	= 9.2741×10^{-21} erg gauss^{-1}
velocity of light (vacuum)	c	2.9979×10^{8} m s^{-1}	
Avogadro number	N	6.0222×10^{23} mol^{-1}	
gas constant	R	8.3143 J K^{-1} mol^{-1}	= 1.9872 cal K^{-1} mol^{-1}
Planck constant	h	6.6262×10^{-34} J s	= 6.6262×10^{-27} erg s
Boltzmann constant	k	1.3806×10^{-23} J K^{-1}	
Faraday constant	F	9.6487×10^{4} C mol^{-1}	
Rydberg constant (H atom)	R_H	1.09737×10^{5} cm^{-1}	
Stefan constant	S	5.670×10^{-8} J m^{-2} K^{-4} s^{-1}	
pi	π	3.14159	
base, natural logarithms	e	2.71828	
	ln 10	2.3026	

Conversion factors

Multiply	by	to obtain
	Energy	
cm^{-1}	1.1962×10^{-2}	kJ mol^{-1}
kJ mol^{-1}	83.59	cm^{-1}
kJ mol^{-1}	0.239	kcal mol^{-1}
kcal mol^{-1}	4.184	kJ mol^{-1}
ev	96.49	kJ mol^{-1}
kJ mol^{-1}	0.01036	ev
ev	23.06	kcal mol^{-1}
k cal mol^{-1}	0.04336	eV
	Length	
nanometre (nm)	10^{-9}	metre (m)
micron (μm)	10^{-6}	metre (m)
nanometre (nm)	10^{-3}	micron (μm)
nanometre (nm)	10	ångstrom (Å)
angstrom (Å)	100	picometre (pm)
picometre (pm)	10^{-2}	ångstrom (Å)
	Pressure	
atmosphere	1.01325×10^{5}	pascal (Pa)
pascal	9.869×10^{-6}	atmosphere
gigapascal (GPa)	10	kilobar (kb)
pascal	7.501×10^{-3}	torr (mm. Hg)
torr	133.3	pascal
bar	0.9869	atmosphere

Conversion table for energy units used in mineral spectroscopy

	10,000 cm^{-1}	1 kJ	1 kcal	1 eV	1 μm
cm^{-1}	10^4	83.59	349.5	8,066	10,000
kJ	119.66	1	4.1835	96.49	119.66
kcal	28.59	0.239	1	23.06	28.59
eV	1.24	0.01036	0.04336	1	1.24
μm	1.0	119.66	28.59	1.24	1

References

Abs-Wurmbach, I. & Langer, K. (1975) Synthetic Mn^{3+}-kyanite and viridine, $(Al_{2-x}Mn^{3+}_x)SiO_5$, in the system Al_2O_3–MnO–MnO_2–SiO_2. *Contrib. Mineral. Petrol.*, **49**, 21–38.

Abs-Wurmbach, I., Langer, K. & Oberhänsli, R. (1985) Polarized absorption spectra of single crystals of the chromium-bearing clinopyroxenes kosmochlor and Cr–aegerine–augite. *Neues Jahrb. Mineral. Abh.*, **152**, 293–319.

Abs-Wurmbach, I., Langer, K. & Tillmans, E. (1977) Structure and polarized absorption spectra of Mn^{3+}-substituted andalusites (viridine). *Naturwiss.*, **64**, 527–8.

Abs-Wurmbach, I., Langer, K., & Schreyer, W. (1983) The influence of Mn^{3+} on the stability relations of the Al_2SiO_5 polymorphs with special emphasis on manganoan andalusites (viridines)$(Al_{1-x}Mn^{3+}_x)_2(O|SiO_4)$: An experimental investigation. *J. Petrol.*, **24**, 48–75.

Abs-Wurmbach, I., Langer, K., Seifert, F. & Tillmanns, E. (1981) The crystal chemistry of (Mn^{3+},Fe^{3+})-substituted andalusites (viridines and kanonaite), $(Al_{1-x-y}Mn^{3+}_xFe_y)_2(O|SiO_4)$: crystal structure refinements, Mössbauer, and polarized optical absorption spectra. *Zeit. Krist.*, **155**, 81–113.

Abu-Eid, R. M. (1976) Absorption spectra of transition metal-bearing minerals at high pressures. In *The Physics and Chemistry of Minerals and Rocks*. (R. G. J. Strens, ed.; J. Wiley, New York), pp. 641–75.

Abu-Eid, R. M. & Burns, R. G. (1976) The effect of pressure on the degree of covalency of the cation–oxygen bond in minerals. *Amer. Mineral.*, **61**, 391–7.

Abu-Eid, R. M. & Langer, K. (1978) Single-crystal uv-spectra of olivine (fayalite) up to 37,000 cm^{-1} at elevated pressures. *Naturwiss.*, **65**, 256–7.

Abu-Eid, R. M., Langer, K. & Seifert, F. (1978) Optical absorption and Mössbauer spectra of purple and green yoderite, a kyanite-related mineral. *Phys. Chem. Minerals*, **3**, 271–89.

Abu-Eid, R. M., Mao, H.-K. & Burns, R. G. (1973) Polarized absorption spectra of gillespite at high pressure. *Ann. Rept. Geophys. Lab., Yearb.* **72**, 564–7.

Adams, J. B. (1968) Lunar and martian surfaces: petrologic significance of absorption bands in the near-infrared. *Science,* **159,** 1453–5.

Adams, J. B. (1974) Visible and near-infrared diffuse reflectance: spectra of pyroxenes as applied to remote sensing of solid objects in the solar system. *J. Geophys. Res.*, **79,** 4829–36.

Adams, J. B. (1975) Interpretations of visible and near-infrared diffuse reflectance spectra of pyroxenes and other rock-forming minerals. In *Infrared and Raman*

Spectroscopy of Lunar and Terrestrial Minerals. (C. Karr Jr, ed.; Academic Press, New York), pp. 91–116.

Adams, J. W. (1965) The visible region absorption spectra of rare earth minerals. *Amer. Mineral.*, **50**, 356–66.

Ahrens, L. H. (1952) The use of ionization potentials. 1. Ionic radii of the elements. *Geochim. Cosmochim. Acta*, **2**, 155–69.

Ahrens, L. H. (1953) The use of ionization potentials. 2. Anion affinity and geochemistry. *Geochim. Cosmochim. Acta*, **3**, 1–29.

Ahsbahs, H., Dehnicke, G., Dehnicke, K. & Hellner, E. (1974) Infra-red spectra of different transition metal complexes under high pressure. *Conf. High Pressure Res.*, Marburg, Germany, p.26.

Aikawa, N., Kumazawa, M. & Tokonami, N. (1985) Temperature dependence of intersite distribution of Mg and Fe in olivine and the associated change of lattice parameters. *Phys. Chem. Minerals*, **12**, 1–8.

Aines, R. D. & Rossman, G. R. (1986) Relationships between radiation damage and trace water in zircon, quartz, and topaz. *Amer. Mineral.*, **71**, 1186–93.

Alberti, A. & Vezzalini, G. (1978) Madelung energies and cation distribution in olivine type structures. *Zeit. Krist.*, **147**, 167–75.

Aldridge, L. P., Tse, J. S. & Bancroft, G. M. (1982) The identification of Fe^{2+} in the M4 site of calcic amphiboles: Discussion. *Amer. Mineral.*, **67**, 340–2 [see Goldman & Rossman (1977b, 1982)].

Allen, G. C. & Warren, K. D. (1971) Electronic spectra of hexafluoride complexes. *Struct. Bonding*, **9**, 49–138.

Amthauer, G. (1976) Crystal chemistry and colour of chrome-bearing garnets. *Neues Jahrb. Mineral. Abh.*, **126**, 158–86.

Amthauer, G. & Rossman, G. R. (1984) Mixed-valence of iron in minerals with cation clusters. *Phys. Chem. Minerals*, **11**, 37–51.

Amthauer, G., Langer, K. & Schliestedt, M. (1980) Thermally activated electron delocalization in deerite. *Phys. Chem. Minerals*, **6**, 19–30.

Anders, E. & Grevesse, N. (1989) Abundances of the elements: Meteoritic and solar. *Geochim. Cosmochim. Acta*, **53**, 197–214.

Andersen, K.L. (1978) *Reflectance spectroscopy as a remote sensing technique for the identification of porphyry copper deposits*. Ph.D. Thesis, MIT.

Anderson, D. L. & Anderson, O. L. (1970) The bulk modulus–volume relationship for oxide compounds and related geophysical problems. *J. Geophys. Res.*, **75**, 3494–500.

Anderson, O. L. (1972) Patterns in elastic constants of minerals important in geophysics. In *The Nature of the Solid Earth*. (E. C. Robertson, ed.; McGraw-Hill, New York), pp. 575–613.

Angel, R. J., Gasparik, T. & Finger, L. W. (1989) Crystal structure of a Cr^{2+}-bearing pyroxene. *Amer. Mineral.*, **74**, 599–603.

Annersten, H. & Eckstrom, T. (1971) Distribution of major and minor elements in coexisting minerals from a metamorphosed iron formation. *Lithos*, **4**, 185–204.

Annersten, H. & Hålenius, U. (1976) Ion distribution in pink muscovite: A discussion. *Amer. Mineral.*, **61**, 1045–50 [see Richardson (1975].

Annersten, H., Adetunai, J. & Filippidis, A. (1984) Cation ordering in Fe–Mn silicate olivines. *Amer. Mineral.*, **69**, 1110–15.

Annersten, H., Ericsson, T. & Filippidis, A. (1982) Cation disordering in Ni–Fe olivines. *Amer. Mineral.*, **67**, 1212–17 [see Ribbe & Lumpkin (1984)].

Anovitz, L. M., Essene, E. J. & Dunham, W. R. (1988) Order–disorder experiments in orthopyroxenes: Implications for the orthopyroxene geospeedometer. *Amer. Mineral.*, **73**, 1060–73.

Arni, R., Langer, K. & Tillmanns, E. (1985) Mn^{3+} in garnets. III. Absence of Jahn–Teller distortion in synthetic Mn^{3+}-bearing garnet. *Phys. Chem. Minerals,* **12**, 279–82.

Aronson, J. R., Bellotti, L. H., Eckroad, S. W., Emslie, A. G., McConnell, R. K. & von Thuna, P. C. (1970) Infrared spectra and radiative thermal conductivity of minerals at high temperatures. *J. Geophys. Res.*, **75**, 3443–56.

Åsbrink, S. & Norrby, L.-J. (1970) A refinement of the crystal structure of copper(II) oxide with a discussion of some exceptional E.S.D.'s. *Acta Cryst.*, **B26**, 8-15.

Bacon, G. E. & Curry, N. A. (1962) The water molecules in $CuSO_4.5H_2O$. *Proc. Royal Soc. (London),* **A266**, 95.

Bailey, S. W. (1984) Review of cation ordering in micas. *Clays & Clay Minerals*, **32**, 81–92.

Balchan, A. S. & Drickamer, H. G. (1959) Effect of pressure on the spectra of olivine and garnet. *J. Appl. Phys.*, **30**, 1446–7.

Balitsky, V. S. & Balitskaya, O. V. (1986) The amethyst–citrine dichromatism in quartz and its origin. *Phys. Chem. Minerals*, **13**, 415–21.

Ballet, O., Fuess, H. & Fritsche, T. (1987) Magnetic structure and cation distribution in $(Fe,Mn)_2SiO_4$ (olivine) by neutron diffraction. *Phys. Chem. Minerals*, **15**, 54–8.

Ballhausen, C. J. (1962) *Introduction to Ligand Field Theory*. (McGraw-Hill, New York), 298 pp.

Bancroft, G. M. & Brown, J. R. (1975) A Mössbauer study of coexisting hornblendes and biotites: Quantitative Fe^{3+}/Fe^{2+} ratios. *Amer. Mineral.*, **60**, 265–72.

Bancroft, G. M. & Burns, R. G. (1967a) Interpretation of the electronic spectra of iron in pyroxenes. *Amer. Mineral.,* **52,** 1278–87 [see White & Keester (1966)].

Bancroft, G. M. & Burns, R. G. (1967b) Distribution of iron cations in a volcanic pigeonite from Mössbauer spectroscopy. *Earth Planet. Sci. Lett.*, **3**, 125–7.

Bancroft. G. M. & Burns, R. G. (1969) Mössbauer and absorption spectral study of alkali amphiboles. *Mineral. Soc. Amer., Spec. Pap.* **2**, 137–48.

Bancroft, G. M., Burns, R. G. & Howie, R. A. (1967) Determination of the cation distribution in the orthopyroxene series by the Mössbauer effect. *Nature*, **213**, 1221–3.

Banks, R. J. (1972) The overall electrical conductivity distribution of the Earth. *J. Geomagn. Geoelectr.*, **24**, 337–51.

Banin, A., Margulies, L. & Chen, Y. (1985) Iron montmorillonite: A spectral analog of Martian soil. *Proc. 15th Lunar Planet. Sci. Conf., J. Geophys. Res.*, **90**, C771–4.

Banno, S. & Matsui, Y. (1973) On the formation of partition coefficients for trace element distribution between minerals and magma. *Chem. Geol.*, **11**, 1–15.

Bargeron, C. B., Avinor, M. & Drickamer, H. G. (1971) The effect of pressure on the Mössbauer resonance for ^{57}Fe in MnS_2. *Inorg. Chem.*, **10**, 1338–9.

Barnes, S. J. (1986) The distribution of chromium among orthopyroxene, spinel and silicate liquid at atmospheric pressure. *Geochim. Cosmochim. Acta*, **50**, 1889–909.

Basaltic Volcanism Study Project (1981) *Basaltic Volcanism on the Terrestrial Planets*. (Pergamon, New York), 1286 pp..

Basolo, F. D. & Pearson, R. G. (1967) *Mechanisms of Inorganic Reactions: A Study of Metal Complexes in Solution, 2nd edn.* (J. Wiley & Sons, New York), 701 pp.

Bass, J. D. (1986) Elasticity of uvarovite and andradite garnets. *J. Geophys. Res.*, **91**, 7505–16.

Basso, R., Dal Negro, A. & Rossi, G. (1979) Fe/Mg distribution in the olivines of ultramafic nodules from Assab (Ethiopia). *Neues Jahrb. Mineral. Mh.*, pp. 197–202.

Baur, W. H. (1976) Rutile-type compounds: V. Refinement of MnO_2 and MgF_2. *Acta Cryst.*, **B32**, 2200–4.

Beckwith, P. J. & Troup, G. J. (1973) The optical and infrared absorption of V^{3+} in beryl ($Be_3Al_2Si_6O_{18}$) *phys. stat. solidi (a)* **16**, 181–6.

Bell, J. F. & Keil, K. (1988) Spectral alteration effects in chondritic gas-rich breccias: implications for S-class and Q-class asteroids. *Proc. 19th Lunar Planet. Sci. Conf.* (Cambridge Univ. Press), pp. 573–80.

Bell III, J. F., McCord, T. B. & Lucey, P. G. (1990a) Imaging spectroscopy of Mars (0.4–1.1 μm) during the 1988 opposition. *Proc. 20th Lunar Planet. Sci. Conf.* (Lunar Planet. Inst., Houston), pp. 479–86.

Bell III, J. F, McCord, T. B. & Owensby, P. D. (1990b) Observational evidence of crystalline iron oxides on Mars. *J. Geophys. Res.*, **95**, 14447–61.

Bell, P. M. (1970) Analyses of olivine crystals in Apollo 12 rocks. *Ann. Rept. Geophys. Lab., Yearb.* **69**, 228–9.

Bell, P. M. & Mao, H.-K. (1969) Crystal field spectra at high pressure. *Ann. Rept. Geophys. Lab., Yearb.* **68**, 253–6.

Bell, P. M. & Mao, H.-K. (1972a) Apparatus for measurement of crystal field spectra of single crystals. *Ann. Rept. Geophys. Lab., Yearb.* **71**, 608–11.

Bell, P. M. & Mao, H.-K. (1972b) Crystal field effects of iron and titanium in selected grains of Apollo 12, 14, and 15 rocks, glasses and fine fractions. *Proc,.3rd Lunar Sci. Conf., Geochim. Cosmochim. Acta, Suppl. 3* (MIT Press), pp. 545–53.

Bell, P. M. & Mao, H.-K. (1972c) Crystal field determinations of Fe^{3+}. *Ann. Rept. Geophys. Lab., Yearb.* **71**, 531–8.

Bell, P. M. & Mao, H.-K. (1973a) Measurements of the polarized crystal-field spectra of ferrous and ferric iron in seven terrestrial plagioclases. *Ann. Rept. Geophys. Lab., Yearb.* **72**, 574–6.

Bell, P. M. & Mao, H.-K. (1973b) Optical and chemical analysis of iron in Luna 20 plagioclase. *Geochim. Cosmochim. Acta*, **37**, 755–9.

Bell, P. M. & Mao, H.-K. (1974) Pressure-effect on charge-transfer processes in minerals. *Ann. Rept. Geophys. Lab., Yearb.*, **73**, 507–10.

Bell, P. M. & Mao, H.-K. (1976) Crystal field spectra of fassaite from the Angra dos Reis meteorite. *Ann. Rept. Geophys. Lab., Yearb.* **75**, 701–5.

Bell, P. M., Mao, H.-K. & Rossman, G. R. (1975) Absorption spectroscopy of ionic and molecular units in crystals and glasses. In *Infrared and Raman Spectroscopy of Lunar and Terrestrial Minerals.* (C. Karr Jr, ed.; Academic Press, New York), pp. 1–38.

Bell, P. M., Mao, H.-K. & Weeks, R. A. (1976) Optical spectra and electron paramagnetic resonance of lunar and synthetic glasses: a study of the effects of controlled atmosphere, composition and temperature. *Proc. 7th Lunar Sci. Conf., Geochim. Cosmochim. Acta, Suppl. 7* (Pergamon Press, New York), pp. 2543–59.

Bell, P. M., Yagi, T. & Mao, H.-K. (1979) Iron–magnesium distribution coefficients between spinel [$(Mg,Fe)_2SiO_4$], magnesiowüstite [(Mg,Fe)O], and perovskite [$(Mg,Fe)SiO_3$]. *Ann. Rept. Geophys. Lab., Yearb.* **78**, 618–21.

Belsky, H. L., Rossman, G. R., Prewitt, C. T. & Gasparik, T. (1984) Crystal structure and optical spectroscopy (300 to 2200 nm) of $CaCrSi_4O_{10}$. *Amer. Mineral.*, **69**, 771–6.

Bernstein, L. W. (1982) Monazite from North Carolina having the alexandrite effect. *Amer. Mineral.*, **67**, 356–9.

Besancon, J. R. (1981) Rate of cation disordering in orthopyroxene. *Amer. Mineral.*, **66**, 965–73.

Bethe, H. (1929) Splitting of terms in crystals. (Termsaufspaltung in Kristallen.) *Ann. Phys.*, **3**, 133–206. [*Transl.*: Consultants Bureau, New York.]

Birle, J. D., Gibbs, G. V., Moore, P. B. & Smith, J. V. (1968) Crystal structures of natural olivines. *Amer. Mineral.*, **53**, 807–24.

Bish, D. L. (1977) A spectroscopic and X-ray study of the coordination of Cr^{3+} ions in chlorites. *Amer. Mineral.*, **62**, 385–9.

Bish, D. L. (1981) Cation ordering in synthetic and natural Ni–Mg olivine. *Amer. Mineral.*, **66**, 770–6.

Bish, D. L. & Burnham, C. W. (1984) Structure energy calculations on optimum distance model structures: application to the silicate olivines. *Amer. Mineral.*, **69**, 1102–9.

Bishop, J. L., Pieters, C. M. & Burns, R. G. (1992) Ferrihydrite found in Fe-rich montmorillonite and its relationship to the reflectance spectra of Mars. *Lunar Planet. Sci.*, **XXIII**, 111–12.

Bjorlykke, H. & Jarp, S. (1950) The content of cobalt in some Norwegian sulphide deposits. *Norsk. Geol. Tids.*, **28**, 151-6.

Blain, C. F. & Andrew, R. L. (1977) Sulfide weathering and the evaluation of gossans in mineral exploration. *Minerals Sci. Engng*, **9**, 119–50.

Blak, A. R., Isotani, S. & Watanabe, S. (1982) Optical absorption and electron spin resonance in blue and green beryl. *Phys. Chem. Minerals*, **8**, 161–5.

Blake, R. L., Hessevick, R. E., Zoltai, T. & Finger, L. W. (1966) Refinement of the hematite structure. *Amer. Mineral.*, **51**, 123–9.

Blazey, K. W. (1977) Optical absorption of MgO:Fe. *J. Phys. Chem. Solids*, **38**, 671–5.

Boon, J. R. & Fyfe, W. S. (1972) The coordination number of ferrous ions in silicate glasses. *Chem. Geol.*, **10**, 287–98.

Boström, D. (1987) Single-crystal X-ray diffraction studies of synthetic Ni–Mg olivine solid solutions. *Amer. Mineral.*, **72**, 965–72.

Bowen, N. L. & Schairer, J. F. (1935) The system MgO–FeO–SiO_2. *Amer. J. Sci.*, **229**, 152–217.

Brimhall, G. H. (1987) Preliminary fractionation patterns of ore metals through earth history. *Chem. Geol.*, **64**, 1–16.

Brookins, D. G. (1990) *Eh–pH Diagrams for Geochemistry.* (Springer–Verlag, New York), 176 pp.

Brown Jr, G. E. (1982) Olivines and silicate spinels. In *Orthosilicates, 2nd edn.* (P. H. Ribbe, ed.; Mineral. Soc. Amer.), *Rev. Mineral.*, **5**, 275–381.

Brown Jr, G. E. & Prewitt, C. T. (1973) High temperature crystal chemistry of hortonolite. *Amer. Mineral.*, **58**, 577–87.

Brown Jr, G. E., Calas, G., Waychunas, G. A. & Periau, J. (1988) X-ray absorption spectroscopy and its applications in mineralogy and geochemistry. In *Spectroscopic Methods in Mineralogy and Geology*. (F. C. Hawthorne, ed.; Mineral. Soc. Amer. Publ.), *Rev. Mineral.*, **18**, 431–512.

Brown Jr, G. E., Prewitt, C. T., Papike, J. J. & Sueno, S (1972) A comparison of the structures of low and high pigeonite. *J. Geophys. Res.*, **77**, 5778–89.

Burnham, C. W., Ohashi, Y., Hafner, S. S. & Virgo, D. (1971) Cation distribution and atomic thermal vibrations in an iron-rich orthopyroxene. *Amer. Mineral.*, **56**, 850–76.

Burns, R. G. (1965a) *Electronic Spectra of Silicate Minerals: Applications of Crystal Field Theory to Aspects of Geochemistry*. Ph.D. Diss., Univ. Calif. Berkeley, California.

Burns, R. G. (1965b) Formation of cobalt(III) in the amorphous FeOOH.nH_2O phase of manganese nodules. *Nature*, **205**, 99.

Burns, R. G. (1966a) Origin of optical pleochroism in orthopyroxenes. *Mineral. Mag.*, **35,** 715–19.

Burns, R. G. (1966b) Apparatus for measuring polarized absorption spectra of small crystals. *J. Sci. Instruments,* **43,** 58–60.

Burns, R. G. (1968a) Enrichments of transition metal ions in silicate structures. In *Symposium on the Origin and Distribution of Elements. Section V. Terrestrial Abundances.* (L. H. Ahrens, ed.; Pergamon Press, New York), pp. 1151–64.

Burns, R. G. (1968b) Optical absorption in silicates. In *The Application of Modern Physics to the Earth and Planetary Interiors.* (S. K. Runcorn, ed.; J. Wiley & Sons, New York), pp. 191–211.

Burns, R. G. (1970) Crystal field spectra and evidence for cation ordering in olivine minerals. *Amer. Mineral.*, **55,** 1608–32.

Burns, R. G. (1972a) Mixed valencies and site occupancies of iron in silicate minerals from Mössbauer spectroscopy. *Canad. J. Spectr.*, **17**, 51–9.

Burns, R. G. (1972b) Site preferences of Ni^{2+} and Co^{2+} in clinopyroxene and olivine: Limitations of the statistical approach. *Chem. Geol.*, **9**, 67–73.

Burns, R. G. (1973) The partitioning of trace transition elements in crystal structures: A provocative review with applications to mantle geochemistry. *Geochim. Cosmochim. Acta,* **37**, 2395–403.

Burns, R. G. (1974) The polarized spectra of iron in silicates: olivine. A discussion of neglected contributions from Fe^{2+} ions in M(1) sites. *Amer. Mineral.*, **59**, 625–9 [see Runciman *et al.* (1973a)].

Burns, R. G. (1975a) On the occurrence and stability of divalent chromium in olivines included in diamonds. *Contrib. Mineral. Petrol.*, **51**, 213–21.

Burns, R. G. (1975b) Crystal field effects in chromium and its partitioning in the mantle. *Geochim. Cosmochim. Acta,* **39**, 857–64.

Burns, R. G. (1976a) Partitioning of transition metal ions in mineral structures of the mantle. In *The Physics and Chemistry of Minerals and Rocks.* (R. G. J. Strens, ed.; J. Wiley, New York), pp. 555–72.

Burns, R. G. (1976b) The uptake of cobalt into ferromanganese nodules, soils, and synthetic manganese(IV) oxides. *Geochim. Cosmochim. Acta,* **40**, 95–102.

Burns, R. G. (1977) Transition element metallogenesis and plate tectonics. In *Geochemistry.* (A. J. Ellis, ed.) *New Zealand DSIR Bulletin,* **218**, 76–82.

Burns, R. G. (1980) Does feroxyhyte occur on the surface of Mars? *Nature,* **285**, 647 [see also *Nature,* **288**, 196 (1980)].

Burns, R. G. (1981) Intervalence transitions in mixed-valence minerals of iron and titanium. *Ann. Rev. Earth Planet. Sci.*, **9**, 345–83.

Burns, R. G. (1982) Electronic spectra of minerals at high pressures: How the mantle excites electrons. In *High-Pressure Researches in Geoscience.* (W. Schreyer, ed.; E. Schweizerhart'sche Verlagsbuchhandlung, Stuttgart), pp. 223–46.

Burns, R. G. (1983) Colours of gems. *Chem. Britain,* (12), 1004–7

Burns, R. G. (1985a) Thermodynamic data from crystal field spectra. In *Microscopic to Macroscopic: Atomic Environments to Mineral Thermodymanics.* (S. W. Kieffer & A. Navrotsky, eds; Mineral. Soc. Amer. Publ.). *Rev. Mineral.*, **14,** 277–316.

Burns, R. G. (1985b) Electronic spectra of minerals. In *Chemical Bonding and Spectroscopy in Mineral Chemistry.* (F. J. Berry & D. J. Vaughan, eds; Chapman & Hall, London), pp. 63–101.

Burns, R. G. (1986) Terrestrial analogues of the surface rocks on Mars. *Nature,* **320,** 55–6.

Burns, R. G. (1987a) Polyhedral bulk moduli from high-pressure crystal field spectra. In *High-Pressure Research in Mineral Physics.* (M. H. Manghnani & Y. Syono, eds; AGU, Washingon, D.C.), *Geophys. Mono.*, **39**, 361–9.

Burns, R. G. (1987b) Ferric sulfates on Mars. *Proc. 17th Lunar Planet. Sci. Conf., J. Geophys. Res.*, **92**, E570–4.

Burns, R. G. (1988) Gossans on Mars. *Proc. 18th Lunar Planet. Sci. Conf.* (Cambridge Univ. Press), pp. 713–21.

Burns, R. G. (1989a) Spectral mineralogy of terrestrial planets: scanning their surfaces remotely. *Mineral. Mag.*, **53**, 135–51.

Burns, R. G. (1989b) Mössbauer spectra of ^{57}Fe in rare earth perovskites: Applications to the electronic states of iron in the Mantle. In *Perovskite: A Structure of Great Interest to Geophysics and Materials Science.* (A. Navrotsky & D. J. Weidner, eds; AGU), *Geophys. Mono.*, **45**, 81–90.

Burns, R. G. (1991) Mixed valency minerals: influences of crystal structures on optical and Mössbauer spectra. In: *Mixed Valence Systems: Applications in Chemistry, Physics and Biology.* (K. Prassides, ed.; Plenum Press) *NATO ASI–C Ser., Math. Phys. Sci.*, **343**, 175–200.

Burns, R., G. (1992a) Origin of electronic spectra of minerals in the visible–near infrared region. In *Remote Geochemical Analysis: Elemental and Mineralogical Composition.* (C. Pieters & P. A. J. Englert, eds; Cambridge Univ. Press), ch. 1.

Burns, R. G. (1992b) Mössbauer spectral characterization of planetary surface materials. In *Remote Geochemical Analysis: Elemental and Mineralogical Composition.* (C. Pieters & P. A. J. Englert, eds; Cambridge Univ. Press), ch. 26.

Burns, R. G. & Burns, V. M. (1977) The mineralogy and crystal chemistry of deep-sea manganese nodules, a polymetallic resource of the 21st century. *Phil. Trans. Royal Soc. (London)*, **A286**, 283–301.

Burns, R. G. & Burns, V. M. (1979) Manganese oxides. In *Marine Minerals.* (R. G. Burns, ed.; Mineral Soc. Amer. Publ.), *Rev. Mineral.*, **6**, 1–46.

Burns, R. G. & Burns, V. M. (1981) Authigenic oxides. In *The Sea, Vol. 7.* (C. Emiliani, ed.; J. Wiley & Sons, Inc.), pp. 875–914.

Burns, R. G. & Burns, V. M. (1982) Crystal chemistry and Mössbauer spectroscopy of the ludwigite–vonsenite series, $(Mg,Fe^{2+})Fe^{3+}BO_5$. *Geol. Soc. Amer., Ann. Meet., Abstr.*, **14**, 457.

Burns, R. G. & Burns, V. M. (1984a) Optical and Mössbauer spectra of transition metal-doped corundum and periclase. In *Structure and Properties of MgO and Al_2O_3 Ceramics.* (W. G. Kingery, ed.; Amer. Ceram. Soc. Inc.), *Adv. Ceramics*, **10**, 46–61.

Burns, R. G. & Burns, V. M. (1984b) Crystal chemistry of meteoritic hibonite. *Proc. 15th Lunar Planet. Sci. Conf., J. Geophys. Res.*, **89**, C313–21.

Burns, R. G. & Dyar, M. D. (1991) Crystal chemistry and Mössbauer spectra of babingtonite. *Amer. Mineral.*, **76**, 892–99.

Burns, R. G. & Fisher, D. S (1990) Evolution of sulfide mineralization on Mars; Iron sulfide mineralogy of Mars: Magmatic evolution and chemical weathering products. *J. Geophys. Res.*, **95**, 14169–73 and 14415–21.

Burns, R. G. & Fuerstenau, D. W. (1966) Electron-probe determination of inter-element relationships in manganese nodules. *Amer. Mineral.*, **51**, 895–902.

Burns, R. G. & Fyfe, W. S. (1964) Site preference energy and selective uptake of transition metal ions during magmatic crystallization. *Science*, **144**, 1001–3.

Burns, R. G. & Fyfe, W. S. (1966a) Distribution of elements in geological processes. *Chem. Geol.*, **1**, 49–56.

Burns, R. G. & Fyfe, W. S. (1966b) The behaviour of nickel during magmatic crystallization. *Nature*, **220**, 1147–8.

Burns, R. G. & Fyfe, W. S. (1967a) Crystal field theory and the geochemistry of transition elements. In *Researches in Geochemistry, Vol. 2.* (P. H. Abelson, ed.; J. Wiley & Son, New York), pp. 259–85.

Burns, R. G. & Fyfe, W. S. (1967b) Trace element distribution rules and their significance. *Chem. Geol.*, **2**, 89–104.

Burns, R. G. & Greaves, C. (1971) Correlations of infrared and Mössbauer site population measurements of actinolites. *Amer. Mineral.*, **56**, 2010–33.

Burns, R. G. and Huggins, F. E. (1973) Visible-region absorption spectra of a Ti^{3+} fassaite from the Allende meteorite: A discussion. *Amer. Mineral.*, **58**, 955–61 [see Dowty and Clark (1973)].

Burns, R. G. and Prentice, F. J. (1968) Distribution of iron cations in the crocidolite structure. *Amer. Mineral.*, **53**, 770–6.

Burns, R. G. and Solberg, T. C. (1990) Crystal structure trends in Mössbauer spectra of ^{57}Fe-bearing oxide, silicate, and aluminosilicate minerals. In *Structures of Active Sites in Minerals.* (L. M. Coyne, S. W. S. McKeever & D. F. Drake eds; Amer. Chem. Soc. Publ.), *ACS Symp. Ser.*, **415**, 262–83.

Burns, R. G. & Straub, D. W. (1992) Mixed-valence iron minerals on Venus: Fe^{2+}–Fe^{3+} oxides and oxysilicates formed by surface–atmosphere interactions. *Intern. Conf. Venus, Pasadena, Abstr.*, pp. 15–7 [see also *EOS, Trans. Amer. Geophys. Union*, **73**, 332–3 (1992)].

Burns, R. G. & Strens, R. G. J. (1966) Infrared study of the hydroxyl bands in clinoamphiboles. *Science*, **153**, 890–2.

Burns, R. G. & Strens, R. G. J. (1967) Structural interpretation of polarized absorption spectra of the Al–Fe–Mn–Cr epidotes. *Mineral. Mag.*, **36**, 204–26.

Burns, R. G. & Sung, C.-M. (1978) The effect of crystal field stabilization energy on the olivine → spinel transition in the system Mg_2SiO_4–Fe_2SiO_4. *Phys. Chem. Minerals*, **2**, 349–64 [see also *Phys. Chem. Minerals*, **2**, 177–97 (1978)].

Burns, D. J. & Vaughan, D. J. (1970) Interpretation of the reflectivity behavior of ore minerals. *Amer. Mineral.*, **55**, 1576–86.

Burns, R. G. & Vaughan, D. J. (1975) Polarized electronic spectra. In *Infrared and Raman Spectroscopy of Lunar and Terrestrial Minerals.* (C. Karr Jr, ed.; Academic Press, New York), pp. 39–76.

Burns, R. G., Abu-Eid, R. M. & Huggins, F. E. (1972a) Crystal field spectra of lunar pyroxenes. *Proc. 3rd Lunar Sci. Conf., Geochim. Cosmochim. Acta, Suppl. 3* (MIT Press), pp. 533–43.

Burns, R. G., Burns, V. M. & Stockman, H. W. (1983) A review of the todorokite–buserite problem: implications to the mineralogy of marine manganese nodules. *Amer. Mineral.*, **68**, 972-80.

Burns, R. G., Burns, V. M. & Stockman, H. W. (1985) The todorokite–buserite problem: further considerations. *Amer. Mineral.*, **70**, 205–8.

Burns, R. G., Clark, M. G. & Stone, A. J. (1966) Vibronic polarization in the electronic spectra of gillespite, a mineral containing iron (II) in square planar coordination. *Inorg. Chem.*, **5**, 1268–72.

Burns, R. G., Clark, R. H. & Fyfe, W. S. (1964) Crystal field theory and applications to problems in geochemistry. In *Chemistry of the Earth's Crust.* (L Vinogradov, A. P., ed.; Acad. Sci. USSR Publ.) *Proc. Vernadsky Centen. Symp.*, **2**, 88–106. [*Transl.*: Israel Sci. Progr. Sci. Transl., Jerusalem, pp. 93–112 (1967).]

Burns, R. G., Huggins, F. E. & Abu–Eid, R. M. (1972b) Polarized absorption spectra of single crystals of lunar pyroxenes and olivines. *Moon,* **4,** 93–102.

Burns, R. G., Parkin, K. M., Loeffler, B. M., Leung, I. S. & Abu-Eid, R. M. (1976) Further characteristics of spectral features attributable to titanium on the moon. *Proc. 7th Lunar Sci. Conf., Geochim. Cosmochim. Acta, Suppl. 7* (Pergamon Press, New York), pp. 2561–78.

Burns, R. G., Tossell, J. A. & Vaughan, D. J. (1972c) Pressure-induced reduction of a ferric amphibole. *Nature,* **240**, 33–5.

Burns, R. G., Vaughan, D. J., Abu-Eid, R. M., Witner, M. & Morawsky, A. (1973) Spectral evidence for Cr^{3+}, Ti^{3+}, and Fe^{2+} rather than Cr^{2+} and Fe^{3+} in lunar ferromagnesian silicates. *Proc. 4th Lunar Sci. Conf., Geochim. Cosmochim. Acta, Suppl. 4* (Pergamon Press, New York), pp. 983–94.

Burns, V. M. & Burns, R. G. (1978) Authigenic todorokite and phillipsite inside deep-sea manganese nodules. *Amer. Mineral.*, **63**, 827–831.

Bush, W. R., Hafner, S. S. & Virgo, D. (1970) Some ordering of iron and magnesium at the octahedrally coordinated sites in a magnesium-rich olivine. *Nature*, **227**, 1339–41.

Byström, A. M. (1949) The crystal structure of ramsdellite, an orthorhombic modification of MnO_2. *Acta Chem. Scand.*, **3**, 163–73.

Byström, D. (1987) Single-crystal X-ray diffraction studies of synthetic Ni–Mg olivine solid solutions. *Amer. Mineral.*, **72**, 965–72.

Calas, G. (1982) Spectroscopic properties of transition elements in glasses of geological interest. *J. de Phys.*, **43** (C9), 311–14.

Calas, G. (1988) Electron paramagnetic resonance. In *Spectroscopic Methods in Mineralogy and Geology*. (F. C. Horthorne, ed.; Mineral. Soc. Amer. Publ.), *Rev. Mineral.*, **18**, 513–71.

Calas, G. & Petiau, J. (1983) Structure of oxide glasses: Spectroscopic studies of local order and crystallochemistry. Geochemical implications. *Bull. Minéral.*, **106**, 33–55.

Calas, G., Manceau, A., Novikoff, A. & Boukili, H. (1984) Comportement du chrome dans les minéraux d'altération du gisement de Campo Formoso (Bahia, Brésil). *Bull. Minéral.*, **107**, 755–66.

Cameron, M. & Papike, J. J. (1981) Structural and chemical variations in pyroxenes. *Amer. Mineral.*, **66**, 1–50.

Carlson, W. D. & Rossman, G. R. (1988) Vanadium- and chromium-bearing andalusite: Occurrence and optical-absorption spectroscopy. *Amer. Mineral.*, **73**, 1366–9.

Carmichael, I. S. E., Fyfe, W. S. and Machin, D. J. (1960) Low-spin ferrous iron in the iron silicate deerite. *Nature*, **211**, 1389.

Carter, J. L. (1970) Mineralogy and chemistry of the Earth's upper mantle based on the partial fusion–partial crystallization model. *Geol. Soc. Amer., Bull.* **81**, 2021–34.

Casey, W. H. & Westrich, H. R. (1991) General relations among weathering rates of some silicate minerals. *Geol. Soc. Amer., Ann. Meet., Abstr.*, **23**, A258.

Cech, F., Povondra, P. & Vrana, S. (1981) Cobaltian staurolite from Zambia. *Bull. Minéral.*, **104**, 526–9.

Cervelle, B. (1991) Application of mineralogical constraints to remote-sensing. *Eur. J. Mineral.*, **3**, 677–88.

Cervelle, B. & Maquet, M. (1982) Cristallochimie des lizardites substitutées Mg–Fe–Ni par spectrométrie visible et infrarouge proche. *Clay. Minerals*, **17**, 377–92.

Charette, M. P., McCord, T. B., Pieters, C. & Adams, J. B. (1974) Application of remote spectral reflectance measurements to lunar geology classification and determination of titanium content of lunar soils. *J. Geophys. Res.*, **79**, 1605–13.

Chenavas, J., Joubert, J. C. & Marezio, M. (1971) Low-spin – high-spin state transition in high-pressure cobalt sequioxide. *Solid State Comm.*, **9**, 1057–60.

Chesnokov, B. V. (1959) Spectral absorption curves of some minerals coloured by titanium. *Dokl. Akad. Sci., SSSR*, **129**, 647–9.

Chesnokov, B. V. (1961) Spectral absorption curves of glaucophane from eclogite. *Zap. Vses. Mineral., Obshchv., SSSR*, **90**, 700–3.

Chopin, C. & Langer, K. (1988) Fe^{2+}–Ti^{4+} charge transfer between face–sharing octahedra: polarized absorption spectra and crystal chemistry of ellenbergerite. *Bull. Minéral.*, **111**, 17–27.

Chukhrov, F. V., Gorshkov, A. I., Drits, L. E., Shterenberg, A. V. & Sakharov, B. A. (1983) Mixed-layer asbolane–buserite minerals and asbolane in oceanic iron–manganese concretions. *Izvest. Akad. Nauk, SSSR, Ser. Geol.*, **5**, 91–9. [*Transl.*: *Intern. Geol. Rev.*, **25**, 838–47 (1983).]

Chukhrov, F. V., Gorshkov, A. I., Vitovskaya, I. V., Drits, L. E., Sivtsov, A. V. & Rudnitskaya, E. S. (1980) Crystallochemical nature of Co–Ni asbolane. *Izvest. Akad. Nauk, S.S.S.R., Ser. Geol.*, **6**, 73–81 [*Transl.*: *Intern. Geol. Rev.*, **24**, 598–604 (1982).]

Cid-Dresdner, H. (1965) Determination and refinement of crystal structure of turquoise, $CuAl_6(PO_4)_4(OH)_8.4H_2O$. *Zeit. Krist.*, **121**, 87–113.

Clark, B. C., Baird, A. K., Weldon, R. J., Tsusaki, D. M., Schnabel, L. & Candelaria, M. P. (1982) Chemical composition of the martian fines. *J. Geophys. Res.*, **87**, 10059–67.

Clark, J. R., Appleman, D. E. & Papike, J. J. (1969) Crystal-chemical characterization of clinopyroxenes based on eight new structure refinements. *Miner. Soc. Amer., Spec. Pap.* **2**, 31–50.

Clark, J. R., Ross, M. & Appleman, D. E. (1971) Crystal chemistry of a lunar pigeonite. *Amer. Mineral.*, **56**, 888–908.

Clark, M. G. & Burns, R. G. (1967) Electronic spectra of Cu^{2+} and Fe^{2+} square planar coordinated by oxygen in $BaXSi_4O_{10}$. *J. Chem. Soc.* pp. 1034–8.

Clark S. P. (1957) Absorption spectra of some silicates in the visible and near infrared. *Amer. Mineral.*, **42**, 732–42.

Clark, T. & Naldrett, A. J. (1978) The distribution of Fe and Ni between synthetic olivine and sulfide at 900 °C. *Econ. Geol.*, **67**, 939–52.

Clarke, F. W. (1924) *The Data of Geochemistry. US Geol. Surv., Bull.* **770**.

Cloutis, E. A. & Gaffey, M. J. (1991) Pyroxene spectroscopy revisited: Spectral compositional correlations and relationship to geothermometry. *J. Geophys. Res.*, **96**, 22809–26.

Cloutis, E. A., Gaffey, M. J., Jackowski, T. L. & Reed, K. L. (1986) Calibrations of phase abundance, composition, and particle size distribution for olivine–orthopyroxene mixtures from reflectance spectra. *J. Geophys. Res.*, **91**, 11641–53.

Coey, J. M. D. (1985) Mössbauer spectroscopy of silicate minerals. In *Mössbauer Spectroscopy Applied to Inorganic Chemistry, Vol. 2.* (G. J. Long, ed; Plenum, New York), pp. 443–509.

Coey, J. M. D., Allan, J., Kan, X., Nguyen, V. D. & Ghose, S. (1984) Magnetic and electrical properties of ilvaite. *J. Appl. Phys.*, **55**, 1963–5.

Coey, J. M. D., Moukarika, A. & McDonagh, C. M. (1982) Electron hopping in cronstedtite. *Solid State Comm.*, **41**, 797–800.

Cohen, A. J. & Hassan, F. (1974) Ferrous and ferric ions in synthetic alpha-quartz and natural amethyst. *Amer. Mineral.*, **59**, 719–28

Cohen, A. J. & Janezic, G. G. (1983) The crystal field spectra of the $3d^3$ ions, Cr^{3+} and Mn^{4+} in green spodumene. In *The Significance of Trace Elements in Solving Petrogenetic Problems and Controversies.* (S. S. Augustithis, ed.; Theophrastus Publ., Athens, Greece), pp. 899–904.

Cohen, A. J. & Makar, L. N. (1985) Dynamic biaxial absorption spectra of Ti^{3+} and Fe^{2+} in a natural rose quartz crystal. *Mineral Mag.*, **49**, 709–15.

Colsen, R. O., McKay, G. A. & Taylor, L. A. (1988) Temperature and composition dependencies of trace element partitioning: Olivine/melt and low-Ca pyroxene/melt. *Geochim. Cosmochim. Acta*, **52**, 539–53.

Cooney, T. F. & Sharma, S. K. (1990) Structure of glasses in the systems: Mg_2SiO_4–Fe_2SiO_4, Mn_2SiO_4–Fe_2SiO_4, Mg_2SiO_4–$CaMgSiO_4$, and Mn_2SiO_4–$CaMnSiO_4$. *J. Non-Cryst. Solids*, **122**, 10–32.

Cotton, F. A. (1964) Ligand field theory. *J. Chem. Educ.*, **41**, 466–76.

Cotton, F. A. (1990) *Chemical Applications of Group Theory, 3rd edn.* (Wiley–Interscience, New York), 579 pp.

Cotton, F. A. & Meyers, M. D. (1960) Magnetic and spectral properties of the spin-free $3d^6$ systems Fe(II) and Co(III) in cobalt(III) hexafluoride ion: Probable observation of dynamic Jahn–Teller effects. *J. Amer. Chem. Soc.*, **82**, 5023–6.

Cotton, F. A. & Wilkinson, G. (1988) *Advanced Inorganic Chemistry, 4th edn.* (J. Wiley & Sons New York), 1396 pp.

Cotton, F. A., Wilkinson, G. & Gaus, P. L. (1988) *Basic Inorganic Chemistry, 2nd edn.* (J. Wiley & Sons, New York).

Cox, R. T. (1977) Optical absorption of the d^4 ion Fe^{4+} in pleochroic amethyst quartz. *J. Phys. C*, **10**, 4631–43.

Crane, S. E. (1981) *Structural Chemistry of the Marine Manganese Minerals.* Ph.D. thesis, Univ. California, San Diego.

Cronan, D. S. (1980) *Underwater Minerals.* (Academic Press, London), 362 pp.

Cruikshank, D. P. & Hartmann, W. K. (1984) The meteorite–asteroid connection: Two olivine–rich asteroids. *Science,* **223,** 281–2.

Curtis, C. D. (1964) Applications of the crystal–field theory to the inclusion of trace transition elements in minerals during magmatic differentiation. *Geochim. Cosmochim. Acta*, **28**, 389–403.

Curtis, L., Gittens, J., Kocman, V., Rucklidge, J. C., Hawthorne, F. C. & Ferguson, R. B. (1975) Two crystal structure refinements of a *P2/n* titanian ferro–omphacite. *Canad. Mineral.*, **13**, 62–7.

Dachs, H. (1963) Neutronen– und Rontgenuntersuchungen am Manganit, MnOOH. *Zeit. Krist.*, **118,** 303–26.

Dahl, P. S., Wehn, D. C. & Feldman, S. G. (1993) The systematics of trace–element partitioning between coexisting muscovite and biotite in pelitic schists from the Black Hills, South Dakota. *Geochim. Cosmochim. Acta*, **57,** (in press).

Dale, I. M. & Henderson, P. (1972) The partition of transition elements in phenocryst–bearing basalts and the implications about melt structure. *24th Intern. Geol. Congr., Sect.* **10**, 105–11.

Day, P. (1976) Mixed valence chemistry and metal chain compounds. In *Low Dimensional Cooperative Phenomena.* (H. J. Keller, ed.; Plenum, New York), pp. 191–214.

Dent–Glasser, L. S. & Ingram, L. (1968) Refinement of the crystal structure of groutite, α–MnOOH. *Acta Cryst.*, **B24,** 1233–6.

Dickson, B. L. & Smith, G. (1976) Low temperature optical absorption and Mössbauer spectra of staurolite and spinel. *Canad. Mineral.*, **14**, 208–15.

Dillard, J. G., Crowther, D. L. & Murray, J. W. (1982) The oxidation states of cobalt and selected metals in Pacific ferromanganese nodules. *Geochim. Cosmochim. Acta*, **46,** 755–9.

Dollase, W. A. (1968) Refinement and comparison of the structures of zoisite and clinozoisite. *Amer. Mineral.*, **53**, 1882–98.

Dollase, W. A. (1969) Crystal structure and cation ordering of piemontite. *Amer. Mineral.*, **54**, 710–7.

Dollase, W. A. (1971) Refinement of the crystal structures of epidote, allanite and hancockite. *Amer. Mineral.*, **56**, 447–64.

Dowty, E. (1978) Absorption optics of low–symmetry crystals – applications to titanian clinopyroxene spectroscopy. *Phys. Chem. Minerals*, **3**, 173–81.

Dowty, E. & Clark, J. R. (1973) Crystal structure refinement and visible–region absorption spectra of a Ti^{3+}-fassaite from the Allende meteorite. *Amer. Mineral.*, **58**, 230–42 and 962–4 [see Burns & Huggins (1973)].

Drake, M. J. & Holloway, J. R. (1981) Partitioning of Ni between olivine and silicate melt: the 'Henry's Law problem' re-examined. *Geochim. Cosmochim. Acta*, **45**, 431–7.

Drake, M. J., Newsom, H. & Capobianco, C. (1989) V, Cr, and Mn in the Earth, Moon, EPB and SPB and the origin of the Moon. *Geochim. Cosmochim. Acta*, **53**, 2101–11.

Drickamer, H. G. (1965) The effect of high pressure on the electronic structure of solids. *Solid State Phys.*, **17**, 1–133.

Drickamer, H. G. & Frank, C. W. (1973) *Electronic Transitions and the High Pressure Chemistry and Physics of Solids.* (Chapman & Hall, London), 220 pp.

Drickamer, H. G., Bastron, V. C., Fisher, D. C. & Grenoble, D. C. (1970) The high pressure chemistry of iron. *J. Solid State Chem.*, **2**, 94-104.

Drifford, M. & Charpin, P. (1967) Variation experimentale des parametres du champ cristallin dans la structure spinelle. *Compte Rendu Acad. Sci., Ser. B.* **264,** 64–6.

Duba, A. & Nicholls, I. A. (1973) The influence of oxidation state on the electrical conductivity of olivine. *Earth Planet. Sci. Lett.*, **18**, 59–64.

Duke, J. M. (1976) Distribution of the period four transition elements among olivine, calcic clinopyroxene and mafic silicate liquid: experimental results. *J. Petrol.*, **17**, 499–521.

Dunitz, J. D. & Orgel, L. E. (1957) Electronic properties of transition element oxides. II. Cation distribution amongst octahedral and tetrahedral sites. *J. Phys. Chem. Solids*, **3**, 318–33.

Dunn, T. M. (1960) The visible and ultraviolet spectra of complex compounds. In *Modern Coordination Chemistry.* (J. Lewis & R. G. Wilkins, eds; Interscience Publ., New York), pp. 229–300.

Dunn, T. M., McClure, D. S. & Pearson, R. G. (1965) *Some Aspects of Crystal Field Theory.* (Harper & Row, New York), 115 pp.

Dupuy, C., Dostal, J., Liotard, J. M. & Leyreloup, A. (1980) Partitioning of transition elements between coexisting clinopyroxene and garnet. *Earth Planet. Sci. Lett.*, **48**, 303–10.

Dyar (1987) A review of Mössbauer data on trioctahedral micas: evidence for tetrahedral Fe^{3+} and cation ordering. *Amer. Mineral.*, **72**, 101–12.

Dyar, M. D. & Burns, R. G. (1981) Coordination chemistry of iron in glasses contributing to remote-sensed spectra of the Moon. *Proc. 12th Lunar Planet. Sci. Conf., Geochim. Cosmochim. Acta, Suppl. 12* (Pergamon Press, New York), pp. 695–702.

Dyar, M. D. & Burns, R. G. (1986) Mössbauer spectral study of ferruginous one-layer trioctahedral micas. *Amer. Mineral.*, **71**, 955–65.

Eigenmann, K. & Gunthard, H. H. (1971) Valence states, redox reactions, and biparticle formation of Fe and Ti doped sapphire. *Chem. Phys. Lett.*, **13**, 58–61.

Eigenmann, K., Kurtz, K. & Gunthard, H. H. (1971) The optical spectrum of α-Al_2O_3:Fe^{3+}. *Chem. Phys. Lett.*, **13**, 54–7.

Eigenmann, K., Kurtz, K. & Gunthard, H. H. (1972) Solid state reactions and defects in doped Verneuil sapphite. III. Systems α-Al_2O_3; Fe, α-Al_2O_3:Ti, and α-Al_2O_3:(Fe,Ti). *Helv. Chim. Acta*, **45**, 452–80.

Engel, A. E. J. & Engel, C. G. (1960) Progressive metamorphism and granitization of the major paragenesis, Northwest Adirondack Mountains, New York. Part 2. Mineralogy. *Geol. Soc. Amer., Bull.* **71**, 1–58.

Engel, A. E. J. & Engel, C. G. (1962) Hornblendes formed during progressive metamorphism of amphibolites, Northwest Adirondack Mountains. *Geol. Soc. Amer., Bull.* **73**, 1499–514.

Ericsson, T. & Filippidis, A. (1986) Cation ordering in the limited solid solution Fe_2SiO_4–Zn_2SiO_4. *Amer. Mineral.*, **71**, 1502–9.

Evans, B. J. & Sergent Jr, E. W. (1975) ^{57}NGR of Fe phases in 'magnetic cassiterites'. I. Crystal chemistry of dodecahedral Fe^{2+} in pyralspite garnets. *Contrb. Mineral. Petrol.*, **53**, 183–94.

Fackler Jr, J. P. & Holah, D. G. (1965) Properties of chromium(II) complexes. I. Electronic spectra of the simple salt hydrates. *Inorg Chem.*, **4**, 954–8.

Farmer, V. C. (1974) Layer silicates. In *The Infrared Spectra of Minerals.* (V. C. Farmer, ed.; Mineral. Soc., London), pp. 331–63.

Farr, T. G., Bates, B. A., Ralph, R. L. & Adams, J. B. (1980) Effects of overlapping optical absorption bands of pyroxene and glass on the reflectance spectra of lunar soils. *Proc. 11th Lunar Planet. Sci. Conf., Geochim. Cosmochim. Acta, Suppl. 11* (Pergamon Press, New York), pp. 19–29.

Farrell, E. F. & Newnham, R. E. (1965) Crystal-field spectra of chrysoberyl, alexandrite, peridot and sinhalite. *Amer. Mineral.*, **50**, 1972–81.

Farrell, E. F. & Newnham, R. E. (1967) Electronic and vibrational absorption spectra of cordierite. *Amer. Mineral.*, **52**, 380–8.

Farrell, E. F., Fang, J. H. & Newnham, R. E. (1963) Refinement of the chrysoberyl structure. *Amer. Mineral.*, **48**, 804–10.

Faye, G. H. (1968a) The optical absorption spectra of certain transition metal ions in muscovite, lepidocrocite and fuchsite. *Canad. J. Earth Sci.*, **5**, 31–8.

Faye, G. H. (1968b) The optical absorption spectra of iron in six-coordinate sites in chlorite, biotite, phlogopite and vivianite. Some aspects of pleochroism in the sheet silicates. *Canad. Mineral.*, **9**, 403–25.

Faye, G. H. (1969) The optical absorption spectrum of tetrahedrally coordinated Fe^{3+} in orthoclase. *Canad. Mineral.*, **10**, 112–17.

Faye, G. H. (1971a) A semi-quantitative microscope technique for measuring the optical absorption spectra of mineral and other powders. *Canad. Mineral.*, **10**, 889–95.

Faye, G. H. (1971b) On the optical spectra of di- and trivalent iron in corundum: A discussion. *Amer. Mineral.*, **56**, 344–8 [see Lehmann & Harder (1970)].

Faye, G. H. (1972) Relationship between crystal-field splitting parameter, 'Δ_{VI}' and M_{host}–O bond distance as an aid in the interpretation of absorption spectra of Fe^{2+} minerals. *Canad. Mineral.*, **11**, 473–87.

Faye, G. H. (1974) Optical absorption spectrum of Ni^{2+} in garnierite: a discussion. *Canad. Mineral.*, **12**, 389–93 [see Lakshman and Reddy (1973b)].

Faye, G. H. (1975) Spectra of shock-affected rhodonite: A discussion. *Amer. Mineral.*, **60**, 939–41 [see Gibbons *et al.* (1974)].

Faye, G. H. & Harris, D. C. (1969) On the origin and pleochroism in andalusite from Brazil. *Canad. Mineral.*, **10**, 47–56.

Faye, G. H. & Hogarth, D. D. (1969) On the reverse pleochroism of a phlogopite. *Canad. Mineral.*, **10**, 25–34.

Faye, G. H. & Nickel, E. H. (1969) On the origin of colour and pleochroism of kyanite. *Canad. Mineral.*, **10**, 35–46.

Faye, G. H. & Nickel, E. H. (1970a) The effect of charge transfer processes on the color and pleochroism of amphiboles. *Canad. Mineral.*, **10**, 616–35.

Faye, G. H. & Nickel, E. H. (1970b) On the pleochroism of vanadium-bearing zoisite from Tanzania. *Canad. Mineral.*, **10**, 812–21.

Faye, G. H., Manning, P. G. & Nickel, E. H. (1968) The polarized optical absorption spectra of tourmaline, cordierite, chloritoid, and vivianite: ferrous–ferric electronic interaction as a source of pleochroism. *Amer. Mineral.*, **53**, 1174–201.

Faye, G. H., Manning, P. G., Gosselin, J. R. & Tremblay, R. J. (1974) Optical absorption spectra of tourmaline: importance of charge transfer processes. *Canad. Mineral.*, **12**, 370–80.

Fegley Jr, B., Treiman, A. H. & Sharpton, V. 1. (1992) Venus surface mineralogy: Observational and theoretical constraints. *Proc. 22nd Lunar Planet. Sci. Conf.* (Lunar Planet. Inst., Houston), pp. 3-19.

Fei, Y., Mao, H.-K. & Mysen, B. O. (1991) Experimental determination of element partitioning and calculation of phase relations in the MgO–FeO–SiO_2 system at high pressure and high temperature. *J. Geophys. Res.*, **96**, 2157–69.

Ferguson, J. & Fielding. P. E. (1971) The origins of the colours of yellow, green and blue sapphires. *Chem. Phys. Lett.*, **10**, 262–5.

Ferguson, J. & Fielding, P. E. (1972) The origins of the colours of natural yellow, blue and green sapphires. *Austral. J. Chem.*, **25**, 1371–85..

Figgis, B. N. (1966) *Introduction to Ligand Fields.* (Interscience Publ., New York), 351 pp.

Finch, J. Gainsford, A. R. & Tennant, W. C. (1982) Polarized optical absorption and ^{57}Fe Mössbauer study of pegmatitic muscovite. *Amer. Mineral.*, **67**, 59–68.

Finger, L. W. (1970) Fe/Mg ordering in olivines. *Ann. Rept. Geophys. Lab.*, Yearb. **69**, 302–5.

Finger, L. W. & Hazen, R. M. (1980) Crystal structure and isothermal compression of Fe_2O_3, Cr_2O_3, and V_2O_3 to 50 kbar. *J. Appl. Phys.*, **51**, 5362–7.

Finger, L. W. & Hazen, R. M. (1987) Crystal structure of monoclinic ilvaite and the nature of the monoclinic–orthorhombic transition at high pressure. *Zeit. Krist.*, **179**, 415–30.

Finger, L.W. & Virgo, D. (1971) Confirmation of Fe/Mg ordering in olivines. *Ann. Rept. Geophys. Lab., Yearb.* **70**, 221–5.

Finger, L. W., Hazen, R. M. & Hemley, R. J. (1989) $BaCuSi_2O_6$: A new cyclosilicate with four-membered tetrahedral rings. *Amer. Mineral.*, **74**, 952–5.

Finger, L. W., Hazen, R. M. & Yagi, T. (1979) Crystal structures and electron densities of nickel and iron silicate spinels at elevated temperatures and pressures. *Amer. Mineral.*, **64**, 1002–9.

Fleet, S. G. & Megaw, H. D. (1962) The crystal structure of yoderite. *Acta Cryst.*, **15**, 721–8.

Fleet, M. E., MacRae, N. D. & Osborne, M. D. (1981) The partition of nickel between olivine, magma and immiscible sulfide liquid. *Chem. Geol.*, **31**, 119–27.

Fleischer, M. (1955) Minor elements in some sulphide minerals. *Econ. Geol., 50th anniv. vol.*, pp. 970–1024.

Fleisher, M., ed. (1962 onwards) *The Data of Geochemistry, 6th edn.* (US Geol. Survey Publ.), *Prof. Pap.* **440**.

Foit, F. F. & Rosenberg, P. E. (1979) The structure of vanadium-bearing tourmaline and its implications regarding tourmaline solid solutions. *Amer. Mineral.*, **64**, 788–98.

Forbes, C. E. (1983) Analysis of the spin hamiltonian parameters for Cr^{3+} in mirror and inversion symmetry sites of alexandrite ($Al_{2-x}Cr_xBeO_4$). Determination of the relative site occupancy by EPR. *J. Chem. Phys.*, **79**, 2590–5.

Ford, R. J. & Hitchman, M. A. (1979) Single crystal electronic and EPR spectra of $CaCuSi_4O_{10}$, a synthetic silicate containing copper(II) in a four-coordinate planar ligand environment. *Inorg. Chim. Acta*, **33**, L167–70.

Fox, K. E., Furukawa, T. & White, W. B. (1982) Transition metal ions in silicate melts. Part 2: Iron in sodium silicate glasses. *Phys. Chem. Glasses*, **23**, 169–78.

Francis, C. A. & Ribbe, P. H. (1980) The forsterite–tephroite series. I. Crystal structure refinements. *Amer. Mineral.*, **65**, 1263–9.

Frentrup, K. R. & Langer, K. (1981) Mn^{3+} in garnets: optical absorption spectrum of a synthetic Mn^{3+}-bearing silicate garnet. *Neues Jahrb. Mineral. Mh.*, pp. 245–56.

Frendrup, K. R. & Langer, K. (1982) Microscope absorption spectrometry of silicate

microcrystals in the range 40,000–5,000 cm^{-1} and its application to garnet end-members synthesized at high pressures. In *High -Pressure Researches in Geoscience.* (W. Schreyer, ed., E. Schweizerbart'sche Verlagsbuchhandlung, Stuttgart), pp. 247–58.
Fritsch, E. & Rossman, G. R. (1987) An update on color in gems. Part 1: Introduction and colors produced by dispersed metal ions. *Gems Gemol.*, **23**, 126–39.
Fritsch, E. & Rossman, G. R. (1988a) An update on color in gems. Part 2: Colors involving multiple atoms and color centers. *Gems Gemol.*, **24**, 3–15.
Fritsch, E. & Rossman, G. R. (1988b) An update on color in gems. Part 3: Colors caused by band gaps and physical phenomena. *Gems Gemol.*, **24**, 81–102.
Fukao, Y., Mizutani, H. & Uyeda, S. (1968) Optical absorption spectra at high temperatures and radiative thermal conductivity of olivines. *Phys. Earth Planet. Interiors*, **1**, 57–62.
Fujino, K., Sasaki, S., Takéuchi, Y. & Sadanaga, R. (1981) X-ray determination of electron distributions in forsterite, fayalite and tephroite. *Acta Cryst.*, **B37**, 513–8.
Fyfe, W. S. (1951) Isomorphism and bond type. *Amer. Mineral.*, **36**, 538–42.
Fyfe, W. S. (1954) The problem of bond type. *Amer. Mineral.*, **39**, 991–1004.
Fyfe, W. S. (1960) The possibility of *d*-electron coupling in olivine at high pressures. *Geochim. Cosmochim. Acta*, **19**, 141–3.
Gabe, E. J., Portheine, J. C. & Whitlow, S. H. (1973) A reinvestigation of the epidote structure: Confirmation of the iron location. *Amer. Mineral.*, **58**, 218–23.
Gaddis, L. R., Pieters, C. M. & Hawke, B. R. (1985) Remote sensing of lunar pyroclastic mantling deposits. *Icarus*, **61**, 461–89.
Gaffey, M. J. (1976) Spectral reflectance characteristics of the meteorite classes. *J. Geophys. Res.*, **81**, 905–20.
Gaffey, M. J. & McCord, T. B. (1978) Asteroid surface materials: Mineralogical characterization from reflectance spectroscopy. *Space Sci. Rev.*, **21**, 555–628.
Gaffney, E. S. (1972) Crystal field effects in mantle minerals. *Phys. Earth Planet. Interiors*, **6**, 385–90.
Gaffney, E. S. (1973) Spectra of tetrahedral Fe^{2+} in $MgAl_2O_4$. *Phys. Rev.*, **B8**, 3384–6.
Gaffney, E. S. & Ahrens, T. J. (1973) Optical absorption spectrum of ruby and periclase at high shock pressures. *J. Geophys. Res.*, **78**, 5942–53.
Gaffney, E. S. and Anderson, D. L. (1973) Effect of low-spin Fe^{2+} on the composition of the lower mantle. *J. Geophys. Res.*, **78**, 7005–14.
Galoisy, L. & Calas, G. (1991) Spectroscopic evidence for five-coordinated Ni in $CaNiSi_2O_6$ glass. *Amer. Mineral.*, **76**, 1777–81.
Ganguli, D. (1977) Crystal chemical aspects of olivine structures. *Neues Yahrb. Mineral. Abh.*, **130**, 303–18.
Gattow, V. & Zemann, J. (1958) Neubestimung der Kristallstruktur von Azurit. *Acta Cryst.*, **11**, 866–72.
Geller, S. (1971) Structures of α-Mn_2O_3, $(Mn_{0.983}Fe_{0.017})_2O_3$ and $(Mn_{0.37}Fe_{0.63})_2O_3$, and relation to magnetic ordering. *Acta Cryst.*, **B27**, 821–8.
George, P. & McClure, D. S. (1959) The effect of inner orbital splitting on the thermodynamic properties of transition metal compounds and coordination complexes. *Progr. Inorg. Chem.*, **1**, 382–463.
Gerloch, M. & Slade, R. C. (1973) *Ligand-Field Parameters*. (Cambridge Univ. Press, Cambridge), 235 pp.
Geschwind, S., Kisliuk, P. Klein,, M. P., Remeika, J. R. & Wood, D. L. (1962) Sharp-line fluorescence, electron paramagnetic resonance, and thermoluminescence of Mn^{4+} in α-Al_2O_3. *Phys. Rev.*, **136**, 1684–6.
Ghazi-Bayat, B., Amthauer, G. & Hellner, E. (1989) Synthesis and characterization of

Mn–bearing ilvaite $CaFe^{2+}_{2-x}Mn_xFe^{3+}[Si_2O_7|O|(OH)]$. *Mineral. Petrol.*, **40**, 101–9 (1987).

Ghazi-Bayat, B., Amthauer, G., Schürmann, K & Hellner, E. (1987) Synthesis and characterization of the mixed valent iron silicate ilvaite, $CaFe_3[Si_2O_7|O|(OH)]$. *Mineral. Petrol.*, **37**, 97–108.

Ghera, A. & Lucchesi, S. (1987) An unusual vanadium-beryl from Kenya. *Neues Jahrb. Mineral. Mh.*, pp. 263–74.

Ghera, A., Graziani, G. & Lucchesi, S. (1986) Uneven distribution of blue color in kyanite. *Neues Jahrb. Mineral. Abh.*, **155**, 109–27.

Ghose, S. (1961) The crystal structure of a cummingtonite. *Acta. Cryst.*, **14**, 622–7.

Ghose, S. (1965) Mg^{2+}–Fe^{2+} order in an orthopyroxene, $Mg_{0.93}Fe_{1.07}Si_2O_6$. *Zeit. Krist.*, **122**, 81–99.

Ghose, S. & Ganguly, J. (1982) Mg–Fe order–disorder in ferromagnesian silicates. In *Advances in Physical Geochemistry, Vol. 2.* (S. K. Saxena, ed.; Springer–Verlag, New York), pp. 1–57.

Ghose, S. & Wan, C. (1973) Luna 20 pyroxenes: evidence for a complex thermal history. *Proc. 4th Lunar Sci. Conf., Geochim. Cosmochim. Acta, Suppl. 4* (Pergamon Press, New York), pp. 901–7.

Ghose, S. & Wan, C. (1974) Strong site preference of Co^{2+} in olivine, $Co_{1.10}Mg_{0.90}SiO_4$. *Contrib. Mineral. Petrol.*, **47**, 131–40.

Ghose, S., Hewat, A. W. & Marenzio, M. (1984) A neutron powder diffraction study of the crystal and magnetic structure of ilvaite from 305 K to 5 K – a mixed valence iron silicate with an electronic transition. *Phys. Chem. Minerals*, **11**, 67–74.

Ghose, S., Kersten, M., Langer, K., Rossi, G. & Ungretti, L. (1986) Crystal field spectra and Jahn–Teller effect of Mn^{3+} in clinopyroxene and clinoamphiboles from India. *Phys. Chem. Minerals*, **13**, 291–305.

Ghose, S., Wan, C., Okamura, F., Ohashi, H. & Weidner, J. R. (1975) Site preferences and crystal chemistry of transition metal ions in pyroxenes and olivines. *Acta Cryst.*, **A31**, S76.

Ghosh, D., Kunda, T., Dasgupta, S. & Ghose, S. (1987) Electron delocalization and magnetic behavior in a single crystal of ilvaite, a mixed valence iron silicate. *Phys. Chem. Minerals*, **14**, 151–5.

Gibbons, R. V., Ahrens, T. J. & Rossman, G. R. (1974) A spectroscopic interpretation of the shock-produced color change in rhodonite ($MnSiO_3$): The shock-induced reduction of Mn(II) to Mn(III). *Amer. Mineral.*, **59**, 177–82 [see also *Amer. Mineral.*, **60**, 942-3 (1975); and Faye (1975)].

Giovanoli, R., Bürki, P., Giuffredi, M. & Stumm, W. (1975) Layer structure manganese oxide hydroxides. IV: The buserite group: structure stabilization by transition elements. *Chimia,* **29,** 110–13.

Glasby, G. P. (1975) Limitations of crystal field theory applied to sedimentary systems. *Geoderma,* **13,** 363–7 [see McKenzie (1970)].

Glassley, W. E. & Piper, D. Z. (1978) Cobalt and scandium partitioning versus iron content for crystalline phase in ultramafic nodules. *Earth Planet. Sci. Lett.*, **39**, 173-8.

Glasstone, S., Laidler, K. J. & Eyring, H. (1941) *The Theory of Rate Processes* (McGraw-Hill Book Co., New York), 611 pp.

Glidewell, C. (1976) Cation distribution in spinels: Lattice energy *versus* crystal field stabilization energy. *Inorg. Chim. Acta*, **19**, L45–7.

Goldman, D. S. & Berg, J. I. (1980) Spectral study of ferrous iron in Ca–Al–borosilicate glass at room temperature and melt temperatures. *J. Non-Cryst. Solids*, **38/39**, 183–8.

Goldman, D. S. & Rossman, G. R. (1976) Identification of a mid–infrared electronic absorption band of Fe^{2+} in the distorted M2 site of orthopyroxene $(MgFe)SiO_3$. *Chem. Phys. Lett.*, **41**, 474–5.

Goldman, D. S. & Rossman, G. R. (1977a) The spectra of iron in orthopyroxene revisited: The splitting of the ground state. *Amer. Mineral.*, **62**, 151–7.

Goldman, D. S. & Rossman, G. R. (1977b) The identification of Fe^{2+} in the M(4) site of calcic amphiboles. *Amer. Mineral.*, **62**, 205–16 [see also *Amer. Mineral.*, **67**, 340–2 (1982); and Aldridge *et al.* (1982)].

Goldman, D. S. & Rossman, G. R. (1978) The site distribution of iron and anomalous biaxiality in osumilite. *Amer. Mineral.*, **63**, 490–8.

Goldman, D. S. & Rossman, G. R. (1979) Determination of quantitative cation distribution in orthopyroxenes from electronic absorption spectra. *Phys. Chem. Minerals*, **4**, 43–53.

Goldman, D. S. & Rossman, G. R. (1982) The identification of Fe^{2+} in the M(4) site of calcic amphiboles: Reply. *Amer. Mineral.*, **67**, 340-2 [see also *Amer. Mineral.*, **62**, 205-16 (1977); and Aldridge *et al.* (1982)].

Goldman, D. S., Rossman, G. R. & Dollase, W. A. (1977) Channel constituents in cordierite. *Amer. Mineral.*, **62**, 1144–57.

Goldman, D. S., Rossman, G. R. & Parkin, K. M. (1978) Channel constituents in beryl. *Phys. Chem. Minerals*, **3**, 225–35.

Goldschmidt, V. M. (1937) The principles of distribution of chemical elements in minerals and rocks. *J. Chem. Soc.*, pp. 655–72.

Goldschmidt, V. M. (1954) *Geochemistry.* (Oxford Univ. Press, London), 730 pp.

Gonschorek, W. (1986) Electron density and polarized absorption spectra of fayalite. *Phys. Chem. Minerals*, **13**, 337–9.

Goodgame, M. & Cotton, F. A. (1961) Magnetic investigation of spin-free cobaltous complexes: IV. Magnetic properties and spectrum of cobalt(II) orthosilicate. *J. Phys. Chem. Solids,* **65**, 791–2.

Gooding, J. L. (1978) Chemical weathering on Mars: thermodynamic stabilities of primary minerals (and their alteration products) from mafic igneous rocks. *Icarus*, **33**, 483–513.

Gooding, J. L. & Muenow, D. W. (1986) Martian volatiles in shergottite EETA 79001: New evidence from oxidized sulfur and sulfur-rich aluminosilicates. *Geochim. Cosmochim. Acta,* **50,** 1049–59.

Gooding, J. L., Wentworth, S. J. & Zolensky, M. E. (1988) Calcium carbonate and sulfate of possible extraterrestrial origin in the EETA 79001 meteorite. *Geochim. Cosmochim. Acta,* **52,** 909–15.

Goto, T., Ahrens, T. J. & Rossman, G. R. (1979) Absorption spectra of Cr^{3+} in Al_2O_3 under pressure. *Phys. Chem. Minerals*, **4**, 253–63.

Goto, T., Ahrens, T. J., Rossman, G. R. & Syono, Y. (1980) Absorption spectrum of shock-compressed Fe^{2+}-bearing MgO and the radiative conductivity of the Lower Mantle. *Phys. Earth Planet. Interiors*, **22**, 277–88.

Goto, T., Sato, J. & Syono, Y. (1982) Shock-induced spin-pairing transition in Fe_2O_3 due to the pressure effect on the crystal field. In *High-Pressure Research in Geophysics.* (S. Akimoto & M. H. Manghani, eds; D. Reidel Publ. Co., Dortrecht), pp. 595–609.

Gray, H. B. (1964) Molecular orbital theory for transition metal complexes. *J. Chem. Educ.*, **41**, 1–12.

Greaves, C. (1983) A powder neutron diffraction investigation of vacancy and covalencein γ-ferric oxide. *Solid State Comm.*, **30**, 257-63.

Greskovich, C. & Stubican, V. S. (1966) Divalent chromium in magnesium chromium spinels. *J. Phys. Chem. Solids*, **27**, 1379–84.

Grover, J. E. & Orville, P. M. (1969) The partitioning of cations between coexisting single- and multi-site phases with application to the assemblages: orthopyroxene–clinopyroxene, and orthopyroxene–olivine. *Geochim. Cosmochim. Acta*, **33**, 205–26.

Grum-Grzhimailo, S. V. (1940) The colouring of minerals produced by chromium. *Trav. Lab. Crist., Acad. Sci.*, **2**, 73–86.

Grum-Grzhimailo, S. V. (1945) The effect of structure of crystals on the colour produced by Mn, Cr, Fe, Ni, Co and Cu. *Acta Phys.*, **20**, 933–46.

Grum-Grzhimailo, S. V. (1947) 'Alexandrite colour' of crystals. *Mem. Soc. Russe Mineral.*, **75**, 253–6.

Grum-Grzhimailo, S. V. (1953) Nature and colour of rose and yellow topaz. *Zap. Vses. Mineral. Obshch.*, SSSR, **82**, 142–6.

Grum-Grzhimailo, S. V. (1954) Curves of spectral absorption as one of the diagnostic characterics of garnets. *Mineral. Sbornik. Lvov Geol. Obshch.*, SSSR, **8**, 281–94.

Grum-Grzhimailo, S. V. (1956) Tourmaline, their examination in polarized ultraviolet light. *Trudy Inst. Krist., Akad. Nauk. SSSR*, **12**, 79–84.

Grum-Grzhimailo, S. V. (1958) The colour of idiochromatic minerals. *Zap. Vses. Mineral. Obshch.*, SSSR, **87**, 129–50.

Grum-Grzhimailo, S. V. (1960) The colour of diamond accessory minerals. *Mat. Vses. Nauk. Issled., Geol. Inst.*, **40**, 57–64.

Grum-Grzhimailo, S. V. (1961) Light absorption surfaces of blue topaz. *Kristallogr.*, **6**, 67–71. [*Transl.*: *Soviet Phys. Cryst.*, **6**, 54–7 (1961).]

Grum-Grzhimailo, S. V. , Anikina, L. I., Belova, E. N. & Tolstikhina, K. I. (1955) Curves of spectral absorption and other physical constants of natural micas. *Mineral. Sbornik. Lvov Geol. Obshch.*, **9**, 90–119.

Grum-Grzhimailo, S. V., Brilliantov, N. A., Sviridova, R. K., Sukhanova, O. N. & Kapitonova, M. M. (1962) Absorption spectra of iron–coloured beryls from 290 to 1.7 K. *Opt. Spektr.*, **13**, 133–4. [*Transl.*: *Optics & Spectr.*, **13**, 72 (1962).]

Grum-Grzhimailo, S. V., Brilliantov, N. A., Sviridov, D. T., Sviridova, R. K. & Sukhanova, O. N. (1963) Absorption spectra of crystals containing Fe^{3+} at temperatures down to 1.7 K. *Opt. Spektr.*, **14**, 228–33. [*Transl.*: *Optics & Spectr.*, **14**, 118–20 (1963).]

Grum-Grzhimailo, S. V. & Klimusheva, G. V. (1960) Temperature dependence of the broad absorption bands in the spectra of crystals with different structures which have been coloured by isotropic impurities. *Opt. Spektr.*, **8**, 342–51. [*Transl.*: *Optics & Spectr.*, **8**, 179–83 (1960).]

Grum-Grzhimailo, S. V. & Perneva, L. A. (1956) Absorption spectra of coloured beryls and topazes. *Trudy Inst. Krist., Akad. Nauk., S.S.S.R.*, **12**, 85–92.

Gurney, J. J. (1990) The diamondiferous roots of our wandering continent. *Sth Afr. J. Geol.*, **93**, 423–37.

Guthrie Jr, G. D. & Bish, D. L. (1991) Refinement of the turquoise structure and determination of the hydrogen positions. *Geol. Soc. Amer., Ann. Meet., San Diego, Abstr.*, **23**, A158.

Güttler, B., Salje, E. & Ghose, S. (1989) Polarized single crystal absorption spectroscopy of the *Pnam–P2₁/a* transition of ilvaite $Ca(Fe^{2+}Fe^{3+})Fe^{2+}Si_2O_8(OH)$ as measured between 300 K and 450 K. *Phys. Chem. Minerals*, **16**, 606–13.

Güven, N. (1988) Smectites. In *Hydrous Phyllosilicates*. (S. W. Bailey, ed.; Mineral. Soc. Amer. Publ.), **19**, 497–559.

Hakli, A. (1963) Distribution of nickel between silicate and sulphide phases in some basic intrusions in Finland. *Bull. Comm. Geol. Finlande*, **209**.

Hakli, T. & Wright, T. L. (1967) The fractionation of nickel between olivine and augite as a geothermometer. *Geochim. Cosmochim. Acta,*, **31**, 877–84.

Halbach, P., Friedrich, G. & von Stackelberg, U. (eds) (1988) *The Manganese Nodule Belt of the Pacific Ocean*. (Ferdinand Enke Verlag, Stuttgart), 254 pp.

Hålenius, U. (1978) A spectroscopic investigation of manganese andalusite. *Canad. Mineral.*, **16**, 567–75.

Hålenius, U. (1979) State and location of iron in sillimanite. *Neues Jahrb. Mineral. Mh.*, pp. 164–9.

Hålenius, U. & Langer, K. (1981) Microscopic–photometric methods for non-destructive Fe^{2+}–Fe^{3+} determination in chloritoids $(Fe^{2+},Mn^{2+},Mg)_2(Al,Fe^{3+})_4Si_2O_{10}(OH)_4$. *Lithos*, **13**, 291–305.

Hålenius, U., Annersten, H. & Langer, K. (1981) Spectroscopic studies on natural chloritoids. *Phys. Chem. Minerals*, **7**, 117–23.

Hargraves, R. B., Collinson, D. W., Arvidson, R. E. & Cates, P. M. (1979) Viking magnetic properties experiment: extended mission results. *J. Geophys. Res.*, **84**, 8379–84.

Harley, S. L. (1984) An experimental study of the partitioning of Fe and Mg between garnet and orthopyroxene. *Contrib. Mineral. Petrol.*, 86, 359–73.

Hart, S. R. & Davis, K. E. (1978) Nickel partitioning between olivine and silicate melt. *Earth Planet. Sci. Lett.*, **40**, 203–19.

Hawkins, D. B. & Roy, R. (1963) Distribution of trace elements between clays and zeolites formed by hydrothermal alteration of synthetic basalts. *Geochim. Cosmochim. Acta*, **27**, 785–95.

Hawthorne, F. C. (1981a) Crystal chemistry of the amphiboles. In *Amphiboles and Other Hydrous Pyriboles – Mineralogy*. (D. R. Veblen, ed.; Mineral. Soc. Amer.Publ.), *Rev. Mineral.*, **9A**, 1–102.

Hawthorne, F. C. (1981b) Amphibole spectroscopy. In *Amphiboles and Other Hydrous Pyriboles – Mineralogy*. (D. R. Veblen, ed.; Mineral. Soc. Amer. Publ.), *Rev. Mineral.*, **9A**, 103–39.

Hawthorne, F. C. (1983) The crystal chemistry of the amphiboles. *Canad. Mineral.*, **21**, 173–480.

Hawthorne, F. C. (1988) Mössbauer spectroscopy. In *Spectroscopic Methods in Mineralogy and Geology*. (F. C. Hawthorne, ed.; Mineral. Soc. Amer.), *Rev. Mineral.*, **18**, 255–340.

Hawthorne, F. C. & Ito, J. (1977) Synthesis and crystal structure refinement of transition metal orthopyroxenes. I. Orthoenstatite and (Mg,Mn,Co) orthopyroxene. *Canad. Mineral.*, **15**, 321–38.

Hazen, R. M. (1976a) Effects of temperature and pressure on the cell dimension and X-ray temperature factors of periclase. *Amer. Mineral.*, **61**, 266–71.

Hazen, R. M. (1976b) Effects of temperature and pressure on the crystal structure of forsterite. *Amer. Mineral.*, **61**, 1280–93.

Hazen, R. M. (1977) Effect of temperature and pressure on the crystal structure of ferromagnesian olivine. *Amer. Mineral.*, **62**, 286–95.

Hazen, R. M. & Burnham, C. W. (1974) The crystal structure of gillespite I and II: A structure determination at high pressure. *Amer. Mineral.*, **59**, 1166–76. [Correction and addendum: *Amer. Mineral.*, **60**, 937–38 (1975).]

Hazen, R. M. & Finger, L. W. (1977) Crystal structure and compositional variation of Angra dos Reis fassaite. *Earth Planet. Sci. Lett.*, **35**, 357–62.

Hazen, R. M. & Finger, L. W. (1978) Crystal structures and compressibilities of pyrope and grossular to 60 kbar. *Amer. Mineral.*, **63**, 297–303.

Hazen, R. M. & Finger, L. W. (1979) Bulk modulus–volume relationships for cation–anion polyhedra. *J. Geophys. Res.*, **B84**, 6723–8.

Hazen, R. M. & Finger, L. W. (1982) *Comparative Crystal Chemistry*. (Wiley–Interscience, New York), 231 pp.

Hazen, R. M. & Finger, L. W. (1982) A program to calculate polyhedral volumes and polyhedral distortion parameters from a set of atomic coordinates and a unit cell. In *Comparative Crystal Chemistry.* (Wiley–Interscience, New York), pp. 103–11.

Hazen, R. M. & Finger, L. W. (1983) High-pressure and high-temperature crystallographic study of the gillespite I–II phase transition. *Amer. Mineral.*, **68**, 595–603.

Hazen R. M., Bell, P. M. & Mao, H.-K. (1977a) Polarized absorption spectra of Angra dos Reis fassaite to 52 kbar. *Ann. Rept. Geophys., Yearb.* **76**, 515–16.

Hazen, R. M., Bell, P. M. & Mao, H.-K. (1978) Effects of compositional variation on absorption spectra of lunar pyroxenes. *Proc. 9th Lunar Planet. Sci. Conf., Geochim. Cosmochim. Acta, Suppl. 9* (Pergamon Press, New York), pp. 2919–34.

Hazen, R. M., Mao, H.-K. & Bell, P. M. (1977b) Effects of compositional variation on absorption spectra of lunar olivines. *Proc. 8th Lunar Sci. Conf., Geochim. Cosmochim. Acta, Suppl. 8* (Pergamon Press, New York), pp. 1081–90.

Heide, H. G. & Boll-Dornberger, K. (1955) Die Struktur des Dioptase, $Cu_6Si_6O_{18}.6H_2O$. *Acta Cryst.*, **8**, 425–30.

Henderson, B. & Imbusch, G. F. (1989) *Optical Spectroscopy of Inorganic Solids.* (Oxford Univ. Press), 645 pp.

Henderson, P. (1979) Irregularities in patterns of element partition. *Mineral. Mag.*, **43**, 399–404.

Henderson, P. (1982) *Inorganic Geochemistry.* (Pergamon Press, New York), 353 pp.

Henderson, P. & Dale, I. M. (1969) The partitioning of selected transition element ions between olivine and groundmass of oceanic basalts. *Chem. Geol.*, **5**, 267–74.

Hendricks, R. C. & Dahl, P. S. (1987) Trace-element partitioning between coexisting metamorphic garnets and clinopyroxenes: crystal field, compositional, and thermal controls. *Geol. Soc. Amer., Ann. Meet., Abstr.*, **19**, 700.

Higgins, J. B., Ribbe, P. H. & Nakajima, Y. (1982) An ordering model for the commensurate antiphase structure of yoderite. *Amer. Mineral.*, **67**, 76–84.

Higuchi, H. & Nagasawa, H. (1969) Partition of trace elements between rock-forming minerals and the host volcanic rocks. *Earth Planet. Sci. Lett.*, **7**, 281–7.

Hill, R. J., Craig, J. R. & Gibbs, G. V. (1979) Systematics of the spinel structure. *Phys. Chem. Minerals*, **4**, 317–39.

Hirsch, L. M. & Shankland, T. J. (1991) Point defects in $(Mg,Fe)SiO_3$ perovskite. *Geophys. Res. Lett.*, **18**, 1305–8.

Hitchman, M. A. (1985) Chemical information from the polarized crystal spectra of transition metal complexes. In *Transition Metal Chemistry, Vol. 9.* (G. A. Melson & B. N. Figgis, eds; Marcel Dekker, Inc., New York), pp. 1–223.

Hitchman, M. A. & Waite, T. D. (1976) Electronic spectrum of the $Cu(H_2O)_6^{2+}$ ion. *Inorg. Chem.*, **15**, 2150–4.

Hofmeister, A. M. & Rossman, G. R. (1983) Color in Feldspars. In *Feldspars, 2nd edn.* (P. H. Ribbe, ed.; Min. Soc. Amer. Publ.) *Rev. Mineral.*, **2**, 271–80.

Hofmeister, A. M. & Rossman, G. R. (1985) A spectroscopic study of irradiation coloring of amazonite: structurally hydrous, Pb-bearing feldspar. *Amer. Mineral.*, **70**, 794–804.

Hofmeister, A. M. & Rossman, G. R. (1986) A spectroscopic study of blue radiation coloring in plagioclase. *Amer. Mineral.*, **71**, 95–8.

Hogarth, D. D., Brown, F. F. & Pritchard, A. M. (1970) Biabsorption, Mössbauer spectra and chemical investigation of five phlogopite samples from Quebec. *Canad. Mineral.*, **10**, 710–22.

Holmes, O. G. & McClure, D. S. (1957) Optical spectra of hydrated ions of the transition metals. *J. Chem. Phys.*, **26**, 1686–94.

Howie, R. A. & Woolley, A. R. (1968) The role of titanium and the effect of TiO_2 on the cell size, refractive index and specific gravity in the andradite–melanite–schorlomite series. *Mineral. Mag.*, **36**, 775–90.

Hu, N., Langer, K. & Boström, K. (1990) Polarized electronic spectra and Ni–Mg partitioning in olivines $(Mg_{1-x}Ni_x)_2[SiO_4]$. *Eur. J. Mineral.*, **2**, 29–41.

Huggins, F. E. (1973) Cation order in olivines: Evidence from vibrational spectra. *Chem. Geol.*, **11**, 99–109.

Huggins, F. E. (1975) The 3*d* levels of ferrous ions in silicate garnets. *Amer. Mineral.*, **60**, 316–9.

Huggins, F. E. (1976) Mössbauer studies of iron minerals at 25 °C and under pressures of up to 200 kilobars. In *The Physics and Chemistry of Minerals and Rocks.* (R. G. J. Strens, ed.; J. Wiley & Sons, New York), pp. 613–40.

Huggins, F. E., Mao, H.-K. & Virgo, D. (1976) Gillespite at high pressure: results of a detailed Mössbauer study. *Ann. Rept. Geophys. Lab., Yearb.* **75**, 756–8.

Huggins, F. E., Virgo, D. & Huckenholtz, H. G. (1977) Titanium-containing silicate garnets. II. The crystal chemistry of melanites and schorlomites. *Amer. Mineral.*, **62**, 646–65.

Huheey, J. E. (1983) *Inorganic Chemistry: Principles of Structure and Reactivity, 3rd edn.* (Harper & Row, New York), 936 pp.

Hunt, G. R. & Ashley, R. P. (1979) Spectra of altered rocks in the visible and near infrared. *Econ. Geol.*, **74**, 1613–29.

Hunt, G. R. & Salisbury, J. W. (1971) Visible and near-infrared spectra of minerals and rocks I. Carbonates. *Modern. Geol.*, **2**, 23–30.

Hunter, J. C. (1981) Preparation of a new crystal form of manganese dioxide: λ-MnO_2. *J. Solid State Chem.*, **39**, 142–7.

Hurlbut Jr, C. S. (1969) Gem zoisite from Tanzania. *Amer. Mineral.*, **54**, 702–9.

Hush, N. S. (1958) Crystal field stabilization and site deformation in crystals and complexes containing transition ions. *Disc. Faraday Soc.*, **26**, 145–56.

Hush, N. S. & Pryce, M. H. L. (1957) Radii of transition ions in crystal fields. *J. Chem. Phys.*, **26**, 143–4.

Hush, N. S. & Pryce, M. H. L. (1958) Influence of the crystal field potential on interatomic separation in salts of divalent iron-group ions. *J. Chem. Phys.*, **28**, 424–9.

Hutton, D. R. (1971) Paramagnetic resonance of VO^{2+}, Cr^{3+} and Fe^{3+} in zoisite. *J. Phys. C*, **4**, 1251–7.

Ihinger, P. D. & Stolper, E. (1986) The color of meteoritic hibonite: an indicator of oxygen fugacity. *Earth Planet. Sci. Lett.*, **78**, 67–79.

Ikeda, K. & Yagi, K. (1977) Experimental study on the phase equilibria in the join $CaMgSi_2O_6$–$CaCrCrSiO_6$ with special reference to the blue diopside. *Contrib. Mineral. Petrol.*, **61**, 91–106 [see also *Contrib. Mineral. Petrol.*, **66**, 343–4 (1978); and Schreiber (1978)].

Ikeda, K. & Yagi, K. (1982) Crystal-field spectra for blue and green diopsides synthesized in the join $CaMgSi_2O_6$–$CaCrAlSiO_6$. *Contrib. Mineral. Petrol.*, **81**, 113–18.

Ikeda, K., Schneider, H., Akasaka, M. & Rager, H. (1992) Crystal-field spectroscopic study of Cr–doped mullite. *Amer. Mineral.*, **77**, 251-7.

Irvine, T. N. (1974) Chromitite layers in stratiform intrusions. *Ann. Rept. Geophys. Lab., Yearb.* **73**, 300–10.

Irvine, T. N. (1975) Crystallization sequences in the Muskox intrusion and other layered intrusions. II. Origin of chromitite layers and similar deposits of other magmatic ores. *Geochim. Cosmochim. Acta*, **39**, 991-1020.

Irvine, T. N. & Kushiro, I. (1976) Partitioning of Ni and Mg between olivine and silicate liquids. *Ann. Rept. Geophys. Lab., Yearb.* **75**, 668–75.

Irving, A. J. (1978) A review of experimental studies of crystal/liquid trace element partitioning. *Geochim. Cosmochim. Acta*, **42**, 743-70.

Ito, E. & Takahashi, T. (1987) Ultrahigh-pressure phase transformations and the constitution of the deep mantle. In *High-Pressure Research in Mineral Physics*. (M. H. Manghnani & Y. Syono, eds; AGU, Washingon, D.C.), *Geophys. Mono.*, **39**, 221–9.

Jackson, I. & Ringwood, A. E. (1981) High pressure polymorphism of the iron oxides. *Geophys. J. Royal Astron. Soc.*, 61, 767–83.

Jackson, W. E., Knittle, E., Brown Jr, G. E. & Jeanloz, R. (1987) Partitioning of Fe within high-pressure silicate perovskite: Evidence for unusual geochemistry in the lower mantle. *Geophys. Res. Lett.*, **14**, 224–6.

Jahn, H. A. & Teller, E. (1937) Stability of polyatomic molecules in degenerate electronic states. I. Orbital degeneracy. *Proc. Royal Phys. Soc. (London)*, **A161**, 220–35.

Jarosch, D. (1987) Crystal structure refinement and reflectance measurements of hausmannite, Mn_3O_4. *Mineral. Petrol.*, **37**, 15–23.

Jeanloz, R. & Ahrens, T. J. (1980) Equation of state of FeO and CaO. *Geophys. J. Royal Astron. Soc.*, **62**, 505–28.

Jeanloz, R. & Knittle, E. (1989) Density and Composition of the lower mantle. *Phil. Trans, Royal Soc. Lond.*, **A328**, 377–89.

Jeanloz, R. & Thompson, A. B. (1983) Phase transitions and mantle discontinuities. *Rev. Geophys. Space Phys.*, **21**, 51-74.

Jensen, B. B. (1973) Patterns of trace element partitioning. *Geochim. Cosmochim. Acta*, **37**, 2227–42.

Johnson, K. H. & Smith Jr, F. C. (1972) Chemical bonding of a molecular transition metal ion in a crystalline environment. *Phys. Rev. B*, **5**, 831-43.

Jones, G. D. (1967) Jahn–Teller splittings in the optical absorption spectra of divalent iron compounds. *Phys. Rev.*, **155**, 259–61.

Kai, A. T., Larsson, S. & Hålenius, U. (1980) The electronic structure and absorption spectra of MnO_6^{9-} octahedra in manganian andalusite. *Phys. Chem. Minerals*, **6**, 77–84.

Kan, X. & Coey, J. M. D. (1985) Mössbauer spectra, magnetic and electrical properties of laihunite, a mixed valence iron olivine mineral. *Amer. Mineral.*, **70**, 576–80.

Karickhoff, S. W. & Bailey, G. W. (1973) Optical absorption spectra of clay minerals. *Clays & Clay Minerals*, **21**, 59–70.

Katzin, L. I. (1961) Energy value of the octahedral–tetrahedral change. *J. Chem. Phys.*, **35**, 467–72.

Keester, K. L. & White, W. B. (1968) Crystal field spectra and chemical bonding in manganese minerals. In *Intern. Mineral. Assoc., Pap. Proc. 5th General Meet., Cambridge, 1966*. (P. Gay, A. F. Seager, H. F. W. Taylor & J. Zussman, eds; Mineral. Soc., London), pp. 22–35.

Keppler, H. (1992) Crystal field spectra and geochemistry of transition metal ions in silicate melts and glasses. *Amer. Mineral.*, **77**, 62–75.

Keppler, H. & Rubie, D. C. (1992) Coordination change of Co and Ni in silicate melts at high pressure – evidence from crystal field spectra. *EOS, Trans. Amer. Geophys. Union*, **73**, 598.

Kerrick, D. M. (1990) *The Al_2SiO_5 polymorphs*. (Mineral. Soc. Amer. Publ.), *Rev. Mineral.*, **22**, 406 pp.

Kersten, M., Langer, K., Almen, H. & Tillmanns, E. (1987) Kristallchemie von Piemontiten: Strukturverfeinerungen und polarisierte Einkristallspektren. *Zeit. Krist.*, **178**, 121

Khomenko, V. M. & Platonov, A. N. (1985) Electronic absorption spectra of Cr^{3+} ions in natural clinopyroxenes. *Phys. Chem. Minerals*, **11**, 261–5.
King, T. V. V. & Ridley, W. I. (1987) Relation of the spectroscopic reflectance of olivine to mineral chemistry and some remote sensing implications. *J. Geophys. Res.*, **92**, 11457–69.
Kinzler, R. J., Grove, T. L. & Recca, S. I. (1990) An experimental study on the effect of temperature and melt composition on the partitioning of nickel between olivine and silicate melt. *Geochim. Cosmochim. Acta*, **54**, 1255–65.
Kleim, W. & Lehmann, G. (1979) A reassignment of the optical absorption bands in biotites. *Phys. Chem. Minerals*, **4**, 65–75.
Klose, D. B., Wood, J. A. & Hashimoto, A. (1992) Mineral equilibria and the high radar reflectivity of Venus mountain tops. *J. Geophys. Res.*, **97**, 16353-69.
Kohn, S. C., Charnock, J. M., Henderson, C. M. B. & Greaves, G. N. (1990) The structural environment of trace elements in dry and hydrous silicate glasses: A manganese and strontium K-edge X-ray absorption spectroscopic study. *Contrib. Mineral. Petrol.*, **105**, 359–68.
Kortum, G. (1969) *Reflectance Spectroscopy : Principles, Methods, Applications.* (Springer–Verlag, New York), 366 pp.
Krebs, J. J. & Maisch, W. G. (1971) Exchange effects in the optical absorption spectra of Fe^{3+} in Al_2O_3. *Phys. Rev.*, **B4**, 757–69.
Kudoh, Y., Ito, E. & Takeda, H. (1987) Effect of pressure on the crystal structure of the perovskite-ype $MgSiO_3$. *Phys. Chem. Minerals*, **14**, 350–4.
Kudoh, Y., Prewitt, C. T., Finger, L. W., Darovskikh, A. & Ito, E. (1990) Effect of iron on the crystal structure of $(Mg,Fe)SiO_3$ perovskite. *Geophys. Res. Lett.*, **17**, 1481–4.
Lakshman, S. V. J. & Reddy, B. J. (1973a) Optical absorption spectrum of Mn^{2+} in rhodonite. *Physica*, **66**, 601–10.
Lachsman, S. V. J. & Reddy, B. J. (1973b) Optical absorption spectrum of Ni^{2+} in garnierite. *Proc. Indian Acad. Sci.*, **77A**, 269–79 [see Faye (1974)].
Lakshman, S. V. J. & Reddy, B. J. (1973c) Optical absorption spectra of Cu^{2+} in chalcanthite and malachite. *Canad. Mineral.*, **12**, 207–10.
Langer, K. (1988) UV to NIR spectra of silicate minerals obtained by microscope spectrometry and their use in mineral thermodynamics and kinetics. In *Physical Properties and Thermodynamic Behaviour of Minerals*. (E. K. H. Salje, ed.; D. Reidel Publ. Co.), pp. 639–85.
Langer, K. (1990) High pressure spectroscopy. In *Absorption Spectroscopy in Mineralogy*. (A. Mottana and F. Burragato, eds; Elsevier Science Publ., Amsterdam), pp. 227–84.
Langer, K. & Abu-Eid, R. M. (1977) Measurements of the polarized absorption spectra of synthetic transition metal-bearing silicate microcrystals in the spectral range 44,000–4,000 cm^{-1}. *Phys. Chem. Minerals*, **1**, 273–99.
Langer, K. & Frentrup, K. R. (1979) Automated microscope-absorption-spectrophotometry of rock-forming minerals in the range 40,000–5,000 cm^{-1} (250–2000 nm). *J. Microscopy*, **116**, 311–20.
Langer, K. & Lattard, D. (1984) Mn^{3+} in garnets. II: Optical absorption spectra of blythite-bearing, synthetic calderites, $Mn_3^{2+[8]}(Fe^{3+}_{1-x}Mn^{3+}_{x})_2^{[6]}[SiO_4]_3$. *Neues Jahrb. Mineral. Abh.*, **149**, 129–41.
Langer, K., Abu-Eid, R. M. & Anastasiou, P. (1976) Absorptionspektren synthetischer Piemontite in den Bereichen 43,000–11,000 cm^{-1} (232.6–909.1 nm) and 4,000–250 cm^{-1} (2.5–40 μm). *Zeit. Krist.*, **144**, 434–6.
Langer, L., Smith, G. & Hålenius, U. (1982) Reassignment of the absorption spectra of purple yoderite. *Phys. Chem. Minerals*, **8**, 143–5.

Langer, K., Hålenius, U. & Fransolet, A.-M. (1984) Blue andalusite from Ottré, Venn–Stavelot Massif, Belgium: a new example of intervalence charge-transfer in the aluminium silicate polymorphs. *Bull. Minéral.*, **107**, 587–96.

Laughon, R. R. (1971) The crystal structure of kinnoite. *Amer. Mineral.*, **56**, 193–200.

Leeman, W. P. & Lindstrom, D. J. (1978) Partitioning of Ni^{2+} between basaltic and synthetic melts and olivine – an experimental study. *Geochim. Cosmochim. Acta*, **42**, 801–16.

Lehmann, G. (1975) On the color centers of iron in amethyst and synthetic quartz: a discussion. *Amer. Mineral.*, **60**, 335–7.

Lehmann, G. & Bambauer, H. U. (1973) Quartz crystals and their colours. *Angewandte Chem. Intern.*, **12**, 283–91.

Lehmann, G. & Harder, H. (1970) Optical spectra of di- and trivalent iron in corundum. *Amer. Mineral.*, **55**, 98–105 [see Faye (1971)].

Lehmann, G. & Harder, H. (1971) On the optical spectra of di- and trivalent iron in corundum: A reply. *Amer. Mineral.*, **56**, 348–50.

Lenglet, M., Guillamet, B., D'Huysser, A., Durr, J. & Jørgensen, C. K. (1986) Fe,Ni coordination and oxidation states in chromites and cobaltites by XPS and XANES. *J. de Phys.*, **47** (C8), 765–9.

Lepicard, G. & Protas, J. (1966) Étude structurale de l'oxyde double de manganese et de calcium orthorhombique $CaMn_2O_4$ (marokite). *Bull. Soc. Franc. Min. Crist.*, **89**, 318–24.

Lever, A. B. P. (1984) *Inorganic Electronic Spectroscopy, 2nd edn.* (Elsevier, Amsterdam). 863 pp.

Levien, L. & Prewitt, C. T. (1981) High-pressure structural study of diopside. *Amer. Mineral.*, **66**, 315–23.

Lewis, G. N. & Randall, M (1961) *Thermodynamics., 2nd edn.* (Revised by K. S. Pitzer & L. Brewer; McGraw-Hill Book Co., New York), 723 pp.

Li, X. & Jeanloz, R. (1990) Laboratory studies of the electrical conductivity of silicate perovskites at high pressures and temperatures. *J. Geophys. Res.*, **95**, 5067–78.

Li, X. & Jeanloz, R. (1991a) Effect of iron content on the electrical conductivity of perovskite and magnesiowüstite assemblages at lower mantle conditions. *J. Geophys. Res.*, **96**, 6113–20.

Li, X., & Jeanloz, R. (1991b) Phases and electrical conductivity of a hydrous silicate assemblage at lower-mantle conditions. *Nature*, **350**, 332–4.

Lin, C. (1981) Optical absorption spectra of Fe^{2+} and Fe^{3+} in garnet. *Bull. Minéral.*, **104**, 218–22.

Lindstrom, D. J. & Weill, D. F. (1978) Partitioning of transition metals between diopside and coexisting silicate liquids. I. Nickel, cobalt, and manganese. *Geochim. Cosmochim. Acta*, **42**, 817–32.

Litterst, F, J. & Amthauer, G. (1984) Electron delocalization in ilvaite: a re-interpretation of its ^{57}Fe Mössbauer spectrum. *Phys. Chem. Minerals*, **10**, 250–5.

Littler, J. G. F. & Williams, R. J. P. (1965) Electrical and optical properties of crocidolite and some other iron compounds. *J. Chem. Soc.*, pp. 6368–71.

Liu, L.-G. & Bassett, W. A. (1986) Chemical and mineral composition of the Earth's interior. In *Elements, Oxides, and Silicates*. (Oxford Univ. Press), pp. 234–44.

Liu, X. & Prewitt, C. T. (1991) High-temperature diffraction study of $LnCoO_3$ perovskites: a high–order electronic phase transition. *J. Phys. Chem. Solids*, **52**, 441–8.

Live, D., Beckett, J. R., Tsay, F.-D., Grossman, L & Stolper, E. (1986) Ti^{3+} in meteoritic and synthetic hibonite: A new oxygen barometer. *Lunar Planet. Sci.*, **XVII**, 488–9.

Loeffler, B. M. & Burns, R. G. (1976) Shedding light on the color of gems and minerals. *Amer. Sci.*, **64**, 636–47.

Loeffler, B. M., Burns, R. G. & Tossell, J. A. (1975) Metal–metal charge transfer transitions: interpretation of visible–region spectra of the moon and lunar materials, *Proc. 6th Lunar Sci. Conf., Geochim. Cosmochim. Acta, Suppl. 6.* (Pergamon Press, New York), pp. 2663–76.
Loeffler, B. M., Burns, R. G., Tossell, J. A., Vaughan, D. J. & Johnson, K. H. (1974) Charge transfer in lunar materials: interpretations of ultraviolet–visible spectral properties of the Moon. *Proc. 5th Lunar Sci. Conf., Geochim. Cosmochim. Acta, Suppl. 5.* (Pergamon Press, New York), pp. 3007–16.
Low, W. (1957) Paramagnetic resonance and optical absorption spectra of Cr^{3+} in MgO. *Phys. Rev.*, **105**, 801–5.
Low, W. (1958a) Paramagnetic and optical spectra of divalent nickel in cubic crystalline fields. *Phys. Rev.*, **109**, 247–55
Low, W. (1958b) Paramagnetic and optical spectra of divalent cobalt in cubic crystalline fields. *Phys. Rev.*, **109**, 256–85.
Low, W. & Dvir, M. (1960) Paramagnetic resonance and optical spectrum of Fe in beryl. *Phys. Rev.*, **119**, 1587–91.
Lyubutin, I. S. & Dodokin, A. P. (1970) Temperature of the Mössbauer effect for Fe^{2+} in dodecahedral coordination in garnet. *Kristallogr.*, **15**, 1249–50.
MacCarthy, G. R. (1926) Colors produced by iron in minerals and the sediments. *Amer. J. Sci.*, **12**, 16–36.
Macfarlane,.R. M. (1963) Analysis of the spectrum of d^3 ions in trigonal crystal fields. *J. Chem. Phys.*, **39**, 3118–26.
Macfarlane, R. M. (1964) Optical and magnetic properties of trivalent vanadium complexes. *J. Chem. Phys.*, **40**, 373–7.
Mackay, D. J., McMeeking, R. F. & Hitchman, M. A. (1979) Magnetic anisotropy and electronic structure of gillespite, a mineral containing planar, four co-ordinate, high-spin iron(II). *J. Chem. Soc., Dalton*, pp. 299–305.
MacLean, W. H. & Shimazaki, H. (1976) The partition of Co, Ni, Cu, and Zn between sulfide and silicate liquids. *Econ. Geol.*, **71**, 1049-57.
McCammon, C. A. & Burns, R. G. (1980) The oxidation mechanism of vivianite as studied by Mössbauer spectroscopy. *Amer. Mineral.*, **65**, 361–6.
McClure, D. S. (1957) The distribution of transition metal cations in spinels. *J. Phys. Chem. Solids*, **3**, 311–17.
McClure, D. S. (1959) Electronic spectra of molecules in crystals. Part II: Spectra of ions in crystals. *Solid State Phys.*, **9**, 399–425.
McClure, D. S. (1962) Optical spectra of transition-metal ions in corundum. *J. Chem. Phys.*, **36**, 2757–79.
McClure, D. S. (1963) Comparison of the crystal field and optical spectra of Cr_2O_3 and ruby. *J. Chem. Phys.*, **38**, 2289–94.
McCord, T. M., ed. (1988) *Reflectance Spectroscopy in Planetary Science: Review and Strategy for the Future.* (NASA Planet. Geol. Geophys. Progr.), *NASA Spec. Rept. 493,* 37pp.
McCord, T. B. & Adams, J. B. (1969) Spectral reflectivity of Mars. *Science,* **163,** 1058–60.
McCord, T. B. & Clark, R. N. (1979) The Mercury soil: presence of Fe^{2+}. *J. Geophys. Res.,* **84,** 7664–8 [see Vilas (1985)].
McCord, T. B. & Adams, J. B. (1977) Use of ground-based telescopes in determining the composition of the surfaces of solar system objects. In *The Soviet–American Conference on Cosmochemistry of the Moon and Planets.* (J. H. Pomeroy & N. J. Hubbard, eds), *NASA Spec. Publ. SP–370, Part II,* pp. 893–922.
McCord, T. B., Clark, R. N., Hawke, B. R., McFadden, L. A., Owensby, P. H.,

Pieters, C. M. & Adams, J. B. (1981) Moon: Near-infrared spectral reflectance, a first good look. *J. Geophys. Res.*, **86,** 10883–92.
McCord, T. B., Clark, R. N. & Singer, R. B. (1982) Mars: Near-infrared spectral reflectance of surface regions and compositional impications. *J. Geophys. Res.*, **87**, 3021-32.
McCormick, T. C., Smyth, J. R. & Lofgren, G. E. (1987) Site occupancies of minor elements in synthetic olivines as determined by channeling-enhanced X-ray emission. *Phys. Chem. Minerals,* **14**, 368–72.
McFadden, L. A. (1989) Remote sensing and the Shergottite–Nakhlite–Chassignite meteorite parent body. *Bull. Amer. Astronom. Soc.*, **21**, 967.
McFadden, L. A., Gaffey, M. J. & McCord, T. B. (1984) Mineralogical–petrological characterization of near-Earth asteroids. *Icarus,* **59,** 25–40.
McFadden, L. A., Gaffey, M. J., Takeda, H., Jackowski, T. L. & Reed, K. L. (1982) Reflectance spectroscopy of diogenite meteorite types from Antarctica and their relationship to asteroids. *Mem. Nat. Inst. Planet. Res.*, **25,** 188–206.
McKenzie, R. M. (1970) The reaction of cobalt with manganese dioxide. *Aust. J. Soil Res.*, **8,** 97–106 [see also *Geoderma*, **13**, 363–7 (1975); and Glasby (1975)].
Manceau, A. & Calas, G. (1985) Heterogeneous distribution of nickel in hydrous silicates from New Caledonia ore deposits. *Amer. Mineral.*, **70**, 549–58.
Manceau, A. & Calas, G. (1986) Nickel-bearing clay minerals: 2. Intracrystalline distribution of nickel: An X-ray absorption study. *Clay Minerals*, **21**, 341–60.
Manceau, A. & Combes, J. M. (1988) Structure of Mn and Fe oxides and oxyhydroxides: A topological approach by EXAFS. *Phys. Chem. Minerals,* **15**, 283–5.
Manceau, A., Llorca, S. & Calas, G. (1987) Crystal chemistry of cobalt and nickel in lithiophorite and asbolane from New Caledonia. *Geochim. Cosmochim. Acta*, **51**, 105–13.
Manning, P. G. (1967a) The optical absorption spectra of some andradites and the identification of the $^6A_1 \rightarrow {}^4A_1,{}^4E(G)$ transition in octahedrally bonded Fe^{3+}. *Canad. J. Earth Sci.*, **4,** 1039–47.
Manning, P. G. (1967b) The optical absorption spectra of the garnets almandine–pyrope, pyrope, and spessartine, and some structural interpretations of mineralogical significance. *Canad. Mineral.*, **9**, 237–51.
Manning, P. G. (1968a) Optical absorption studies of chrome-bearing tourmaline, black tourmaline and buergerite. *Canad. Mineral.*, **9**, 57–70.
Manning, P. G. (1968b) Optical absorption spectra of octahedrally bonded Fe^{3+} in vesuvianite. *Canad. J. Earth Sci.*, **5**, 89–92.
Manning, P. G. (1969a) Absorption spectra of the manganese-bearing chain silicates pyroxmangite, rhodonite, bustamite and serandite. *Canad. Mineral.*, **9**, 348–57.
Manning, P. G. (1969b) On the origin of colour and pleochroism of astrophyllite and brown clintonite. *Canad. Mineral.*, **9**, 663–77.
Manning, P. G. (1969c) Optical absorption studies of grossular, andradite (var colophonite) and uvarovite. *Canad. Mineral.*, **9**, 723–30.
Manning, P. G. (1970) Racah parameters and their relationships to lengths and covalencies of Mn^{2+} and Fe^{3+} oxygen bonds in silicates. *Canad. Mineral.*, **10,** 677–87.
Manning, P. G. (1973) Effect of second-nearest-neighbour interaction on Mn^{3+} absorption in pink and black tourmalines. *Canad. Mineral.*, **11**, 971–7.
Manning, P. G. (1976) Ferrous–ferric interaction on adjacent face-sharing antiprismatic sites in vesuvianites: evidence for ferric ion in eight coordination. *Canad. Mineral.*, **14**, 216–20.
Manning, P. G. & Nickel, E. H. (1969) A spectral study of the origin of color and

pleochroism of a titanaugite from Kaiserstuhl and of a riebeckite from St. Peter's Dome, Colorado. *Canad. Mineral.*, **10**, 71–83.

Manning, P. G. & Tricker, M. J. (1975) Optical absorption and Mössbauer spectral studies of iron and titanium site-populations in vesuvianites. *Canad. Mineral.*, **13**, 110–16.

Manson, N. B., Gourley, J. Y., Vance, E. R., Sengupta, D. & Smith, G. (1976) The $^5T_{2g} \rightarrow {}^5E_g$ absorption in MgO. *J. Phys. Chem. Solids*, **37**, 1145–8.

Mao, H.-K. (1973) Electrical and optical properties of the olivine series at high pressure. Observations of optical absorption and electrical conductivity in magnesiowüstite at high pressures. Thermal and optical properties of the Earth's mantle. *Ann. Rept. Geophys. Lab., Yearb.* **72**, 52–64.

Mao, H.-K. (1976) Charge transfer processes at high pressure. In *The Physics and Chemistry of Minerals and Rocks.* (R. G. J. Strens, ed.; J. Wiley & Sons, New York), pp. 573–81.

Mao, H.-K. & Bell, P. M. (1971) Crystal field spectra. *Ann. Rept. Geophys, Lab., Yearb.* **70**, 207–15.

Mao, H.-K. & Bell, P. M. (1972a) Optical and electrical behaviour of olivine and spinel (Fe_2SiO_4) at high pressure. Interpretation of the pressure effect on the olivine absorption bands of natural fayalite to 20 kb. Crystal field stabilization of the olivine–spinel transition. *Ann. Rept. Geophys. Lab., Yearb.* **71**, 520–8

Mao, H.-K. & Bell, P. M. (1972b) Electrical conductivity and the red-shift of absorption in olivine and spinel at high pressure. *Science*, **176**, 403–5.

Mao, H.-K. & Bell, P. M. (1973a) Polarized crystal-field spectra of microparticles of the Moon. In *Analytical Methods Developed for Application to Lunar-Sample Analysis*. (Amer. Soc. Testing Materials Publ.), *STP* **539**, 100–19.

Mao, H.-K. & Bell, P. M. (1973b) Luna 20 plagioclase: crystal field effects and chemical analysis of iron. *Ann. Rept. Geophys. Lab., Yearb.* **72**, 662–5.

Mao, H.-K. & Bell, P. M. (1973c) Observations of optical absorption and electrical conductivity in magnesiowüstite at high pressure. *Ann. Rept. Geophys. Lab., Yearb.* **72**, 554–7.

Mao, H.-K. & Bell, P. M. (1974a) Crystal field effects of trivalent titanium in fassaite from the Pueblo de Allende meteorite. *Ann. Rept. Geophys. Lab., Yearb.* **73**, 488–92.

Mao, H,-K. & Bell, P. M. (1974b) Crystal field effects of ferric iron in goethite and lepidocrocite: Band assignment and geochemical applications at high pressure. *Ann. Rept. Geophys. Lab., Yearb.* **73**, 502–7.

Mao, H.-K. & Bell, P. M (1975a) Crystal-field effects in spinel: oxidation states of iron and chromium. *Geochim. Cosmochim. Acta*, **39**, 865–74 [see also *Ann. Rept. Geophys. Lab., Yearb.* **73**, 332–41 (1974)].

Mao, H.-K. & Bell, P. M. (1975b) Contribution of anionic complexes to charge-transfer and associated optical, electrical, and thermal effects at high pressure. *Ann. Rept. Geophys. Lab., Yearb.*, **74**, 559–61].

Mao, H.-K. & Bell, P. M. (1978) High-pressure physics: Sustained static generation of 3.16 to 1.72 megabars. *Science*, **200**, 1145–47.

Mao, H.-K. & Seifert, F. (1974) A study of crystal field effects of iron in the amphiboles anthophyllite and gedrite. *Ann. Rept. Geophys. Lab., Yearb.* **73**, 500–3.

Mao, H.-K., Bell, P. M. and Dickey, J. S. (1972) Comparison of the crystal-field spectra of natural and synthetic chrome diopside. *Ann. Rept. Geophys. Lab., Yearb.* **71**, 538–41.

Mao, H.-K., Bell, P. M. & Virgo, D. (1977) Crystal field spectra of fassaite from the Angra dos Reis meteorite. *Earth Planet. Sci. Lett.*, **35**, 353–6.

Mao, H.-K., Bell, P. M. & Yagi, T. (1982) Iron–magnesium fractionation model for the Earth. In *High–Pressure Research in Geophysics*. (S. Akimoto & M. H. Manghnani, eds; Acad. Publ., Tokyo, Japan), pp. 319–25.

Mao, H.-K., Bell, J. W., Shaner, J. W. & Steinberg, D. J. (1978) Specific volume measurements of Cu, Mo, Pd, and Ag and calibration of the ruby R_1 fluorescence pressure gauge from 0.06 to 1 Mbar. *J. Appl. Phys.*, **49**, 3276–83.

Mao, H.-K. , Hemley, R. J., Fei, Y., Shu, J. F., Chen, L. C., Jephcoat, A. P. & Wu, Y. (1991) Effect of pressure, temperature, and composition on lattice parameters and density of (Fe,Mg)SiO_3-perovskites to 30 GPa. *J. Geophys. Res.*, **96**, 8069–79.

Mao, H.-K., Xu, J. & Bell, P. M. (1986) Calibration of the ruby pressure gauge to 800 kbar under quasihydrostatic conditions. *J. Geophys. Res.*, **91**, 4673–6.

Marfunin, A. S. (1979a) *Physics of Minerals and Inorganic Materials*. (Springer–Verlag, New York), 340 pp.

Marfunin, A. S. (1979b) *Spectroscopy, Luminescence and Radiation Centers in Minerals*. (Springer–Verlag, New York), 352 pp.

Marshall, M. & Runciman, W. A. (1975) The absorption spectrum of rhodonite. *Amer. Mineral.*, **60**, 88–97.

Martinez, J. & Martinez, A. (1952) Pleochroism and structure of natural silicates. *Bull. Soc. Chim. France*, **19**, 563–5.

Marusak, L. A., Messier, R. & White, W. B. (1980) Optical absorption spectrum of hematite, α-Fe_2O_3, near IR to near UV. *J. Phys. Chem. Solids*, **41**, 981–4.

Mashimo, T., Kondon, K., Sawaoka, A., Syono, Y., Takei, H. & Ahrens, T. J. (1980) Electrical conductivity measurements of fayalite under shock compression up to 56 GPa. *J. Geophys. Res.*, **85**, 1876–81.

Mason, B. & Moore, C. (1982) *Principles of Geochemistry, 4th edn.* (J. Wiley & Sons, New York), 344 pp.

Matsui, Y. & Banno, S. (1969) Partition of divalent transition metals between coexisting ferromagnesian minerals. *Chem. Geol.*, **5**, 259-65.

Matsui, Y. & Nishizawa, O. (1974) Iron (II)–magnesium exchange equilibrium between olivine and calcium–free pyroxene over a temperature range 800 °C to 1300 °C. *Bull. Soc. Franc. Mineral.*, **97**, 122–30.

Mattson, S. M. & Rossman, G. R. (1984) Ferric iron in tourmaline. *Phys. Chem. Minerals*, **11**, 225–34.

Mattson, S. M. & Rossman, G. R. (1987a) Identifying characteristics of charge transfer transitions in minerals. *Phys. Chem. Minerals*, **14**, 94–9.

Mattson, S. M. & Rossman, G. R. (1987b) Fe^{2+}–Fe^{3+} interactions in tourmalines. *Phys. Chem. Minerals*, **14**, 163–71.

Mattson, S. M. & Rossman, G. R. (1988) Fe^{2+}–Ti^{4+} charge transfer in stoichiometric Fe^{2+},Ti^{4+}-minerals. *Phys. Chem. Minerals*, **16**, 78–82.

Meagher, E. P. (1982) Silicate garnets. In *Orthosilicates, 2nd edn.* (P. H. Ribbe, ed.; Mineral. Soc. Amer. Publ.), *Rev. Mineral.*, **5**, 25–66.

Menchetti, S. & Sabelli, C. (1976) Crystal chemistry of the alunite series: Crystal structure refinement of alunite and synthetic jarosite. *Neues Jahrb. Mineral. Mh.*, pp. 406-17.

Mercy, E. I. P. & O'Hara, M. J. (1967) Distribution of Mn, Cr, Ti, and Ni in coexisting minerals of ultramafic rocks. *Geochim. Cosmochim. Acta*, **31**, 2331–41.

Mero, J. L. (1965) *The Mineral Resources of the Sea.* (Elsevier Publ. Co., New York), 312 pp.

Meyer, H. O. A. & Boyd, F. R. (1972) Composition and origin of crystalline inclusions in natural diamonds. *Geochim. Cosmochim. Acta*, **36**, 1255–73.

Mocala, K., Navrotsky, A. & Sherman, D. M. (1992) High-temperature heat capacity of Co_3O_4 spinel: Thermally induced spin-unpairing transition. *Phys. Chem. Minerals*, **19**, 88–95.
Modine, F. A., Sonder, E. & Weeks, R. A. (1977) Determination of the Fe^{2+} and Fe^{3+} concentrations in MgO. *J. Appl. Phys.*, **48**, 3514–8.
Moore, P. B. & Araki, T. (1976) Braunite: Its structure and relationships to bixbyite, and some insights on the genealogy of fluorite derivative structures. *Amer. Mineral.*, **61**, 1226–40.
Morre, R. K. & White, W. B. (1971) Intervalence electron transfer effects in the spectra of melanaite garnets. *Amer. Mineral.*, **56**, 826–40.
Moore, R. K. & White, W. B. (1972) Electronic spectra of transition metal ions in silicate garnets. *Canad. Mineral.*, **11**, 791–811.
Moore, W. S., Ku, T.-L., MacDougall, J. D., Burns, V. M., Burns, R. G., Dymond, J., Lyle, M. W. & Piper, D. Z. (1981) Fluxes of metals to a manganese nodule: Radiochemical, chemical, structural, and mineralogical studies. *Earth Planet. Sci. Lett.*, **52**, 151–71.
Morris, R. V. & Lauer Jr, H. V. (1990) Matrix effects for reflectivity spectra of dispersed nanophase (superparamagnetic) hematite with applications to martian spectral data. *J. Geophys Res.*, **95,** 5101–9.
Morris, R. V., Agresti, D. G., Lauer Jr, H. V., Newcomp, J. A., Shelfer, T. D. & Murali, A. V. (1989) Evidence for pigmentary hematuite on Mars based on optical, magnetic, and Mössbauer studies of superparamagnetic (nanocrystalline) hematite. *J. Geophys. Res.*, **94**, 2760–78.
Morris, R. V., Gooding, J. L., Lauer Jr, H. V. & Singer, R. B. (1990) Origins of Marslike spectral and magnetic properties of a Hawaiian palagonitic soil. *J. Geophys. Res.*, **95,** 14427–34.
Morris, R. V., Lauer Jr, H. V., Lawson, C. A., Gibson Jr., E. K., Nace, G. A. & Stewart, C. (1985) Spectral and other physicochemical properties of submicron powders of hematite (α-Fe_2O_3), maghemite (γ-Fe_2O_3), magnetite (Fe_3O_4), goethite (α-FeOOH), and lepidocrocite (γ–FeOOH). *J. Geophys. Res.*, **90**, 3126–44.
Morris, R. V., Neely, S. C. & Mendell, W. W. (1982) Application of Kebulka–Munk theory of diffuse reflectance to geologic problems: The role of scattering. *Geophys. Res. Lett.*, **9**, 113-16.
Muller, R. & Gunthard, H. H. (1966) Spectroscopic study of the reduction of nickel and cobalt ions in sapphire. *J. Chem. Phys.*, **44**, 365–73.
Murphy, W. M. & Helgeson, H. C. (1987) Thermodynamic and kinetic constraints on reaction rates among minerals and aqueous solution. III. Activated complexes and the pH-dependence of the rates of feldspar, pyroxene, wollastonite, and olivine hydrolysis. *Geochim. Cosmochim. Acta*, **51**, 3137–53.
Mustard, J. F. (1992) Chemical analysis of actinolite from reflectance spectra. *Amer. Mineral.*, **77,** 345–58.
Mustard, J. F. & Pieters, C. M. (1987) Quantitative abundance estimates from bidirectional reflectance measurements. *J. Geophys. Res.*, **92,** E617–26.
Mustard, J. F. & Pieters, C. M. (1989) Photometric phase functions of common geologic minerals and applications to quantitative analysis of mineral mixture reflectance spectra. *J. Geophys. Res.*, **94,** 13619–34.
Mysen, B. O. (1978) Experimental determination of nickel partition coefficients between liquid, pargasite, and garnet peridotite minerals and concentration limits of behavior according to Henry's Law at high pressure and temperature. *Amer. J. Sci.*, **278**, 217–43.
Mysen, B. O. & Virgo, D. (1989) Redox equilibria, structure, and properties of Fe-

bearing aluminosilicate melts: Relationships among temperature, composition, and oxygen fugacity in the system Na_2O–Al_2O_3–SiO_2–Fe-O. *Amer. Mineral.*, **74**, 58–76.

Nabelek, P. I. (1980) Nickel partitioning between olivine and liquid in natural basalts: Henry's law behavior. *Earth Planet. Sci. Lett.*, **48**, 293–302.

Nash, D. B. & Conel, J. E. (1974) Spectral reflectance systematics for mixtures of powdered hypersthene, labradorite, and ilmenite. *J. Geophys. Res.*, **79**, 1615–21.

Nassau, K. (1978) The origins of color in minerals. *Amer. Mineral.*, **63**, 219–29.

Nassau, K. (1980) The causes of color. *Sci. Amer.*, October, 124–54

Nassau, K. (1983) *The Physics and Chemistry of Color: The Fifteen Causes of Color.* (Wiley, New York), 454 pp.

Navrotsky, A. & Kleppa, O. J. (1967) The thermodynamics of cation distribution in simple spinels. *J. Inorg. Nucl. Chem.*, **29**, 2701–14.

Nelson, C. & White, W. B. (1980) Transition metal ions in silicate melts: I. Manganese in sodium silicate melts. *Geochim. Cosmochim. Acta*, **44**, 887–93.

Nelson, C. & White, W. B. (1986) Transition metal ions in silicate melts: IV. Cobalt in sodium silicate and related glasses. *J. Mat. Sci.*, **1**, 130–8.

Nelson, C., Furukawa, T., and White, W. B. (1983) Transition metal ions in glasses: Network modifiers or quasi-molecular complexes? *Mat. Res. Bull.*, **18**, 959–66.

Newnham, R. E., Santoro, R., Pearson, J. & Jansen, C. (1964) Ordering of Fe and Cr in chrysoberyl. *Amer. Mineral.*, **49**, 427–30.

Neuhaus, A. (1960) Uber die Ionenfarben der Kristalle und Minerale am Beispiel der Chrommfarbungen. *Zeit. Krist.*, **113**, 195–233.

Newman, D. J., Price, D. C. & Runciman, W. A. (1978) Superposition model analysis of the near-infrared spectrum of Fe^{2+} in pyrope–almandine garnets. *Amer. Mineral.*, **63**, 1278–81.

Newnham, R. E. and de Haan, Y. M. (1962) Refinement of the α-Al_2O_3, Ti_2O_3, V_2O_3, and Cr_2O_3 structures. *Zeit. Krist*, **117**, 235–7.

Nickel, E. H. (1954) The distribution of iron, manganese, nickel and cobalt between coexisting pyrite and biotite in wall-rock alteration. *Amer. Mineral.*, **39**, 494–503.

Nickel, E. H. (1968a) Structural stability of of minerals with the pyrite, marcasite, arsenopyrite and löllingite structures. *Canad. Mineral.*, **9**, 311–21.

Nickel, E.H. (1968b) The crystal chemistry of the skutterudite minerals. *Canad. Mineral.*, **9**, 578–9.

Nitsan, U. & Shankland, T. J. (1976) Optical properties and electronic structure of mantle silicates. *Geophys. J. Royal Astron. Soc.*, **45**, 59–87.

Nockolds, S. R. (1966) The behaviour of some elements during fractional crystallization of magma. *Geochim. Cosmochim. Acta*, **30**, 267–78.

Nolet, D. A. & Burns, R. G. (1979) Ilvaite: a study of temperature dependent electron delocalization by the Mössbauer effect. *Phys. Chem. Minerals*, **4**, 221–34.

Nolet, D. A., Burns, R. G., Flamm, S. L. & Besancon, J. R. (1979) Spectra of Fe–Ti silicate glasses: implications to remote-sensing of planetary surfaces. *Proc. 10th Lunar Planet. Sci. Conf., Geochim. Cosmochim. Acta, Suppl. 10* (Pergamon Press, New York), pp. 1775–86.

Nord, A. G., Annersten, H. & Filippidis, A. (1982) The cation distribution in synthetic Mg–Fe–Ni olivines. *Amer. Mineral.*, **67**, 1206–11.

Nordstrom, D. K. & Munoz, J. L. (1986) *Geochemical Thermodynamics.* (Blackwell Sci. Publ.), 477 pp.

Novak, G. A. & Gibbs, G. V. (1971) The crystal chemistry of silicate garnets. *Amer. Mineral.*, **56**, 791–825.

Nover, G. & Will, G. (1981) Structure refinements of seven natural olivine crystals

and the influence of the oxygen partial pressure on cation distribution. *Zeit. Krist.*, **155**, 27–45.
Nuber, B. & Schmetzer, K. (1979) The lattice position of Cr^{3+} in tourmaline: structural refinement of a chromium-rich Mg–Al tourmaline. *Neues Jahrb. Mineral., Abh.*, **137**, 184–97.
O'Neill, H. St C. (1985) Thermodynamics of Co_3O_4: a possible electron spin-unpairing transition in Co^{3+}. *Phys. Chem. Minerals*, **12**, 149–54.
O'Neill, H. St C. & Navrotsky, A. (1983) Simple spinels: crystallographic parameters, cation radii, lattice energies, and cation distribution. *Amer. Mineral.*, **68**, 181–94.
O'Neill, H. St C. & Wood, B. J. (1979) An experimental study of Fe–Mg partitioning between garnet and olivine and its calibration as a geothermometer. *Contrib. Mineral. Petrol.*, **75**, 291–300.
O'Nions, R. K. & Smith, D. G. W. (1973) Bonding in silicates: An assessment of bonding in orthopyroxene. *Geochim. Cosmochim. Acta*, **37**, 249–57.
Obata, M., Banno, S. & Mori, T. (1974) The iron–magnesium partitioning between naturally occurring coexisting olivine and Ca-rich clinopyroxene: an application of the simple mixture model to olivine solid solution. *Bull. Soc. Franc. Mineral.*, **97**, 101–7.
Oles, A., Szytula, A. & Wanic, A. (1970) Neutron diffraction study of γ-FeOOH. *phys. stat. solidi (a)* **41**, 173–7.
Olsen, R. O., McKay, G. A. & Taylor, L. A. (1988) Temperature and composition dependencies of trace element partitioning: Olivine/melt and low-Ca pyroxene/melt. *Geochim. Cosmochim. Acta*, **52**, 539–53.
Onuma, N., Higuchi, H., Wakita, H. & Nagasawa, H. (1968) Trace element partition between two pyroxenes and the host lava. *Earth Planet. Sci. Lett.*, **5**, 47–51.
Orgel, L. E. (1952) The effects of crystal fields on the properties of transition metal ions. *J. Chem. Soc.*, pp. 4756–61.
Orgel, L. E. (1966) *An Introduction to Transition Metal Chemistry: Ligand-Field Theory, 2nd edn.* (Methuen, London), 186 pp.
Osborne, M. D., Parkin, K. M. & Burns, R. G. (1978) Temperature-dependence of Fe–Ti spectra in the visible region: implications to mapping Ti concentrations on hot planetary surface. *Proc. 9th Lunar Planet. Sci. Conf., Geochim. Cosmochim. Acta, Suppl. 9* (Pergamon Press, New York), pp. 2949–60.
Oshnishi, S. (1978) A theory of the pressure-induced high-spin to low-spin transition of transition metal oxides. *Phys. Earth Planet. Interiors*, **17**, 130–39.
Oshnishi, S. & Sugano, S. (1981) Strain interaction effects on the high-spin – low-spin transition of transition metal compounds. *J. Phys. C.*, **14**, 39–55.
Ottonello, G. (1987) Energies and interactions in binary (*Pbnm*) orthosilicates: A Born parameterization. *Geochim. Cosmochim. Acta*, **51**, 3119–35.
Ottonello, G., Della Giusta, A. & Molin, G. M. (1989) Cation ordering in Ni–Mg olivines. *Amer. Mineral.*, **74**, 411–21.
Ottonello, G., Princivalle, F. & Giusta, A. D. (1990) Temperature, composition and f_{O2} effects on intersite distribution of Mg and Fe^{2+} olivines. *Phys. Chem. Minerals*, **17**, 301–12.
Pabst, A. (1959) Structures of some tetragonal sheet silicates. *Acta Cryst.*, **12**, 733–9.
Pappalardo, R. Wood, D. L. & Linares Jr, R. C. (1961a) Optical absorption study of Ni-doped oxide systems. *J. Phys. Chem.*, **35**, 1460–78.
Pappalardo, R. Wood, D. L. & Linares, Jr., R. C. (1961b) Optical absorption study of Co-doped oxide systems. *J. Phys. Chem.*, **35**, 2041–59.
Papike, J. J. & Clark, J. R. (1968) The crystal structure and cation distribution of glaucophane. *Amer. Mineral.*, **53**, 1156–73.
Parkin, K. M. & Burns, R. G. (1980) High temperature crystal field spectra of

transition metal-bearing minerals: Relevance to remote-sensed spectra of planetary surfaces. *Proc. 11th Lunar Planet. Sci. Conf., Geochim. Cosmochim. Acta, Suppl. 11* (Pergamon Press, New York), pp. 731–55.

Parkin, K. M., Loeffler, B. M. & Burns, R. G. (1977) Mössbauer spectra of kyanite, aquamarine and cordierite showing intervalence charge transfer. *Phys. Chem. Minerals*, **1**, 301–11.

Pauling, L. (1960) *The Nature of the Chemical Bond, 3rd edn.* (Cornell Univ. Press, New York), 644 pp.

Pauling, L. & Kamb, B. (1982) The crystal structure of lithiophorite. *Amer. Mineral.*, **67**, 817–21.

Petrov, I., Schmetzer, K. & Eysel, H. H. (1977) Absorption spectrum of chromium in topaz. *Neues Jahrb. Mineral. Mh.*, pp. 365–71.

Pettengill, G. H., Ford, P. G. & Wilt, R. J. (1992) Venus surface radiothermal emission as observed by Magellan. *J. Geophys. Res.*, **97**, 13091-102.

Peyronneau, J. & Poirier, J. P. (1989) Electrical conductivity of the Earth's lower mantle. *Nature*, **342**, 537–9.

Phillips, C. S. G. & Williams, R. J. P. (1966) *Inorganic Chemistry* (Clarendon Press, Oxford), 2 vols.

Philpotts, J. A. (1978) The law of constant rejection. *Geochim. Cosmochim. Acta*, **42**, 909–20.

Pieters, C. M. (1978) Mare basalt types on the front side of the Moon: A summary of spectral reflectance data. *Proc. 9th Lunar Planet. Sci. Conf., Geochim. Cosmochim. Acta, Suppl. 9* (Pergamon Press, New York), pp. 2825–50.

Pieters, C. M. (1982) Copernicus crater central peak: Lunar mountain of unique composition. *Science*, **215**, 59–61.

Pieters, C. M. (1983) Strength of mineral absorption features in the transmitted component of near-infrared reflected light: First results from RELAB. *J. Geophys. Res.*, **88**, 9534–44.

Pieters, C. M. (1986) Composition of the lunar highland crust from near-infrared spectroscopy. *Rev. Geophys.*, **24**, 557–78.

Pieters, C. M. (1989) Seeing through the dust and alteration products of Mars. *Lunar Planet. Sci.*, **XX**, 850–1.

Pieters, C. M., Adams, J. B., Mouginis-Marx, P. J., Zisk, S. H., Smith, M. G., Head, J. W. & McCord, T. B. (1985) The nature of crater rays: The Copernicus example. *J. Geophys. Res.*, **90**, 12393–413.

Pieters, C. M., Hawke, B. R., Gaffey, M. & McFadden, L. A. (1983) Possible lunar source areas of meteorite ALHA81005: Geochemical remote sensing information. *Geophys. Res. Lett.*, **10**, 813–16.

Pieters, C. M., Head, J. W., Patterson, W., Pratt, S., Garvin, J., Barsukov, V. I., Basilevsky, A. T., Khodakovsky, I. L., Panfilov, A. S., Gektin, Yu. M. & Narayeva, Y. M. (1986) The color of the surface of Venus. *Science*, **234**, 1379–83.

Pinet, P. & Chevrel, S. (1990) Spectral identification of geological units on the surface of Mars related to the presence of silicates from Earth-based near-infrared telescopic charge-coupled device imaging. *J. Geophys. Res.*, **95**, 14435–46.

Pitt, G. D. & Tozer, D. C. (1970) Radiative heat transfer in dense media and its magnitude in olivines and some other ferromagnesian minerals under typical upper mantle conditions. *Phys. Earth Planet Interiors*, **2**, 189–99.

Pol'shin, E. V., Platonov, A. N., Borutzky, B. E., Taran, M. N. & Rastsvetaeva, R. K. (1991) Optical and Mössbauer study of minerals of the eudialyte group. *Phys. Chem. Minerals*, **18**, 117–25.

Pollak, H. (1976) Charge transfer in cordierite. *phys. stat. solidi (a)* **74**, K31–4.

Pollak, H., Quartier, R. & Bruyneel, W. (1981) Electron relaxation in deerite. *Phys. Chem. Minerals*, **7**, 10-14.
Poole, Jr., C. P. (1964) The optical spectra and color of chromium-containing solids. *J. Phys. Chem. Solids*, **25**, 1169–82..
Post, J. E. & Appleman, D. E. (1988) Chalcophanite, $ZnMn_3O_7.3H_2O$: New crystal-structure determination. *Amer. Mineral.*, **73**, 1401–4.
Post, J. E. & Bish, D. L. (1988) Rietveld refinement of the todorokite structure. *Amer. Mineral.*, **73**, 861–9.
Post, J. E. & Bish, D. L. (1989) Rietveld refinement of the coronodite structure. *Amer. Mineral.*, **74**, 913–17.
Post, J. E. & Buchwald, V. F. (1991) Crystal structure refinement of akaganeite. *Amer. Mineral.*, **76**, 272–7.
Post, J. E. & Veblen, D. R. (1990) Crystal structure determinations of synthetic sodium, magnesium, and potassium birnessite using TEM and the Rietveld method. *Amer. Mineral.*, **75**, 477–89.
Post, J. E., Von Dreele, R. B. & Buseck, P. R. (1982) Symmetry and cation displacements in hollandites: structure refinements of hollandite, cryptomelane and priderite. *Acta Cryst.*, **B38**, 1056–65.
Pratt, Jr., G. W. & Coelho, R. (1959) Optical absorption of CoO and MnO above and below the Néel temperature. *Phys. Rev.*, **116**, 281–6.
Prewitt, C. T., Shannon, R. D. & White, W. B. (1972) Synthesis of a pyroxene containing trivalent titanium. *Contrib. Mineral. Petrol.*, **35**, 77–82.
Price, D. C., Vance, E. R., Smith, G., Edgar, A. & Dickson, B. L. (1976) Mössbauer effect studies of beryl. *Journ. de Phys.*, **12**, (C6) 811-17.
Princivalle, F. (1990) Influence of temperature and composition on Mg–Fe^{2+} intracrystalline distribution in olivines. *Mineral. Petrol.*, **43**, 121–9.
Princivalle, F. & Secco, L. (1985) Crystal structure refinement of 13 olivines in the forsterite–fayalite series from volcanic rocks and ultramafic nodules. *Tschermaks Mineral. Petrol. Mitt.*, **34**, 105–15.
Rager, H. (1977) Electron spin resonance of trivalent chromium in forsterite. *Phys. Chem. Minerals*, **1**, 371–8.
Rager, H. and Weiser, G. (1981) Polarized absorption spectra of trivalent chromium in forsterite, Mg_2SiO_4. *Bull. Minéral.*, **104**, 603–9.
Rager, H., Hosoya, S. & Weiser, G. (1988) Electron paramagnetic resonance and polarized absorption spectra of Ni^{2+} in synthetic forsterite. *Phys. Chem. Minerals*, **15**, 383–9.
Rahman, H. U. & Runciman, W. A. (1971) Energy levels and *g* values of vanadium corundum. *J. Phys. C*, **4**, 1576–90.
Rajamani, V. (1976) Distribution of iron, cobalt and nickel between synthetic sulfide and orthopyroxene at 900 °C. *Econ. Geol.*, **71**, 795-802.
Rajamani, V. & Naldrett, A. J. (1978) Partitioning of Fe, Co, Ni, and Cu between sulfide liquid and basaltic melts and the composition of Ni–Cu sulfide deposits. *Econ. Geol.*, **67**, 939-52.
Rajamani, V., Brown, G. E. & Prewitt, C. T. (1975) Cation ordering in Ni–Mg olivine. *Amer. Mineral.*, **60**, 292–9.
Ramberg, H. & DeVore, G. W. (1951) The distribution of Mg^{2+} and Fe^{2+} ions in coexisting olivines and pyroxenes. *J. Geol.*, **59**, 193–210.
Rankama, K. and Sahama, Th.G. (1950) *Geochemistry*. (Chicago Univ. Press), 912 pp.
Rastsvetaeva, R. K. & Borutzky, B. E. (1988) Crystal–chemical peculiarities of eudialyte proceeding from new structural data. *Mineral. Zhurnal*, **10**, 48–57.
Reinen, D. (1964) Farbe und Komstitution bei anorganischen Festoffen. II. Die

Lightabsorption des okraedrisch koordinierten Co^{2+} ions in der Mischkristallriehe $Mg_{1-x}Co_xO$ und anderen oxidischen Wirtsgittern. *Monat. Chem.*, **96**, 730–9.
Reinen, D. (1969) Ligand-field spectroscopy and chemical bonding in Cr^{3+}-containing oxidic solids. *Struct. Bonding*, **6**, 30–51.
Reinen, D. (1970) Kationenverteilung zweiwertiger $3d^n$-Ionen in oxidischen Spinell-, Granat- und anderen Structuren. *Struct. Bonding*, **7**, 114–54.
Ribbe, P. H. & Lumpkin, G. R. (1984) Cation ordering in Ni–Fe olivines: corrections and discussion. *Amer. Mineral.*, **69**, 161–3 [see Annersten *et al.* (1982)].
Richardson, S. M. (1975) A pink muscovite with reverse pleochroism from Archer's Post, Kenya. *Amer. Mineral.*, **60**, 73–8 [see Annersten & Hålenius (1976)].
Richardson, S. M. (1976) Ion distribution in pink muscovite: A reply. *Amer. Mineral.*, **61**, 1051–2.
Ringwood, A. E. (1955) The principles governing trace element distribution during magmatic crystallization. *Geochim. Cosmochim. Acta*, **7**, 242–55.
Ringwood, A. E. (1956) Melting relations in the system Mg_2SiO_4–Ni_2SiO_4. *Geochim. Cosmochim. Acta*, **8**, 297–303.
Ringwood, A. E. (1969) Composition and evolution of the upper mantle. In *The Earth's Crust and Upper Mantle. AGU Geophys. Mono.*, **13**, 1–17.
Ringwood, A. E. (1975) *Composition and Petrology of the Earth's Mantle.* (McGraw-Hill, New York), 618 pp.
Ringwood, A. E., Kato, T., Hibbeson, W. & Ware, N. (1991) Partitioning of Cr, V, and Mn between mantles and cores of differentiated planetesimals: Implications for giant impact hypothesis of lunar origin. *Icarus*, **89**, 122–8.
Robbins, D. W. & Strens, R. G. J. (1972) Charge transfer ferromagnesian silicates: The polarized electronic spectra of trioctahedral micas. *Mineral. Mag.*, **38**, 551–63.
Robin, M. B. & Day, P. (1967) Mixed valence chemistry – a survey and classification. *Adv. Inorg. Chem. Radiochem.*, **10**, 247–63.
Robinson, K., Gibbs, G. V. & Ribbe, P. H. (1971) Quadratic elongation: A quantitative measure of distortion in coordination polyhedra. *Science*, **172**, 567–70.
Roeder, P. L. (1974) Activity of iron and olivine solubility in basaltic liquides. *Earth Planet. Sci. Lett.*, **23**, 397–410.
Rosenberg, P. E. & Foit Jr, F. J. (1977) Fe^{2+}–F avoidance in silicates. *Geochim. Cosmochim. Acta*, **41**, 345–6.
Roslar, H. J. & Beuge, P. (1983) Geochemistry of trace elements during regional metamorphism. In *The Significance of Trace Elements in Solving Petrogenetic Problems and Controversies*. (S. S. Augustithis, ed.; Theophrastus Publ., Athens, Greece), pp. 407–30.
Ross II, C. R., Rubie, D. C. & Paris, E. (1990) Rietveld refinement of the high-pressure polymorph of Mn_3O_4. *Amer. Mineral.*, **75**, 1249–52.
Rossman, G. R. (1974) Optical spectroscopy of green vanadium apophyllite from Poona, India. *Amer. Mineral.*, **59**, 621–2.
Rossman, G. R. (1975) Spectroscopic and magnetic studies of ferric iron hydroxy sulfates: intensification of color in ferric iron clusters bridged by a single hydroxide ion. *Amer. Mineral.*, **60**, 698–704.
Rossman, G. R. (1976) Spectroscopic and magnetic studies of ferric iron hydroxy sulfates: the series $Fe(OH)SO_4.nH_2O$ and the jarosites. *Amer. Mineral.*, **61**, 398–404.
Rossman, G. R. (1980) Pyroxene spectroscopy. In *Pyroxenes*. (C. T. Prewitt, ed.; Mineral. Soc. Amer. Publ.). *Rev. Mineral.*, **7**, 91–115.
Rossman, G. R. (1984) Optical spectra of micas. In *Micas*. (S, W, Bailey, ed.; Mineral Soc. Amer. Publ.), *Rev. Mineral.*, **13**, 145–81.
Rossman, G. R. (1988) Optical spectroscopy. In *Spectroscopic Methods in*

Mineralogy and Geology. (F. C. Hawthorne, ed.; Mineral. Soc. Amer. Publ.), *Rev. Mineral.*, **18**, 207–54.

Rossman, G. R. & Mattson, S. M. (1986) Yellow, Mn-rich elbaite with Mn–Ti intervalence charge transfer. *Amer. Mineral.*, **71**, 599–602.

Rossman,G. R., Grew, E. S. & Dollase, W. A. (1982) The colors of sillimanite. *Amer. Miner.*, **67**, 749–61.

Rossman, G. R., Shannon, R. D. & Waring, R. K. (1981) Origin of the yellow color of complex nickel oxides. *J. Solid State Chem.*, **39**, 277–87.

Rost, F. & Simon, E. (1972) Zur geochemie und Färbun des cyanits. *Neues Jahrb. Mineral. Mh.*, pp. 383–95.

Roush, T. L. & Singer, R. B. (1986) Gaussian analysis of temperature effects on the reflectance spectra of mafic minerals in the 1-μm region. *J. Geophys. Res.*, **91**, 10301–8.

Runciman, W. A. & Sengupta, D. (1974) The spectrum of Fe^{2+} ions in silicate garnets. *Amer. Mineral.*, **59**, 563–6.

Runciman, W. A., Sengupta, D. & Gurley, J. T. (1973a) The polarized spectra of iron in silicates. II. Olivine. *Amer. Mineral.*, **58**, 466–70 [see also *Amer. Mineral.*, **59**, 630–1 (1974); and Burns (1974)].

Runciman, W. A., Sengupta, D. & Marshall, M. (1973b) The polarized spectra of iron in silicates. I. Enstatite. *Amer. Mineral.*, **58**, 444–50.

Sack, R. O. (1980) Some constraints on the thermodynamic mixing properties of Fe–Mg orthopyroxenes and olivines. *Contrib. Mineral. Petrol.*, **71**, 257–69.

Samoilovich, M. I., Isinober, L. I. & Dunin-Barkovskii, R. L. (1971) Nature of the coloring in iron-containing beryl. *Kristallogr.*, **16**, 186–9. [*Transl.*: *Sov. Phys. Crystallog.*, **16**, 147-50 (1971).]

Sanz, J. & Stone, W. E. (1979) NMR studies of micas. II. Distribution of Fe^{2+}, F^- and OH^- in the octahedral sites of phlogopites. *Amer. Mineral.*, **64**, 119–26.

Sasaki, S., Takeuchi, Y., Fujiko, K. & Akimoto, S. (1982) Electron-density distributions of three orthopyroxenes, $Mg_2Si_2O_6$, $Co_2Si_2O_6$ and $Fe_2Si_2O_6$. *Zeit. Krist.*, **158**, 279–97.

Sawamoto, H. & Horiuchi, H. (1990) β-$(Mg_{0.9}Fe_{0.1})_2SiO_4$: Single crystal structure, cation distribution, and properties of coordination polyhedra. *Phys. Chem. Minerals*, **17**, 293–300.

Sawkins, F. J. (1990) *Metal Deposits in Relation to Plate Tectonics, 2nd edn.* (Springer–Verlag, New York), 461 pp.

Saxena, S. K. & Ghose, S. (1971) Mg^{2+}–Fe^{2+} order–disorder and the thermodynamics of the orthopyroxene crystalline solution. *Amer. Mineral.*, **56**, 532–59.

Schaefer, M. W. (1985) Site occupancy and two-phase character of 'ferrifayalite'. *Amer. Mineral.*, **70**, 729–36 [see also *Nature*, **303**, 325–7].

Schärmeli, G. H. (1979) Identification of radiative thermal conductivity of olivine to 25 kb and 1500 K. In *High-Pressure Science and Technology. Sixth AIRAPT Conf.* (K. D. Timmershauf & M. S. Barber, eds; Plenum, New York), **2**, 60–74

Schatz, J. F. & Simmons, G. (1972) Thermal conductivity of earth materials at high temperatures. *J. Geophys. Res.*, **77**, 6966–83.

Scheetz, B. E. & White, W. B. (1972) Synthesis and optical absorption spectra of Cr^{2+}-containing silicates. *Contrib. Mineral. Petrol.*, **37**, 221–7.

Schmetzer, K. (1978) Der Alexandrit-Effekt in Festkorpern. *Naturwiss.*, **65**, 592.

Schmetzer, K. (1982) Absorption spectroscopy and colour of V^{3+}-bearing natural oxides and silicates – a contribution to the crystal chemistry of vanadium. *Neues Jahrb. Mineral. Abh.*, **144**, 73–126.

Schmetzer, K. & Bank, H. (1980a) The colour of natural corundum. *Neues Jahrb. Mineral. Abh.*, **139**, 58–68.

Schmetzer, K, & Bank, H. (1980b) Explanations of the absorption spectra of natural and synthetic Fe- and Ti-containing corundums. *Neues Jahrb. Mineral. Abh.*, **139**, 216–25.

Schmetzer, K. & Bank, H. (1981) An unusual pleochroism in Zambian emeralds. *J. Gemol.*, **37**, 443–6.

Schmetzer, K. & Berdesinski, W. (1978) The absorption spectrum of Cr^{3+} in zoisite. *Neues Jahrb. Mineral. Mh.*, pp. 197–202.

Schmetzer, K. & Gübelin, E. (1980) Alexandrite-like spinel from Sri Lanka. *Neues Jahrb. Mineral. Mh.*, pp. 28–32.

Schmetzer, K. & Ottemann, J. (1979) Crystal chemistry and colour of vanadium-bearing garnets. *Neues Jahrb. Mineral. Abh.*, **136**, 146–68.

Schmetzer, K., Bank, H. & Gübelin, E. (1980) The alexandrite effect in minerals; chrysoberyl, garnet, corundum and fluorite. *Neues Jahrb. Miner. Abh.*, **138**, 147–64.

Schmetzer, K., Berdesinski, W. & Traub, I. (1975) Vanadium-bearing grossular from Kenya. *Z. Dt. Gemmol. Ges.*, **24**, 229–34.

Schmitz-DuMont, O. & Friebel, C. (1967) Farbe und Konstitution bei anorganischen Festoffen. 15. Die Lichtabsorption des zweiwertigen Kobalts in Silikaten vom Olivintypus. *Monat. Chem.*, **98**,1583-602

Schmitz-Dumont, O., Gossling, H..& Brokopf, H. (1959) Die Lichtabsorption des zweiwertigen Nickels in oxydischen Koordinationsgittern. *Z. anorg. allgem. Chemie*, **300**, 159-74.

Schorin, H. (1983) Behaviour of the trace elements Zr, Ga, Zn, Cu, Ni, Mn, Cr, Ca, Sr and Ba during the lateritic weathering of a diabase sill from the Serrania de los Guaicas, Venezuela. In *The Significance of Trace Elements in Solving Petrogenetic Problems and Controversies.* (S. S. Augustithis, ed.; Theophrastus Publ., Athens, Greece), pp. 695–730.

Schreiber, H. D. (1977) On the nature of synthetic blue diopside crystals: The stabilization of tetravalent chromium. *Amer. Mineral.*, **62**, 522–7.

Schreiber, H. D. (1978) Chromium, blue diopside, and experimental petrology. *Contrib. Mineral. Petrol.*, **66,** 341–2 [see Ikeda & Yagi (1977, 1982)].

Schreiber, H. D. & Haskin, L. A. (1976) Chromium in basalts: Experimental determination of redox states and partitioning among synthetic silicate glasses. *Proc. 7th Lunar Sci. Conf., Geochim. Cosmochim. Acta, Suppl.* 7 (Pergamon Press, New York), pp. 1221–59.

Schulien, S., Hornemann, U. & Stoffler, D. (1978) Electrical conductivity of dunite during shock compression from 12.5 to 45 GPa. *Geophys. Res. Lett.*, **5**, 345–8.

Schumm Jr, B., Middaugh, R. L., Grotheer, M. P. & Hunter, J. C. (eds) (1985) *Manganese Dioxide Electrode Theory and Practice for Electrochemical Applications.* (Electrochemical Soc., Pennington, NJ), *Proc. vol.* **85-4**, 685 pp.

Schwartz, H. P. (1967) The effect of crystal field stabilization on the distribution of transition metals between metamorphic minerals. *Geochim. Cosmochim. Acta*, **31**, 503–17.

Schwartz, K. B., Nolet, D. A. & Burns, R. G. (1980) Mössbauer spectroscopy and crystal chemistry of natural Fe–Ti garnets. *Amer. Mineral.*, **65**, 142–53.

Shankland, T. J. (1968) Pressure shift of absorption bands in MgO:Fe^{2+}; the dynamic Jahn–Teller effect. *J. Phys. Chem. Solids*, **29**, 1907–9.

Shankland, T. J. (1970) Pressure shift of infrared absorption bands in minerals and the effect on radiative heat transport. *J. Geophys. Res.*, **75**, 409–13.

Shankland, T. J. (1975) Electrical conduction in rocks and minerals: Parameters for interpretation. *Phys. Earth Planet. Interiors*, **10**, 209–19.

Shankland, T. J., Duba, A. G. & Woronow, A. (1974) Pressure shifts of optical

absorption bands in iron-bearing garnet, spinel, olivine, pyroxene and periclase. *J. Geophys. Res.*, **79**, 3273–82.
Shankland, T. J., Nitsan, U. & Duba, A. G. (1979) Optical absorption and radiative transport in olivine and high temperature. *J. Geophys. Res.*, **84**, 1603–10.
Shannon, R. D. (1976) Revised effective ionic radii and systematic studies of interatomic distances in halides and chalcogenides. *Acta Cryst.*, **A32**, 751–7.
Shannon, R. D. & Prewitt, C. T. (1969) Effective ionic radii in oxides and fluorides. *Acta Cryst.*, **B25**, 925–46.
Shannon, R. D., Gumerman, P. S. & Chenavas, J. (1975) Effect of octahedral distortion on mean Mn^{3+}–O distances. *Amer. Mineral.*, **60,** 714–16.
Shaw, D. M. (1953) The camouflage principle of trace element distribution in magmatic minerals. *J. Geol.*, **61,** 142–51.
Sherman, D. M. (1984) The electronic structures of manganese oxide minerals. *Amer. Mineral.*, **69**, 788–99.
Sherman, D. M. (1985a) The electronic structures of Fe^{3+} coordination sites in iron oxides: Applications to spectra, bonding, and magnetism. *Phys. Chem. Minerals*, **12**, 161–75.
Sherman, D. M. (1985b) SCF–Xα–SW MO study of Fe–O and Fe–OH chemical bonds: Applications to the Mössbauer spectra and magnetochemistry of hydroxyl-bearing Fe^{3+} oxides and silicates. *Phys. Chem. Minerals*, **12**, 311–14.
Sherman, D. M. (1987a) Molecular orbital (SCF–Xα–SW) theory of metal–metal charge transfer processes in minerals. I. Applications to $Fe^{2+} \rightarrow Fe^{3+}$ charge transfer and 'electron delocalization' in mixed-valence iron oxides and silicates. *Phys. Chem. Minerals,* **14**, 355–64.
Sherman, D. M. (1987b) Molecular orbital (SCF–Xα–SW) theory of metal–metal charge transfer processes in minerals. II. Applications to $Fe^{2+} \rightarrow Ti^{4+}$ charge transfer transitions in oxides and silicates. *Phys. Chem. Minerals*, **14**, 364–7.
Sherman, D. M. (1988) High-spin to low-spin transition of iron(II) oxides at high pressures: Possible effects on the physics and chemistry of the lower mantle. In *Structural and Magnetic Phase Transitions in Minerals.* (S. Ghose, J. M. D. Coey & E. Salje, eds; Springer–Verlag, New York), *Adv. Geochem.*, **7**, 113–28.
Sherman, D. M. (1989) The nature of the pressure-induced metallization of FeO and its implications to the core–mantle boundary. *Geophys. Res. Lett.*, **16**, 515–18.
Sherman, D. M. (1990) Molecular orbital (SCF–Xα–SW) theory of Fe^{2+}–Mn^{3+}, Fe^{3+}–Mn^{2+}, and Fe^{3+}–Mn^{3+} charge transfer and magnetic exchange in oxides and silicates. *Amer. Mineral.*, **75**, 256–61.
Sherman, D. M. (1991) The high pressure electronic structure of magnesiowüstite (Mg,Fe)O: Applications to the physics and chemistry of the Lower Mantle. *J. Geophys. Res.*, **96**, 14299–312.
Sherman, D. M. & Vergo, N. (1988a) Optical spectrum, site occupancy, and oxidation state of Mn in montmorillonite. *Amer. Mineral.*, **73**, 140–4.
Sherman, D. M. & Vergo, N. (1988b) Optical (diffuse reflectance) and Mössbauer spectroscopic study of nontronite and related iron smectites. *Amer. Mineral.*, **73**, 1346–54.
Sherman, D. M. & Waite, T. D. (1985) Electronic spectra of Fe^{3+} oxides and oxide hydroxide in the near-IR to near-UV. *Amer. Mineral.*, **70**, 1262–9.
Sherman, D. M., Burns, R. G. & Burns, V. M. (1982) Spectral characteristics of the iron oxides with application to the Martian bright region mineralogy. *Proc. 12th Lunar Planet. Sci. Conf., J. Geophys. Res.*, **87,** 10169–80.
Shimizu, N. & Allegre, C. J. (1978) Geochemistry of transition elements in garnet lherzolite nodules in kimberlites. *Contrib. Mineral. Petrol.*, **67**, 41–50.
Shinno, I. (1974) Mössbauer study of ferric iron in olivine. *Phys. Chem. Minerals*, **7,** 91–5.

Siegel, M. D. & Turner, S. (1983) Crystalline todorokite associated with biogenic debris in manganese nodules. *Science*, **219**, 172–4.

Sillitoe, R. H. (1972) A plate tectonic model for the origin of porphyry copper deposits. *Econ. Geol.*, **67**, 184–97.

Sillitoe, R. H. (1973) The tops and bottoms of porphyry copper deposits. *Econ. Geol.*, **68**, 799–815.

Simons, B. (1980) Composition–lattice parameter relationships of the magnesiowüstite solid solution series. *Ann. Rept. Geophys. Lab., Yearb.* **79**, 376–80.

Singer, R. B. (1981) Near-infrared spectral reflectance of mineral mixtures: Systematic combinations of pyroxenes, olivines, and iron oxides. *J. Geophys. Res.*, **86,** 7967–82.

Singer, R. B. (1982) Spectral evidence for the mineralogy of high-albedo soils and dust on Mars, *J. Geophys. Res.*, **87**, 10159–68.

Singer, R. B. (1985) Spectroscopic observations of Mars. *Adv. Space Res.*, **5,** 59–68.

Singer, R. B. & Roush, T. L. (1985) Effects of temperature on remotely sensed mineral absorption features. *J. Geophys. Res.*, **90,** 12434–44.

Singer, R. B., McCord, T. B., Clark, R. N., Adams, J. B. & Huguenin, R. L. (1979) Mars surface composition from reflectance spectroscopy: A summary. *J. Geophys. Res.*, **84,** 8415–26.

Singer, R. B. & McSween Jr, H. Y. (1992) The composition of the Martian crust: Evidence from remote sensing and SNC meteorites. In *Resources of Near-Earth Space*. (J. S. Lewis, ed.; Univ. Arizona Press), *in press*.

Skogby, H. and Rossman, G. R. (1991) The intensity of amphibole OH bands in the infrared absorption spectrum. *Phys. Chem. Minerals*, **18**, 64–8.

Smith, G. (1977) Low temperature optical studies of metal–metal charge transfer transitions in various minerals. *Canad. Mineral.*, **15**, 500–7.

Smith, G. (1978a) A re-assessment of the role of iron in the 5,000–30,000 cm^{-1} region of the electronic absorption spectra of tourmaline. *Phys. Chem. Minerals*, **3**, 343–73.

Smith, G. (1978b) Evidence for absorption by exchange coupled Fe^{2+}–Fe^{3+} pairs in the near-infrared spectra of minerals. *Phys. Chem. Minerals*, **3**, 373–83.

Smith, G. (1980) Evidence for optical absorption by Fe^{2+}–Fe^{3+} interactions in MgO:Fe. *phys. stat. solidi (a)*, **61**, K191–5.

Smith, G. & Langer, K. (1982a) Single crystal spectra of olivines in the range 40,000 – 5,000 cm^{-1} at pressures up to 200 kbars. *Amer. Mineral.*, **67**, 343–8.

Smith, G. & Langer, K. (1982b) High pressure spectra of olivines in the range 40,000 – 11,000 cm^{-1}. In *High -Pressure Researches in Geoscience*. (W. Schreyer, ed.; E. Schweizerbart'sche Verlagsbuchhandlung, Stuttgart), pp. 259–68.

Smith, G. H. & Langer, K. (1983) High pressure spectra up to 120 kbars of the synthetic garnet end-members spessartine and almandine. *Neues Jahrb. Mineral. Mh.*, pp. 541–55.

Smith, G. & Strens, R. G. J. (1976) Intervalence transfer absorption in some silicate, oxide and phosphate minerals. In *The Physics and Chemistry of Minerals and Rocks*. (R. G. J. Strens, ed.; J. Wiley & Sons, New York), pp. 583–612..

Smith, G., Hålenius, U. & Langer, K. (1982) Low temperature spectral studies of Mn^{3+}-bearing andalusite and epidote type minerals in the range 30,000–5,000 cm^{-1}. *Phys. Chem. Minerals*, **8**, 136–42.

Smith, G., Hålenius, U., Annersten, H. & Ackermann, L. (1983) Optical and Mössbauer spectra of manganese-bearing phlogopites: Fe^{3+}_{IV}–Mn^{2+}_{VI} pair absorption as the origin of reverse pleochroism. *Amer. Mineral.*, **68**, 759–68.

Smith, G., Howes, B. & Hasan, Z. (1980) Mössbauer and optical spectra of biotite: A case for Fe^{2+}–Fe^{3+} interactions. *phys. sat. solidi (a)*, **57**, K187–92.

Smith, J. V. (1971) Minor elements in Apollo 11 and Apollo 12 olivine and plagioclase. *Proc. 2nd Lunar Sci. Conf., Geochim. Cosmochim. Acta, Suppl. 2* (MIT Press), pp. 143–50.

Smith, M. R., Laul, J. C., Ma, M. S., Huston, T., Verkouteren, R. M., Lipschutz, M. E. & Schmitt, R. A. (1984) Petrogenesis of the SNC (Shergottites, Nakhlites, Chassignites) meteorites. Implications for their origin from a large dynamic planet, possibly Mars. *J. Geophys. Res.*, **89**, B612–30.

Smyth, J. R. & Bish, D. L. (1988) *Crystal Structures and Cation Sites of the Rock-Forming Minerals.* (Allen & Unwin, Boston), 332 pp.

Smyth, J. R. & Hazen, R. M. (1973) The crystal structure of forsterite and hortonolite at several temperatures up to 900 °C. *Amer. Mineral.*, **58**, 588–93.

Smyth, J. R. & Taftø, J. (1982) Major and minor element site occupancies in heated natural forsterite. *Geophys. Res. Lett.*, **9**, 1113–16.

Solberg, T. C. & Burns, R. G. (1989) Iron Mössbauer spectral study of weathered Antarctic and SNC meteorites. *Proc. 19th Lunar Planet. Sci. Conf.* (Cambridge Univ. Press), pp. 313–22.

Stacey, F. D. (1969) *Physics of the Earth.* (J. Wiley & Sons, New York), 324 pp.

Steffen, G., Langer, K. & Seifert, F. (1988) Polarized electronic absorption spectra of synthetic (Mg–Fe)-orthopyroxenes, ferrosilite and Fe^{3+}-bearing ferrosilite. *Phys. Chem. Minerals,* **16**, 120–9.

Stephens, D. R. & Drickamer, H. G. (1961) Effect of pressure on the spectrum of ruby. *J. Chem. Phys.*, **35**, 427–9.

Stergiou, A. C. & Rentzeperis, P. J. (1987) Refinement of the crystal structure of a medium iron epidote. *Zeit. Krist.*, **178**, 297–305.

Stouff, P. & Boulégue, J. (1988) Synthetic 10-Å and 7-Å phyllomanganates: Their structures as determined by EXAFS. *Amer. Mineral.*, **73**, 1162–9.

Straub, D. W. & Burns, R. G. (1992) A kinetic study of the conversion of hematite to magnetite with applications to the metastability of hematite on Venus. *Lunar Planet. Sci.*, **XXIII**, 111–2.

Straub, D. W., Burns, R. G. & Pratt, S. F. (1991) Spectral signature of oxidized pyroxenes: Implications to remote-sensing of terrestrial planets. *J. Geophys. Res.*, **96**, 18819–30.

Strens, R. G. J. (1966a) The axial-ratio-inversion effect in Jahn–Teller distorted ML_6 octahedra in the epidote and perovskite structures. *Mineral. Mag.*, **35**, 777–80.

Strens, R. G. J. (1966b) Pressure-induced spin-pairing in gillespite $BaFeSi_4O_{10}$. *Chem. Comm.*, p. 777.

Strens, R. G. J. (1968) Stability of the Al_2SiO_5 solid solutions. *Mineral Mag.*, **36**, 839–49.

Strens, R. G. J. (1969) The nature and geophysical importance of spin-pairing in mineral of iron(II). In *The Applications of Modern Physics to the Earth and Planetary Interiors*. (S. K. Runcorn, ed.; Wiley–Interscience, London), pp. 213–22.

Strens, R. G. J. (1974) The common chain, ribbon, and ring silicates. In *The Infrared Spectra of Minerals.* (V. C. Farmer, ed.; Mineralogical Soc. Publ.), pp. 305–30.

Strens, R. G. J. (1976) Behaviour of iron compounds at high pressure, and the stability of Fe_2O in planetary mantles. In *The Physics and Chemistry of Minerals and Rocks*. (R. G. J. Strens, ed.; J. Wiley, New York), pp. 545–54.

Strens, R. G. J., Mao, H.-K. & Bell, P. M. (1982) Quantitative spectra and optics of some meteoritic and terrestrial titanian clinopyroxenes. In *Advances in Physical Geochemistry, Vol 2.* (S. K. Saxena, ed.; Springer–Verlag, New York), pp. 327–46.

Sugano, S. & Oshnishi, S. (1984) Electron theory of transition-metal compounds under high pressure. In *Material Science of the Earth's Interior.* (I. Sunagawa, ed; Terra Sci. Publ. Co., Tokyo), pp. 173–89.

Sung, C.-M. (1976) New modification of the diamond anvil press: A versatile apparatus for research at high pressure and temperature. *Rev. Sci. Instrum.*, **47**, 1133-6.

Sung, C.-M., Abu-Eid, R. M. & Burns, R. G. (1974) Ti^{3+}/Ti^{4+} ratios in lunar pyroxenes: Implications to depth and origin of mare basalt magma. *Proc. 5th Lunar Sci. Conf., Geochim. Cosmochim. Acta, Suppl. 5* (Pergamon Press, New York), pp. 717–26.

Sung, C.-M., Singer, R. B., Parkin, K. M. & Burns, R. G. (1977) Temperature dependence of Fe^{2+} crystal field spectra: implications to mineralogical mapping of planetary surfaces. *Proc. 8th Lunar Planet. Sci. Conf.*, Geochim. Cosmochim. Acta, Suppl. 5 (Pergamon Press, New York), pp. 1063–79.

Sunshine, J. M., Pieters, C. M. & Pratt, S. F. (1990) Deconvolution of mineral absorption bands: an improved approach. *J. Geophys. Res.*, **95**, 6955–66.

Surkov, Y. A., Barsukov, V. L., Moskalyeva, L. P. & Kharyukova, V. P. (1984): New data on the composition, structure, and properties of Venus rock obatined by Venera 13 and Venera 14. *J. Geophys. Res.*, **89**, B393–402.

Surkov, Y. A., Moskalyeva, L. P., Kharyukova, V. P., Dudin, A. D., Smirnov, S. Y., and Zaitseva, S. Y. (1986) Venus rock composition and the Vega 2 landing site. *J. Geophys. Res.*, **91**, E215–18.

Süsse, V. (1967) Verfeinerung der Kristallstruktur des malachits. *Acta. Cryst.*, **22**, 146–51.

Sviridov, D. T., Sevastyanov, B. K., Orekhova, V. P., Sviridova, R. K. & Veremeichik, T. F. (1973) Optical absorption spectra of excited Cr^{3+} ions in magnesium spinel at room and liquid nitrogen temperatures. *Opt. Spektrosk.*, **35**, 102–7.

Swinnea, J. S. & Steinfink H. (1983) Crystal structure and Mössbauer spectrum of vonsenite, $2FeO.FeBO_3$. *Amer. Mineral.*, **68**, 827–32.

Syono, Y., Ito, A. & Morimoto, S. (1984) Mössbauer spectra of the high pressure phase of Fe_2O_3. *Solid State Chem.*, **50**, 97–100.

Syono, Y., Tokonami, M. & Matsui, Y. (1971) Crystal field effect on the olivine–spinel transformation. *Phys. Earth Planet. Ineriors*, **4**, 347–52.

Szytula, A., Burewicz, A., Dimitrijevic, A., Krasnicki, S., Rzany, H., Todorovic, J., Wanic, A. & Wolski, W. (1968) Neutron diffraction studies of α-FeOOH. *phys. stat. solidi (a)* **26**, 429–34.

Taftø, J. (1982) The cation–atom distribution in a $(Cr,Fe,Al,Mg)_3O_4$ spinel as revealed from the channeling effect in electron-induced X-ray emission. *J. Appl. Crystallogr.*, **15**, 378–81.

Taftø, J. & Spence, J. C. H. (1982) Crystal lattice location of iron and trace elements in a magnesium–iron olivine by a new crystallographic technique. *Science*, **218**, 49–51.

Takahashi, E. (1978) Partitioning of Ni^{2+}, Co^{2+}, Fe^{2+}, Mn^{2+} and Mg^{2+} between olivine and silicate melts. *Geochim. Cosmochim. Acta*, **42**, 1829–44.

Takeuchi, Y., Haga, N. & Bunno, M. (1983) X-ray study on polymorphism of ilvaite, $HCaFe^{2+}{}_2Fe^{3+}O_2[Si_2O_7]$. *Zeit. Krist.*, **163**, 267–83.

Takeda, H., Miyamoto, M. & Reid, A. M. (1974) Crystal-chemical control on element partitioning for coexisting chromite–ulvospinel and pigeonite–augite in lunar rocks. *Proc. 5th Lunar Sci. Conf., Geochim. Cosmochim. Acta, Suppl. 5* (Pergamon Press, New York), pp. 727–41.

Tamada, O., Fujino, K. & Sasaki, S. (1983) Structures and electron distributions of α-Co_2SiO_4 and α-Ni_2SiO_4 (olivine structure). *Acta Cryst.*, **B39**, 692–7.

Taylor, L. A., Mao, H.-K. & Bell, P. M. (1974) Identification of the hydrated iron oxide mineral akaganeite in sample 66095,85. *Ann. Rept. Geophys. Lab., Yearb.* **73**, 477–80.

Taylor, S. R. (1982) *Planetary Science: A Lunar Perspective.* (Lunar & Planet, Inst., Houston.), 481 pp.

Tejedor-Tejedor, M. I., Anderson, M. A. & Herbillon, A. J. (1983) An investigation of the coordination number of Ni^{2+} in nickel-bearing phyllosilicates using diffuse reflectance spectroscopy. *J. Solid State Chem.*, **50**, 153–62 [see Faye (1974)].

Thonis, M. & Burns, R. G. (1975) Manganese ore deposits and plate tectonics. *Nature*, **253**, 614–15.

Tossell, J. A. (1976) Electronic structures of iron-bearing oxidic minerals at high pressures. *Amer. Mineral.*, **61**, 130–44.

Tossell, J. A. (1978) Self-consistent-field–Xα study of one-electron energy levels in Fe_3O_4. *Phys. Rev.*, **B17**, 484–7.

Tossell, J. A. (1985) Quantum mechanical models in mineralogy. In *Chemical Bonding and Spectroscopy in Mineral Chemistry.* (F. J. Berry & D. J. Vaughan, eds; Chapman and Hall, London), pp. 1–30.

Tossell, J. A. & Vaughan, D. J. (1991) *Theoretical Geochemistry: Applications of Quantum Mechanics in the Earth and Mineral Science.* (Oxford Univ. Press), 416 pp.

Tossell, J. A., Vaughan, D. J., Burns, R. G. & Huggins, F. E. (1972) Do ferric ions occur in the Mantle? *EOS, Trans. Amer. Geophys. Union*, **53**, 1130.

Tossell, J. A., Vaughan, D. J. & Johnson, K. H. (1973a) Electronic structure of ferric iron octahedrally coordinated to oxygen. *Nature Phys. Sci.*, **244**, 42–5.

Tossell, J. A., Vaughan, D. J. & Johnson, K. H. (1973b) The electronic structure of rutile, wüstite and hematite from molecular orbital calculations. *Amer. Mineral.*, **59**, 319–34.

Toulmin III, P. A., Baird, A. K., Clark, B. C., Keil, K., Rose Jr, H. J., Christian, R. P., Evans, P. H. & Keliher, W. C. (1977) Geochemical and mineralogical interpretation of the Viking inorganic chemical results. *J. Geophys. Res.*, **82**, 4625–34.

Townsend, M. G. (1968) Visible charge transfer band in blue sapphire. *Solid State Comm.*, **6**, 81–3.

Townsend, M. G. (1970) On the dichroism of tourmaline. *J. Phys. Chem. Sol.*, **31**, 2481–8.

Townsend, M. G. & Faye, G. H. (1970) Polarized electronic absorption spectrum of vivianite. *phys. stat. solidi (a)*, **38**, K57–60.

Townsend, T. E. (1987) Discrimination of iron alteration minerals in visible and near-infrared reflectance data. *J. Geophys. Res.*, **92**, 1441–54.

Tsang, T. & Ghose, S. (1971) Electron parmagnetic resonance of V^{2+}, Mn^{2+}, Fe^{3+}, and optical spectra of V^{3+} in blue zoisite, $Ca_2Al_3Si_3O_{12}(OH)$. *J. Chem. Phys.*, **54**, 856–62.

Turner, S. & Buseck, P. R. (1979) Manganese oxide tunnel structures and their intergrowths. *Science*, **203**, 456–8.

Turner, S. & Buseck, P. R. (1981) Todorokites: a new family of naturally occurring manganese oxides. *Science*, **212**, 1024–7.

Turner, S. & Buseck, P. R. (1983) Defects in nsutite (γ-MnO_2) and dry-cell battery efficiency. *Nature*, **304**, 143–6.

Turner, S. & Post, J. E. (1988) Refinement of the substructure and superstructure of romanechite. *Amer. Mineral.*, **73**, 1155–61.

Turner, S., Siegel, M. D. & Buseck, P. R. (1982) Structural features of todorokite intergrowths in manganese nodules. *Science*, **296**, 841–2.

Tyrna, P. L. & Guggenheim, S. (1991) The crystal structure of norrishite, $KLiMn^{3+}{}_2Si_4O_{10}$: An oxygen-rich mica. *Amer. Mineral.*, **76**, 266–71.
Ulmer, G. & White, W. B. (1966) Existence of chromous ion in the spinel solid solution series $FeCr_2O_4$–$MgCr_2O_4$. *J. Amer. Ceram. Soc.*, **49**, 50–1.
Vassilikou-Dova, A. B. & Lehmann, G. (1987) Investigations of minerals by electron paramagnetic resonance. *Fortschr. Mineral.*, **65**, 173–202.
van Santen, J. H. & van Wieringen, J. S. (1952) Some remarks on the ionic radii of iron-group elements. *Rec. Trav. Chem. Pay.-Bas.*, **71**, 420–40.
Vance, E. R. & Price, D. C. (1984) Heating and radiation effect on the optical and Mössbauer spectra of Fe-bearing cordierites. *Phys. Chem. Minerals*, **10**, 200–8.
Vaughan, D. J. & Burns, R. G. (1973) Low-oxidation states of Fe and Ti in the Apollo 17 orange soil. *EOS, Trans. Amer. Geophys. Union,* **54,** 618–20.
Vaughan, D. J. & Burns, R. G. (1977) Electronic absorption spectra of lunar minerals. *Phil. Trans. Royal Soc. Lond.* **A285,** 249–58.
Vaughan, D. J. & Craig, J. R. (1985) The crystal chemistry of iron–nickel thiospinels. *Amer. Mineral.*, **70**, 1036–43.
Vaughan, D. J. & Tossell, J. A. (1978) Major transition metal oxide minerals: Their electronic structures and interpretations of mineralogical properties. *Canad. Mineral.*, **16**, 159–68.
Vaughan, D. J. & Tossell, J. A. (1980a) The chemical bond and the properties of sulfide minerals: I. Zn, Fe and Cu in tetrahedral and triangular coordinations with sulfur. *Canad. Mineral.*, **18**, 157–63.
Vaughan, D. J. & Tossell, J. A. (1980b) Electronic structures of thiospinel minerals: Results from MO calculations. *Amer. Mineral.*, **66**, 1250–3.
Vaughan, D. J. & Tossell, J. A. (1983) Electronic structures of sulfide minerals – theory and experiment. *Phys. Chem. Minerals*, **9**, 253–62.
Vaughan, D. J., Burns, R. G. & Burns, V. M. (1971) Geochemistry and bonding of thiospinel minerals. *Geochim. Cosmochim. Acta*, **35**, 365-81.
Vaughan, D. J., Tossell, J. A. & Johnson, K. H. (1974) The bonding of ferrous iron to sulfur to sulfur and oxygen in tetrahedral coordination: A comparative study using SCF–Xα–scattered wave molecular orbitals calculations. *Geochim. Cosmochim. Acta*, **38**, 993–1005.
Verble, J. L. (1974) Temperature-dependent lightscattering studies of the Verwey transition and electron disorder in magnetite. *Phys. Rev. (B)*, **9**, 5236–48.
Verwey, E. J. & Haayman, P. W. (1941) Electronic conductivity and transition point of magnetite. (Fe_3O_4). *Physica*, **8**, 979–87.
Vilas, F. (1985) Mercury: Absence of crystalline Fe^{2+} in the regolith. *Icarus*, **64**, 133–8 [see McCord & Clark (1979)].
Vinogradov, A. P. (1959) *The Geochemistry of Rare and Dispersed Chemical Elements in Soils, 2nd edn.* (Consult. Bureau. Inc., New York), 209 pp.
Virgo, D. & Hafner, S. S. (1969) Fe^{2+},Mg order–disorder in heated orthopyroxenes. *Mineral. Soc. Amer., Spec. Pap.* **2**, 67–81.
Virgo, D. & Hafner, S. S. (1972) Temperature-dependent Mg,Fe distribution in a lunar olivine. *Earth Planet. Sci. Lett.*, **14**, 305–12.
Waddington, T. C. (1959) Lattice energies and their significance in inorganic chemistry. *Adv. Inorg. Chem. Radiochem.*, **1**, 157–221.
Wager, L. R. & Mitchell, R. L. (1951) Distribution of trace elements during strong fractionation of basic magma – a further study of the Skaergaard intrusion. *Geochim. Cosmochim. Acta*, **1**, 129–208.
Walsh, D., Donnay, G. & Donnay, J. D. H. (1974) Jahn–Teller effects in ferro–magnesian minerals: pyroxenes and olivines. *Bull. Soc. Franc. Mineral. Cristallogr.*, **97**, 170–83.

Walsh, D., Donnay, G. & Donnay, J. D. H. (1976) Ordering of transition metal ions in olivine. *Canad. Mineral.*, **14**, 149–50.

Wan, C., Ghose, S. & Rossman, G. R. (1978) Guildite, a layer structure with a ferric hydroxy-sulphate chain and its optical absorption spectrum. *Amer. Mineral.*, **63**, 478–83.

Wang, P. J. & Drickamer, H. G. (1973) Reduction of Cu(II) at high pressure. *J. Chem. Phys.*, **59**, 713–17.

Watanabe, H. (1982) Thermodynamic properties of synthetic high-pressure compounds relevant to the Earth's mantle. In *High-Pressure Research in Geophysics*. (M. H. Manghnani & S. Akimoto, eds; Center Acad. Publ.,Tokyo), pp. 441–64.

Watson, T. L. (1918) The color change in vivianite and its effect on the optical properties. *Amer. Mineral.*, **3**, 159–61.

Watson, E. B. (1977) Partitioning of manganese between forsterite and silicate liquid. *Geochim. Cosmochim. Acta,* **41**, 1363–74.

Waychunas, G. A. (1987) Synchrotron radiation XANES spectroscopy of Ti in minerals: Effects of Ti bonding differences, Ti valence and site geometry on absorption edge structure. *Amer. Mineral.*, **72**, 89–101.

Waychunas, G. A. (1991) Crystal chemistry of oxides and oxyhydroxides. In *Oxide Minerals: Petrologic and Magnetic Significance*. (D. H. Lindsley, ed.; Mineral Soc. Amer. Publ.), *Rev. Mineral.*, **25**, 11–68.

Waychunas, G. A. & Rossman, G. R. (1983) Spectroscopic standard for tetrahedrally coordinated ferric iron: γ-$LiAlO_2$: Fe^{3+}. *Phys. Chem. Minerals*, **9**, 212–5.

Waychunas, G. A., Apted, M. J. & Brown Jr, G. E. (1983) X-ray K-edge absorption spectra of Fe minerals and model compounds: I. Near-edge structure. *Phys. Chem. Minerals*, **10**, 1–9.

Waychunas, G. A., Brown Jr, G. E. & Apted, M. J. (1986) X-ray K-edge absorption spectra of Fe minerals and model compounds: II. EXAFS. *Phys. Chem. Minerals*, **13**, 31–47.

Waychunas, G. A., Brown Jr, G. E., Ponader, C. W. & Jackson, W. E. (1988) Evidence from X-ray absorption for network-forming Fe^{2+} in molten alkali silicates. *Nature*, **332**, 251–3.

Wedepohl, K. H., exec. ed. (1969–78) *Handbook of Geochemistry.* (Springer–Verlag), 5 vols.

Weeks, R. A., Pigg, J. C. & Finch, C. B. (1974) Charge transfer spectra of Fe^{3+} and Mn^{2+} in synthetic forsterite (Mg_2SiO_4). *Amer. Mineral.*, **59**, 1259–66.

Weiss, Z, Bailey, S. W. & Rieder, M.(1981) Refinement of the crystal structure of kanonaite, $(Mn^{3+},Al)^{[6]}(Al,Mn^{3+})^{[5]}O[SiO_4]$. *Amer. Mineral.*, **66**, 561–7.

Wells, A. F. (1984) *Structural Inorganic Chemistry, 5th edn.* (Oxford Univ. Press), 1382 pp.

Wendlandt, W. & Hecht, H. G. (1966) *Reflectance Spectroscopy* (Springer–Verlag, New York), 298 pp.

Weyl, W. A. (1951) Light absorption as a result of the interaction of two states of valency of the same element. *J. Phys. Colloid. Chem.*, **55**, 507–12.

Wenk, H. R. & Raymond, K. N. (1973) Four new structure refinements of olivine. *Zeit. Krist.*, **137**, 86–195.

Wherry, E. T. (1918) Note on iron as a cause of blue color in minerals. *Amer. Mineral.*, **3**, 161.

White, E. W. & White, W. B. (1967) Electron microprobe and optical absorption study of colored kyanites. *Science*, **158**, 915–17.

White, W. B. & Keester, K. L. (1966) Optical absorption spectra of iron in the rock-forming silicates. *Amer. Mineral.*, **51,** 774–91 [see Bancroft & Burns (1967a)].

White, W. B. and Moore, R. K. (1972) Interpretation of the spin-allowed bands of Fe^{2+} in silicate garnets. *Amer. Mineral.*, **57**, 1692–710.

White, W. B., McCarthy, G. J. & Scheetz, B. E. (1971) Optical spectra of chromium, nickel and cobalt-containing pyroxenes. *Amer. Mineral.*, **56**, 72–89.

White, W. B., Roy, R. & Crichton, J. M. (1967) The 'alexandrite effect': An optical study. *Amer. Mineral.*, **52**, 867–71.

Whittaker, E. J. W. (1967) Factors affecting element ratios in the crystallization of minerals. *Geochim. Cosmochim. Acta*, **31**, 2275–88.

Whittaker, E. J. W. & Muntus, R. (1970) Ionic radii for use in geochemistry. *Geochim. Cosmochim. Acta*, **34**, 945–56.

Wilkins, R. W. T. (1967) The hydroxyl stretching region of the spectrum of biotite mica. *Mineral. Mag.*, **36**, 325–33.

Wilkins, R. W. T., Farrell, E. F. & Naiman, C. S. (1969) The crystal field spectra and dichroism of tourmaline. *J. Phys. Chem. Sol.*, **30**, 43–56.

Will, G. & Nover, G. (1979) Influence of oxygen partial pressure on the Mg/Fe distribution in olivine. *Phys. Chem. Minerals*, **4**, 199–208.

Williams, Q., Knittle, E. & Jeanloz, R. (1989) Geophysical and crystal chemical significance of (Mg,Fe)SiO_3 perovskite. In *Perovskite: A Structure of Great Interest to Geophysics and Materials Science.* (A. Navrotsky and D. J. Weidner, eds; Amer. Geophys. Union), *Geophys. Mono.* **45**, 1–12.

Williams, R. J. P. (1959) Deposition of trace elements in a basic magma. *Nature*, **184**, 44.

Winter, J. K. & Ghose, S. (1979) Thermal expansion and high-temperature crystal chemistry of the Al_2SiO_5 polymorphs. *Amer. Mineral.*, **64**, 573–86.

de Wolff, P. M., Visser, J. W., Giovanoli, R. & Brütsch., R. (1978) Über ε-· mangandioxid. *Chimia*, **32**, 257–9.

Wood, B. J. (1974) Crystal field spectrum of Ni^{2+} in olivine. *Amer. Mineral.*, **59**, 244–8.

Wood, B. J. (1981) Crystal field electronic effects on the thermodynamic properties of Fe^{2+} minerals. In *Thermodynamics of Minerals and Melts*. (R. C. Newton, A. Navrotsky & B. J. Wood, eds; Springer–Verlag, New York), *Adv. Phys. Geochem.*, **1**, 63–84.

Wood, B. J. & Strens, R. G. J. (1972) Calculation of crystal field splittings in distorted coordination polyhedra: Spectra and thermodynamic properties of minerals. *Mineral. Mag.*, **38**, 909–17.

Wood, D. L. & Nassau, K. (1968) The characterization of beryl and emerald by visible and infrared absorption spectroscopy. *Amer. Mineral.*, **53**, 777–800.

Wood, M. M. (1970) The crystal structure of ransomite. *Amer. Mineral.*, **55**, 727–34.

Xu Zitu, Zheng Chusheng & Peng Mingsheng (1982) Calculation of crystal field theory of Mn^{3+} in piemontite. *Kexue Tongbao*, **27**, 1199–203.

Yagi, T. & Mao, H.-K. (1977) Crystal field spectra of the spinel polymorph of Ni_2SiO_4 at high pressure. *Ann. Rept. Geophys. Lab., Yearb.* **76**, 505–8.

Yagi, T., Bell, P. M. & Mao, H.-K. (1979) Phase relations in the system MgO–FeO–SiO_2 between 150 and 700 kbar at 1000 °C. *Ann. Rept. Geophys. Lab., Yearb.* **78**, 614–18.

Zhao, S.-B., Wang, H,-S., Zhou, K. W. & Xiao, T. B. (1986) The spin-forbidden absorption spectrum of Fe^{2+} in orthopyroxenes. *Phys. Chem. Minerals*, **13**, 96–101.

Subject index

A crystal field state, 56, 467
a_{1g} orbital, 11
absorbance,
  see optical density, 47
absorption bands,
  effects of temperature, 82, 120, 362–4, 412–15
  effects of pressure, 91, 94. 120, 126, 360–74
  energies, 50–64
  intensities, 64–80, 91, 93
  number, 60
  shape, 47, 80, 82
  widths, 47, 80–2, 120
absorption coefficients, 46, 47
  and radiative transfer in the mantle, 390
absorption spectra, polarized,
  of actinolite, 196
  of blue sapphire, 127
  of calcic amphibole, 196
  of calcic pyroxene, 196, 402
  of cummingtonite, 192
  of enstatite, 182
  of epidote, 109
  of fayalite, 104, 162
  of forsterite under pressure, 372
  of gillespite, 49
  of glaucophane, 125
  of grunerite, 193
  of hedenbergite, 181, 402
  of knebellite, 104
  of liebenbergite, 116
  of lusakite, 102
  of manganiferous olivines, 104
  of Ni^{2+}–Mg olivines, 166
  of olivine under pressure, 104
  of orthoferrosilite, 183
  of piemontite, 96
  of plagioclase, 207
  of ruby, 111
  of sapphires, 127
  of staurolite, 100
  of tanzanite, 114
  of tephroite, 104
  of Ti-doped corundum, 127
  of vanadium zoisite, 114
  of vivianite, 121
  of yellow sapphire, 127
absorption spectra, unpolarized,
  of almandine, 157
  of andradite, 109
  of hexahydrated Co^{2+}, 102
  of hexahydrated Fe^{2+}, 48
  of hexahydrated Ti^{3+}, 48
  of spinel containing Co^{2+}, 102
abundances, of transition elements, 1, 301, 460
acceptor ligands, 438–9
acentric coordination site, 21
  in epidote, 95–7
  in olivine, 103
acentric vibrational modes, 68
  and intensities of CF spectra, 72
actinolite,
  cation ordering, 258, 281
  CF parameters for Fe^{2+}, 189, 228, 229, 288
  IVCT energies, 118
  polarized spectra, 196
activation energies,
  in substitution reactions, 331, 334
admitted trace element, 303
aegerine,
  CF parameters for cations, 189
akaganeite,
  crystal chemical data, 418–19
  CF parameters for Fe^{3+}, 418–19
  effects of pressure on CF spectra, 365
albite–diopside glasses,
  CF spectra, 316
  CFSE of cations, 316

alexandrite,
  birthstone, 106
  colour due to Cr^{3+}, 106, 112
  CF parameters for Cr^{3+}, 176, 213
  gem chrysoberyl, 105
  site occupancy of Cr^{3+}, 269
alexandrite effect, 112, 115
  in chrysoberyl, 112
  in coquimbite, 112
  in corundum, 112
  in fluorite, 112
  in garnets, 112
  in V^{3+}-doped Al_2O_3, 113
alkali amphiboles,
  see glaucophane; riebeckite
Allende meteorite Ti^{3+} pyroxene, 93
  crystal chemical data, 179
  IVCT energy, 119
  polarized spectra, 94
  pressure effects on optical spectra, 94, 363, 366
almandine,
  absorption spectrum, 157
  colour due to Fe^{2+}, 110
  coordination site properties, 470
  energy level diagram for Fe^{2+}, 158
  CF parameters for Fe^{2+}, 228, 229, 288
  CFSE of Fe^{2+}, 157, 288
  effect of pressure on CF spectra, 364, 371–2
  effect of temperature on CF spectra, 364, 371–2
  polyhedral bulk modulus of Fe^{2+}, 376
α-spin configuration, 10
  see also spin-up configuration
alpine peridotites, Ni in, 328
aluminosilicates,
  cation ordering, 250, 261, 281
  CF spectra, 172–4
  CFSE of Fe^{2+}, 288
  oxygen ligands, 30
amazonite, colour of, 207
amethyst,
  birthstone, 106
  gem quartz, 88
  optical spectra, 207
  cause of colour, 106, 207
amphiboles,
  cation ordering, 258
  CF parameters for Fe^{2+}, 228, 230, 288
  CF spectra, 189, 190–8
  crystal structure, 191
  effect of pressure on optical spectra, 366
  fluorine avoidance, 260
  polarized spectra, 192, 193, 196
andalusite,
  coordination site properties, 471
  CF parameters
    for Cr^{3+}, 213
    for Mn^{3+}, 218
  influence of Mn^{3+} on phase equilibria, 100, 294
  IVCT energies, 117, 119
  stability of Mn^{3+}, 34, 39, 99, 294
  structure, 99, 172
andradite,
  absorption spectrum, 109
  coordination site properties, 470
  CF parameters for Fe^{3+}, 159, 224
Angra dos Reis meteorite pyroxene, 126
  colour due to IVCT, 126
  coordination site properties, 179
  crystal chemical data, 179
  effect of pressure on optical spectra, 366
  IVCT energy, 118
  optical spectra, 126
angular functions, 8
angular momentum,
  orbital, 9
  spin, 10
  total orbital, 52
anions, 3, 7
anion radii, 7, 465
annite,
  coordination site properties, 473
  mixed-valence properties, 141
anthophyllite
  cation ordering, 258, 279
antibonding orbitals, 15, 435, 437, 446
antisymmetric wave functions, 13, 65
  see also ungerade states
apatite
  colours due to lanthanide elements, 115
apophyllite
  optical spectra of VO_2^+, 211
April birthstone, 106
aquamarine
  birthstone, 106
  colour due to IVCT, 106
  gem beryl, 105
  IVCT energy, 117
argon core, 14
arsenides
  MO diagrams, 442
  π-bonding, 442
asbolane
  crystal chemical data, 339
  in manganese nodules, 346
asteroids
  spectra and surface features, 399
astrophyllite
  valence of Ti, 126
atmophile elements, 302
atomic orbitals, 2, 12
augite
  crystal chemical data, 178
  CF spectra of Fe^{2+}, 181
  IVCT energy, 117
August birthstone, 107

axial spectrum, 74
axinite
CF parameters for V^{3+}, 210
azimuthal quantum number, *l*, 9, 66
azurite
colour due to Cu^{2+}, 115
CF parameters for Cu^{2+}, 238
interatomic distances, 245
Jahn–Teller distortion, 243, 245

B crystal field state, 56, 63, 467
babingtonite
CF parameters for Fe^{2+}, 188, 228, 230, 288
intensity enhancement by IVCT, 454
IVCT energies, 117, 119
band-gap transitions, 108
baricentre, 14, 16, 26, 99, 319
basalt
CF spectra of glasses, 67, 316
CFSE of cations in glasses, 316
reflectance spectra, 413
bastnaesite
colours due to lanthanide elements, 115
Beer–Lambert law, 91, 102
deviations from, 92
beryl
coordination site properties, 473
CF parameters for cations, 198, 200, 210, 213, 218, 225, 230
CF parameters for Fe^{2+}, 228, 230, 288
structure, 198
β-spin configuration, 10
see also spin-down configuration
biaxial minerals, 50, 75
biophilic elements, 302
biotite
cation ordering, 259
coordination site properties, 473
CF parameters for Fe^{2+}, 228, 230, 288
crystal structure, 203
effect of temperature on optical spectra, 366
fluorine avoidance, 260
IVCT energies, 118
optical spectra, 204
OMCT, 133
Mössbauer spectra, 259
birnessite
crystal chemical data, 339
in manganese nodules, 347
structural stability from MO theory, 456
birthstones, 106–7
bixbyite
interatomic distances, 244
Jahn–Teller distortion by Mn^{3+}, 244
blanfordite
CF parameters for Mn^{3+}, 188, 218
blue sapphire
IVCT energies, 118
effect of temperature on IVCT, 366
polarized spectra, 127
see also corundum and sapphire
blue-shifts, 361
Born equation, 276, 304
botryogen
CF parameters for Fe^{3+}, 225
braunite
interatomic distances, 244
Jahn–Teller distortion by Mn^{3+}, 244
bridging oxygens in silicates, 30
bunsenite
CF parameters for Ni^{2+}, 151, 236
buserite
interatomic distances in, 246
Jahn–Teller distortion by Mn^{3+}, 246
in manganese nodules, 347
structural stability from MO theory, 456
Bushveld Complex, chromite in, 324
bustamite
CF parameters for Mn^{2+}, 220
butlerite
CF parameters for Fe^{3+}, 225

calcic amphiboles
coordination site properties, 473
polarized spectra, 196
see also actinolite *and* tremolite
calcic pyroxenes
coordination site properties, 472–3
CF parameters for cations, 189, 225
CF spectra, 180–2
crystal structure, 177
polarized spectra, 181, 402
reflectance spectra, 402, 406, 410
temperature effects on spectra, 413
see also diopside *and* hedenbergite
calcite
colours due to lanthanide elements, 115
calderite
CF parameters for Mn^{3+}, 159, 218
camouflaged trace element, 303
carbonaceous chondrites,
composition of Mantle, 354
hibonite in, 129, 292
cation, 14
sequence of CF splittings, 28
nephelauxitic series, 434
cation ordering, 39, 250
from absorption spectra, 92, 102, 254
in Mg–Fe^{2+} orthopyroxenes, 97, 103
in Mg–Ni^{2+} olivines, 168
Mn^{2+} in olivines 103
Mn^{3+} in epidotes, 96
cation radii, 464
see ionic radii

cation valences from absorption spectra, 93
cattierite
  interatomic distance, 442
  π-bonding, 442
central field covalency, 433
centre of gravity
  d orbitals, 14
  rule, 16
centrosymmetric coordination site, 22, 55
Ceres
  spectra and surface features, 399
CFSE, definition, 3, 4, 17
  in aluminosilicates, 172
  in amphiboles, 189
  in beryls, 200
  in clay silicates, 200
  in corundum, 154
  in feldspars, 201
  in garnets, 159
  in hexahydrated cations, 29
  in micas, 200–1
  in olivines, 170
  in periclase, 151
  in pyroxenes, 188
  in tourmalines, 200
  see also individual transition elements
CFSE of mixing, 281
  in Cr^{3+} oxides, 284
  in Mg–Fe^{2+} olivines, 281
  in Mg–Ni^{2+} olivines, 282
  in Mn^{3+} minerals, 283
chalcanthite
  CF parameters for Cu^{2+}, 238
  interatomic distances, 245
  Jahn–Teller distortion by Cu^{2+}
chalcophanite
  crystal chemical data, 339
  crystal structure, 342
  structural stability from MO theory, 456
chalcophilic elements, 302, 429, 440
character tables, 75
charge transfer spectra, 27
  see IVCT transitions *and* OMCT transitions
chemical isomer shift, 11, 17
chlorite
  CF parameters for Cr^{3+}, 213
  CF parameters for Fe^{2+}, 201, 228, 230, 288
  IVCT energy, 118
chloritoid
  CF parameters for Fe^{2+}, 201, 228, 230, 288
  IVCT energy, 118
chromite
  coordination site properties, 470
  CF parameters for
    Cr^{2+}, 155, 216
    Cr^{3+}, 155, 213
  effects of pressure on CF spectra, 362
  in layered intrusions, 3, 301, 324
  polyhedral bulk modulus of Cr^{3+}, 376
chromium
  abundance data, 301, 460
  in lherzolites, 325
  in metamorphic processes, 350
  oxidation states in Eh–pH diagrams, 336–7
  in Skaergaard intrusion, 312
chromium (II),
  CF activation energies, 334
  CF parameters for compounds, 212–16
  CF spectra, 214–16
  CF splitting
    in hexahydrated cation, 29, 216
    in periclase, 29
  CF states in octahedral sites, 215
  CFSE
    in compounds, 216
    in hexahydrated cation, 29
    in monoclinic sites, 264
    in octahedral sites, 19, 249, 264
    in tetragonal sites, 38, 264
    in tetrahedral sites, 23, 249
    in trigonal sites, 264
  electronic configurations, 15, 19, 23
  electronic entropy, 285
  energy level diagrams, 60, 215
  intersite partition coefficient for
      orthopyroxene, 256
  ionic radii, 464
  in magmatic crystallization, 317
  in Moon rocks, 14, 397
  octahedral site preference energy, 249
  Orgel diagram, 60
  spectroscopic terms, 53
  stability in olivines, 34, 39, 329
  Tanabe–Sugano diagram, 215
chromium(III),
  cation ordering of Cr^{3+}
    in alexandrite, 260
    in epidote, 258
    in tourmaline, 260
  colours in minerals due to Cr^{3+}, 88, 105,
      106–17, 115
  CF activation energies, 334
  CF parameters for minerals, 212–14
  CF spectra, 212–14
  CF splitting
    in corundum or ruby, 29, 153–4, 213
    in hexahydrated cation, 28, 29, 213
    in periclase, 29, 151, 213
  CF states in octahedral sites, 212
  CFSE
    in hexahydrated cation, 29, 32
    in minerals, 213
    in monoclinic site, 264
    in octahedral sites, 19, 249, 264
    sequence in minerals, 214
    in tetragonal sites, 38, 264

chromium(III) (*cont.*)
CFSE (*cont.*)
in tetrahedral sites, 23, 103, 216, 249
in trigonal sites, 264
on Earth, 14, 397
electronic configurations, 15, 19, 23
electronic entropy, 285
energy level diagrams, 61, 202
in gems, 88, 105–17
interphase partitioning, 293
ionic radii, 464
in magmatic crystallization, 317
in Moon rocks, 14, 397
in normal spinels, 248
octahedral site preference energy, 249
Orgel diagram, 61
Racah parameters, 214, 432
in silicate glasses, 315
in silicate melts, 306
spectroscopic terms, 53
Tanabe–Sugano diagram, 202
chromium(VI),
on Earth, 14, 397
electronic configuration, 15
spectroscopic term, 53
stability in oxidizing environments, 340
chrysoberyl,
alexandrite effect, 112
cation ordering, 259–60
colour due to Cr^{3+} in alexandrite, 112
colour due to Fe^{3+} in sinhalite, 115
CF parameters for Cr^{3+}, 112, 176, 213
CF parameters for V^{3+}, 210
chrysocolla,
CF parameters for Cu^{2+}, 238
clinochlore,
CF parameters for Cr^{3+}, 200, 213
clusters,
coordination, 5, 67
MO diagrams, 17, 436, 441, 444, 452, 455
cobalt,
abundance data, 301, 460
oxidation states in Eh–pH diagrams, 337
in manganese nodules, 346–9
in metamorphic processes, 350
cobalt(II),
colours in minerals due to Co^{2+}, 115
CF activation energies, 334
CF parameters for minerals, 232–4
CF spectra, 234
CF splitting
in corundum, 29
in hexahydrated cation, 29
in oxides
in periclase, 29, 151, 232
CF states in octahedral sites, 231, 234
CFSE
in hexahydrated cation, 29
in minerals, 232–3
in monoclinic sites, 264
in octahedral sites, 19, 249, 264
in oxides, 276
in sulphides, 276
in tetragonal sites, 38, 264
in tetrahedral sites, 23, 249
in trigonal sites, 264
on Earth, 14, 397
electronic configurations, 15, 19, 23
electronic entropy, 285
energy level diagrams, 61, 231
interphase partitioning, 290–2
intersite partition coefficients for olivine and orthopyroxene, 256
ionic radii, 464
in magmatic crystallization, 317
in Moon rocks, 14, 397
octahedral site preference energy, 249
Orgel diagram, 61
π-bond formation, 440
spectroscopic terms, 53
Tanabe–Sugano diagram, 231
in tetrahedral sites in staurolite, 22
cobalt(III),
CF activation energies, 334
CF spectra of compounds, 227–31
CF splitting
in corundum, 29, 154, 231
in hexahydrated cation, 29, 231
CF states in octahedral sites, 227
CFSE
in compounds, 231
in hexahydrated cation, 29
in octahedral sites, 19, 264
in oxides, 276
in sulphides, 276
in tetragonal sites, 38, 264
in tetrahedral sites, 23, 249 264
in trigonal sites, 264
on Earth, 14, 397
electronic configurations, 15, 19, 23
electronic configurations of CF states, 58
electronic entropy, 285
energy level diagrams, 55, 231
ionic radii, 464
low-spin configuration in oxides, 30, 243, 379
in magmatic crystallization, 317
in Mn(IV) oxides, 348
octahedral site preference energy, 249
Orgel diagram, 60
π-bond formation, 440
spectroscopic terms, 53
Tanabe–Sugano diagrams, 57, 231
cobalt diopside,
coordination site properties, 473
CF parameters for Co^{2+}, 188, 232

cobalt enstatite,
  coordination site properties, 188, 472
  CF parameters,
cobalt olivine, 171, 232
cobalt silicate spinel,
  coordination site properties, 472
colour,
  concept, 105
  in minerals, 30, 39, 88, 106–7
  origins, 108–133
colour centres,
  in birthstones, 106–7
composition determinative curves,
  of Mg–Fe^{2+} olivines, 163
  of Mg–Fe^{2+} pyroxenes, 184
  of Mg–Ni^{2+} olivines, 168
Condon–Shortley parameters, 53
configurational entropy, 278–81
  ferromagnesian silicates, 282
  liebenbergite, 279, 280
  lunar olivine, 281
  orthopyroxenes, 278–9
coordination cluster, 17, 67
  octahedral $[FeO_6]^{-9}$, 444
  octahedral $[FeO_6]^{-10}$, 444
  octahedral $[ML]_6$, 17
  octahedral $[MnO_6]^{-9}$, 455
  shared octahedra in $[Fe_2O_{10}]^{-15}$
  tetrahedral $[MnO_4]^{-4}$, 455
coordination sites, 4
  in minerals hosting transition elements, 470–4
Copernicus,
  reflectance spectrum of Moon, 412
copper,
  abundance data, 301, 460
  in manganese nodules, 346–8
  in metamorphic processes, 350
  oxidation states in Eh–pH diagrams, 337
  in porpyhry copper deposits, 1, 301, 322, 397
copper(I),
  on Earth, 14
  electronic configurations, 15, 19, 23
  spectroscopic term, 53
copper(II),
  colours in minerals due to Cu^{2+}, 115
  CF activation energies, 334
  CF parameters for minerals, 235, 238–9
  CF spectra, 235–9
  CF splitting
    in hexahydrated cation, 29, 52
    in oxides, 29
  CF states in octahedral sites, 235
  CFSE
    in hexahydrated cation, 29
    in monoclinic sites, 264
    in octahedral sites, 19, 249, 264
    in oxides, 276
    in sulphides, 276
    in tetragonal sites, 38, 264
    in tetrahedral sites, 23, 249
    in trigonal sites, 264
  on Earth, 14, 397
  electronic configurations, 15, 19, 23
  electronic entropy, 285
  energy level diagrams, 51, 60
  ionic radii, 464
  in magmatic crystallization, 317
  in Moon rocks, 14, 397
  octahedral site preference energy, 249
  Orgel diagram, 60
  π-bond formation, 440
  spectroscopic term, 53
coquimbite,
  alexandrite effect in, 112
  CF parameters for Fe^{3+}, 225
cordierite,
  coordination site properties, 473
  CF parameters for Fe^{2+}, 228, 230, 288
  effect of temperature on IVCT, 366
  IVCT energy, 117
corundum, alexandrite effect in, 112
  CF parameters
    for cations, 29, 154, 200, 208, 210, 213, 216, 218, 224, 229, 231, 232, 236
    for Fe^{2+}, 228, 229, 288
  crystal structure, 151
  IVCT, 129
  effect of pressure on CF spectra, 369
  polarized spectra, 70, 111, 127
  polyhedral bulk moduli of cations, 376
  see also blue sapphire, ruby *and* yellow sapphire
cosmic abundances, 1, 460
coupling interactions, 71
  see also pair-transitions
covalent bonding, 3, 5, 28, 53
  and Goldschmidt's rules, 303
  influence on CF spectra, 67, 428
  at high pressures, 385, 435
  and intersite partitioning, 268
  in Mn(IV) oxides, 454–7
  and sulphide mineralogy, 429–39, 440–2
crocidolite, *see* riebeckite
crocoite,
  OMCT, 132
  effect of pressure on IVCT, 367
cronstedtite,
  electron delocalization, 133
  mixed-valence effects, 141
crustal abundances, 1, 301, 460
crystal field spectra (CF spectra),
  classification, 148
  intensities, 72
  of mineral groups, 146–208
  of silicate glasses, 314, 320

crystal field spectra (CF spectra) (*cont.*)
of transition metal ions, 208–39
crystal field (CF) splitting, Δ, 2, 5, 15, 27
in corundum, 29
in cubic coordination, 22–3
dependence on interatomic distance, 27
effect of temperature, 30
effect of pressure, 31
in dodecahedral coordination, 22–3
evaluation of, 27, 273–5
in hexahydrated cations, 29
in octahedral coordination, 15–21
in periclase, 29
in tetrahedral coordination, 21–2
and type of cation, 27
and type of ligand, 28
crystal field stabilization energy (CFSE), 3, 17
crystal field states, 50, 53
Fe^{2+} in coordination sites, 65
crystal field theory, 1, 3, 7–43
history, 4
origin of name, 5
scope, 5
relation to MO theory, 429, 437, 448
crystalline field, 5
cubic coordination, 20
CF splitting parameter, Δ_c, 21–3
Orgel diagrams, 60–2
cuboctahedral coordination, 20
see dodecahedral coordination
cummingtonite,
CF parameters for Fe^{2+}, 189, 228, 230, 288
crystal structure, 191
polarized spectra, 192
cuprorivaite,
colour due to Cu^{2+}, 115
CF spectra, 79, 115, 201, 206, 238
interatomic distances, 246
cyanide complexes,
and low-spin states, 28, 228, 383

d orbitals, 2, 9, 10, 12
$d \rightarrow d$ spectra,
see absorption spectra; CF spectra
D spectroscopic term, 50, 466
December birthstones, 107
deerite,
electron delocalization, 133
mixed-valence properties, 141
structure, 141
delocalized electrons 133
see electron delocalization
Δ, 5, 15, 437
see crystal field splitting
demantoid,
birthstone, 106
gem garnet, 106
see andradite
diagenesis, 349
diamond,
birthstone, 106
Cr olivine inclusions, 329
dichroism, 108
diffusion, 349
diopside,
blue due to Cr, 215
coordination site properties, 472
crystal chemical data, 178
CF parameters
for cations, 188, 210, 213, 216, 229, 232, 236
for Fe^{2+}, 228, 229, 288
crystal structure, 177, 178
interatomic distances at high pressures, 359
diopside–albite glass,
CF parameters for cations, 210–37
CFSE of cations, 316
dioptase,
colour due to Cu^{2+}, 115
interatomic distances, 246
Jahn–Teller distortion by Cu^{2+}, 246
dipolar ligands, 3, 7
dipole moment, 66, 99
direct product, 75, 78
disproportionation,
of Mn(III) compounds, 18, 95
in substitution reactions, 333
to post-spinel phases in Lower Mantle, 388
distorted sites,
CFSE, 36, 38, 264
effects on cation distribution, 263
and electronic entropy, 40, 286
energy level diagrams, 39, 62, 263
and interphase partitioning, 291
see also Jahn–Teller effect
distribution coefficients, 295, 305
for crystal–melt systems, 295
and ionic radii, 305
and OSPE, 296
for Skaergaard intrusion, 313
divalent cations,
CFSE
in oxides, 276
in sulphides, 276
distribution in spinels, 248
intersite partitioning in silicates, 262–8
interphase partitioning, 290–2
magmatic crystallization, 296, 313
octahedral site preference energy, 249
dodecahedral coordination, 20
CF splitting parameter, Δ_d, 21–3
Orgel diagrams, 60–2
donor ligands, 438–9
doublet state, 54, 59, 466
Dq parameter, 15, 24–27
in MO diagrams, 437

dynamic Jahn–Teller effect, 36, 81
 in Co^{3+} CF spectra, 227
 in Fe^{2+} CF spectra, 223
 in magnesiowüstite spectra, 150

E crystal field state, 54, 467
e and e_g-group *d* orbitals, 12, 13
 in MO diagrams, 445
Earth,
 chemical fractionation, 1
 composition, 1, 301, 460
 structure, 353, 354
Egyptian blue,
 see cuprorivaite
Eh–pH diagrams, 336–7
electric dipole, 65, 74, 75, 79, 99
 antisymmetric property, 75
electric vector, 50, 73, 75, 78
electrical conductivity,
 in mixed-valence minerals, 134, 135, 142
electromagnetic radiation, 65
electromagnetic spectrum, 27
electron,
 angular momentum, 9
 polar coordinates, 8
 spin, 10
electron delocalization, 120, 133, 135
 effects in Mössbauer spectra, 133, 135, 137–9, 140
 mixed-valence minerals, 120, 133–42
electron density, 11, 13
electron hopping,
 see electron delocalization
electron paramagnetic resonance (EPR),
 and Jahn–Teller distortion, 36
 and spin–orbit coupling, 11
electron spin resonance (ESR), 3
 see also electron paramagnetic resonance
electronegativity,
 and trace element distribution, 304
electronic axes, 63–4, 77
electronic configurations of first transition series, 15
 in octahedral coordination, 19
 in tetrahedral coordination, 23
 of Fe^{2+} CF states, 55
 of Fe^{3+} CF states, 58
electronic entropy, 40, 285
 contribution to thermodynamic properties, 287
 effect on olivine → spinel transitions, 287
 Fe^{2+} in octahedral sites, 286
 spinel crystal chemistry, 249, 287
 spin-state transitions, 287
electrostatic field, 7
 non-spherical, 14
ellenbergerite,
 IVCT energy, 117 129, 131
 structure, 131
emerald,
 birthstone, 106
 colour due to Cr^{3+}, 105, 106, 115
 CF parameters for Cr^{3+}, 199, 200, 213
 gem beryl, 88, 105
energy level diagrams,
 crystal field, 16, 21, 39
 molecular orbital, 17, 436, 438, 444, 452, 455
 Orgel 59–61
 Tanabe–Sugano, 56–9
energy units, 45, 475–6
enrichment principle, 303
enstatite,
 coordination site properties, 472
 crystal chemical data, 178
 effect of pressure on CF spectra, 363
 CF parameters for cations, 188, 213, 229, 232, 236, 238
 crystal structure, 185
 polarized spectra, 182
 reflectance spectra, 406, 411
entropy,
 electronic, 40, 285
 of fusion, 309, 322
 of mixing, *see* configurational entropy
epidote,
 cation ordering of Fe^{3+}, 175
 coordination site properties, 471
 CF parameters for Fe^{3+}, 225
 CF spectra, 109
 colour due to Fe^{3+}, 109, 115
 effect of temperature on CF spectra, 365
 oxygen ligands, 30
 pleochroism, 109
 polarized spectra, 109
 site occupancies of cations, 259
 see also piemontite, tawmawite *and* zoisite
eskolaite,
 colour due to Cr^{3+}, 30, 101
 coordination site properties, 470
 CF parameters for Cr^{3+}, 154, 213
 Racah *B* parameter, 432
euclase,
 CF parameters for Fe^{2+} 228, 229, 288
 CF spectra of Cr^{3+}, 176
 IVCT energy, 118
eudialyte,
 CF spectra of Fe^{2+}, 203
 structure, 203
 trace elements, 306
EXAFS, 3
 see X-ray absorption spectra
excess CFSE of mixing, 281
 Mg^{2+}–Fe^{2+} olivines, 281–2
 Mg^{2+}–Ni^{2+} olivines, 282–3
 Cr^{3+} in Al_2O_3–Cr_2O_3 solid solutions, 283–4
excited state, 68

extinction coefficient,
see molar extinction coefficient, 47
extraterrestrial pyroxenes,
Ti^{3+} in, 93

f orbitals, 9, 10, 13
F spectroscopic term, 54, 466
face-shared octahedra,
in corundum, 128, 132, 151, 223
in ellenbergerite, 129, 132
in hematite, 420
in hibonite, 129
IVCT, 131
fassaite,
crystal chemical data, 179
IVCT energies, 118–19
polyhedral bulk moduli, 376–7
see also Allende meteorite pyroxene *and* Angra dos Reis meteorite pyroxene
fayalite,
coordination site properties, 472
CF spectra of Fe^{2+} olivines, 161–5
crystal structure, 160
effects of pressure on CF spectra, 363
effects of pressure on OMCT spectra, 367
effects of temperature on CF spectra, 363
effect of temperature on OMCT spectra, 367
electrical conductivity at elevated pressures, 392
interatomic distances, 160
interatomic distances at high pressures, 358
polarized spectra, 104, 162
polyhedral bulk moduli of Fe^{2+}, 377
^{57}Fe isotope, 3,
see also Mössbauer spectroscopy
February birthstone, 106
feldspar,
birthstone, 106
CF parameters for Fe^{3+}, 201, 206–7
iridescence in gem moonstone, 106, 108
system: $NaAlSi_3O_8$–$KAlSi_3O_8$, 303
feroxyhyte,
in manganese nodules, 347
on Mars, 421
ferrifayalite, *see* laihunite
ferrihydrite,
in Fe^{3+}-exchanged clays, 421
in manganese nodules, 347
on Mars, 421
ferromagnesian silicates,
cation ordering, 250–69
CFSE of Fe^{2+}, 229–30, 288
CF parameters for Fe^{2+}, 228
dissolution, 332
oxidation on Mars, 421
phase changes in Mantle, 355–7
reflectance spectra of lunar minerals, 406
ferromanganese oxide deposits,
see manganese nodules
ferrosilite,
coordination site properties, 472
crystal chemical data, 178
effect of pressure on CF spectra, 363
polarized spectra, 183
polyhedral bulk moduli of Fe^{2+}, 377
see also orthoferrosilite
ferric,
see iron(III)
ferrous,
see iron(II)
fibroferrite,
CF parameters for Fe^{3+}, 226
first transition series, 1, 2, 14, 15
fitzroyite,
CF spectra of Ti^{3+} in phlogopite, 205
five-fold coordination in silicate melts, 315, 317, 326
five-fold degeneracy, of *d* orbitals, 3, 10
fluorides,
CFSE and lattice energies, 274
and high-spin configurations, 28
Jahn–Teller distortion, 243
nephelauxitic series for cations, 434
spectrochemical series for cations, 28
fluorine avoidance, 260
fluorite,
colours due to lanthanide elements, 125
alexandrite effect, 112, 125
forsterite,
coordination site properties, 472
CF parameters for cations, 170–1, 213, 229, 236, 232
CF spectra
of Fe^{2+} olivines, 161–5
of Ni^{2+} olivines, 165–8
effect of pressure on CF spectra, 363, 372
effect of temperature on CF spectra, 363, 372
electrical conductivity at high pressures, 392
interatomic distances at high pressures, 358
phase changes in Mantle, 355
polarized spectra, 372
reflectance spectra, 406, 411
spin-pairing of Fe^{2+}, 380, system: Mg_2SiO_4–Ni_2SiO_4, 305, 320, 322
fractional crystallization of magma, 4, 312
framework silicates,
CF parameters for cations, 201
CF spectra, 206–7
Frank–Condon principle, 67
free energy of fusion, 309
free energy of mixing, 278
free ion, 3

friedelite,
CF parameters for Mn^{2+}, 220
fuchsite,
CF parameters for Cr^{3+}, 213

g orbitals, 9–10
G spectroscopic term, 54, 466
gadolinite,
colours due to lanthanide elements, 115
ganphyllite,
CF parameters for Mn^{2+}, 220
garnets,
absorption spectra, 109, 157
alexandrite effect, 112
birthstones, 106
coordination site properties, 470
CF parameters for cations, 159
CF spectra, 156–7
crystal structure, 155
high pressure CF spectra, 364, 371
radiative heat transfer in, 391
see also almandine, andradite, goldmanite, tsavorite *and* uvarovite
garnierite,
colour due to Ni^{2+}, 115
CF parameters for Ni^{2+}, 201, 206, 236
laterite deposits, 329
mis-identification of tetrahedral Ni^{2+}, 101
gas-phase deposition, 310
Gaussian profiles, 47, 80, 82
gems, 88, 106–7
gerade state, *g*, 11, 65, 66
gillespite,
colour due to Fe^{2+}, 110, 115
CF parameters for Fe^{2+}, 228, 229, 288
CF spectra, 49–50, 73, 76–9, 201, 206, 215
crystal structure, 76
effects of pressure on CF spectra, 206
interatomic distances at high pressure, 359
pleochroism, 110
polarized spectra, 49
spin-pairing of Fe^{2+}, 381
glaucochroite,
CF parameters for Mn^{2+}, 220
glaucophane,
coordination site properties, 473
CF parameters for Fe^{2+}, 189, 230
crystal structure, 123
IVCT energies, 117
mixed-valence properties, 134
pleochroism, 124
polarized spectra, 125, 197
goethite,
crystal chemical data, 418–19
CF parameters for Fe^{3+}, 224, 418
effect of pressure on optical spectra, 365, 367
in manganese nodules, 347
reflectance spectra, 403
goldmanite,
CF parameters for V^{3+}, 159, 210
colour due to V^{3+}, 115

Goldschmidt rules, 302
gossans, 417
on Mars, 421
granite,
CF spectra of glasses, 316,
CFSE of cations in glasses, 316
Great Dyke,
chromite in, 324
grossular,
coordination site properties, 470
CF parameters for cations, 159, 210–24
interatomic distances at high pressures, 358
see also goldmanite *and* tsavorite
ground-state, 54, 56, 68
group theory, 3, 11, 34, 54, 59, 62, 75
and MO theory, 435, 445
groutite,
interatomic distances, 244
Jahn–Teller distortion by Mn^{3+}, 244
grunerite,
CF parameters for Fe^{2+}, 189, 228
CF spectra of Mg–Fe^{2+} amphiboles, 193, 194–5, 228
CFSE of Fe^{2+}, 195, 230, 288
hydroxyl stretching frequencies, 253
polarized spectra, 193
guildite,
CF parameters for Cu^{2+}, 226
CF parameters for Fe^{3+}, 238
interatomic distances, 245
Jahn–Teller distortion by Cu^{2+}, 245
gypsum,
on Mars, 424

H spectroscopic term, 54, 466
Halon,
standard in reflectance spectra, 380
hauerite,
spin-pairing of Fe^{2+}, 380
hausmannite,
interatomic distances, 244
Jahn–Teller distortion by Mn^{3+}, 234, 243–4
heats of hydration,
and CFSE, 4, 275, 330
hedenbergite, coordination site properties, 472
crystal chemical data, 178
CF parameters for Fe^{2+}, 229
CF spectra of Fe^{2+} in calcic pyroxenes, 180–2, 402
CFSE of Fe^{2+}, 180
reflectance spectrum, 402

hematite,
  coordination site properties, 470
  crystal chemical data, 418
  CF parameters for Fe^{3+}, 154, 224, 418–19
  intensification of spin-forbidden transitions, 420
  magnetic ordering, 152
  nanophase on Mars, 417
  paired transitions in, 420
  reflectance spectra, 403
  structure, 152, 420
hemimorphite,
  tetrahedral Zn^{2+} in, 304
Henry's law,
  and ideal solid-solutions, 278, 300
Hermann–Mauguin symbols, 62, 468–9
heterovalent IVCT, 116, 126
  MO calculations, 454
hexahydrated cations,
  absorption spectra of Ti^{3+}, 48
  absorption spectra of Fe^{2+}, 48
  CF parameters, 29
  CFSE, 29
hibonite,
  IVCT energy, 118, 129
  optical spectra, 129
  containing Ti^{3+}, 129, 216, 292, 310
  containing V^{3+}, 129, 293, 310
high-spin configurations, 17
  octahedral coordination, 19
  tetrahedral coordination, 23
hollandite,
  crystal chemical data, 338
  lattice images, 344
  crystal structure, 343
homonuclear IVCT, 116, 126
  MO calculations, 451–4
hornblende,
  OMCT in, 132
Hund's rules, 14, 17, 380
hydroxyl stretching frequencies,
  in spectra of garnierite, 101
  in spectra of Mg–Fe amphiboles, 194
  cation site occupancies from, 253
  in amphiboles, 253
  in micas, 253
hydroxysilicates, 30, 37, 332

I spectroscopic term, 54, 466
ilesite,
  CF parameters for Mn^{2+}, 221
ilmenite,
  coordination site properties, 470
ilvaite,
  coordination site properties, 472
  crystal structure, 138
  electron delocalization, 133
  IVCT energies, 117
  Mössbauer spectra, 137–9
  optical spectra, 140
  properties of mixed-valence properties, 134, 137–40
  on Venus, 122, 134, 422
indicolite,
  birthstone, 107
  see tourmaline
inesite,
  CF parameters for Mn^{2+}, 221
insolubility,
  of high valence oxides, 340
integrating sphere, 89
intensities,
  of absorption bands, 64–72, 93, 449–50
interatomic distances,
  and crystal field splitting, 30
  in high-pressure phases, 358–9
  and interphase partitioning, 291
  and lattice energies, 276
  in olivines, 243
  in pyroxenes, 178–9
intercrystalline partitioning,
  see interphase partitioning
interelectronic repulsion, 14, 52
  and Hund's rule, 17
interphase partitioning,
  crystal–melt systems, 295
  divalent cations in ferromagnesian silicates, 290–2
  iron:magnesium ratios in coexisting silicates, 288–90
  trivalent cations in aluminosilicates, 292–4
intersite partition coefficients, 250
  and ionic radii, 256
  in olivines, 255
  in orthopyroxenes, 255
intervalence charge transfer (IVCT) transitions, 108, 116
  in blue sapphire, 128
  energies in mixed-valence minerals, 117–9
  in ilvaite, 140
  influence on CF spectra, 147
  influence of pressure, 120, 361, 366
  influence of temperature, 120, 361, 390
  intensities, 72
  and interatomic distances, 130
  and MO diagrams, 4514
  in kyanite, 129
  in vivianite, 130
intracrystalline partitioning,
  see intersite partitioning coefficients *and* cation ordering
inverse spinel, 248
ionic radii,
  and distribution coefficients, 305
  first-series transition elements, 242, 464
  and Goldschmidt's rules, 302

ionic radii (*cont.*)
  influence on cation ordering, 260–2
  and intersite partition coefficients, 256
  lanthanide elements, 241
  and trace element distribution, 307
ionic structures, 5
ionization energy, 9
iridescence, 108
  opal, 107, 108
  moonstone, 106, 108
iron: magnesium ratios,
  in silicates, 288–90
iron,
  abundance data, 301, 460
  oxidation states in Eh–pH diagrams, 337–8
  in manganese nodules, 347
  in seafloor deposits, 346
iron(II),
  absorption spectra of hexahydrated cation, 48
  cation ordering
    in amphiboles, 258
    in biotites, 259
    in olivines, 254
    in orthopyroxenes, 257
    in pigeonites, 258
  colours in minerals due to Fe^{2+}, 106–7, 115
  CF activation energies, 334
  CF parameters for Fe^{2+} minerals, 229
  CF spectra of minerals, 148–207, 223–7
  CF splitting
    in hexahydrated cation, 29, 56, 69, 71
    and interatomic distances, 228
    in oxides, 276
    in periclase, 29, 151
    in sulphides, 276
  CF states in octahedral sites, 55–9, 223
  CFSE
    in aluminosilicates, 227, 288
    in distorted sites, 265
    in ferromagnesian silicates, 223, 288
    in hexahydrated cation, 29, 229
    in minerals, 229–30
    in monoclinic sites, 264
    in octahedral sites, 19, 249, 264
    in oxides, 276
    in sulphides, 276
    in tetragonal sites, 38, 264
    in tetrahedral sites, 23, 249
    in trigonal sites, 264
  dynamic Jahn–Teller effect, 223
  on Earth, 14, 397
  effective charge in $[FeO_6]^{-10}$ cluster, 449
  electronic configurations, 15, 19, 23, 63
  electronic configurations of CF states, 55
  electronic entropy, 284–8
  energy level diagrams, 55, 57, 60, 150, 158, 164, 186, 227
  interphase partitioning, 288
  intersite partition coefficients in olivine and orthopyroxene, 256
  ionic radii, 464
  low-spin, 228
  in Lower Mantle, 379–83
  low-spin state, 56
  in magmatic crystallization, 317
  in metamorphic processes, 350
  in Moon rocks, 14, 397
  octahedral site preference energy, 249
  Orgel diagram, 60
  pressure-induced reduction, 385
  in silicate glasses, 315
  spectroscopic terms, 53
  spin-pairing transitions, 56, 379–81
  substitution reactions, 325
  Tanabe–Sugano diagram, 57, 227
iron(II) → iron(III) IVCT, 116
  in aquamarine, 198
  in augite, 180
  in blue sapphire, 128
  in calcic amphiboles, 197
  calculated energies, 118, 454
  in chlorite, 205
  in chloritoid, 205
  in cordierite, 199
  effect of pressure, 366
  energies computed for dimeric clusters, 451–4
  energies in minerals, 117, 130
  in glaucophane, 124–5, 197
  in ilvaite, 140
  and interatomic distances, 130
  in kyanite, 129, 173
  in orthopyroxene, 184
  in osumilite, 199
  in sillimanite, 174
  in vivianite, 121–4
iron(II) → titanium(IV) IVCT, 95, 116
  in Angra dos Reis fassaite, 126
  in biotite, 205
  in blue sapphire, 128–9
  calculated energies, 118, 454
  effect of pressure, 126, 366
  energies in minerals, 117
  in ellenbergerite, 129
  in hibonite, 129
  in kyanite, 129, 173
  in neptunite, 129
  in pyroxenes, 131
  in taramellite, 129
iron(III),
  cation ordering
    in epidote, 259
    in olivine, 257
    in sinhalite, 260
  colours in minerals due to Fe^{3+}, 106–7, 115
  CF activation energies, 334

CF energies in octahedral *vs* tetrahedral sites, 62
CF parameters for minerals, 224–6
CF spectra, 222–3
CF splitting
in corundum, 29, 154
in hexahydrated cation, 29
CF states in octahedral sites, 58–9
CFSE
in hexahydrated cation, 29
in monoclinic sites, 264
in octahedral sites, 19, 249, 264
in tetragonal sites, 38, 264
in tetrahedral sites, 23, 249
in trigonal sites, 264
on Earth, 14, 397
effective charge in $[FeO_6]^{-9}$ cluster, 449
electronic configurations, 15, 19, 23
of CF states, 58
electronic entropy, 285
energy level diagrams, 57–8, 70
ionic radii, 464
in magmatic crystallization, 317
in Moon rocks, 14, 397
octahedral site preference energy, 249
pair-excitation in corundum, 71, 223
pair-excitation in hematite, 420
Racah parameters, 224–5, 431
in silicate glasses, 315
spectroscopic terms, 53
spin-forbidden transitions, 59, 69–71, 222
iron(III) clay silicates, on Mars, 416, 421
iron(III) → manganese(III),
IVCT in babingtonite, 119
iron(III) oxides, 438–9
antiferromagnetic ordering, 449
CF parameters for Fe^{3+}, 438–9
interatomic distances, 438–9
and isostructural Mn(IV) oxides, 338, 438–9
on Mars, 397, 417, 421
reflectance spectra, 403
iron(III) sulphates, on Mars, 416, 421
CF parameters for Fe^{3+}, 225–6
irreducible representations, 75
isotopes,
stable, of transition elements, 462–3
itinerant electrons,
see electron delocalization

jadeite,
CF parameters for Cr^{3+}, 189
Jahn–Teller effect and distortion, 33–6, 243–7
in Cr^{2+} compounds, 215, 243, 266, 329
in Cu^{2+} minerals, 243–5
influence on cation ordering, 266
influence on CF spectra of Cu^{2+} in minerals, 235–8
influence on crystallization from magma, 318
in Mn^{3+} minerals, 205, 217, 243, 266, 259, 294, 344
see also dynamic Jahn–Teller effect
January birthstone, 106
jarosite,
CF parameters for Fe^{3+}, 225, 418–19
crystal chemical data, 418–19
johannsenite,
colour due to Mn^{2+}, 115
juddite,
CF parameters for Mn^{3+}, 189, 198, 218
Jahn–Teller distortion, 247, 268
Mn^{3+} ordering, 259, 268
July birthstone, 107
June birthstone, 106

kaersutite,
colour due to OMCT, 132
kammererite,
colour due to Cr^{3+}, 115
CF parameters for Cr^{3+}, 200, 213
kanonaite,
CF parameters for Mn^{3+}, 172, 218
enrichment of Mn^{3+}, 293
interatomic distances, 244
Jahn–Teller distortion by Mn^{3+}, 172, 246, 247
stability of Mn^{3+}, 99
structure, 99, 172
karelianite,
coordination site properties, 470
Kebulka–Munk function, 401, 403
kerolite,
CF parameters for Ni^{2+} in talc, 201, 237
kimberlite,
Cr olivines in diamonds, 329
kinetics, 4, 331–5
kinnoite,
interatomic distances, 246
Jahn-Teller distortion by Cu^{2+}, 246
knebellite,
polarized spectrum, 104
kornuperine,
CF parameters for cations, 200, 210, 213
kosmochlor,
coordination site properties, 473
CF parameters for cations, 95, 189, 213
kyanite,
causes of colour, 129
coordination site properties, 471
CF parameters
for cations, 173, 210, 213, 225
for Fe^{2+}, 228, 229, 288
effect of temperature on IVCT, 366
IVCT energies, 117, 129
mixed-valence properties, 133, 173
structure, 129, 173

L or total orbital angular momenta, 52, 466
labile, 331
laihunite,
electron delocalization, 133
Fe^{3+} site occupancy, 269
mixed-valence properties, 141–2
structure, 142
on Venus, 134–5, 422
lanthanide elements, 1, 14
causes of colour in minerals, 115
crystal field effects, 240
ionic radii, 241
Laporte selection rule, 65–9, 79
intensities, 72
laterites, 329, 335, 346
lattice energies,
and CFSE, 274, 319, 330
and trace element distributions, 311
lattice images,
of Mn(IV) oxides, 343
lazulite,
IVCT energy, 117
layer silicates,
CF parameters of cations, 201
optical spectra, 204–6
layered intrusions,
chromite, 301, 324
leaching, 330–5
lepidocrocite,
crystal chemical data, 418–19
CF parameters for Fe^{3+}, 224, 418
effect of pressure on spectra, 365, 367
reflectance spectra, 403
leucophanite,
CF parameters for Mn^{2+}, 220
$LiAlO_2$, synthetic γ,
CF parameters for Fe^{3+}, 224
liebenbergite,
assignments of CF spectra, 167
coordination site properties, 472
CF parameters for Ni^{2+}, 170, 236
CF spectra, 165–8
CFSE of Ni^{2+}, 280
Ni^{2+} site occupancies, 256
lattice energy and CFSE, 274
polarized spectra, 166
thermodynamic data, 280
lifetime,
of electronic transitions, 36, 67, 81
of Mössbauer transition in ^{57}Fe, 134
of vibrational transitions, 68
ligands, 3, 7
nephelauxitic series, 434
spectrochemical series, 28
ligand field theory, 3, 82–4
light scattering, 88
lithiophilite,
CF parameters for Mn^{2+}, 220
lithiophorite,
crystal chemical data, 339
Ni in laterites, 346
structure, 346
lithophilic elements, 302, 355
lizardite,
CF parameters for Ni^{2+}, 201, 237
Lower Mantle,
composition of, 355
low-spin configurations, 17
characterization 17, 28
in Co(III) oxides, 30, 348, 379
in Fe(II) minerals, 56
interatomic distances, 242, 464
in Mantle, 379–83
in octahedral coordination, 19
and π-bonds, 243, 440
in pyrite, 441
and spectrochemical series, 28
in tetrahedral coordination, 23
low-symmetry sites, 33, 37
effects on CF spectra, 60, 147
energy level diagrams, 62–4
L–S terms, 52
see spectroscopic terms
ludwigite,
electron delocalization, 140
mixed-valence properties, 140
structure, 140
Luna missions, 397, 422
lunar meteorites,
reflectance spectra, 424
lunar minerals, 14
reflectance spectra, 406
lusakite,
colour due to Co^{2+}, 115
CF parameters for Co^{2+}, 175, 234
polarized spectra, 102

maghemite,
crystal chemical data, 418–19
CF parameters for Fe^{3+}, 224, 418
reflectance spectra, 403
magnesiocopiapite,
CF parameters for Fe^{3+}, 225
magnesiowüstite, 149
CF spectrum of Fe^{2+}, 150
effect of pressure on absorption spectra, 362, 367
electrical conductivity at high pressures, 392
energy level diagram for Fe^{2+}, 150
Fe^{2+} partitioning in post-spinel phases, 388
spin-pairing of Fe^{2+}, 382
transition element partitioning in Lower Mantle, 384
see also periclase
magnetic dipole, 74

magnetic field,
orbiting electrons, 10
spinning electrons, 10
magnetic quantum number, m_l, 9
magnetic susceptibility, 3
magnetic vector, 73, 74
magnetite,
coordination site properties, 470
crystal structure, 136
electron delocalization, 133
mixed-valence properties, 134
Mössbauer spectra, 137
on Venus, 134, 422
Verwey transition, 137
majorite, 356
in Mantle, 356
malachite,
colour due to Cu^{2+}, 115
CF parameters for Cu^{2+}, 238
interatomic distances, 245
Jahn–Teller distortion by Cu^{2+}, 243, 245
manganese,
abundance data, 301, 460
oxidation states in Eh–pH diagrams, 337
in porphyry copper systems, 327
in seafloor hydrothermal deposits, 2, 346, 397
in manganese nodules, 347
in metamorphic processes, 350
manganese(II),
cation ordering in olivines, 102–4, 257, 279
colours in minerals due to Mn^{2+}, 115
CF activation energies, 334
CF energies in octahedral *vs* tetrahedral sites, 62
CF parameters for minerals, 217, 220–2
CF spectra, 217, 222
CF splitting in hexahydrated cation, 29
CF states in octahedral sites, 58
CFSE
in hexahydrated cation, 29
in monoclinic sites, 264
in octahedral sites, 19, 249, 264
in tetragonal sites, 38, 264
in tetrahedral sites, 23, 249
in trigonal sites, 264
on Earth, 14, 397
electronic configurations, 15, 19, 23
of CF states, 58
electronic entropy, 285
energy level diagram, 58
interphase partitioning, 290–2
intersite partition coefficients for olivine and orthopyroxene, 256
ionic radii, 464
in magmatic crystallization, 317
in Moon rocks, 14, 397
octahedral site preference energy, 249
Orgel diagram, 58
Racah parameters, 217, 431
spectroscopic terms, 53
manganese(II) → manganese(III) IVCT, 116
energies in minerals, 119
in yoderite, 119
manganese(II) → iron(III) IVCT, 116, 454
calculated, 119, 454
manganese(II) → titanium(IV) IVCT,
in tourmaline, 119
manganese(III),
cation ordering in epidotes, 98, 259
colours in minerals due to Mn^{3+}, 115
CF activation energies, 334
CF parameters for minerals, 217–19
CF spectra, 217
CF splitting
in corundum, 29
in hexahydrated cation, 29
in piemontite, 98
CF states in octahedral sites, 215
CFSE
in hexahydrated cation, 29
in minerals, 217–8
in monoclinic sites, 264
in octahedral sites, 19, 249, 264
in piemontite, 99
in tetragonal sites, 38, 264
in tetrahedral sites, 23, 249
in trigonal sites, 264
on Earth, 14, 397
disproportionation, 18, 95
effective charge in $[MnO_6]^{-9}$ cluster, 457
electronic configurations, 15, 19, 23
electronic entropy, 285
energy level diagrams, 60, 98, 215
interatomic distances in minerals, 244
interphase partitioning, 293
ionic radii, 464
Jahn–Teller effect, 217, 243, 344
in magmatic crystallization, 317
in Moon rocks, 14, 397
octahedral site preference energy, 249
Orgel diagram, 60
pressure-induced reduction, 385
spectroscopic terms, 53
stability in andalusites, 34, 99
stability in epidotes, 34, 39, 95–9
stability in tetragonally distorted sites, 34
in substitution reactions, 335
Tanabe–Sugano diagram, 215
manganese(III) → iron(III) IVCT, 116,
in babingtonite, 119
calculated, 119, 454
manganese(IV),
CF activation energies, 334
CF parameters for Mn^{4+} in corundum, 154
CF splitting in hexahydrated cation, 29

manganese(IV) (*cont.*)
CFSE
in octahedral sites, 19, 264
in tetragonal sites, 38, 264
in tetrahedral sites, 23
in trigonal sites, 264
on Earth, 14, 397
effective charge in $[MnO_6]^{-8}$ cluster, 456
electronic configurations, 15, 19, 23
electronic entropy, 285
ionic radii, 464
spectroscopic terms, 53
in spodumene, 188
manganese(IV) oxides,
crystal chemical data for, 338
enrichments of Co, 263, 348
in hydrothermal crusts, 346
in porphyry copper systems, 328
structural stabilities from MO calculations, 454–7
manganese nodules, 328, 346–9
manganite, 345
interatomic distances, 244
Jahn–Teller distortion by Mn^{3+}, 244
manganophyllite,
CF parameters for Mn^{3+} in phlogopite, 200, 205, 218
Jahn–Teller effect in Mn^{3+}, 205
manganosite,
CF parameters for Mn^{2+}, 151, 220
manjiroite, 343
see hollandite
Mantle,
composition, 354, 460
influence of CFSE on olivine → spinel transition, 386
low-spin Fe^{2+}, 231, 381
mineralogy, 355
pressures, 353–4
temperatures, 353, 390
March birthstone, 106
marokite,
interatomic distances, 244
Jahn–Teller distortion by Mn^{3+}, 244
Mars,
bright-region spectra, 416
composition, 397, 415, 460
dark-region spectra, 416
ferric oxides, 417
gossans, 417
oxidation states of cations, 14, 397
reflectance spectra, 416
spectra and surface features, 398
martian meteorites,
see SNC meteorites
May birthstone, 106
melanite garnet,
Ti^{3+} in, 93
melting points,
in Lower Mantle phases, 382
and trace element distribution, 308, 322
Mercury,
oxidation states of cations, 397
reflectance spectra, 414
spectra and surface features, 398
metalliferous deposits, 328
metal → metal charge transfer,
see intervalence charge transfer
metamorphic minerals,
partitioning of transition elements, 349–50
glaucophane in blueschists, 124
meteorites,
Allende, 93, 126, 179, 208
Angra dos Reis, 126, 179
carbonaceous chondrites, 129, 292, 354
lunar, 424
martian, 424
pyroxenes in, 93, 95, 126, 179, 208, 423
reflectance spectra, 422–4
SNC, 424, 460–1
Toluca, 95
mica,
CF parameters for cations in, 200, 209
crystal structure, 203
preferred orientation in spectra, 89
microscopes,
in mineral spectroscopy, 89–91
mid-infrared region, 46
mixed-mineral assemblages,
reflectance spectra, 411
mixed-valence minerals,
classification, 135
IVCT energies, 117–19
Mn oxides, 243
optical spectra, 115–33
opaque, 133–42, properties, 133
phase D in Mantle, 393
on Venus, 422
MnO_2,
synthetic, crystal chemical data, 338
structure of λ-MnO_2, 340, 456
modified spinel, β-phase,
see wadsleyite, 355
molar extinction coefficient, 47
calculation, 49, 92
for IVCT transitions, 120
values, 72
molecular orbital (MO) diagrams, 5, 17, 319
for $[FeL_6]$, 17, 436
for $[FeO_6]^{-9}$, 444
for $[FeO_6]^{-10}$, 444
for $[Fe_2O_{10}]^{-15}$, 452
for FeS_2, 441
for $[MnO_4]^{-4}$, 455
for $[MnO_6]^{-8}$, 455
octahedral Fe^{3+}, 436

molecular orbital (MO) theory, 3
molybdenum(III),
  CF splitting in hexahydrated ion, 28
monazite,
  alexandrite effect, 113
  colours due to lanthanide elements, 115
monoclinic distorted site, 263
  energy level diagram, 263
  CFSE, 264
montmorillonite,
  CF parameters of Mn^{3+}, 201, 219
Moon,
  composition, 397, 460
  oxidation states of cations, 397
  reflectance spectra, 408–14
  spectra and surface features, 398
morganite,
  colour due to Mn, 105
  CF parameters for Mn^{3+}, 199, 200, 218
  gem beryl, 88, 105
Mössbauer spectroscopy, 3, 148
  of biotite, 259
  chemical isomer shift, 11, 17
  of deerite, 141, 381
  and electron delocalization, 133
  of epidote, 259
  of fayalite, 163
  of garnet, 157
  of gillespite, 381
  at high pressures, 385
  identification of low-spin Fe^{2+}, 28
  of ilvaite, 139
  of kyanite, 173
  of laihunite, 142
  lifetime of transition in ^{57}Fe, 134
  of magnetite, 137
  in ^{61}Ni, 3, 462
  of orthopyroxenes, 257
  and *s*-electron density, 11
  of silicate glasses, 314
  site populations from, 252
  of SNC meteorites, 424
  of vonsenite, 140
  in ^{67}Zn, 3, 462
mullite,
  CF parameters for Cr^{3+}, 176, 213
muscovite,
  coordination site properties, 473
  CF spectra of cations, 200, 204, 210, 213, 218, 225
Muskox intrusion,
  chromite, 324

nanophase iron(III) oxides, 421
near-infrared region, 27, 46
nephelauxetic series, 28, 434
nepouite,
  CF parameters for Ni^{2+} in serpentine, 201, 237
neptunite,
  IVCT energies, 118, 131
  structure, 131
Nernst distribution coefficient, 295
nickel,
  abundance data, 301, 460
  geochemistry, 320–3
  in layered intrusions, 324
  in magmatic crystallization, 320–2
  in manganese nodules, 346–8
  in metamorphic processes, 350–1
  Mössbauer effect in ^{61}Ni, 3
  oxidation states in Eh–pH diagrams, 337
  partitioning in crystal–melt systems, 297, 305, 322
  in silicate glasses, 315
  in Skaergaard intrusion, 312
  in ultramafic rocks, 2, 325, 329, 335, 346, 397
nickel(I),
  in Lower Mantle, 385
nickel(II),
  cation ordering in ferromagnesian silicates, 261
  cation ordering in olivines, 256, 279
  colours in minerals due to Ni^{2+}, 115
  CF activation energies, 334
  CF parameters for minerals, 234, 236–7
  CF spectra, 234–5
  CF splitting
    in corundum, 29
    in hexahydrated cation, 29
    in oxides
    in periclase, 29
  CF states in octahedral sites, 234
  CFSE
    in hexahydrated cation, 29
    in minerals, 236–7
    in monoclinic sites, 264
    in octahedral sites, 19, 249, 264
    in oxides, 276
    in sulphides, 276
    in tetragonal sites, 38, 264
    in tetrahedral sites, 23, 249
    in trigonal sites, 264
  on Earth, 14, 397
  electronic configurations, 15, 19, 23
  electronic entropy, 285
  energy level diagrams, 61, 235
  interphase partitioning, 290–1
  intersite partition coefficients for olivine and orthopyroxene, 256
  ionic radii, 464
  in magmatic crystallization, 317
  in Moon rocks, 14, 397
  octahedral site preference energy, 249
  Orgel diagram, 61
  π-bond formation, 440

nickel(II) (*cont.*)
partitioning in crystal–melt systems, 322
phase equilibria in Ni–Mg olivines, 305, 321
site occupancy in spinels, 249
spectroscopic terms, 53
in sulphides, 440, 442
in substitution reactions, 335
Tanabe–Sugano diagram, 235
thermodynamic data for Mg–Ni^{2+} olivines, 280
in Upper Mantle minerals, 290
nickel(III),
CF activation energies, 334
CF parameters in corundum, 29, 154
CF splitting in hexahydrated cation, 29
CF states in octahedral sites, 231
CFSE
in hexahydrated cation, 29
in corundum, 29, 154
in octahedral sites, 19, 264
in tetragonal sites, 38, 264
in tetrahedral sites, 23
in trigonal sites, 264
electronic configurations, 15. 19, 23
electronic entropy, 285
energy level diagrams, 231
ionic radii, 464
Jahn–Teller effect in, 247
low-spin configuration in oxides, 30
π-bond formation, 440
spectroscopic terms, 53
Tanabe–Sugano diagram, 231
nickel diopside,
coordination site properties, 473
nickel olivines,
assignments of absorption bands, 167
coordination site properties, 472
CFSE of Ni^{2+}, 280
polarized spectra, 166
thermodynamic data, 280
see also liebenbergite
nickel silicate spinel,
coordination site properties, 472
CF parameters of Ni^{2+}, 169, 236
non-bonding oxygens,
in silicates, 30
non-crossing rule, 59
non-cubic coordination sites, 25, 33
nontronite,
CF parameters for Fe^{3+}, 201, 225, 418–19
crystal chemical data, 418
normal spinel, 247
norrishite,
interatomic distances, 245
Jahn–Teller distortion by Mn^{3+}, 245
November birthstone, 107
nsutite,
crystal chemical data, 338
structure, 342

oceanic crust,
transition elements, 327, 462
octahedral coordination, 15
CF splitting parameter, Δ_o, 15
CFSE, 19, 249, 264
electronic configurations, 19
energy levels for *d* orbitals, 16, 21, 263
in magma, 312, 317
in silicate melts, 314, 323–4
orientation of ligands, 16
Schöenflies symbol, 63
octahedral crystal field splitting, Δ_o, 15
octahedral site preference energy (OSPE), 249
and cation distribution in spinels, 248
and distribution coefficients, 296
and fractional crystallization of magma 314, 317, 320, 322, 326
October birthstones, 107
olivine,
α-phase in Upper Mantle, 158, 355
cation ordering, 103, 162, 166, 267
colour due to Fe^{2+}, 107, 110, 115
configurational entropy, 279
coordination site properties, 472
covalent bonding, 268
CF parameters
for cations, 170–1, 232, 236
for Fe^{2+}, 228, 229, 288
CF spectra, 158–72
CFSE
of cations in distorted sites, 264
of Fe^{2+}, 165, 288
of Ni^{2+}, 167, 236
crystal structure, 160
in diamonds, 329
effects of pressure on optical spectra, 362–3, 367, 373
effects of temperature on optical spectra, 362, 373
electrical conductivity at high pressures, 392
energy level diagrams for Fe^{2+}, 164
extinction coefficient calculation, 92
gem peridot, 107, 110, 117
interatomic distances, 261
intersite partitioning, 254–7
in oxidation products on Venus, 415
pleochroism, 110, 161
polarized spectra, 104, 162, 372
polyhedral bulk moduli, 377
reflectance spectra, 406, 411, 413
in reflectance spectra of the Moon, 408–12, 424
radiative heat transfer, 390
spin-pairing of Fe^{2+}, 380

structure, 159–60
temperature variation of reflectance spectra, 413
on terrestrial planets, 400
trace elements, 306
omphacite,
IVCT energy, 118
effect of pressure on optical spectra, 363
opal, birthstone, 107
iridescence, 107, 108
ophiolites,
Ni in, 228
and Upper Mantle, 354
optical density, 46, 47
optical spectra, 44
see also absorption spectra, CF spectra *and* IVCT transitions
orange glass,
spectra of Fe and Ti in samples from the Moon, 410
orbital angular quantum number, *l*, 9
orbitals, 7–14
antibonding, 435
a_{1g}, 11
a_{2u}, atomic, 12
d, 2, 9–12
e_g, 12, 13
f, 6, 13
molecular, 445
non-bonding, 445
overlap of, 3
p, 9–12
π, 437
s, 9–12
shape of, 12
σ, 435
σ^*, 434
symmetry, 11–14
t_{1u}, 13
t_{2u},13
t_{2g}, 12, 13
valence band, 445
ore deposits,
and chalcophilic properties, 440–2
Orgel diagrams, 59–61
for *D* spectroscopic terms, 60
for *F* spectroscopic terms
and widths of absorption bands, 80
orthoclase,
CF parameters for Fe^{3+}, 201, 225
orthoferrosilite,
energy level diagram for Fe^{2+}, 186
polarized spectra, 183
see also ferrosilite
orthopyroxene,
cation ordering, 103, 180, 257–9, 267
configurational entropy, 278
coordination site properties, 473
covalent bonding, 268
CF parameters
for cations, 182–7, 188
for Fe^{2+}, 228, 229, 288
CF spectra, 183–7
CFSE
in distorted sites, 264
of Fe^{2+}, 187, 288
effects of pressure on CF spectra, 363
effects of temperature on CF spectra, 363
crystal structure, 185
energy level diagrams for Fe^{2+}, 186
interatomic distances, 178
intersite partition coefficients, 256
IVCT energy, 117
polarized spectra, 182–3
polyhedral bulk moduli, 377
reflectance spectra, 406
in reflectance spectra of the Moon, 408–12
reflectance spectra at elevated temperatures, 411, 413
trace elements, 306
orthosilicates,
CF parameters for cations, 170–1
CF spectra, 155–76
CFSE and lattice energies, 274, 276
oscillator strength, 47
osumilite,
CF spectra of Fe^{2+}, 199
IVCT energy, 117
oxidation,
in blue sapphire, 128
in sedimentary environments, 330, 335–40
in Tanzanian zoisite, 113–15
in vivianite, 122
oxidation states of transition elements, 14, 15
on Earth, 14, 397, 400
in Moon rocks, 14, 397, 400
on Mars, 14, 397, 460
on Mercury, 397
pressure-induced changes, 385
and Racah parameters, 431
on Venus, 397, 460
oxides,
CFSE and lattice energies, 274, 276
see also corundum, periclase *and* Mn(IV) oxides
oxybiotite,
colour due to OMCT, 133
on Venus, 134, 422
oxygen ligands in minerals, 30, 33, 37
influence on colour, 30
in MO diagrams, 449
oxygen → metal charge transfer (OMCT), 108, 132
effect of pressure on spectra, 133, 361, 367, 369
effect of temperature on spectra, 367, 369

oxygen → metal charge transfer (OMCT) (*cont.*)
and electrical conductivity at high pressure, 133, 392
intensities, 72
and MO theory, 132, 450
and radiative heat transfer at high pressure, 133, 390
oxyhornblende,
colour due to OMCT, 133
on Venus, 134, 422

p orbitals, 9, 12
in molecular orbital theory, 435, 445
P spectroscopic term, 54, 46
pair-excitations, 71, 115
in babingtonite, 71, 124
in blue sapphire, 120, 124, 128
in corundum, 70–1, 223
IVCT influencing CF transitions, 71
in iron(III) oxides, 418
in hematite, 71, 223, 420
in kyanite, 120, 124, 130, 173
in lazulite, 124
in sillimanite, 124, 174
in tourmaline, 71, 202
in vivianite, 71, 120, 124, 128
in yellow sapphire, 71, 223
pairing energies, 29, 380
Pallas,
spectra and surface features, 399
parity,
of orbital wave functions, 13, 66
selection rule, 65
and OMCT transitions, 450
Pauli exclusion principle, 13
pelagic sediments,
transition elements in, 327, 340
pentagonal bipyramidal coordination,
and *d* orbital energy levels, 20
in S_N2 reactions, 333–5
periclase,
coordination site properties, 470
CF parameters
for cations, 29, 151, 213, 224, 232, 236
for Fe^{2+}, 228, 229, 288
CFSE of cations, 29, 151
CF spectrum of Fe^{2+}, 150
crystal structure, 149
energy level diagram for Fe^{2+}, 150
effect of pressure on optical spectra, 362, 368–9
interatomic distances at high pressure, 359
in Lower Mantle, 357
polyhedral bulk moduli of cations, 376
reflectance spectra standard, 403
see also magnesiowüstite
peridot, birthstone, 107
colour due to Fe^{2+} in gem olivine, 107, 110, 115
peridotite,
and Upper Mantle composition, 354
periodic table, 1, 2, 304
perovskite ($CaTiO_3$),
crystal structure, 357
on Venus, 422
see also silicate perovskite
perovskites,
synthetic,
CF parameters of cations, 213, 233
phase D,
mixed-valence and electrical conductivity, 393
phase equilibria, in systems:
Al_2SiO_5–Mn_2SiO_5, 294
MgO–FeO–SiO_2, 289
Mg_2SiO_4–Ni_2SiO_4, 305, 321
$(Mg,Ni)_2SiO_4$–$K_2O.6SiO_2$, 323
$NaAlSi_3O_8$–$KAlSi_3O_8$, 303
phase transitions,
electronic in ilvaite, 137
in Mantle, 355–7
olivine → spinel, 386
post-spinel, 388
phlogopite,
coordination site properties, 473
CF parameters for cations, 200, 205
structure, 204
Phobus mission,
transition element data for Mars, 415
π-bonds and π-molecular orbitals, 437, 438–9
in minerals, 440
π spectrum, 74
piemontite,
colour due to Mn^{3+}, 110, 115
coordination site properties, 471
CF parameters for Mn^{3+}, 95–9, 218
CFSE of Mn^{3+}, 99, 218
crystal structure, 97
effect of pressure on CF spectra, 365
energy level diagram for Mn^{3+}, 98
interatomic distances, 97, 244
Jahn–Teller distortion by Mn^{3+}, 244, 247
pleochroism, 99, 110
polyhedral bulk modulus of, 377
polarized spectra, 96
stability of Mn^{3+}, 95–9, 217
pigeonite,
cation ordering, 258, 281
crystal chemical data, 178
CF parameters for Fe^{2+}, 187, 228, 230, 288
CFSE of Fe^{2+}, 187, 230, 288
effect of temperature on CF spectra, 363
plagioclase,
CF parameters for Fe^{2+}, 201, 228, 230, 288
in lunar meteorites, 424
polarized spectra, 207
reflectance spectrum, 406

plate tectonics,
and transition element metallogenesis, 325
pleochroism, 105, 108
causes in minerals, 108
in epidotes, 110
in gillespite, 110
in glaucophane, 124
invisible, 108
in tanzanite, 113
point negative charges, 7, 28, 428
polar coordinates, 8
polarization,
of absorption bands, 72–80
polarized absorption spectra,
see absorption spectra, polarized
polarized light, 50, 73
polarizers, 89–90
polished crystals,
spectra of, 88
polyhedral bulk modulus, 31, 374–9
and high pressure CF spectra, 376
polyhedral volumes, 291, 470–4
porphyry copper deposits, 2, 301, 327, 397
powdered minerals,
spectra, 89
post-spinel phases, 388
see silicate perovskite *and* periclase
potential energy diagram,
electronic transitions, 68
preferred orientation,
effects on spectra, 89
pressure,
effect on covalent bonding, 435
effects on CF spectra, 362–5
effect on CF splittings, 30
effects on IVCT transitions, 366
effects on mineral spectra, 91, 94, 360–74
effects on OMCT transitions, 367
probability,
of electronic transition, 64,66
see also selection rules
propagation direction, 74
principal quantum number, *n*, 8
psilomelane, 338, 343
see romanechite
pyrite,
diamagnetism, 441
interatomic distances, 440
low-spin Fe^{2+}, 379–80
MO diagram, 441
structure, 440
transition element distribution, 442
on Venus, 422
pyrolusite,
crystal chemical data, 338
crystal structure, 341
pyrope,
coordination site properties, 470
CF parameters
for cations, 159, 210, 213
for Fe^{2+}, 157, 228, 229, 288
crystal structure, 156
interatomic distances at high pressures, 358
pyroxenes,
cation ordering, 103, 180, 257–9, 267
coordination site properties, 472
CF parameters
for cations, 188
for Fe^{2+}, 228, 229, 288
CF spectra, 176–90
CFSE in distorted sites, 264
crystal chemical data, 178–9
crystal structures, 177, 185
effects of pressure on CF spectra, 363
effects of pressure on optical spectra, 363, 366
interatomic distances, 178
oxidation products on Venus, 415
polarized spectra, 94, 181, 182, 183, 402
radiative heat transfer, 391
reflectance spectra, 402, 406, 411
reflectance spectra of lunar samples, 406
reflectance spectra of Mars, 417
in SNC meteorites, 417
in remote-sensed spectra of Moon, 406
spectral determinative curve, 407
temperature variations of reflectance spectra, 413
trace elements, 306
pyroxmangite,
colour due to Mn^{2+}, 115
CF parameters for Mn^{2+}, 220
pyrrhotite,
transition element distribution, 442

quantum numbers, 8
azimuthal, *l*, 9
magnetic, m_l, 9
orbital angular momentum, *l*, 9
principal, *n*, 8
spin, m_s,10
quartet state, 59, 466
quartz,
as birthstone amethyst, 106
CF parameters for Fe^{3+}, 201, 206, 225
quintet state, 55, 466

Racah *B* parameter, 3, 52–3, 56, 59, 381, 430–5
of Co^{2+} minerals, 232–3
of Co^{3+} minerals, 231
of Cr^{3+} minerals, 213, 431, 432
effect of pressure, 362–4, 385, 434–5
evaluation, 432
of Fe^{3+} minerals, 224–6, 432–3
for field-free cations, 430

Racah *B* parameter (*cont.*)
of Mn^{2+} minerals, 220–1, 432–3
and nephelauxetic ratio, 433–4
of Ni^{2+} minerals, 236–7
of Ni^{2+}–Mg^{2+} olivines, 168, 269
of V^{3+} minerals, 210
and Tanabe–Sugano diagrams, 56
radial integral, 26, 82–4,
radial wave function, 8
radiative heat transfer,
in mantle minerals, 389
radius ratio, 307, 383, 455
Raman spectra,
of silicate glasses, 314
ramsdellite,
crystal chemical data, 338
crystal structure, 343
ransomite,
interatomic distances, 245
Jahn–Teller distortion by Cu^{2+}, 245
rate constant,
for substitution reactions, 333
red-shift, 111, 360
reflectance spectra,
of basalt, 413
of calcic pyroxenes, 402, 406, 413
diffuse, 401
effect of temperature, 413
in future space missions, 425
of iron(III) oxides, 403
of lunar meteorites, 424
of lunar minerals, 406
measurement through telescopes, 404
of meteorites, 422–4
of mineral mixtures, 409–11
of Mars, 416
of Moon, 405, 408
of olivine, 406, 413
origin on terrestrial planets, 401
of orthopyroxenes, 406, 413
of SNC meteorites, 422
specular, 401
relaxation,
of selection rules, 66
remote-sensed spectra, 39, 89
Moon's surface, 408, 412
surface of Mars, 406
telescopic, 404–5
rhabdophane,
colours due to lanthanide elements, 115
rhodocrosite,
CF parameters for Mn^{2+}, 220
rhodolite,
birthstone, 106
gem garnet, 106
see almandine *and* pyrope
rhodonite,
colour due to Mn^{2+}, 115
CF parameters for Mn, 218, 220
riebeckite,
electron delocalization, 132
IVCT energies, 117
Mn^{3+} in juddite, 189, 198
ring silicates,
optical spectra, 198–203
CF parameters for cations, 201
ringwoodite,
γ-phase in Mantle, 158, 355, 386
coordination site properties, 472
CF parameters for Fe^{2+}, 228, 229, 288
CF spectra of Fe^{2+}, 169, 288
CFSE of Fe^{2+}, 169, 229
effects of pressure on optical spectra, 362, 367
interatomic distances at high pressures, 358
in olivine $\rightarrow$ spinel transition, 386–8
polyhedral bulk modulus of Fe^{2+}, 376
see also silicate spinel
Ringwood's rules, 304
rockbridgeite,
IVCT energy, 117
romanechite (psilomelane),
crystal chemical data, 338
interatomic distances, 244
Jahn–Teller distortion by Mn^{3+}, 244
lattice images, 344
structural stability from MO theory, 456–7
structure, 343
roscoelite,
CF parameters for V^{3+} in, 200, 210
rubellite,
birthstone, 107
see tourmaline
ruby, 88
birthstone, 107
colour, 30, 111–12
coordination site properties, 470
CF parameters for Cr^{3+}, 153–4, 213
effects of pressure on CF spectra, 362, 370
effect of pressure on Racah parameters, 435
effect of temperature on CF spectra, 362
interatomic distances at high pressures, 358
polarized spectra, 111
polyhedral bulk modulus of Cr^{3+}, 376
Russell–Saunders terms, 52, 466
see also spectroscopic terms
rutile structure, 342

s orbitals 9, 12
in MO diagrams, 445
S spectroscopic term, 54, 466
S or total spin angular momenta, 52, 287
sanidine,
CF parameters for Fe^{3+}, 206, 225
sapphire,
birthstone, 107
causes of colour, 127–9

effect of temperature on optical spectra, 366
IVCT energies, 118, 127–9
mixed-valence properties, 134
pair-excitations, 70–1, 223
polarized spectra
of blue, 127
of yellow, 70, 128
see also blue sapphire, corundum *and* yellow sapphire
scandium,
abundance data, 301, 460
in metamorphic processes, 350
scandium(III),
CF activation energies, 334
electronic configurations, 15, 19, 23
electronic entropy, 285
ionic radii, 464
in magmatic crystallization, 317
in Moon rocks, 460
octahedral site preference energy, 249
spectroscopic term, 53
scattering coefficient,
in reflectance spectra, 401
scheelite,
colours due to lanthanide elements, 68
Schöenflies symbols, 62, 468–9
schorlomite,
Ti^{3+} in garnets, 93
Schrödinger wave equation, 7–8, 443
seawater,
and manganese nodules, 346
second transition series, 1, 14
sedimentary geochemistry,
of transition elements, 32, 310, 330–49
seismic data,
and structure of Earth, 354
selection rules, 64–71
Laporte, 65–8
parity, 65–8
spin-multiplicity, 69–71
self-consistent field Xα scattered wave calculations, 443
September birthstone, 107
serandite,
CF parameters for Mn^{2+}, 220
sextet state, 59, 466
shergottites,
and composition of Mars' surface, 460
Mössbauer spectra, 424
reflectance spectra, 417, 424
siderophilic elements, 302
σ-bonds and σ-molecular orbitals, 435–8
σ spectrum, 74
silicate glasses,
CF parameters, 316
CFSE of cations, 316
silicate melts,
coordination sites, 33, 314–20
interphase partition coefficients, 296, 302, 355
spectroscopic measurements, 314
transition metal ions, 314
silicate perovskite,
Cr^{2+} in, 330
coordination site properties, 474
crystal structure, 357
effect of pressure on optical spectra, 374
interatomic distances at high pressures, 359
in Lower Mantle, 357
polyhedral bulk modulus of Fe^{2+}, 377
spin-pairing of Fe^{2+}, 382
silicate spinel,
γ-phase in Mantle, 158, 386
coordination site properties, 472
CF parameters
for Fe^{2+}, 169, 229,
for Ni^{2+}, 169, 236
disproportionation in Lower Mantle, 388
effect of pressure on CF spectra, 362, 367, 371
electrical conductivity at high pressures, 393
polyhedral bulk moduli, 376
radiative heat transfer, 391
sillimanite,
coordination site properties, 471
CF parameters
for cations, 174, 213, 225
for Fe^{2+}, 228, 229, 288
structure, 173.
singlet state, 52, 379, 466
sinhalite,
CF parameters of Fe^{3+} in chrysoberyl, 176
site occupancy of Fe^{3+}, 259
site distortion, and cation ordering, 263
influence on interphase partitioning, 291
see also Jahn–Teller distortion
site incompressibility,
see polyhedral bulk modulus, 31, 374
site occupancies,
in amphiboles, 258
in biotites, 259
in epidotes, 259
in olivines, 254–7
in pyroxenes, 257–8
in spinels, 247–9
techniques for measuring, 251–4
site preferences,
see cation ordering, 250
site symmetries,
in minerals, 65, 470–4
Skaergaard intrusion, 4
trace transition element fractionation, 312
skutterudite,
MO diagrams, 442

smectite,
  CF parameters for Cr^{3+}, 213
sodium chloride structure,
  lattice energies of divalent oxides, 274
spectra,
  see absorption spectra; CF spectra; IVCT spectra; Mössbauer spectra; optical spectra; reflectance spectra
spectrochemical series, 28, 260, 434
spectrophotometer, 89, 90, 108
spectroscopic terms, 50–3, 466
spessartine,
  colour due to Mn^{2+}, 115
  coordination site properties, 470
  CF parameters for Mn^{2+}, 220
  effect of temperature on CF spectra, 364
sphalerite,
  lattice energy, 274
spherically symmetrical,
  s orbitals, 11
  electronic configurations, 56, 64
spin-allowed transitions, 57, 69, 98, 116
  intensities, 72
  and OMCT transitions, 450
spin-down configuration, 10, 55
  in MO diagrams, 444, 448
  in optically induced IVCT transitions, 453
spinel,
  cation distribution, 247–50
  cations in tetrahedral sites, 22, 100
  coordination site properties, 470
  CF parameters
    for cations, 155, 210, 213, 216, 232, 237
    for Fe^{2+}, 228, 229, 288
  CFSE of cations, 249
  crystal structure, 136
  effect of pressure on CF spectra, 371
  polyhedral bulk moduli, 376
spinel,
  modified or β-phase, 158
  see wadsleyite
spin-forbidden transitions, 57, 69, 70
  in Fe^{3+}, 59, 70, 222
  intensification by magnetic dipole interactions, 74
  intensities, 72
  in Mn^{2+}, 59, 222
  in Ni^{2+}, 234
spin function, 8
spin multiplicity, 52, 54
spin-multiplicity selection rule, 69–71
spin–orbit coupling, 11
  and EPR spectra, 11
  in CF spectra, 11
spin-pairing transitions, 18
  in deerite, 381
  energies 29, 231, 380
  in Earth's Mantle, 18, 31, 379–83
  electrical conductivity at high pressures, 393
  Fe^{2+} partitioning in Lower Mantle, 388
  in gillespite, 381
  in hematite, 381
  interphase partitioning in Lower Mantle, 384
  pressure-induced in Fe^{2+} in hauerite (MnS_2), 379
  in silicate perovskites, 382
  site occupancies of transition elements, 384
  in wüstite, 381
spin quantum number, m_s, 10, 52
spin-unrestricted formalism,
  and MO diagrams 444
spin-up configuration, 10, 55, 58
  in MO diagrams, 444, 448
spin wave function, 8
spodumene,
  coordination site properties, 473
  CF parameters for cations, 188, 189, 210, 213
square antiprism coordination,
  and *d* orbital energy levels, 20
square planar coordination,
  CF states of Fe^{2+}, 65
  d orbital energy levels 25
  in eudialyte, 202
  in gillespite, 20
  vibrational modes, 79
square pyramidal coordination,
  and *d* orbital energy levels, 25
  in eudialyte, 20, 203
  in S_N1 reactions, 332, 334–5
staurolite,
  coordination site properties, 471
  CF parameters for Fe^{2+}, 175, 228, 229, 288
  Fe^{2+} in tetrahedral coordination, 100–1
  polarized spectra, 100
  see also lusakite
Stefan's constant, 390
stewartite,
  CF parameters for Fe^{3+}, 226
stishite,
  CF parameters for Cr^{3+}, 226
substitution reactions, 332
substitutional blocking,
  influence on IVCT, 120
  in glaucophane, 124
sulphides, 5
  covalent bonding, 5
  CFSE and lattice energies, 274, 276
  element distribution in coexisting, 442
  interatomic distances, 442
  MO diagrams, 441, 442
  π-bonding, 440
  and transition element mineralogy, 440
symmetric wave functions, 13, 65
  see gerade state

symmetry,
of crystal field states, 54, 68
of electric dipole vector, 75
orbitals, 3, 68
of vibrational modes, 68
symmetry-allowed transitions,
see Laporte selection rule 65–8
symmetry restricted covalency, 433

T crystal field state, 54, 467
t_2 and t_{2g}-group *d* orbitals, 12, 13
in MO diagrams, 445
talc, CF parameters for Ni^{2+}, 201, 237
Tanabe–Sugano diagrams, 56–9
for Cr^{2+}, 215; for Cr^{3+}, 212; for Co^{2+}, 231; for Co^{3+}, 227; for Fe^{2+}, 57, 227; for Fe^{3+}, 58, 70; for Mn^{2+}, 58; for Mn^{3+}, 215; for Ni^{2+}, 235; for Ni^{3+}, 231; for V^{2+}, 212; for V^{3+}, 209
widths of absorption bands, 80
tanzanite,
colours caused by oxidation of vanadium, 113–14, 115
CF parameters for V^{3+}, 210
gem zoisite, 88
optical spectra, 114, 175, 211
polarized spectra, 114
taramellite,
IVCT energies, 118, 131
tawmawite,
colours caused by Cr^{3+}, 115
CF parameters for Cr^{3+}, 175, 213
temperature,
influence on band shape, 82
CF spectra of silicate melts, 319
effects on CF splitting parameter, 31, 68, 82
effects on CF spectra, 362–4
effects on reflectance spectra, 412–15
effects on optical spectra of influence on IVCT, 120
in Mantle, 390
variation of CFSE, 289, 319
tenorite,
interatomic distances, 245
Jahn–Teller distortion by Cu^{2+}, 243, 245
tephroite,
cation ordering, 102–4, 257
coordination site properties, 472
CF spectra of Mn^{2+}, 103–4
CF parameters for Mn^{2+}, 220
polarized spectra, 104
terrestrial planets,
compositions of, 396, 460
tetragonally distorted octahedral sites, 34–6, 64, 263
CFSE, 38
CF states of Fe^{2+}, 63
electronic configurations of Fe^{2+}, 63
energy level diagrams, 35, 39, 263
ligand field energy levels, 83–4
Schöenflies symbol, 63
stability of Mn^{3+}, 34
tetrahedral coordination, 20
Co^{2+} in staurolite, 22, 100–2
CF splitting, 21–2
CFSE, 23, 249
electronic configurations, 23
energy levels for *d* orbitals, 21
in magma, 312, 317
mis-identification of cation sites, 101, 216
Orgel diagrams, 60–2
in silicate melts, 314, 323–4
Schöenflies symbol, 63
in staurolite, 100–2
thermal expansion,
influence on CF splitting, 81
influence on CF spectra, 81–2
thermochromic effect, 112
thermodynamic data,
from CF spectra, 3, 275–6
influence on trace element distribution, 309
thermodynamic properties, 272–98
thiospinels,
MO diagrams, 442
third transition series, 1, 14
three-fold degeneracy,
of *p* orbitals, 11
of t_{2g}-group *d* orbitals, 13
thulite,
colour due to Mn^{3+}, 115
tirodite,
cation ordering of Mn^{2+}, 258
colour due to Mn^{2+}, 115
titanaugite,
colours caused by IVCT, 95, 126
titanium,
abundance data, 301, 460
in metamorphic reactions, 350
oxidation states in Eh–pH diagrams, 336, 340
titanium(III),
absorption spectrum of hexahydrated cation, 48
in Allende meteorite, 93, 292
colours in minerals due to Ti^{3+}, 115
in corundum, 94, 128
CF activation energies, 339
CF parameters for minerals, 208–9
CF spectra, 208–9
CF splitting
in corundum, 29, 154, 208
in hexahydrated cation, 29, 52, 69, 208
in periclase, 29, 151
CF states in octahedral sites, 50–2
CFSE
in hexahydrated cation, 29, 208

titanium(III) (*cont.*)
  CFSE (*cont.*)
    in minerals, 208
    in monoclinic sites, 264
    in octahedral sites, 19, 249, 264
    in tetragonal sites, 38, 264
    in tetrahedral sites, 23, 249
    in trigonal sites, 264
  on Earth, 14, 397
  electronic configurations, 15, 19, 23
  electronic entropy, 40, 285
  energy level diagrams, 51, 60
  in extraterrestrial minerals, 88, 93, 94, 292, 310
  in garnets, 93, 95
  interphase partitioning, 292
  ionic radii, 464
  in kosmochlor, 95
  in magmatic crystallization, 317
  in micas, 93
  in Moon rocks, 14, 397
  octahedral site preference energy, 249
  Orgel diagram, 60
  polarized spectra of Allende pyroxene, 94
  spectroscopic term, 53
  stability at high pressures, 385
  in terrestrial minerals, 95, 125–6
titanium(IV),
  CF activation energies, 334
  CFSE
    in octahedral sites, 19, 249, 264
    in tetragonal sites, 38, 264
    in tetrahedral sites, 23, 249
    in trigonal sites, 264
  on Earth, 14, 397
  effective charge in $[TiO_6]^{-8}$ cluster, 456
  electronic configuration, 15
  electronic entropy, 285
  ionic radii, 464
  in magmatic crystallization, 317
  in Moon rocks, 14, 397
  octahedral site preference energy, 249
  spectroscopic term, 53
titanium(III) → titanium (IV) IVCT, 94, 116
  in Allende meteorite pyroxene, 94, 119
  in andalusite, 119
  in corundum, 119, 128
  effect of pressure, 94, 366
  energies in minerals, 119
todorokite,
  crystal chemical data, 338
  crystal structure, 343, 347
  interatomic distances, 244
  Jahn–Teller distortion by Mn^{3+}, 244
  lattice images, 344
  in manganese nodules, 344, 347
  structural stability from MO theory, 456–7
topaz,
  birthstone, 107
  coordination site properties, 471
  CF paramaters for cations, 175, 213
tourmaline,
  coordination site properties, 473
  CF parameters
    for cations, 200, 202, 210, 213, 218, 230
    for Fe^{2+}, 228, 230, 288
  gem indicolite, 107
  gem rubellite, 107
  IVCT energies, 117–18
  structure, 198
trace elements,
  crustal abundances, 301, 460
  definition, 300
  distribution rules, 301
  and ionic radii, 307
transition elements,
  definition 14
  abundances, 460–1
  electronic configurations, 15
  first series, 1, 2, 14, 15
  ionic radii, 464–5
  isotopes, 462–3
  second series, 1, 14
  spectroscopic terms, 53
  third series, 1, 14
  lanthanide series, 1, 14
transition metal compounds,
  heats of hydration, 275
  lattice energies, 274–5
transition moment, 66
transition probability, 66, 75, 99
transition state,
  in substitution reactions, 333
  in MO diagrams, 444
Transition Zone, 354, 388
  mineralogy, 355
tremolite,
  coordination site properties, 472
  CF parameters for cations, 189, 197, 210, 213
  hydroxyl stretching frequencies, 253
triangular coordination,
  and *d* orbital energy levels 25
trigonal bipyramidal coordination,
  and *d* orbital energy levels 25
  in andalusite, 172
  in hibonite, 129
trigonally distorted octahedral site, 37
  CF states of Fe^{2+}, 65
  CFSE, 265
  energy level diagrams, 39, 263
  Schöenflies symbol, 63
triplet state, 52, 379, 466
trivalent cations,
  CFSE in oxides, 276
  distribution in spinels, 248

intersite partitioning in silicates, 262–8
interphase partitioning, 292–4
magmatic crystallization, 296, 313
octahedral site preference energy, 249
tsavorite,
gem garnet, 88, 105
colour due to V^{3+}, 105, 115
CF parameters for V^{3+}, 159, 210
tunnel structures, 342, 343
turquoise,
birthstone, 107
colour due to Cu^{2+}, 115
interatomic distances, 245
Jahn–Teller distortion, by Cu^{2+}, 245
two-fold degeneracy, of e_g-group *d* orbitals, 13

ultraviolet region, 46
ungerade state, *u*, 13, 65, 66
electromagnetic radiation, 89–90
uniaxial minerals, 50, 74
universal stage,
use in mineral spectroscopy, 89–90
unpaired electrons, 14
Upper Mantle, 354
abundance data, 460–1
element distribution, 302, 327
transition elements, 355
polyhedral bulk moduli, 376–8
ureyite,
see kosmochlor
uvarovite,
colour due to Cr^{3+}, 115
coordination site properties, 470
CF parameters for Cr^{3+}, 159, 213
effect of pressure on CF spectra, 159, 364
Racah *B* parameter, 432

vacancies,
in chalcophanite, 342
in laihunite, 142, 269
in Mn(IV) oxides, 345, 347, 456–7
vaesite,
bond-distance, 442
π-bonding, 442
valence of cations from spectra, 93
valence orbitals,
energies for $[FeO_6]$ clusters, 446–7
vanadinite,
OMCT, 132
effect of pressure on OMCT, 367
vanadium,
abundance data, 301, 460
oxidation states in Eh–pH diagrams, 336, 340
in metamorphic reactions, 350
vanadium(II),
CF splitting
in hexahydrated cation, 29
in oxides, 276
in sulfides, 276
CFSE
in hexahydrated cation, 29
in monoclinic site, 264
in octahedral sites, 19, 264
in oxides, 276
in sulphides, 276
in tetragonal sites, 38, 264
in tetrahedral sites, 23
in trigonal sites, 264
electronic configuration, 15
electronic entropy, 285
ionic radii, 464
in unheated Tanzanian zoisite, 115, 211
vanadium(III),
cation ordering in epidote, 260
cation ordering in tourmaline, 260
causes of colour in tanzanite, 113–15
colours in minerals due to V^{3+}, 115
CF activation energies, 334
CF parameters for minerals, 210
CF spectra, 209–12
CF splitting
in corundum, 29
in hexahydrated cation, 29
in periclase, 29
CF states in octahedral sites, 209
CFSE
in hexahydrated cation, 29
in minerals, 210
in monoclinic sites, 264
in octahedral sites, 19, 249, 264
in tetragonal sites, 38, 264
in tetrahedral sites, 23, 249
in trigonal sites, 264
on Earth, 14, 397
electronic configuration, 15, 19, 23
electronic entropy, 285, 249, 287
energy level diagrams, 61, 209
in gems, 88
interphase partitioning, 293
ionic radii, 464
in magmatic crystallization, 317
in Moon rocks, 14, 397
octahedral site preference energy, 249
Orgel diagram, 61
in silicate glasses, 315
site occupancy in spinels, 248, 249, 287
spectroscopic terms, 53
Tanabe–Sugano diagram, 209
in tanzanite, 113–15, 211
vanadium(IV),
in apophyllite, 211
CFSE
in monoclinic sites, 264
in octahedral sites, 19, 264
in tetragonal sites, 38, 264

vanadium(IV) (*cont.*)
CFSE (*cont.*)
in tetrahedral sites, 23
in trigonal sites, 264
in Eh–pH diagrams, 336
electronic configuration, 15
in heated Tanzanian zoisite, 115
spectroscopic term, 53
stability at low pH, 340
in vanadyl ions, 115
vanadium(V),
on Earth, 14
in Eh–pH diagrams, 336
electronic configuration, 15
spectroscopic term, 53
stability at high pH, 340
vanadyl ions,
in apophyllite, 211
in tanzanite, 115
Vega missions,
transition metal data for Venus, 415, 422, 460
Venera missions,
transition metal data for Venus, 415, 422, 460
Venus,
composition, 397, 415, 460
dehydroxylated silicates, 422
Fe(III) minerals on surface, 422
high-radar reflectivity, 422
surface features, 398
vernadite,
crystal chemical data, 338
in manganese nodules, 347
structural stability from MO theory, 456
structure, 345
Verwey transition,
in magnetite, 137, 142
Vesta,
source of basaltic achondrites, 397
spectra and surface features, 399
vesuvianite,
CF parameters for Cr^{3+} 176
CF parameters for Fe^{3+}, 225
vibrational modes, 66–9, 79, 80
octahedral site, 69
square planar site, 77–9
vibronic coupling, 67–9
CF spectra
of gillespite, 78–9
of Mg–Ni^{2+} olivines, 167
of ruby, 154
Viking missions,
transition metal data for Mars, 397, 415, 421, 460
violarite,
low-spin Fe^{2+}, 379
viridine,
colour due to Mn^{3+} in andalusite, 115
CF parameters for Mn^{3+}, 218
CF spectra, 99, 172
stability of Mn^{3+}, 99–100
structure, 172
visible region, 46, 105
visible-region spectra,
see CF spectra, IVCT transitions *and* OMCT transitions
vivianite,
colour due to IVCT, 87, 122
crystal structure, 122
effect of pressure on IVCT, 366
intensity enhancement by IVCT, 454
IVCT energy, 117, 119
mixed-valence properties, 134–5
polarized absorption spectra, 121
volume of mixing, 278
vonsenite,
electron delocalization, 133
mixed-valence effects, 140
structure, 140

wadsleyite,
β-phase in Mantle, 355, 358, 386
cation ordering, 356
coordination site properties, 474
interatomic distances, 358
oxygen ligands, 30
structure, 355
wave functions, 7, 11
wavelength units, 45, 46
wavenumber, 45
weathering, 329, 330, 335
widths,
of absorption bands, 47, 80–2
of IVCT bands, 120
willemite,
tetrahedral Zn^{2+} in, 304
winchite,
cation ordering of Mn^{3+}, 259, 268
colour due to Mn^{3+}, 115
CF parameters for Mn^{3+}, 189, 198
Jahn–Teller effect, 198, 247
wulfenite,
effect of pressure on OMCT, 367
wurtzite structure, 275
wüstite,
spin-pairing of Fe^{2+} in, 381

XANES,
see X-ray absorption spectroscopy
XPS,
see X-ray photoelectron spectroscopy
X-ray absorption spectroscopy, 148
characterization of low-spin configurations, 28
of silicate glasses, 314, 317, 320
and site occupancies 254

X-ray amorphous,
Fe(III) oxides on Mars, 421
X-ray diffraction,
and crystal structure, 193, 302, 307, 317, 381
difficulty in distinguishing Fe and Mn, 103, 251
difficulty in distinguishing mixed-valence cations, 251
and Jahn–Teller distortion, 36
site populations, 251, 257, 258, 302
structures of glasses, 317
X-ray photoelectron spectroscopy, 3
characterization of low-spin configurations, 28
xenoliths,
and composition of Mantle, 354
xenotime,
colour due to lanthanide elements in, 115

yellow sapphire,
colour due to Fe^{3+}, 128
CF parameters of Fe^{3+}, 154, 224
pair-excitations, 70–1, 223
polarized spectra, 70, 128
yoderite,
coordination site properties, 471
IVCT energy, 117
colour due to $Mn^{2+} \rightarrow Mn^{3+}$ IVCT, 119, 174
CF parameters for Mn^{3+}, 218
structure, 174

zinc,
abundance data, 301, 460
in manganese nodules, 346
Mössbauer effect in ^{67}Zn, 3, 462
zinc(II),
cation ordering,
in olivine, 257
in chalcophanite, 342, 345
CF activation energies, 334
CFSE
in monoclinic sites, 264
in octahedral sites, 19, 249, 264
in oxides, 276
in sulphides, 276
in tetragonal sites, 38, 264
in tetrahedral sites, 23, 249
in trigonal sites, 264
electronic configurations, 15, 19, 23
electronic entropy, 285
intersite partition coefficients,
for olivine, 256, 279
for orthopyroxene, 256, 258
ionic radii, 464
in magmatic crystallization, 317
octahedral site preference energy, 249
zincite,
CF parameters for Co^{2+}, 232
CF parameters for Ni^{2+}, 237
zircon,
birthstone, 107
zirconia,
cubic, CF parameters for Co^{2+}, 232
CF parameters for Ni^{2+}, 237
zoisite,
coordination site properties, 471
CF parameters for Cr^{3+}, 213
CF parameters for V^{3+},113–5, 210
gem tanzanite, 88, 105
origin of blue color when heated, 113–5
polarized spectra of specimen from Tanzania, 114